Organofluorine Chemistry

Principles and Commercial Applications

TOPICS IN APPLIED CHEMISTRY

Series Editors: **Alan R. Katritzky, FRS**
Kenan Professor of Chemistry
University of Florida, Gainesville, Florida

Gebran J. Sabongi
Laboratory Manager, Encapsulation Technology Center, 3M
St. Paul, Minnesota

BIOCATALYSTS FOR INDUSTRY
Edited by Jonathan S. Dordick

CHEMICAL TRIGGERING
Reactions of Potential Utility in Industrial Processes
Gebran J. Sabongi

THE CHEMISTRY AND APPLICATION OF DYES
Edited by David R. Waring and Geoffrey Hallas

HIGH-TECHNOLOGY APPLICATIONS OF ORGANIC COLORANTS
Peter Gregory

INFRARED ABSORBING DYES
Edited by Masaru Matsuoka

LEAD-BASED PAINT HANDBOOK
Jan W. Gooch

ORGANOFLUORINE CHEMISTRY
Principles and Commercial Applications
Edited by R. E. Banks, B. E. Smart, and J. C. Tatlow

POLY(ETHYLENE GLYCOL) CHEMISTRY
Biotechnical and Biomedical Applications
Edited by J. Milton Harris

RADIATION CURING
Science and Technology
Edited by S. Peter Pappas

RESORCINOL
Its Uses and Derivatives
Hans Dressler

STRUCTURAL ADHESIVES
Edited by S. R. Hartshorn

TARGET SITES FOR HERBICIDE ACTION
Edited by Ralph C. Kirkwood

A Continuation Order Plan is available for this series. A continuation order will bring delivery of each new volume immediately upon publication. Volumes are billed only upon actual shipment. For further information please contact the publisher.

Organofluorine Chemistry

Principles and Commercial Applications

Edited by

R. E. Banks

*The University of Manchester Institute
of Science and Technology (UMIST)
Manchester, United Kingdom*

B. E. Smart

*DuPont Central Research & Development
Wilmington, Delaware*

and

J. C. Tatlow

*Editorial Office of the Journal
of Fluorine Chemistry
Birmingham, United Kingdom*

Plenum Press • New York and London

ACz-3252

Library of Congress Cataloging-in-Publication Data

Organofluorine chemistry : principles and commercial applications /
  edited by R.E. Banks, B.E. Smart, and J.C. Tatlow.
       p.   cm. -- (Topics in applied chemistry)
    Includes bibliographical references and index.
    ISBN 0-306-44610-3
    1. Organofluorine compounds.   I. Banks, R. E. (Ronald Eric),
  1932-   . II. Smart, B. E. (Bruce E.)   III. Tatlow, J. C.
  IV. Series.
  QD305.H15074   1994
  661'.891--dc20   94-26835
                                                                  CIP

ISBN 0-306-44610-3

©1994 Plenum Press, New York
A Division of Plenum Publishing Corporation
233 Spring Street, New York, N.Y. 10013

Printed in the United States of America

Contributors

Y. W. Alsmeyer, Fluorochemical Process Technology Center, 3M Company, St. Paul, Minnesota 55144-1000

Madhu Anand, Air Products and Chemicals, Inc., Allentown, Pennsylvania 18195-1501

Wolfgang K. Appel, Hauptlaboratorium, Hoechst AG, D-6230 Frankfurt-am-Main 80, Germany

Bruce E. Baker, DuPont Speciality Chemicals, Jackson Laboratory, Chambers Works, Deepwater, New Jersey 08023

R. E. Banks, Department of Chemistry, The University of Manchester Institute of Science and Technology, Manchester M60 1QD, England

Bernd A. Blech, RLMTC, Hoechst Celanese Corporation, Summit, New Jersey 07901

Ian J. Brass, Air Products PLC, Walton-on-Thames, Surrey KT12 4RZ, England

David Cartwright, Zeneca Agrochemicals, Jeallott's Hill Research Station, Bracknell, Berkshire RG12 6EY, England

W. V. Childs, Fluorochemical Process Technology Center, 3M Company, St. Paul, Minnesota 55144-1000

Richard A. Du Boisson, PCR Inc., Gainesville, Florida 32602

Philip Neil Edwards, Zeneca Pharmaceuticals, Alderley Park, Macclesfield, Cheshire SK10 4TG, England

Arthur J. Elliott, Halocarbon Products Corporation, North Augusta, South Carolina 29841-0369

A. Engel, Bayer AG, Research and Development—Dyes and Pigments, D-51368 Leverkusen 1, Germany

Andrew E. Feiring, DuPont Central Research & Development Department, Experimental Station, Wilmington, Delaware 19880-0328

Peter Field, PGF Associates Ltd., Buxton, Derbyshire SK17 6EW, England

R. M. Flynn, Fluorochemical Process Technology Center, 3M Company, St. Paul, Minnesota 55144-1000

S. W. Green, Rhône-Poulenc Chemicals Ltd., Avonmouth, Bristol BS11 9HP, England; *present address*: 69 South Road, Kingswood, Bristol BS15 2JF, England

Donald F. Halpern, Ohmeda, P.P.D., Murray Hill, New Jersey 07076; *present address*: 44 Murray Hill Square, Murray Hill, New Jersey 07974.

K. J. Herd, Bayer AG, Research and Development—Dyes and Pigments, D-51368 Leverkusen 1, Germany

J. P. Hobbs, Air Products and Chemicals, Inc., Allentown, Pennsylvania 18195-1501

Takeshi Inoi, Technical Division, Chisso Corporation, Tokyo, Japan; *present address*: 8-8-6 Sugita, Isogo-ku, Yokohama-shi, Kanagawa-ken 236, Japan

†Nobuo Ishikawa, F & F Research Center, Shizushin Building, Minato-ku, Tokyo 107, Japan

Bernard Langlois, Laboratoire de Chimie Organique III, Université Claude Bernard-Lyon I, 69622 Villeurbanne Cedex, France

André Lantz, Atochem, Centre de Recherche Rhône-Alpes, 69310 Pierre-Bénite, France

Anestis L. Logothetis, DuPont Elastomers, Experimental Station, Wilmington, Delaware 19880-0328

Kenneth C. Lowe, Mammalian Physiology Unit, Life Science Department, University of Nottingham, University Park, Nottingham NG7 2RD, England

Ganpat Mani, Allied-Signal Inc., Fluorine Products Division, Morristown, New Jersey 07962-1139

Guiseppe Marchionni, Ausimont/Montefluos, Montedison/Montefluos Group, 20021 Bollate, Milano, Italy

Haruhisa Miyake, Research Center, Asahi Glass Co. Ltd., Kanagawa-ku, Yokohama 221, Japan

†Deceased.

J. S. Moilliet, Zeneca Agrochemicals, Huddersfield HD2 1FF, England; *present address*: BNFL Fluorochemicals, Springfields, Salwick, Preston PR4 OXJ, England

G. G. I. Moore, Fluorochemical Process Technology Center, 3M Company, St. Paul, Minnesota 55144-1000

Yohnosuke Ohsaka, Daikin Industries, Ltd., Chemical Division, 1-1 Nishi Hitotsuya, Settsu-shi, Osaka 566, Japan; *present address*: 1-16-5 Shirakawa, Ibaraki, Osaka, Japan

Ralph J. De Pasquale, Ausimont/Montefluos, Montedison/Montefluos Group, 20021 Bollate, Milano, Italy; *present adddress*: 5500 Atlantic View, St. Augustine, Florida 32084

Richard L. Powell, ICI Klea, Runcorn Technical Centre, Runcorn, Cheshire WA7 4QD, England

Nandakumar S. Rao, DuPont Speciality Chemicals, Jackson Laboratory, Chambers Works, Deepwater, New Jersey 08023

V. N. M. Rao, DuPont Central Research & Development, Experimental Station, Wilmington, Delaware 19880-0262

George A. Shia, Allied-Signal Inc., Fluorine Products Division, Buffalo Research Laboratory, Buffalo, New York 14210

Dario Sianesi, Ausimont/Montefluos, Montedison/Montefluos Group, 20021 Bollate, Milano, Italy

R. N. F. Simpson, Rhône-Poulenc Chemicals Ltd., Avonmouth, Bristol BS11 9HP, England; *present address*: Fluoro Systems Ltd., Abbots Leigh, Bristol BS8 3RP, England

D. S. L. Slinn, Rhône-Poulenc Chemicals Ltd., Avonmouth, Bristol BS11 9HP, England; *present address*: Fluoro Systems Ltd., Abbots Leigh, Bristol BS8 3RP, England

Bruce E. Smart, DuPont Central Research & Development, Experimental Station, Wilmington, Delaware 19880-0328

J. C. Smeltzer, Fluorochemical Process Technology Center, 3M Company, St. Paul, Minnesota 55144-1000

J. Hugo Steven, ICI Klea, Runcorn Technical Centre, Runcorn, Cheshire WA7 4QD, England

Michael Stöbbe, Hauptlaboratorium, Hoechst AG, D-6230 Frankfurt-am-Main 80, Germany

J. C. Tatlow, Editorial Office of the Journal of Fluorine Chemistry, 30 Grassmoor Road, Kings Norton, Birmingham B38 8BP, England

Claude Wakselman, CNRS-CERCOA, 94320 Thiais, France

A. J. Woytek, Air Products and Chemicals Inc., Industrial Gas Division, Allentown, Pennsylvania 18105

Masaaki Yamabe, Research Center, Asahi Glass Co. Ltd., Kanagawa-ku, Yokohama 221, Japan

Preface

During the past fifteen years commercial interest in compounds containing carbon–fluorine bonds has burgeoned beyond all expectations, mainly owing to business opportunities arising from work on biologically active fluoroorganics—particularly agrochemicals, the relentless search for new markets for fluoropolymers and fluorocarbon fluids, developments in the field of medical diagnostics, and the drive to find replacements for ozone-depleting CFCs and Halon fire-extinguishing agents. Judging the situation to warrant the publication of a comprehensive collection of up-to-date reviews dealing with commercial organofluorine compounds within a *single* volume of *manageable* size (and hence reasonable cost), we were delighted to be invited by Plenum Publishing Corporation to produce a suitable book.

In order to provide an authentic and wide-ranging account of current commercial applications of fluoroorganic materials, it clearly was necessary to assemble a sizeable team of knowledgeable contributing authors selected almost entirely from industry. Through their efforts we have been able to produce an almost complete coverage of the modern organofluorochemicals business in a manner designed to attract a readership ranging from experts in the field, through chemists and technologists currently unaware of the extent of industrial involvement with fluoroorganics, to students of applied chemistry. Promised chapters dedicated to perfluoroolefin oxides and ^{18}F labeling of radiopharmaceuticals failed to materialize. This is somewhat unfortunate in view of our aim to achieve comprehensive coverage of the subject. However, not only do other chapters touch on these topics, but an excellent 149-page review, *Fluorine-18 Labeling of Radiopharmaceuticals* (National Academy Press, Washington DC, 1990) by Michael R. Kilbourn is available gratis from The Board on Chemical Sciences and Technology, National Research Council, 2101 Constitution Avenue NW, Washington, D.C. 20418, U.S.A. Finally we must emphasize that the production of this book has been a team effort and our sincere thanks go to all those who have contributed: librarians, typists, the staff at Plenum, co-workers and colleagues of authors, and—preëminently—the collaborating authors themselves. Sadly, one of our authors, Professor Nobuo Ishikawa, died in 1991, and we respectfully dedicate this book to his memory.

Employment of such a large number of contributing authors, coupled with the busy lives they lead and the need to have manuscripts cleared for publication by the companies involved, has resulted in an unusually wide range of receipt dates for typescripts. In fact, chapters started to arrive in the summer of 1990 and continued to appear until February 1992! Not wishing to leave too many gaps, we just held on as long as seemed reasonable. Our apologies for the delay in publication are offered to those authors who were within six months of the original deadline. Our own efforts toward the completion of this book are dedicated severally to our wives, who seem to possess endless patience.

R. E. Banks
Manchester, England
Bruce Smart
Wilmington, Delaware
J. C. Tatlow
Birmingham, England

Contents

1. Organofluorine Chemistry: Nomenclature and Historical Landmarks

R. E. Banks and J. C. Tatlow

1.1. Preamble 1
1.2. Nomenclature 2
    1.2.1. Highly Fluorinated Compounds 2
    1.2.2. Fluorocarbon Code Numbers 4
1.3. Historical Landmarks 5
    1.3.1. The HF Problem 5
    1.3.2. Synthesis of C—F Bonds 6
    1.3.3. Aliphatic Fluorides Go Commercial 12
    1.3.4. Perfluorocarbons 14
    1.3.5. Wartime Advances: The Manhattan Project 17
    1.3.6. Postwar Progress 17
1.4. References 21

2. Synthesis of Organofluorine Compounds

R. E. Banks and J. C. Tatlow

2.1. Introduction 25
2.2. Synthesis Methodology 26
    2.2.1. The Building-Block Approach 27
    2.2.2. The Formation of Carbon–Fluorine Bonds 27
2.3. The Current Position 29
2.4. Tabular Summary of Fluorination Methods and Reagents 29
2.5. References 53

3. **Characteristics of C–F Systems**

   Bruce E. Smart

3.1. Introduction 57
3.2. Physical Properties 58
   3.2.1. General 58
   3.2.2. Solvent Polarity 64
   3.2.3. Lipophilicity 66
   3.2.4. Acidity and Basicity 67
   3.2.5. Hydrogen Bonding 69
3.3. Chemical Properties 70
   3.3.1. Bond Strengths and Reactivity 70
   3.3.2. Reactive Intermediates 74
   3.3.3. Steric Effects 80
3.4. References 82

4. **Perfluorocarbon Fluids**

   S. W. Green, D. S. L. Slinn, R. N. F. Simpson, and A. J. Woytek

4.1. Introduction 89
   4.1.1. Perfluorocarbon Gases 89
   4.1.2. Perfluorocarbon Liquids 90
4.2. Production of Perfluorocarbons 91
   4.2.1. Processes for Perfluorocarbon Gases 91
   4.2.2. Processes for Perfluorocarbon Liquids 92
4.3. Physical Properties 93
4.4. Applications of Perfluorocarbons 95
   4.4.1. Perfluorocarbon Gases as Etchants 95
   4.4.2. Perfluorocarbon Liquids as Alternatives to Chlorofluorocarbons . . . 97
   4.4.3. Perfluorocarbon Liquids as Heat Transfer Agents 104
   4.4.4. Miscellaneous Applications of Perfluorocarbon Liquids 114
4.5. References 118

5. **Electrochemical Fluorination and Its Applications**

   Y. W. Alsmeyer, W. V. Childs, R. M. Flynn,
   G. G. I. Moore, and J. C. Smeltzer

5.1. Coverage 121
5.2. Simons Electrochemical Fluorination 121
   5.2.1. Introduction 121

    5.2.2. Experimental Technique 122
    5.2.3. Anode Film   123
    5.2.4. Mechanisms 124
    5.2.5. Scope 126
    5.2.6. Commercial Utility of ECF-Derived Materials 129
5.3. Cave–Phillips Process for Electrochemical Fluorination 133
    5.3.1. Systems 133
    5.3.2. Feeds 135
    5.3.3. Nonwetting Behavior 135
    5.3.4. Polarization 137
    5.3.5. Brief History of an Anode 138
    5.3.6. Fluorination Results 138
    5.3.7. Summary of the CAVE ECF Process 140
5.4. References 140

6. Chlorofluorocarbons

Arthur J. Elliott

6.1. Introduction 145
6.2. Production 147
    6.2.1. Halogen Exchange 147
    6.2.2. Chlorotrifluoroethylene Telomerization 149
6.3. Properties and Specifications 150
6.4. Applications 151
    6.4.1. General 151
    6.4.2. Declining Applications 151
    6.4.3. Continuing Applications 153
6.5. References 156

7. Alternatives to Chlorofluorocarbons (CFCs)

V. N. M. Rao

7.1. Introduction 159
7.2. Synthesis 162
    7.2.1. 1,1,1, 2-Tetrafluoroethane, CF_3CH_2F (HFC-134a) 162
    7.2.2. 2,2-Dichloro-1,1,1-trifluoroethane, CF_3CHCl_2 (HCFC-123) 165
    7.2.3. 2-Chloro-1,1,1,2-tetrafluoroethane, CF_3CHFCl (HCFC-124) 167
    7.2.4. Pentafluoroethane, CF_3CHF_2 (HFC-125) 168
    7.2.5. 1,1-Dichloro-1-fluoroethane, $CFCl_2CH_3$ (HCFC-141b) 169
    7.2.6. 1,1-Difluoroethane, CHF_2CH_3 (HFC-152a) 169

7.2.7. Dichloropentafluoropropanes 170
7.3. Commercial Aspects 171
7.4. Properties 171
7.5. Applications 172
    7.5.1. Refrigeration 172
    7.5.2. Foaming Agents 172
    7.5.3. Solvents 173
7.6. References 173

8. Perfluoroalkyl Bromides and Iodides

Claude Wakselman and André Lantz

8.1. Introduction 177
8.2. Preparation 178
8.3. Reactions 178
    8.3.1. "Classical" Radical Reactions 178
    8.3.2. Reactions with Nucleophiles 180
    8.3.3. Reactions with Metals 183
    8.3.4. Reactions with Acidic Reagents 184
    8.3.5. Electrochemical Transformations 185
8.4. Applications 185
    8.4.1. Halons 185
    8.4.2. Perfluoroalkyl Iodides 188
8.5. References 190

9. Industrial Routes to Ring-Fluorinated Aromatic Compounds

J. S. Moilliet

9.1. Introduction 195
9.2. "Halex" Fluorinations 196
    9.2.1. General Considerations 196
    9.2.2. Examples 198
    9.2.3. Fluoride Sources 199
    9.2.4. Solvents 200
    9.2.5. Procedure for the Halex Process 201
9.3. Diazotization Methods 203
    9.3.1. The Balz–Schiemann Reaction 203
    9.3.2. HF-Diazotization/Dediazoniation 207
9.4. Other Fluorination Methods 211

9.4.1. Cobalt Fluoride Fluorination 211
9.4.2. Using Hydrogen Fluoride 212
9.4.3. Direct Fluorination 213
9.5. Comparison of the Three Principal Methods 213
9.6. The Industrial Scene 214
9.6.1. HF-Diazotization/Dediazoniation 214
9.6.2. Halex Fluorination 217
9.6.3. Balz–Schiemann Methodology 217
9.7. Concluding Remarks 217
9.8. References 218

10. Side-Chain Fluorinated Aromatic Compounds Routes to Benzotrifluorides

Bernard Langlois

10.1. Introduction 221
10.2. Synthesis of Benzotrifluorides from the
    Corresponding Toluenes: Liquid-Phase Methods 222
10.2.1. General Considerations 222
10.2.2. Fluorination of Benzotrichlorides 224
10.3. Vapor-Phase Routes to Benzotrifluorides and
    (Trifluoromethyl)pyridines 225
10.4. Synthetic Manipulation of Benzotrifluorides 225
10.4.1. Electrophilic Substitution 225
10.4.2. Nucleophilic Displacement of Nuclear Halogen from
    4-Halogenobenzotrifluorides 227
10.4.3. Hydrolysis of Trifluoromethyl Groups 228
10.5. Newer Commercially Interesting Methods for the
    Trifluoromethylation of Aromatic Compounds 229
10.5.1. Electrophilic Trifluoromethylation 229
10.5.2. "Nucleophilic" Trifluoromethylation 230
10.5.3. Radical Trifluoromethylation 230
10.6. Concluding Remarks 232
10.7. References 232

11. Recent Developments in Fluorine-Containing Agrochemicals

David Cartwright

11.1. Introduction 237
11.2. Occurrence of Fluorine in Agrochemicals 239

11.3. Fluorine-Containing Herbicides 240
    11.3.1. Herbicides Containing an Aromatic-Type CF₃ Group 240
    11.3.2. Herbicides Containing a Fluoroaromatic Group:
        Pyridyloxyacetic Acids 244
    11.3.3. Herbicides Containing a Fluoroalkoxy Group:
        Sulfonylureas 245
    11.3.4. Herbicides Containing a Trifluoromethanelsulfonyl Group:
        Trifluoromethanesulfoanilides 245
11.4. Fluorine-Containing Insecticides 246
    11.4.1. Compounds Affecting Insect Growth 246
    11.4.2. Pyrethroid Insecticides 247
    11.4.3. Others 252
11.5. Fluorine-Containing Fungicides 252
    11.5.1. Sterol Biosynthesis Inhibitors 252
    11.5.2. Amide Fungicides 255
11.6. Fluorine-Containing Plant Growth Regulators 255
    11.6.1. Compounds that Interfere with Gibberellin Biosynthesis 255
    11.6.2. Benzylamine Derivatives 256
11.7. Rodenticides 257
11.8. References 257

12. Fluorinated Liquid Crystals

Takeshi Inoi

12.1. Introduction 263
12.2. Properties and Structural Classification of Liquid Crystals 263
12.3. Applications 267
12.4. Molecular Design 267
12.5. Fluorinated Liquid Crystals 270
    12.5.1. Semifluorinated Alkanes 271
    12.5.2. Schiff Bases (Azomethines) 271
    12.5.3. Benzoates 271
    12.5.4. Biphenyls 275
    12.5.5. Cyclohexanecarboxylates 276
    12.5.6. Liquid Crystals with Hybridized
        Structures and Multiring Systems 276
12.6. Ferroelectric Liquid Crystals 283
12.7. Conclusions 284
12.8. References 285

13. Fluorine-Containing Dyes
A. Reactive Dyes

K. J. Herd

13A.1. Introduction 287
13A.2. Dyestuffs with One Reactive System 288
    13A.2.1. General Information on Reactive Dyes 288
    13A.2.2. Heterocyclic Carrier Systems with
                 Fluorine as the Leaving Group 289
    13A.2.3. Preparation of the Reactive Compounds 297
    13A.2.4. Preparation of the Reactive Dyes 299
13A.3. Dyes with Two or More Reactive Systems 302
    13A.3.1. Homobifunctional Reactive Dyes 303
    13A.3.2. Heterobifunctional Reactive Dyes 303
    13A.3.3. Polyfunctional Reactive Dyes 306
13A.4. Conclusion 307
13A.5. References 307

13.B. Other Fluorinated Dyestuffs

A. Engel

13B.1. Introduction 315
13B.2. Properties of Fluorine-Containing Dyes 315
13B.3. Summary and Outlook 320
13B.4. References 320

14. Textile Finishes and Fluorosurfactants

Nandakumar S. Rao and Bruce E. Baker

14.1. Textile Repellent Finishes 321
    14.1.1. Intrinsic Repellency and Fluorocarbon Structure 321
    14.1.2. Synthesis of Fluoroalkyl Intermediates 325
    14.1.3. Synthesis of Fluorochemical Repellents 329
    14.1.4. Soil-Release Finishes 331
    14.1.5. Future Developments 332
    14.1.6. Major Manufacturers 332
14.2. Fluorosurfactants 333
    14.2.1. Fluorosurfactant Synthesis 333
    14.2.2. Aqueous Solutions of Fluorosurfactants 334

14.2.3. Properties and Uses 336
14.3. References 336

15. Fluoroplastics

Andrew E. Feiring

15.1. Introduction 339
15.2. Fluorinated Vinyl Monomers 341
15.3. Crystalline Perfluoroplastics 342
    15.3.1. Poly(tetrafluoroethylene) (PTFE) 342
    15.3.2. Perfluorinated Copolymers (FEP and PFA) 344
    15.3.3. Properties of the Perfluoroplastics 346
    15.3.4. Applications and Commercial Aspects 348
15.4. Amorphous Perfluoroplastics 349
15.5. Poly(chlorotrifluoroethylene) 350
    15.5.1. Production 351
    15.5.2. Properties 351
    15.5.3. Applications and Commercial Aspects 352
15.6. Partially Fluorinated Plastics 352
    15.6.1. Ethylene–Tetrafluoroethylene Copolymer 352
    15.6.2. Ethylene–Chlorotrifluoroethylene Copolymer 354
    15.6.3. Poly(Vinylidene Fluoride) 356
    15.6.4. Poly(Vinyl Fluoride) 358
15.7. Other Fluorine-Containing Plastics 360
    15.7.1. Addition Polymers 360
    15.7.2. Condensation Polymers 362
    15.7.3. Surface-Fluorinated Plastics 363
15.8. Outlook and Conclusions 364
15.9. References 364

16. Fluoroelastomers

Anestis L. Logothetis

16.1. Introduction 373
    16.1.1. Chemical Compositions 374
    16.1.2. Structural Considerations 374
16.2. Fluoroelastomers Based On Vinylidene Fluoride Copolymers 376
    16.2.1. General Description 376
    16.2.2. Production 377

16.2.3. Curing (Vulcanization) of Fluoroelastomers Containing
Vinylidene Fluoride 380
16.2.4. Properties 383
16.2.5. Processing 385
16.3. Fluoroelastomers Based on Tetrafluoroethylene/Propylene Copolymers 387
16.3.1. Introduction 387
16.3.2. Polymerizations 387
16.3.3. Curing Chemistry and Properties 388
16.3.4. Processing 388
16.4. Perfluoroelastomers Based on
Tetrafluoroethylene/Perfluoro(alkyl vinyl ether) Copolymers 389
16.4.1. Polymer Description 389
16.4.2. Polymerizations 390
16.4.3. Curing Chemistry 390
16.4.4. Properties 392
16.4.5. Processing 393
16.5. Uses of Fluorinated Elastomers 393
16.6. References 394
16.6.1. General References 394
16.6.2. Specific References 394

17. Fluoropolymer Coatings

Masaaki Yamabe

17.1. Introduction 397
17.2. Thermoplastic Fluoropolymer Coatings 397
17.2.1. Tetrafluoroethylene Polymers 397
17.2.2. Poly(Vinylidene Fluoride) 398
17.3. Curable Fluoropolymer Coatings 399
17.4. References 401

18. Fluorinated Membranes

Masaaki Yamabe and Haruhisa Miyake

18.1. Introduction 403
18.2. Structure and Properties 404
18.3. Preparation 405
18.4. Applications 407
18.4.1. Chlor-Alkali Process 407
18.4.2. Outlook 409
18.5. References 410

19. Monomers and Polymers from Hexafluoroacetone

Wolfgang K. Appel, Bernd A. Blech, and Michael Stöbbe

19.1. Introduction 413
19.2. Synthesis of Hexafluoroacetone (HFA) 413
19.3. Synthesis of Intermediates Containing the
    Hexafluoroisopropylidene Group 414
    19.3.1. The Overall Reaction Sequence 414
    19.3.2. Formation of Arylbis(trifluoromethyl)carbinols 414
    19.3.3. Properties of Arylbis(trifluoromethyl)carbinols 417
    19.3.4. Formation of Diphenylhexafluoropropanes 418
19.4. Directly Accessible HFIP-Bridged Monomers and Intermediates 419
19.5. Polymers Containing Hexafluoroisopropylidene Groups 421
    19.5.1. Aromatic 6F-Polyesters 421
    19.5.2. Aromatic 6F-Polyamides 422
    19.5.3. Aromatic 6F-Polyimides 423
19.6. Conclusion 427
19.7. References 427

20. Perfluoropolyethers (PFPEs) from Perfluoroolefin Photooxidation: Fomblin® and Galden® Fluids

Dario Sianesi, Guiseppe Marchionni, and Ralph J. De Pasquale

20.1. Introduction 431
20.2. The Photooxidation Route to Neutral PFPEs 432
    20.2.1. Oxidative Photopolymerization of Fluoroolefins 432
    20.2.2. Removal of Peroxide Linkages 435
    20.2.3. End-Group Modification to Give Neutral PFPEs 437
20.3. Mechanism of HFP and TFE Oxidative Photopolymerization 438
20.4. Chemical and Physical Properties of
    Fomblin® and Galden® Fluids 440
    20.4.1. Chemical Properties and Solubilities 440
    20.4.2. Thermal Properties—Influence of Metals 442
    20.4.3. Oxidative Stability, Radiation Resistance, and Bioinertness . . . 443
    20.4.4. Physical Properties 443
    20.4.5. Greases 446
    20.4.6. Tribology 446
    20.4.7. Rheology 448
20.5. Functional PFPEs from Photooxidation 450
20.6. Applications of PFPEs 453
    20.6.1. Testing of Electronic Devices and Equipment 453
    20.6.2. Polymer Curing 454
    20.6.3. Lubrication 454

20.6.4. Surfactants, Emulsions, and Cosmetics 454
20.6.5. Polymer Modification 455
20.6.6. Conductive Films 455
20.6.7. Mass Spectrometry 455
20.6.8. Environmental Applications 456
20.7. Future Prospects 456
20.8. References 457

21. Perfluoropolyether Fluids (Demnum®) Based on Oxetanes

Yohnosuke Ohsaka

21.1. Introduction 463
21.2. Preparation 463
21.2.1. Monomers 463
21.2.2. Oligomers 464
21.3. Properties 464
21.3.1. General 464
21.3.2. Toxicity 465
21.3.3. Comparative PFPE Properties 465
21.4. Applications 466
21.4.1. Base Oil for Greases 466
21.4.2. Semiconductor Manufacturing 466
21.4.3. Miscellaneous 467
21.5. References 467

22. Surface Fluorination of Polymers

Madhu Anand, J. P. Hobbs, and Ian J. Brass

22.1. Introduction 469
22.2. Reactions of Elemental Fluorine with Organic Polymers 469
22.3. Properties and Applications of Surface-Fluorinated Polymers 471
22.3.1. Barrier Properties 471
22.3.2. Adhesion and Surface Energy 473
22.3.3. Other Properties of Fluorinated Surfaces 474
22.4. Techniques for the Direct Fluorination of Polymer Surfaces 475
22.4.1. Post-Treatment 475
22.4.2. *In Situ* Treatment 477
22.5. Safety and Environmental Concerns 478
22.6. Conclusions 479
22.7. References 479

23. Fluorinated Carbon

George A. Shia and Ganpat Mani

23.1. Introduction 483
23.2. Physical Properties 484
23.3. Chemical Properties 488
23.4. Manufacture 489
23.5. Applications 490
    23.5.1. Batteries 490
    23.5.2. Lubrication 492
    23.5.3. Imaging 494
23.6. Fluorine Intercalation Compounds of Graphite 495
23.7. Health/Safety 495
    23.7.1. Toxicity 495
    23.7.2. Dermal and Ocular Effects 496
    23.7.3. Cytotoxicity 496
    23.7.4. Handling of Fluorinated Carbons 496
23.8. References 496

24. Uses of Fluorine in Chemotherapy

Philip Neil Edwards

24.1. General Introduction 501
24.2. Fluoroquinolone Antibacterial Agents 502
    24.2.1. Introduction 502
    24.2.2. Early 6-Fluoro-7-piperazino-quinolones 503
    24.2.3. Mode of Action 505
    24.2.4. New Structural Types 506
24.3. Electronegativity and Its Consequences 509
    24.3.1. What Is Electronegativity and Can "It" Be Quantified? 509
24.4. Dipole Moments and Electron Density
    Distributions in Halogeno Compounds 511
24.5. Fluorine Substitution in Acid-Sensitive Compounds 512
24.6. Hydrogen Bonding in Fluoro Compounds 514
    24.6.1. Fluorine as Hydrogen-Bond Acceptor 515
    24.6.2. Halogeno-Substituent Effects on
        Hydrogen-Bond Donor and Acceptor Ability 520
24.7. Fluoroketone Enzyme Inhibitors 525
24.8. Inhibition of Pyridoxal-Dependent Enzymes 527
24.9. Drug Absorption and Distribution 530
    24.9.1. Fluorine Substitution in Relation to log $P_{octanol}$ Values 531

24.9.2. Solubility Effects 532
24.10. Fluorination Effects on
Metabolism and Elimination 533
24.11. The Size and Shape of Fluoro Substituents 536
24.12. References 539

25. Fluorinated Inhalation Anesthetics

Donald F. Halpern

25.1. Introduction 543
25.2. The Fluorine Revolution 544
25.2.1. Halothane 546
25.2.2. Methoxyflurane 546
25.2.3. Enflurane 547
25.2.4. Isoflurane 547
25.2.5. Sevoflurane 550
25.2.6. Desflurane 551
25.3. Conclusions and Acknowledgments 552
25.4. References 553

26. Properties and Biomedical Applications of
Perfluorochemicals and Their Emulsions

Kenneth C. Lowe

26.1. Introduction 555
26.2. Properties of PFCs 556
26.3. Synthesis of PFCs 558
26.4. Emulsification of PFCs 558
26.4.1. "First-Generation" Emulsions 558
26.4.2. Improved Emulsions 560
26.4.3. Surfactants 561
26.5. Biocompatibility Assessment 562
26.5.1. Tissue Uptake and Excretion 563
26.5.2. Effects on Lymphoid Tissues and the Reticuloendothelial System 563
26.5.3. Effects on Immunological Competence 563
26.5.4. Effects on Tissue Biochemistry 565
26.5.5. Other in vivo Responses 565
26.5.6. Cellular Effects in vitro 566
26.5.7. Effects of Surfactant(s) 567
26.6. Biomedical Applications 568

26.6.1. Microcirculatory Support in Ischemic Tissues 568
26.6.2. Coronary Angioplasty 569
26.6.3. Cancer Therapy 570
26.6.4. Contrast Media and Diagnostic Imaging 570
26.6.5. Respiratory Distress Syndrome 570
26.6.6. Decompression Sickness 571
26.6.7. Lung Damage and Respiratory Failure 571
26.6.8. Blood Diseases 572
26.6.9. Applications in Ophthalmology 572
26.6.10. Organ Perfusion and Preservation 572
26.6.11. Cell Culture Studies 572
26.7. Concluding Remarks 573
26.8. References 573

27. The Fluorochemical Industry
A. The Fluorochemical Industry in the United States

Richard A. Du Boisson

27A.1. The Manufacture of Hydrogen Fluoride in the United States 579
27A.1.1. Industrial Preparation of Hydrogen Fluoride 579
27A.1.2. U.S. Manufacturers of Hydrogen Fluoride 580
27A.1.3. Hydrogen Fluoride Production Statistics 580
27A.2. Industrial Uses of Hydrogen Fluoride 582
27A.2.1. Introduction 582
27A.2.2. Chlorofluorocarbons and Related Halogenoalkanes 583
27A.2.3. Aluminum Production 590
27A.2.4. Petroleum Alkylation 590
27A.2.5. Aqueous Hydrofluoric Acid 591
27A.2.6. Fluorine 591
27A.2.7. Uranium 591
27A.2.8. Chemical Intermediates 592
27A.3. References 592

27. B. Organofluorine Products and Companies in Western Europe

Peter Field

27B.1. Introduction 595
27B.2. Fluoroaliphatics 595
27B.2.1. Fluorocarbons 595
27B.2.2. Halogenofluorocarbons 596
27B.2.3. Trifluoroacetic Acid and Its Derivatives 597
27B.2.4. Trifluoromethanesulfonic Acid and Its Derivatives 598

27B.2.5. Higher (>C₂) Fluoroalkane Derivatives 599
27B.3. Fluoroaromatics 600
    27B.3.1. General Comments 600
    27B.3.2. Nuclear-Fluorinated Aromatics 601
    27B.3.3. Production of Nuclear-Fluorinated Aromatics in Europe 604
    27B.3.4. Side-Chain Fluorinated Aromatics 605
27B.4. Fluoropolymers 607
27B.5. Note Added in Proof: Developments Concerning Companies 607
27B.6. References 608

27.C. Manufacturers of Organic Fluoro Compounds in Japan 609

    Nobuo Ishikawa

28. CFCs and the Environment: Further Observations

    Richard L. Powell and J. Hugo Steven

28.1. Introduction 617
28.2. Refrigeration 618
28.3. The Ozone Layer Problem 621
28.4. Replacements for the Chlorofluorocarbons 622
28.5. Discovery of Antarctic Ozone Depletion 624
28.6 The Montreal Protocol 625
28.7. Replacements for the Chlorofluorocarbons: A Second Look 626
28.8. Commercial Production of Chlorofluorocarbon Replacements:
    The Third Period of Refrigeration 629
28.9. References 629

Index 631

1

Organofluorine Chemistry: Nomenclature and Historical Landmarks

R. E. BANKS and J. C. TATLOW

1.1. PREAMBLE

Fluorine—the *super*halogen—is by no means a rare element: found only in the form of its mononuclidic $^{19}_{9}F^-$ ion, it lies 13th in order of abundance of the elements in Earth's crust, and therefore outranks chlorine and the other members of the "salt-forming" family (Cl, 20th; Br, 46th; I, 60th; Ar, the rarest element).[1,2a] It is the most electronegative of all the chemical elements (Pauling values: F, 4.0; O, 3.4; Cl, 3.2; C, 2.6: H, 2.2) and easily the most reactive—a fact superbly underscored in the early 1960s by the direct synthesis of the noble-gas fluorides XeF_x ($x = 2,4,6$),[1] the difluoride being obtainable simply by exposing a mixture of xenon and fluorine to sunlight.[2b]

How to control the violence with which fluorine reacts with hydrocarbon material, and so provide a generally applicable route to compounds containing C—F bonds, became one of the master problems of organic chemistry once the French chemist Moissan had isolated fluorine in 1886.[2a] Even now only limited success in this quest can be claimed, and the vast edifice of modern organofluorine chemistry owes its existence mainly to indirect fluorination techniques based on the availability of hydrogen fluoride. As explained and exemplified in Chapter 3, fluorination of organic materials by whatever means often causes dramatic changes in physical properties and chemical reactivities, as today's amazingly diverse and constantly expanding commercial range of fluoroorganic products bears witness. Just how, by which companies, and for what purposes those products are manufactured comprise the main theme of this

R. E. BANKS • Department of Chemistry, The University of Manchester Institute of Science and Technology, Manchester M60 1QD, England. J. C. TATLOW • Editorial Office of the Journal of Fluorine Chemistry, 30 Grassmoor Road, Kings Norton, Birmingham B38 8BP, England.

Organofluorine Chemistry: Principles and Commercial Applications, edited by R. E. Banks *et al.* Plenum Press, New York, 1994.

book. The main objective of this first chapter is to provide a concise historical background to modern commercial organofluorine chemistry.

1.2. NOMENCLATURE

1.2.1. Highly Fluorinated Compounds

Naturally, organic compounds containing fluorine are named as hydrocarbon derivatives according to universal codes of practice. For highly fluorinated molecules, however, certain conventions have been adopted to facilitate rapid comprehension of structures. This section deals primarily with that aspect and also with the numerical codes used internationally to define simple chlorofluorocarbons (CFCs) and related compounds. Giving special attention to the nomenclature of organic compounds with a high fluorine content, i.e., per- and poly-fluorinated materials, stems from the fact that such compounds comprise a massively important group within the subject.

1.2.1.1. Perfluoro Nomenclature[2c,3,4]

Conventional use of Greek or Latin numeral roots to indicate the number of fluorine atoms present in a completely fluorinated organic compound is quite acceptable for C_1–C_4 aliphatics [e.g., tetrafluoroethene ($CF_2{=}CF_2$), pentafluoropropionic acid ($C_2F_5CO_2H$), octafluoropropane (C_3F_8)] or monocyclic aromatics [e.g., hexafluorobenzene (C_6F_6), pentafluoropyridine (C_5F_5N)]. With larger molecular systems, however, such names become cumbersome and do not reveal immediately that the compounds are fully fluorinated, e.g., dodecafluorocyclohexane (c-C_6F_{12}) or pentadecafluorooctanoic acid (n-$C_7F_{15}CO_2H$). In such cases, therefore, it is advantageous to use the prefix *perfluoro* in conjunction with standard hydrocarbon nomenclature (e.g., perfluorocyclohexane and perfluorooctanoic acid). This has been common practice for 40 years, and many authors use perfluoro nomenclature even for quite simple molecules, e.g., perfluoropropene ($CF_3CF{=}CF_2$).

No attempt has been made to standardize on either type of nomenclature in this book and, as often found, our authors have made judicious use of the perfluoro system. Note carefully, however, that the term *perfluoro* denotes substitution of *all* hydrogen atoms attached to carbon atoms *except* those whose substitution would affect the nature of the functional groups present. For example, perfluorobutyraldehyde is C_3F_7CHO (not C_3F_7COF which clearly is an acyl fluoride, perfluorobutyryl fluoride), perfluoropropionic acid is $C_2F_5CO_2H$ [not $C_2F_5C(O)OF$, an acyl hypofluorite], and perfluorodimethylamine is $(CF_3)_2NH$ [not $(CF_3)_2NF$, perfluoro-N-fluorodimethylamine].

Another point to watch is that perfluoro may refer to the whole word or to part of the word to which it is attached, but not to more than one word. Parentheses should be used where necessary to avoid any ambiguity; for example, cyclo-$C_6F_{11}CF_3$ must be named perfluoro(methylcyclohexane) not perfluoromethylcyclohexane, and $C_6F_5CF{=}CF_2$ (perfluorovinyl)benzene not perfluorovinylbenzene. Furthermore, the use of perfluoro excludes names in which this prefix is preceded by other prefixes. Thus, the compound $CF_3CFBrCF_3$ is called perfluoro-2-bromopropane (or 2-bromoheptafluoropropane) and not 2-bromoperfluoropropane. This limitation is imposed to avoid the implication that some atom other than hydrogen has been substituted. For this reason also, the convenient term fluorocarbon hydride strictly should not be used to define compounds of the type R_FH, where R_F = perfluoroalkyl (e.g., CF_3) or -cycloalkyl (e.g., cyclo-C_6F_{11}).

A convenient way to define the structure of a polyfluoro compound containing hydrogen *not* associated with a functional group is to use a prefix consisting of the numbers or symbols allocated to the carbon atoms bearing the hydrogen atoms, each followed by an italicized letter H. For example, $CF_3CHFCHF_2$ and $CHF_2CF_2CF_2CF_2CO_2H$ are named 1H,2H-hexafluoropropane and δH-octafluorovaleric acid (or 5H-octafluoropentanoic acid), respectively; this avoids having to use the cumbersome names 1,1,1,2,3,3-hexafluoropropane and 2,2,3,3,4,4,5,5-octafluoropentanoic acid. Compounds named according to this method should comply with the rule that ordinarily the number of hydrogen atoms not associated with the functional group must not exceed four, and the ratio of such hydrogen atoms to fluorine of atoms likewise not part of a functional group (e.g., COF or SO_2F) must not be greater than 1:3.

1.2.1.2.   *F*-Nomenclature

In the early 1970s, an alternative system was authorized by the American Chemical Society in which the symbol F conveys the sense of perfluoro.[5] According to "*F*-nomenclature" C_3F_7CHO is *F*-butyraldehyde, $C_2F_5CO_2H$ is *F*-propionic acid, and $(CF_3)_2NH$ is di(*F*-methyl)amine.   The prefix *hydryl* is used to denote replacement of fluorine by hydrogen in a perfluorocarbon system; for example, CF_3CHFCF_3 is 2-hydryl-*F*-propane. Other examples are as follows: $CF_3CF_2CF_2CF_2CF_2CF{=}CF_2$, *F*-hept-1-ene; $CH_3OCF_2CF_2CF_2CF_3$, *F*-butyl methyl ether; $CF_3OCF_2CF_2CF_2CF_3$, *F*-butyl *F*-methyl ether; $C_6F_5NO_2$, nitro-*F*-benzene. This system of nomenclature has not proved popular enough to be used in this book, except in an isolated case (Table 4, Chapter 25).

J.H. Simons, who invented the electrochemical fluorination process for the production of perfluorinated compounds (Chapter 5), disliked perfluoro terminology. Unsuccessfully, he advocated the adoption of a separate and distinct nomenclature for the field of perfluorocarbon chemistry based on placement of the syllable *for* before the final syllable in an otherwise standard name, e.g., methforane (CF_4) or ethforene ($CF_2{=}CF_2$).[6] Other important pioneers of fluorine chemistry proposed the forerunner of the *F*-system described above but used the Greek letter phi (e.g., φ-heptane, C_7F_{16}) instead of the Latin F.[7]

The practice of using a small subscript F (or F) in conjunction with the conventional symbols R and Ar to represent general cases of monovalent perfluoroalkyl (R_F) or perfluoroaryl (Ar_F) groups is very well established. Another useful notation is a placement of a large F in the center of ring structures to symbolize that all substituents not shown are fluorine.

Finally, the original practice of referring to compounds containing only carbon and fluorine as fluorocarbons has lapsed. Such materials are now called perfluorocarbons (PFCs) to make it quite clear that they are composed of only carbon and fluorine. The former term, fluorocarbons, is often used nowadays as an all-embracing name for organic fluorides.

1.2.2. Fluorocarbon Code Numbers

To avoid confusion and errors arising from the use of systematic names, manufacturers of relatively simple compounds of the CFC (chlorofluorocarbon), HCFC (hydrochlorofluorocarbon), and HFC (hydrofluorocarbon) classes decided, long ago, to adopt a naming system based on a numerical code. Initially this caused more confusion, because different companies often used different codes for the same compound. For example, the well-known refrigerant dichlorodifluoromethane, CF_2Cl_2, was referred to as Freon® 12 by Du Pont, Arcton® 6 by ICI, and Isceon® 122 by ISC. Eventually (1957), however, a common set of rules was adopted,[8] based on the so-called Du Pont system, so nowadays the numbers derived are prefixed either by a tradename or by single-letter alphabetic prefixes which reveal the applications area for a batch of material (R for refrigerant, P for aerosol propellant, S for solvent).

Devised originally in 1929 by the founding fathers of the CFC refrigeration industry (Henne, Midgley, and McNary), extended by another pioneer, Park,[9] standardized by the American Society of Refrigeration Engineers (ASRE Standard 34; 1957) for methane, ethane, and cycloalkane derivatives,[10] and still evolving unofficially, the coding system now includes fluoropropanes, fluorobutanes, fluoroalkenes, and fluoroethers (styled E-compounds).[10,11]

In brief, when reading a code from right to left the first digit defines the number of fluorine atoms; the second is the number of hydrogen atoms *plus one*; and the third is the number of carbon atoms *minus one*, and is omitted if zero (e.g., CF_2Cl_2 is coded 12 not 012); any discrepancy with a valency of four for carbon is made up with chlorine atoms after allowance has been made for any bromine present, which is indicated by a suffix made up of the letter B followed by the number of such atoms per molecule (e.g., CF_3Br is coded 13B1; note again the deletion of the zero for carbon). For cyclic derivatives the letter C is used before the numerical code, e.g., perfluorocyclobutane's number is C-318 (C_4F_8).

When positional (constitutional) isomers exist, deviation from symmetry is indicated by an alphabetical code. For example, there are two quite distinct compounds of molecular formula $C_2H_2F_4$, namely CHF_2CHF_2 (1,1,2,2-tetrafluoroethane) and CF_3CH_2F (1,1,1,2-tetrafluoroethane). The former is coded 134 according to ASRE nomenclature, and the latter 134a, the subrule being that when isomers exist, the most symmetrical one takes precedence over the others, which are allocated the letters a,b,c, ... as they become more unsymmetrical. Symmetry is determined by adding the atomic weights of the substituents attached to each carbon atom and subtracting one sum from the other (H = 1, F = 19, Cl = 35.5). In the case of trichlorotrifluoroethane ($C_2F_3Cl_3$), for example, $CF_2ClCFCl_2$ is CFC-113 and CF_3CCl_3 is CFC-113a; and for CH_3CFCl_2

which has three possible isomers, $CH_2ClCHFCl$ is HCFC-141, CH_2FCHCl_2 is HCFC-141a, and CH_3CFCl_2 is 141b. All of these ethanes have become important in the search for so-called ozone-friendly alternatives to CFCs (see Chapters 7 and 28).

Obviously, the situation is more complex when one considers fluoroalkanes containing three or more carbon atoms per molecule. The HCFC propane isomers $CF_3CF_2CHCl_2$ and CF_2ClCF_2CHFCl, for example, which are possible substitutes for S-113 ($CF_2ClCFCl_2$), are represented by the codes 225ca and 225cb, respectively. Derivation of the numerals follows the standard rule, while the second alphabetical suffix is derived from the symmetry rule applied only to the terminal carbons (C-1 and C-3). The first appended letter refers to the central carbon (C-2), the greater the total atomic mass of the substituents, the nearer the letter is to the beginning of the alphabet (CCl_2 = a, $CFCl$ = b, CF_2 = c, $CHCl$ = d, CHF = e, CH_2 = f). The 225cc isomer is $CFCl_2CF_2CHF_2$, and 2,3-dichloropentafluoropropane, $CF_3CHClCF_2Cl$, is 225da.

References 10 and 11 are replete with further examples and more detailed information on codes, and readers will be able to gain further expertise here by reading other chapters (6–8, 27, and 28). Note that a complication exists with bromofluorocarbons used as fire-extinguishing (FE) agents because the fire-fighting industry uses a numbering system [Halon (short for halogenated hydrocarbon) FE numbers] which reveal, reading from left to right, the number of carbon, fluorine, chlorine, and bromine atoms, respectively, in the molecules. For example, the most important extinguishants, CF_3Br and CF_2ClBr, are Halons 1301 and 1211, respectively (see Chapter 8).

1.3. HISTORICAL LANDMARKS

Detailed accounts of the development of organofluorine chemistry during the pioneering years 1835–1940,[2d] and of work done during World War II when fluorine chemistry as a whole came of age,[2e] can be found in a book published in 1986 to commemorate the centenary of Moissan's isolation of fluorine.[2] To provide continuity with other chapters in the present monograph, the slightly longer period 1835–1960 is covered here, albeit in briefer style, with emphasis on matters directly related to commercial developments.

1.3.1. The HF Problem

Hydrogen fluoride is the life blood of the fluorochemicals industry,[12] and from it, principally via halogen-exchange reactions, stems the vast array of today's commercial organic fluorides. Studies on the reaction used to produce anhydrous hydrogen fluoride (AHF) commercially [CaF_2 (native fluorspar) + $H_2SO_4 \rightarrow 2\ HF + CaSO_4$] date from 1764.[2a] In principle, today's manufacturing process is the same (though substantially improved in design and engineering terms) as that employed in 1931 to service the emerging CFC refrigeration industry in the U.S.[2f] Clearly, mastery of the preparation and handling of this highly dangerous chemical has been the key to the development of organofluorine chemistry, and rightly deserves attention here.

Benchmark events[2a] in hydrogen fluoride's early history are (i) Scheele's demonstration (1771) that sulfuric acid liberates a peculiar acid (Flussäure) from the metallurgical flux fluorospar (Latin *fluo*, I flow) which destroys glass; (ii) the preparation of fairly pure aqueous hydrogen fluoride (hydrofluoric acid) by Meyer and Wenzel (*ca.* 1780) using metal apparatus; (iii) the production of tolerably pure anhydrous hydrogen fluoride (AHF) by Gay-Lussac and Thénard (1808) and their observation of the extraordinary burns it inflicts on the skin; (iv) Ampère's deduction that hydrofluoric acid is analogous to hydrochloric acid (1810); and (v) the introduction of Frémy's method for making pure AHF, namely, thermal decomposition of potassium bifluoride (KHF_2) in platinum apparatus (1856). Moissan used Frémy's method to prepare the AHF used in his famous experiment on June 26, 1886, when he produced fluorine by electrolyzing hydrogen fluoride made conducting with potassium fluoride in a platinum–iridium cell cooled to $-50°C$.[2g]

That Moissan succeeded where so many other eminent 19th century chemists (including Humphry Davy and Michael Faraday) failed over a period of 75 years bears witness to the difficulties and serious hazards associated with the use of AHF, and to the great reactivity of fluorine once liberated.[2a]

Nowadays, of course, AHF (bp 19.5°C) is rarely made in the laboratory because samples of 99.8% purity can be purchased in mild steel cylinders, from whence it can be transferred via flexible copper tubing to chilled poly(ethylene) or poly(propylene) receptacles or metal reactors (mild steel, nickel or nickel-based alloys). Aqueous material (< 70% HF) can be purchased in poly(ethylene) bottles or drums (with steel overpacks). Neither form of HF should be used in any laboratory or location where adequate safety precautions cannot be exercised. It is absolutely essential that the protocol established includes arrangements for the special medical treatment required for "HF burns" and inhalation of "HF vapors."[2a]

1.3.2. Synthesis of C–F Bonds

Not only does fluorine form the strongest single bond to carbon encountered in organic chemistry, but also its spatial requirements are very moderate—especially when compared with the other halogens (F is the smallest substituent after H) (see Chapter 3). Only in the case of fluorine, therefore, can total substitution of hydrogen by halogen in *any* hydrocarbon or functionalized derivative thereof be contemplated. Furthermore, when the opportunities for stepwise conversion of hydrocarbon compounds to perfluorocarbon analogs are considered (e.g., $CH_3CO_2H \rightarrow CH_2FCO_2H \rightarrow CHF_2CO_2H \rightarrow CF_3CO_2H$), it is abundantly clear that considerably more organofluorine compounds are capable of existence than there are organic substances known today. This situation was perceived by pioneering fluorine chemists about 50 years ago, when it was estimated that a trillion (US) fluoroorganic compounds could be patterned structurally on the organic substances known at the time. With appropriate investment in the C—F area, it was argued, the chances of discovering materials of commercial value would be unprecedented.

Writing at the end of World War II, when publications dealing with the enormous advances made in fluorine chemistry under the sponsorship of the U.S. Government (the Manhattan Project) were being prepared, McBee pronounced: "Peacetime products containing fluorine are expected to include new and useful dyes, plastics, pharmaceuticals, lubricants, tanning agents, metal fluxes, fumigants, insecticides, fungicides, germicides, fire extinguishers, solvents, fireproofing compounds, heat transfer media, and other products of benefit to society".[13] How right he was!

1.3.2.1. Aliphatic Fluorides: The First Seventy Years

Most of the synthesis methodology required to underpin the modern organofluorochemicals industry was established before 1950. Progress was very slow initially, a gap of about 20 years separating the seminal synthesis of the first (and simplest) organic fluoride, methyl fluoride, by Dumas and Péligot in 1835[14] and the second, ethyl fluoride, by Frémy (1854).[2d] Both compounds were obtained by heating sulfates of the corresponding alcohols with potassium fluoride, e.g., $(CH_3)_2SO_4 + 2$ $KF \rightarrow 2\ CH_3F + K_2SO_4$.

Progress was not accelerated by Moissan's preparation of elemental fluorine because he failed to develop methods for controlling its reaction with hydrocarbon-based substrates. Attempts were made to fluorinate more than 40 compounds (including methane, ethene, chloroform, ethanol, benzene, pyridine, and even strychnine); the reactions needed no initiation and proceeded in catastrophic fashion, with flames and detonation, causing the formation of carbon and hydrogen fluoride.[15] Owing to the difficulty and expense of producing fluorine using a Moissan-type cell (corrosion resulted in a Pt loss-rate of 5 g per g of utilizable F_2)[2c] and the general fear—certainly not unwarranted—of handling AHF, studies on elemental fluorine were few in number for some years after Moissan's death in 1907. Significant progress in the utilization of fluorine to produce C—F bonds was not made until the 1930s. This contrasts markedly with the chlorocarbon situation, for by the time Moissan had isolated fluorine, direct chlorination as a route to organic chlorides was textbook stuff.[2d]

In collaboration with Meslans, however, Moissan did contribute significantly to C—F bond synthesis methodology through pioneering studies on halogen exchange between organic bromides or iodides and silver fluoride.[15,16] Meslans also used AgF, AsF_3, ZnF_2, and SbF_3 to prepare acid fluorides via halogen exchange. This extended the seminal work of the Russian composer-chemist Borodin who, in the early 1860s, carried out the first ever halogen-exchange ("halex") fluorination when he treated benzoyl chloride with Frémy's salt (KHF_2) in a platinum retort and obtained benzoyl fluoride.[17] Nowadays, halex fluorination is easily the most important general method for the synthesis of C—F bonds, and not least because of the priceless contributions of the Belgian chemist Frédéric Swarts during the period 1890–1930 (see Section 1.3.3).

1.3.2.2.   Ring-Fluorinated Aromatic Compounds

a.   The First Phase (1870–1925).   Thanks to the discovery of aromatic diazo compounds by Griess in the 1850s, aromatic fluorine chemistry was in far better shape than its aliphatic counterpart by the time Moissan isolated fluorine (1886). The first of numerous reports of successful diazonium-mediated syntheses of ring-fluorinated arenes began to appear in 1870 (Schmitt and von Gehren).[2d]

Fluorobenzene, the simplest and nowadays the most important (tonnagewise) nuclear fluoroaromatic, was first reported in 1883 by the Italians Paternò and Oliveri;[18] they used Lenz's method (1877)[2d] to procure *p*-fluorobenzenesulfonic acid then desulfonated it in now classical fashion: $4\text{-}^-\text{O}_3\text{SC}_6\text{H}_4\text{N}_2^+ \to$ (heat with conc. HF aq.) $4\text{-HO}_3\text{SC}_6\text{H}_4\text{F} \to 4\text{-KO}_3\text{SC}_6\text{H}_4\text{F} \to$ (heat with conc. HCl aq.) $\to \text{C}_6\text{H}_5\text{F}$.[18] The methodology then took a quantum leap in 1886 with the introduction of Wallach's method, namely, the diazopiperidide route,[19,20] which involves *in situ* thermal decomposition of arenediazonium fluorides generated by adding strong hydrofluoric acid to diazopiperidides, e.g., $\text{C}_6\text{H}_5\text{NH}_2 \to$ (with HCl aq. + NaNO_2 at 0°C) $\text{C}_6\text{H}_5\text{N}_2^+\text{Cl}^-$ aq. $\to$ (with piperidine + KOH aq. at 0°C) cyclo-$\text{C}_5\text{H}_{10}\text{N}$—N$=N\text{C}_6\text{H}_5 \to$ (with HF aq.) cyclo-$\text{C}_5\text{H}_{10}\overset{+}{\text{N}}\text{H}_2\text{F}^- + {}^+\text{C}_6\text{H}_5\text{N}_2{}^+\text{F}^- \to$ (heat) $\text{C}_6\text{H}_5\text{F} + \text{N}_2$. This method, occasionally still used today, was extended (with Heusler,[20] and independently by others) to produce numerous fluoroaromatics. Fluoroaromatic chemistry never looked back, and from 1900 onward the area started to become thoroughly systematized as studies on electrophilic substitution in fluorobenzene by Holleman and his group in Holland got underway.[21]

What is often overlooked is that, during his initial work on the diazopiperidide route, Wallach carried out the first HF-diazotization (as it is now called) of an aromatic amine when he added sodium nitrite to aniline in cold fuming hydrofluoric acid, and heated the mixture to give fluorobenzene in 20% yield (the diazopiperidide route afforded a 50% yield). This "direct" method was scaled up during the early 1940s in Germany (see later), and nowadays lies at the heart of major production facilities worldwide (see Chapters 9 and 27).

Note that by the time Wallach's work was carried out, hydrofluoric acid was an article of commerce, being shipped in lead or (for small quantities) gutta percha bottles (40–60% HF) or wooden barrels ($\leq 35\%$ HF); high-quality samples were transported in ceresine wax or platinum bottles. The English chemist George Gore who, during the period 1860–69, made a thorough study of AHF and hydrofluoric acid, regularly purified crude ("ordinary") hydrofluoric acid on a multigallon scale by distilling it in lead retorts fitted with platinum condensers and receivers. Such crude hydrofluoric acid was manufactured in England at the time chiefly for glass-etching and related purposes by heating fluorspar with sulfuric acid in iron retorts attached to water-filled leaden receivers (Meyer's method).[2a]

b.   The Balz–Schiemann Method.   The problem with Otto Wallach's method was that it involved using strong hydrofluoric acid. Clearly an alternative type of reaction was needed which could be carried out in standard laboratory glassware and did not involve the same HF-burn risk. Finding one was not easy. Eventually, in 1927, the German chemists Balz and Schiemann published a fine solution to the HF-problem, namely, a method based on diazonium tetrafluoroborates.[22,23] This brought aromatic fluorides well within the reach of any competent experimentalist, and fluoroaromatic chemistry took on a new lease of life.

The Balz–Schiemann method, one of the most important fluorination procedures ever discovered, involves two steps: preparation and isolation of a dry arenediazonium tetrafluoroborate ($ArNH_2 + HNO_2 + BF_4^- \rightarrow ArN_2^+BF_4^-$), followed by controlled thermolysis of that salt ($ArN_2^+BF_4^- \rightarrow ArF + N_2 + BF_3$), as described in detail in Chapter 9. Its virtues are extolled in excellent reviews,[23–25] one of which states: "The phenomenon which makes possible the Schiemann reaction is the remarkable stability of the dry diazonium fluoborates (sometimes called diazonium borofluorides). These salts, almost alone amongst the diazonium salts, are quite stable and insensitive to shock, and many can be handled safely in quantities of several kilograms. Most of them have definite decomposition temperatures, and the rates of decomposition, with few exceptions, are easily controlled. The overall yields in general are satisfactory. No special apparatus is required, and the inorganic fluoborates necessary as intermediates may be purchased or easily prepared".[24] Note the omission of Balz's name—a malpractice often encountered.

During World War II, IG Farben used the Balz–Schiemann reaction to manufacture the pharmaceutical intermediate 3-fluoro-4-methoxytoluene [en route to 3-fluorotyrosine ("Pardinon Bayer" for treatment of hyperthyreosis)[26]] and 4,4'-difluorobiphenyl, a component of the burn/wound ointment "Epidermin".[27]

c. HF-Diazotization Goes Commercial. Wallach's one-pot method for the synthesis of fluorobenzene, i.e., treatment of aniline with sodium nitrite in cold strong hydrofluoric acid, followed by *in situ* thermal decomposition of the benzenediazonium fluoride formed, was utilized on occasion by other workers prior to the introduction of the Balz–Schiemann reaction. The Russians Chichibabin and Rjazancev (1915), for example, used it to convert 2-aminopyridine into 2-fluoropyridine, the first ring-fluorinated heteroaromatic compound.[28] Industrial interest in the method was revealed in 1934 by the publication of an IG Farben patent in which the inventors Oswald and Scherer claimed the production of fluorobenzene in 87% via diazotization of aniline in AHF.[29]

During World War II, IG Farben built a developmental plant at Hoechst designed to produce 20 tonnes of fluorobenzene per month exclusively for use in the production of the DDT analog GIX [$2 C_6H_5F + CCl_3CHO\text{-}ClSO_3H \rightarrow (4\text{-}FC_6H_4)_2CHCCl_3$]. Owing to excessive corrosion of the cast iron reaction vessels, fluorobenzene production never exceeded 12 tpm, the average yield being 76% based on the feedstock aniline hydrochloride. Full details of the plant and its operation can be found in official British Intelligence Reports published in 1947.[30]

Nowadays, considerable industrial importance attaches to the Lenz–Wallach HF-diazotization route to ring-fluorinated aromatics (see Chapters 9 and 27). The

Balz–Schiemann method has not yet achieved the same prominence in commercial circles, but it is still the preferred method on laboratory or "multikilo" scales.

   d. Halex Fluorination. The first report on the use of anhydrous potassium fluoride to effect replacement of ring Cl by F in aryl chlorides suitably activated toward nucleophilic attack came in 1936 from Gottlieb (Victor Chemical Works, Illinois, U.S.).[31] He used dry KF in a solvent-free system at 205°C to convert 1-chloro-2,4-dinitrobenzene into its fluoroanalog, a compound prepared previously via nitration of 4-fluoronitrobenzene produced from 4-nitroaniline by Wallach's diazopiperidide technique (Holleman and Beekman; 1904).[32] During World War II Cook and Saunders (1947) improved Gottlieb's method by using nitrobenzene as a suspending agent for the KF, and more than doubled the yield of 1-fluoro-2,4-dinitrobenzene (from 30% to 71%).[33] Known nowadays as Sanger's reagent for the determination of *N*-terminal amino-acid residues in peptide chains,[34] 1-fluoro-2,4-dinitrobenzene was of interest to Bernard Saunders as part of his wartime studies on the physiological action of fluoroorganic compounds (see Section 1.3.6.2).[35]

   The appearance in 1956 of a benchmark paper from Finger and Kruse on Halex fluorination of chloroaromatic substrates (which included information on replacement of NO_2 by F) in hot dipolar "aprotic" (nonhydrogen-bond-donor) solvents, such as DMF and DMSO, heralded the move toward commercialization of aromatic halex fluorination (see Chapter 9) and the development of routes to *per*fluorinated aromatics (see Section 1.3.2.2.f). Finger's group at the Illinois State Geological Survey made outstanding contributions to fluoroaromatic chemistry over many years.[36b]

   e. Side-Chain Fluoroaromatics. Benzotrifluoride is easily the most extensively studied fluoroaromatic compound,[37] having been used commercially since the 1930s as an intermediate for the production of dyestuffs, agrochemicals, or pharmaceuticals. Nearly a century ago,[38] Swarts showed how easy it is to effect side-chain (benzylic) halogen exchange in aromatic systems, benzotrichloride being found to react sequentially with antimony(III) fluoride: $C_6H_5CCl_3 \rightarrow C_6H_5CFCl_2 \rightarrow C_6H_5CF_2Cl \rightarrow C_6H_5CF_3$; and through electrophilic nitration of benzotrifluoride he established the *meta*-directing effect of the CF_3 group in aromatic systems. Reduction of the nitro derivative gave 3-trifluoromethylaniline, while catalytic hydrogenation of benzotrifluoride gave (trifluoromethyl)cyclohexane, revealing the reductive stability of the CF_3 group.[39]

   Patents dealing with the use of AHF to effect the conversion $ArCCl_3 \rightarrow ArCF_3$ began to appear in the early 1930s,[40] when IG Farben began to invest in the development of trifluoromethylated intermediates for the production of dyestuffs with improved light and wash color clarity.[41] By the time World War II commenced, four vat dyes had been approved for marketing, including Indanthren® Blue CLB—known as airforce gray.[42]

   Processes for the production of bis(trifluoromethyl)benzenes required as feedstocks for the manufacture of saturated perfluorocarbons by the cobalt fluoride process (see Section 1.3.5) were developed during World War II by McBee's Manhattan Project team.[43,44] AHF at 110–120°C and 90–100 atm pressure in a 110-gallon steel autoclave was used to effect the conversion o/p-$C_6H_4(CCl_3)_2 \rightarrow o/p$-$C_6H_4(CF_3)_2$. Milder reaction conditions and shorter times could be used when the AHF was used in conjunction with Swarts-type catalysts (SbF_3 or $SbCl_5$), but corrosion problems increased. Fluorination of 1,3,5-tris(trichloromethyl)benzene with AHF at 200°C gave the corresponding tris(trifluoromethyl) compound in 49% yield.[45] McBee's group also prepared the first side-chain fluorinated heterocycles [including 5-chloro-2-(trifluoromethyl)pyridine and 2,4,6-tris(trifluoromethyl)-1,3,5 triazine] via halex fluorination of the corresponding trichloromethyl compounds with AHF or AHF–$SbCl_5$.[46]

f. Perfluorinated Aromatics. The chemistry of perfluorinated aromatic compounds has been developed almost entirely since 1955, the year in which the first practical synthesis of hexafluorobenzene was disclosed, namely, pyrolysis of tribromofluoromethane obtained by Swarts-fluorination of tetrabromomethane: $CBr_4 \rightarrow$ (with SbF_3–Br_2 at 125°C) $CFBr_3 \rightarrow$ (flow pyrolysis in Pt at 635°C) C_6F_6.[47] Amazingly, this method was discovered by Mlle. Désirant in the mid-1930s while studying in Swarts's laboratory at the University of Ghent (Gand), and a preliminary communication was deposited "sous plis cacheté" at the offices of the Belgian Royal Academy on 15 December 1936.[47] A brief footnote in the full paper, which appeared in 1958,[48] states simply that the work could not be continued. Swarts retired from his teaching post in 1936 but worked in his laboratory until a few months before his death on September 6, 1940.

Before authorizing publication of her work, Désirant revealed the discovery to Leo Wall, Chief of the Polymer Chemistry Section at the U.S. National Bureau of Standards; his group proceeded to show that the yield of C_6F_6 can be increased from 45% to 55% by conducting the pyrolysis of $CFBr_3$ at elevated pressure (4.5 atm) rather than atmospheric.[49] The method is no longer used, having been rendered obsolete at the end of the 1950s when, independently, Banks and Tatlow discovered that octafluorocyclohexadienes produced via CoF_3-fluorination of benzene can be aromatized with hot iron or nickel to give hexafluorobenzene in good yield.[50] Subsequently (1963), Yakobson's group in Russia introduced halex fluorination of hexachlorobenzene with KF as a route to hexafluorobenzene and chloropentafluorobenzene.[51] Both methods have been commercialized (see Chapter 9).

For the record, however, the first literature on aromatic perfluorocarbons appeared in 1947, when McBee's group reported that hexafluorobenzene and octafluorotoluene can be obtained in very poor yields (*ca.* 5% overall) by tedious and hazardous stepwise fluorination of hexachlorobenzene and pentachlorobenzotrifluoride, respectively, followed by dehalogenation of the products, e.g., $C_6Cl_6 \rightarrow$ with (BrF_3) $C_6Br_2Cl_4F_6 \rightarrow$ (with SbF_5) $C_6BrCl_4F_7 \rightarrow$ (with Zn) C_6F_6.[52] Also, the first example of a cobalt fluoride-based route to highly fluorinated aromatics was reported in 1956 by Tatlow's group. The CoF_3 was used to convert benzene to polyfluorocyclohexanes, which were dehydrofluorinated to re-create aromatic systems: $C_6H_6 \rightarrow C_6H_3F_9 \rightarrow$ (with KOH) C_6F_6; $C_6H_6 \rightarrow C_6H_4F_8 \rightarrow$ (with KOH) C_6HF_5.[53]

1.3.3. Aliphatic Fluorides Go Commercial

1.3.3.1. The Work of Frédéric Swarts

The first ventures into the subdiscipline of aliphatic fluorine chemistry (summarized in Section 1.3.2.1) gave no hint of the novel and extensive chemistry awaiting discovery. The pioneer who laid firm foundations in this area was F. Swarts, who systematically developed routes to polyfluorides at the University of Ghent (Belgium).[21] Starting around 1890, he carried out halogen exchange on a number of polychlorides and polybromides, using as his halex reagent antimony trifluoride with added bromine (SbF_3Br_2). Typical early conversions were $CCl_4 \rightarrow CFCl_3$; $CHCl_3 \rightarrow CHFCl_2$; and $CHBr_2Cl \rightarrow CHFBrCl$. Bromofluoroethanes such as $CFBr_2CHFBr$, CF_2BrCH_2Br, and CHF_2CH_2Br were made from appropriate bromoethanes. A mixture of SbF_3 and $SbCl_5$ (behaving as SbF_3Cl_2) later became his standard exchange reagent, the first sequence reported being $CHCl_2CHCl_2 \rightarrow CHFClCHCl_2 \rightarrow CHF_2CHCl_2 \rightarrow CHF_2CHFCl$. Two very important general reactions of these fluorohalogenoethanes were discovered, namely, olefin formation via zinc dehalogenation or base-induced dehydrohalogenation. Preferential elimination of a halogen other than F was shown to occur, e.g., $CHFClCHCl_2 + Zn \rightarrow CHF{=}CHCl$, and $CF_2BrCH_2Br + K_2CO_3 \rightarrow CF_2{=}CHBr$ (alkoxides, such as $NaOCH_3$, were often used as bases, but gave ethers as well as olefins). Swarts made mono-, di-, and trifluoroethylene by these methods. Fluorinated carboxylic acids RCO_2H, where R = CHF_2, CF_2Br, CF_2Cl, $CFBr_2$, $CFClBr$, CH_2F, $CHFBr$, and $CHFI$, were synthesized, either by oxidative fluorination of haloacyl halides or by direct chlorination/bromination of C—H bonds. This work established the principle that clusters containing fluorine are fairly stable and that, in competitive situations, C—F bonds do not usually break as readily as do other halogen–carbon bonds.

Using halex reagents then available, no aliphatic precursor was found to give a product containing a CF_3 group. By contrast benzotrifluoride ($C_6H_5CF_3$) was easily made from the corresponding chloroarene using SbF_3 alone (Section 1.3.2.2.e); clearly, the arene ring promoted the exchange of side-chain halogen. After World War I, Swarts carried out further work in this area and synthesized, for example, trifluoromethylcyclohexane and m-aminobenzotrifluoride. These were oxidatively degraded to produce trifluoroacetic acid, effectively demonstrating the stability of the CF_3 group. Trifluoroacetic acid was the strongest organic acid known at that time. It was the first true functionalized perfluorocarbon system, and via reactions of the carboxyl group Swarts converted it into a range of derivatives, including salts, esters, amide, anhydride, nitrile, and acid fluoride. Further sequences afforded products CF_3X, where X = $C(OH)(CH_3)_2$, $CH(OH)CH_3$, CH_2OH, $C(O)CH_3$, and $C(O)CH_2CO_2CH_3$. All had unusual properties and enhanced acidities.[21]

Not only was Swarts undoubtedly a synthetic chemist *par excellence*, but also he made precise physicochemical measurements on his products, thus providing thermochemical data and information on intermolecular forces. His outstanding solo contribution to organofluorine chemistry will surely never be surpassed. He always worked alone, hence none of his papers carry the name of a co-author. He received honors from his contemporaries,[2d] but his best memorial is the extent to which the foundations he laid have been extended and built on, as exemplified in many places in this book.

1.3.3.2. Chlorofluorocarbons (CFCs)

Prior to World War II, many leading organic chemists published work involving fluoroarenes, more so after the introduction of the Balz–Schiemann reaction. In contrast, the isolated work of Swarts in the aliphatic area was regarded as very specialized and distinct from mainstream organic chemistry—an attitude that plagues fluorocarbon chemistry to this day! It is ironic therefore that this little-recognized area was the starting point for a large-scale industrial effort that provided an answer to a serious technological deficiency.

The situation in the refrigeration industry in the 1920s is summarized later (Chapter 28). While searching for a new and inert working fluid, Midgley, Henne, and McNary of the Frigidaire Company (General Motors) deduced that the answer might lie among the aliphatic fluorides, despite a general belief that such compounds were toxic. Having no faith in the then recorded boiling point of −15 °C for carbon tetrafluoride (see Section 1.3.4.1) they preferred to attempt the synthesis of dichlorodifluoromethane by a Swarts-type reaction, $CCl_4 + SbF_3–SbCl_5 \rightarrow CF_2Cl_2$. They acquired all the SbF_3 to be found in the U.S. at that time (5 × 1 oz bottles), and used the material from one bottle to make a sample of the desired product, which proved to be inert and nontoxic. Subsequent reactions, however, produced toxic material. Fortunately, this was shown to be due to the presence of phosgene, which arose because all the other bottles were contaminated with a double salt having water of crystallization! Once phosgene had been washed out, the excellent properties of CF_2Cl_2 (later called Freon® 12 by Du Pont) were revealed. The basic idea was sound, but it was by good luck that it was not stultified, and from this chance the massive CFC industry arose.[2d,10,54] (Further developments are covered in Chapters 6 and 7.) Interestingly, Swarts himself did not report this exact fluorination, but characterized CF_2Cl_2 as a minor product of a reaction between CHF_2CH_2OH and Cl_2. However, Midgley and Henne clearly acknowledged that his pioneering work was the basis for the manufacture of CF_2Cl_2.[54]

1.3.3.3. Another Fortunate Accident: The Discovery
of Poly(tetrafluoroethylene) (PTFE)

In 1938, R. J. Plunkett, a Du Pont chemist, was using tetrafluoroethylene (TFE) as an intermediate in a program aimed at the synthesis of new refrigerants.[55] About 45 kg of TFE had been made via the reaction $CF_2ClCF_2Cl + Zn \rightarrow CF_2{=}CF_2$ (bp −76°C) and then stored in several dozen small steel cylinders. However, when the valve of one of these was opened, no gas emerged. Dismantling of the valve revealed that a white powder had been formed; samples were present also in the other cylinders, one of which was sawn in half to recover the contents. The powder was identified as the homopolymer of TFE and found to possess remarkable properties. Further development and commercialization of the polymer ("Teflon") is discussed later in this Chapter (Section 1.3.4.2) and in Chapter 15. Plunkett was not expecting the TFE to polymerize, nor did he know then of the inherent dangers associated with the handling and storage of this monomer (under certain circumstances it decomposes explosively to $C + CF_4$).[55] His discovery—of such great importance to mankind, for PTFE now touches peoples' everyday lives in numerous ways—is one of the best known examples of scientific serendipity.

1.3.4. Perfluorocarbons

1.3.4.1. The Preliminaries (1890–1937)

Prior to 1937, only three properly characterized perfluorocarbons had been reported in the literature: carbon tetrafluoride, hexafluoroethane, and tetrafluoroethylene[4]; this situation stemmed mainly from the difficulties, costs, and hazards associated with the generation and use of fluorine at the time.[2d] Three perfluorocarbon derivatives: CF_3CO_2H, CF_3NO, and CF_3NF_2, were also known, the first having been made by Swarts (1920) during his work on benzotrifluoride (see Section 1.3.3.1), and the last two in Otto Ruff's famous German school of inorganic fluorine chemistry via the action of fluorine on silver cyanide contaminated with silver nitrate (1936).[56] As explained earlier (Section 1.3.2.2.f), the aromatic perfluorocarbon C_6F_6 had been synthesized by Désirant, but that information remained unpublished until 1955.

Not unexpectedly, the first report of a perfluorocarbon came from Moissan (1890),[57] who thought that he had separated carbon tetrafluoride from volatile material produced by igniting powdered carbon in fluorine. However, the boiling point ($-15°C$) he quoted for CF_4 is hopelessly incorrect (today's value is $-129°C$), and the credit for the isolation of this, the simplest perfluorocarbon, belongs to his compatriots Lebeau and Damiens, who reinvestigated the fluorination of carbon in the 1920s.[58] Ruff and Keim carried out similar work in Germany, and both sets of researchers believed that the CF_4 was accompanied by its higher homologs. However, neither group was able to prepare sufficient material for proper investigation owing to the frequent explosions they encountered when treating carbon with fluorine. Eventually, Ruff's group discovered that graphite absorbs fluorine under certain conditions to give so-called carbon monofluoride $[CF]_n$, an intercalation compound that decomposes violently when heated rapidly to produce clouds of soot, carbon tetrafluoride, and higher perfluorocarbons.[59]

Just after World War II, the Rüdorffs and Wadsworth of ICI (UK) revealed independently that they had confirmed Ruff's work; methods for the preparation of "graphite fluoride" ranging in composition from $CF_{0.678}$ to $CF_{1.04}$ were described, and X-ray structural data were presented.[60] The Rüdorffs also showed that in the presence of hydrogen fluoride, fluorine combines with carbon at room temperature to give black material of composition C_4F (tetracarbon monofluoride).[60] So-called super-stoichiometric graphite fluoride of composition $[CF_x]_n$, where $x > 1$, is snow-white in color. Commercial samples of graphite fluorides first became readily available in the 1970s (see Chapter 23).

Swarts obtained the first pure sample of hexafluoroethane (bp $-78°C$) via electrolysis of aqueous trifluoroacetic acid (1930).[61] Shortly afterward (1933),[62] Ruff and Bretschneider reported that decomposition of carbon tetrafluoride in an electric arc struck between carbon electrodes had given tetrafluoroethylene (bp $-76°C$), which they separated from the co-products (mainly C_2F_6) by converting it to the easily isolated dibromide CF_2BrCF_2Br and then regenerating it with zinc suspended in ethanol. This work provided the first authentic sample of a perfluorinated olefin. Almost immediately (1934) a paper appeared from Henne's group in the U.S. describing the synthesis of tetrafluoroethylene by zinc dechlorination of Freon® 114 (CF_2ClCF_2Cl) produced via a Swarts-type fluorination of hexachloroethane with SbF_3Cl_2.[63] The stage was now set for the serendipitous discovery of poly(tetrafluoroethylene) by Plunkett in 1938 (see Section 1.3.3.3).

1.3.4.2. "Joe's Stuff"

In 1937,[64] the Americans Simons and Block reported briefly how, through inadvertent use of an amalgamated copper tube as a reactor, they had discovered that mercury(II) fluoride can promote a smooth reaction between carbon and fluorine to give a range of perfluorocarbons; full details appeared in 1939.[65,66] Passage of fluorine over Norite or sugar charcoal impregnated with catalytic amounts of mercury(I) or (II) chloride at temperatures just below red heat was bound to give—"steadily and without explosions"—a complex mixture of saturated perfluorocarbons boiling over the range -128 to $160°C$. From this Simons and Block isolated CF_4, C_2F_6, the new compound C_3F_8 (bp $-38°C$), and fractions with analyses corresponding to C_4F_{10} (at least two isomers, bp -4.7 and $3°C$), C_5F_{10} ($23°C$), C_6F_{12} ($51°C$), and C_7F_{14} ($80°C$); the last three fractions were thought to comprise cyclic compounds. All the materials possessed great chemical and thermooxidative resistance.

Thus was established the benchmark fact that both open and closed chains of CF_2 groups are perfectly stable—a conclusion confirmed emphatically in 1938 by Plunkett's discovery of poly(tetrafluoroethylene) (Section 1.3.3.3). The high thermooxidative stability of Simon's perfluorocarbons, coupled with their unusual physical properties (see Chapter 3), clearly differentiated them from their hydrocarbon analogs.

In July 1940, Joe Simons sent almost his entire stock (*ca.* 2 cm^3) of liquid perfluorocarbon material to Columbia University, where Manhattan Project scientists established its ability to resist attack by the highly reactive, volatile uranium derivative UF_6.[2e,67] All materials tested previously had been found to undergo degradation. Consequently, a huge secret effort was mounted to produce perfluorocarbons (codenamed "Joe's Stuff") useful as buffer gases, coolants, lubricants, and sealants in gas-diffusion plants for the concentration of the U-235 isotope required for the development of an atomic bomb.

Development work at Du Pont on poly(tetrafluoroethylene) (PTFE) required for the fabrication of gaskets, valve packings, reactor linings, and pipes inert toward fluorine and UF_6 also forged ahead; and methods for the production of both low and high polymers of chlorotrifluoroethylene were researched by W.T. Miller in a successful drive to discover inert oils and thermoplastics from which seals, gaskets, and tubing could be more easily fabricated than from PTFE.[2e]

1.3.4.3. Use of Elemental Fluorine

In the mid-1930s Miller had studied under L. A. Bigelow, who became renowned for his work on the so-called direct fluorination of organic compounds, i.e., using F_2.[2d] Following important work by German chemists [Bockemüller (1933), Fredenhagen and Cadenbach (1934)],[2d] Bigelow's group devised methods for controlling the action of fluorine on hydrocarbons in the vapor phase; benchmark achievements around the end of the decade were the synthesis of the first perfluorinated ketone, hexafluoroacetone, from acetone at 60°C, the smooth conversion of ethane at room temperature to a mixture of C_2F_6, CF_3CHF_2, CH_3CHF_2, and CHF_2CH_2F, and the production of perfluorocyclohexane from benzene at 90°C. By 1940, Bigelow was writing modern free-radical chain mechanisms to explain his fluorination results.[68]

Overcoming the fluorine-ignition phenomenon which had so frustrated Moissan (and others) was clearly shown by Bigelow's group to depend crucially on controlling the rate at which the conversion of hydrocarbon material was allowed to proceed, and on efficient dissipation of the large amounts of heat generated. The German chemist Bockemüller had argued in 1933 that provision of adequate heat sinks was the key to controlling direct fluorination reactions; having computed that the heats of reaction for the highly exothermic changes $C{-}H + F_2 \rightarrow C{-}F + HF$ and $C{=}C + F_2 \rightarrow CF{-}CF$ are -102.5 and -107.2 kcal mol^{-1}, respectively, he pointed out that such amounts of energy are much higher than those encountered in chlorination and ample to disrupt C—C linkages. Bückemüller used the "old" value of 66.8 kcal mol^{-1} for the dissociation energy of F_2; "modern" values became available in the 1950s, but use of today's datum (38 kcal mol^{-1} for $\Delta H_{dissoc.}$ of F_2) and other contemporary bond-energy terms hardly affects the outcome of Bockemüller's calculations.[2c]

Based on the work done by Bigelow's group, a team assembled at Columbia University's War Research Laboratories in January 1942 developed technology for the preparation of perfluorocarbons in moderate-to-high yield by allowing hydrocarbon vapors and fluorine, each diluted with nitrogen, to mix gradually in heated vertical metal reactors packed with silver-plated copper turnings.[70] This relatively uncomplicated method proved valuable in providing perfluorocarbon samples for testing in the early stages of the Manhattan Project, and showed promise for the production of fluorinated lubricants. Nevertheless, the simultaneous development of the two-stage vapor-phase cobalt trifluoride process by researchers at Johns Hopkins University in Baltimore indicated that this was the method of choice for the production of $C_6{-}C_8$ perfluorocarbons in quantity: fluorine + CoF_2 (from $CoCl_2$ and HF) $\rightarrow CoF_3$; hydrocarbon + $CoF_3 \rightarrow$ perfluorocarbon + CoF_2 (reconverted to CoF_3).[71] The enthalpy change in the actual hydrocarbon fluorination step (C—H + 2 CoF$_3$ → C—F + HF + 2 CoF$_2$) was estimated to be *ca* –52 kcal mol^{-1}, i.e., roughly half of that involved in direct fluorination. Herein lies the success of the cobalt fluoride method, which nowadays is a well-known commercial method for manufacturing perfluorocarbons (Chapter 4).

1.3.5. Wartime Advances: The Manhattan Project[2e]

In the summer of 1942 the U.S. Army Corps of Engineers organized the Manhattan District as the unit which would control the procurement and engineering aspects of the American nuclear weapons program. Commanded by General Leslie Groves, the enterprise became known in the annals of warfare as the Manhattan Project. Britain's wartime nuclear effort was organized under the code name Tube Alloys, a body managed for the Government by Wallace Akers, the Research Director of ICI.[72]

The remarkable feats of engineering that led to the use of atomic bombs with such devastating effect against Japan in August 1945 have been thoroughly documented. Much less attention has been paid to the key role played by fluorine chemistry in the realization of nuclear weaponry, as emphasized in Goldwhite's recent review devoted to that aspect of the Manhattan Project.[2e] Correspondingly, fluorine chemistry's contribution to the development of peaceful applications of nuclear power have been absolutely indispensable. Meshri estimated in 1986 that 45,000 tonnes of uranium will be needed in 1995 to satisfy global demand for UF$_6$.[2f]

That fluorine chemistry came of age during the course of the Manhattan Project is clearly revealed in the famous March 1947 issue of *Industrial and Engineering Chemistry*, which was devoted to wartime developments in the subject; the contents of Volume I of Simon's *Fluorine Chemistry* serve to emphasize that statement.[66] In particular, the generation and handling of fluorine on an industrial scale had been mastered; large-scale perfluorocarbon production through direct or indirect (via CoF$_3$) fluorination of hydrocarbons had been developed; and Simon's electrochemical fluorination route (ECF) to perfluorocarbon derivatives had been invented [e.g., RCOF → R$_F$COF, RSO$_2$F → R$_F$SO$_2$F, R$_3$N → (R$_F$)$_3$N]. Quite obvious was the fact that a new and potentially very large branch of organic chemistry based on C—F rather than C—H bonds had been discovered—and one which offered glittering prizes in both commercial and academic circles.

1.3.6. Postwar Progress

1.3.6.1. Developments in Synthesis Methodology

Immediately postwar, a considerable effort got underway, mainly in the U.K., U.S., and U.S.S.R., to develop the chemistry of perfluorocarbon derivatives, i.e., compounds of type R$_F$X where R$_F$ is a perfluoroalkyl group and X is a familiar functional group of "normal" (hydrocarbon-based) organic chemistry, e.g., I, CHO, CO$_2$H, CN, OH, SH, SO$_3$H, NH$_2$, NO$_2$, MgI. Direct functionalization of saturated perfluorocarbons was not a synthesis option, of course, owing to the inertness of these compounds. Hence progress had to be based firmly on the Swarts–Henne halogen-exchange methodology developed prior to World War II, coupled with wartime extensions [e.g., $C_6Cl_6 \rightarrow$ (with SbF_5) $CF_2(CF_2)_3CCl{=}CCl \rightarrow$ (with $KMnO_4$) $HO_2C(CF_2)_4CO_2H \rightarrow$ $\rightarrow \rightarrow NC(CF_2)_4CN$].

Simons AHF-based ECF method also played a major role. Discovered at the beginning of World War II, this technique was not reported in the open literature until 1949 for reasons of national security, by which time the 3M Company in the U.S. had long since acquired the commercial rights and developed the technique to pilot plant level.[73] Importantly, ECF was used commercially to prepare trifluoroacetic, pentafluoropropionic, heptafluorobutyric, and perfluorooctanoic acid from the corresponding alkanoyl chlorides or fluorides: $C_nH_{2n+1}COX \rightarrow C_nF_{2n+1}COF \rightarrow C_nF_{2n+1}CO_2H$ ($n = 1,2,3,8$; X = Cl,F).

Synthesis of the first perfluoroalkyl iodides CF_3I (from CI_4 and IF_5) and C_2F_5I (from $CI_2{=}CI_2$ and IF_5) by Emeléus and his co-workers at Cambridge University (UK) in the late 1940s unlocked the door to the chemistry of perfluorinated organometallic and organometalloidal compounds—a very rich field of study. The first compounds synthesized were mercurials (R_FI + Hg/UV light or heat $\rightarrow R_FHgI$; $R_F = CF_3$, C_2F_5),[74] and many more types were to follow from the Cambridge laboratories [e.g., $(CF_3)_3P$, $CF_3S_2CF_3$, CF_3SeCF_3].[4] Progress accelerated when, independently, Henne and Haszeldine found that perfluoroalkyl iodides could be synthesized via the Hunsdiecker reaction [$R_FCO_2Ag + I_2$ (heat) $\rightarrow R_FI + CO_2 + AgI$].[75] These workers prepared the first Grignard reagents of the perfluoroalkyl class: R_FI + Mg (in Et_2O at $-30°C$) $\rightarrow R_FMgI$ ($R_F = CF_3$, C_2F_5, n-C_3F_7).[76]

Haszeldine went on to develop perfluoroalkyl iodides widely as tools for the synthesis of numerous functionalized fluorocarbons [e.g., R_FI + CH≡CH (heat) $\rightarrow$ $R_FCH{=}CHI \rightarrow$ (with KOH) $R_FC{≡}CH$]; as part of this work he pioneered in the commercially useful area of free-radical addition of the elements of R_F—I across carbon–carbon double bonds, using a variety of olefin types, including perfluorocarbon [e.g., CF_3I (UV light or 220°C) $\rightarrow CF_3{\cdot} + I{\cdot}$; $CF_3{\cdot} + CF_2{=}CF_2 \rightarrow CF_3CF_2CF_2{\cdot}$ $\rightarrow$ (chain transfer with CF_3I) $CF_3CF_2CF_2I + CF_3{\cdot} \rightarrow$ cycle repeats; *or* with an excess of tetrafluoroethylene, $CF_3CF_2CF_2{\cdot} + n$-1 $CF_2{=}CF_2 \rightarrow CF_3[CF_2CF_2)_n{\cdot}$ (propagation) $\rightarrow$ (chain transfer with CF_3I) $CF_3[CF_2CF_2]_nI + CF_3{\cdot} \rightarrow$ etc.].[77]

Perfluoroalkanecarboxylic acids and perfluoroalkyl iodides were soon firmly established as premier building blocks for the synthesis of molecules containing perfluorocarbon groups; and with fluorinated alkenes—already established as intermediates of immense potential through wartime work in the U.S. and Germany—they rapidly became the "Blessed Trinity" of perfluorocarbon synthesis.[4] Of particular historical note is the early (1940s) work on the thermal homo- and co-dimerization of tetrafluoroethylene [e.g., $C_2F_4 \rightarrow$ perfluorocyclobutane; $C_2F_4 + CH_2{=}CHR$ (R = alkyl, aryl, vinyl) $\rightarrow CF_2CF_2CH_2CHR$],[78] and on base-catalyzed addition of alcohols to this and other 1,1-difluorinated ethenes [e.g., $CF_2{=}CFX + C_2H_5OH$—C_2H_5ONa $\rightarrow C_2H_5OCF_2CHFX$ (X = F, Cl)].[79]

The considerable postwar attention paid to the development of chain-growth (addition) fluoropolymers, particularly in the U.S., led to much interest in free-radical attack on fluoroolefins. Du Pont dubbed its PTFE "Teflon" in 1945, and in 1950 brought on stream the world's first full-scale plant near Parkersburg, West Virginia, where the Washington works is still the major source of PTFE. Top-secret wartime negotiations between Kinetic Chemicals, Du Pont, and ICI led to the transfer of scientific and technological information which enabled the British firm to embark on the manufacture of tetrafluoroethylene (TFE monomer) and its Fluon-brand homopolymer. Almost 5 tonnes of Fluons® were produced in a semitechnical plant at ICI's General Chemicals Division (Widnes, Cheshire) during the period April 1947–August 1948; Fluon® production at the present Hillhouse site (near Blackpool, Lancashire) commenced in the 1950s, a 200-tpa plant being commissioned there in 1956.[55] Du Pont researchers had invented the only significant commercial route to TFE monomer during the war, namely, pyrolysis of the refrigerant HCFC-22 prepared using Swarts–Henne halogen-exchange methodology: $CHCl_3 \rightarrow$ (with $AHF/SbCl_5$) $CHF_2Cl \rightarrow$ (at 700°C) $HCl + :CF_2 \rightarrow CF_2{=}CF_2$.[80] This is carbene chemistry on a truly grand scale!

By the end of the 1950s, publications detailing progress in the chemistry of polyfluoroaliphatic compounds were becoming a regular feature of the chemical literature. Not so with polyfluoroaromatic compounds—a situation arising from Mlle. Désirant's failure to take action concerning publication of her note describing the synthesis of hexafluorobenzene from $CFBr_3$, and the unattractiveness of McBee's wartime routes to hexafluorobenzene and octafluorotoluene (Section 1.3.2.2.f). In fact, the chemistry of perfluorinated aromatic compounds has been developed almost entirely since 1955, when the "$CFBr_3$ route" was published. Progress was slow at first, but commercialization of the new Banks–Tatlow fluorination–defluorination route to fluoroaromatics[50] by the Imperial Smelting Corporation (UK)[81] at the end of the 1950s led to a massive surge of activity in both academic and industrial circles. The method was used independently by the Manchester and Birmingham groups in 1959 to synthesize pentafluoropyridine [ECF of pyridine $\rightarrow CF_2(CF_2)_4NF \rightarrow$ (with hot Fe) C_5F_5N].[82] Not long afterward (1963), the KF-halex routes to pentafluoropyridine [$C_5Cl_5N + 5KF \rightarrow$ (at 500°C + 5 KCl]^{83} and polyfluorobenzenes ($C_6Cl_6 \rightarrow C_6F_6 + C_6F_5Cl$)[51] were developed.

1.3.6.2.   Biologically Active Compounds

Hexafluorobenzene (bp 80.5°C), the first commercial perfluoroaromatic compound, was shown in the early 1960s to possess several good features when used as an inhalation anesthetic in large animals (dogs, sheep, pigs, ponies). Unfortunately, however, its biodegradative behavior and low flammability limits in oxygen and nitrous oxygen precluded further developments.[84] By that time, however, several inhalation anesthetics of the organofluorine class had already been investigated clinically in man, and their nonflammability under conditions likely to be encountered in modern operating theaters had begun to revolutionize the field of anesthesiology (see Chapter 25). Fluroxene ($CF_3CH_2OCH{=}CH_2$) was the first fluorinated anesthetic to be tested clinically (1953), closely followed (early 1956) by the famous ICI compound $CF_3CHClBr$, (Fluothane®), which was first marketed in 1956.[84,85]

Commercial interest in the biological properties of compounds containing C—F bonds was stimulated by discoveries in IG Farben laboratories in the early 1930s. Work by Ufer and Schrader resulted in the introduction of 2-fluoroethanol and certain of its derivatives as pesticides in the mid-1930s, but these were soon displaced by organophosphorus compounds.[86] The key to the action of 2-fluoroethanol and its derivatives lies in the *in vivo* oxidation of this alcohol to monofluoroacetic acid, CH_2FCO_2H, a potent mammalian poison which owes its legendary lethality to an ability to disrupt the central nervous system and heart. The well-known, though controversial rodenticide CH_2FCO_2Na (compound 1080) depends on this property.[86]

During World War II, Saunders and others at Cambridge University (UK) carried out extensive work on the toxicity of fluoroacetate sources [and on phosphorus-based chemical warfare agents, e.g., the "nerve gas" $[(CH_3)_2CHO]_2P(O)F$, which is appreciably less toxic to rats than $CH_2FCO_2^-$]. In 1948, it was established that citric acid accumulates in the tissues of animals treated with sources of monofluoroacetate (MFA, $CH_2FCO_2^-$); this led to the demonstration by the famous Oxford biochemist Sir Rudolph Peters and his collaborators that MFA has no toxic effect on cells until it is converted into the actual culprit "fluorocitrate," $^-O_2CCH_2C(OH)(CO_2^-)CHFCO_2^-$; Peters dubbed the conversion "lethal synthesis."[87] Exactly how fluorocitrate produces its toxic effect is still under discussion.[88] Recent studies indicate that fluorocitrate is a "suicide" substrate for aconitase, rather than a competitive inhibitor as originally suggested. Note that only one diastereomer of fluorocitrate is toxic, namely, $2R,4R$-2-fluorocitrate.

Fluoroacetate was the first of the very few naturally occurring molecules containing C—F bonds to be discovered, its potassium salt having been isolated from the dangerous South African plant *Dichapetalum Cymosum* (gifblaar) by Marais in the early 1940s.[89]

Despite the paucity of organofluorine natural products (see Chapter 2), study of the effects of fluorine substitution for hydrogen or hydroxyl groups in biochemically active molecules has become phenomenally important, particularly in medical circles where the premise that correct placement of fluorine ought to block metabolism originated.[90] The first significant report on the successful application of this vital strategy came in 1953 from the Americans Fried and Sabo (Squibb Company), who found that replacement of hydrogen by fluorine at the 9α-position of cortisone dramatically enhanced the glucocorticoid activity.[91] Since then, the introduction of fluorine into steroidal hormones has become one of the most successful ways in which medicinal chemists have manipulated these molecules.[92–94]

Shortly after Fried's publications on fluorosteroids appeared, Heidelberger's group in the U.S. described the tumour-inhibitory properties, biochemical mechanisms of action, and initial clinical trial of 5-fluoroacil (5-FU) in patients with advanced cancer.[95] The landmark development in cancer chemotherapy led to much study of 5-FU and its nucleosides, and to the successful development of a commercial process for producing 5-FU using fluorine (see Chapter 9).

As the 1950s ended, the trifluoromethylated 2,6-dinitroaniline derivative trifluralin, one of the most successful fluorinated herbicides ever discovered, was introduced by the Eli Lilly Company (US). Much progress in the agrochemicals area of organofluorine chemistry has been made since then (see Chapter 11)—a situation true of many other commercial fluorochemical fields of endeavor since 1960, as the contents of this book clearly reveal.

1.4. REFERENCES

1. N. N. Greenwood and A. Earnshaw, *Chemistry of the Elements*, 1st edn., Chapter 17 ("The Halogens"), Pergamon Press, Oxford (1984).
2. R. E. Banks, D.W.A. Sharp, and J. C. Tatlow (eds.), *Fluorine: The First Hundred Years (1886–1986),* Elsevier Sequoia, Lausanne and New York (1986) [reproduced in J. Fluorine Chem., *33* (1986)]: (a) R. E. Banks, p. 3; (b) J. M. Holloway, p. 749; (c) R. E. Banks and J. C. Tatlow, p. 285; (d) R. E. Banks and J. C. Tatlow, p. 71; (e) H. Goldwhite, p. 109; (f) D. T. Meshri, p. 195; (g) J. Flahaut and C. Viel, p. 27.
3. *Handbook for Chemical Society Authors*, Special Publication No. 14, p. 191 (1960).
4. R. E. Banks, *Fluorocarbons and Their Derivatives*, 2nd edn., Macdonald, London (1970).
5. J. A. Young, *J. Chem. Documentation 14*, 98 (1974).
6. J. H. Simons, in: *Fluorine Chemistry* (J. H. Simons, ed.), Vol. I, p. 404, Academic Press, New York (1950).
7. A. V. Grosse and G. H. Cady, *Ind. Eng. Chem. 39*, 367 (1947).
8. H. Steinle , *Kältetech 12*, 392 (1960); *Refrig. Eng. 65*, 49 (1957).
9. J. D. Park, see Ref. 6, p. 528.
10. J. M. Hamilton, in: *Advances in Fluorine Chemistry* (M. Stacey, J. C. Tatlow, and A. G. Sharpe, eds.), Vol. 3, p. 117, Butterworths, London (1963).
11. *PCR Research Chemicals Catalog (1990–91)*, pp. 176–181 (available from PCR Inc., P.O. Box 1466, Gainesville, Florida 32602, USA).
12. A. K. Barbour, in: *Organofluorine Chemicals and Their Industrial Applications* (R. E. Banks, ed.), p. 44, Horwood, Chichester (1979).
13. E. T. McBee, *Ind. Eng. Chem. 39*, 236 (1947).
14. R. E. Banks and J. C. Tatlow, *J. Fluorine Chem. 35*, 3 (1987).
15. H. Moissan, *Le Fluor et ses Composés*, Steinheil, Paris (1900).
16. M. Meslans, *Ann. Chim. Phys. 1*, 346 (1894); M. Meslans and L. Giradet, *C.R. Acad. Sci. Paris 122*, 239 (1986).
17. A. Borodin, *Chemical News 7*, 267 (1862); *Justus Liebigs Ann. Chem. 126*, 58 (1863); R. E. Banks and J. C. Tatlow, *Chem. Eng. News 65* (June 15), 3 (1987); see C. B. Hunt, *Chem. Br. 23*, 547 (1987) for a succinct account of Aleksandr Borodin's achievements as a chemist and a musician.
18. E. Paternò and V. Oliveri, *Gazzetta 13*, 533 (1883).
19. O. Wallach, *Justus Liebigs Ann. Chem. 235*, 233, 255 (1886).
20. O. Wallach and F. Heusler, *Justus Liebigs Ann. Chem. 243*, 219 (1888).
21. See references cited in Ref. 2d.
22. G. Balz and G. Schiemann, *Ber. 60*, 1186 (1927).
23. G. Schiemann, *Chem.-Ztg. 54*, 269 (1930); *J. Prakt. Chem. 140*, 97 (1934).
24. A. Roe, *Org. React. 5*, 193 (1949).
25. H. Suschitzky, in: *Advances in Fluorine Chemistry* (M. Stacey, J. C. Tatlow, and A. G. Sharpe, eds.), Vol. 4, p. 1, Butterworths, London (1965).
26. Miscellaneous Pharmaceuticals and Pharmaceutical Intermediates Manufactured at I.G. Farbenindustrie A.G., Elberfeld, F.I.A.T. Final Report No. 1014 (1974).
27. G. C. Finger, *Chem. Met. Eng. 51*(6), 101 (1944).

28. A. E. Chichibabin and M. D. Rajazancev, *J. Russ. Phys. Chem. Soc.* 46, 1571 (1915) [CA *10*, 2898 (1916)].
29. P. Oswald and O. Scherer, Ger. Patent 600,706 (to IG Farben) [*CA 28*, 7260 (1934)].
30. *Fluorobenzene Manufacture*, FIAT Final Report No. 998 (1947); *The Manufacture, Formulation and Application of the Major Pest Control Products in the British, U.S. and French Zones of Germany*, B.I.O.S. Final Report No. 1480, Item 22, p. 125 (1947).
31. H. B. Gottlieb, *J. Am. Chem. Soc.* 58, 532 (1936).
32. A. F. Holleman and J. W. Beekman, *Recl. Trav. Chim. Pays-Bas 23*, 240 (1904).
33. H. G. Cook and B. C. Saunders, *Biochem. J. 41*, 558 (1947).
34. F. Sanger, *Biochem. J. 39*, 507 (1945).
35. B. C. Saunders, in: *Advances in Fluorine Chemistry* (M. Stacey, A. G. Sharpe, and J. C. Tatlow, eds.), Vol. 2, p. 183, Butterworths, London (1961).
36. (a) G. C. Finger and C. W. Kruse, *J. Am. Chem. Soc.* 78, 6034 (1956); (b) R. H. Shiley, D. R. Dickerson, and G. C. Finger, *Aromatic Fluorine Chemistry at the Illinois State Geological Survey: Research Notes, 1934–1976*, Circular 501/1978.
37. A. K. Barbour, L. J. Belf, and M. W. Buxton, in: *Advances in Fluorine Chemistry* (M. Stacey, A. G. Sharpe, and J. C.Tatlow, eds.), Vol. 3, p. 181, Butterworths, London (1963).
38. F. Swarts, *Bull. Classe Sci., Acad. Roy. Belg. 35*, 375 (1898); *113*, 241 (1913).
39. F. Swarts, *Bull. Acad. Roy. Belg.*, 389, 399 (1920); 331 (1922); 343 (1922); *Bull. Soc. Chim. Belg. 32*, 367 (1923).
40. See, for example, French Patent 745293 (to IG Farben [*CA 27*, 4414 (1933)]; P. Osswald, F. Müller, and F. Steinhaüser, Ger. Patent 575593 (to IG Farben [*CA 27*, 4813 (1933)]; L. C. Holt and E. L. Mattison, U.S. Patent 2 005 712 (to Kinetic Chemicals) [*CA 29*, 5123 (1935)].
41. G. Wolfrum, in: *Organofluorine Chemicals and Their Industrial Applications* (R.E. Banks, ed.), p. 208, Horwood, Chichester (1979).
42. P. Scherer, in: *Research Work on Fluorine and Fluorine Compounds* (O.G. Direnga, ed.), F.I.A.T. Final Report No. 1114, p. 42 (1947).
43. E. T. McBee, H. B. Hass, P. E. Weiner, G. M. Rothrock, W. E. Burt, R. M. Robb, and A. R. Van Dyken, *Ind. Eng. Chem. 39*, 298 (1947).
44. R. L. Murray, W. S. Beanblossom, and B. H. Wojcik, *Ind. Eng. Chem. 39*, 303 (1947); E. T. McBee, H. B. Haas, P. E. Weiner, W. E. Burt, Z. D. Welch, R. M. Robb, and F. Speyer, *Ind. Eng. Chem. 39*, 387 (1947).
45. E. T. McBee and R. E. Leech, *Ind. Eng. Chem. 39*, 393 (1947).
46. E. T. McBee, H. B. Hass, and E. M. Hodnett, *Ind. Eng. Chem. 39*, 389 (1947); E. T. McBee, O.R. Pierce, and R. O. Bolt, *Ind. Eng. Chem. 39*, 391 (1947).
47. Y. Désirant, *Bull. Acad. Roy. Belg., Classe Sci. 41*, 759 (1955).
48. Y. Désirant, *Bull. Soc. Chim. Belg. 67*, 676 (1958).
49. M. Hellmann, E. Peters, W. J. Pummer, and L. A. Wall, *J. Am. Chem. Soc. 79*, 5654 (1957).
50. R. E. Banks, A. K. Barbour, A. E. Tipping, B. Gething, C. R. Patrick, and J. C. Tatlow, *Nature 183*, 586 (1959).
51. N. N. Vorozhtsov, V. E. Platonov, and G. G. Yakobson, *Izv. Akad. Nauk SSSR, Ser. Khim.* 1524 (1963); *Zh. Obshch. Khim. 35*, 1158 (1965).
52. E. T. McBee, V. V. Lindgren, and W. B. Ligett, *Ind. Eng. Chem. 39*, 378 (1947).
53. J. A. Godsell, M. Stacey, and J. C. Tatlow, *Nature 178*, 199 (1956); R. Stephens and J. C. Tatlow, *Chem. Ind. (London)* 821 (1957).
54. T. Midgley and A. L. Henne, *Ind. Eng. Chem. 22*, 542 (1930).
55. R. E. Banks, *Chem. Br. 24*, 453 (1988).
56. O. Ruff and M. Giese, *Ber. 69B*, 598, 604, 684 (1936).
57. H. Moissan, *Compt. Rend. 110*, 951 (1890).
58. P. Lebeau and A. Damiens, *C.R. Acad. Sci. Paris 182*, 1340 (1926); *191*, 939 (1930).
59. O. Ruff, O. Bretschneider, and F. Ebert, *Z. Anorg. Allg. Chem. 217*, 1 (1934).

60. See N. Watanabe and T. Nakajima, in: *Preparation, Properties and Industrial Applications of Organofluorine Compounds* (R.E. Banks, ed.), p. 297, Wiley (Halstead Press), New York (1982) for more detailed historical information and references.
61. F. Swarts, *Bull. Sci. Acad. Roy. Belg. 17*, 27 (1931).
62. O. Ruff and O. Bretschneider, *Z. Anorg. Allg. Chem. 210*, 173 (1933).
63. E. G. Locke, W. R. Brode, and A. L. Henne, *J. Am. Chem. Soc. 56*, 1726 (1934).
64. J. H. Simons and L. P. Block, *J. Am. Chem. Soc. 59*, 1407 (1937).
65. J. H. Simons and L. P. Block, *J. Am. Chem. Soc. 61*, 2962 (1939).
66. See also J. H. Simons, in: *Fluorine Chemistry* (J. H. Simons, ed.), Vol. I, p. 401, Academic Press, New York (1950); J. H. Simons Memorial Issue, *J. Fluorine Chem. 32*, 7 (1986).
67. T. J. Brice, in: *Fluorine Chemistry* (J. H. Simons, ed.), Vol. I, p. 423, Academic Press, New York (1950).
68. For a review, see L. A. Bigelow, *Chem. Rev. 40*, 51 (1947).
69. W. Bockemüller, *Justus Liebigs Ann. Chem. 506*, 20 (1933); W. Bockemüller, Organische Fluorverbindungen, Enke, Stuttgart (1936).
70. G. H. Cady, A. V. Grosse, E. J. Barber, L. L. Burger, and Z. D. Sheldon, *Ind. Eng. Chem. 39*, 290 (1947).
71. R. D. Fowler, W. B. Burford, J. M. Hamilton, R. G. Sweet, C. E. Weber, J. S. Kasper, and I. Litant, *Ind. Eng. Chem. 39*, 292 (1947).
72. R. W. Clark, *The Greatest Power on Earth*, Sidgwick and Jackson, London (1980).
73. T. Abe and S. Nagase, in: *Preparation Properties, and Industrial Applications of Organofluorine Compounds* (R. E. Banks, ed.), p. 19, Halstead Press (John Wiley), New York (1982).
74. A. A. Banks, H. J. Eméleus, R. N. Haszeldine, and V. Kerrigan, *J. Chem. Soc.* 2188 (1948); H. J. Eméleus and R. N. Haszeldine, *J. Chem. Soc.*, 2948 (1949).
75. A. L. Henne and W. G. Finnegan, *J. Am. Chem. Soc. 72*, 3806 (1950); R. N. Haszeldine, *J. Chem. Soc.* 584 (1951).
76. A. L. Henne and W. C. Francis, *J. Am. Chem. Soc. 75*, 992 (1953); R. N. Haszeldine, *J. Chem. Soc.* 1273 (1954); 3423 (1952); 1748 (1953).
77. R. N. Haszeldine, *J. Chem. Soc.* 3761 (1953).
78. A. F. Benning, F. B. Downing, and J. D. Park, U.S. Patent 2,394,581 (to Du Pont) [*CA 40*, 3460 (1946)]; J. D. Harmon, U.S. Patent 2,404,374 (to Du Pont) [*CA 40*, 7234 (1946)]; A. L. Henne and R. P. Ruh, *J. Am. Chem. Soc. 69*, 279 (1947).
79. W. E. Hanford and G. W. Rigby, U.S. Patent 2,409,274 [*CA 41*, 982^b (1947)]; W. T. Miller, E. W. Fager, and P. H. Griswold, *J. Am. Chem. Soc. 70*, 431 (1948).
80. J. D. Park, A. F. Benning, F. B. Downing, J. F. Laucius, and R. C. McHarness, *Ind. Eng. Chem. 39*, 354 (1947).
81. R. E. Banks, A. K. Barbour, C. R. Patrick, and J. C. Tatlow, U.S. Patent 3,004,007/1961 (to ISC).
82. R. E. Banks, A. E. Ginsberg, and R. N. Haszeldine, *Proc. Chem. Soc. (London)* 211 (1960); *J. Chem. Soc.* 1740 (1961); R. N. Haszeldine, R. E. Banks and A. E. Ginsberg, U.S. Patent 3, 232, 946/1966; J. Burdon, D. J. Gilman, C. R. Patrick, M. Stacey, and J. C. Tatlow, *Nature 186*, 232 (1960).
83. R. E. Banks, R. N. Haszeldine, J. V. Latham, and I. M. Young, *Chem. Ind. (London)* 835 (1964); R. D. Chambers, J. Hutchinson, and W. K. R. Musgrave *Proc. Chem. Soc.* 83 (1964).
84. W. G. M. Jones, in: *Preparation, Properties, and Industrial Applications of Organofluorine Compounds* (R.E. Banks, ed.), p. 157, Halstead Press (John Wiley), Chichester (1982).
85. E. R. Larsen, in: *Fluorine Chemistry Reviews* (P. Tarrant, ed.), Vol. 3, p. 1, Dekker, New York (1969).
86. G. T. Newbold, in: *Organofluorine Chemicals and Their Industrial Applications* (R. E. Banks, ed.), p. 169, Horwood, Chichester (1979); R. L. Metcalf, in: *Pharmacology of Fluorides*, Part 1 (F. A. Smith, ed.), p. 354, Springer-Verlag, Berlin (1966).
87. R. A. Peters, in: *Carbon-Fluorine Compounds: Chemistry, Biochemistry, and Biological Activities*, p. 55, Elsevier, Amsterdam (1972).
88. D. D. Clarke, *Neurochem. Res. 16*(9), 1055 (1991).
89. J. C. S. Marais, *Onderstepoort J. Vet. Sci. Anim. Ind. 18*, 203 (1943); *20*, 67 (1944).

90. C. Walsh, in: *Advances in Enzymology and Related Areas of Molecular Biology* (A. Meisters, ed.), p. 197, Wiley, New York (1983).

91. J. Fried and E. F. Sabo, *J. Am. Chem. Soc. 75*, 2273 (1953); *76*, 1455 (1954).

92. N. F. Taylor and P. W. Kent, in: *Advances in Fluorine Chemistry* (M. Stacey, J. C. Tatlow, and A. G. Sharpe, eds.), Vol. 4, p. 113, Butterworths, London (1965).

93. A. Wettstein, in: *Carbon–Fluorine Compounds: Chemistry, Biochemistry, and Biological Activities*, Elsevier, Amsterdam (1972).

94. R. Filler, in: *Organofluorine Chemicals and Their Industrial Applications* (R. E. Banks, ed.), p. 123, Horwood, Chichester (1979).

95. C. Heidelberger, N. Chaudhur, P. Danneberg, D. Mooren, L. Griesbach, R. Duschinsky, R. J. Schnitzer, E. Pleven, and J. Scheiner, *Nature 179*, 663 (1957); for a review, see C. Heidelberger, in: *Carbon–Fluorine Compounds: Chemistry, Biochemistry, and Biological Activities*, p. 125, Elsevier, Amsterdam (1972).

2

Synthesis of Organofluorine Compounds

R. E. BANKS and J. C. TATLOW

2.1. INTRODUCTION

Fluoroorganic chemistry is virtually a man-made subject. Indeed, perfluorocarbon chemistry—which straddles organic and inorganic chemistry—is sometimes facetiously referred to as unnatural product chemistry. Fluorometabolites do occur in nature, but they are few in number and only monofluorides have been detected so far.[1] The best known is the plant metabolite CH_2FCO_2H (see Chapter 1); the other nine include the related compounds CH_2FCOCH_3, 2-fluorocitric acid, $F(CH_2)_nCO_2H$ ($n = 9,13,15$), and $F(CH_2)_8CH{=}CH(CH_2)_7CO_2H$. The most intriguing by far is nucleocidin (1), an antibiotic fluoro-sugar derivative produced by the microorganisms *Streptomyces calvus*.

Since Dumas and Péligot prepared the first organic fluoride, CH_3F, from dimethyl sulfate and potassium fluoride 160 years ago (Chapter 1), an impressive array of reagents for the synthesis of C—F bonds has been assembled. And the list is still growing through continuing efforts to increase yields, eliminate handling problems, achieve better selectivity (regio and stereo), and find ways of avoiding environmental problems. The result is that newcomers to the field of C—F bond synthesis often find difficulty in gaining an adequate working knowledge of the subject as quickly as they would like. The problem is compounded by the pathetic coverage of organofluorine chemistry in current student texts devoted to organic chemistry.

The object of the necessarily brief discussion here is to point out and exemplify some important ground rules of fluoroorganic synthesis. C—F Bond construction, which lies at the heart of the matter, is emphasized at the expense of the building block

R. E. BANKS • Department of Chemistry, The University of Manchester Institute of Science and Technology, Manchester M60 1QD, England.   J. C. TATLOW • Editorial Office of the Journal of Fluorine Chemistry, 30 Grassmoor Road, Kings Norton, Birmingham B38 8BP, England.

Organofluorine Chemistry: Principles and Commercial Applications, edited by R. E. Banks *et al.* Plenum Press, New York, 1994.

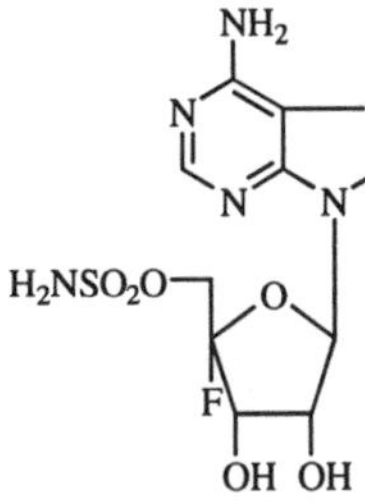

1

or disconnection/synthon approach. Both aspects are well illustrated throughout the rest of this book. Readers seeking detailed and wide-ranging discussions about the synthesis of fluoroorganic compounds are advised to consult initially a 1983 review[2] and two relatively recent books.[3,4] Preparative details for more than 300 selected organofluorine compounds, mainly per- and poly-fluoro in type, are available in a book published in 1985 and based on a two-volume Russian edition covering 500 procedures.[5]

The last edition (1976) of Milŏs Hudlický's famous book *Chemistry of Organic Fluorine Compounds* is still so much in demand by practitioners of the subject that it has been reissued recently;[6] it emphasizes laboratory aspects and contains numerous detailed preparative procedures.

Like all organic synthesis nowadays, work in the fluorine field is a highly sophisticated pursuit which demands considerable knowledge of reaction mechanisms, stereochemical principles, analytical methods, separation techniques, and selective reagents. Reaction mechanism underpins synthetic strategy to such a degree that these two cornerstones of organic chemistry are inseparable. Chapter 3 deals with mechanistic principles associated with fluoroorganic systems, and illustrates useful synthesis aspects.

2.2. SYNTHESIS METHODOLOGY

Clearly, two strategies are available when planning the synthesis of a fluorinated molecule: (1) purchase a starting material already containing the C—F bond(s) needed (the "building-block" approach); or (2) insert the C—F bond(s) required at a convenient stage using a fluorinating agent. Depending on the target molecule, either or both approaches may have to be employed.[7]

The indirect method (1) is preferred by researchers who are not experts in fluorine chemistry, because it often avoids dealing with hazardous chemicals and using unfamiliar techniques or special equipment. When there is no alternative to a direct method (2), the usual stratagem is to construct the C—F bond(s) at as late a stage in the route as possible; this minimizes loss of fluorinated material through side-reactions in transformation still required to reach target molecules. Nowadays, fortunately, truly impressive ranges of both fluoroorganic intermediates and fluorinating agents (including fluorine) are available commercially.

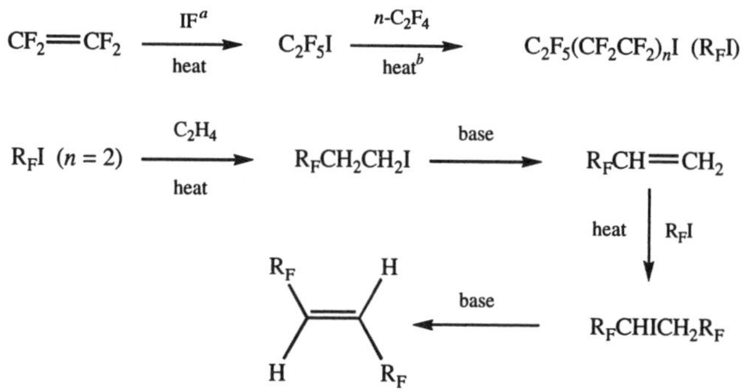

$$CF_2\!=\!CF_2 \xrightarrow[\text{heat}]{IF^a} C_2F_5I \xrightarrow[\text{heat}^b]{n\text{-}C_2F_4} C_2F_5(CF_2CF_2)_nI \ (R_FI)$$

$$R_FI \ (n=2) \xrightarrow[\text{heat}]{C_2H_4} R_FCH_2CH_2I \xrightarrow{\text{base}} R_FCH\!=\!CH_2$$

$$R_FCHICH_2R_F \xleftarrow{\text{base}} \qquad\qquad R_FCH\!=\!CH_2 \xrightarrow[R_FI]{\text{heat}} R_FCHICH_2R_F$$

(trans alkene: R_F and H on one carbon, H and R_F on the other)

Scheme 1. Example of the building-block approach to fluoroorganic targets. [a]Generated *in situ* from $I_2 + 2IF_5$. [b]Usually in the presence of a peroxide.

2.2.1. The Building-Block Approach

Per- and poly-fluorinated alkenes, alkanecarboxylic acids, and alkyl iodides are just as significant today (see Chapter 27) as synthesis intermediates as they were 30 years ago (see Chapter 1); a "modern" example is the preparation of the second-generation oxygen-transport agent *trans*-1,2-bis(perfluoro-*n*-hexyl)ethene (Scheme 1), which is of interest in blood substitute research (see Chapter 26). Notable commercially available additions to this group nowadays are fluorinated oxiranes, particularly perfluoropropene oxide, hexafluoroacetone, and certain fluoroalkyl bromides (such as CF_3Br, CF_2Br_2, and CF_2ClBr) which, though less reactive than iodides, are less expensive.

In the fluoroaromatics area, monofluorobenzene and benzotrifluoride still rank as the most important commercial intermediates, but numerous others are available (see Chapters 9, 10, and 27). Hexa- and pentafluorobenzene play important roles in researches on polyfluorinated compounds. For example, a plethora of pentafluorophenyl derivatives—organic and organoelemental—have been prepared via nucleophilic displacement of ring fluorine: $C_6F_6 + Nu^- \rightarrow C_6F_5Nu + F^- \{Nu^- = H^-$, HO^-, HS^-, CH_3^-, H_2N^-, $[\pi\text{-}C_5H_5Fe(CO)_2]^-$ etc.$\}$.[7]

2.2.2. The Formation of Carbon–Fluorine Bonds

2.2.2.1. Classification of Reagents

Thirty years ago,[8] it was customary to divide fluorinating agents into two major groups according to their activity and hence the degree of control which could at that time be exercised over them: (I) fluorine, high-valency fluorides of certain metals (such as CoF_3 or AgF_2) and halogen fluorides (such as ClF_3 or BrF_3); (II) hydrogen fluoride and derived compounds (such as KF, SbF_3, or SF_4). Group (I) reagents were classed as *vigorous fluorinating agents*, because in the main they had been found difficult to control and capable of replacing *all* hydrogen by fluorine, and of saturating all multiple C—C bonds and aromatic systems (so-called exhaustive fluorination). They were prized, therefore, for the synthesis of perfluorocarbons from hydrocarbons. By contrast, Class (II) reagents were categorized as *mild fluorinating agents* used for introducing fluorine at selected sites in organic molecules and not normally capable of replacing hydrogen by fluorine. Naturally, Simons electrochemical fluorination (ECF), which utilizes AHF yet achieves complete fluorination of all C—H bonds, was accorded Class I status.

Inevitably, the science of C—F bond synthesis methodology has advanced appreciably since that necessarily simple classification of fluorinating agents was recommended: adequate small-scale methods for controlling the degree of fluorination with aggressive reagents (especially F_2) have been developed[7]; easier-to-handle modifications of hazardous reagents, notably AHF and SF_4,[2] have been introduced for research and developmental purposes; alternatives to the potentially dangerous electrophilic fluorinating agent perchloryl fluoride have been discovered, e.g., XeF_2 and numerous compounds containing N—F bonds[3]; ways of using fluorine in an electrophilic rather than a radical mode have successfully been researched[9]; and two more electrochemical methods of fluorination have been described (see Chapter 5 and Tables 1 and 2).

Important reaction types (e.g., C—H → C—F; C—Cl → C—F; C—OH → C—F; C=C → CH—CF) used for the synthesis of C—F bonds are presented and exemplified in Tables 1 and 2. Mechanistic categories of reagents are designated by the symbols F^-, $F\cdot$, and F^+; these indicate in which form (actual for fluorine atom; actual or incipient for fluoride ion; apparent in the case of fluoronium ion) fluorine is "delivered" to a suitably activated carbon site.

2.2.2.2. [^{18}F]Fluorination

Fluorine-18, the longest-lived of the five artificial isotopes of the monoisotopic natural element (^{19}F), has played a major role in the development[10] of one of the most significant advances in modern medicine: positron emission transaxial tomography (PET). This technique is a powerful noninvasive diagnostic method that allows the circulatory and metabolic behavior of small, well-defined volumes of any body tissue to be monitored and measured without trauma in living patients. For measuring function, it is superior to other imaging methods such as magnetic resonance imaging (MRI) or computed axial X-ray tomography (CAT), which mainly provides anatomical data. Almost entirely (97%; 3% electron capture) a positron emitter ($^{18}_{9}F \rightarrow {}^{18}_{8}O + \beta^+$), fluorine-18 is a particularly attractive candidate radionuclide for use in PET because its half-life is sufficiently long (110 minutes) to allow complex or multistep syntheses to be carried out on sites moderately distant from medical premises. Fluorine-18 can be prepared by a number of nuclear reactions, using either charged particle sources (linear accelerators or cyclotrons) or nuclear reactors.

The necessary reagents are produced either from ^{18}F-labelled F_2 or [^{18}F]fluoride. The element is usually generated by bombarding neon-20 with deuterons [$^{20}_{10}$Ne(d, α)$^{18}_{9}$F] while the most successful method for the preparation of ^{18}F$^-$ appears to be proton irradiation of oxygen-18 enriched water targets [$^{18}_{8}$O(p,n)$^{18}_{9}$F].

2.3. THE CURRENT POSITION

If the option exists, cost/convenience considerations normally make fluorinations involving AHF or its derivatives (fluorides; KF, for example, is particularly important) much more attractive propositions in both academia and industry than those needing fluorine or one of its aggressive derivatives. To date, relatively little use has been made of fluorine to produce fluoroorganic products in commercial circles, the main applications being the production of perfluorocarbons via the cobalt fluoride process (Chapter 4), fluorine-finishing (end-capping) of perfluoropolyethers (PFPEs) (Chapters 20 and 21), postpolymerization (surface) fluorination of hydrocarbon polymers (Chapter 22), and the conversion of uracil into 5-fluorouracil (Chapters 1 and 9). Industrial interest is mounting, however, in the fluorination of biologically important molecules, notably steroids, with selective electrophilic reagents of the N—F class that require fluorine for their production (see Table 1). Also, new techniques for effecting high-yield exhaustive fluorination (perfluorination) of organic molecules with fluorine are emerging, e.g., temperature-programmed LaMar solid-phase fluorination[11]; Adcock aerosol fluorination,[12] and liquid-phase thermal fluorination or photofluorination in inert solvents such as CFCs or PFPEs.[13] These hold considerable promise for the future, especially if partial fluorination of substrate molecules can first be achieved economically using building-block or fluoride-based techniques, a ploy which has been known for some time to minimize C—C bond fission in direct,[14] cobalt fluoride,[15] or electrochemical[16] fluorination reactions (e.g., see Scheme 6, Chapter 5).

During the first half of the present century, organofluorine chemistry acquired quite a reputation as a particularly problematic area of organic synthesis. Despite the remarkable expansion of the subject since then, difficulties in the synthesis of C—F bonds persist—even for "experts". This makes all the more commendable the achievements of the technologists who have forged today's organofluorochemicals industry.

2.4. TABULAR SUMMARY OF FLUORINATION METHODS AND REAGENTS

Table 1 contains summaries of the principal general methods available for producing C—F bonds at specific molecular sites; the list of fluorinating agents is not comprehensive, priority having been given to those most widely used or of interest commercially. Exhaustive fluorination is summarized in Table 2. In addition to sources of information on C—F bond synthesis already mentioned,[2–8] much valuable information can be found in *Organic Reactions* Vol. 21, which is devoted to fluorination with sulfur tetrafluoride[17a] and methods for the preparation of monofluoroaliphatic

Table 1. An Introductory Guide to the Selective Synthesis of C—F Bonds: Summary of Important Fluorination Methods

Common reagents	Typical conversions	Comments	Mechanistic category	Specific references [this volume]
A. *Halogen exchange*: C—X → C—F	[X=Cl (main preoccupation commercially), Br or I]			
1. In polychloroalkanes				
AHF/SbCl$_5$[a] (liquid phase)	CCl$_4$ → CFCl$_3$, CF$_2$Cl$_2$, CF$_3$Cl CHCl$_3$ → CHFCl$_2$, CHF$_2$Cl CCl$_3$CH$_3$ → CF$_3$CH$_3$	Product ratios are determined by the reaction conditions. The order of reactivity for Sb halides is SbF$_5$ > SbF$_3$Cl$_2$ > SbF$_3$ / SbCl$_5$ mixtures > SbF$_3$. Antimony(V) chlorofluorides, SbCl$_x$F$_{5-x}$ are produced *in situ* (AHF + SbCl$_5$) industrially.	C$^+$ (incipient) + F$^-$	20 [Chapters 1,6,7]
AHF/Cr-based catalysts (vapor phase)	CCl$_3$CCl$_3$ → CFCl$_2$CCl$_3$ + CFCl$_2$CFCl$_2$ + CF$_2$ClCFCl$_2$ + CF$_2$ClCF$_2$Cl + CF$_3$CF$_2$Cl			
SbF$_3$/SbCl$_5$ (laboratory scale)	CCl$_3$CHClCCl$_3$ → CFCl$_2$CHClCCl$_3$ + CFCl$_2$CHClCFCl$_2$ + CF$_2$ClCHClCFCl$_2$ CCl$_3$CH$_2$CH$_2$Cl → CF$_3$CH$_2$CH$_2$Cl			
2. In *gem*-dihalogeno-alkanes[b] (or -cycloalkanes)				
HgO/AHF or HgF$_2$	CH$_2$ClCCl$_2$CH$_3$ → CH$_2$ClCF$_2$CH$_3$		C$^+$ (incipient) + F$^-$	20, 21
SbF$_3$	CH$_2$BrCBr$_2$CH$_3$ → CH$_2$BrCF$_2$CH$_3$			20
AgBF$_4$	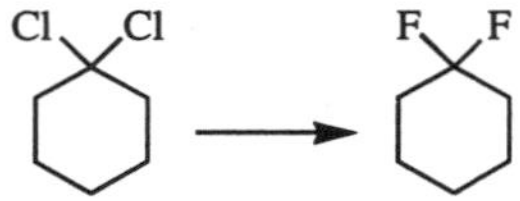 1,1-dichlorocyclohexane → 1,1-difluorocyclohexane			22

3. In monohalogenoalkanes

Reagent	Reaction	Comments	Ref.
KF	$Cl(CH_2)_4Cl \rightarrow F(CH_2)_4Cl,\ F(CH_2)_4F$ $Br(CH_2)_6CO_2C_2H_5 \rightarrow F(CH_2)_6CO_2C_2H_5$	F^- in S_N2 mode (KF, TBAF) or with electrophilic assistance (incipient C^+; AgF or CuF). Nonsolvated F^- is not only a good nucleophile but also rather a strong base, hence elimination reactions tend to occur with secondary halides and often exclusively so with tertiary substrates. Anhydrous KF (preferably spray-dried) is normally used in conjunction with nonhydrogen-bonding ("dipolar aprotic") solvents, sometimes with added crown ethers or cryptands to provide truly "naked fluoride." Since milder reaction conditions can be used, AgF (highly active but expensive) in conjunction with organic bromides or iodides is often preferred to KF when "sensitive" molecules (e.g., steroids, carbohydrates) need to be fluorinated.	20,23
AgF	$(CH_3)_2CHI \rightarrow (CH_3)_2CHF$		24
"CuF" (from Cu_2O + AHF)	$n\text{-}C_8H_{17}Br \rightarrow n\text{-}C_8H_{17}F$		25
$n\text{-}Bu_4N^+F^-$ (TBAF)	$C_6H_5CH_2Br \rightarrow C_6H_5CH_2F$		26

4. In allylic or benzylic ("activated") positions

Reagent	Reaction	Comments	Ref.
AHF AHF/SbCl$_5$ SbF$_3$/SbCl$_5$	$CCl_3CCl=CCl_2 \rightarrow CF_3CCl=CCl_2$ $CCl_2=CClCCl=CCl_2 \rightarrow CF_3CCl=CClCF_3$	$C^+ + F^-$ Allylic Cl is exchanged more readily than chlorine in polychloroalkanes. Vinylic chlorine is seldom exchanged directly. 1,4-Addition of chlorine occurs first in the C_4Cl_6/SbF_3Cl_2 reaction.	20
AHF	$C_6H_5CCl_3 \rightarrow C_6H_5CF_3$ $C_6H_5CCl_2CCl_3 \rightarrow C_6H_5CF_2CCl_3 + C_6H_5CF_2CFCl_2 + C_6H_5CF_2CF_2Cl$		[Chapters 1, 10, 11, 27]
SbF$_3$	$p\text{-}ClC_6H_4CCl_3 \rightarrow p\text{-}ClC_6H_4CF_3$	Nuclear chlorine remains unaffected in homoaromatics	

(Continued)

Table 1. (Continued)

Common reagents	Typical conversions	Comments	Mechanistic category	Specific references [this volume]
5. In carbonyl compounds				
KF	$CH_2ClCO_2CH_3 \rightarrow CH_2FCO_2CH_3$		F^- in S_N2 mode	20
	$C_6H_5COCl \rightarrow C_6H_5COF$			
	$CCl_3COCl \rightarrow CCl_3COF$			
AgF	$n\text{-}C_6H_{13}CHBrCO_2CH_3 \rightarrow$ $n\text{-}C_6H_{13}CHFCO_2CH_3$		F^- in S_N2 mode	27
AHF/catalyst (e.g., Cr salts)	$CCl_3COCl \rightarrow CF_3COF$ $(CCl_3)_2CO \rightarrow (CF_3)_2CO$		F^- (S_N2) and/or $C^+ + F^-$	[Chapters 19, 27]
6. In "activated" nuclear halogeno-aromatics				
KF	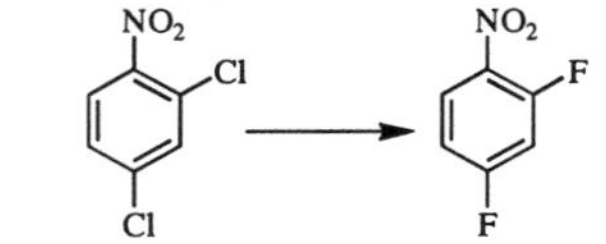 $C_6Cl_6 \rightarrow C_6F_{6-x}Cl_x$ ($x = 0,1,2,3$)	An important commercial method in which a variety of activating (electron-withdrawing) groups can be used (NO_2, CN, CF_3, CHO, Cl, ring N, etc.). KF is sometimes activated by addition of expensive CsF (order of activity CsF > RbF > KF > NaF > LiF), and is normally suspended in hot dipolar aprotic solvents. Phase-transfer reagents are sometimes needed. The last two examples shown are high-temperature (450–500 °C) solvent-free reactions.	F^- in S_NAr mode	[Chapters 1, 9, 13, 27]

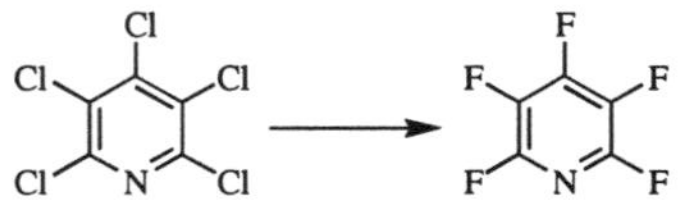

HF, NaF

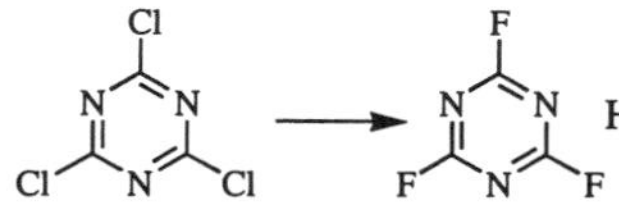

Highly reactive substrates are required.

B. *Oxygen replacement:*   C—OX → C—F [OX = OH, OSO$_2$R, OC(O)F, OC (epoxide components)]
C=O → CF$_2$
C(=O)OH → CF$_3$

1. Fluorodehydroxylation

AHF/pyridine/NaF

n-C$_8$H$_{17}$OH → n-C$_8$H$_{17}$F

(CH$_3$)$_3$CCH$_2$OH → (CH$_3$)$_3$CCH$_2$F

In general, the classical route ROH + HX → RX + H$_2$O (X = halogen) fails or gives low yields of RF in complex tarry mixtures when X = F. Use of AHF in conjunction with a Lewis base (e.g., THF, pyridine, melamine, urea) improves matters remarkably, not only in terms of yield but ease of experimentation (AHF boils at 19.5 °C).

F$^-$/C—$^+$OH$_2$
(S_{N2} . . . S_{N1} continuum)

28, 29

AHF/pyridine

CH$_3$CH$_2$CH(OH)CH$_3$ → CH$_3$CH$_2$CHFCH$_3$

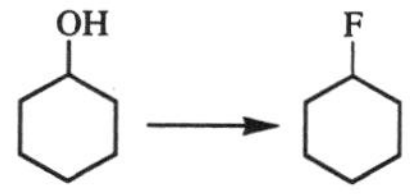

(CH$_3$)$_3$COH → (CH$_3$)$_3$CF

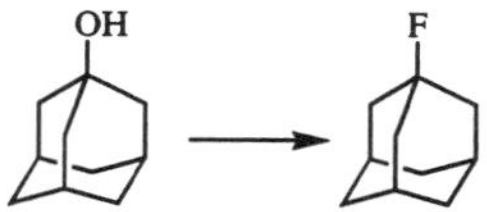

The most popular variant is Olah's Reagent—pyridinium poly(hydrogen fluoride) [PPHF, C$_5$H$_5$NH$^+$(HF)$_x$F$^-$] known as "tamed HF" and normally encountered as a 1C$_5$H$_5$N:9HF mixture (30:70 wt.%). This suffers no appreciable loss of HF at temperatures up to 50 °C, but must be handled with care (HF burns!) in poly(ethylene) or similar equipment. 3° Alcohols react more readily (≤0 °C) than 2° (20–50 °C); 1° alcohols require added F$^-$ (NaF) for the reaction to proceed smoothly.

F$^-$/C—$^+$OH$_2$
(S_{N2} . . . S_{N1} continuum)

28, 29

(Continued)

Table 1. (Continued)

Common reagents	Typical conversions	Comments	Mechanistic category	Specific references [this volume]
$(C_2H_5)_2NCF_2CHFX$ [X = Cl (Yarovenko's reagent), CF$_3$ (Ishikawa's reagent)]	$CH_3CHBrCH_2CH_2OH$ $\rightarrow$ $CH_3CHBrCH_2CH_2F$ $CH_2BrCH_2CH(OH)CH_3$ $\rightarrow$ $CH_2BrCH_2CHFCH_3$ $C_6H_5CH(OH)COC_6H_5$ $\rightarrow$ $C_6H_5CHFCOC_6H_5$	Fluorodehydroxylation of alcohols with these so-called FARs (fluoroalkylamine reagents) is one of the simplest, most convenient, and safe fluorination techniques; in use, the reagents, become converted into amides. Easily prepared by treating the corresponding fluoroolefins with diethylamine [CF$_2$=CFX + (C$_2$H$_5$)$_2$NH], they usually react fairly cleanly with 1° and 2° alcohols (cyclic 2° alcohols tend to dehydrate extensively; 3° alcohols prove troublesome, too). Yarovenko's reagent, which has the shorter shelf-life, has been used extensively to fluorinate steroidal alcohols.	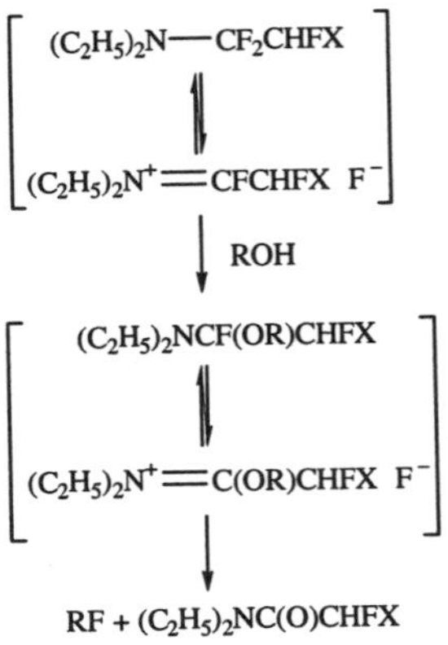 $(C_2H_5)_2N\!-\!CF_2CHFX$ $\rightleftharpoons$ $(C_2H_5)_2N^+\!=\!CFCHFX \; F^-$ $\xrightarrow{ROH}$ $(C_2H_5)_2NCF(OR)CHFX$ $\rightleftharpoons$ $(C_2H_5)_2N^+\!=\!C(OR)CHFX \; F^-$ $\rightarrow$ $RF + (C_2H_5)_2NC(O)CHFX$ F$^-$ delivered via an $S_N2\ldots S_N1$ continuum	17b, 29
SF$_4$	$CHF_2(CF_2)_5CH_2OH \rightarrow$ $CHF_2(CF_2)_5CH_2F$	Sulfur tetrafluoride reacts best with acidic alcohols; being a toxic gas (bp −38 °C) which readily hydrolyzes to HF and SOF$_2$, it is inconvenient to use in the laboratory (metal autoclaves are required). Its liquid derivative (C$_2$H$_5$)$_2$NSF$_3$ (see below) is much more convenient to handle.	F$^-$ delivered via an $S_N2\ldots S_N1$ continuum in a mechanism analogous to that shown for the SF$_4$ derivative DAST (see immediately below).	17a, 31, 32
SF$_4$/AHF	$(NO_2)_3CCH_2OH \rightarrow (NO_2)_3CCH_2F$ $CH_2(OH)CH(NH_2)CO_2H \rightarrow$ $CH_2FCH(NH_2)CO_2H$			

$(C_2H_5)_2NSF_3$ (DAST)	n-$C_8H_{17}OH \rightarrow n$-$C_8H_{17}F$ $CH_2BrCH_2OH \rightarrow CH_2BrCH_2F$ $C_6H_5CH(OH)CO_2H \rightarrow$ $C_6H_5CHFCO_2H$ 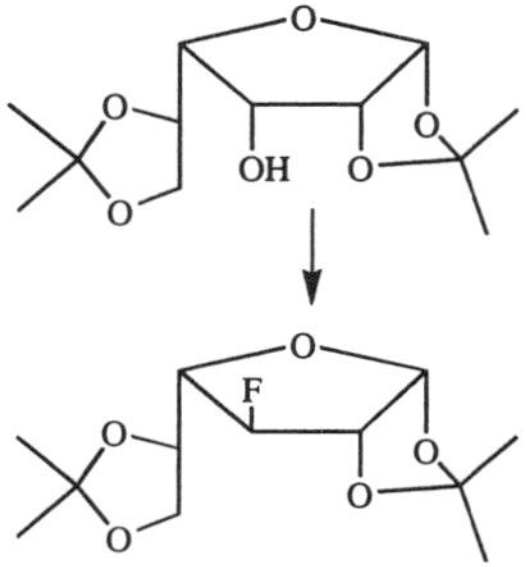 	Diethylaminosulfur trifluoride [$(C_2H_5)_2NSi(CH_3)_3 + SF_4 \rightarrow$ $(C_2H_5)_2NSF_3$ (DAST; bp 46–47 °C at 10 mm Hg]) is much more convenient to handle but has the potential to decompose violently when heated above 50 °C; it is usually employed at ≥20 °C in solution (CH_2Cl_2, THF, toluene) in glass apparatus, but is moisture-sensitive (store in polyethylene). Morpholinosulfur trifluoride (MOST) is much safer to use.[30] DAST and its analogs react with most 1°, 2°, and 3° alcohols to give high yields of the corresponding fluorides. DAST has proved to be very useful in the carbohydrate field for OH $\rightarrow$ F conversions.	Many fluorinations involving DAST proceed with inversion of configuration, as expected from an S_N2 mechanism: 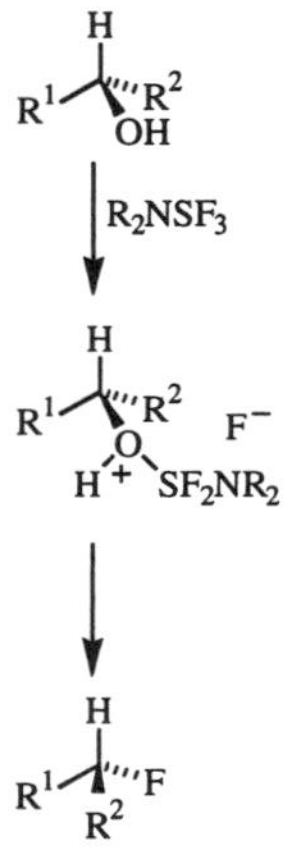 	33, 34
2. Ester fluorolysis (Ts = 4-$CH_3C_6H_4SO_2$; Tf = CF_3SO_2)			
KF	$CH_3OTs \rightarrow CH_3F$ $CH_2ClCH_2OTs \rightarrow CH_2ClCH_2F$  	An extension of the seminal 1830s Dumas–Péligot experiment (see Chapter 1) this method constitutes a convenient and reliable laboratory synthesis of 1° and 2° alkyl fluorides. Depending on the thermal stability of the tosylate, reactions are run either "dry," or in hot diethylene glycol or dipolar aprotic solvents.	F^- in S_N2 mode	4, 35

(Continued)

Table 1. (*Continued*)

Common reagents	Typical conversions	Comments	Mechanistic category	Specific references [this volume]
CsF or $R_4N^+F^-$ ($R = CH_3, C_4H_9$)	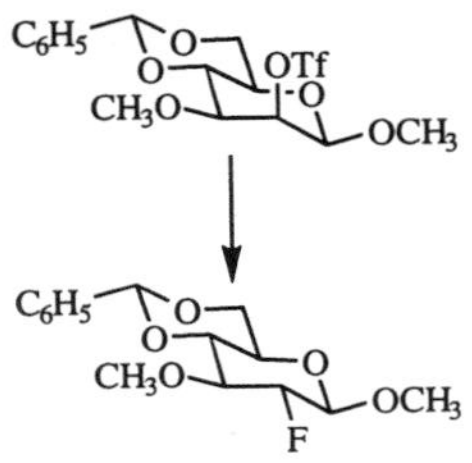 [carbohydrate structure: C_6H_5–, CH_3O, OTf, OCH_3 → C_6H_5–, CH_3O, F, OCH_3]	Useful for "sensitive" substrates, the triflate ($CF_3SO_3^-$) "super" leaving-group enabling conversion to proceed under mild conditions in CH_3CN or THF.	F^- in S_N2 mode	4

3. Thermal fluorodecarboxylation of halogenoformates

Synthesis of fluoroformates

Common reagents	Typical conversions	Comments	Mechanistic category	Specific references [this volume]
CaF_2	$COCl_2 \rightarrow COFCl$	Alkali metal fluorides can be used.		36
COFCl	$ROH/(C_4H_9)_3N \rightarrow ROC(O)F$			6
	$ArOH/(C_4H_9)_3N \rightarrow ArOC(O)F$			36
KF	$ROC(O)Cl \rightarrow ROC(O)F$			37
AHF/pyridine	$ROC(O)NH_2/NaNO_2 \rightarrow ROC(O)F$			2

Decomposition to fluorides

Common reagents	Typical conversions	Comments	Mechanistic category	Specific references [this volume]
$BF_3 \cdot O(C_2H_5)_2$	$ROC(O)F \rightarrow RF$ [$R = C_2H_5$, $(CH_3)_2CH$, *cyclo*-C_6H_{11} etc.]			6

Pt/Al_2O_3	$C_6H_5OC(O)F \rightarrow C_6H_5F + CO_2$	Clearly of commercial interest. Aryl chloroformates can be used as feedstock with the AHF/AlF_3 systems; *in situ* halogen exchange takes place initially		36
AHF/AlF_3	$4\text{-}FC_6H_4OC(O)F \rightarrow 4\text{-}FC_6H_4F + CO_2$			38
4. Ring-opening of epoxides				
$(i\text{-}C_3H_7)_2NH \cdot 3HF$	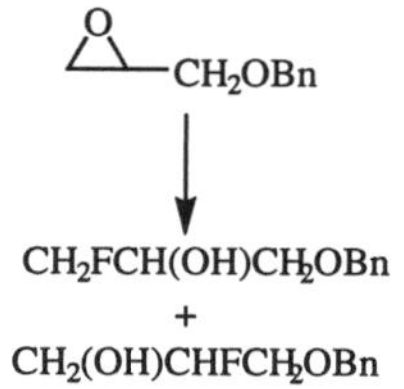O CH₂OBn CH₂FCH(OH)CH₂OBn + CH₂(OH)CHFCH₂OBn (product ratio 15:1; Bn = benzyl)	Many other sources of HF have been used; di-isopropylamine trishydrofluoride, like its congener $(C_2H_5)_3N \cdot 3HF$,[40] are excellent sources of "tamed" HF and, unlike HF/pyridine, can be used in standard *glass* apparatus. KHF_2 in ethylene glycol is popular for opening epoxide rings in carbohydrates.[41]	F^- Delivery in acid-catalyzed modes of ring-opening.	39
AHF/pyridine	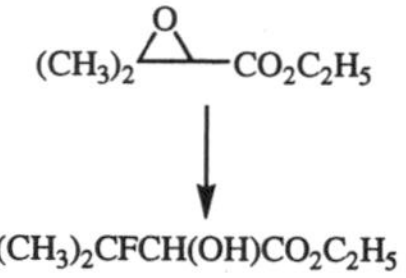O (CH₃)₂ CO₂C₂H₅ (CH₃)₂CFCH(OH)CO₂C₂H₅			
$AHF/BF_3 \cdot O(C_2H_5)_2$	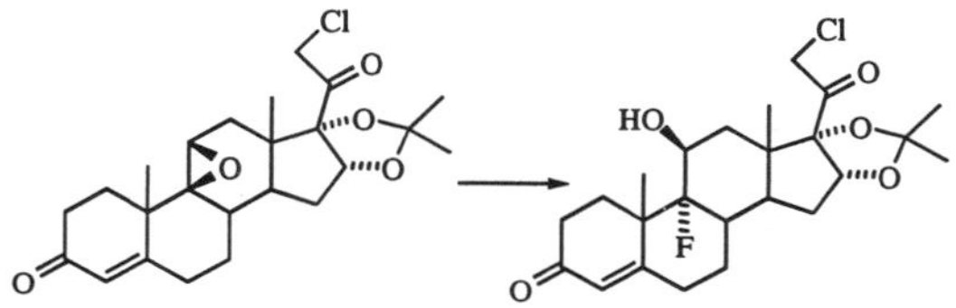Cl O O O O O Cl O HO O O F O	The method has been widely used to produce steroidal fluorohydrins.[17b]		

(Continued)

Table 1. (Continued)

Common reagents	Typical conversions	Comments	Mechanistic category	Specific references [this volume]
5. Replacement of oxygen in aldehydes, ketones, and carboxylic acids				
SF_4 [often with HF or BF_3 (catalysts)]	$n\text{-}C_6H_{13}CHO \rightarrow n\text{-}C_6H_{13}CHF_2$ $C_6H_5CHO \rightarrow C_6H_5CHF_2$ $C_6H_5COC_6H_5 \rightarrow C_6H_5CF_2C_6H_5$ $C_6H_5OC(O)F \rightarrow C_6H_5OCF_3$ $CH_3COCOCH_3 \rightarrow CH_3CF_2CF_2CH_3$ $C_6H_5CO_2H \rightarrow [C_6H_5COF] \rightarrow C_6H_5CF_3$ $HO_2CC\equiv CCO_2H \rightarrow CF_3C\equiv CCF_3$ $H_2NCH_2CO_2H$ (in AHF) $\rightarrow H_2NCH_2CF_3$ 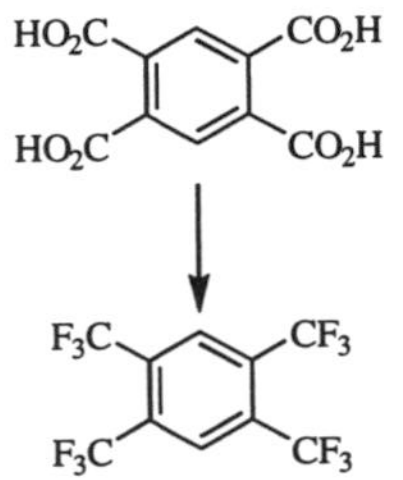	Use of SF_4 (which is expensive) requires special precautions and equipment (see Section B.1). Since its introduction as a fluorinating agent in 1960 by Du Pont chemists, SF_4 has been studied virtually *ad infinitum*; thus considerably more is known than indicated here, and knowledge of the area is a specialty in its own right. Carbonyl compounds are ranked in order of ease of conversion as follows: ROH $>$ RCHO $\approx$ R_2CO $>$ RCO_2H $\approx$ $RCONR'_2$ $>$ RCO_2R $\approx$ $(RCO_2)_2O$.	Mechanisms are still being debated.[42] HF-catalyzed reactions seemingly are driven by initial formation of $SF_3^+HF_2^-$; conversions then proceed at the carbonyl group via the sequence: $R_2CO \rightarrow R_2C^+\!-OSF_3$ $\quad\quad\quad\quad\quad\downarrow F^-$ $R_2CF\!-OSF_2^+ \leftarrow R_2CF\!-OSF_3$ $\quad\downarrow F^-$ $R_2CF_2 + SOF_2$ The last step can lie on an $S_N2 \ldots S_N1$ continuum.	17a, 31, 32

Reagent	Examples	Comments	Mechanism	Ref.
$(C_2H_5)_2NSF_3$ (DAST)	$(CH_3)_3CCHO \rightarrow (CH_3)_3CCHF_2$ $C_6H_5COCO_2C_2H_5 \rightarrow$ $C_6H_5CF_2CO_2C_2H_5$ $4\text{-}C_6H_5COC_6H_4CHO \rightarrow$ $4\text{-}C_6H_5COC_6H_4CHF_2$	Widely used in research on bioactive fluorides (steroids, carbohydrates, etc.), DAST is much easier to handle than SF_4 (see Section B.1). Milder in action than SF_4 and not often used with catalysts, the order of reactivity is $ROH > RCHO > R_2CO > RCO_2H > RCO_2R'$. Thus OH can be replaced by F selectively (e.g. see Section B.1). DAST analogs are available (e.g. MOST; see Section B.1). Carboxylic acids yield acyl fluorides with DAST under "safe" conditions; SF_4 needs to be used to acquire trifluoromethyl derivatives.	Similar to reactions involving SF_4.	33

C. Nitrogen replacement: $C\text{—}NH_2 \rightarrow C\text{—}N_2^+ \rightarrow C\text{—}F$; $C\text{—}N\text{—}C \rightarrow CF\text{—}N\text{—}C$

1. Deaminative fluorination of α-amino acids

Reagent	Examples	Comments	Mechanism	Ref.
AHF/pyri- dine/$NaNO_2$	$C_2H_5CH(NH_2)CO_2H \rightarrow$ $C_2H_5CHFCO_2H$ $(CH_3)_2CHCH(NH_2)CO_2H$ $\rightarrow$ $(CH_3)_2CHCHFCO_2H$ + $(CH_3)_2CFCH_2CO_2H$ (ratio 2:1)	Provides a simple and convenient route to α-fluorocarboxylic acids from readily available α-amino acids. Skeletal rearrangements (cationic) occur in some cases prior to fluoride capture, giving β-fluorocarboxylic acids. High pyridine:HF ratios (ca 50:50 wt%; cf. Section B.1) favor the formation of α-fluoro products.	Nucleophilic $(C^+ + F^-)$ dediazoniation following in situ diazotization of the NH_2 group.	4

2. Fluorodediazoniation of aromatic diazonium salts

Reagent	Examples	Comments	Mechanism	Ref.
AHF/$NaNO_2$	$C_6H_5NH_2 \rightarrow C_6H_5N_2^+F^-$ (heat in situ) $\downarrow$ C_6H_5F	A commercial method ("HF-diazotization;" Lenz–Wallach Reaction)—see Chapter 9 for full details. Metal equipment is essential.	$Ar^+ + F^-$ (extracted from BF_4^- in the Balz–Schiemann Reaction).	[Chapters 9, 27]

(Continued)

Table 1. (*Continued*)

Common reagents	Typical conversions	Comments	Mechanistic category	Specific references [this volume]
HBF$_4$ aq. or NaBF$_4$	$C_6H_5NH_2 \rightarrow C_6H_5N_2^+Cl^- \rightarrow$ $C_6H_5N_2^+ BF_4^-$ (isolated) $\downarrow$ (heat) $C_6H_5F + N_2 + BF_4$ $(4\text{-}H_2NC_6H_4)_2CH_2 \rightarrow$ $(4\text{-}FC_6H_4)_2CH_2$	Used commercially for certain products (see Chapter 9). An invaluable laboratory method (the Balz–Schiemann Reaction) since standard glassware can be used.		43 [Chapters 1,9]
3. Ring-opening of azirine and aziridines				
AHF/pyridine	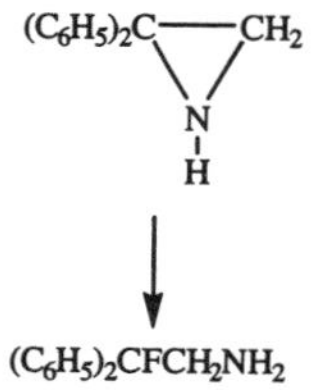 $(C_6H_5)_2C\!\!-\!\!CH_2$ ring with N–H $\longrightarrow$ $(C_6H_5)_2CFCH_2NH_2$	The stereochemistry and complexity of this reaction can be highly dependent on the substrate's structure and the actual composition of the reagents used. Remember that Olah's reagent is much easier to handle than AHF on a laboratory scale (simple polyolefin apparatus is used at atmospheric pressure; see Section B.1).	F^- in $S_N1 \ldots S_N2$ continuum. The aziridine shown reacts via the S_N1 mode.	4, 44
D. Replacement of hydrogen: C—H → C—F				
1. Electrophilic fluorination with fluorine				
a. Directly in aliphatic systems				
F$_2$/N$_2$	$n\text{-}C_6H_{13}CH(CH_3)C_2H_5 \rightarrow$ $n\text{-}C_6H_{13}CF(CH_3)C_2H_5$ 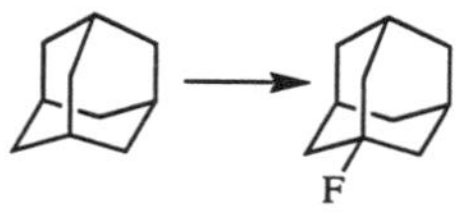 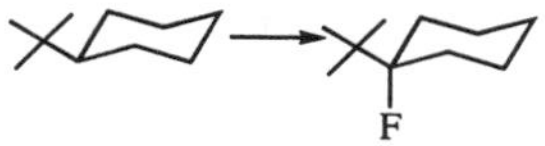	Use of fluorine in an electrophilic mode demands the rigorous suppression of radical reactions. Typically, 1–10% v/v F$_2$ in N$_2$ (or other inert gas, e.g., He, Ar) is bubbled slowly through efficiently stirred dilute solutions of substrates in CFCl$_3$, CFCl$_3$–CHCl$_3$ blends, or CH$_3$CN at −40 to −78 °C. Nobody believes that even a solvated fluoronium ion (F$^+$) is involved. Commercial F$_2$–N$_2$ blends are widely used. C—H bonds with the highest σ-orbital contribution are	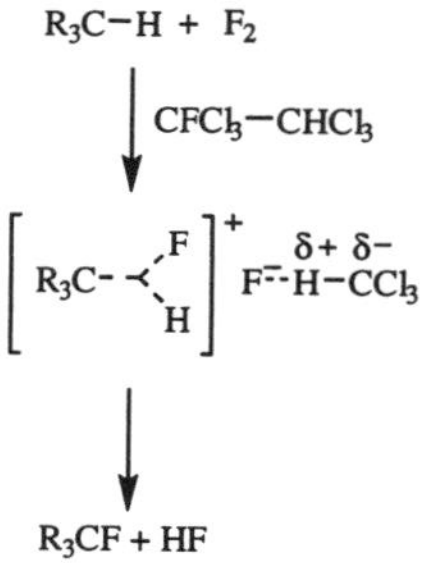 $R_3C\!-\!H + F_2 \xrightarrow{\ CFCl_3-CHCl_3\ } \left[R_3C\!\cdots\!\overset{F}{\underset{H}{Y}} \right]^{+} \overset{\delta+}{F}\cdots\overset{\delta-}{H}\!-\!CC_3 \longrightarrow R_3CF + HF$	45

b. Directly in aromatic systems

F_2/N_2

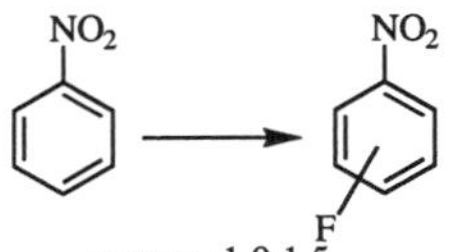

$o:m:p = 1:9:1.5$

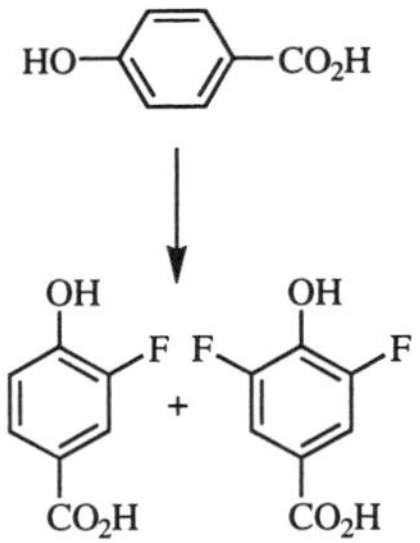

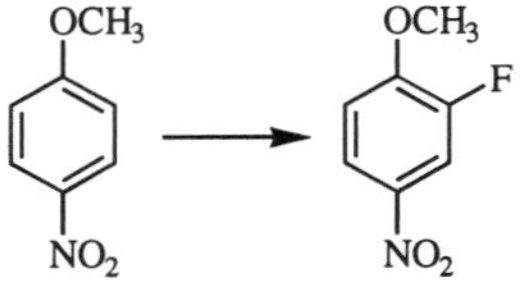

See comments immediately above (1a) regarding techniques. The phenolic substrates were fluorinated in AHF (polyolefin reaction vessels) to prevent attack at the heteroatoms (OH → OF; NH → NF).[48] Note the radiolabeling application[10,49]; fortuitously, production of fluorine 18 involves neon (see main chapter text, Section 2.2.2.2), an excellent inert diluent for F_2. Commercial ^{19}F fluorination of aromatics via F_2 appears to be confined to the production of 5-fluorouracil at present (see Chapter 9). Catalysis with Lewis acids (BCl_3, $AlCl_3$) has been studied.[48] At both plant and laboratory level, the method is considerably more costly and inconvenient than methods based on F^- sources (Halex and Balz–Schiemann methods; Sections A.6 and C.2 above).

Simplified approach:

(i) $ArH + "F^{+"} \rightarrow [ArHF]^+$

(ii) $[ArHF]^+ + F^- \rightarrow ArF + HF$

46–48
[Chapter 9]

(Continued)

Table 1. (Continued)

Common reagents	Typical conversions	Comments	Mechanistic category	Specific references [this volume]

^{18}F[F$_2$]/Ne

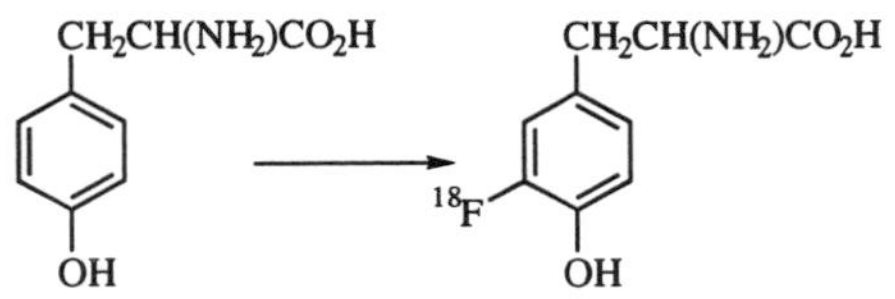

CH$_2$CH(NH$_2$)CO$_2$H   CH$_2$CH(NH$_2$)CO$_2$H

2. Electrophilic fluorination with "fluorine carriers" (usually prepared from F$_2$)^3

Oxidative fluorination via [ArH]$\cdot^+$ produced by SET (single electron transfer to O—F, Xe—F or N—F bonds) (cf. Section D.2.b below).

50

a. In aromatic systems

CF$_3$OF

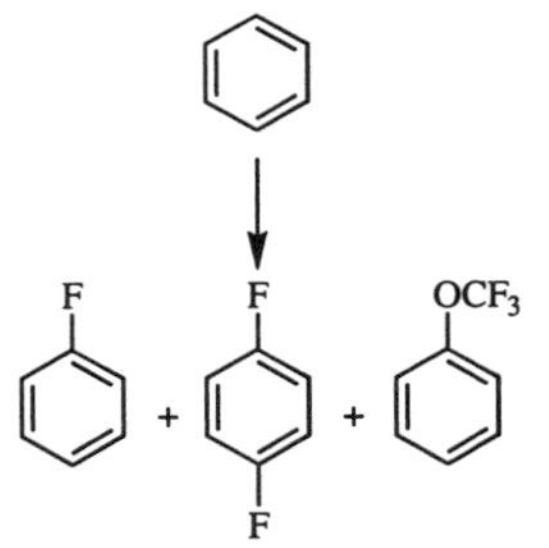

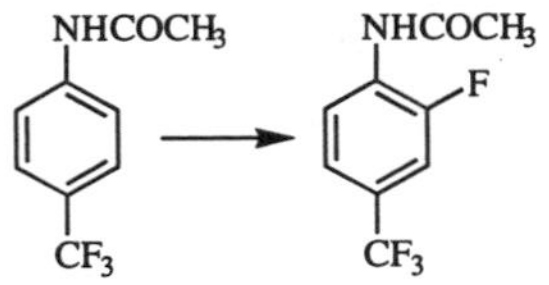

Trifluoromethyl hypofluorite is a hazardous gas (bp $-95\,°C$; $D[O\!-\!F] \sim 180$ kJ mol^{-1}) which must be manipulated with great care. It is made via treatment of CO or COF$_2$ with F$_2$. Cesium fluoroxysulfate (Cs$^+$FOSO$_3^-$, made from F$_2$ and CsSO$_4$ aq.) and acetyl hypofluorite (CH$_3$CO$_2$F, from, e.g., CH$_3$CO$_2$H–CH$_3$CO$_2$K + F$_2$) have been extensively investigated as sources of "F$^+$"; both are hazardous reagents useful only on a small scale.^{7,51}

52

XeF$_2$/HF

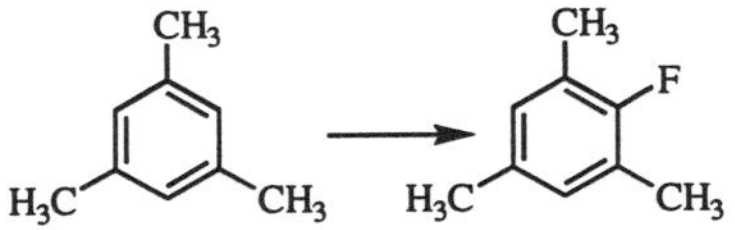

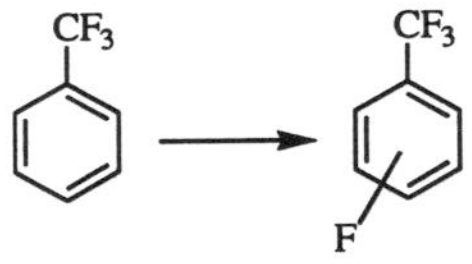

For comments on XeF$_2$ see Section E below.

m:p = 72:4

53

(C$_6$H$_5$SO$_2$)$_2$NF

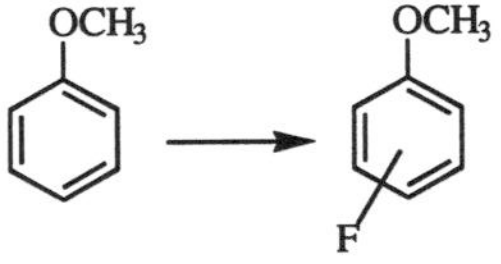

For information on *N*-fluoro(benzene sulfonimide), see Section b immediately below.

(*o:m:p* ratio = 58:5:37)

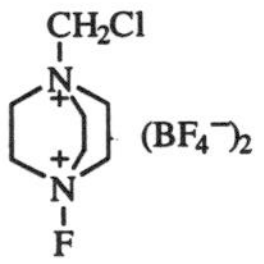

54

(Selectfluor™)

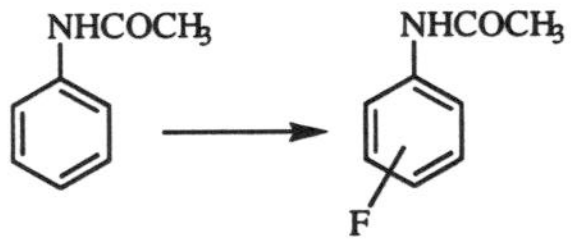

For information on Selectfluor™ reagents, see Section b immediately below.

(*o:p* ratio = 62:38)

(Continued)

Table 1. (*Continued*)

Common reagents	Typical conversions	Comments	Mechanistic category	Specific references [this volume]
b. Via fluorination of overt or masked carbanions				
FClO$_3$	C$_2$H$_5$CNa(CO$_2$Et)$_2$ → C$_2$H$_5$CF(CO$_2$Et)$_2$	Except for perchloryl fluoride (KClO$_4$ + FSO$_3$H → FClO$_3$), a hazardous gas (bp −46.7 °C), these reagents are made using F$_2$. The last two are easily handled, commercially available solids; the diazabicyclo[2.2.2]octane derivative is a member of the Selectfluor™ family (Air Products, USA), developed particularly for fluorosteroid synthesis.	S$_N$2 (Halophilic)...SET continuum (see footnote *c*).	17b, 55
(CF$_3$SO$_2$)$_2$NF	(CH$_3$CO)$_2$CHCH$_3$ → (CH$_3$CO)$_2$CFCH$_3$			56
(C$_6$H$_5$SO$_2$)$_2$NF	K$^+$ $^-$CH(CH$_3$)P(O)(OEt)$_2$ → CHF(CH$_3$)P(O)(OEt)$_2$ Na$^+$ $^-$C(CN)$_2$C$_6$H$_5$ → FC(CN)$_2$C$_6$H$_5$			53
Selectfluor™ (see Section a, p. 42)	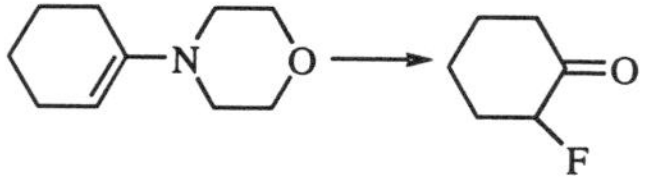			54
3. Anodic fluorination of aromatic hydrocarbons (Knunyants-Rozhkov ECF)				
(C$_2$H$_5$)$_3$N·3HF or (C$_2$H$_5$)$_4$N$^+$F$^-$	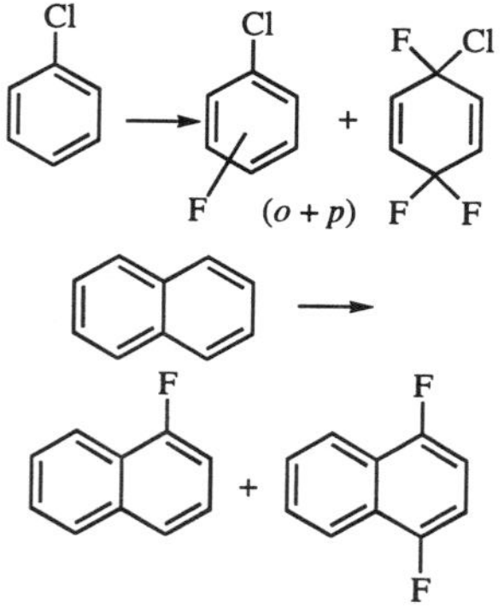	Solutions of substrates in CH$_3$CN containing tri- or tetraalkylammonium fluorides are electrolyzed in single-compartment cells with Pt anodes. The method is perhaps rather too "specialized" to be included here, except that a complete coverage of electrochemical fluorination (ECF) methods is warranted. (See Table 2 for other electrochemical processes.)	Electrochemical oxidation (see footnote *d*).	57, 58

E. Addition to C=C or C≡C bonds

			$C^+ + F^-$ Electrophilic addition of H^+ or Hal^+ followed by F^- attack.	
AHF/FSO₃H	$HC\equiv CH \rightarrow CH_2{=}CHF \rightarrow$ CH_3CHF_2			[Chapter 15]
AHF/pyridine	$C_4H_9C\equiv CH \rightarrow C_4H_9CF_2CH_3$	See Section B.1 (Olah's Reagent).		2, 3, 28
AHF/pyridine/NBA or NBS	$C_4H_9CH{=}CH_2 \rightarrow C_4H_9CHFCH_2Br$ $C_6H_5C\equiv CC_6H_5 \rightarrow$ $C_6H_5CF{=}CBrC_6H_5$ $CH_2{=}CHCO_2H \rightarrow$ $CH_2FCHBrCO_2H \downarrow$ (with NH₃) $CH_2FCH(NH_2)CO_2H$	Indirect ClF (via *N*-chloroamides or Cl₂) and IF (see below) additions are also well known. Other types of "positive halogen" sources (e.g., AgNO₂/Br₂) and of F⁻ have been used [e.g., (C₂H₅)₃N·3HF, which is less corrosive than AHF/pyridine]. Note that the halogen fluorides BrF and IF are unstable compounds, hence the need to use a "stoichiometric equivalent" source. ClF can be isolated, but it is highly reactive and not convenient to prepare in a non-specialist laboratory (Cl₂ + F₂ or ClF₃). Extensive study has been made of addition of preformed ClF across heteroatomic multiple bonds, e.g. CF₃CN + 2ClF → CF₃CF₂NCl₂; SO₃ + ClF → ClOSO₂F.		

AHF/pyridine/NBS or NIS then AgF/AHF/pyridine

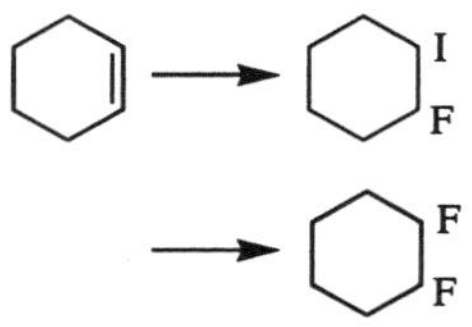

$C_6H_5CH{=}CHC_6H_5 \xrightarrow{\text{Steps}} \xrightarrow{} C_6H_5CHFCHFC_6H_5$

(Continued)

Table 1. (Continued)

Common reagents	Typical conversions	Comments	Mechanistic category	Specific references [this volume]
AHF or AHF/pyridine	$(CH_3)_2C=CH_2 \rightarrow (CH_3)_3CF$	Yields vary depending on the olefin and the reaction conditions. Low temperatures are used to suppress polymerization. Olah's reagent is used in laboratory circumstances (cf. Section B.1).		
HF/SbCl$_5$	$CCl_2=CHCl \rightarrow [CFCl_2CH_2Cl] \rightarrow CF_3CH_2Cl$	Forcing conditions are required with electron-poor C=C bonds. With chloro compounds subsequent halogen exchange is utilized commercially (see Section A.1).		[Chapter 7]
KF/H$_2$NCONH$_2$ or (i) AgF/CH$_3$CN (ii) H$_2$O	$CF_3CF=CF_2 \rightarrow CF_3CHFCF_3$	Susceptibility to nucleophilic attack is a hallmark of perfluoro-olefins; F$^-$ addition to give perfluorinated carbanions is of enormous importance in the building block approach to synthesis in organofluorine chemistry.	(i) $C=C \rightarrow FC-C^-$ (ii) $FC-C^- + [H^+] \rightarrow FC-CH$	7
IF$_5$/I$_2$	$CF_2=CF_2 \rightarrow CF_3CF_2I$ $CF_2=CH_2 \rightarrow CF_3CH_2I$	Mixtures of stoichiometric quantities of halogen fluoride and halogen ($IF_5 + 2I_2 \rightarrow 5IF$; $BrF_3 + Br_2 \rightarrow 3BrF$) are employed at elevated temperatures. IF addition to perfluoroalkenes is very important commercially. Note that the halogen exchange reaction $CI_4 + IF_5 \rightarrow CF_3I$ and the addition/exchange $CI_2=CI_2 + IF_5 \rightarrow C_2F_5I$ played important roles in the pioneering studies on organoelemental aspects of perfluorocarbon chemistry (see Chapter 1).	Thought to involve "electrophilic mechanisms."	3, 7 [Chapter 8]
BrF$_3$/Br$_2$	$CF_3CF=F_2 \rightarrow CF_3CFBrCF_3$			

XeF_2	$(C_6H_5)_2C{=}CH_2 \rightarrow (C_6H_5)_2CFCH_2F$ 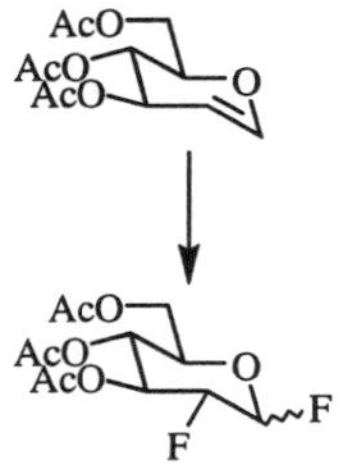 (Ac=CH_3CO)	Xenon difluoride (a solid, subl.p. 114 °C; commercially available, but costly) is normally used under mild conditions (≤25 °C) in CH_3CN (no catalyst) or CH_2Cl_2 (often in the presence of HF or BF_3-etherate as catalyst). Suitable only for small-scale work in view of the cost, it can be used in standard glassware when no catalyst is required.	$C{=}C + XeF_2/HF$ (or BF_3) → $[C{=}C]^{+\cdot}\ \dot{X}eF\ HF_2^-$ (or BF_4^-) → $CF{-}C^+\ Xe\ HF_2^-$ (or BF_4^-) → $CF{-}CF + Xe + HF$ (or BF_3)	4, 52

[a]AHF = anhydrous hydrogen fluoride.

[b]The adjectival prefix *halogeno* excludes fluorine.

[c]
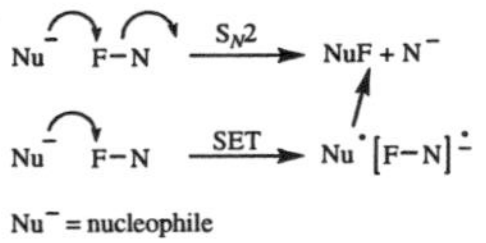
Nu⁻
F–N
SN2
NuF + N⁻
SET
Nu˙ [F–N]⁻˙
Nu⁻ = nucleophile

[d](i) $ArH - e^- \rightarrow [ArH]^{+\bullet}$
(ii) $[ArH]^{+\bullet} + F^- \rightarrow [ArFH]^\bullet$
(iii) $[ArFH]^\bullet - e^- \rightarrow [ArFH]^+$
(iv) $[ArFH]^+ - H^+ \rightarrow ArF$ and $[ArFH]^+ + F^- \rightarrow ArF_2H$

Table 2. Examples of Exhaustive (Per) Fluorination of Hydrocarbon-Type Feedstocks^a

Fluorinating agent	Typical conversions	Comments	Mechanisms	Specific references [this volume]
F_2	$C \rightarrow CF_4$	A commercial route, included here for the record; CF_4 is used as a plasma etchant and CFC replacement.		59, 7 [Chapter 4]
F_2	C_3H_8 $\rightarrow\uparrow$ C_3F_8 $CH_3CH{=}CH_2$ $(CH_3)_2CO \rightarrow (CF_3)_2CO$	Vapor-phase techniques ("Jet Fluorination" or "Porous Tube" Fluorination) are used for small molecules; C_3F_8 is utilized in the electronics industry (plasma etchant) and as a CFC replacement. Commercially, $(CF_3)_2CO$ is made from $CF_3CF{=}CF_2$ or $(CCl_3)_2CO/HF$ (Chapters 19, 27).	Reactions proceed via radical-chain mechanisms.^{b} Control of fluorine concentrations, rates of conversions of feedstocks, and efficient dissipation of the large amount of heat generated is absolutely crucial if skeletal fragmentation, and even catastrophic phenomena (charring, fires, and detonation), are to be avoided. Thermodynamic data for radical fluorinations (based on values for CH_4) are given in footnote b. Note the possibility for initiation via $F_2\cdots$ HR collisions (molecule–molecule initiation); since that step (ii) and propagation steps (iii,iv) have negative free-energy changes, the reactions can occur spontaneously even in cooled systems, despite the negligible dissociation of F_2 under such conditions.	11
F_2	$(CH_3)_2CHOCH(CH_3)_2$ $\downarrow$ LaMar $(CF_3)_2CFOCF(CF_3)_2$ $-[CH_2CH_2O]_x-$ $\downarrow$ LaMar $-[CF_2CF_2O]_x-$ 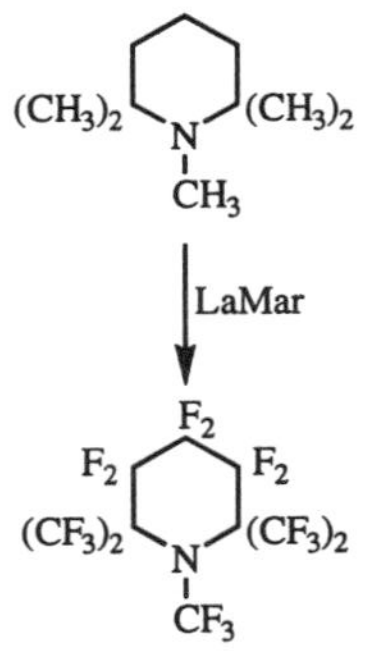 	Quite complicated molecules can often be perfluorinated successfully by a solid-phase low-temperature gradient (LTG) technique known as the LaMar slow-batch process, after its inventors (Lagow/Margrave). Yields vary considerably and conversion rates need to be kept low. Scale-up for liquid perfluoropolyether production ("gallons per hour") has been claimed by the Exfluor Research Corporation (USA).		

F2	$(CH_3)_4C$ $\downarrow AF$ $(CF_3)_4C$ $CH_3COC(CH_3)_3$ $\downarrow AF$ $CF_3COC(CF_3)_3$	A flow version of the LaMar LTG slow-batch process known as aerosol direct fluorination (AF) has been developed by Adcock, who is championing its scale-up.	12
F2	$-[CH_2CH(CH_3)]_n-$ $\downarrow LaMar$ $-[CF_2CF(CF_3)]_n-$	LaMar fluorination has been developed for converting finely divided solid hydrocarbon polymers into their perfluorocarbon analogs (cf. surface fluorination of polymers—Chapter 22).	
CoF3	$C_5H_{12} \rightarrow C_5F_{12}$ 	CoF$_3$ is the most frequently used HVMF (high-valency metal fluoride) reagent; AgF$_2$, MnF$_3$, CeF$_4$, PbF$_4$, and BiF$_5$ and related complexes (notably KCoF$_4$ and CsCoF$_4$, which are milder reagents than CoF$_3$) have also been utilized. AgF$_2$ has proved valuable for liquid-phase reactions, while CoF$_3$ is normally employed for vapor-phase work (and has been commercialized for such—see Chapter 4).	15, 60 [Chapter 4]
		Oxidative fluorination via radical-cation intermediates has been proposed [see footnote c].	

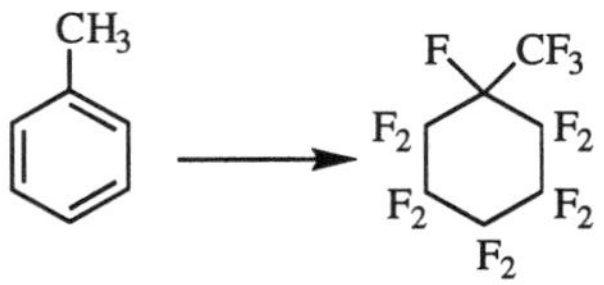

(Continued)

Table 2. (Continued)

Fluorinating agent	Typical conversions	Comments	Mechanisms	Specific references [this volume]
$KCoF_4$	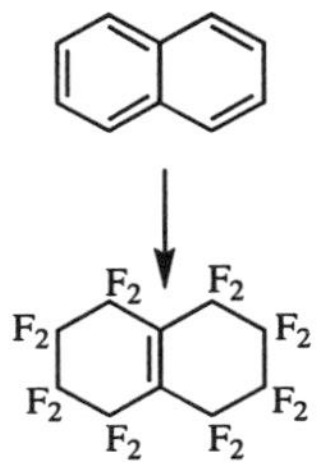	F_2 is needed to prepare HVMFs. Products are usually multicomponent (per and partial fluorination, degradation and skeletal rearrangement occur), the actual composition depending on the HVMF used and reaction temperature/contact time. HF is a byproduct so work-up needs care. Partially fluorinated feedstocks can be used advantageously, and in simple systems have proved valuable when H retention is sought, e.g., CHF_2CH_3 or $CF_2{=}CH_2 \rightarrow CF_3CH_2F$ (HFC-134a).		
AHF (Simons ECF)	$CH_3COF \rightarrow CF_3COF$ $C_7H_{15}COF \rightarrow C_7F_{15}COF +$ 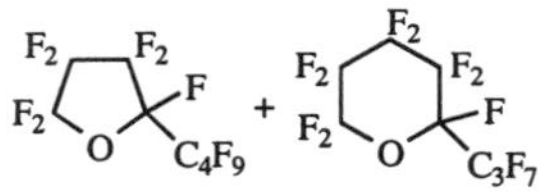 $(C_3H_7)_2O \rightarrow (C_3F_7)_2O$ $C_4H_9OCH_2CH_2OH \rightarrow C_4F_9OC_2F_5$ $(C_2H_5)_3N \rightarrow (C_2F_5)_3N$ $CS_2 \rightarrow CF_3SF_5$	Simons electrochemical fluorination is distinguished from exhaustive fluorination with F_2 or CoF_3 by the ability to retain important functional groups. Like CoF_3 fluorinations, perfluorination is accompanied by skeletal fragmentation, rearrangement, and cyclization processes. With pyridine, for example, the main products are C_5F_{12} [normal (mainly) and iso] and NF_3. Note the now-classical cyclization of the C_8 skeleton in octanoic acid; this gives useful yields of the two ethers shown. Octanesulfonyl chloride gives a *ca* 4:1 mixture of normal and branched $C_8F_{17}SO_2F$, plus mixed perfluorocarbons.	The primary act seems to be formation of a film of anodic nickel fluoride, probably followed (in part, at least) by oxidative fluorination via radical cations (cf. CoF_3 above). Much still remains to be discovered about the mechanism of this old technique.	16 [Chapter 5]

AHF (Simons ECF)	$CH_3SO_2F \rightarrow CF_3SO_2F$
	$CH_2(SO_2F)_2 \rightarrow CF_2(SO_2F)_2$
	$C_8H_{17}SO_2Cl \rightarrow C_8F_{17}SO_2F$
KF·2HF (Phillips ECF)	$C_2H_6 \rightarrow C_2F_xH_{6-x}$ (x = 1–6)	Gaseous or vaporized liquid substrates are fed into (and confined to) porous carbon anodes operating in a medium-temperature (100 °C) fluorine generator (electrolyte:molten KF·2HF).	[Chapter 5]

[a] The most important methods for converting hydrocarbon-based feedstocks directly into perfluorocarbon entities involve using F_2, CoF_3 (generated using F_2), or Simon's electrochemical fluorination (requiring AHF). Safe techniques for manipulating F_2 and AHF are well documented; and with the former reagent in particular there is no reason why any skillful organic chemist should not be able to undertake perfluorination experiments (neat F_2 and F_2—N_2 blends are available commercially). Metal equipment is required for all the techniques covered here, which adds considerably to the cost; this is particularly true of Simon's ECF. Nickel or nickel-based alloys are preferred materials of construction, but often copper or mild steel will suffice; the major problem is usually establishing safe working procedures (a dedicated laboratory, etc.).

[b] Reaction steps	ΔH_{298K} (kJ mol^{-1})	ΔG_{298K} (kJ mol^{-1})
Initiation		
(i) $F_2 \rightarrow 2F\cdot$	158	124
(ii) $F_2 + RH \rightarrow R\cdot + HF + F\cdot$	16	−24
Propagation		
(iii) $RH + F\cdot \rightarrow R\cdot + HF$	−141	−151
(iv) $R\cdot + F_2 \rightarrow RF + F\cdot$	−289	−285
Termination		
(v) $R\cdot + F\cdot \rightarrow RF$	−447	−408
(vi) $R\cdot + R\cdot \rightarrow R_2$	−351	−294
Overall change		
$F_2 + RH \rightarrow RF + HF$	−430	−435

Table 2. (*Continued*)

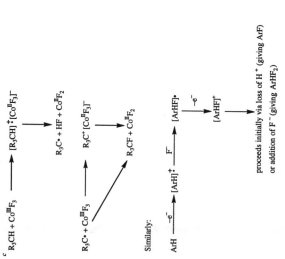

^{c} $R_3CH + Co^{III}F_3 \longrightarrow [R_3CH]^{\cdot +} [Co^{II}F_3]^-$

$\longrightarrow R_3C\cdot + HF + Co^{II}F_2$

$R_3C^+ [Co^{II}F_3]^-$

$R_3C\cdot + Co^{III}F_3 \longrightarrow$

$R_3CF + Co^{II}F_2$

Similarly:

$ArH \xrightarrow{-e^-} [ArH]^{\cdot +} \xrightarrow{F^-} [ArHF]^{\cdot}$

$\xrightarrow{-e^-} [ArHF]^+$

proceeds initially via loss of H^+ (giving ArF)
or addition of F^- (giving $ArHF_2$)

compounds.[17b] Older, but still useful book sources of data are provided in Ref. 18; and readers should also note that a supplement to Hudlický's book[6] will soon be published (ACS Books), and that a Houben-Weyl volume on Organofluorine Chemistry has already been planned. The compact, near-exhaustive, easily accessible early-1980s review of methods of fluorination in organic chemistry by Gerstenberger and Haas is recommended reading[19a]; so is Wilkinson's 1992 review "Recent Advances in the Selective Formation of the C—F Bond",[19b] in which the author points out that more than 6% of the 10 million or so compounds registered in the American Chemical Society's *Chemical Abstracts* possess one or more C—F bonds.

Readers seeking further information concerning the preparation, properties, and uses of inorganic fluorides employed as reagents in organofluorine chemistry will find useful the appropriate section (Fluorine Compounds: Inorganic) of the *Kirk-Othmer Encyclopedia of Chemical Technology*; organic fluorine compounds are also reviewed there, of course.

Interestingly, most of the techniques used commercially to create C—F bonds were well known prior to 1970, and many before 1950.

2.5. REFERENCES

1. S. L. Neidleman and J. Geigert, *Biohalogenation: Principles, Basic Roles and Applications*, Ellis, Horwood, Chichester (1986); M. Meyer and D. O'Hagan, *Chem. Br.* 28, 785 (1992).
2. M. Hudlicky and T. Hudlicky, in: *The Chemistry of Functional Groups, Supplement D: The Chemistry of Halides, Pseudohalides and Azides* (S. Patai and Z. Rappoport, eds.), Part 2, p. 1021, Wiley, Chichester (1983).
3. L. German and S. Zemskov (eds.), *New Fluorinating Agents in Organic Synthesis*, Springer-Verlag, Berlin (1989).
4. J. T. Welch and S. Eswarakrishnan, *Fluorine in Bioorganic Chemistry*, Wiley, New York (1991).
5. I. L. Knunyants and G. G. Yakobson (eds.), *Syntheses of Fluoroorganic Compounds*, Springer-Verlag, Berlin (1985).
6. M. Hudlický, *Chemistry of Organic Fluorine Compounds: A Laboratory Manual with Comprehensive Literature Coverage*, Ellis Horwood, Chichester (1976; second revised edn.; reissued 1984 and 1992).
7. For a recent overview, see R.E. Banks and J.C. Tatlow, in: *Fluorine: The First Hundred Years (1886–1986)* (R. E. Banks, D. W. A. Sharp, and J. C. Tatlow, eds.), pp. 251–272, Elsevier Sequoia, Lausanne (1986) [reproduced in *J. Fluorine Chem.* 33 (1986)].
8. R. Stephens and J. C. Tatlow, *Quart. Rev. Chem. Soc.* 16, 44 (1962); R. E. Banks and H. Goldwhite, *Handbook of Experimental Pharmacology*, Vol. XX: *Pharmacology of Fluorides, Part 1* (F. A. Smith, ed.), p. 1, Springer-Verlag, Berlin (1966).
9. S. T. Purrington, B. S. Kagen, and T. B. Patrick, *Chem. Rev.* 86, 997 (1986).
10. M. R. Kilbourn, *Fluorine-18 Labeling of Radiopharmaceuticals*, National Academy Press, Washington DC (1990) (Nuclear Medicine Series; NAS-NS-3203).
11. R. J. Lagow, in Ref. 7, p. 321.
12. J. L. Adcock, in Ref. 7, p. 327.
13. K. V. Scherer, K. Yamanouchi, and T. Ono, *J. Fluorine Chem.* 50, 47 (1990); A. C. Sievert, W. R. Tong, and M. J. Nappa, *J. Fluorine Chem.* 53, 397 (1991).
14. J. M. Tedder, in: *Advances in Fluorine Chemistry* (M. Stacey, J. C. Tatlow, and A. G. Sharpe, eds.), Vol. 2, p. 104, Butterworths, London (1961).
15. M. Stacey and J. C. Tatlow, in Ref. 14, Vol. 1, p. 166 (1960); Ref. 7, p. 267.

16. T. Abe and S. Nagase, in: *Preparation, Properties, and Industrial Applications of Organofluorine Compounds* (R. E. Banks, ed.), p. 19, Ellis Horwood, Chichester (1982).

17. (a) G. A. Boswell, W. C. Ripka, R. M. Scribner, and C. W. Tullock, *Org. React. 21*, 1 (1974); (b) C. M. Sharts and W. A. Sheppard, *Org. React. 21*, 125, (1974).

18. A. M. Lovelace, D. A. Rausch, and W. Postelnek, *Aliphatic Fluorine Compounds*, Reinhold, New York (1958); W. A. Sheppard and C. M. Sharts, *Organic Fluorine Chemistry*, Benjamin, New York (1969); R. D. Chambers, *Fluorine in Organic Chemistry*, Wiley-Interscience, New York (1973).

19. (a) M. R. Gerstenberger and A. Haas, *Angew. Chem., Int. Ed. Engl. 20*, 647 (1981); (b) J. A. Wilkinson, *Chem. Rev. 92*, 505 (1992).

20. A. K. Barbour, L. J. Belf, and M. W. Buxton, in Ref. 14, Vol. 3, p. 181 (1963).

21. J. S. Filippo and L. J. Romano, *J. Org. Chem. 40*, 782 (1975).

22. A. J. Bloodworth, K. J. Bowyer, and J. C. Mitchell, *Tetrahedron Lett. 28*, 5347 (1987).

23. G. G. Yakobson, and N. E. Akhmetova, *Synthesis* 169 (1983).

24. T. Ando, D. G. Cork, M. Fujita, T. Kimura, and T. Tatsuno, *Chem. Lett.* 1877 (1988).

25. N. Yoneda, T. Fukuhara, K. Yamagishi, and A. Suzuki, *Chem. Lett.* 1675 (1987).

26. D. P. Cox, J. Terpinski, and W. Lawrynowicz, *J. Org. Chem. 49*, 3216 (1984).

27. S. A. Pogany, G. M. Zentner, and C. D. Ringeisen, *Synthesis* 718 (1987).

28. G. A. Olah, J. G. Shih, and G. K. S. Prakash, in Ref. 7, p. 377.

29. R. A. Du Boisson, "Fluorodehydroxylation", PCR Report for September 1988 (available PCR Incorporated, PO Box 1466, Gainesville, FL 21602, USA).

30   P. A. Messina, K. C. Mange, and W. J. Middleton, *J. Fluorine Chem. 42*, 137 (1989); K. C. Mange and W. J. Middleton, *J. Fluorine Chem. 43*, 405 (1989).

31. C.-L. J. Wang, *Org. React. 34*, 319 (1987).

32. A. I. Burmakov, B. V. Kunshenko, A. Alekseeva, and L. M. Yagupolskii, in Ref. 3, p. 197.

33. M. Hudlicky, *Org. React. 35*, 513 (1988).

34. L. N. Markovskii and V. E. Pashinnik, in Ref. 3, p. 254.

35. W. F. Edgell and L. Parts, *J. Am. Chem. Soc. 77*, 4899 (1955).

36. D. P. Ashton, T. A. Ryan, B. R. Webster, and B. A. Wolfindale, *J. Fluorine Chem. 27*, 263 (1985).

37. J. Cuomo and R. A. Olofson, *J. Org. Chem. 44*, 1016 (1979); V. A. Dang, R. A. Olafson, P. R. Wolf, M. D. Piteau, and J.-P. G. Senet, *J. Org. Chem.* 1847 (1990).

38. H. Garcia, M. C. Perrod, L. Gilbert, S. Ratton, and C. Rochin, *J. Fluorine Chem. 54*, 117 (1991).

39. M. Muehlbacher and C. D. Poulter, *J. Org. Chem. 53*, 1026 (1988) and references cited therein.

40. R. Franz, *J. Fluorine Chem. 15*, 423 (1980).

41. See, for example, A. D. Borthwick, S. Butt, K. Biggadike, A. M. Exall, S. M. Roberts, P. M. Youds, B. E. Kirk, B. R. Booth, J. M. Cameron, S. W. Cox, C. L. P. Marr, and M. D. Shill, *J. Chem. Soc., Chem. Commun.* 656 (1988).

42. Yu. M. Pustovit and V. P. Nazaretian, *J. Fluorine Chem. 55*, 29 (1991).

43. H. Suschitzky, in Ref. 14, Vol. 4, p. 1 (1965).

44. T. N. Wade, *J. Org. Chem. 45*, 5328 (1980); T. N. Wade and R. Khéribet, *J. Org. Chem. 45*, 5333 (1980); G. M. Lavernhe, C. M. Ennakoua, S. M. Lacombe, and A. J. Laurent, *J. Org. Chem. 46*, 4938 (1981).

45. S. Rozen and C. Gal, *J. Org. Chem. 52*, 2769 (1987); *53*, 2803 (1988); S. Rozen, *Acc. Chem. Res. 21*, 307 (1988).

46. V. Grakauskas, *Intra-Sci. Chem. Rep. 5* (1), 85 (1971); *J. Org. Chem. 35*, 723 (1970).

47. S. Misaki, *Chem. Express 1* (11), 683 (1986).

48. S. T. Purrington and D. L. Woodard, *J. Org. Chem. 56*, 142 (1991).

49. R. Chirakal, E. S. Garnett, G. J. Schrobilgen, C. Nahmias, and G. Firnau, *Chem. Br. 27*, 47 (1991).

50. F. M. Mukhametshin, in Ref. 3, p. 69.

51. G. G. Furin, in Ref. 3, p. 35.

52. V. V. Bardin and Y. L. Yagupolskii, in Ref. 3, p. 1.

53. E. Differding and H. Ofner, *Synlett.* 187, 395 (1991); F. A. Davis and W. Han, *Tetrahedron Lett. 32*, 1631 (1991).

54. R. E. Banks, S. N. Mohialdin-Khaffaf, G. S. Lal, I. Sharif, and R. G. Syvret, *J. Chem. Soc., Chem. Commun.* 595 (1992).
55. Y. Takeuchi, K. Nagara, and R. Koizumi, *J. Org. Chem. 52*, 5062 (1987).
56. Z-.Q. Xu, D. D. Desmarteau, and Y. Gotoh, *J. Chem. Soc., Chem. Commun*, 179 (1991).
57. I. N. Rozhkov, *Russ. Chem. Rev. 45*, 615 (1976).
58. J. H. H. Meurs, D. W. Sopher, and W. Eilenberg, *Angew. Chem., Int. Ed. Engl. 28*, 927 (1989); J. H. H. Meurs and W. Eilenberg, *Tetrahedron 47*, 705 (1991).
59. A. J. Woytek, in Ref. 7, p. 331.
60. B. D. Joyner, in Ref. 7, p. 337.

3

Characteristics of C–F Systems

BRUCE E. SMART

3.1. INTRODUCTION

The physical properties and chemical reactivities of organic molecules can be dramatically affected by fluorination. Today's diverse commercial applications of organo-fluorine materials clearly manifest the possible beneficial effects of fluorination, and the numerous ways in which industry has taken practical advantage of these effects are the subjects of the following chapters in this book. This chapter focuses on the characteristic substituent effects that underlie the physicochemical properties of organofluorine compounds, especially those important to the design of commercial products. Fortunately, advances in both experimental and theoretical physical organo-fluorine chemistry over the past two decades have made the "unusual" behavior of fluorinated compounds much more understandable and predictable. General principles that govern the characteristic effects of fluorination will be presented, but important exceptions also will be noted. Misleading generalizations like "fluorination increases lipophilicity" and myths about fluorine steric effects will be dispelled, for example.

Actually, many of the characteristic effects of fluorination can be anticipated simply by comparing some fundamental atomic properties of fluorine with those of other elements (Table 1). The high ionization potential of fluorine[1] and its relatively low polarizability[2] imply very weak intermolecular interactions, low surface energies, and low refractive indexes for perfluorocarbons. The extreme electronegativity of fluorine[1] insures that it always will be electron-withdrawing inductively when bonded to carbon, and the bonds will be strongly polarized $\delta^+C \rightarrow F\delta^-$. Consequently, C—F bonds will have relatively high ionic character and will be stronger than other C—X bonds. (Bond polarity in *molecules* can be derived explicitly from electronegativity differences.[3]) Another consequence of the C—F bond dipole is that partially fluorinated materials can have significant polar character, and their physical properties

BRUCE E. SMART • DuPont Central Research and Development, Experimental Station, Wilmington, Delaware 19880-0328.

Organofluorine Chemistry: Principles and Commercial Applications, edited by R. E. Banks *et al.* Plenum Press, New York, 1994.

Table 1. Atomic Physical Properties[a]

Atom	IP (kcal mol^{-1})	EA (kcal mol^{-1})	α_v (Å^3)	r_v (Å)	χ_p
H	313.6	17.7	0.667	1.20	2.20
F	401.8	79.5	0.557	1.47	3.98
Cl	299.0	83.3	2.18	1.75	3.16
Br	272.4	72.6	3.05	1.85	2.96
I	241.2	70.6	4.7	1.98	2.66
C	240.5	29.0	1.76	1.70	2.55
N	335.1	[−6.2]	1.10	1.55	3.04
O	314.0	33.8	0.82	1.52	3.44

[a]IP, ionization potential [Ref. 1]; EA, electron affinity [Ref. 1]; α_v, atom polarizability [Ref. 2]; r_v, van der Waals' radius [Ref. 4]; χ_p, electronegativity (Pauling) [Ref. 1].

therefore often are quite different from those of either their hydrocarbon or perfluoro-carbon counterparts.

In short, the electronic effects of fluorination on molecular properties can be attributed to the unique combination of the atom's properties: its high electronegativity and moderately small size,[4] its three tightly bound, nonbonding electron pairs, and the excellent match between its $2s$ or $2p$ orbitals with the corresponding orbitals of carbon.

3.2. PHYSICAL PROPERTIES

3.2.1. General

Saturated perfluorocarbons (PFCs) are characterized by an unusual combination of physical properties relative to their analogous hydrocarbons (HCs).[5,6] Some specific comparative data for perfluoro-n-hexane versus n-hexane are given in Table 2.[5] Their densities are typically about 2.5 times those of the corresponding HCs, and they have significantly greater compressibilities and viscosities but lower internal pressures. Saturated PFCs have the lowest dielectric constants, refractive indexes, and surface tensions of any liquids at room temperature, which reflects their nonpolar character and extremely low polarizabilities. Many of these properties also are shared by perfluorinated ethers and tertiary amines.[7]

These physical properties, coupled with their outstanding chemical and thermal stabilities, make perfluorinated materials ideal candidates for several commercial applications. They are excellent convective coolants owing to their high densities, viscosities, and coefficients of expansion. Their chemical inertness combined with low dielectric constants, low dielectric losses, and high dielectric strengths and resistance makes them superior insulating materials, especially for electronics applications (see Chapters 4 and 5).

Although many physical properties of partially fluorinated alkanes (HFCs) fall between those of PFCs and HCs, there can be pronounced differences. For example,

Table 2. Comparative Physical Properties of *n*-Hexanes[5,8]

Property[a]	C_6F_{14}	$F(CF_2)_3(CH_2)_3H$	C_6H_{14}
bp (°C)	57	64	69
ΔH_v (kcal mol^{-1})	6.7	7.9	6.9
T_c (°C)	174	200	235
d^{25} (g cm^3)	1.672	1.265	0.655
η^{25} (cP)	0.66	0.48	0.29
γ^{25} (dyn cm^{-1})	11.4	14.3	17.9
β (10^{-6} atm^{-1})	254	198	150
n_D^{25}	1.252	1.290	1.372
ε	1.69	5.99	1.89

[a]ΔH_v, heat of vaporization; T_c, critical temperature; d, density; η, viscosity; γ, surface tension; β, compressibility at 1 atm; n, refractive index; ε, dielectric constant.

the dielectric constant of *n*-$CF_3(CF_2)_2(CH_2)_2CH_3$ is much higher than the values for either *n*-C_6F_{14} or *n*-C_6H_{14}; and *n*-$C_6H_{13}F$ (bp 91.5°C) boils much higher and has a considerably higher surface tension (19.8 dyn cm^{-1}) than either *n*-C_6F_{14} or *n*-C_6H_{14}.[8] These "anomalies" point to the importance of polar effects in HFCs owing to net C—F or C—C dipoles that are absent in PFCs or HCs.

There are many excellent reviews of the physical properties of fluorocarbons,[5–7,9] and only the boiling points, surface energies, and solvent/solubility properties are chosen to illustrate the characteristic effects of fluorination in the following subsections. The influence of fluorination on acid and base strengths is covered in Section 3.2.4, and the related effects on hydrogen bonding, including the role of C—F bonds as hydrogen-bond acceptors, are discussed in the final section.

3.2.1.1. Boiling Points

The boiling points (bps) of homologous linear PFCs and HCs closely track each other (Table 3). This is remarkable considering the much higher molecular weights of the PFCs. *n*-Hexane and CF_4, for example, have nearly the same molecular weights, 86 and 88, respectively, yet their bps differ by nearly 200°C! The bps of saturated PFCs clearly are much lower than expected based on their molecular weights; in fact, they

Table 3. Boiling Points of Homologous Perfluoroalkanes and Alkanes

	bp (°C)									
	n = 1	2	3	4	5	6	7	8	9	10
n-C_nF_{2n+2}	−128	−78	−38	−1	29	57	82	104	125	144
n-C_nH_{2n+2}	−161	−88	−42	−0.5	36	69	98	126	151	174
c-C_nF_{2n}	—	—	−32	−6	23	53	81	102	—	—
c-C_nH_{2n}	—	—	−34	13	50	81	118	151	—	—

boil only 25–30°C higher than noble gases with similar molecular weights. Moreover, branching has a negligible effect on the bps of perfluorinated compounds, in marked contrast with the behavior of the corresponding hydrocarbon systems. Some comparisons are given below.[6,10]

The lower bps of perfluoro-ethers and -ketones compared to their HC analogs also belie their much higher molecular weights, and this holds for tertiary amines as

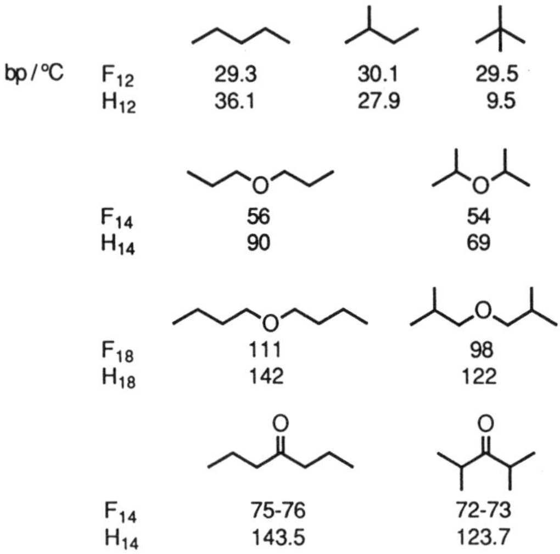

bp / °C				
	F_{12}	29.3	30.1	29.5
	H_{12}	36.1	27.9	9.5
	F_{14}	56	54	
	H_{14}	90	69	
	F_{18}	111	98	
	H_{18}	142	122	
	F_{14}	75-76	72-73	
	H_{14}	143.5	123.7	

well—bp $(n\text{-}C_4F_9)_3N$ = 174–178°C versus bp $(n\text{-}C_4H_9)_3N$ = 216.5°C, for example. These characteristic boiling-point trends reflect extremely low intermolecular interactions in perfluorinated compounds, which behave as nearly ideal liquids.[6]

Partially fluorinated hydrocarbons (HFCs) display entirely different trends indicative of substantial intermolecular interactions. The effects of increasing halogen content on methane bps is illustrative (Table 4). As expected, the bps of chloro- and bromomethanes progressively increase with halogen content (and molecular weight), but for the fluoromethanes, the bps increase from –161°C for CH_4 to a maximum of –51°C for CH_2F_2, and then decrease upon further fluorination. This peculiar trend indicates CH_2F_2 is rather polar and it is not coincidental that the bps of the fluoromethanes parallel their dipole moments[11]: CH_3F (μ = 1.85 D), CH_2F_2 (1.97 D),

Table 4.   Boiling Points of Halomethanes

$CH_{4-n}X_n$	bp (°C)			
X =	n = 1	2	3	4
F	–78.6	–51.6	–82.2	–128
Cl	–24.2	40.1	61.3	76.8
Br	3.6	98.2	149.5	189.5

CHF$_3$ (1.65 D), and CF$_4$ (0.0 D). (An associated, intermolecular hydrogen-bonded structure for CH$_2$F$_2$ also has been proposed to account for its high bp.[2]) Other HFCs show some correlation between bp and dipole moment, but there are many exceptions. For example, CH$_3$CF$_3$, with a highly polar C—C bond (μ = 2.32 D) boils at –47°C, considerably higher than either C$_2$H$_6$ (bp –88°C) or C$_2$F$_6$ (bp –78°C), but CH$_3$CHF$_2$ (μ = 2.27 D, bp –24.7°C) has nearly the same dipole moment as CH$_3$CF$_3$ and yet it boils about 22°C higher. The relationship is also somewhat irregular for fluoroalkenes, in particular, the bp of CH$_2$=CF$_2$ is unexpectedly low: CH$_2$=CH$_2$ (bp –102°C), CH$_2$=CHF (bp –72°C, μ = 1.43 D), CH$_2$=CF$_2$ (bp –83°C, μ = 1.38 D), CHF=CF$_2$ (bp –51°C, μ = 1.40 D), and CF$_2$=CF$_2$ (bp –76°C). For cis/trans isomers, however, the isomer with the higher dipole moment normally has the higher boiling point. Compare, for instance, cis-CHF=CHF (bp –25°C, μ = 2.42 D) versus trans-CHF=CHF (bp –51°C) and cis-CF$_3$CH=CHCF$_3$ (bp 33°C) versus trans-CF$_3$CH=CHCF$_3$ (bp 9°C).[13] Although boiling points vary considerably depending upon the degree and positions of fluorination, fluorination actually has only a small, usually less than 1 kcal mol^{-1}, and unpredictable effect on heats of vaporization (and sublimation).[14,15]

The importance of C—F or C—C bond dipoles to the properties of partially fluorinated materials will be a leitmotif in the following sections. From a practical standpoint, the contrasting piezoelectric character of the isomeric polymers poly(vinylidene fluoride) (PVF$_2$) and poly(ethylene-co-tetrafluoroethylene) (ETFE) provides a particularly striking example of the key role polar C—F bonds can have (see Chapter 15). Poled PVF$_2$ adopts a zigzag chain conformation in which the C—F bond dipoles are all aligned in the same direction, both micro- and macroscopically.[16] A highly anisotropic polar structure results, and this is the source of the material's useful piezoelectric property. By contrast, ETFE cannot similarly adopt a polar conformation and its piezoelectric character is negligible.

PVF$_2$ 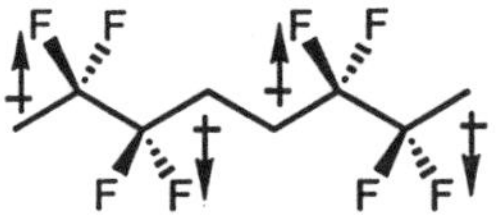   ETFE

3.2.1.2. Surface Energies

a. Surface Tension. Since surface tension (γ) is a measure of molecular forces per unit length (in dyn cm^{-1} = mN m^{-1}) on a liquid surface that oppose expansion of its surface area, PFCs not surprisingly have the lowest γ values of any organic liquids (Table 5) and will completely wet practically any solid surface. (The only exceptions are graphite fluorides and Langmuir–Blodgett films of certain fluorinated monolayers,

Table 5.   Surface Tensions of Perfluorocarbons[5,6] vs Hydrocarbons[17]

Structure	γ (dyn cm^{-1})[a]	
	PFC	HC
n-Pentane	9.4	15.2
n-Hexane	11.4	17.9
n-Octane	13.6	21.1
Methyl-*c*-hexane	15.4	23.3
Decalin	17.6	29.9[b,c]
Benzene	22.6[d]	28.9[b]

[a]At 25°C unless noted otherwise.
[b]20°C.
[c]*trans* isomer.
[d]23°C.

vide infra.) Perfluorinated ethers and amines also have low surface tensions, typically 15–16 dyn cm^{-1},[7] which reflect their PFC-like character.

The surface tensions of HFCs are always greater than those of the corresponding PFCs but can be greater, smaller, or equal to those of their HC counterparts. For instance, the γ values for the hexanes $H(CH_2)_n(CF_2)_{6-n}F$ progressively increase from 12.6 dyn cm^{-1} for $H(CF_2)_6F$ to 17.9 dyn cm^{-1} for $H(CH_2)_5CF_3$, which has the same value as n-C_6H_{14}.[8] For n-$C_6H_{13}F$, however, $\gamma = 19.8$ dyn cm^{-1} which implies more polar character than found with the other hexanes in this series. With longer-chain homologs, the hydrocarbon character often dominates the polar contributions. Thus, n-$C_{12}H_{26}$, n-$C_{12}H_{25}F$, and n-$C_{11}H_{23}CF_3$ all have about the same surface tensions (24.9, 24.6, and 24.1 dyn cm^{-1}, respectively). Further fluorination then progressively reduces the surface tensions but, interestingly, γ levels off at about 17.9 dyn cm^{-1} for $H(CH_2)_n(CF_2)_{12-n}F$ ($n = 2$–6).[8]

The effect of fluorination on solid surface free energies parallels the liquid trends.[18,19] The PFC polymers have the lowest γ_C values, 16.2 dyn cm^{-1} for poly[(CF_2CF(CF_3))] and 18.5 for poly(CF$_2$CF$_2$), which directly relate to their useful antistick and low frictional properties (see Chapters 7 and 10), whereas HFC polymers have substantially higher values [γ_C for poly(CH$_2$CF$_2$) = 25; 28 for poly(CH$_2$CHF)] approaching that for polyethylene ($\gamma_C = 31$ dyn cm^{-1}). Substituting the more polarizable Cl atom for F also markedly increases the surface free energy [$\gamma_C = 31$ dyn cm^{-1} for poly(CF$_2$CFCl)].

The lowest γ_C value reported is 6 dyn cm^{-1} for a $CF_3(CF_2)_{10}CO_2H$ monolayer on platinum.[19,20] This surface comprises closely packed CF_3 groups and is the least wettable surface known. Substituting but one H for F, however, markedly increases the surface free energy to 15 dyn cm^{-1} [$H(CF_2)_{10}CO_2H$ monolayer].[19] Fluorinated graphites $(C_2F)_n$ and $(CF)_n$ also have extremely low γ_C values, approaching 6 dyn cm^{-1}.[21]

Perfluorinated materials are *not* required to achieve low surface energies, which is important to the economical design of commercial water/oil repellants and other surfactants (Chapter 14). Only the outermost surface groups need to be perfluorinated. Monolayers of $CF_3(CF_2)_n(CH_2)_{16}CO_2H$ have values comparable to those for $CF_3(CF_2)_nCO_2H$ (6–9 dyn cm^{-1}) when $n \geq 6$.[19] The acrylate and methacrylate polymers of n-$C_7F_{15}CH_2O_2C(R)$=CH_2 (R = H; CH$_3$) have γ_C values of 10.4 and 10.6 dyn cm^{-1},[19] respectively, and several methyl(ethyleneperfluoroalkyl)siloxane polymers have similarly low surface energies.[22] A terminal CF$_3$ group alone, however, is not sufficient to produce particularly low γ_C values ($\gamma_C \cong 18$ dyn cm^{-1} for a $CF_3(CH_2)_{16}CO_2H$ monolayer) and a fairly long attached $(CF_2)_n$ chain ordinarily is necessary to insure proper molecular alignment at the surface. The fascinating relationships among head-group structure and chain composition, orientation and order, and monolayer surface properties have been reviewed[19] and continue to be actively investigated.[23–26]

b. Surface Activity. When a hydrophilic functional group is attached to a PFC, an extremely surface-active material results. Fluorocarbon surfactants can commonly reduce the surface tension of water (72 dyn cm^{-1}) to 15–20 dyn cm^{-1}, compared with 25–35 dyn cm^{-1} for HC surfactants.[27,28] The longer the PFC chain, the higher is its surface activity. For example, aqueous solutions of n-$C_nF_{2n+1}CO_2Li$ above their critical micelle concentrations (cmcs) have $\gamma = 27.8$, 24.6, and 20.5 dyn cm^{-1} for $n = 6$, 8, and 10, respectively.[28] Both nonionic and polymeric materials also can be very effective surfactants. Above its cmc, $\gamma = 16$ dyn cm^{-1} for solutions of n-$C_6F_{15}CH_2C(O)N[(C_2H_4O)_2CH_3]_2$[29]; and block copolymers of 2-(perfluoroethyl)- or 2-(perfluoropropyl)-oxazoline with 2-ethyloxazoline can lower γ to 12.8 dyn cm^{-1} despite their very short PFC chains.[30] As expected, more polar or polarizable head groups on the surfactant reduce its surface activity. The minimum γ values for aqueous $HCF_2(CF_2)_5CO_2H$ and $Cl(CF_2CFCl)_3CF_2CO_2H$ are 21.8 and 24.0 dyn cm^{-1}, respectively, compared with 15.3 dyn cm^{-1} for $CF_3(CF_2)_6CO_2H$.[20]

Among the many applications for fluorosurfactants (Chapter 14), they have proved particularly useful for microemulsifying PFCs. Infinitely stable aqueous microemulsions for potential artificial blood applications have been prepared with the very active n-$C_nF_{2n+1}CH_2(OC_2H_4)_nOH$ and n-$C_nF_{2n+1}CH_2C(O)N[(C_2H_4O)_mCH_3]_2$ classes of surfactants.[29,31] A combination of n-$C_8F_{17}CH_2CH_2CO_2K$ and n-$C_4F_9CH_2CH_2OH$ surfactants has been reported recently to microemulsify n-$C_8F_{17}CH$=CH_2 in $HC(O)NH_2$.[32] This is the first example of a nonaqueous microemulsion.

Perhaps the most unusual class of surfactants are the diblock amphiphiles $F(CF_2)_m(CH_2)_nH$. They can behave like prototypical surfactants[33] and form micelles in both PFC and HC media.[34]

c. Micellization. The more surface active the amphiphile, the greater is its tendency to form micelles. The cmc of $CF_3(CF_2)_6CO_2Na$ (0.036 M) is ten times smaller than that for $CH_3(CH_2)_6CO_2Na$ (0.4 M), and in general the cmcs of linear C_6–C_{12} perfluorocarboxylates are about equal to those of their HC counterparts with 50% longer chains.[27,28] Nonionic, fluorinated poly(oxyethylene) surfactants similarly have cmcs equal to those of their hydrocarbon analogs with lipophilic chains 1.5–1.8 times longer (cmc = 4.8×10^{-5} M for $n\text{-}C_7F_{15}CH_2CH_2(OC_2H_4)_5OH$ and 5.8×10^{-5} M for $n\text{-}C_{10}H_{15}CH_2CH_2(OC_2H_4)_5OH$).[29] Perfluoropolyether surfactants, however, behave much like their PFC counterparts. The cmc of $CF_3[OCF_2CF(CF_3)]_{3-4}OCF_2CO_2NH_4$ (2×10^{-4} M),[35] for instance, compares to the values of $2–4 \times 10^{-4}$ M for $n\text{-}C_8F_{17}CO_2M$ salts.[27,36] Again, the properties of lightly fluorinated surfactants can be markedly different. The cmcs of $CF_3(CH_2)_nCO_2Na$ actually are about twice those of $CH_3(CH_2)_nCO_2Na$,[37] which imply that the polar CF_3—CH_2 group imparts increased hydrophilicity (see Section 3.2.3).

Fluorination also can significantly affect the size and shape of micelles.[38–41] Linear PFC surfactants form larger aggregates than their HC counterparts, and rod-like, cylindrical micelles are produced by $n\text{-}C_7F_{15}CO_2NH_4$[39] and $n\text{-}C_8F_{17}SO_3NR_4$ (R = Me, Et) salts,[41] for example, in marked contrast to their HC analogs which give only spherical micelles. (The counterion in the PFC cases can have an influence too. Unlike the NH_4 salt, $n\text{-}C_7F_{15}CO_2Na$ forms smaller, spherical micelles.[38]) Even certain mixed micelles of $n\text{-}C_7F_{15}CO_2NH_4$ and $n\text{-}C_{10}H_{21}CO_2NH_4$ are reported to be cylindrical.[40] Other aggregate structures have been reported. Disc-like micelles are claimed for $n\text{-}C_8F_{17}CO_2NHEt_2$,[40] and liposomes can be formed from multifluorocarbon-tailed surfactants.[42]

The tendency of PFC amphiphiles to produce rod-like rather than spherical aggregates has been attributed to two important characteristics of the $(CF_2)_n$ chain: its conformation and its stiffness. Unlike the zigzag $(CH_2)_n$ chain, $(CF_2)_n$ adopts a twisted, helical structure. This is true in the solid state,[33,43] in mono- and multilayers,[25,26,44] and even in the gas phase.[45] There is cogent evidence that the helical $(CF_2)_n$ chain is much stiffer than $(CH_2)_n$. Results of fluorescence-quenching experiments in solution,[46] and the temperature-dependent behavior of $F(CF_2)_n(CH_2)_mH$ diblock[47] and $F(CF_2)_n(CH_2)_m(CF_2)_nF$ triblock[48] materials, $[F(CF_2)_8(CH_2)_{10}Co_2]_2Cd$ multilayers,[44] and polymers with pendant PFC groups[25] all reveal greater conformational flexibility of $(CH_2)_n$ versus $(CF_2)_n$ chains. Moreover, the unusual liquid crystalline properties of $F(CF_2)_n(CH_2)_mH$ diblock alkanes[49] and polymers containing $(CF_2)_n$ blocks[50] imply prototypical, rigid-rod mesogen character in the $(CF_2)_n$ chain. The relatively stiff, helical chains in PFC surfactants thus are better accommodated by cylindrical rather than spherical packing in the larger aggregates.[39,40]

A recent report on the aqueous emulsion polymerization of $CF_2{=}CF_2$ (TFE) using $n\text{-}C_8F_{17}CO_2Li$ raises the intriguing possibility that the size and structure of the surfactant aggregate may control polymer morphology.[51] Normal spherical poly(TFE) particles result when the surfactant concentration is below its cmc, but above its cmc long rod-shaped particles are produced!

3.2.2. Solvent Polarity

Organofluorine compounds represent some of the least and most polar solvents known. Perfluorinated compounds are extremely nonpolar and are poor solvents for all materials except those with very low cohesive energies, such as gases, PFCs, or some inorganic fluorides like WF_6. Saturated PFCs are practically insoluble in water, HF, and usually only slightly soluble in HCs, but can dissolve low-molecular-weight HCs to an appreciable extent.[6,52] The cohesive pressures of PFCs are only about half those of their corresponding HCs,[52] heats of solution of PFCs are much different from those of HCs,[52,53] and enthalpies of interaction between PFCs and HCs are smaller than interactions between HCs.[53] In terms of solvent–solute interactions, PFCs are more like Ar and Kr than HCs.[5] The profound difference in interaction energies of PFCs versus HCs revealed earlier by their boiling-point trends also is clearly manifested by the grossly nonideal behavior of their mixtures, which is a subject of long-standing interest to physical chemists.[9,54–56]

Among several empirical solvent polarity scales,[57] one recently introduced by Middleton and co-workers[58] based on solvatochromism is particularly useful for ranking fluorinated solvents. Some comparative values are shown in Table 6. The values emphasize the nonpolar character of PFCs but also reveal the increased polarity of HFCs ($P_S = 7.52$ for C_6H_5F versus 6.95 for C_6H_6) and the extraordinarily high polarity of fluorinated alcohols. The values of 10.2 and 11.08 for CF_3CH_2OH and $(CF_3)_2CHOH$ compare to 10.64 and 12.1 for 50% aqueous HCO_2H and H_2O, respectively, on this scale. This reflects the strong hydrogen-bonding character of the alcohols (see Section 3.2.5).

A peculiar and useful property of PFCs is their ability to dissolve oxygen and other gases[59]: PFCs dissolve about three times more oxygen than their analogous HCs, and about ten times more than water. This property underlies their use as oxygen carriers for artificial blood and organ perfusion applications (Chapter 26). The higher solubility of O_2 is not due to any specific attractive interaction between PFCs and O_2,[60–62] but rather results from the existence of large cavities (free volume) in PFC liquids that can accommodate the gaseous molecules. The solubility of O_2 and other gases directly correlates with the isothermal compressibility of the PFC.[63]

Table 6. Solvent Spectral Polarity Index[58]

Solvent	P_S	Solvent	P_S
n-C_6F_{14}	0.00	n-C_6H_{14}	2.56
c-$C_6F_{11}CF_3$	0.46	c-$C_6H_{11}CH_3$	3.34
n-C_8F_{18}	0.55	n-C_8H_{18}	2.86
$(n$-$C_4F_9)_3N$	0.68	$(n$-$C_4H_9)_3N$	3.93
c-$C_{10}F_{18}{}^a$	0.99	c-$C_{10}H_{18}{}^b$	4.07
$CF_2ClCFCl_2$	3.22	$CHCl_2CHCl_2$	9.23
$CFCl_3$	3.72	CCl_4	4.64
C_6F_6	4.53	C_6H_6	6.95
CF_3CO_2Et	6.00	CH_3CO_2Et	6.96
C_6H_5F	7.52	C_6H_5Cl	8.30
o-$C_6H_4F_2$	7.86	o-$C_6H_4Cl_2$	8.94
CF_3CH_2OH	10.2	CH_3CH_2OH	8.05
$(CF_3)_2CHOH$	11.08	$(CH_3)_2CHOH$	7.85

[a] Perfluorodecalin.
[b] Decalin.

3.2.3. Lipophilicity

Lipophilicity is an important consideration in the design of biologically active compounds since it often controls absorption, transport, or receptor binding. It is commonly held that fluorination increases lipophilicity, but this is true only for aromatic fluorination and fluorination adjacent to most atoms or groups with π-electrons. Monofluorination and trifluoromethylation of saturated aliphatic groups normally *decrease* lipophilicity.

Octanol/water partition coefficients (P) are the most common quantitative measure of lipophilicity, and several Hansch–Leo π substituent values derived from substituted benzenes[64] ($\pi_X = \log P_{C_6H_5X} - \log P_{C_6H_6}$) are listed in Table 7. Fluorination clearly increases lipophilicity and its effect can be very large when heteroatoms are present. The sulfone $C_6H_5SO_2CF_3$, for instance, is 150 times more lipophilic than $C_6H_5SO_2CH_3$ (The relative π values vary somewhat from system to system; cf. $\pi_{CH_3} = 0.54$ and $\pi_{CF_3} = 1.16$ for 4-substituted phenylacetic acids.[65])

Aromatic fluorination invariably increases lipophilicity, even when it significantly increases acidity and hydrogen-bonding capability (see Sections 3.2.4 and

Table 7. Hydrophobic[a] and Electronic Substituent Constants[b]

Substituent	π	σ_I	σ_R^0
F	0.14	0.52	–0.34
Cl	0.71	0.46	–0.23
OH	–0.67	0.29	–0.43
NO_2	–0.27	0.56	0.22
CH_3	0.56	–0.04	–0.15
CF_3	0.88	0.39	0.10
CH_3CH_2	1.02	–0.02	–0.14
CF_3CH_2	—	0.14	–0.05
CF_3CF_2	1.89	0.41	0.11
OCH_3	–0.02	0.27	–0.43
OCF_3	1.04	0.39	–0.04
SCH_3	0.61	0.23	–0.17
SCF_3	1.44	0.42	0.06
SO_2F	0.05	0.75	0.26
SO_2CH_3	–1.63	0.48	0.16
SO_2CF_3	0.55	0.73	0.31
$NHSO_2CH_3$	–1.18	0.42	–0.21
$NHSO_2CF_3$	0.92	0.49	–0.10
C_6H_5	1.96	0.08	–0.09
C_6F_5	—	0.25	0.02

[a]Ref. 64.
[b]Values from ^{19}F NMR data [Refs. 64, 69].

3.2.5). For example, $\log P = 3.23$ and 1.48 for C_6F_5OH and C_6H_5OH,[64] respectively. For partitioning between solvents with markedly different hydrogen-bonding basicities, C_6F_5OH is still measurably more lipophilic [$\log P(c\text{-}C_6H_{12}/H_2O) = -0.52$ for C_6F_5OH and -1.00 for C_6H_5OH].[64]

Fluorination of C adjacent to atoms with π-electrons also increases lipophilicity, and this seems to hold for $C{=}C$ bonds as well ($\log P = 1.13$, 1.24, and 2.00 for $CH_2{=}CH_2$, $CH_2{=}CF_2$, and $CF_2{=}CF_2$, respectively[66]). The situation is more complex for fluorinated aliphatic carbonyl compounds, however. α-Fluorinated ketones or aldehydes commonly form stable hydrates and ketals or acetals[67] and their *apparent* lipophilicities depend on the choice of partitioning solvents.[64] For $n\text{-}C_9H_{19}OH/H_2O$ partitioning, $\log P = 0.26$ and 0.59 for $CH_3C(O)CH_2C(O)CH_3$ and $CF_3C(O)CH_2C(O)CH_3$, respectively, which deceptively implies that fluorination increases lipophilicity. For $n\text{-}C_6H_{14}/H_2O$ partitioning, however, $\log P = 0.02$ and -0.50 for the respective diketones, which more properly reflects decreased lipophilicity upon fluorination. Similarly for aliphatic carboxylic acids, the effects of fluorination on acidity and hydrogen bonding become important. The respective values of $\log P$ (C_6H_6/H_2O) = -1.74 and -1.89 for CH_3CO_2H and CF_3CO_2H, and $\log P$ ($CHCl_3/H_2O$) = -1.52 and -1.96 for CH_3CO_2H and FCH_2CO_2H, imply fluorination increases hydrophilicity, but for partitioning between solvents that are both good hydrogen-bond acceptors, the opposite conclusion is reached, namely, $\log P$ (Et_2O/H_2O) = -0.36 for CH_3CO_2H and -0.27 for CF_3CO_2H. Obviously any unqualified generalization about the lipophilicities of α-fluorinated carbonyl compounds would be misleading.

Consistent with their more polar character, HFCs are often more hydrophilic than their corresponding HCs. Notably, $\log P = 1.09$, 0.51, 0.20, 0.64, and 1.18 for CF_nH_{n-4} ($n = 0$–4, respectively),[66] which is precisely the same relative order of their bps and dipole moments. Similarly, $\log P = 0.75$ for CH_3CHF_2 versus 1.81 for CH_3CH_3, and 2.33 for $n\text{-}C_5H_{11}F$ versus 3.11 for $n\text{-}C_5H_{12}$.[64,66] The fluorinated alcohols $CF_3(CH_2)_mOH$ are significantly less lipophilic than $CH_3(CH_2)_mOH$ ($m = 4,5$)[68] and the earlier mentioned relatively high cmcs for $CF_3(CH_2)_nCO_2H$ surfactants indicate trifluoromethylation increases hydrophilicity. Even higher levels of fluorination do not necessarily increase lipophilicity. Compared with $n\text{-}C_6H_{14}$ and $n\text{-}C_{12}H_{26}$, the fluorinated analogs $F(CF_2)_n(CH_2)_{6-n}H$ ($n = 0$–4) and $F(CF_2)_m(CH_2)_{12-m}H$ ($m = 0$–3) are also completely miscible with $n\text{-}C_6H_{14}$ but are more soluble in 10% aqueous methanol.[8] From the limited available data, it seems safe to conclude that monofluorination and trifluoromethylation decrease lipophilicity, provided the site of fluorination is separated from any heteroatom or π-bond by at least three $C{-}C$ bonds.

3.2.4. Acidity and Basicity

Since F and fluorinated substituents are very electron-withdrawing inductively[64,69] (Table 7), the acidities of acids, alcohols, and amides are always increased by fluorination, and the effects can be impressive (Table 8). For example, $(CF_3)_3COH$ is over 10^{13} times more acidic than $t\text{-}BuOH$. As FSO_2 and CF_3SO_2 are two of the most electron-withdrawing groups known,[69] their effects are especially pronounced. [Note

Table 8. Acidities and Hydrogen Bonding Parameters[71,72]

Acid	pK_a	α_2^H	β_2^H
CH_3CO_2H	4.76	0.550	0.42
CF_3CO_2H	0.52	0.951	—
$C_6H_5CO_2H$	4.21	0.588	0.42
$C_6F_5CO_2H$	1.75	0.889	—
CH_3CH_2OH	15.9	0.33	0.44
CF_3CH_2OH	12.4	0.567	0.18
$(CH_3)_2CHOH$	16.1	0.32	0.47
$(CF_3)_2CHOH$	9.3	0.771	0.03
$(CH_3)_3COH$	19.0	0.32	0.49
$(CF_3)_3COH$	5.4	0.862	—
C_6H_5OH	10.0	0.596	0.22
C_6F_5OH	5.5[a]	0.763	0.02
	9.61	0.493	—
	2.1[b]	0.86[c]	—

[a]Ref. 73.
[b]Ref. 74.
[c]Estimated from data in Ref. 74.

that $CF_3SO_2N{=}S(O)CF_3$, whose electron-withdrawing power is about equal to *two* NO_2 groups, is the strongest electron-withdrawing substituent so far reported.[70]] Thus the phenol $2,4,6\text{-}(CF_3SO_2)_3C_6H_2OH$ is about 20 times more acidic than picric acid[75]; $(CF_3SO_2)_2NH$ is reportedly the most acidic amide known[76]; $(CF_3SO_2)_2CH_2$ ($pK_a =$ -1.0) is more acidic than CF_3CO_2H[77]; and $(FSO_3)_3CH$ is about as acidic as H_2SO_4[78] ($H_0 = -11.9$) (see Section 3.3.2.2 for more about C—H acidity). At the extreme in acidity are the Brønsted superacids[79,80]: CF_3SO_3H ($H_0 = -13.8$), FSO_3H (-15.1), and FSO_3H–20% SbF_5 (-20). Anhydrous FSO_3H and HF ($H_0 = -15.1$) are the strongest pure acids known, and the acidity of HF–0.5% SbF_5 ($H_0 = -21$) is the highest so far measured for any solution. (The predominant species in HF–SbF_5 is $H_3F_2^+$ up to 40 mol% SbF_5, and H_2F^+ is observed at higher concentrations.[81]) The superacidic tri-fluoromethanesulfonic acid and its esters ("triflates") are some of the most important reagents in modern synthetic and mechanistic organic chemistry.[80,82]

Fluorination has an equally pronounced effect of decreasing the basicity of amines, ethers, and carbonyl compounds. 2,2,2-Trifluoroethylamine ($pK_b = 3.3$)[83] and $C_6F_5NH_2$ ($pK_b = -0.36$)[84] are about 10^5 times less basic than $CH_3CH_2NH_2$ and

$C_6H_5NH_2$, respectively. Perfluoro secondary amines, $(R_F)_2NH$, and perfluoropyridine (C_5F_5N) do not react with HCl or BF_3, and $(R_F)_3N$ amines have no basic character at all.[85] Some relative gas-phase proton affinity values (in kcal mol^{-1}) illustrate the powerful influence of fluorination[86]: CF_3NMe_2 (193.8) versus Me_3N (225.1), C_5F_5N (177.5) versus C_5H_5N (213.7), CF_3CH_2OEt (186.4) versus Et_2O (200.2), and $CF_3(CO)CF_3$ (161.4) versus $CH_3C(O)CH_3$ (196.7). Hexafluoroacetone is so nonbasic that it cannot be protonated in solution by superacids.[87] The perfluorinated amines $(R_F)_3N$ and ethers $(R_F)_2O$ are more like PFCs than typical amines and ethers, and this chemical inertness is crucial to their special commercial applications.

3.2.5. Hydrogen Bonding

Table 8 lists some values of hydrogen-bond acidity (basicity) parameters derived by Abraham *et al.*[72,73] from free energies of complexation with bases (acids) in CCl_4 (for monomeric water, $\alpha_2^H = 0.33$ and $\beta_2^H = 0.38$). As expected, fluorination increases H-bond acidity but decreases basicity, and at least within each family of compounds, H-bonding ability is proportional to pK_a. The fluorinated alcohols are notably strong H-bond donors and relatively nonnucleophilic, poor H-bond acceptors, which makes them excellent solvents for polar materials, including polymers,[88] and useful media for solvolysis studies.[57,89]

Fluorohalocarbons which have acidic C—H bonds also are reasonably good H-bond donors,[72] and it is known, for example, that the H-bond acidity of halocarbons is important to their anesthetic properties[90,91] (Chapter 25). Some representative data (pK_a, α_2^H) for known anesthetics are[72,92]: $CF_3CHClBr$ (23.8, 0.22), $CHCl_3$ (24.1, 0.20), $CH_3OCF_2CHCl_2$ (26.1, 0.17), and CHF_2OCF_2CHFCl (26.7, 0.19). Fluoroform, which is a weaker acid and poorer H-bond donor than $CHCl_3$, has no anesthetic potency.[92]

Much like C—OR bonds, the highly polar C—F bond with negatively charged F can act as a proton acceptor in H-bonding, but the effect is often unpredictable. Theoretical calculations show that hydrofluoromethanes can form weak H-bonded dimers[93] and NH_4^+ prefers to bind to fluorobenzene as a C_6H_5F---HNH_3^+ H-bonded complex,[94] but the most compelling evidence comes from experimental studies of intramolecular H-bonding in fluoro-alcohols and -phenols, and from X-ray crystal structure data. 2-Fluoroethanol is the most intensively studied system and in both the gas phase and solution it highly favors a *gauche* conformation (**1**) with a bent OH---F intramolecular hydrogen bond whose strength is about 2 kcal mol^{-1}.[95] An intramolecular H-bonded conformer also exists in $FCH_2CH_2CH_2OH$, but it is not the most stable one.[96] For 2-halophenols, the intramolecular H-bond strengths increase in the order F = Cl > Br > I in the vapor phase, but in solution the order is Cl > Br > I $\cong$ F.[97] Neither 2-fluorobenzyl alcohol[98] nor 2-fluoroaniline[99] show any evidence for intramolecular H-bonding in solution, however. A survey of crystal structures of fluorinated carboxylic acids and related compounds revealed a few examples of both inter- and intramolecular C—F---HO and C—F---HN H-bonding, albeit the interactions were deemed rather weak.[100] Perhaps most remarkable is the reported crystal structure of **2**, which has two F---H bond distances well below the sum of H and F van der Waals' radii (2.67 Å), implying substantial intramolecular H-bonding.[101]

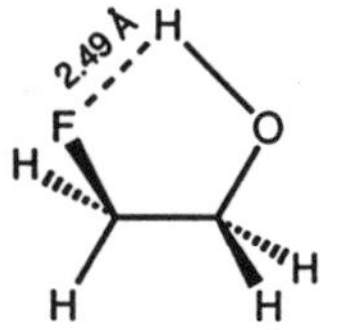

1 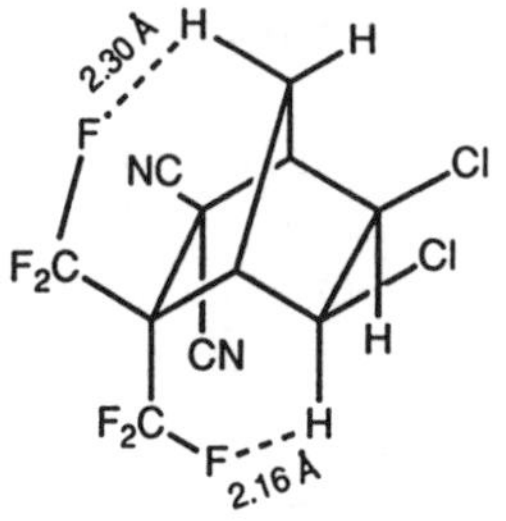   **2**

Although H-bonding to O is comparatively stronger, F in a C—F bond unquestionably can be a H-bond acceptor and this possibility should be considered, particularly when assessing the effects of fluorination on biological activity (see Chapter 24).

3.3. CHEMICAL PROPERTIES

The effects of fluorination on bond strengths, stabilities of reactive intermediates, and steric interactions that provide predictive insight into the reactivities of fluorocarbons are highlighted in the following sections. Much more thorough treatments, including detailed discussions of the electronic effects that govern structure, bonding, and reactivity, can be found in several reviews[102–105] and texts.[85,106,107]

3.3.1. Bond Strengths and Reactivity

3.3.1.1. Saturated Systems

Fluorine forms the strongest single bond with carbon. In monohaloalkanes, the C—F bond is about 25 kcal mol^{-1} stronger than the C—Cl bond[102,108] and the difference in *heterolytic* bond dissociation energies is even greater—about 30 kcal mol^{-1}.[109] As a consequence of the relatively strong C—F bond, and the poor leaving-group ability of fluoride ion,[110] alkyl fluorides are 10^2–10^6 times less reactive than the corresponding chlorides in typical S_N1 solvolysis or S_N2 displacement reactions.[106,107] Their displacements, however, can be strongly catalyzed by acid where H-bonding assists the departure of fluoride.[111,112] For $C_6H_5CH_2X$ solvolysis in 10% aqueous acetone, $k_F/k_{Cl} = 3.2 \times 10^{-2}$, but with 6 M $HClO_4$, $k_F/k_{Cl} = 2.6 \times 10^3$.[112] The decomposition of benzyl fluoride autocatalyzed by HF can be violent and cause storage problems.[113]

Fluorination has pronounced, characteristic effects on adjacent bond strengths (See Tables 9 and 10; the average values listed reflect the latest revised heats of formation of simple alkyl radicals[114,115] and differ somewhat from previous compila-

Table 9. Bond Dissociation Energies of Ethanes[102,113]

X	D° (C—X) (kcal mol^{-1})			
	CH$_3$CH$_2$—X	CH$_3$CF$_2$—X	CF$_3$CH$_2$—X	CF$_3$CF$_2$—X
H	100.1	99.5	106.7	102.7
F	107.9	124.8	109.4	126.8
Cl	83.7	—	—	82.7
Br	69.5	68.6	—	68.7
I	55.3	52.1	56.3	52.3

tions,[102] but their accuracies are seldom better than ±2 kcal mol^{-1}.) *α-Fluorination always markedly increases C—F and C—O bond strengths* but does not significantly affect C—H, C—Cl, or C—Br bonds. The increase in C—F bond energies from 108.3 kcal mol^{-1} in CH$_3$F to 130.5 kcal mol^{-1} in CF$_4$ shows how extraordinarily large the effect can be.[102] By contrast, *β-fluorination significantly increases C—H bond strengths,* but has little effect on C—F bonds. The tertiary C—H bond in (CF$_3$)$_3$CH is estimated to be at least 15 kcal mol^{-1} stronger than that in (CH$_3$)$_3$CH,[116] which makes it stronger than the C—H bond in CH$_4$!

Both α- and β-fluorination decrease the reactivity of saturated systems toward nucleophilic displacement. Compared with CH$_3$CH$_2$Br, CF$_3$CH$_2$Br reacts about 10^5 times slower with NaI in acetone,[117] and RCF$_2$Br compounds are inert to halide exchange under identical conditions. These diminished reactivities are attributed mainly to the shielding of the carbon center by F and to inductive effects, but the strong C—F bonds also contribute to the unreactivities of alkyl CF$_3$ and CF$_2$H groups toward F$^-$ displacement or hydrolysis. Although polyfluorohaloalkanes resist direct attack on carbon, they can react with nucleophiles, but these reactions invariably involve initial attack on halogen by either one- or two-electron transfer processes.[118] The two reactions below, for example, are formal displacements, but the first (Eq. 1) involves a radial-anion mechanism (S$_{RN}$1),[119] and the second (Eq. 2) is a carbene-mediated ion-chain process.[118]

(1) $\text{NO}_2\text{C(CH}_3)_2^- \text{ Li}^+ \; + \; \text{n-C}_8\text{F}_{17}\text{I} \quad \xrightarrow{\text{DMF}} \quad \text{NO}_2\text{C(CH}_3)_2\text{C}_8\text{F}_{17} \; + \; \text{Li}^+\text{I}^-$

(2) $\text{C}_6\text{H}_5\text{O}^-\text{Na}^+ \; + \; \text{CF}_2\text{Br}_2 \quad \xrightarrow{\text{DMF}} \quad \text{C}_6\text{H}_5\text{OCF}_2\text{Br} \; + \; \text{Na}^+\text{Br}^-$

Table 10. C—C and C—O Bond Dissociation Energies[102]

Ethane	D° (C—C)^a	Ether	D° (C—O)^a
CH$_3$CH$_3$	88.8	CH$_3$OCH$_3$	83.2
CH$_3$CF$_3$	101.2	CF$_3$OCF$_3$	105.2
CF$_3$CF$_3$	98.7	—	—

^{a}In kcal mol^{-1}.

Aliphatic C—C bonds are usually strengthened by fluorination (cyclopropanes and epoxides are notable exceptions wherein fluorination increases their strain energies[102,120]). The CF_3—CF_3 bond is 10 kcal mol^{-1} stronger than the CH_3—CH_3 bond, and the C—C bonds in poly(CF_2CF_2) are some 8 kcal mol^{-1} stronger than those in poly(CH_2CH_2). Geminal, partially fluorinated alkanes can have even stronger C—C bonds. For example, the CF_3—CH_3 bond, with its considerable ionic character, is 2.5 kcal mol^{-1} stronger than the CF_3—CF_3 bond. Fluorination increases the C—C bond strengths in four- and larger-ring cycloalkanes, and is believed to actually reduce the strain energy of cyclobutane.[102,120,121]

The very high C—F and C—C bond strengths in PFCs contribute to their characteristic, outstanding thermal and chemical stabilities. PFC thermal stabilities are limited only by the strengths of their C—C bonds, which decrease with increasing chain length or chain branching.[85] The most robust PFC is CF_4, whose C—F bonds measurably homolyze only above 2000°C. Temperatures approaching 1000°C are required to pyrolyze n-C_2F_6 or n-C_3F_8,[85] but poly(CF_2CF_2) rapidly decomposes above 500°C and its copolymers with perfluoroalkenes are significantly less stable.[122] PFCs with tertiary C—C bonds thermolyze around 300°C and highly branched systems can be more labile than their HC analogs (see Section 3.3.3). Perfluorocyclopropanes are a uniquely different class, however. Compared with c-C_4F_8, which undergoes homolysis at a rate of <5% hr^{-1} at 500°C,[123] c-C_3F_6 extrudes CF_2: at about 170°C.[124]

Perfluoroethers often are more thermally stable than PFCs owing to their especially strong C—O bonds. At 585°C, poly(CF_2CF_2O) decomposes about 10 times slower than poly(CF_2CF_2),[122] and CF_2CF_2/perfluoro(vinyl ether) copolymers are as stable as poly(CF_2CF_2), despite their branched structures (see Chapter 15). Three-membered ring ethers, which are unusually thermolabile[125] and chemically reactive,[126] are again special exceptions: tetrafluoroethylene and hexafluoropropylene oxides smoothly eject CF_2: at 165°C.

HFCs are less stable thermally than their PFC counterparts, but they decompose primarily by HF elimination rather than by C—C bond rupture. Even though poly(CH_2CF_2) has stronger C—C bonds than poly(CF_2CF_2), it is unstable above about 350°C and starts to lose HF rapidly.[122] Similarly, CF_3CF_2H exclusively loses HF at 925°C and C—C bond scission becomes significant only around 1125°C.[127] The eliminations of HF from HFCs of course are greatly accelerated by bases, which *inter alia* provide useful means to cure partially fluorinated elastomers (Chapter 16) and to functionalize the surfaces of hydrofluorinated plastics.[128]

The chemical unreactivity commonly associated with saturated PFCs also has its exceptions. They are susceptible to defluorination by reducing agents. Fusion with alkali metals, which is used for elemental analysis, converts most PFCs to carbon and metal fluorides.[129] Sodium naphthalide is employed commercially to etch perfluoropolymers,[130] and dipotassium benzoin dianion rapidly defluorinates poly(CF_2CF_2) at 50°C.[131] The high electron affinities of PFCs account for their reactivity toward reducing agents,[132] and defluorination proceeds by a sequential one-electron transfer mechanism (Eq. 3) whereby the inherently strong C—F bond becomes exceptionally labile toward dissociation to a carbon-centered radical and F^- when an electron is added

(3)

$$-CF_2CF_2- \xrightarrow{e^-} [-CF_2CF_2-]^{\cdot -} \xrightarrow{-F^-} [-CF_2\dot{C}F-]$$

$$\xrightarrow{e^-} [-CF_2\bar{C}F-] \xrightarrow{-F^-} -CF=CF- \xrightarrow{e^-} \text{etc.}$$

(4) [structure: perfluorodecalin] $+ \ C_6H_5S^-Na^+ \xrightarrow{\text{DMF}}$ [product: octakis(phenylthio)naphthalene with SC_6H_5 groups]

to its σ^* antibonding orbital. Nonetheless, several PFC radical anions can be observed in the gas phase[132,133] or matrix isolated.[134] PFCs with relatively low energy tertiary C—F σ^* orbitals are especially easy to reduce. MacNicol and Robertson[135] have reported a striking example of a PFC reaction that is triggered by one-electron transfer[136] (Eq. 4).

3.3.1.2. Unsaturated Systems

The C—F bonds in fluoroalkenes (and fluorobenzenes) are quite strong, 116 kcal mol^{-1} in C_2H_3F (125 kcal mol^{-1} in C_6H_5F),[114] but their C=C π-bond strengths vary considerably with the degree of fluorination. The π-bond dissociation energies (D_π) of $CH_2=CH_2$ (64–65 kcal mol^{-1})[137] and $CH_2=CF_2$ (62.8 ± 2 kcal mol^{-1})[138] are about the same, but $D_\pi \cong 53$ kcal mol^{-1} for $CF_2=CF_2$.[139] Experimental D_π values for other fluoroethylenes are not known, but related thermodynamic data indicate monofluorination stabilizes double bonds whereas vicinal difluorination and trifluorination are destabilizing.[102] Fluorination of acetylenes, however, is always highly destabilizing.[140] Both $HC \equiv CF$ and $FC \equiv CF$ are dangerously explosive and $CF_3C \equiv CCF_3$ is an extraordinarily reactive dienophile and enophile.[104,141]

Addition and polymerization reactions of *gem*-difluoroalkenes are more exothermic than the corresponding HC reactions, but this can reflect the greater stability of saturated over vinylic CF_2 groups rather than just relative double bond stabilities.[102] The π-bond character of fluoroalkenes, however, underlies in part their well-known propensity to undergo thermal [2+2] cyclodimerizations and cycloadditions via biradical mechanisms to provide a wide variety of fluorocyclobutanes and -cyclobutenes.[104,142] Vinyl CF_2 groups are usually necessary but not sufficient for biradical reactivity, and $CF_2=CF_2$, $CF_2=CFCl$, and $CF_2=CCl_2$ are especially reactive in this context. The low π-bond energies of these olefins provide the kinetic driving force for their participation in biradical additions.

Fluorination markedly increases C=O π-bond energies.[102] The π-bond in $CF_2=O$ is 25–35 kcal mol^{-1} stronger than that in $CH_2=O$.[102,143] The strong C=O bonds in acid fluorides and stabilization by $F—C=O \leftrightarrow {}^+F=C—O^-$ resonance impart greater chemical stability over their analogous acid chlorides, and the mechanisms of hydrolysis or ammonolysis of acid fluorides and chlorides can be quite different.[144] By contrast, fluoroalkylation significantly *destabilizes* C=O bonds.[102] The C=O bond in $(CF_3)_2C$=O is at least 21 kcal mol^{-1} weaker than that in acetone[145]; consequently, $(CF_3)_2C$=O is very susceptible to addition reactions[146] and can even undergo free-radical copolymerization and grafting reactions.

(5)

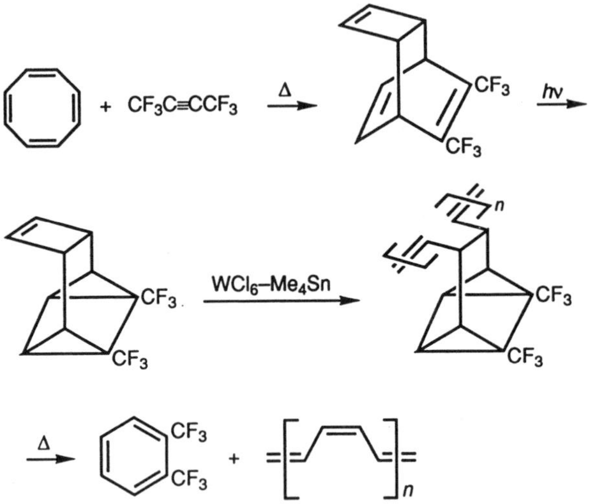

Although perfluoroalkylation thermodynamically destabilizes double bonds and small rings,[102] it has a remarkable tendency to *kinetically* stabilize highly strained ring compounds.[147] This so-called *"perfluoroalkyl effect"*[148] has enabled the isolation of diverse types of compounds, especially valence-bond isomers of aromatics and heteroaromatics,[149] that are uncommon in HC chemistry. This effect has been put to practical use in an improved synthesis of poly(acetylene) (Eq. 5).[150]

3.3.2. Reactive Intermediates

3.3.2.1. Carbocations

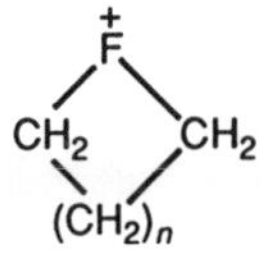

α-Fluorine stabilizes carbocations by resonance (3) which dominates its opposing inductive effect in fully-developed ions. The gas-phase ion stabilities increase in the orders $^+CH_3 < {}^+CF_3 < {}^+CH_2F < {}^+CHF_2$ and $^+CH_2CH_3 << {}^+CF_2CH_3 \cong {}^+CHFCH_3$,[151,152] and in solution α-halogens stabilize carbocations in the order F > Cl > Br > I.[80,153] Alkyl groups, however, stabilize carbocations better than does F.[154,155]

Although α-F stabilizes carbocations relative to H, the influence of F on reaction rates is more complicated since its opposing inductive effect makes fluoro-alkenes and -aromatics more electron deficient than their HC analogs. Fluorine will be activating only in electrophilic reactions with very late transition states. The faster rates of addition of CF_3CO_2H[156] or H_2SO_4[157] to $CH_2{=}CFCH_3$ versus $CH_2{=}CHCH_3$ are often cited, but examples of enhanced fluoroalkene reactivity in solution are actually rare,[158] and there are no reports of $CH_2{=}CF_2$ being more reactive than $CH_2{=}CH_2$ toward electrophilic addition. Much more common are instances of electrophilic aromatic substitution where F is *ortho-para* directing like other halogens but also activating relative to H.[159]

β-Fluorine strongly destabilizes carbocations inductively, and no simple β-fluoroalkyl carbocations have been observed in the gas phase or in solution.[160] The $FCH_2CH_2^+$ ion is calculated to be 29.6 kcal mol^{-1} less stable than CH_3CHF^+, and even 5.4 kcal mol^{-1} less stable than the ethylene fluoronium ion (**4**, $n = 0$).[161] (Actually, $FCH_2CH_2^+$ is not a stable species but instead is a transition state for proton scrambling in CH_3CHF^+.) Unlike the other halogens, however, there is no evidence for alkylfluoronium ion formation in solution,[154,162,163] although in the gas phase $CH_3FCH_3^+$ is long-lived[164] and **4** ($n = 0$) is a short-lived intermediate,[165] while ions of type **4** where n is 2 or greater are predicated to be kinetically stable to ring opening.[166]

Highly fluorinated olefins are relatively resistant to electrophilic attack, particularly when one or more perfluoroalkyl groups are present so that the β-fluorine destabilization is most pronounced. The HF/FSO_3H superacid does not react with $CF_2{=}CF_2$ or $CF_3CF{=}CF_2$,[154] and even $CF_3CH{=}CH_2$ is not protonated by FSO_3H but instead ionizes to the 1,1-difluoroallyl cation.[167] Hydrofluoro- and halofluoroethylenes have been shown to react with several species of electrophile, but their reactivity decreases with increasing F content.[168] For instance, $CF_2{=}CH_2$ is nitrofluorinated by HNO_3/HF 6 times faster than $CF_2{=}CHF$ at –60°C, and much higher temperatures (10–20°C) are required for addition to $CF_2{=}CF_2$. The same trend holds for fluoroaromatic substitutions.[153,154,169] Pentafluorobenzene reacts with some electrophiles, but only under forcing conditions, and C_6F_6 resists electrophilic substitution, which would require elimination of F^+.

The destabilization effects of β- or γ-fluorination on solvolysis or hydrolysis reactions also can be enormous.[160] For example, $C_6H_5C(CF_3){=}CH_2$ undergoes acid-catalyzed hydration 10^{10} times slower than $C_6H_5(CH_3){=}CH_2$[170]; triflate **5** (X = F) solvolyzes 10^{11} times slower than its H-analog **5** (X = H) in aqueous dioxan,[171] and $PhC(CF_3)_2OTs$ is a phenomenal 10^{18} times less reactive than $PhC(CH_3)_2OTs$ toward solvolysis.[172] This effect has been used to overcome the intrinsic instability of oxetane acetals.[173,174] Notably, Fried *et al.* have prepared the chemically stable, potent thromboxane A_2 agonist **6** (X = F) that is about 10^8 times more stable than TXA_2 itself (**6**, X = H) toward aqueous hydrolysis.[174,175]

The combined α- and β-fluorine effects imply that *fluoroolefins normally will react regiospecifically with electrophiles so as to minimize the number of fluorines β*

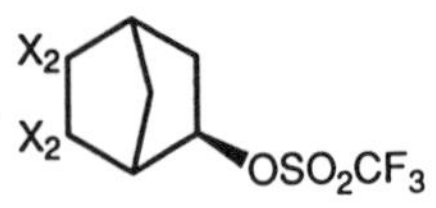

5

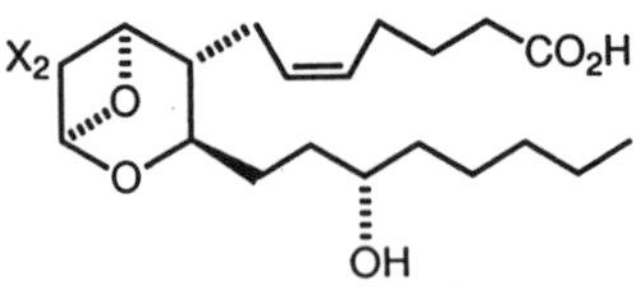

6

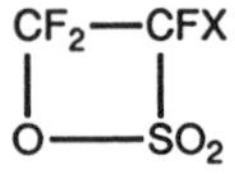

7   (X = CF₃, Cl, H)   **8**

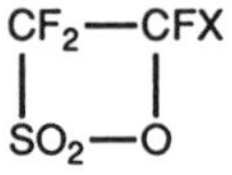

to the electron-deficient carbon in the transition state. For example, $CH_2{=}CF_2$ reacts with FSO_3H to give $CH_3CF_2OSO_2F$ exclusively,[154] $CF_2{=}CHF$ and $NO_2^+F^-$ give only CF_3CHFNO_2,[168] and the olefins $CF_2{=}CFX$ (X = CF_3, Cl or H) give **7**, not **8**, with SO_3.[176] There are some exceptions to this rule, however, particularly for Friedel-Crafts additions of halomethanes to polyfluoroalkenes.[177]

3.3.2.2. Carbanions

The effects of α- and β-fluorination on carbanions are essentially opposite to their effects on carbocations. α-Halogens stabilize carbanions in the order Br > Cl > F, a fact which can be adduced from C—H acidity data for haloalkanes.[178,179] For the haloforms, pK_{CsCHA} = 30.5 (CHF_3), 24.4 ($CHCl_3$), 22.7 ($CHBr_3$); and the acidities of CF_3CHX_2 and $(CF_3)_2CHX$ likewise increase in the order X = F < Cl < Br.[179,180] Although α-F is inductively stabilizing, its opposing I_π electron-pair repulsion (**9**) predominates, and carbanion stability is determined primarily by the degree of this repulsion, which increases in the order Br < Cl < F. Since I_π repulsion is maximized in planar systems, α-fluorocarbanions strongly prefer to be pyramidal. The pyramidal CF_3^- anion, for instance, is calculated to have an inversion barrier of over 100 kcal mol^{-1} compared with <2 kcal mol^{-1} for CH_3^-,[181,182] and even polyfluoroallyl anions are predicted to have pyramidal instead of classical, planar delocalized structures.[182]

Even though α-F is always less stabilizing than other halogens, it is still highly stabilizing relative to α-H *in pyramidal anions* (CF_3H is some 10^{40} times more acidic than CH_4).[178,179] In planar anions, however, α-F can be destabilizing. The greater acidity of fluorene itself versus 9-fluorofluorene, and the comparative acidities of esters (CH_3CO_2Et > CHF_2CO_2Et) and nitromethanes (CH_3NO_2 > CH_2FNO_2; $CH_2(NO_2)_2$ > $CHF(NO_2)_2$), illustrate the role of I_π repulsion in planar carbanions.[179]

9    **10**

CF$_3$O$^-$ TAS$^+$

11

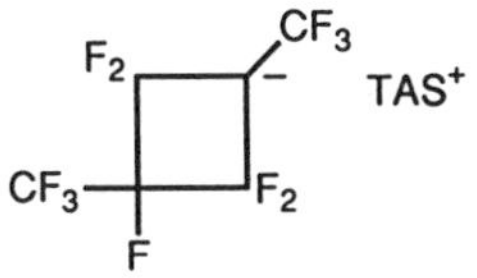

12

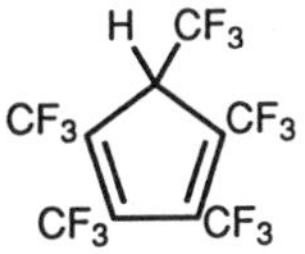

13

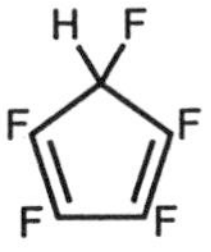

14

β-Fluorination highly stabilizes anions, both inductively and by negative (anionic) hyperconjugation (**10**). The latter has been a very controversial issue for several decades, but more recent theoretical[183] and experimental[184] data strongly support its importance. The X-ray crystal data for anion salts **11**[185] and **12**[186] [TAS$^+$ = (Me$_2$N)$_3$S$^+$] provide particularly compelling structural evidence for anionic hyperconjugation.

The relative approximate acidities of the HFCs (pK_a in MeOH) CF$_3$H (31), CF$_3$(CF$_2$)$_5$CF$_2$H (30), (CF$_3$)$_2$CFH (20), (CF$_3$)$_3$CH (11)[178,179] and the observation that **13** (pK_a ≤–2 in H$_2$O)[187] is about 10^{18} times more acidic than cyclopentadiene itself, whereas **14** is only slightly more acidic,[188] indicate β-F is much more stabilizing than α-F in both pyramidal and planar carbanions. Notably, stable salts of several tertiary perfluoroalkyl[189] and -allyl[190] carbanions, as well as R$_F$O$^-$[186,191] and R$_F$S$^-$[192] anions, have been isolated, but to date no long-lived α-F carbanion has even been detected spectroscopically in solution.

The combined effect of F to both increase the electrophilicity of double bonds and stabilize carbanions insures that fluoroalkenes (and fluoroaromatics) will be much more susceptible to nucleophilic attack than their HC counterparts. *Fluoroalkenes react with nucleophiles so as to maximize the number of fluorines β to the electron-rich carbon in the transition state.* In general, the relative stabilities of the intermediate carbanions govern both relative reactivity and orientation of attack. For example, fluoroalkene reactivities increase in the orders CF$_2$=CF$_2$ < CF$_2$=CFCF$_3$ << CF$_2$=C(CF$_3$)$_2$ and CF$_2$=CF$_2$ < CF$_2$=CFCl < CF$_2$=CFBr, and nucleophilic attack occurs exclusively at the CF$_2$= sites of the unsymmetrical olefins.[178,193] Although the orientation of attack is usually very predictable, the product distribution is much less so. Partitioning between addition and addition–elimination pathways depends upon the specific olefin and nucleophile, and the reaction conditions. For instance, NaOMe reacts with CF$_2$=CFCl in MeOH to give mostly MeOCF$_2$CHFCl, but in THF, the product is CH$_3$OCF=CFCl.[193] Perfluorocyclobutene with NaOMe, by contrast, gives 1-methoxy-pentafluorocyclobutene (addition–elimination) in either proton-donating or aprotic solvents.[194] The various factors that control the course of a fluoroalkene's reactions with nucleophiles have been extensively reviewed.[107,194–197]

Since base-promoted eliminations of hydrogen halides from polyfluoroalkanes typically proceed via carbanionic intermediates ($E1cb$ mechanism),[195,198] their regiochemistry also is governed by the characteristic effects of halogen on carbanion stability or, equivalently, C—H acidity. For example, $CF_3CH_2CHBrCH_3$ with KOH/EtOH gives $CF_3CH=CHCH_3$ via $CF_3\bar{C}HCHBrCH_3$ (not $CF_3CH_2CH=CH_2$ via $CF_3CH_2CHBr\bar{C}H_2$), and $HCCl_2CCl_2CHF_2$ gives $Cl_2C=ClCHF_2$ via $\bar{C}Cl_2CCl_2CHF_2$ (not $CHCl_2CCl=CF_2$ via $CHCl_2CCl_2\bar{C}F_2$).[107]

The rates and orientation of nucleophilic substitution in fluorinated aromatics and heterocycles show their own interesting characteristic trends. For C_6F_5X derivatives, nucleophilic attack occurs mainly *para* to X for most substituents, but their reactivities can vary enormously depending upon how the substituent X affects a planar carbanion's stability. The relative rates of $NaOC_6H_5$ attack on C_6F_5X (in dimethyl acetamide at 106°C), for instance, increase in the order X = F (0.9) < H (1.0) < Cl (32) < CF_3 (2.4 × 10^4).[199] Extensive surveys of the regiochemistry and kinetics of nucleophilic substitutions of polyfluoroaromatics are available.[200]

The characteristic stabilizing effects of F on anions correctly imply that anionic processes should be much more prevalent in fluorocarbon than hydrocarbon chemistry. The apt analogy drawn between the role of F^- in FC chemistry and H^+ in HC chemistry is particularly illustrative.[178] Like the familiar electrophilic processes in HC chemistry, there are analogous F^--promoted isomerizations and dimerizations (Eq. 6), oligomerizations (Eq. 7), additions (Eq. 8), and even Friedel-Crafts alkylations (Eq. 9) that all proceed via carbanionic intermediates.[178]

(6) $2\ CF_3CF=CF_2 \xrightarrow{F^-} (CF_3)_2CFCF=CFCF_3 \rightleftharpoons (CF_3)_2C=CFC_2F_5$

(7) $5\ CF_2=CF_2 \xrightarrow{F^-} CF_3CF=CF(CF_3)C(C_2F_5)_2CF_3$

(8) $CF_3CF=CF_2 \xrightarrow[F^-]{HCONH_2} CF_3CFHCF_3$

(9) $(CF_3)_2C=CF_2 \ +$ $\xrightarrow{F^-}$

3.3.2.3. Free Radicals

α-Fluororadicals prefer to be pyramidal to minimize I_π repulsion, but the effect is much smaller than in carbanions. The calculated inversion barrier in $\cdot CF_3$ is about 25 kcal mol^{-1}.[201] β-Fluorination does not significantly affect geometry and, unlike other halogens, bridging or σ-delocalization by F is negligible.[201,202] The effect of F on radical stability is somewhat ambiguous, although most evidence suggests it is very small.[104,201–204] Fluorination nonetheless can greatly influence the rates and orientation of free-radical abstraction and addition reactions, but this chemistry is controlled by polar, bond-strength, or steric factors, and radical stability plays a minor role.[205]

Fluorination inductively deactivates C—H abstraction by *electrophilic* radicals (Cl·, Br·, HO·, etc.). For example, the relative rates of tertiary CH abstraction by Br· in $XCH_2C(CH_3)_2H$ are X = F (1.0) < Cl (11) < CH_3 (27.5) < Br (220).[206] When the pronounced C—H bond-strengthening by β-fluorination (see Section 3.3.1.1) is superimposed, the effect can be quite large, namely, $k(CH_3CH_3) / k(CH_3CF_3) \cong 2.2 \times 10^4$ for H-abstraction by Cl·.[207] For abstraction by *nucleophilic* radicals, however, fluorination can be activating. For instance, CF_3H is less reactive than CH_4 toward Br·, but the opposite is true for attack by the nucleophilic ·CH_3 radical.[205] This reflects the much more favorable polarization in the transition state $[\delta^-CF_3\text{---}H\text{---}CH_3\delta^+]·$ versus $[\delta^+CF_3\text{---}H\text{---}Br\delta^-]·$.

The reactions of HFCs and HCFCs with HO· provide a mechanism for their removal from the atmosphere (Chapters 7 and 28), and there is now a vast body of kinetic data available[208] for calculating ozone and greenhouse warming potentials. The approximate relative rates of reaction of some commercial CFC-replacement candidates with HO· at 25°C are: CF_3CHF_2 (1.0) < CH_3CF_2Cl (1.4) < CH_3CFCl_2 (2.4) < CF_3CH_2F (3.4) < CF_3CHFCl (4.1) < CF_3CHCl_2 (13.4) where $k(CH_3CH_3)/k(CF_3CHF_2)$ $\cong 110.$[208] Obviously both polar and bond-strength effects underlie this reactivity order, and correlations of hydrohaloalkane reaction rates have been developed to help eludicate the trends.[208,209]

Free-radical additions to fluoroalkenes involve a complex interplay of substituent effects. Rules for predicting rates and regioselectivity have been developed,[205] and only a couple of illustrations of polar effects are presented here. The $k(CF_2{=}CF_2)$ / $k(CH_2{=}CH_2)$ rate ratio for addition of ·CH_3 is 9.5, but is 0.1 for addition of ·CF_3 at 164°C.[210] This reflects the preferred attack by the nucleophilic ·CH_3 radical on the electron-deficient olefin $[\delta^+CH_3\text{---}CF_2\text{---}CF_2\delta^-]$, whereas the highly electrophilic ·$CF3$ radical prefers the more electron-rich olefin $[\delta^-CF_3\text{---}CH_2\text{---}CH_2\delta^+]$. These polar effects account for the tendency of perfluoroalkenes and alkenes to combine to produce highly regular, alternating copolymers (Chapter 15). The orientation of addition to a fluoroalkene also can depend critically on the electronic character of the radical. The contrasting product distributions for additions of $CF_3S·$ and $CH_3S·$ to $CF_2{=}CFCF_3$ (Eq. 10)[211] are consistent with the preferred orientation of attack by electrophiles and nucleophiles (cf. Sections 3.3.2.1 and 3.3.2.2).

(10)

$$RSH + CF_2{=}CFCF_3 \xrightarrow{h\nu} RSCF_2CHFCF_3 + RSCF(CF_3)CHF_2$$

| | R = CF_3 | 45% | 55% |
| | R = CH_3 | 92% | 8% |

Besides their electrophilic character, fluoroalkyl radicals differ from their HC counterparts in other respects. Fluorine-atom migration is very uncommon, although there is recent evidence for the $CH_2FCF_2· \rightarrow ·CH_2CF_3$ rearrangement,[212] but disproportionation by F-atom transfer never occurs under normal reaction conditions. Disproportionation between PFC and HC radicals ($R_FCF_2CF_2· + CH_3CH_2· \rightarrow R_FCF_2CHF_2 + CH_2{=}CH_2$) is well-known,[213] but $R_FCF_2CF_2· + CH_3CH_2· \rightarrow$

$R_FCF{=}CF_2 + CH_3CH_2F$ or $2 R_FCF_2CF_2 \cdot \rightarrow R_FCF{=}CF_2 + R_FCF_2CF_3$ are not observed. Because the latter disproportionation is not available to terminate the free-radical polymerization of $CF_2{=}CF_2$, extraordinarily high molecular weights can be achieved (Chapter 15).

3.3.2.4. Carbenes

Fluorination markedly affects the structure, stability, and reactivity of carbenes. In contrast to CH_2, all α-F carbenes are ground state singlets[214] owing to resonance stabilization analogous to that in α-F carbocations (3). The electrophilicity of singlet CX_2 carbenes increases in the order $X = F < Cl < Br < H$, and $CF_2 < CHF < CH_2$. α-F carbenes display characteristic singlet chemistry, i.e., they add stereospecifically to simple alkenes to yield cyclopropanes and do not insert into C—H bonds competitively with C=C addition.[107,215] Experimentally, CF_2 is the most selective electrophilic carbene known. The $(CF_3)_2C$ carbene in marked contrast is an exceedingly electrophilic triplet species that indiscriminately adds to C=C and inserts into CH bonds.[107] A quantitative scale of carbene selectivities has been developed,[216] although selectivity order can be temperature-dependent.[217]

By far the most commercially important role of fluorocarbenes is in the manufacture of $CF_2{=}CF_2$ and $CF_2{=}CFCF_3$ from CF_2 (Chapter 15). The thermodynamics and kinetics of the pyrolysis reaction $2 CHF_2Cl \rightarrow [2 CF_2] + 2 HCl \rightarrow CF_2{=}CF_2$ have been quantified experimentally.[218] The high-temperature, thermal decompositions of linear PFCs also are believed to involve CF_2.[85]

3.3.3. Steric Effects

The old Pauling value of 1.35 Å for the van der Waals' radius of F compared to 1.2 Å for H is still commonly cited in the literature, along with the conclusion that F and H are about the same size. The conclusion is incorrect and the data upon which it is based have been superceded. Bondi's accepted values of 1.47 Å and 1.52 Å for the van der Waals' radii of F and O, respectively (Table 1), more recent corroborating X-ray crystal structure data,[219] and the modified Taft steric parameters[64] ($E_S°(F) = -0.46$, $E_S°(OH) = -0.55$) all in fact indicate that *F and O, not H, are very nearly isosteric*. It actually has been correctly recognized for some time now that F and OH are "chemical isosteres".[220]

Examples of bona-fide F steric effects are most commonly encountered in dynamic processes. The barriers to *i*-Pr rotation in **15**[221] and ring inversion in **16**[222] are both about 5 kcal mol^{-1} higher for X = F than X = H, while the barriers to biphenyl rotation in **17** ($\Delta G^{\ddagger} = 14.2 \pm 0.7$ kcal mol^{-1} for X = F, Y = Me; 16.1 ± 0.6 kcal mol^{-1} for X = OH, Y = Me)[223] reflect the similar size of F and OH. The largest F steric effect on record is the k_H/k_F rate ratio of 10^{11} at 25°C for the *meta* ring-flip in **18**.[224] A notable instance where F steric effects control stereochemistry is in free-radical additions to norbornenes **19**, where *trans* addition is highly favored with X = H, but *cis* addition predominates for X = F.[225]

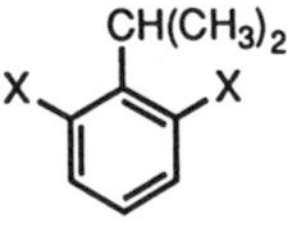

15 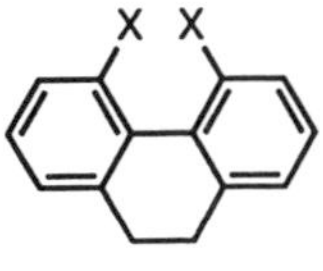   **16** 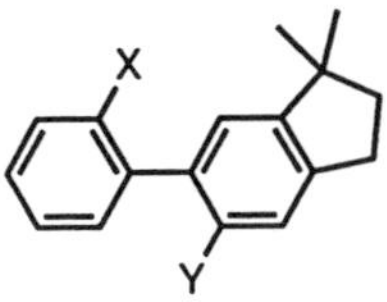   **17**

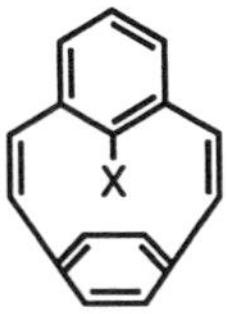

18 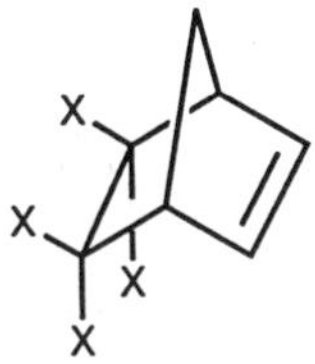   **19**

The CF_3 group is much larger than CH_3 by any measure of steric size. Comparative Taft E_S^o values, conformational A values,[226] and Charton v steric parameters[227] are listed in Table 11. These data, combined with the observation that the rotation barriers for **17** (X = CF_3, Y = Me) and **17** (X = i-Pr, Y = Me) are identical within experimental error ($\Delta G^{\ddagger} \cong 22$ kcal mol^{-1}),[223] indicate that CF_3 is *sterically at least as large as* $CH(CH_3)_2$.

The activation energy for *syn–anti* isomerization of $(CF_3)_2C{=}NCF(CF_3)_2$ implies $CF(CF_3)_2$ and t-Bu are about the same size,[228] but there are no quantitative data on the steric size of other perfluoroalkyl groups. The unusual thermolability of highly branched PFCs, however, does suggest considerable steric strain. For example, $E_a =$ 36 kcal mol^{-1} for C—C bond rupture in $(i$-$C_3F_7)_2C(CF_3)CF_2CF_3$[229] compared with 51.3 kcal mol^{-1} for the more branched HC $(t$-Bu)$(i$-Pr)C(CH$_3$)CH$_2$CH$_3$.[230]

The unprecedented stability of the PFC radicals C_2F_5 $(i$-$C_3F_7)_2$C· and $(i$-$C_3F_7)_3$C· is one of the most striking manifestations of a fluorine steric effect.[231] Their persistence is a kinetic effect that arises from shielding of the trivalent carbon by the steric bulk of the perfluoroalkyl groups.[232]

Table 11.   Alkyl Group Steric Parameters

Group	E_S^{oa}	A^b	v^c
CH$_3$	−1.24	1.70	0.52
(CH$_3$)$_2$CH	−1.71	2.20	0.76
(CH$_3$)$_2$CHCH$_2$	−2.17	—	0.98
CF$_3$	−2.40	2.4–2.5	0.91

^{a}Modified Taft values [Ref. 64].
^{b}Cyclohexane ΔG^o (axial-equatorial) values [Ref. 226].
^{c}Charton steric parameter [Ref. 227].

3.4. REFERENCES

1. K. D. Sen and C. K. Jorgensen, *Electronegativity*, Springer-Verlag, New York (1987).
2. J. K. Nagel, *J. Am. Chem. Soc. 112*, 4740 (1990).
3. L. C. Allen, D. A. Egolf, E. T. Knight, and C. Liang, *J. Phys. Chem. 94*, 5602 (1990).
4. A. Bondi, *J. Phys. Chem. 68*, 441 (1964).
5. A. Maciejewski, *J. Photochem. Photobio., A: Chemistry 51*, 87 (1990).
6. T. M. Reed III, in: *Fluorine Chemistry* (J. H. Simons, ed.), Vol. 5, Chapter 2, Academic Press, New York (1964); H. G. Bryce, in: *ibid.*, Vol. 5, Chapter 4 (1964); T. J. Brice, in: *ibid.*, Vol. 1, Chapter 13 (1950).
7. D. S. L. Slinn and S. W. Green, in: *Preparation, Properties, and Industrial Applications of Organofluorine Compounds* (R. E. Banks, ed.), Chapter 2, Ellis Horwood, Chichester (1982).
8. W. Mahler (DuPont), personal communication.
9. C. R. Patrick, in: *Preparation, Properties, and Industrial Applications of Organofluorine Compounds* (R. E. Banks, ed.), Chapter 10, Ellis Horwood, Chichester (1982).
10. D. F. Persico, H.-N. Huang, R. J. Lagow, and L. C. Clark, Jr., *J. Org. Chem. 50*, 5156 (1985).
11. R. D. Nelson, Jr., D. R. Lide, Jr., and A. A. Maryott, *Selected Values of Electric Dipole Moments in the Gas Phase*, NSRD-NBS 10, Government Printing Office, Washington, DC (1967); A. L. McClellan, *Tables of Experimental Dipole Moments*, Vol. 1, W. H. Freeman, San Francisco (1963).
12. M. Hudlicky, *Chemistry of Organofluorine Compounds*, 2nd edn., Chapter 7, Wiley, New York (1976); W. A. Sheppard and C. M. Sharts, *Organic Fluorine Chemistry*, pp. 17, 42–43, W. A. Benjamin, New York (1969).
13. A. M. Lovelace, D. A. Rausch, and W. Postelnek, *Aliphatic Fluorine Compounds*, Reinhold, New York (1958).
14. V. Majer and V. Svoboda, *Enthalpies of Vaporization of Organic Compounds*, Blackwell, Boston (1985).
15. J. S. Chickos, in: *Molecular Structure and Energetics* (J. F. Liebman and A. Greenburg, eds.), Vol. 2, VCH Publishers, New York (1987).
16. A. J. Lovinger, *Science 220*, 1115 (1983); A. J. Lovinger, in: *Developments in Crystalline Polymers* (D. C. Bassett, ed.), Applied Science Publishers, London (1982).
17. J. A. Dean (ed.), *Lange's Handbook of Chemistry*, 13th edn., McGraw-Hill, New York (1985).
18. S. Wu, *Polymer Interface and Adhesion*, Marcel Dekker, New York (1982).
19. A. G. Pittman, in: *Fluoropolymers* (L. A. Wall, ed.), High Polymers Vol. XXV, Chapter 13, Wiley, New York (1972).
20. E. F. Hare, E. G. Shafrin, and W. A. Zisman, *J. Colloid. Sci. 58*, 236 (1954).
21. N. Watanabe, T. Nakajima, and H. Touhara, *Graphite Fluorides*, Studies in Inorganic Chemistry 8, Elsevier, Oxford (1988).
22. M. M. Doeff and E. Lindler, *Macromolecules 22*, 2951 (1989).
23. A. J. Arduengo, III, J. R. Moran, J. Rodriguez-Parada, and M. D. Ward, *J. Am. Chem. Soc. 112*, 6153 (1990).
24. A. Sekiya and M. Tamura, *Chem. Lett.* 707 (1990).
25. J. Schneider, C. Erdelen, H. Ringsdorf, and J. F. Rabolt, *Macromolecules 22*, 3475 (1989)
26. S. G. Wolf, M. Deutsch, E. M. Landau, M. Lahav, L. Leiserowitz, K. Kjaer, and J. Als-Nielsen, *Science 242*, 1286 (1988); I. Weissbuch, F. Frolow, L. Addadi, M. Lahav, and L. Leiserowitz, *J. Am. Chem. Soc. 112*, 7718 (1990).
27. K. Shinoda, M. Hato, and T. Hayaski, *J. Phys. Chem. 76*, 909 (1972).
28. H. Kuneida and K. Shinoda, *J. Phys. Chem. 80*, 2468 (1976); W. Guo, T. A. Brown, and B. M. Fung, *J. Phys. Chem. 95*, 1829 (1991).
29. G. Mathis, P. Leempoel, J.-C. Ravey, C. Selve, and J.-J. Delpuech, *J. Am. Chem. Soc. 106*, 6162 (1984); L. Matos, J.-C. Ravey, and G. Serratrice, *J. Colloid. Interface Sci. 128*, 341 (1989); C. Selve, J.-C. Ravey, M.-J. Stebe, C. E. Moudjahid, and J.-J. Delpuech, *Tetrahedron 47*, 411 (1991).
30. M. Miyamoto, K. Aoi, and T. Saegusa, *Macromolecules 22*, 3540 (1989).
31. C. Selve, E. M. Moumni, J.-J. Delpuech, *J. Chem. Soc., Chem. Commun.* 1437 (1987).

32. I. Rico, A. Lattes, K. P. Das, and B. Lindman, *J. Am. Chem. Soc. 111*, 7267 (1989).
33. T. P. Russell, J. F. Rabolt, R. J. Twieg, R. L. Siemens, and B. L. Farmer, *Macromolecules 19*, 1135 (1986); R. J. Twieg, T. P. Russell, R. Siemens, and J. F. Rabolt, *Macromolecules 18*, 1361 (1985).
34. M. P. Turnberg and J. E. Brady, *J. Am. Chem. Soc. 110*, 7797 (1988).
35. G. Martini, M. F. Ottaviani, S. Ristori, D. Lenti, and A. Sanguineti, *Colloids Surf. 45*, 177 (1990).
36. K. Kalyanasundaram, *Langmuir 4*, 942 (1988).
37. N. Muller and R. H. Birkhahn, *J. Phys. Chem. 71*, 957 (1967).
38. S. S. Berr and R. R. M. Jones, *J. Phys Chem. 93*, 2555 (1989).
39. S. J. Burkitt, R. H. Ottewill, J. B. Hayter, and B. T. Ingram, *Colloid Polym. Sci. 265*, 619 (1987).
40. S. J. Burkitt, R. H. Ottewill, J. B. Hayter, and B. T. Ingram, *Colloid Polym. Sci. 265*, 628 (1987).
41. H. Hoffman, J. Klaus, K. Reizlein, W. Ulbricht, and K. Ibel, *Colloid Polym. Sci. 260*, 435 (1982).
42. R. Elbert, T. Folda, and H. Ringsdorf, *J. Am. Chem. Soc. 106*, 7687 (1984); T. Kunitake, Y. Okahata, and S. Yasunami, *J. Am. Chem. Soc. 104*, 5547 (1982).
43. W. P. Zhang and D. L. Dorset, *Macromolecules 23*, 4322 (1990); C. W. Bunn and E. R. Howells, *Nature 174*, 549 (1954); E. S. Clark and L. T. Muus, *Z. Kristallogr. 117*, 119 (1962).
44. C. Naselli, J. D. Swalen, and J. F. Rabolt, *J. Chem. Phys. 90*, 3855 (1989).
45. D. A. Dixon and F. A. Van-Catledge, *Int. J. Supercomputer Appl. 2*, 62 (1988); D. A. Dixon, F. A. Van-Catledge, and B. E. Smart, *Abstracts of the 9th IUPAC Conf. on Phys. Org. Chem.* A15 (1988).
46. D. F. Eaton and B. E. Smart, *J. Am. Chem. Soc. 112*, 2821 (1990).
47. R. J. Twieg and J. F. Rabolt, *J. Polym. Sci., Polym. Phys. Ed. 21*, 901 (1983).
48. R. J. Twieg and J. F. Rabolt, *Macromolecules 21*, 1806 (1988).
49. J. Höpken and M. Möller, *Macromolecules 25*, 2482 (1992); W. Mahler, D. Guillon, and A. Skoulios, *Mol. Cryst. Liq. Cryst., Lett. Sect. 2*, 111 (1985).
50. F. Tournilhac, L. Bosio, J. F. Nicoud, and J. Simon, *Chem. Phys. Lett. 145*, 452 (1988).
51. A. E. Feiring, in: *Frontiers of Macromolecular Science* (T. Saegusa, T. Higashimura, and A. Abe, eds.), Blackwell, Oxford (1989).
52. A. M. F. Barton, *Handbook of Solubility Parameters and Other Cohesive Parameters*, CRC Press, Boca Raton (1983).
53. R. Fusch, E. J. Chambers, and W. K. Stephenson, *Can. J. Chem. 65*, 2624 (1987); R. Fusch, L. A. Peacock, and K. Das, *Can. J. Chem. 58*, 2301 (1980).
54. J. H. Hildebrand, J. M. Prausnitz, and R. L. Scott, *Regular and Related Solutions*, Van Nostrand-Reinhold, New York (1970); J. S. Rowlinson and F. L. Swinton, *Liquids and Liquid Mixtures*, 3rd edn., Butterworth, London (1982).
55. T. Handa and P. Mukerjee, *J. Phys. Chem. 85*, 3916 (1981); P. Mukerjee, A. Y. S. Yang, *J. Phys. Chem. 80*, 1388 (1976); J. Carlfors and P. Stibbs, *J.Phys. Chem. 88*, 4410 (1984).
56. D. L. Dorset, *Macromolecules 23*, 894 (1990).
57. C. Reichardt, *Solvents and Solvent Effects in Organic Chemistry*, 2nd edn., VCH Publishers, New York (1988).
58. B. K. Freed, J. Biesecker, and W. J. Middleton, *J. Fluorine Chem. 48*, 63 (1990).
59. R. Battino and H. L. Clever, *Chem. Rev. 66*, 395 (1966); E. Wilhelm and R. Battino, *Chem. Rev. 73*, 11 (1973).
60. M'H. A. Hamza, G. Serratrice, M.-J. Stébé, and J. J. Delpuech, *J. Am. Chem. Soc. 103*, 3733 (1981).
61. H. G. Mack and H. Oberhammer, *J. Chem. Phys. 87*, 2158 (1987); S. Oikawa, M. Tsuda, and N. Nagayama, *Theor. Chim. Acta 64*, 403 (1984).
62. J. Afzal, S. R. Ashlock, B. M. Fung, and E. A. O'Rear, *J. Am. Chem. Soc. 90*, 3019 (1986).
63. G. Serratrice, J. J. Delpuech, and R. Diguet, *Nouv. J. Chem. 6*, 489 (1982).
64. C. Hansch and A. Leo, *Substituent Constants for Correlation Analysis in Chemistry and Biology*, Wiley, New York (1979).
65. T. Fujita, *Prog. Phys. Org. Chem. 14*, 75 (1983).
66. J. Sangster, *J. Chem. Phys. Ref. Data 18*, 3 (1989).
67. A. E. Feiring and B. E. Smart, in: *Ullmann's Encyclopedia of Industrial Chemistry*, Vol. A11, pp. 367–370, VCH Publishers, New York (1988).
68. N. Muller, *J. Pharm. Sci. 75*, 987 (1986).

69. L. M. Yagupol'skii, A. Ya. Il'chenko, and N. V. Kondratenko, *Russ. Chem. Rev. 43*, 32 (1974).
70. L. M. Yagupol'skii, *J. Fluorine Chem. 36*, 1 (1987).
71. M. H. Abraham, P. L. Grellier, D. V. Prior, P. P. Duce, J. J. Morris, and P. J. Taylor, *J. Chem. Soc., Perkin Trans. 2*, 699 (1989).
72. M. H. Abraham, P. L. Grellier, D. V. Prior, J. J. Morris, and P. J. Taylor, *J. Chem. Soc., Perkin Trans. 2*, 521 (1990).
73. J. M. Birchall and R. M. Haszeldine, *J. Chem. Soc.* 3653 (1959).
74. J. Hine, S. Hahn, and J. Hwang, *J. Org. Chem. 53*, 884 (1988).
75. J. M. Carpentier, F. Ferrier, R. Schall, N. V. Ignatev, V. N. Boiko, and L. M. Yagupol'skii, *Bull Soc. Chim. Fr.* 150 (1985).
76. J. Foropoulous, Jr. and D. D. DesMarteau, *Inorg. Chem. 23*, 3720 (1984).
77. R. J. Koshhar and R. A. Mitsch, *J. Org. Chem. 38*, 3358 (1973).
78. G. Klöter, H. Pritzkow, and K. Seppelt, *Angew Chem., Int. Ed. Engl. 19*, 942 (1980).
79. R. J. Gillespie and J. Liang, *J. Am. Chem. Soc. 110*, 6053 (1988).
80. G. A. Olah, G. K. S. Prakash, and J. Sommer, *Superacids*, Wiley, New York (1985).
81. D. Moutz asnd K. Bartmann, *Angew. Chem., Int Ed., Engl. 27*, 391 (1988).
82. P. J. Stang, M. Hanak, and L. R. Subramanian, *Synthesis* 85 (1982); F. Effenberger, *Angew Chem., Int. Ed. Engl. 19*, 751 (1980).
83. D. D. Perrin, *Dissociation Constants of Organic Bases in Aqueous Solution*, Butterworths, London (1965), Suppl. (1972).
84. R. Filler, *Fluorine Chem. Rev. 8*, 1 (1977).
85. R. E. Banks, *Fluorocarbons and Their Derivatives*, Macdonald, London (1970).
86. J. F. Liebman, in: *Fluorine Containing Molecules* (J. F. Liebman, A. Greenberg, and W. R. Dolbier, Jr., eds.), VCH Publishers, New York (1988); R. W. Taft, *Prog. Phys. Org. Chem. 14*, 247 (1983).
87. G. A. Olah and Y. R. Mo, *Adv. Fluorine Chem. 7*, 69 (1973).
88. M. Goodman, I. G. Rosen, and M. Safdy, *Biopolymers 2*, 503, 519, 537 (1964); W. J. Middleton, U.S. Patent 3,418,337 (to DuPont) [*CA 60*, 7000 (1964)].
89. T. W. Bentley, C. T. Bowen, D. H. Morten, and P. v. R. Schleyer, *J. Am. Chem. Soc. 103*, 5466 (1981); T. W. Bentley and G. E. Carter, *J. Am. Chem. Soc. 104*, 5741 (1982).
90. A. L. Brown, Y. Chiang, A. J. Kresge, Y. S. Tang, and W.-H. Wang, *J. Am. Chem. Soc. 111*, 4918 (1989).
91. C. Sandorfy, R. Buchet, L. S. Lussier, P. Ménassa, and L. Wilson, *Pure Appl. Chem. 58*, 1115 (1986).
92. P. Hobza, F. Mulder, and C. Sandorfy, *J. Am. Chem. Soc. 104*, 925 (1982).
93. A. A. Hasanein, *J. Comput. Chem. 5*, 528 (1984); A. Popowicz and T. Ishida, *Chem. Phys. Lett. 83*, 520 (1981).
94. C. A. Deakyne and M. Meot-Ner Mautner, *J. Am. Chem. Soc. 107*, 474 (1985).
95. J. Huang and K. Hedberg, *J. Am. Chem. Soc. 111*, 6909 (1989); D. A. Dixon and B. E. Smart, in: *Selective Fluorination in Organic and Bioorganic Chemistry*, ACS Symposium Series 456 (J. T. Welch, ed.), American Chemical Society, Washington, D.C. (1991).
96. W. Caminati, *J. Mol. Spectrosc. 92*, 101 (1982).
97. J.-M. Dumas, M. Gomel, and M. Guerin, in: *The Chemistry of Functional Groups, Supplement D* (S. Patai and Z. Rappoport, eds.), Chapter 21, Wiiey, Chichester (1983).
98. G. A. Crowder, *J. Fluorine Chem. 14*, 77 (1979).
99. D. Christen, D. Damiani, and D. G. Lister, *J. Mol. Struct. 41*, 315 (1977).
100. P. Murray-Rust, W. C. Stallings, C. T. Monti, R. K. Preston, and J. P. Glusker, *J. Am. Chem. Soc. 105*, 3206 (1983).
101. A. N. Chekhlov, V. L. Fetisov, S. 1. Kolbasenko, and I. V. Martynov, *Dokl. Akad. Nauk SSSR 297*, 1177 (1987).
102. B. E. Smart, in: *Molecular Structure and Energetics* (J. F. Liebman and A. Greenberg, eds.), Vol. 3, Chapter 4, VCH Publishers, Deerfield Beach, FL (1986).
103. R. E. Banks and J. C. Tatlow, in: *Fluorine: The First Hundred Years (1886–1986)* (R. E. Banks, D. W. A. Sharp, and J. C. Tatlow, eds.), Chapter 11, Elsevier, New York (1986).

104. B. E. Smart, in: *The Chemistry of Functional Groups, Supplement D*, Part 2 (S. Patai and Z. Rappoport, eds.), Chapter 14, Wiley, New York (1983).

105. J. F. Liebman, A. Greenberg, and W. R. Dolbier, Jr. (eds.), *Fluorine Containing Molecules*, VCH Publishers, New York (1988).

106. M. Hudlický, *Chemistry of Organofluorine Compounds*, Ellis Horwood, Chichester (1976).

107. R. D. Chambers, *Fluorine in Organic Chemistry*, Wiley, New York (1973).

108. D. F. McMillen and D. M. Golden, *Ann. Rev. Phys. Chem. 33*, 493 (1982).

109. S. K. Shin and J. L. Beauchamp, *J. Am. Chem. Soc. 111*, 900 (1989).

110. H. F. Koch, *Acc. Chem. Res. 17*, 137 (1984); C. J. Stirling, *Acc. Chem. Res. 12*, 198 (1974); R. E. Parker, *Adv. Fluorine Chem. 3*, 63 (1963).

111. C. G. Swain and R. E. T. Spalding, *J. Am. Chem. Soc. 82*, 6104 (1960).

112. M. Namavari, N. Satayamurthy, M. E. Phelps, and J. R. Barrio, *Tetrahedron Lett. 31*, 4973 (1990).

113. S. S. Szucs, *Chem. Eng. News 68*, 4 (1990).

114. D. Griller, J. M. Kanabus-Kaminska, and A. Maccol, *J. Mol. Struct. (Theochem) 163*, 125 (1988).

115. J. A. Seetula, J. J. Russell, and D. Gutman, *J. Am. Chem. Soc. 112*, 1347 (1990).

116. B. S. Evans, I. Weeks, and E. Whittle, *J. Chem. Soc., Faraday Trans. 1, 79*, 147 (1983).

117. F. G. Bordwell and W. Brannen, Jr., *J. Am. Chem. Soc. 86*, 4645 (1964).

118. C. Wakselman, *J. Fluorine Chem. 59*, 367 (1992); C. Wakselman and C. Kaziz, in: *Fluorine: The First Hundred Years (1886–1986)* (R. E Banks, D. W. A. Sharp, and J. C. Tatlow, eds.), Chapter 12, Elsevier, New York (1986).

119. A. E. Feiring, *J. Org. Chem. 50*, 3269 (1985); A. E. Feiring, *J. Org. Chem. 48*, 347 (1983).

120. J. F. Liebman, W. R. Dolbier, Jr., and A. Greenberg, *J. Phys. Chem. 90*, 394 (1986).

121. M. M. Rahman, B. A. Secor, K. M. Morgan, P. R. Schafer, and D. M. Lemal, *J. Am. Chem. Soc. 112*, 5986 (1990).

122. L. A. Wall, in: *Fluoropolymers* (L. A. Wall, ed.), High Polymers Vol. XXV, Chapter 12, Wiley, New York (1972).

123. J. N. Butler, *J. Am. Chem. Soc. 84*, 1393 (1962).

124. M. J. Birchall, R. Fields, R. N. Haszeldine, and R. J. McClean, *J. Fluorine Chem. 15*, 487 (1980).

125. R. C. Kennedy and J. B. Levy, *J. Fluorine Chem. 7*, 101 (1976); W. Mahler and P. R. Resnick, *J. Flourine Chem. 3*, 451 (1973).

126. H. Millauer, W. Schwertfeger, and G. Siegemund, *Angew. Chem., Int Ed. Engl. 24*, 161 (1985); P. Tarrant, C. G. Allison, K. P. Barthold, and E. C. Stump, Jr., *Fluorine Chem. Rev. 5*, 77 (1971).

127. V. Aviyente and Y. Inel, *Can. J. Chem. 68*, 1332 (1990).

128. A. J. Dias and T. J. McCarthy, *Macromolecules 17*, 2529 (1984); J. V. Brennan and T. J. McCarthy, *Polym. Prep., Am. Chem. Soc., Div. Polym. Chem. 29*, 338 (1988).

129. C. Y. Wang and J. A. Tarter, *Anal Chem. 55*, 1775 (1983); L. Kavan, F. P. Dousek, and K. Micka, *J. Phys. Chem. 94*, 5127 (1990).

130. R. C. Benning and T. J. McCarthy, *Macromolecules 23*, 2648 (1990); S. B. Dake, S. V. Bhoraskar, S. V. Patil, and N. S. Narashiman, *Polymer 27*, 910 (1986).

131. C. A. Costello and T. J. McCarthy, *Macromolecules 20*, 2819 (1987).

132. A. A. Christodoulides, D. L. McCorkle, and L. G. Christophorou, in: *Electron Molecule Interactions and Their Applications* (K. G. Christophorou, ed.), Academic, New York (1984); P. Kebarle and S. Chowdhury, *Chem. Rev. 87*, 513 (1987).

133. S. M. Spyrou, S. R. Hunter, and L. G. Christophorou, *J. Chem. Phys. 83*, 641 (1985).

134. A. Hasegawa, M. Shiotani, and F. Williams, *J. Chem. Soc., Faraday Discuss. 63*, 157 (1977); M. Hasegawa and F. Williams, *Chem. Phys. Leff. 45*, 275 (1977).

135. D. D. MacNicol and C. D. Robertson, *Nature (London) 332*, 59 (1988).

136. D. L. Cooper, N. L. Allan, and R. L. Powell, *J. Fluorine Chem. 49*, 421 (1990); J. A. Marsella, A. G. Gilicinski, A. M. Coughlin, and G. P. Pez, *J. Org. Chem. 57*, 2856 (1992).

137. S. Y. Wang and W. T. Borden, *J. Am. Chem. Soc. 111*, 7282 (1989).

138. J. M. Pickard and A. S. Rodgers, *J. Am. Chem. Soc. 99*, 695 (1977).

139. E. C. Wu and A. S. Rodgers, *J. Am. Chem. Soc. 98*, 6112 (1976).

140. D. A. Dixon and B. E. Smart, *J. Phys. Chem. 93*, 7772 (1989).

141. M. I. Bruce and W. R. Cullen, *Fluorine Chem. Rev. 4*, 79 (1969).
142. W. H. Sharkey, *Fluorine Chem. Rev. 2*, 1 (1969); P. D. Bartlet, *Quart. Rev. 24*, 473 (1970).
143. L. Batt and R. Walsh, *Int. J. Chem. Kinet. 14*, 933 (1982).
144. B. D. Song and W. P. Jencks, *J. Am. Chem. Soc. 111*, 8470, 8479 (1989).
145. F. E. Rogers and R. J. Rapiejko, *J. Am. Chem. Soc. 93*, 4596 (1971).
146. C. G. Krespan and W. J. Middleton, *Fluorine Chem. Rev. 1*, 145 (1967); W. J. Middleton, in: *Kirk Othmer: Encyclopedia of Chemical Technology*, Vol. 10, Wiley, New York (1980).
147. A. Greenberg, J. F. Liebman, and D. van Vechten, *Tetrahedron 36*, 1161 (1980).
148. D. M. Lemal and L. H. Dunlap, *J. Am. Chem. Soc. 94*, 6562 (1972).
149. Y. Kobayashi and I. Kumadaki, *Top. Curr. Chem. 123*, 103 (1984).
150. W. J. Feast and J. N. Winter, *J. Chem. Soc., Chem. Commun.* 202 (1985).
151. R. J. Blint, T. B. McMahon, and J. L. Beauchamp, *J. Am. Chem. Soc. 96*, 1269 (1974).
152. A. D. Williams, P. R. LeBreton, and J. L. Beauchamp, *J. Am. Chem. Soc. 98*, 2705 (1976).
153. G. A. Olah and Y. K. Mo, in: *Carbonium Ions* (G. A. Olah and P. v. R. Schleyer, eds.), Vol. V, Chapter 36, Wiley, New York (1976).
154. G. A. Olah and Y. K. Mo, *Adv. Fluorine Chem. 7*, 69 (1973).
155. R. W. Taft, R. H. Martin, and F. W. Lampe, *J. Am. Chem. Soc. 87*, 2490 (1965).
156. P. E. Peterson, R. J. Bopp, and M. M. Ajo, *J. Am. Chem. Soc. 92*, 2834 (1970).
157. A. D. Allen and T. T. Tidwell, *J. Am. Chem. Soc. 104*, 3145 (1982).
158. W. J. Johnson, G. W. Daub, T. A. Lyle, and M. Niwa, *J. Am. Chem. Soc. 102*, 7802 (1980).
159. G. Modena and G. Scorrano, in: *The Chemistry of the Carbon–Halogen Bond* (S. Patai, ed.), Chapter 6, Wiley, London (1973).
160. X. Creary, *Chem. Rev. 91*, 1625 (1991); A. D. Allen and T. T. Tidwell, in: *Advances in Carbocation Chemistry* (X. Creary, ed.), Chapter 1, JAI Press, Greenwich (1989).
161. G. P. Ford and K. S. Raghuveer, *Tetrahedon 44*, 7489 (1988).
162. G. F. Koser, in: *The Chemistry of Functional Groups; Supplement D*, Part 2 (S. Patai and Z. Rappoport, eds.), Chapter 25, Wiley, New York (1983).
163. G. A. Olah, G. K. S. Prakash, and V. V. Krishnamurthy, *J. Org. Chem. 48*, 5116 (1983).
164. F. Cacace, *Acc. Chem. Res. 21*, 215 (1988).
165. V. Nguyen, X. Cheng, and T. H. Morton, *J. Am. Chem. Soc. 114*, 7127 (1992).
166. D. A. Stams, T. D. Thomas, D. C. MacLaren, D. Ji, and T. H. Morton, *J. Am. Chem. Soc. 112*, 1427 (1990).
167. P. C. Myhre and G. D. Andrews, *J. Am. Chem. Soc. 92*, 7595, 7596 (1970).
168. B. L. Dyatkin, E. P. Mochalina, and I. L. Knunyants, *Fluorine Chem. Rev. 3*, 45 (1969); G. G. Belen'kii and L. S. German, in: *Soviet Scientific Reviews, Section B, Chemistry Reviews*, Vol. 5 (M. E. Vol'pin, ed.), pp. 183–218, Ellis Horwood, London (1984).
169. G. G. Yakobson and G. G. Furin, in: *Soviet Scientific Reviews, Section B, Chemistry Reviews*, Vol. 5 (M. E. Vol'pin, ed.), pp. 255–296, Ellis Horwood, London (1984).
170. K. M. Koshy, D. Roy, and T. T. Tidwell, *J. Am. Chem. Soc. 101*, 357 (1979).
171. W. Kirmse, U. Mrotzeck, and R. Siegfried, *Angew. Chem., Int. Ed. Engl. 24*, 55 (1985).
172. V. Kanagasabapathy, J. F. Sawyer, and T. T. Tidwell, *J. Org. Chem. 50*, 503 (1985).
173. A. J. Kirby, H. Ryder, and V. Matassa, *J. Chem. Soc., Perkin Trans. 2*, 825 (1990).
174. J. Fried, E. A. Hallinan, and M. J. Swedo, Jr., *J. Am. Chem. Soc. 106*, 3871 (1984).
175. J. Fried, V. John, M. J. Swedo, Jr., C. K. Chen, C. O. Yang, T. A. Morinelli, A. K. Okwu, and P. V. Halushka, *J. Am. Chem. Soc. 111*, 4510 (1989).
176. I. L. Knunyants and G. A. Sokolski, *Angew Chem., Int Ed. Engl. 11*, 583 (1972).
177. O. Paleta, *Fluorine Chem. Rev. 8*, 39 (1977).
178. R. D. Chambers and M. R. Bryce, in: *Comprehensive Carbanion Chemistry*, Part C (E. Buncel and T. Durst, eds.), Chapter 5, Elsevier, Amsterdam (1987).
179. O. A. Reutov, I. P. Beletskaya, and K. P. Butin, *CH-Acids*, Pergamon, Oxford (1978).
180. A. Streitwieser, Jr., D. Holtz, G. R. Ziegler, J. O. Stoffer, M. L. Brokaw, and F. Guibe, *J. Am. Chem. Soc. 98*, 5229 (1976).
181. D. S. Marynick, *J. Mol. Struct. 87*, 161 (1982).

182. D. A. Dixon, T. Fukunaga, and B. E. Smart, *J. Phys. Org. Chem. 1*, 153 (1988); J. H. Hammons, D. A. Hrovat, and W. T. Borden, *J. Phys. Org. Chem. 3*, 635 (1990).
183. A. E. Reed and P. v. R. Schleyer, *J. Am. Chem. Soc. 112*, 1434 (1990): D. A. Dixon, T. Fukunaga, and B. E. Smart, *J. Am. Chem. Soc. 108*, 4027 (1986); D. S. Friedman, M. M. Francl, and L. C. Allen, *Tetrahedron 41*, 499 (1985).
184. J. H. Sleigh, R. Stephens, and J. C. Tatlow, *J. Fluorine Chem. 15*, 411 (1980).
185. W. B. Farnham, B. E. Smart, W. J. Middleton, J. C. Calabrese, and D. A. Dixon, *J. Am. Chem. Soc. 107*, 4565 (1985).
186. W. B. Farnham, D. A. Dixon, and J. C. Calabrese, *J. Am. Chem. Soc. 110*, 2607 (1988).
187. E. D. Laganis and D. M. Lemal, *J. Am. Chem. Soc. 102*, 6633 (1980).
188. G. Paprott and K. Seppelt, *J. Am. Chem. Soc. 106*, 4060 (1984).
189. A. E. Bayliff and R. D. Chambers, *J. Chem. Soc., Perkin Trans. 1* 201 (1988); B. E. Smart, W. J. Middleton, and W. B. Farnham, *J. Am. Chem. Soc. 108*, 4905 (1986).
190. W. B. Farnham, W. J. Middleton, W. C. Fultz, and B. E. Smart, *J. Am. Chem. Soc. 108*, 3125 (1986).
191. D. A. Dixon, W. B. Farnham, and B. E. Smart, *Inorg. Chem. 29*, 3954 (1990).
192. B. E. Smart and W. J. Middleton, *J. Am. Chem. Soc. 109*, 4982 (1987).
193. R. D. Chambers and R. H. Mobbs, *Adv. Fluorine Chem. 4*, 50 (1965).
194. J. D. Park, R. J. McMurtry, and J. H. Adams, *Fluorine Chem. Rev. 2*, 55 (1968).
195. H. F. Koch, in: *Comprehensive Carbanion Chemistry* (E. Buncel and T. Durst, eds.), Chapter 6, Elsevier, Amsterdam (1987); H. F. Koch and J. G. Koch, in: *Fluorine-Containing Molecules* (J. F. Liebman, A. Greenberg, and W. R. Dolbier, Jr., eds.), Chapter 6, VCH Publishers, New York (1988).
196. Z. Rappoport, *Adv. Phys. Org. Chem. 7*, 1 (1969).
197. C. G. Krespan, F. A. Van-Catledge, and B. E. Smart, *J. Am. Chem. Soc. 106*, 5544 (1986); C. G. Krespan and B. E. Smart, *J. Org. Chem. 51*, 320 (1986).
198. J. R. Gandler, J. W. Storer, and D. A. A. Ohlberg, *J. Am. Chem. Soc. 112*, 7756 (1990).
199. R. J. dePasquale and C. Tamborski, *J. Org. Chem. 32*, 2163 (1967).
200. P. P. Rodinov and G. G. Furin, *J. Fluorine Chem. 47*, 361 (1990); L. S. Kobrina, *Fluorine Chem. Rev. 7*, 1 (1974).
201. D. J. Pasto, R. Krasnansky, and C. Zercher, *J. Org. Chem. 52*, 3062 (1987).
202. H. G. Viehe, Z. Janousek, and R. Merenyi (eds.), *Substituent Effects in Radical Chemistry*, Reidel, Dordrecht (1986); L. Kaplan, *Bridged Free Radicals*, Marcel Dekker, New York (1972).
203. X.-K Jiang and G.-Z. Ji, *J. Org. Chem. 57*, 6051 (1992); J. M. Dust and D. R. Arnold, *J. Am. Chem. Soc. 105*, 1221 (1983).
204. G. Leroy, D. Peeters, and C. Wilante, *J. Mol. Struct. 88*, 217 (1982).
205. J. M. Tedder, *Angew. Chem., Int. Ed. Engl. 21*, 401 (1982); J. M. Tedder and J. C. Walton, *Adv. Phys. Org. Chem. 16*, 51 (1978).
206. P. S. Skell and K. J. Shea, *Israel J. Chem. 10*, 493 (1972).
207. E. Tschuikow-Roux, T. Yano, and J. Niedzielski, *J. Chem. Phys. 82*, 65 (1985).
208. R. Atkinson, *Chem. Rev. 85*, 69 (1985); R. Liu, R. E. Huie, and M. J. Kurylo, *J. Phys. Chem. 94*, 3247 (1990).
209. N. Cohen and S. W. Benson, *J. Phys. Chem. 91*, 162, 171 (1987).
210. J. M. Tedder and J. C. Walton, *Adv. Free Radical Chem. 6*, 155 (1980).
211. J. F. Harris and F. W. Stacey, *J. Am. Chem. Soc. 83*, 840 (1961).
212. M. Kotaka, T. Kohida, and S. Sato, *Z. Naturforsch. B, Chem. Sci. 45*, 721 (1990).
213. G. O. Pritchard, S. H. Abbas, J. M. Kennedy, S. J. Paquette, D. B. Hudson, M. A. Meleason, and D. D. Shoemaker, *Int. J. Chem. Kinet. 22*, 1051 (1990).
214. E. A. Carter and W. A. Goddard, III, *J. Chem. Phys. 88*, 1752 (1988); D. A. Dixon, *J. Phys. Chem. 90*, 54 (1986).
215. D. J. Burton and J. L. Hanfeld, *Fluorine Chem. Rev. 8*, 119 (1977); R. A. Moss, in: *Carbenes* (R. A. Moss and M. Jones, Jr., eds.), Vol. I, Wiley, New York (1973).
216. R. A. Moss, *Acc. Chem. Res. 22*, 15 (1989); *13*, 58 (1980).
217. B. Giese, W.-B. Lee, and J. Meister, *Ann. Chem. 725* (1980).
218. P. B. Chinoy and P. D. Sunavala, *Ind. Eng. Chem. Res. 26*, 1340 (1987).

219. D. E. Williams and D. J. Houpt, *Acta Cryst. B42*, 286 (1986).
220. C. W. Thornber, *Chem. Soc. Rev 8*, 563 (1979).
221. T. Schaefer, R. P. Veregin, R. Laatikainen, R. Sebastian, K. Marat, and J. L. Chareton, *Can. J. Chem.* *60*, 2611 (1982).
222. R. Cosmo and S. Sternhell, *Aust. J. Chem. 40*, 35 (1987).
223. G. Bott, L. D. Field, and S. Sternhell, *J. Am. Chem. Soc. 102*, 5618 (1980).
224. S. A. Sherrod, R. L. daCosta, R. A. Barnes, and V. Boekelheide, *J. Am. Chem. Soc. 96*, 1565 (1974).
225. B. E. Smart, *J. Org. Chem. 38*, 2027, 2035, 2039 (1973).
226. J. Hirsh, *Top. Stereochem. 1*, 199 (1967); E. W. Della, *J. Am. Chem. Soc. 89*, 5221 (1967).
227. R. Gallo, *Prog. Phys Org. Chem. 14*, 115 (1983).
228. W. H. Dawson, D. H. Hunter, and C. J. Willis, *J. Chem. Soc., Chem. Commun.* 874 (1980).
229. R. E. Fernandez, *Diss. Abstr. Int. B48*, 3569 (1988).
230. S. Hellmann, H.-D. Beckhaus, and C. Rüchardt, *Chem. Ber 116*, 2238 (1983).
231. K. V. Scherer, Jr., T. Ono, K. Yamanouchi, R. Fernandez, P. Henderson, and H. Goldwhite, *J. Am. Chem. Soc. 107*, 718 (1985); C. Tonelli and V. Tortelli, *J. Chem. Soc., Perkin Trans. 1*, 23 (1990).
232. K. V. Scherer, Jr., in: *Fluorine: The First Hundred Years* (R. E. Banks, D. W. A. Sharp, and J. C. Tatlow, eds.), Appendix 11.5, Elsevier, New York (1986).

4

Perfluorocarbon Fluids

S. W. GREEN, D. S. L. SLINN, R. N. F. SIMPSON, and A. J. WOYTEK

4.1. INTRODUCTION

This chapter deals almost exclusively with the production, properties, and applications of saturated compounds containing only carbon and fluorine, i.e., *saturated perfluorocarbons.*[*] It updates a previous account[1] of a group of commercial perfluorocarbon (PFC) liquids known as Flutec™ Fluids with boiling points lying in the range 29–160°C.[1] This range has now been extended to 260°C. Information on gaseous perfluorocarbons has also been included.

4.1.1. Perfluorocarbon Gases

The major merchant perfluorocarbon gases which have commercial applications are tetrafluoromethane (CF_4; F-14), usually called carbon tetrafluoride, hexafluoroethane (C_2F_6; F-116), and octafluoropropane (C_3F_8; F-C318).[**] These gases have not achieved the commercial significance of chlorofluorocarbons (CFCs), hydrochlorofluorocarbons (HCFCs), or hydrofluorocarbons (HFCs), their total worldwide production being less than 500 tonnes per year (1990). In fact, until the early 1980s there was little commercial interest in these compounds, but with the

---

[*]See Chapter 1 for remarks about fluorocarbon nomenclature.
[**]For an explanation of numerical codes used to specify simple perfluorocarbons and related compounds, such as CFCs, see Chapter 1.

S. W. GREEN • Rhône-Poulenc Chemicals Ltd., Avonmouth, Bristol BS11 9HP, England. D. S. L. SLINN and R. N. F. SIMPSON • Rhône-Poulenc Chemicals Ltd., Avonmouth, Bristol BS11 9HP England. A. J. WOYTEK • Air Products and Chemicals Inc., Industrial Gas Division, Allentown, Pennsylvania 18105. *Present address for S. W. G.:* 69 South Road, Kingswood, Bristol BS15 2JF, England. *Present address for D. S. L. S. and R. N. F. S.:* Fluoro Systems Ltd., Abbots Leigh, Bristol B58 3RP, England.

Organofluorine Chemistry: Principles and Commercial Applications, edited by R. E. Banks *et al.* Plenum Press, New York, 1994.

development of dry etching (plasma) technology in the microelectronics industry, all three gases are in demand today. Carbon tetrafluoride is by far the most important in this respect.

A major factor limiting the commercial use of perfluorocarbon gases is their cost. Selling prices range from \$7 kg^{-1} for CF$_4$ to \$20 kg^{-1} for C$_3$F$_8$ (1990 prices). Major producers are Air Products and Chemicals, Du Pont, Asahi Glass, Daikin, and Kanto Denka. The costs of these gases are related to their high fluorine content and the process technology required for their production. Therefore, they have not been able to compete with the low-cost CFCs for large market applications. However, with the forthcoming ban on CFCs and new taxes in the U.S.A. affecting the economics of their use, opportunities for perfluorocarbons are expected to emerge. Although the production cost of perfluorocarbon gases will always be greater than that of the traditional CFCs, larger volume production could significantly reduce their current price levels.

4.1.2. Perfluorocarbon Liquids

These liquids possess a unique combination of properties which has led to their major uses as inert fluids for electronics testing, the cooling of electronic devices, and vapor-phase soldering. The quantities required for each application are in the approximate ratio of 3:1:1 respectively with a total world market in excess of 1000 tonnes (1990).

In recent years the continuing search for economic processes for the production of higher boiling fluids has met with considerable success; thus high-purity perfluorocarbon liquids are now available (Flutec™ Fluids—see Table 1—produced

Table 1. Current Range of Fluctec™ Fluids^a

Commercial designation	Molecular formula	Main molecular species or mixture of isomers
PP50	C$_5$F$_{12}$	Perfluoropentanes
PP1C	C$_6$F$_{12}$	Perfluoro(methylcyclopentanes)
PP1	C$_6$F$_{14}$	Perfluorohexanes
PP2	C$_7$F$_{14}$	Perfluoro(methylcyclohexane)
PP3	C$_7$F$_{14}$/C$_8$F$_{16}$	Perfluoro(methyl/dimethylcyclohexanes)
PP5^b/PP6	C$_{10}$F$_{18}$	Perfluorodecalin (*cis*- and *trans*- isomers)
PP7/PP9^b	C$_{10}$F$_{18}$/C$_{11}$F$_{20}$	Perfluoro(decalin/methyldecalin) (*cis*- and *trans*- isomers)
PP10	C$_{13}$F$_{22}$	Perfluoroperhydrofluorene
PP11	C$_{14}$F$_{24}$	Perfluoroperhydrophenanthrene
PP24	C$_{16}$F$_{26}$	Perfluoroperhydrofluoranthene
PP25	C$_{17}$F$_{30}$	Perfluoro(cyclohexylmethyldecalin)

^{a}In 1993 Rhône-Poulenc Chemical Ltd sold its "Flutec" Perfluorocarbon Fluid business to BNFL Fluorochemicals Ltd. who will be producing Flutec Fluids in the future.
^{b}Medical grade.

by Rhône-Poulenc Chemicals Ltd and Multifluor Inert Fluids by Air Products and Chemicals Inc.) with boiling points up to 260°C. The complete range of Flutec Fluids is shown in Table 1, and it is the availability of this extended range which has led to developments within established uses and also to new applications since the last review (1982).[1]

A significant requirement which has arisen in recent years is to find nonozone depleting substitutes for CFCs (see Chapter 7). The major applications of CFCs lay in areas concerned with coolants, solvents, and blowing agents for polymeric foams. Perfluorocarbons can be used as substitutes in specialized sectors of these markets, and examples will be quoted.

4.2. PRODUCTION OF PERFLUOROCARBONS

4.2.1. Processes for Perfluorocarbon Gases

Carbon tetrafluoride (F-14) was first produced by Moissan by the direct fluorination of carbon, following his isolation of fluorine in 1886.[2] However, he obtained a mixture of fluorocarbons and was not able to isolate pure carbon tetrafluoride. It was not until the late 1920s that the French chemists Lebeau and Damiens succeeded where Moissan failed.[3] Independently, Ruff and Keim studied the fluorination of wood charcoal,[4] isolated pure carbon tetrafluoride by careful distillation of the products, and fully characterized it. Since that time, numerous investigators have studied the preparation of carbon tetrafluoride. The methods now available can be categorized as follows:

1. Direct fluorination of carbon.
2. Halogen exchange between chlorocarbons and hydrogen fluoride.
3. Electrochemical fluorination utilizing hydrocarbons.
4. Various fluorinations using metal fluorides, such as cobalt trifluoride.

Two of these methods are used commercially, namely, halogen exchange and direct fluorination of carbon. The halogen-exchange route for producing carbon tetrafluoride utilizes a vapor-phase process in which carbon tetrachloride or dichlorodifluoromethane is treated with hydrogen fluoride over solid catalysts at temperatures up to 500°C.[5] This produces a mixture of halocarbons which requires extensive distillation to produce CF_4. Typically, commercial-grade carbon tetrafluoride made by this process is 98% pure, the major contaminants being the chlorofluorocarbons CF_2Cl_2 and CF_3Cl. Refinements to the purification process have upgraded CF_4 purity levels to 99.9+%.

The critical requirements of the microelectronics industry for high-purity carbon tetrafluoride in plasma etching technology led to the development of a new commercial process based on the direct fluorination of carbon.[6] This process,[7] which totally eliminates chlorine contamination, was commercialized almost a century after Moissan performed his initial experiments on fluorine and carbon.

Hexafluoroethane (F-116) is produced in a similar way to carbon tetrafluoride by the halogen-exchange process utilizing hydrogen fluoride. It is also obtained as a byproduct in the manufacture of 1,1,2-trichlorotrifluoroethane (CFC-113), 1,2-dichlorotetrafluoroethane (CFC-114), and chloropentafluoroethane (CFC-115) (see Chapter 6). It can be prepared by the fluorination of ethane or partially fluorinated ethanes with cobalt trifluoride.

Octafluoropropane (F-318) can be prepared by several methods [direct (F_2/N_2),[8] electrochemical,[9] or CoF_3 fluorination[10]] using commercial hexafluoropropene (C_3F_6) as the starting material. Passage of C_3F_6 over cobalt trifluoride is claimed to give a 99% yield of 99.9% pure C_3F_8.[10] Electrochemical fluorination of propane can also be used to prepare C_3F_8.[11]

4.2.2. Processes for Perfluorocarbon Liquids

The higher-boiling perfluoro-alkanes and -cycloalkanes are manufactured predominately either by electrochemical fluorination (Simons Process—see Chapter 5)[12] or by cobalt trifluoride fluorination[13,14] of the corresponding alkanes, alkenes, or aromatic hydrocarbons (C_5–C_{18}). Industrial-scale production of C_6–C_{14} perfluoroalkanes, such as C_7F_{16} (FC™ 84) by electrochemical fluorination and C_7F_{14} (Flutec™ PP2) by cobalt trifluoride fluorination, is quite significant. Inherently, electrochemical fluorination (ECF) should be the lower-cost process for the production of the lower-molecular-weight perfluorocarbon liquids, since it avoids the cost of generating elemental fluorine. However, it does have certain disadvantages when a product of high chemical specification is required because it gives lower yields and selectivity, and often leads to extensive molecular rearrangement. Also, with a limited recycle possibility, higher concentrations of residual hydrogen content are likely to be found. The cobalt fluoride process also leads to rearrangement of the parent hydrocarbon (see, for example, comparative studies on a series of *n*-alkanes[15]). However, by control of process conditions and careful selection of feedstocks, it is practicable to manufacture some perfluorocarbons at the 99% or even 99.9% purity level (for example, as is required in certain tracer applications). Owing to the recycle capability of the cobalt fluoride process, it is also economically feasible to reduce trace impurities such as CH-containing material to extremely low levels (<0.1 ppm w/w).

Comparing C_7 perfluorocarbon samples manufactured by the two industrial processes, perfluoroheptane by ECF contains *ca* 70% of C_7F_{14}, whereas material produced by CoF_3 fluorination is *ca* 90% C_7F_{14}. Technical-grade perfluoro(methylcyclohexane) produced by CoF_3 fluorination for industrial use contains *ca* 90% $C_6F_{11}CF_3$, which can be enhanced to *ca* 99% $C_6F_{11}CF_3$ for tracer applications. The manufacture of perfluoroperhydrophenanthrene (Flutec PP11) is typical of the production of high-molecular-weight cyclic perfluorocarbons:

$$C_{14}H_{10} + 17F_2 \xrightarrow{\text{CoF}_3,\ 350°C} C_{14}F_{24} + 10HF$$

In the finished product about 85% of the material possesses the phenanthrene skeleton, and the rest comprises perfluorobicycloalkanes.

The cobalt trifluoride process used at Rhône-Poulenc Chemicals Ltd was pioneered in the U.S.A by Fowler's Group in the 1940s,[13] developed further in the U.K. at Birmingham University,[14] and is now operated in a continuous stirred reactor, as described in the patent literature.[16] Vaporized hydrocarbon feedstock (e.g., phenanthrene) and fluorine are fed simultaneously to a reactor containing a CoF_2/CoF_3 mixture to produce a crude perfluorocarbon product which is condensed and separated from the HF byproduct. Physical separation techniques are used to recycle underfluorinated material to the reactor, followed by chemical stabilization steps and drying of the highly fluorinated material. A similar sequence is used in the ECF process. The specification of the finished perfluorocarbon may be tailored to meet the physical or chemical requirements of the intended application. Very often physical, thermal, or electrical properties of a final product are more important than its precise chemical composition.

4.3. PHYSICAL PROPERTIES

Physical properties of the gaseous perfluorocarbons carbon tetrafluoride, hexafluoroethane, octafluoropropane, and decafluorobutane are shown in Table 2.[17] Typical properties of Flutec Fluids are given in Table 3.[18] The compositions of these perfluorocarbon liquids can vary during production, but for most applications pure compounds are not required.

Table 2. Physical Properties of Perfluorocarbon Gases

	CF_4	C_2F_6	C_3F_8	C_4F_{10}
Molecular weight	88.00	138.01	188.02	283.02
Boiling point (°C) at 1 atm	−128.0	−78.2	−36.7	−2.2
Freezing point (°C) at 1 atm	−184.0	−100.6	−183	−128.2
Specific volume at 21.1°C and 1 atm (liter kg^{-1})	274.7	174.8	126.1	101.1
Density of gas at bp (kg $liter^{-1}$)	8.04	0.0057^a		0.0246^b
Density of liquid (kg $liter^{-1}$)	1.33^c	1.607^d	1.350^b	1.517^b
Critical temperature (°C)	−45.5	24.3	71.9	113.2
Critical pressure (bar)	37.39	33.03	26.8	23.23
Critical density (kg $liter^{-1}$)	0.64	0.6	0.628	0.629
Latent heat of vaporization at bp (kJ kg^{-1})	136.4	117.1	104.4	96.3
Latent heat of fusion at mp (kJ kg^{-1})	7.95	4.65		

^{a}At 25°C; ^{b}at 20°C; ^{c}at −80°C; ^{d}At −78.2°C.

Table 3. Typical Properties of Flutec™ Liquids

Property[a]	PP50	PP1C	PP1	PP2	PP3	PP5/6	PP7/9	PP10	PP11	PP24	PP25
Molecular weight	288	300	338	350	400	462	512	574	624	686	774
Density (kg liter^{-1})	1.604	1.707	1.682	1.788	1.828	1.917	1.972	1.984	2.03	2.052	2.049
Boiling point (°C) at 1 atm	29	48	57	76	102	142	160	194	215	244	260
Pour point (°C)	−120	−70	−90	−30	−70	−8[b]	−70	−40	−20	0	−10
Viscosity (kinematic) (mm^2 s^{-1})	0.29	0.615	0.39	0.873	1.06	2.66	3.25	4.84	14.0	15.3	56.1
Viscosity (dynamic) (mPa s)	0.465	1.049	0.656	1.561	1.919	5.10	6.41	9.58	28.4	31.5	114.5
Surface tension (mN m^{-1})	9.4	12.6	11.1	15.4	16.6	17.6	18.5	19.7	19	22.2	
Vapor pressure (mbar)	862	368	294	141	48	8.8	2.9	<1	<1	<1	<1
Heat of vaporization at boiling point (kJ kg^{-1})	90.8	75.8[c]	85.5	85.9	82.9	78.7	75.5	71[c]	68[c]	65.8[c]	67.9[c]
Specific heat (kJ kg^{-1} °C^{-1})	1.05	0.878[c]	1.09	0.963	0.963	1.05	1.09	0.92[c]	1.07[c]	0.93[c]	0.957[c]
Critical temperature (°C)	148.7	180.8[c]	177.9	212.8	241.5	292.0	313.4	357.2[c]	377[c]	388.7[c]	400.4[c]
Critical pressure (bar)	20.48	22.64[c]	18.34	20.19	18.81	17.53	16.60	16.2[c]	14.6[c]	15.1[c]	11.34[c]
Critical volume (liter kg^{-1})	1.626	1.567[c]	1.582	1.522	1.520	1.521	1.500	1.59[c]	1.58[c]	1.606[c]	1.574[c]
Thermal conductivity (mW m^{-1} °C^{-1})	64.0	66.4[c]	65.3	59.9	60.4	57.0	57.5	56[c]	52.6	64.6[c]	63.8[c]
Coefficient of expansion at 0 °C	0.00189	0.00167	0.00159	0.00138	0.00123	0.00104	0.000974	0.00078	0.00075	0.00078	0.00084
at bp	0.00213		0.00205	0.00190	0.00178	0.00170	0.00167				
Refractive index n_D^{20}	1.2383	1.2650	1.2509	1.2781	1.2895	1.3130	1.3195	1.3289	1.3348	1.3462	1.3376

[a]Temperature-dependent properties were determined at 25 °C.
[b]−8 °C for a typical *cis/trans* mixture.
[c]Estimated value.

4.4. APPLICATIONS OF PERFLUOROCARBONS

4.4.1. Perfluorocarbon Gases as Etchants

The major application for perfluorocarbon gases is in "dry" etching processes for the microelectronics industry. This "dry" process, commonly referred to as plasma etching, uses a gaseous compound in the presence of an RF (Radio Frequency) glow discharge plasma to generate reactive species which react with the substrate to form volatile etch products.[19-21] For silicon surfaces, fluorine-containing gases are utilized to generate the volatile byproduct SiF_4. Dry etching, which was first commercialized in the microelectronics industry in the late 1970s, has made enormous progress during the 80s and is used almost exclusively for the manufacture of the more sophisticated microelectronic circuits. Dry etching processes offer the advantage of anisotropic etching, which can pattern finer line widths (<1 micron) not obtainable with wet processes utilizing acid etchants. A comparison of the etch profiles typifying the wet and dry processes is shown in Figure 1.

The most common of the gaseous perfluorocarbons used to etch silicon-type materials is carbon tetrafluoride, which is a very stable gas. However, when subjected to a glow discharge in a plasma (where electron temperatures are greater than gas temperatures by several orders of magnitude[21]) it dissociates into a number of species including electrons, positive ions, negative ions, fluorine atoms, and CF_3 radicals: $CF_4 + e^- \rightarrow CF_3 \cdot + F \cdot + e^-$; where the predominant etching species is the fluorine atom. Under these circumstances, the fluorine atoms react with the silicon substrate to produce the volatile species SiF_4.

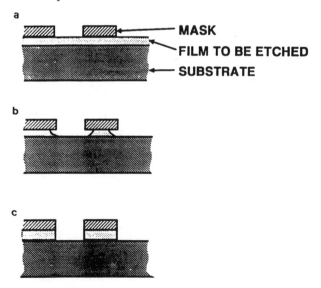

Figure 1. Etching profiles for integrated circuits. (a) Patterned device before etching; (b) liquid etching, isotropic profile; (c) plasma etching, anisotropic profile.

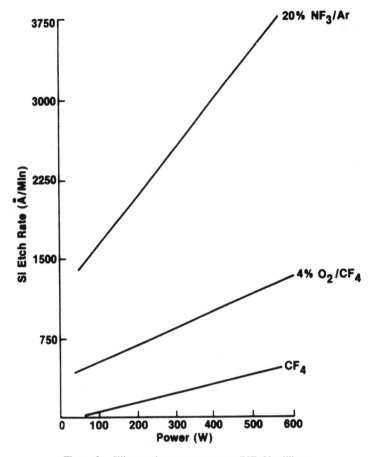

Figure 2.   Silicon etch rate versus power. RIE, 80 millitorr.

In addition to utilizing pure perfluorocarbon gases in plasma etching, extensive work has been done with additives such as CO_2 and, in particular, O_2 which is used extensively to promote etch rates and selectivity.[22]

Other gases such as SF_6, BF_3, and NF_3 are also used in plasma technology. Nitrogen trifluoride is a particularly important gas when high etch rates are desired.[23] It also offers the advantage of producing noncarbon-containing byproducts and liberating only nitrogen and volatile silicon compounds. A comparison of the etching performance on silicon of CF_4, NF_3, and CF_4/O_2 mixtures is shown in Figure 2.

Additional applications for perfluorocarbon gases are constantly being investigated. In particular, the banning of CFCs during the present decade has caused renewed interest in perfluorocarbon gases as possible substitutes. Although these gases cannot be expected to reach the large volume uses CFCs found in refrigeration, thermal insulation, and solvent applications, opportunities do exist for the smaller niche market. Applications under investigation include doping of silica, optical fiber glass cleaning, fire suppression and aerosols. A recent patent which could represent a large market opportunity is based on the use of CF_4 to remove hydrogen during a steel-making process.[24]

4.4.2. Perfluorocarbon Liquids as Alternatives to Chlorofluorocarbons

4.4.2.1. Cooling of Electrical Equipment

The continuing trend for size/cost reduction in the electronics and heavier electrical industries leads to higher packing densities of energy-dissipating devices, and therefore an increased need for effective heat removal. As previously described,[25] this has been achieved in many cases by direct liquid cooling, and 1,1,2-trichloro-1,2,2-trifluoroethane (CFC-113) has been used extensively. However, this situation is beginning to change because CFC-113 is implicated in stratospheric ozone depletion; hence perfluorocarbons are now being studied and used for larger volume applications, e.g., in distribution transformers and large thyristors.

Since heat transfer during boiling increases with pressure, then, in an hermetically sealed system, the vapor pressure/temperature characteristic of CFC-113 (boiling point 47.6°C at 1 atm) minimizes the possibility of thermal damage and gives good heat transfer. It would obviously be advantageous for perfluorocarbon alternatives to have a similar vapor pressure/temperature characteristic, and some potential candidates are listed in Table 4.

Two important considerations for heat transfer under ebullient conditions concern the critical heat flux as a function of saturated boiling temperature (this affects the size and cost of a device), and the temperature difference between the heat emitting surface and the liquid coolant (this may have relevance with temperature-sensitive components). Examination of Figure 3 shows that the critical heat flux for perfluoro(methylcyclopentane) (Flutec PP1C) is very similar to that of CFC-113 and there is an advantage (>30%) compared with perfluoro-n-hexane (Flutec PP1, Fluorinert FC-72) and perfluoro(dimethylcyclobutane) (KCD 9445).[26] However, Figure 4 illustrates that the difference in temperature between the heat-emitting surface (T_p) and the boiling liquid (T_{sat}) is greater and therefore less advantageous for perfluoro(methylcyclopentane), since the component will operate at a higher temperature (about 10°C max.) at

Table 4.   Perfluorocarbons as Potential Alternatives to CFC-113

Fluid	Commercial designation	Boiling point at 1 atm
Perfluoro(methylcyclopentane)	Flutec PP1C (Rhône-Poulenc)	47°C
Perfluoro-n-hexane	Flutec PP1 (Rhône-Poulenc) Fluorinert FC72 (3M) Hostinert S7 (Hoechst)	57°C
Perfluoro(dimethylcyclobutane)	KCD 9445 (Du Pont)	45°C

CRITICAL HEAT FLUX

w/cm²

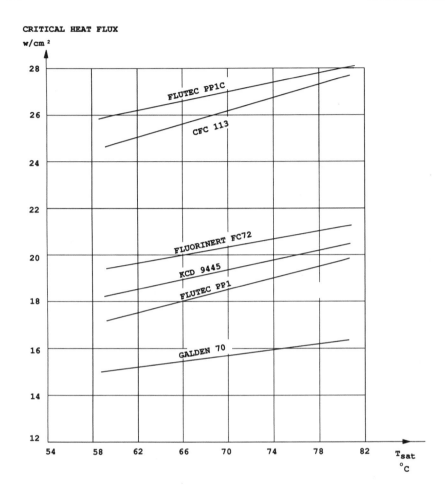

Figure 3. Comparison of critical heat flux for different fluids as a function of T_{sat}.

the same heat flux. It is unlikely that this small difference will have any adverse effect on component reliability.

An example of the advantage of high critical heat flux is a direct CFC-113 liquid-cooled thyristor used in a high-speed electric locomotive.[26] The maximum surface temperature for this device is 125°C and an operating saturated boiling temperature of 80°C. Under these conditions, the heat dissipation is 16–17 W/cm^{-2}, which is comfortably below the critical heat flux of perfluoro(methylcyclopentane) and CFC-113 (Figure 3), but close to that of other fluids.

Other heat transfer properties of perfluoro(methylcyclopentane) are similar to CFC-113, as shown in Table 5.

Although CFC-113 and perfluorocarbons have adequate high-voltage breakdown strength for most applications, perfluoro(methylcyclopentane) has a higher breakdown

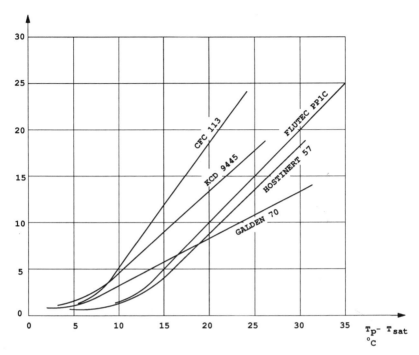

Figure 4. Heat flux versus temperature difference between heated surface and liquid.

strength than either CFC-113 or perfluoro-*n*-hexane, in both the liquid and vapor states. This has obvious advantages under ebullient, high-voltage conditions.

In conclusion, the critical heat flux of perfluoro(methylcyclopentane) is at least as good as that of CFC-113; therefore this perfluorocarbon is a preferred substitute since it allows minimum size of heat exchange surfaces, thereby reducing constructional and installation costs.

Table 5.   Comparison of Heat Transfer Properties of CFC-113 and
Perfluoro(methylcyclopentane)

Property (at 25°C)	CFC-113	Perfluoro(methylcyclopentane)
Density/(kg liter^{-1})	1.561	1.707
Specific heat (kJ kg^{-1})	0.91	0.85[a]
Thermal conductivity (mW m^{-1}°C^{-1})	740	66.4[a]
Latent heat (kJ kg^{-1}) at boiling point	143.9	75.8[a]

[a]Estimated.

4.4.2.2. Solvent Systems and Cleaning

The unique properties of CFC-113, combined with the exponential growth in precision engineering and electronics over the past few decades, has resulted in a growth rate of about 20% for CFC-113 and a peak global usage of about 200,000 tones p.a. Restrictions on CFC-113 now in force (see Chapter 28)[27,28] will result in the cessation of the use of this material for surface treatment in the decade 1990–2000. Industry is responding to this challenge, and aqueous processes seem destined to

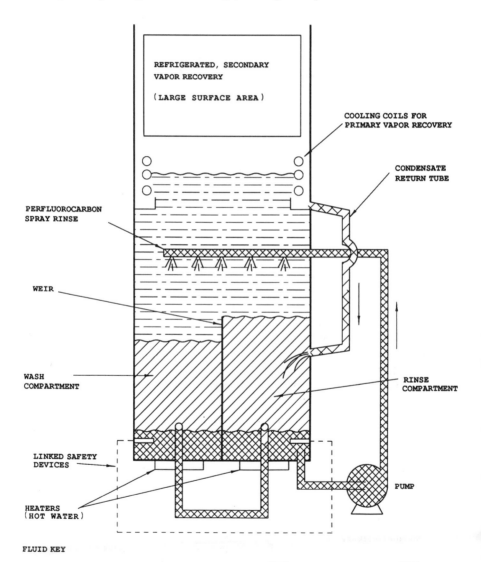

FLUID KEY

Figure 5. Perfluorocarbon/alcohol batch cleaning system. ▨ , Perfluorocarbon liquid; ▨ , alcohol liquid; ▤ , nonflammable perfluorocarbon/alcohol vapor.

replace a large proportion of the previous CFC-113 market. However, there are still severe difficulties with certain high-reliability items, for which manufacturers are reluctant to use water-based formulations because moisture in interstices could cause corrosion or electrical breakdown in the long term. Furthermore, aqueous cleaning systems are more complicated, more expensive, and consume more energy (typically three times); thus, there is a pressing need for an alternative cleaning system.

Although liquid perfluorocarbons are poor solvents, their solubility for oils is still sufficient to remove traces of oil.[29] In addition, their high densities and low surface tensions make them effective for the physical removal of particulate matter by spray/jet cleaning, e.g., surface cleaning of computer-memory disks. Technically, perfluorocarbons such as Flutecs PP1 and PP1C can be used as drop-in replacements for CFC-113 in standard equipment; however, their cost and poor solvent characteristics limit their applications. (A way of overcoming these disadvantages is described below.)

Alternatively, perfluorocarbons can be used in conjunction with solvents such as alcohol to form the basis of a cleaning method in which the perfluorocarbon acts as a heat-transfer agent, forms a nonflammable vapor blanket, reduces the flash point of the solvent, and finally rinses off solvent residues.[30,31]

A typical "perfluorocarbon-inerted" cleaning system is illustrated in Figure 5. Lower layers of perfluorocarbon are heated under a higher-boiling flammable solvent,

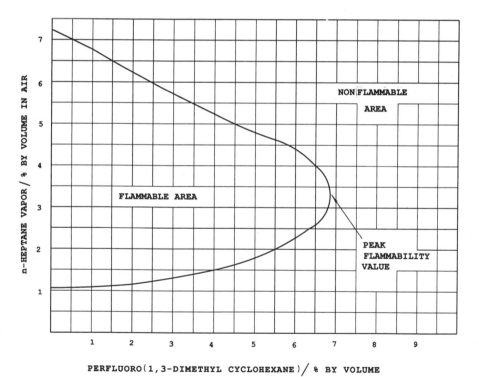

Figure 6. Flammability of n-heptane/perfluoro(1,3-dimethylcyclopentane)/air mixtures.

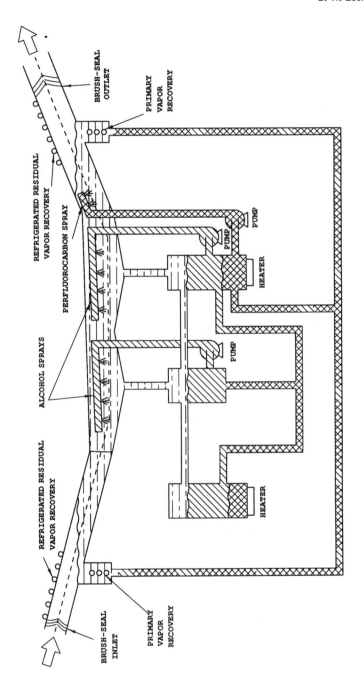

Figure 7. Perfluorocarbon/alcohol in-line cleaning system. ▨, Perfluorocarbon liquid; ▨, alcohol liquid; ▦, nonflammable perfluorocarbon/alcohol vapor.

such as an aliphatic alcohol or a hydrocarbon, to form a nonflammable vapor layer above the flammable solvent. The perfluorocarbon is selected on the basis of its boiling point so that the system is operating well outside the flammable area. An example is illustrated in Figure 6 for a mixture of perfluoro(1,3-dimethylcyclohexane)/n-heptane/air. At a concentration of 6.8 vol% perfluorocarbon, the flame ceases to propagate. A simple means is thus offered whereby the flammable solvent can be purified by distillation under nonflammable conditions and continuously used for rinsing cleaned items before a final rinse with perfluorocarbon. Details of the operation of the equipment are described elsewhere.[32]

Various modifications of the process are possible: for example, Figure 7 illustrates an in-line system having the advantage that only a relatively small alcohol inventory is required.

Because of the high cost of perfluorocarbons and the requirement that large quantities of volatile organic solvents should not be released into the atmosphere, it is envisaged that cleaning plants will be hermetically sealed during both operation and shutdown—a new concept for this type of process. Systems based on the above principles are now used industrially for cleaning such items as computer disk drives and printed circuit boards.

4.4.2.3. Foam Blowing

Environmental pressures to replace chlorofluorocarbon blowing agents, such as trichlorofluoromethane (CFC-11) and dichlorodifluoromethane (CFC-12), have focused attention on perfluorocarbons, e.g., perfluoro-n-pentane and perfluoro-n-hexane for blowing phenolic, polyurethane,[33] and fluoropolymer foams.[34] Perfluorocarbons have low thermal conductivities similar in value to those of CFCs (see Table 6), so that similar insulation thicknesses can still be maintained. This has economic advantages in applications such as refrigeration cabinets.

Perfluorocarbons are very stable, high-value materials and it is therefore necessary to avoid any significant release into the atmosphere during the manufacture, use, or disposal of foamed products. All three modes of loss can be controlled so that the perfluorocarbon is recyclable. The low solvent power of perfluorocarbons[29] ensures

Table 6. Comparison of Vapor Thermal Conductivities of Some Chlorofluorocarbons and Perfluorocarbons

Compound	Vapor thermal conductivity at 30 °C (mW m^{-1} °C)
CFC-11	8.4
CFC-12	9.6
Perfluoro-n-pentane[a] (Flutec PP50)	3.6[b]
Perfluoro-n-hexane[a] (Flutec PP1)	8.8[b]

[a]Main structural isomer.
[b]Estimated value.

negligible permeation losses through closed cellular foam structures, thus retaining the perfluorocarbon and conserving the low conductivity properties. By contrast, the higher solvent power of CFC blowing agents causes significant loss through the cell walls during the product lifetime, with consequent pollution of the environment and reduction of thermal insulative properties.

4.4.2.4. Refrigerants

Refrigeration applications operating in the –20 to –40°C temperature range have increasingly used an azeotropic mixture of R-22 (CF_2HCl) and R-115 (C_2F_5Cl) in place of R-22 alone, known as R-502. The high halogen content of R-115 imparts extra thermal capacity benefits which both decrease the temperature of compression and improve the energy efficiency. However, R-115 is now being banned because of its ozone-depleting effects and it has proved very difficult to find an R-502 substitute. It is possible to substitute perfluoropropane (R-218) for R-115 and obtain similar temperature-reduction benefits, without ozone-depletion problems. Some loss of efficiency occurs where this is done and can be overcome by adding a minor proportion of propane as a third ingredient.[35] The ternary mixture then has better properties than R-502 and is being commercialized as Isceon 69-L by Rhône-Poulenc Chemicals Ltd as an interim solution. Although this mixture has a much reduced ozone-depletion potential, R-22 is not regarded as an acceptable refrigerant for the long term.

4.4.2.5. Fire Extinguishants

The efficacy of perfluorocarbons as fire-extinguishing agents has been known for a long time,[36] but they were historically dismissed in favor of Halon-1211 (CF_2ClBr) and Halon-1301 (CF_3Br) for efficacy/cost effectiveness. Halon-1301 is exceptional in being the only material usable in enclosed spaces where humans are present. Hence its use on oil platforms, in aircraft, etc. Finding an alternative to Halon-1301 is proving difficult.

The very low toxicities of perfluorocarbons allows their use in human-occupied enclosed situations and work is in progress to assess their suitability.[37] One factor which must be considered is the propensity of some perfluorocarbons to form the highly toxic perfluoroisobutene under high-temperature conditions.

4.4.3. Perfluorocarbon Liquids as Heat Transfer Agents

4.4.3.1. Transformers

From a technical viewpoint, perfluorocarbons are ideal candidates for all types of fluid cooled transformers, but applications have been limited to those which justify the high cost of the fluid, e.g., mobile radar.

In the past, nonflammability in transformers for the general distribution of electrical power has been achieved by using such liquids as polychlorinated biphenyl/trichlorobenzene blends (PCBs), perchloroethylene, and 1,1,2-trichloro-1,2,2-trifluoroethane (CFC-113). PCBs and CFC-113 have become environmentally unacceptable, and perchloroethylene is receiving increasing adverse comment regarding its possible carcinogenicity and accumulation in various natural lipophiles.

Power distribution in densely populated areas, or in underground substations, still requires the fluid used to be nonflammable, of low toxicity, and, of more recent importance, environmentally friendly. The need has been most acute in Japan where gas-cooled transformers containing sulfur hexafluoride became generally available around 1980. Because of the poor heat-transfer properties of sulfur hexafluoride, the transformers were considerably larger than their standard liquid-filled counterparts. A reduction in size has been accomplished using a perfluorocarbon liquid/SF6 gas combination.[38-40] In order to achieve acceptable costs, a relatively small quantity of perfluorocarbon liquid (typical boiling point 100°C) is circulated from a lower sump through a heat exchanger, followed by spray delivery onto the transformer coils from where it returns to the sump. The internal free space is filled with SF6 gas at about 2.5 atmospheres pressure to attain a satisfactory dielectric strength. The core and windings are cooled by both direct contact with SF6 gas and the perfluorocarbon liquid, which partly evaporates, to give an enhanced cooling effect. The direct heat-sinks for both the SF6 gas and perfluorocarbon vapor are the container walls and excess perfluorocarbon spray droplets. The latter give very efficient heat transfer because of their large surface area. Perfluorocarbon-cooled transformers now account for a small, but significant proportion of the Japanese market.

One potential problem of the SF6 gas/perfluorocarbon liquid combination is the inferior dielectric strength of a gas compared to that of a liquid. Also, degassing of a liquid could cause bubble formation which, together with fiber contamination, may cause electrical breakdown.[41]

A different approach is currently under development in the U.K. (Rhône-Poulenc Chemicals Ltd) using immersion cooling with liquid perfluorocarbon alone. This avoids SF6 degassing and markedly reduces transformer weight and size. In this design, each of three transformer coils (3-phase) is enclosed in a discrete, closely fitting container, which comprises two concentric epoxy-glass fiber cylinders, with the anulus sealed at both ends except for perfluorocarbon liquid inlets and outlets. The cooling circuit is illustrated schematically in Figure 8. A comparison of the power losses, weight, transformer volume, and liquid content of a Flutec-filled transformer with other types of transformers is given in Table 7. In Table 8 a comparison is made of the test results obtained on a Flutec-filled transformer manufactured to a 500 kVA specification, showing that the required performance is well within the limits set.

It can be seen that nonflammable perfluorocarbon-cooled transformers offer an attractive option where electrical power has to be delivered to high-cost or densely populated areas. The small transformer volume and weight of the liquid-cooled type make the Flutec-filled transformer particularly attractive for reducing building and installation costs. Although the unit requires both a pump and a fan, these items are now highly reliable and should be acceptable to power distribution companies. Using pool-boiling, variations of this design are feasible in which radiators (condensers) are

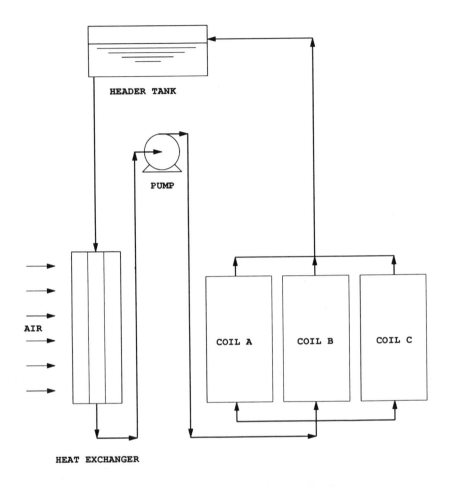

Figure 8. Perfluorocarbon-filled transformer cooling circuit.

Table 7. Comparison of Typical Transformers Filled with Flutec Fluid, Silicone, and Cast Resin

| | Power losses | | Gross weight | Transformer | Liquid |
	Iron core	Copper coil	(kg)	volume (m^3)	content (liter)
Flutec (liquid filled)	880	6860	1500	1.20	50
Silicone/ester (liquid filled)	1030	6860	1970	3.48	600
Cast resin	1750	4500	2120	2.03	0

Table 8. Specification and Test Results for
Flutec-Filled Transformer[a]

	Specification (maximum values)	Test results
Impedance	4.75%	4.28%
No load losses	880 watts	880 watts
Load losses (75%)	6860 watts	6423 watts
Average temperature rise of:		
Windings	70°C	Low voltage 58°C
		High voltage 57°C
Liquid at top	60°C	43°C

[a]The transformer is based on a 500-kVA distribution transformer to ESI 35.1 and BS 171.

situated above the three coils (boilers) and the system operates under a vacuum. Both the pump and fan are thus eliminated, but the top-mounted radiator increases the overall size.

4.4.3.2. Capacitors

These devices have inevitably been included in the modern electronics trend to reduce component size, preferably without comprising performance. Liquid impregnants have a significant role in achieving these objectives, and perfluorocarbons are particularly effective because they have all the desirable properties except high relative permittivity, which tends to be low (typically 1.8–2) compared with that of other organic capacitor impregnants (typically 3–5.5). Thus an adverse effect on power-to-size ratio would be expected according to Coulomb's Law ($Q_1Q_2 = 4\pi kx^2F$; where F = force between Q_1, and Q_2, Q_1Q_2 = product of charges on opposite capacitor plates, k = relative permittivity, and x = distance between capacitors).

Hence a low relative permittivity would be expected to give a low energy storage (Q_1Q_2) for an equivalent potential on the capacitor surfaces.[42] However, other properties of perfluorocarbons overcome this defect, so that capacitors made at Sandia Laboratories had considerably improved energy density and reliability compared with those impregnated with hydrocarbon or silicone.[43] These capacitors were wound at predetermined stresses and hermetically sealed under degassed conditions. It was concluded that the improvement was due to evaporative cooling at hot-spots. The devices were operated at near-atmospheric pressures, where perfluorocarbon vapors have a high enough dielectric strength to reduce the likelihood of electrical breakdown because of the gas bubbles formed during ebullition.

The other properties believed to be important are the low surface tension and viscosities of the perfluorocarbons; these assist in efficient wetting and impregnation of the capacitor windings.

A typical performance improvement was an increase of pulses before failure from about 10^6 to 10^8 at electrical stress levels of about three times those possible with

Table 9. Comparison of Typical Physical Properties of
Perfluorocarbons Relevant to Capacitors

Perfluoro fluid	Main molecular structure	Boiling point (°C)	Breakdown strength (kV/(2.5mm))	Electrical resistivity (ohm m^{-1})	Dissipation factor	Surface tension (mN m^{-1})	Kinematic viscosity (nm^2 s^{-1})
Flutec							
PP1C	Perfluoro(methyl-cyclopentane)	48	33^a	10^{15}	<0.0001	12.6	0.615
PP1	Perfluorohexane	57	33^a	10^{15}	<0.0001	11.1	0.39
PP2	Perfluoro(methyl-cyclohexane)	76	33^a	10^{15}	<0.0001	15.4	0.873
PP3	Perfluoro(1,3-di-methylcyclo-hexane)	102	33^a	10^{15}	<0.0001	16.6	1.06
Fluorinert							
FC72	Perfluorohexanes	56	38	10^{15}	<0.0003	12	0.4
FC75	Perfluorinated cyclic ethers	102	40	8 × 10^{15}	<0.0001	15	0.8
FC40	Perfluoro(tributyl-amine)	155	46	4 × 10^{15}	<0.0003	16	2.2

^{a}Minimum after filtration through 30 μm filter.

silicone or hydrocarbon impregnants. Since energy density is proportional to the square of the stress levels, the improvement in energy storage is an order of magnitude.

The fluids used were based on perfluorotributylamine and perfluorohexanes. Other perfluorocarbons are also obvious candidates; perfluoro(methylcyclohexane) and perfluoro(1,3-dimethylcyclohexane), for example, have some of the highest vapor dielectric strengths known.[44]

Selected relevant typical properties of some perfluoro compounds are given in Table 9.

4.4.3.3. Vapor-Phase Soldering Update

A more correct name for this technique (described previously[45]) is condensation soldering; however, the VPS abbreviation has become widely accepted and therefore will be used here.

Since its initial use in the mid-seventies, VPS has continued to provide a vital means of solder reflow,[46] which is the most widely used procedure for Surface Mount Technology (SMT). Although this method of assembly has been used historically for only a minor proportion of printed circuit board assembly, it allows substantial size reduction and cost savings, therefore SMT is now becoming widely used. Briefly, solder paste is screen-printed onto the circuit substrate followed by accurate placement of components; these are retained in position by surface tension until the solder paste is reflowed and solidification has occurred.

During the early eighties, VPS took an increasing share of the reflow equipment market, mainly at the expense of infrared (IR) systems. The manufacturers of these systems responded by markedly improving the performance of their equipment, which became effectively a forced hot-air/IR hybrid. Other improvements such as selective wavelength and computer-controlled temperature profiling were incorporated. IR soldering thus became applicable to more densely packed SMT assemblies than was previously thought possible. The availability of this improved equipment, together with problems associated with VPS equipment at the time, reversed the trend in favor of IR (IR/VPS:85/15).

However, with the continuing miniaturization trend in electronics which results in very densely packed boards, IR is again reaching its limits; and where this occurs, the VPS technique is being used. Another disadvantage with IR is the time taken to set correct temperature profiles, whereas VPS requires little time. Thus VPS is preferred for contract manufacturing, involving short production runs or the assembly of "one-offs".

The dominance of IR for solder reflow in recent years has been in no small part due to inadequacies in early VPS equipment and the availability of only one fluorocarbon working fluid. Some of these inadequacies, which have now been overcome, are discussed below:

a. Fluid Losses.   Certain types of early equipment (single fluid, in-line) suffered from design defects which caused excessive fluid losses. These have been remedied by the reduction of ventilation flow rates, preheating the workload prior to soldering, and the use of metal carrier belts, refrigerated vapor recovery, and mist filters. The latter retains the aerosol formed when perfluorocarbon vapor contacts cooler air. Figure 9 illustrates the dependence of fluid loss on air flow rate, with and without an aerosol filter. It shows an approximate 50% smaller loss with an aerosol filter. A comparison is shown in Figure 10 of the relative loss rates of two perfluorocarbons, Fluorinert FC70 and Flutec PP11. These perfluorocarbons have identical boiling points (215°C), different molecular weights (820 and 620 respectively), and almost identical vapor pressure/temperature characteristics. Figure 10 shows that the relative loss rate of the two fluids is reversed when an aerosol filter is fitted.

b. CFC-113 Secondary Blanket.   The technique of floating a secondary CFC-113 blanket above the primary perfluorocarbon blanket to reduce perfluorocarbon losses is obsolescent. New equipment is simpler since it uses perfluorocarbon liquid only. Environmental pressures are accelerating a change to single fluid equipment.

c. Thermal Stability.   During the early years of VPS, corrosion of cooling coils was a common feature because both primary and secondary fluids decomposed at vapor soldering temperatures. Newer fluids have improved stability and, in correctly designed equipment, such corrosion is negligible.[47]

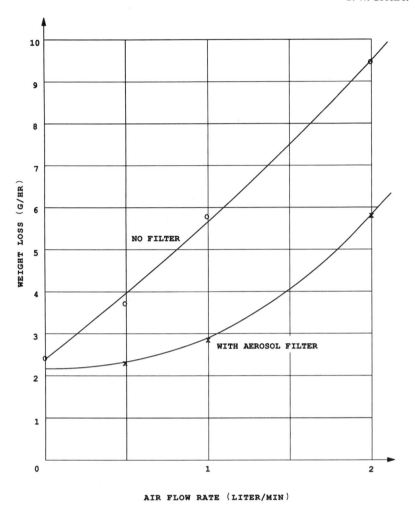

Figure 9. Dependence of Flutec PP11 loss on air flow rate.

d. Temperature Selection and Control. Soft-soldering temperatures range from about 140°C to 320°C and the limited range of fluids available in the early years of VPS was a disadvantage; however, there are now families of suitable fluids (such as Flutec PP10, PP24, and PP25—see Table 1). The general rule is to maintain a minimum temperature compatible with effective soldering, since component failure due to thermal damage increases exponentially with temperature. It is important that the boiling point of the perfluorocarbon does not drift upward excessively during use. Some perfluorocarbon production methods produce mixtures which are distilled to

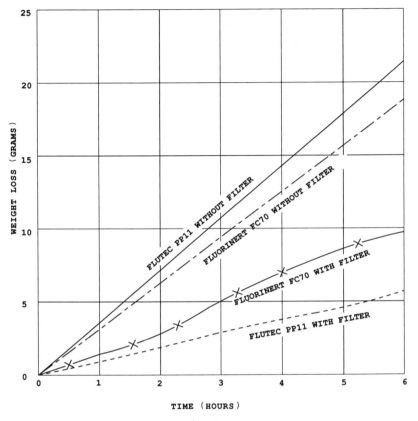

Figure 10. Effect of mist filter on fluid loss.

give various boiling range cuts. These typically give a temperature drift of about 10°C, which is acceptable in the majority of cases.

e. Soldering Defects. Two deleterious phenomena known as "tombstoning" (components stand on end during solder reflow) and "solder-balling" can be amplified by the VPS process, unless the printed circuit assembly is adequately preheated before reflow. This facility was absent on early equipment.

f. Toxicology. Perfluoroisobutene (PFIB) formation continues to be a perceived toxicological problem, and although various commercial fluorocarbons produce PFIB under operating conditions, it is usually nondetectable, or found at levels well below[48] the accepted time-weighted average (TWA), for an 8-hour working day, of 0.01 ppm. The formation of PFIB is temperature-dependent and results for Flutec PP11 are shown in Table 10. PFIB is just detectable at 275°C, i.e., well above the normal process operating temperature of 215°C.

Table 10. Flutec PP11—PFIB Formation at Various
Temperatures

Temperature (°C)	Perfluoroisobutene concentration $(\mu g\ g^{-1}\ hr^{-1})$
<250	nil
275	0.00005
300	0.0026

4.4.3.4. Vapor-Phase Sterilization

Vapor-phase sterilization is an application for perfluorocarbons where advantages accrue from the ability to control temperature precisely under inert, anaerobic conditions and where heat transfer is both rapid and even.

Historically, sterilization has been carried out in small portable units based on fan-assisted ovens or steam autoclaves. Both of these methods have disadvantages and some doubt has been expressed about uniform temperatures being achieved on the variety of profiles presented.[49] In fan-assisted ovens, quiescent zones may exist, giving nonuniform heating; and in steam autoclaves, temperature uniformity is questionable because of the variable internal geometry and the possibility of residual air pockets. Also, the combination of air and steam is aggressive on the sharp edges of instruments. Furthermore, steam sterilizers are subject to pressure vessel regulations and because of their large mass require an extended sterilization cycle. Clearly, there is a need to improve sterilizer technology.

In 1970, R.E. Pickstone described the use of perfluorocarbon vapors (Fluorinert FC-43) for sterilization in an insulated vessel containing boiling perfluorocarbon, and also a means of controlling the level of the vapor.[50] The following potential advantages were apparent: (1) Rapid heat transfer due to vapor condensation can be achieved. This is independent of the geometry and varying thermal masses of the items to be sterilized. (2) The low surface tension of perfluorocarbons assists in the penetration of minute interstices which may contain bacteria. Anaerobic conditions and the inert nature of perfluorocarbons reduce damage to instruments. (3) The process is conducted at atmospheric pressure, thus avoiding the necessity for a pressure vessel. Reductions in both weight and sterilization times are thus achievable.

A sterilizer was constructed which operated successfully for six years in the Ophthalmology Department of a hospital in Hastings, U.K., and, during that time, the equipment was returned only once to the manufacturer for servicing. The serviceable life of instruments with microsharp edges was markedly improved, sterilization times were shortened, and operators found the device very convenient to use.

The fluid originally used was perfluoro(tributylamine) (Fluorinert FC43),[51] but in later work perfluorodecalin, perfluoro(methyldecalin), and perfluoroperhydrofluorene (Flutecs PP5, PP9, and PP10, respectively) were employed.[52] The last perfluorocarbons have improved thermal stability and purity, which are important in this application.

Figure 11. Bench-top perfluorocarbon sterilizer.

Success in applying this sterilization technique relies heavily on suitable equipment being available. In the original Pickstone design losses of perfluorocarbon were small and costs were well within acceptable limits.

In another design (Figure 11), a front-opening door is used for operator convenience. Although each time the door is opened, air saturated with perfluorocarbon vapor is lost, it is normal to cool the equipment at the end of the sterilization cycle so that the perfluorocarbon has a low vapor pressure; the losses are still small enough for the procedure to remain economical.

Kill times in perfluorocarbon vapors using *stearothermophilus* spore strips are shown in Table 11.

Table 11. Kill Times in Perfluorocarbon Vapors for *Stearothermophilus* Spore Strips

Flutec fluid	Boiling point (°C)	Kill time (s)
PP5	140	300
PP9	155	150
FC-43	179	60
PP10	189	16

4.4.4. Miscellaneous Applications of Perfluorocarbon Liquids

4.4.4.1. Tracers

Because of environmental concerns relating to acid rain and urban smog, experiments using perfluorocarbon tracers have been conducted worldwide to identify pollution sources and transport mechanisms.[53] Cyclic perfluorocarbons containing 6–9 carbon atoms are particularly suitable and have strong electron affinities which allow their detection by electron capture techniques down to very low levels (1 pp10^{17}).[54]

More recently, evaluations have been made using perfluorocarbons as tracers for water mixing processes.[55,56] There are essentially two methods of studying these. The first, and up to now the most common, is to make use of naturally occurring physical and chemical variables, for example, temperature, salinity, and trace metals content. The second approach has been to add a suitable tracer deliberately and monitor its progress. An advantage of this method is that it allows the dynamics of the mixing process to be studied.

Tracers added deliberately should fulfill the following criteria:

- possess long lifetimes under aqueous conditions;
- be nontoxic and environmentally acceptable;
- occur naturally in only very low concentrations;
- have low detection limits using readily available equipment;
- undergo rapid dispersion into water after introduction.

Radionuclides such as ^{82}Br fulfill most of these criteria, but their use has been limited for safety reasons. In contrast, perfluorocarbons are virtually inert biologically, but because of their low water solubility and high density (relative to water), injection and dispersion are difficult. However, these difficulties can be overcome by the use of water-based emulsions similar to those already developed for medical applications (see Chapter 26).

An early investigation into the possibility of using perfluorodecalin (Flutec PP5) as a marine tracer was undertaken in the English Channel.[55] The results indicated that a perfluorodecalin/Pluronic F-68/water emulsion gave acceptable mixing characteristics. The perfluorodecalin was detectable down to approximately 4×10^{-15} mol per liter using gas chromatography and an electron capture detector. Analysis times were of the order of a few minutes.

More recently,[56] a study has been made of the discharge of cooling water from a Swedish nuclear power station into a sea-water lagoon employing a perfluoro(1-methyldecalin)/Pluronic F-68/water emulsion together with *iso*-propanol and pentane as co-solvents. The emulsion was stable for two months. Mathematical modeling enabled the mixing process to be determined in detail. Moreover, the results indicated an almost 100% recovery of perfluorocarbon, suggesting that no significant loss from either evaporation or sedimentation occurred. Thus it would appear that perfluorocarbons are suitable tracers for aqueous mixing studies.

Table 12.   Comparison of Boundary Lubricant Properties of
Perfluorocarbons and a Mineral Oil

Perfluorocarbon	Boiling point (°C)	Initial seizure load (kg)	Weld load (kg)
Flutec PP9	160	—	200
Flutec PP11	215	70	225
Flutec PP25	260	90	250
ISO 10 Mineral oil	—	36	126

4.4.4.2.   Lubricants

Perfluorocarbon ethers combine the usual inert, nonflammable characteristics of perfluorocarbons with exceptionally low vapor pressures and wide liquid ranges. Consequently, these fluids have become extensively used as lubricants, both in high vacuum and oxygen handling technologies (see Chapter 20). However, in certain other applications, limitations occur because of the formation of Lewis acids which cause chemical degradation at the C—O—C linkage. Examples of such applications are the handling of corrosive halogen compounds or the lubrication of high-temperature heat pumps. Where such limitations occur, the use of compounds containing carbon and fluorine only (i.e., perfluorocarbons) should be considered.

Because of the volatility of perfluorocarbon lubricants, the low boiling types can only be used under hermetically sealed conditions. However, those having boiling points of about 250°C or greater should have sufficiently low vapor pressures for open lubrication of glands, valves, etc. They could also be suitable for greases incorporating low-molecular-weight fluorocarbon polymers as thickening agents.

Three Flutec perfluorocarbons have been tested for boundary lubrication properties according to the Institute of Petroleum Four Ball Test 239/69 and compared to an ISO 10 mineral oil having an approximately equivalent viscosity.[57] The results (Table 12) show that in all cases, perfluorocarbons required a greater load than the mineral oil to cause both initial seizure and weld. The latter contained performance-enhancing additives, whereas the perfluorocarbon contained none.

The Four Ball Test measures the effectiveness of a lubricant under boundary layer conditions where severe adhesive wear occurs. Both the scar diameters and temperatures that occurred with the perfluorocarbons appeared to be greater at equivalent loads, indicating an increased tendency for mild wear compared to that with mineral oil. However, this disadvantage might be overcome with suitable viscosity modifications of the perfluorocarbons or the inclusion of additives.

4.4.4.3.   Particle Physics

When subatomic particles are accelerated at very high velocities and then caused to collide with one another, mass is converted to energy according to Einstein's equation, $E = mc^2$. The energy then reverts to new mass structures in an analogous way to water droplets condensing to form steam. However, for some unknown reason,

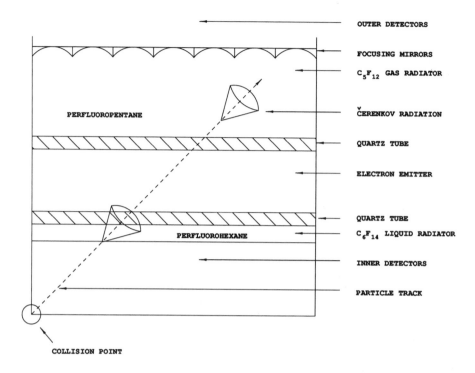

OUTER DETECTORS

FOCUSING MIRRORS

C_5F_{12} GAS RADIATOR

ČERENKOV RADIATION

QUARTZ TUBE

ELECTRON EMITTER

QUARTZ TUBE

C_6F_{14} LIQUID RADIATOR

INNER DETECTORS

PARTICLE TRACK

PERFLUOROPENTANE

PERFLUOROHEXANE

COLLISION POINT

Figure 12. RICH detector cross-section.

specifically structured particles are formed.[58,59] Detection of these particles is a major consideration when building "atom-smashing" machines, and perfluorocarbons have been used in part of the detection assemblies.

Both the accelerators at CERN (Geneva) and SLAC (Stamford University, California) use detectors comprising a space between two concentric cylinders filled with a gaseous reactive medium and a suitable detection system described below. By using multiconcentric cylinders, a variety of media and detection systems can be used to cover the wide range of particles formed. These can be emitted at all angles from the point of collision on the central axis of the concentric assembly. The CERN system is based on four detectors code named Aleph, Delphi, L3, and Opal. The Delphi detector incorporates gaseous perfluoropentanes and liquid perfluorohexanes. The SLAC detector uses a similar system. Both devices are known as ring-imaging Cerenkov detectors (RICH—see Figure 12).

When a charged particle with sufficient velocity passes through the perfluorocarbon, it generates a cone of Č UV photons around its trajectory. This cone (lower wavelength 190 nm) is focused into a ring by means of mirrors within a drift tube, where the photons are converted into electrons by means of a photoelectric gas. The electrons drift parallel to the axis and are read by computerized pulse counter. Gaseous perfluoropentanes are used because of their high refractive index (low Č threshold), high density (high number of photons), and good UV transparency. The pressure is fixed, so the refractive index is kept constant. Liquid perfluorohexanes, used for similar reasons, enable the range of particles which can be detected to be extended.

4.4.4.4. Supplying Oxygen to Aerobic Submerged Cultures

Since perfluorocarbons have a significantly greater capacity to dissolve gases than water (see Chapter 26), they can be used to increase yields in biological cultures requiring oxygen (normally supplied by a combined bubbling and stirring technique) as an essential nutrient. Thus, yield increases up to eight times were noted when air-saturated Flutec PP9 was sprayed into the top of a fermentation column containing *E. coli* cells in an aqueous nutrient medium.[60] Other advantages claimed were:

- Attainment of higher oxygen supply rates owing to the large increase in surface area (the perfluorocarbon droplets are smaller than air bubbles, which tend to coalesce).
- Relatively low energy was required to form small droplets (because of the low surface tension of perfluorocarbons) compared with that required by the combined air-bubbling and stirring method.
- No foaming was observed (this can be a significant problem when bubbling air through submerged cultures).
- Relatively low air-flow rates were required to saturate the perfluorocarbon.

Perfluorocarbon recoveries of 98.8% were reported at low to moderate cell yields[60]; however, at higher cell yields, a greater retention of the perfluorocarbon occurs in the cell mass, hence the recovery is lower. For the process to be optimized, some method of separation needs to be devised. It may be possible, for example, to use a more volatile perfluorocarbon so that it can be recovered by an evaporative process that does not damage the cells thermally.

4.4.4.5. Liquid Barrier Filters for Ultraclean Air

This application also takes advantage of the relatively high solubility of gases in perfluorocarbon fluids. Gases pass through these fluids by initial dissolution followed by diffusion. However, any particulate matter present in the gas will collide with the liquid molecules, lose momentum, and ultimately remain trapped. Consequently, this leads to the possibility of perfluorocarbons being used as liquid barrier filters.

Conventional "mechanical" filters rely on reducing the mean free path of particles and then retaining the particles. There has been considerable interest in the possibility of using a thin liquid barrier on the face of a conventional filter and, with this approach, entrapment of particles approaching 100% has been achieved.[61,62]

Thus liquid barrier filters could have important applications in the semiconductor industry where the ever-decreasing size of device requires even more efficient particle filtration. Work has also been undertaken into the use of liquid barrier filters incorporated into face masks for the removal of potentially dangerous particles.[62]

4.5.   REFERENCES

1. D. S. L. Slinn and S. W. Green, in: *Preparation, Properties and Industrial Applications of Organofluorine Compounds* (R. E. Banks, ed.), p. 45, Ellis Horwood, Chichester (1982).
2. H. Moissan, *Compt. Rend. 110*, 276, 951 (1890).
3. P. Lebeau and A. Damiens, *Compt. Rend. 182*, 1340 (1926); *191*, 939 (1930).
4. O. Ruff and R. Keim, *Z. Anorg. Chem. 192*, 249 (1930).
5. Japanese Patent 55 113, 728 (to Daikin Kogyo Co.) [*CA 95*, 97016 (1981)].
6. A. J. Woytek, in: *Fluorine: The First Hundred Years (1886–1986)* (R. E. Banks, D. W. A. Sharp, and J. C. Tatlow, eds.), p. 331, Elsevier Sequoia, New York (1986).
7. *Chemical Marketing Reporter*, p. 57 (26th April 1982); *Journal of Commerce*, p. 10A (3rd May 1982).
8. S. Fukui and H. Yoneda, Japanese Patent 58 41, 829 (to Asahi Glass Co.) [*CA 100*, 67823 (1984)].
9. Japanese Patent 60 77,983 (to Daikin Kogyo Co.) [*CA 103*, 61499 (1985)].
10. Japanese Patent 60 81, 134 (to Showa Denka K. K.) [*CA 103*, 122980 (1985)].
11. R. A. Paul and M. B. Howard, U. S. Patent 3, 840, 445 (to Phillips Petroleum Co.) [*CA 82*, 49313 (1975)].
12. D. Sato, K. Yamanouchi, and R. Murashima, Japanese Patent 48 76,841 (to Green Cross Corp.) [*CA 80*, 82239 (1974)].
13. R. D. Fowler, W. B. Burford, J. M. Hamilton, R. G. Sweet, C. E. Weber, J. S. Kasper, and I. Litant, *Ind. Eng. Chem. 39*, 292 (1947).
14. M. Stacey and J. C. Tatlow, in: *Advances in Fluorine Chemistry* (M. Stacey, J. C. Tatlow, and A. G. Sharpe, eds.), Vol. 1, p. 166, Butterworths, London (1960).
15. J. Burdon, *Centenary of the Discovers of Fluorine, International Symposium, Abstracts*, p. 9, Paris (1986).
16. R. J. Kingdom, G. D. Bond, and W. L. Linton, British Patent 1,281,822 (to ISC Chemicals). Ger. Patent 2, 000, 830 [*CA 73*, 130690 (1970)].
17. Air Products and Chemicals, Inc., Allentown (U.S.A.).
18. Flutec™ PP Fluorocarbon Liquids, Rhône-Poulenc Chemicals Ltd, Avonmouth, Bristol (U. K.).
19. A. J. Woytek and J. Stach, *Ind. Res. Dev. 24*, No. 7, 107 (1982).
20. B. Chapman, *Glow Discharge Processes*, Chap. 7, Wiley, New York (1980).
21. H. W. Hess, *Chem. Tech. 9*, 432 (1979).
22. A. Jacob, U. S. Patent 3, 795, 557 (to LFE Corp.) [*CA 80*, 114048 (1974)].
23. A. J. Woytek, J. T. Lileck, and J. A. Barkanic, *Solid State Technol. 27*, 172 (1984).
24. K. J. Fioravanti, D. W. Kern, and P. D. Stelts, U.S. Patent 4, 869, 749 (to Air Products and Chemicals) [*CA 112*, 40440 (1990)].
25. D. S. L. Slinn and S. W. Green, *Preparation, Properties and Industrial Applications of Organofluorine Compounds* (R. E. Banks, ed.), p. 69, Ellis Horwood, Chichester (1982).
26. T. Jomard and M. Lallemand, "Transferts de Chaleur en Ébullition Libre de fluides de Remplacement du R 113", Colloque SFT 30, Colloque de thermique: Progrès et défis actuels, Nantes (16/17 Mai 1990).
27. C. Lea, *Circuit World, 14*, No. 4, 4 (1988).
28. C. Lea, *Circuit World 15*, No. 3, 22 (1989).
29. Flutec PP Fluorocarbon Liquids, Rhône-Poulenc Chemicals Ltd, Section E9.
30. D. S. L. Slinn, British Patent Appl. 8915464 (24th January 1990) (to RTZ Chemicals—ISC Division).
31. D. S. L. Slinn, European Patent Appl. 89306894 (6th July 1989) (to ISC Chemicals).
32. D. S. L. Slinn and B. H. Baxter, *Proceedings of the Technical Program NEPCOM West '90*, Annaheim, California, Vol. II, p. 1810 (1990).

33. BASF, European Patent Appl. 0351614 (1st July 1989).
34. D. A. Reed and H. E. Lunk, British Patent 2, 143, 237 (to Raychem Corp.) [*CA 102*, 204942 (1985)].
35. S. F. Pearson and J. Brown, *18th International Congress of Refrigeration*, Montreal, Paper 162 (1991).
36. J. H. Simons (ed.), *Fluorine Chemistry*, Vol. 5, p. 356, Academic Press, New York (1964).
37. J. S. Nimitz, S. R. Skaggs, and R. E. Tapscott, *Proc. 1991 International CFC and Halon Alternatives Conf.*, Baltimore, M.D. (Dec. 1991).
38. Electrical Power Research Institute, (Palo Alto, California), Report, Gas/Vapor and Fire Resistant Transformers, EI-1430 (June 1980).
39. K. Tokoro, Y. Harumoto, Y. Kabayama, A. Yamauchi, T. Sato, and T. Ina, *IEEE Trans. Power Appar. Syst. PAS-101*, 4341 (1982).
40. Y. Harumoto, Y. Kabayama, Y. Kuroda, Y. Yoshida, H. Kan, Y. Miura, E. Tamaki, and T. Hakata, *IEEE Trans. Power Appar. Syst. PAS-104*, 2501 (1985).
41. I. Takagi, M. Higaki, K. Endoo, T. Shirone, K. Hiraishi, and K. Kawashima, *IEEE Trans. Power Delivery 3*, 1809 (1988).
42. R. Kron, *Electrical Review 215*, 30 (1984).
43. G. H. Maudlin, *IEEE Trans. Electr. Insul. EI-20*, 60 (1985).
44. D. R. James, L. G. Christophorou, R. Y. Pai, M. O. Pace, R. A. Mathis, I. Sauers, and C. C. Chan, *Gaseous Dielectr., Proc. Int. Symp. (CONF-780301)*, 224 (1978, Eng.).
45. D. S. L. Slinn and S. W. Green, *Preparation, Properties and Industrial Applications of Organofluorine Compounds* (R. E. Banks, ed.), p. 66, Ellis Horwood, Chichester (1982).
46. S. W. Hinch, *Printed Circuits Handbook*, Third Edition (C. F. Coombs, ed.), p. 31, McGraw-Hill, New York (1988).
47. D. S. L. Slinn, *Proceedings of the Technical Program NEPCOM West 1984*, Annaheim, California, p. 121 (1984).
48. L. J. Turbini and F. M. Zado, *Electron. Packag. Prod. 20*, 49 (1980).
49. Proposed American National Standard/American Dental Association Specification No. 59 for portable steam sterilizers for health care use; AIP (1987).
50. R. E. Pickstone, British Patent 1,195,052 (17th June 1970).
51. M. Klein and E. G. Millwood, *Trans. Ophth. Soc. 88*, 355 (1968); *Ophthalmology 222*, 1501 (1970).
52. D. S. L. Slinn, British Patent 2,196,532 (29th October 1986) (to RTZ Chemicals—ISC Division).
53. J. E. Lovelock and G. J. Ferber, *Atmos. Environ. 16*, 1467 (1982).
54. R. W. Dietz, Paper presented at Ispra Courses, Regional and Longrange Transport of Air Pollution, Ispra (Varese) Italy (September 15–19, 1986).
55. A. J. Watson, M. I. Liddicoat, and J. R. Ledwell, *Deep-Sea Res. 34*, 19 (1987).
56. E. Fogelqvist, M. Krysell, and P. Oehman, *Mar. Chem. 26*, 339 (1989).
57. Unpublished work conducted on behalf of Rhône-Poulenc Chemicals by Swansea Tribology Centre, University College of Swansea, U.K. (1987).
58. J. Hassard, *New Scientist*, 65 (8th July 1989).
59. C. Sutton, *New Scientist*, 1 (11th Feb. 1989).
60. D. Damiano and S. S. Wang, *Biotechnol. Lett. 7*, 81 (1985).
61. W. D. Seufert, F. Bessette, G. Lachiver, and H. Merdy, *Health Phys. 42*, 209 (1982).
62. W. D. Seufert, F. Bessette, and G. Lachiver, *Br. J. Ind. Med. 41*, 1 (1984).

5

Electrochemical Fluorination and Its Applications

Y. W. ALSMEYER, W. V. CHILDS, R. M. FLYNN, G. G. I. MOORE, and J. C. SMELTZER

5.1. COVERAGE

The Simons and CAVE (or Phillips) electrochemical fluorination (ECF) processes are surveyed in this chapter, together with commercial applications of perfluorinated fluids produced by the Simons methods. Applications of ECF-derived *reactive* materials are covered in Chapter 14.

Caution: Any work in this area must be reviewed for safety by qualified professionals. Since the highly dangerous chemical, anhydrous hydrogen fluoride, is used in quantity, competent medical support must always be available (see Chapter 1).

5.2. SIMONS ELECTROCHEMICAL FLUORINATION

5.2.1. Introduction

Electrochemical fluorination (ECF) was invented by the American chemist Joseph Simons at Pennsylvania State University.[1] 3M Company licensed the technology in 1946, and by 1952 had significant sales of fluorochemicals made via ECF. The process has been reviewed previously,[2-14] and recent primary publications and patents have shown that workers in university, government, and commercial laboratories in several countries have a renewed interest in this well established, but still poorly understood, perfluorination process. Hopefully, new knowledge from this activity will lead to improvements in the control of the process, and to new commercial applications.

Y. W. ALSMEYER, W. V. CHILDS, R. M. FLYNN, G. G. I. MOORE, and J. C. SMELTZER • Fluorochemical Process Technology Center, 3M Company, St. Paul, Minnesota 55144-1000.

Organofluorine Chemistry: Principles and Commercial Applications, edited by R. E. Banks *et al.* Plenum Press, New York, 1994.

This survey covers principally the period since the mid-1980s. Chapters 1 and 2 introduce the process.

5.2.2. Experimental Technique

In the Simons process, organic molecules are dissolved in anhydrous hydrogen fluoride (AHF) and fluorinated at the anode, usually nickel, of an electrochemical cell. Essentially all hydrogen atoms are substituted by fluorine and all carbon–carbon multiple bonds are saturated by fluorine. The electrode reactions for the ECF of a saturated hydrocarbon can be expressed as shown in Scheme 1.

A typical cell is shown diagrammatically in Figure 1. The cell body may be of mild steel or other AHF resistant material. Alternating nickel anode and cathode plates are immersed in AHF. Soluble organic substrates such as amines,[15–17] sulfonyl halides,[18,19] and acyl halides[20–22] (among others) are protonated in the AHF and form conductive electrolytes. Slightly soluble substrates may be fluorinated, often with the aid of conductivity additives such as alkali metal and alkaline earth fluorides or organic co-feeds which are protonated and fluorinated themselves.[2] Fluorinated products are generally not soluble in the electrolyte and collect at the bottom of the cell due to their high densities. Volatile products may be collected with the overhead condenser configuration shown in Figure 1. The effluent gas stream, which includes hydrogen generated at the cathode and other noncondensable species, exits from the condenser. This stream may be passed through a bed of NaF or washed with NaOH or KOH to remove HF, and then scrubbed with aqueous KI or Na_2SO_3 to remove oxidants such as OF_2.[8,16,17,22] Cold traps are used to collect volatile fluorinated products.[17] Pre-electrolysis may be used to remove traces of water.[21,22] Cells are operated in batch or continuous modes. The important operational variables include temperature, pressure, reactant concentration, and cell voltage (or current density). ECF processes are ordinarily performed at temperatures below the boiling point of HF, with cell voltages of 4.5 to 7.0 V, current densities up to about 2 A dm^{-2}, and organic substrate concentrations under 16% by weight.[12]

Other reactor configurations used for ECF include rotating disk, rotating ring disk, rotating cylinder, porous, and hollow electrode reactors.[2] An ECF channel-flow reactor was described in a recent patent application.[23] Occasionally platinum is used as anode material.[2] Steel may substitute for nickel as the cathode.[3] Pt,[15] Cu/CuF_2,[21,22] Pd/H_2,[2] and Hg/Hg_2F_2[2] have been used as reference electrodes.

anode: $C_mH_n + nHF \rightarrow C_mF_n + 2nH^+ + 2ne^-$

cathode: $2nH^+ + 2ne^- \rightarrow nH_2$

Scheme 1. Anode and cathode reactions in Simons ECF.

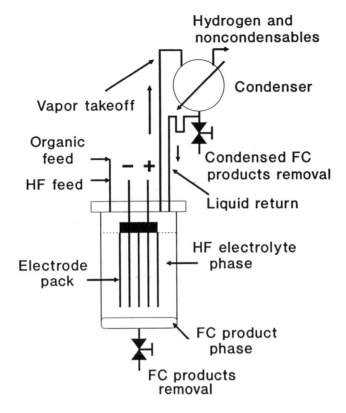

Figure 1. Typical Simons configuration showing cell and low-temperature condenser for return of HF and collection of volatile perfluorocarbons.

5.2.3. Anode Film

During electrochemical fluorination, a film forms on the nickel anode. Its color and thickness vary with the environment and the electrode history.[8] Electron transfer occurs at the interface between the electrode and the electrolyte,[24] so the film significantly affects the ECF process. Without knowledge of the mechanism, we can only argue that the reversible potential for ECF must be less than the 2.85 volts often assigned to the hydrogen/fluorine cell.[13] The cathodic overpotential is about 0.5 volts, so a large fraction of the applied voltage (4.5 to 7.0 V) is anodic overpotential.[22] The role of the anode film can be illustrated by its function in the overlapping stages of the overall process.[2]

The first stage is the rapid formation of a passive nickel fluoride film on the anode. In the absence of an organic substrate the system is poorly conductive[25] and the anode corrodes. When an organic substrate is added, the conductivity of the system increases dramatically and the corrosion rate drops.[22,26] The addition of the organic substrate

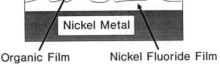

AHF Electrolyte

Figure 2.   Anode film model for Simons ECF.

causes significant, but presently poorly known, changes in the nature of the film. A representation of the anode film model is shown in Figure 2.

The second stage is steady-state fluorination of the organic substrate. This often follows an induction period. The organic substrate diffuses into the film, is adsorbed and fluorinated. The products diffuse back into the bulk electrolyte. The anode film appears to serve as a barrier to electron transfer between the electrode and the bulk electrolyte.[21]

In the third stage, an ECF run ends with a significant drop in the current. This can be associated with a depletion of the organic substrate or the formation of a tarry film on the anode. Such coatings are often observed in the ECF of long-chain alkyl derivatives ($>C_8$)[8] or thiols[27] and under conditions of high cell temperature or exhaustion of the organic substrate.[22] The tar can cover the active area of the electrode surface, decreasing the conductivity dramatically. It has been suggested that the tar formation is from the association of radical species generated on the anode surface.[21]

Because of the difficulties of characterizing anode films, their nature and role in the ECF process remain poorly understood. It has been demonstrated that mass transfer of the organic substrate within the bulk electrolyte, and across the interface between the bulk electrolyte and the organic anodic film, are not rate-determining steps.[28–30] However, mass transfer within the anode film may be rate-limiting.[2]

5.2.4. Mechanisms

The proposed mechanisms for the ECF process have been based mostly on product analyses and hence remain a matter for discussion. Dimerized products suggest free radical intermediates; rearranged and cleavage products suggest that carbocations are involved. Nonelectrochemical reaction of the substrate with or under the influence of HF complicates the interpretations. The proposed mechanisms fall into two groups depending on whether the electrochemical step (electron transfer) is from an inorganic or an organic substrate.

In the first group, fluorination is thought to be accomplished by inorganic fluorinating agents generated at the anode. Those proposed include atomic or molecular fluorine, or its loose complexes with NiF_2 on the anode surface,[9,31,32] reacting via free radicals (Rüdiger *et al.* have renewed this argument for "active fluorine", at least

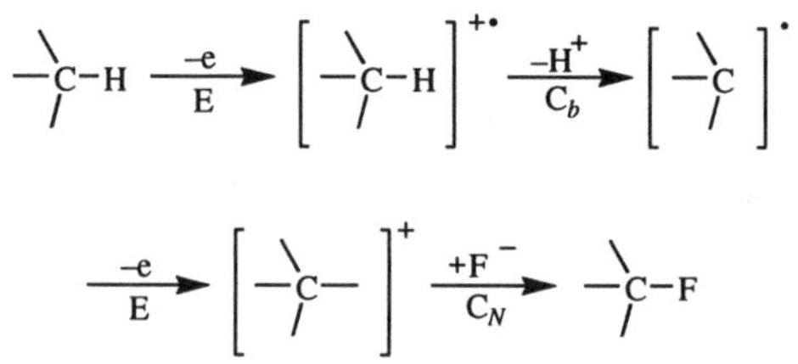

$$\overset{|}{\underset{|}{-C}}-H \xrightarrow[E]{-e} \left[\overset{|}{\underset{|}{-C}}-H\right]^{+\bullet} \xrightarrow[C_b]{-H^+} \left[\overset{|}{\underset{|}{-C}}\right]^{\bullet}$$

$$\xrightarrow[E]{-e} \left[\overset{|}{\underset{|}{-C-}}\right]^{+} \xrightarrow[C_N]{+F^-} \overset{|}{\underset{|}{-C}}-F$$

Scheme 2. EC_bEC_N mechanism.

when the organic substrate is depleted[15]), and simple or complex forms of high valent nickel fluoride, NiF_x or $NiF_6^{-(6-x)}$, [12,21,33–36] which promote carbocation formation. The observed "induction period" has been attributed to the initial electrochemical generation of such fluorinating agents, but product distributions have generally been deemed inconsistent with these, particularly products arising from cyclization, cleavage, and isomerization.

The second group of mechanistic concepts involves the direct anodic oxidation of the organic substrate at the anode.[6,7,26,36–42] A representation of this group of mechanisms is the four-step EC_bEC_N mechanism, proposed independently by two groups in 1960,[12,42] and refined in 1972.[37] In this mechanism (Scheme 2), the adsorbed organic substrate is electrochemically oxidized (E) to form a radical cation; a proton is eliminated in a chemical step (C_b) to form a radical; further electrochemical oxidation (E) generates a carbocation; this is captured by the fluoride ion (C_N). This mechanism is particularly attractive to organic chemists and allows rationalization of both cationic and radical products. It has been used to explain product isomerization,[7,26] cyclization,[20 27,44–46] and the residual hydrides often found on carbon atoms adjacent to nitrogen in the fluorination of trialkylamines,[26,43,47,48] N-alkylmorpholines,[26] and aminoalkyl ethers.[49] Gambaretto et al. viewed the observed tendency of materials to give perfluorinated products as a "zipper mechanism".[26] That is, the substrate was thought to adsorb on the active anode surface until all possible EC_bEC_N events had occurred. Meinert et al. argue that the lower yields from an ethylenediamine than from the analogous propylenediamine reflect poorer contact with the surface by the ethylenediamine.[41]

Some of these arguments are not convincing. Some systems, such as dibutylmethylamine[15–17] and propanesulfonyl fluorides,[51] do yield partially fluorinated species. Rüdiger and co-workers have shown that for an amine feed, roughly one-third of the current passed produces liquid products, one-third gives rise to gases, and one-third goes to form HF-soluble polyfluoro material (the "HF-phase").[16] These materials do not undergo further fluorination to perfluoro(dibutylmethylamine), so "zipper" perfluorination arguments are not affected, but at the same time a "nonzipper" pathway seems to be operative. Gambaretto et al. argue that hydride retention in the α- position of protonated amines supports an EC_bEC_N mechanism,[26] but the α-position would be less reactive in any mechanistic scheme. They also cited isomerization during ECF of

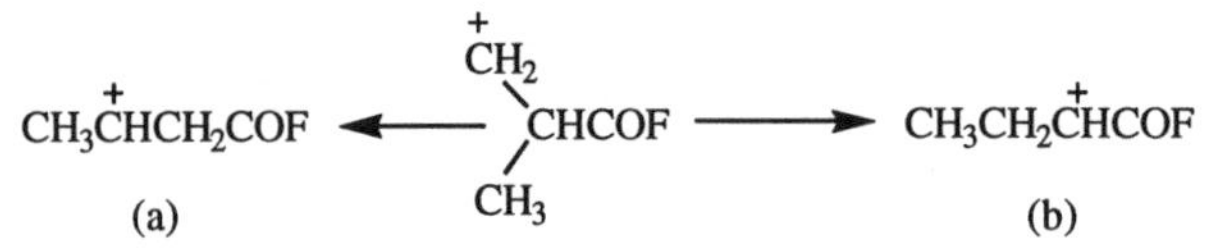

$$CH_3\overset{+}{C}HCH_2COF \longleftarrow \underset{\underset{CH_3}{|}}{\overset{\overset{\overset{+}{C}H_2}{\diagdown}}{}}CHCOF \longrightarrow CH_3CH_2\overset{+}{C}HCOF$$

(a)   (b)

Scheme 3.   (a) COF migration, (b) CH$_3$ migration.

isobutyryl fluoride (giving *n*-perfluorobutyryl fluoride) as support for the involvement of EC$_b$EC$_N$ processes, as shown in Scheme 3, but, as drawn, their carbocation (a) must have resulted from COF group migration; more conventional methyl migration would yield a less stable carbocation (b).

Likewise, the concept has not proven reliable in predicting yields in homologous series.[50] A major complication lies in the multitude of partially fluorinated species possible in any reaction, and the consequent unpredictable relative stabilities of the possible intermediates. Despite these difficulties, the EC$_b$EC$_N$ concept is appealing. It can help to explain the large differences in ECF behavior among various chemical classes, if not the subtle differences within a class.

Understanding of the Simons ECF mechanism clearly is far from complete. The function of the anode film, wide product distribution ranges, strong dependence on the operating conditions and structure of the organic substrate, and relatively poor reproducibility, have all added to the complexity of the problem. It seems likely, too, that the reactions occurring during the ECF process are due to more than just one type of mechanism.

5.2.5. Scope

ECF has been used for the fluorination of hydrocarbons, fluorohydrocarbons, chlorohydrocarbons, amines, N-heteroaromatics, ethers, carboxylic acids and their derivatives, thiols, and alkanesulfonyl fluorides. Products obtained from this process are generally *completely fluorinated*. Thus, all hydrogen is replaced with fluorine, carbon–carbon multiple bonds are saturated with fluorine, and divalent sulfur compounds are oxidized to derivatives of sulfur hexafluoride [R$_f$SF$_5$, (R$_F$)$_2$SF$_4$]. However, the S=O bonds in sulfones, sulfates, and alkanesulfonyl halides, and the C=O bonds in carboxylic acids and their derivatives, remain intact to a useful degree (see Chapter 14).

Much of the ECF work of the last decade has been directed toward the production of inert fluids. For example,[53–55] ECF of a group of *N*-cycloalkylpyrrolidines and piperidines has given the desired tertiary perfluoroamines in 40–57% yield. In a series of octahydro-quinolizidines, the 4-methyl derivative led to ring expansion in addition to the desired material, while the 4-trifluoromethyl analog gave a lower yield due to cleavage[54] (Scheme 4). The enamine 1-morpholinocyclohexene has been fluorinated to provide a saturated perfluorocycloalkylamine (Scheme 4) in 31% yield,[41] and

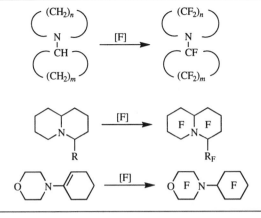

Scheme 4.   Simons ECF of some tertiary cycloalkylamines.

perfluorinated cycloaliphatic aminoethers have been prepared from the nonfluorinated precursors.[52,56]

Tetra-substituted ethylene- and propylene-diamines were fluorinated with the following yield trends[57]: ethylene < propylene; and Me_2N- < Et_2N- < pyrrolidino- < morpholino- < piperidino.

In a related series of α-, ω-dimorpholino-, and dipiperidino-alkanes (Scheme 5), the latter amines again gave higher yields,[41] but note, however, that the yields listed include the F-3-methylpyrrolidines formed by apparent ring contraction of the piperidines. These yields are especially impressive given the tendency of amines to undergo β-cleavage when subjected to ECF.[57,58]

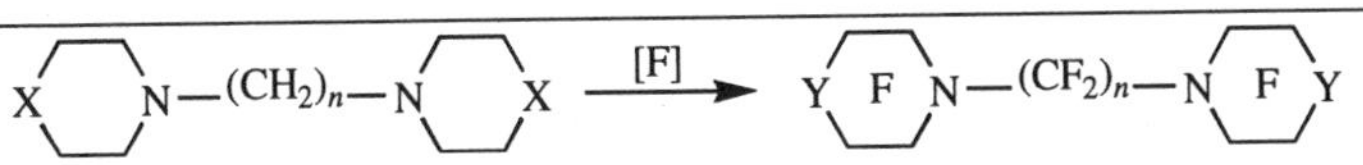

$$X\!\!-\!\!N-(CH_2)_n-N\!\!-\!\!X \xrightarrow{[F]} Y\ F\ N-(CF_2)_n-N\ F\ Y$$

(major product[a])

	X = O, CH_2			Y = O, F_2		
n =	1[b]	2	3	4	5	6
morpholine series:	0	28%	25%	30%	26%	20%
piperidine series:	0	26%	41%	22%	42%	16%

Scheme 5.   Simons ECF of α,ω-dimorpholino- and α,ω-dipiperidino-alkanes.[41] [a]The piperidines led to $CF_2(CF_2)_4N\!\!-\!\!(CF_2)_n\!\!-\!\!NCF_2CF(CF_3)CF_2CF_2$ and $CF_2CF_2(CF_3)CF_2N\!\!-\!\!(CF_2)_n\!\!-\!\!NCF_2CF(CF_3)CF_2CF_2$, as minor products. [b]These substrates were decomposed by AHF.

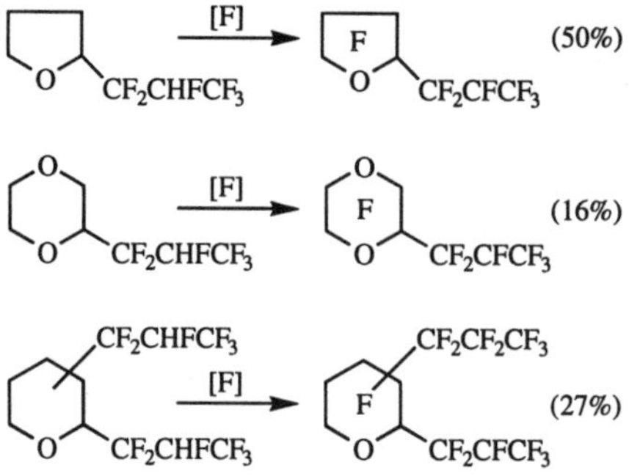

Scheme 6.   Simons ECF of α-fluoroalkylated ethers.[59]

$$FSO_2(CH_2)_nSO_2F \xrightarrow{[F]} FSO_2(CF_2)_nSO_2F$$

Yields: $n = 3, 34\%$
           $n = 4, 28\%$

Scheme 7.   Simons ECF of alkane disulfonyl fluorides.[60]

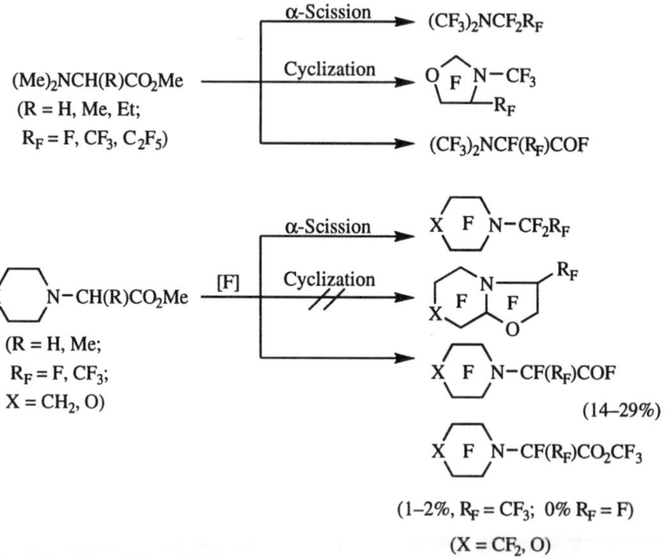

Scheme 8.   Simons ECF of α-(dialkylamino)acid esters.[63,64]

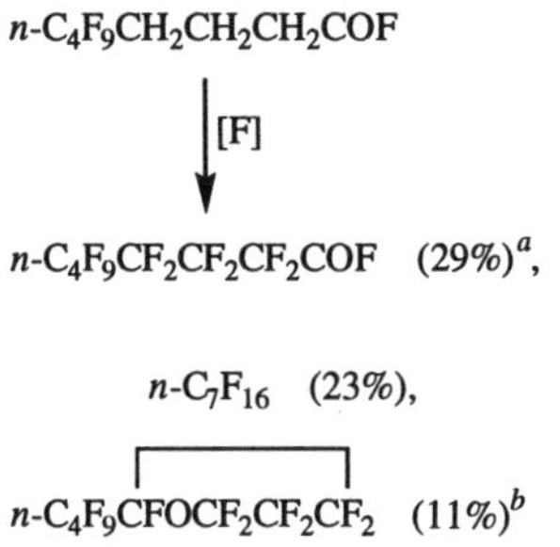

$$n\text{-}C_4F_9CH_2CH_2CH_2COF$$

$$\downarrow [F]$$

$$n\text{-}C_4F_9CF_2CF_2CF_2COF \quad (29\%)^a,$$

$$n\text{-}C_7F_{16} \quad (23\%),$$

$$n\text{-}C_4F_9CFOCF_2CF_2CF_2 \quad (11\%)^b$$

Scheme 9. Simons ECF of nonafluorooctanoyl fluoride.[65] [a]Small amounts of an isomeric alkanoyl fluoride were detected. [b]No $n\text{-}C_3F_7CFOCF_2CF_2CF_2$ was detected.

The well-known technique of partially fluorinating substrates by other techniques in order to enhance ECF yields has been applied successfully to mono- and bis-1,1,2,3,3,3-hexafluoropropyl derivatives of cyclic ethers (Scheme 6)[59]; and the now-classical (Chapter 2) cyclization side reaction of carboxylic acid derivatives has been used to prepare a variety of structurally complex cyclic ethers.[8,60,61,62]

Interest in functional groups continues. α-, ω-perfluoroalkylenedisulfonyl fluorides have been prepared in reasonable yields (Scheme 7),[18,60] and fluorination of α-(dialkylamino)acid esters has given good yields of the corresponding acid fluorides (Scheme 8).[63,64] With these latter substrates, the effects of structure on the side reactions of α-scission and cyclization were reported. Cyclization was seen in the case of dialkylamino derivatives but not for the cyclic amino-esters. This was rationalized in terms of enhanced strain caused by the large covalent bond angles known for perfluoro t-amines. Some of the methyl 2-(cyclic amino) propionates gave significant amounts of the corresponding perfluoroesters. This was attributed to steric protection of the ester link.

The stratagem of partially fluorinated substrates has led to improved yields of perfluoroacyl fluorides (Scheme 9). The use of nonafluorooctanoyl fluoride was found to increase the yield of perfluorooctanoyl fluoride from 13% to 29% because cyclization to perfluoro-(2-n-propyl)pyran was blocked.[65] An earlier example of this strategy gave similar results by blocking cyclization to perfluoro-(2-n-butyl)furan[66]; thus, with $n\text{-}C_4H_9CF_2CH_2CH_2COF$ the yield was 31% as compared to 14% from n-octanoyl fluoride. References 45 and 66 contain many other examples of the benefits of starting with partially fluorinated substrates.

5.2.6. Commercial Utility of ECF-Derived Materials

The use of products from the Simons ECF Process was commercialized by the 3M Company in the U.S. nearly forty years ago.[67,68] 3M remains preëminent in the field, but other companies have patented and in some cases commercialized ECF-based materials. These companies include Tokuyama Soda, Asahi Glass Company,

Daikin Industries, Mitsubishi Metal Corporation, Green Cross Corporation, Kanto Denka Kogyo Co., Dainippon Inc. (all of Japan), Bayer A.G. (Germany), and Miteni (Italy). Green Cross has marketed perfluorinated compounds in emulsion form for use as artificial blood substitutes (see Chapter 26). The best known of these compositions is Fluosol-DA which contains perfluorodecalin (prepared by CoF_3 fluorination; see Chapter 4) and perfluorotripropylamine as the "active" ingredients. Fluosol-DA was recently approved by the United States Food and Drug Administration for use during balloon catheterization angioplasty.

Perfluorinated compounds obtained via ECF can be divided into two general classes. Members of the first class contain a functional group, e.g., a carboxylic or sulfonic acid moiety, which can be elaborated using standard synthetic techniques. The second class comprises the so-called *inert fluids* which have no readily functionalized sites. These fluids include perfluoro-alkanes and perfluoro-cycloalkanes, as well as the tertiary perfluoro-amines and perfluoro-ethers, which are devoid of basic character due to the powerful inductive effects ($-I$) of perfluoro substituents and which share many of the properties of saturated perfluorocarbons (see Chapter 3).

5.2.6.1. Functionalized Products

The fluorinated carboxylic and sulfonic acids (available from hydrolysis of the ECF-derived carbonyl or sulfonyl fluorides) or derivatives thereof find great utility as specialized surfactants (see Chapter 14) and are marketed under the 3M *Fluorad* trademark. Derivatives of the sulfonyl fluorides are also especially useful in fire-fighting foams, and are marketed under the 3M *Light Water* trademark. Materials derived primarily from perfluoroalkanesulfonyl fluorides provide soil and stain resistance to textiles, carpets, paper, and nonwovens. They are marketed under the 3M *Scotchgard* and *Scotchban* trademarks. These properties typically include soiling and stain resistance, as well as the oil and water repellency expected for compounds containing a long-chain perfluoroalkyl group (see Chapters 3 and 14). All these industrial applications have been reviewed elsewhere recently.[68-71]

5.2.6.2. Inert Fluids

Perfluorinated fluids are marketed under the 3M *Fluorinert* Electronic Liquids trademark. The use of these in electronic applications has been reviewed[70] and the aim here is to provide an update in this area. Chapter 4 also contains information on this subject.

a. Properties. The special properties of perfluorinated liquids (see Chapter 4) which make them attractive and useful in the electronics industry are now well known. Briefly, *Fluorinert* fluids are completely nonpolar and are very poor solvents for most organic compounds and water. Hydrocarbon alkanes, low-molecular-weight ethers, and some highly chlorinated solvents such as carbon tetrachloride, are relatively soluble in perfluorocarbons. This solubility decreases markedly as the molecular weight of the perfluorocarbon increases. The perfluorinated liquids are completely miscible with each other and with polyhalogenated solvents such as CFC-113 ($CF_2ClCFCl_2$), HCFC-123 (CF_3CHCl_2), and HCFC-141b (CH_3CFCl_2). The higher-molecular-weight perfluoropolyethers such as Dupont's *Krytox* vacuum pump fluids (from homopolymers of hexafluoropropene oxide), Montefluos's *Fomblin* fluids (photooxidation products of hexafluoropropene or tetrafluoroethene; see Chapter 20), or Daikin's *Demnum* fluids (the perfluorinated homopolymer of tetrafluorooxetane; see Chapter 21) are also freely miscible with perfluorocarbons. More polar organic solvents such as acetone, methanol, or toluene are only slightly soluble.

Fluorinert liquids are dense (1.68–1.94 kg liter^{-1}), nonflammable, have very low toxicities, display very low surface tensions (12–18 mN m^{-1}), and possess low heats of vaporization and boiling points relative to their molecular weights, especially in comparison to their hydrocarbon analogs. Their electrical properties are excellent [dielectric constants of less than 2; dissipation factors generally less than 3×10^{-4} (1 kHz); volume resistivities of about 10^{15} ohm-cm; dielectric strengths of 38–46 kV], and this situation, coupled with their marked thermal and chemical stabilities, has led to their use in the testing and cooling of electronic components.

Tertiary perfluoro-amines and perfluoro-ethers are slightly less stable, both thermally and chemically, than perfluoroalkanes, but in use they can often be thought of as functionally equivalent. Their slightly lower stability is in practice not generally a problem but, at elevated temperatures, Lewis acids such as aluminum chloride can catalyze the decomposition of both perfluoro-amines and perfluoro-ethers (see Chapter 20).[71]

The *Fluorinert* fluids obtained via ECF are normally selected for end-use on the basis of boiling point alone. Some of them such as FC-43, perfluorotributylamine (bp 174°C), are chemically relatively pure, although in the ECF process the occurrence of rearrangement and cleavage reactions typically produces a fairly broad mixture of products. Other *Fluorinert* products are mixtures of several compounds. Lower-boiling (< 110°C) *Fluorinerts* include products consisting mainly of C_6F_{14}, C_7F_{16}, C_8F_{18}, and perfluoro(2-butyltetrahydrofuran), the last being a co-product from the ECF of octanoic acid or octanoyl halides (see Scheme 9). *Fluorinert* liquids generally sell at prices in the range of $20 to $40 per pound.

b.   Cooling, Testing, Reflow Soldering, and CFC Replacement.   From what began as not particularly desirable co-products in the ECF of octanoic acid,[67] *Fluorinert* liquids have come to play a critical role in the electronics industry, with the major uses being cooling, testing, and reflow soldering.

Cooling refers to the broad-based use of perfluorocarbons as direct contact coolants in a variety of applications. With their favorable electrical properties and superior materials compatibility, perfluorinated liquids are ideal for use in the direct contact cooling of radar transmitters, high voltage transformers, radar klystrons, power supplies, lasers, and fuel cells. An excellent discussion of the mechanisms of direct contact cooling can be found in Ref. 67. Since that work was published there have been some significant advances in the cooling area. The first is the use of the *Fluorinert*

Liquid Heat Sink. These heat sinks, available in a variety of shapes and sizes, are special laminated plastic pouches filled with a perfluorinated liquid, which is then degassed, and the pouch sealed. These thin pouches are placed directly on the component to be cooled (for example, a complex circuit board). Heat transfer is effected as a result of natural fluid convection. The *Fluorinert Liquid Heat Sink* can in some cases eliminate the need for forced air cooling and has found specific application in laptop microcomputers and fixed disk drives.

A second application which has achieved commercialization in the past several years is the use of a perfluorinated liquid as the total-immersion coolant for a new generation of supercomputers. The power supplies, memory boards, logic circuits, and main processors of the Cray-2 supercomputer are entirely immersed in a sealed 155-gallon tank containing a 100°C boiling point *Fluorinert* liquid. This mode of cooling allows the Cray-2 to have six to twelve times the output of the earlier Cray-1, and yet be contained in approximately half the space. More recently, Cray has developed a Model Y-MP supercomputer which also is cooled by a perfluorinated liquid, although not in a direct-contact mode. As circuits become smaller and more densely placed, and as the semiconductor industry moves to gallium arsenide chips, cooling with perfluorinated liquids becomes even more of a necessity.

Vapor-phase reflow soldering was first introduced in 1975. In this process, a part to be soldered is introduced into a chamber containing a refluxing perfluorinated liquid. The hot vapor condenses onto the relatively cool surface of the part, releasing the latent heat of vaporization and rapidly raising the temperature of the part to the solder reflow point. The solder used must of course have a flow point lower that the boiling point of the perfluorinated liquid. The use of a perfluorinated liquid provides a controlled processing temperature (the temperature inside the chamber cannot exceed that of the condensing liquid), rapid heat transfer, and a nonoxidizing environment. The liquid leaves no residue and, because of its low surface tension, drains rapidly and completely from the finished part. This process has remained relatively unchanged over the past several years. With the introduction of ever more complicated and crowded circuit boards, the advantages of vapor-phase reflow soldering have become even more apparent.

In some batch-type vapor-phase reflow soldering equipment, it is often desirable to use a second, much lower boiling fluid to provide a blanket on top of the soldering liquid to minimize loss of the expensive perfluorinated liquid. In the past the liquid of choice for this purpose was CFC-113 ($CF_2ClCFCl_2$), but with the implementation of the Montreal Protocol for the protection of the Earth's stratospheric ozone layer, the use of chlorofluorocarbons such as CFC-113 has become tightly regulated under international law (see Chapter 28). Since a lower-boiling "ozone-friendly" perfluorocarbon can be used as the secondary fluid for vapor-phase soldering systems, 3M has introduced fluids suitable as drop-in replacements for CFC-113.

Perfluorocarbons have proved to be viable *substitutes for CFCs* in other applications. Other examples of direct replacement include use as the cooling fluid for ion implanters (replacing CFC-113), the cooling of large transformers found in Japanese trains (another CFC-113 replacement), and application as a solvent for the coating of lubricant on magnetic hard disks. In the last example, a major lubricant in this industry is a perfluorinated polyether (see Chapter 20) which is soluble in perfluorinated alkanes and CFC-113. A recent patent describes the use of perfluoroalkanes as additives in the blowing of polyurethane foams (CFC-11 is the chlorofluorocarbon most used in foam blowing today),[72] and perfluorocarbons are also under active investigation as potential replacements for Halon fire-fighting agents (see Chapter 8). Since perfluorocarbons contain no chlorine or bromine and do not take part in the atmospheric catalytic cycle which destroys ozone, they are exempt from the restrictions of the Montreal Protocol.

The use of perfluorinated liquids for *testing* of electronic components remains important (liquid burn-in testing, gross and fine leak testing, thermal shock testing, hot spot determination, and accelerated aging[71]). Electrical environmental testing is done by using a refluxing liquid to provide a chamber at a uniform temperature. The part to be tested is placed inside the chamber and rapidly brought to thermal equilibrium. The speed at which this can be done by the condensing vapors is a significant advantage. The detection of cracks in ceramic substrates is a novel use for perfluorinated liquids. These cracks are frequently invisible to the eye and can even be missed during fine leak (hermetic seal) testing. If a fine mist of a heated perfluorocarbon impinges on the surface of a ceramic chip to be tested, cracks in the surface became clearly visible; the effect is quite striking and there is an apparent magnification of the crack by as much as a factor of 1000. Adsorption into the crack of the low-viscosity, low-surface-tension liquid from the areas immediately adjacent to the defect has the effect of giving a duller appearance to the crack area when the surface is illuminated. Since this type of defect has always been extremely difficult to detect, and has undoubtedly been responsible for many previously mysterious failures, this new test should find great utility.

New uses for perfluorinated liquids continue to be discovered, and the next decade will see the commercialization of yet more applications.

5.3. CAVE–PHILLIPS PROCESS FOR ELECTROCHEMICAL FLUORINATION

This process was invented and initially developed at Phillips Petroleum Company.[73] At the 3M Company this process is referred to as CAVE (Carbon Anode Vapor phase Electrochemical fluorination). It was reviewed recently.[2]

5.3.1. Systems

5.3.1.1. Electrolyte

The electrolyte, nominally KF·2HF, is that used in medium temperature fluorine cells, i.e., fluorine generators. It is prepared by the careful addition of anhydrous HF vapor to solid KF·HF (potassium bifluoride) until the mixture contains 20.7 to 20.9 meq of HF per gram. (*Caution*: addition of liquid HF to KF·HF, like the addition of

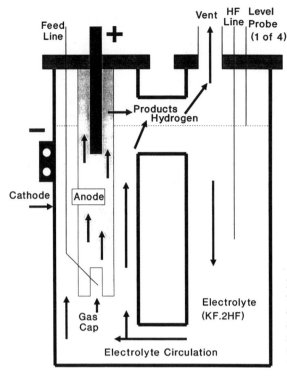

Figure 3. Laboratory CAVE cell with gas cap feeding (not to scale). An anode 3.5 cm in diameter, immersed to a depth of 28 cm and operating at 200 mA cm^{-2}, passes 2 faradays per hour (53.6 amperes). Cells operate at 90 to 100°C.

HF vapors or liquid to KF, leads to an extremely vigorous reaction and should not be attempted.)

5.3.1.2. Cells

Figure 3 shows a cell design used in our laboratory. The cell body can be fabricated from a mild steel that resists hydrogen embrittlement or an equivalent material. Steel corrodes slowly in this environment and a sludge of potassium fluoroferrate will accumulate. The cell case usually serves as the cathode with an anode-to-cathode spacing of 1 to 2 cm. The cell is jacketed (not shown) for circulation of a tempered fluid for temperature control. The cell lids are of poly(chlorotrifluoroethylene) or unfilled poly(tetrafluoroethylene).

Cells passing currents up to 4000 amps were operated at Phillips.

5.3.1.3. Anodes

The anodes are of porous amorphous (nongraphitic) carbon. The Union Carbide products PC25, PC45, and PC60 are satisfactory. PC25 is very friable and requires care in handling, but it polarizes less frequently than do less permeable carbons (see below). An anode 3.5 cm in diameter, 35 cm long, and immersed to a depth of 28 cm will run at two faradays per hour (53.6 amps) at 200 mA cm^{-2}. The current collector is a copper or nickel rod carefully fitted into the porous carbon.

5.3.1.4. Controls

A computer-based system is used for control and data acquisition. This incorporates several stages of safety interlocks and an independent timer that shuts down the system if the computer does not reset it each 90 seconds.

5.3.2. Feeds

5.3.2.1. HF

Special care and attention must be given to materials selection and to the handling and control of AHF. Two air-operated fail-closed bellows valves, "control" and "backup", are used to control the addition of HF to the system. Routine operation is controlled through a computer, but a backup system is in place that is completely independent of the computer.

5.3.2.2. Substrates

In the CAVE process a gas or a volatile liquid substrate that is not particularly soluble in the molten KF·2HF electrolyte is introduced into the network of pores at the bottom of the carbon anode. While it is contained within the anode it reacts with elemental fluorine generated at the anode surface. Substrates include alkanes, chloroalkanes, acyl fluorides, ethers, and esters. Products range from monofluoro to perfluoro compounds.

The substrate is introduced into the network via the feed line into the gas cap as shown in Figure 3. Since the electrolyte does not wet the anode, the substrate passes into the network of pores and moves upward through the anode.[73] Fluorine is generated at the surface of the anode and enters the network of pores also. Whenever the substrate and fluorine meet they react. In most cases this is a very efficient reaction.

5.3.3. Nonwetting Behavior

As shown in Figure 4a, molten KF·2HF wets porous carbon, but if the carbon is made anodic (see Figure 4b) it does not. This behavior is irreversible and is thought to be due to the formation of a film of so-called graphite fluoride "CF$_x$" (see Chapter 23).

Figure 5 shows plots of cell voltage and current density versus time during the startup of a new anode with ethane feed. During the first twelve minutes the current density was controlled at 25 mA cm^{-2}; the initial voltage was a little over three volts. The voltage rose for eight minutes, then flattened out at 5.2 volts. At twelve minutes the current density was increased to 55 mA cm^{-2} and the voltage rose to 6.5 volts. Over

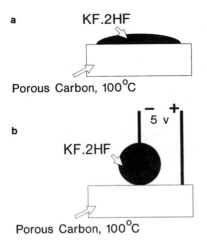

a

KF.2HF

Porous Carbon, 100°C

b

5 v

KF.2HF

Porous Carbon, 100°C

Figure 4. (a) Molten KF·2HF wets porous carbon but (b) not a porous carbon anode.

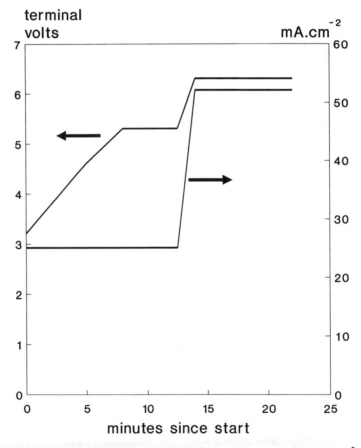

Figure 5. Electrical parameters during start-up of CAVE fluorination of ethane (at 25 mA cm^{-2} the voltage increases due to the irreversible formation of a layer of CF$_x$).

several hours the current density was increased to 200 mA cm^{-2}; the voltage rose to 7.5 volts.

During the first few minutes of startup, the porous carbon is wetted by the electrolyte and the feed bypasses and bubbles up around the anode. Initially, this is a lower energy path than displacing the electrolyte that wets the anode and fills the pores. The bubbles and bypassing stop when the cell voltage flattens out. Clearly, the low energy path for the feed is now to displace the electrolyte which now does not wet the anode, enter the anode, move through the network of pores (where it reacts with fluorine), and escape from the anode into the vapor space above the electrolyte.

The process at the initial voltage is probably production of a CF_x film, following which the voltage increases owing to the energy required to tunnel through that layer, formation of which is essentially irreversible. Lowering the current does not reduce the voltage to the initial value nor does it lead to the resumption of feed bubbles bypassing the anode.

5.3.4. Polarization

Here, and in fluorine generation, polarization has a meaning not closely related to its usual meanings in electrochemistry.

Consider a cell operated at constant current. (A similar treatment can be written for constant voltage operation.) At 200 mA cm^{-2} the cell will operate initially at about 7 volts, but with time the cell voltage will increase to 9 to 10 volts. (We usually set a hardware limit of 12 to 15 volts.) After some time the voltage will rapidly increase to the hardware limit and only a small fraction of the usual current will pass. The cell is then said to be polarized.

The cell can be depolarized by allowing the voltage to rise at constant current. Normally it will rise to 60 to 80 volts. If the cell is turned off, then back on after 90 seconds at the elevated voltage, it will usually operate normally and is said to be depolarized. (If the HF concentration is too low, polarization may soon recur.) Bai and Conway,[74] using a small anode, reported that a high-voltage treatment lasting 60 minutes gave better results than a short treatment. Because of heat-removal considerations, it is difficult to give an extended high-voltage treatment to a large anode.

The frequency of polarization depends on the choice of carbon, the current density, the electrolyte composition, and the feed. The initial polarization will likely occur hours to days after startup; the elapsed time between subsequent polarizations may be measured in days or even weeks.

Polarization is apparently due to the formation of a film of gas (perhaps elemental fluorine) over much of the anode surface. This film blocks the passage of current through the anode/electrolyte interface. It may form because the carbon becomes highly nonwetting (perhaps x in CF_x increases); or it may form because paths for fluorine to pass from the surface into the anode interior are blocked.

The high-voltage treatment depolarizes the cell by stripping off some of the anode surface. Bai and Conway[74,75] liken it to polishing. Considerable energy is dissipated at the anode/electrolyte interface during the treatment.

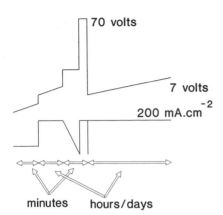

Figure 6. Current and voltage changes during long-term operation of a CAVE cell, emphasizing the polarization–depolarization cycle.

The Rüdorffs,[76] Rudge,[77] and Watanabe[78] were pioneers in the area of fluorine cell polarization, which is still under investigation.[74,75,78–80] (The Rüdorffs called polarization *aging*, so perhaps depolarization should be called *rejuvenation*.)

5.3.5. Brief History of an Anode

Figure 6 is a sketch of voltage and current during the first several hundred hours of operation of a PC25 anode. There are four significant periods. The first few minutes during CF_x formation were discussed in detail above. The period leading up to the first polarization may stretch to one-hundred or more hours with anodes fabricated from PC25, or to only a few hours with PC60. Polarization at constant current is foreshadowed by a voltage increase and erratic voltage swings until the cell reaches the hardware voltage limit, and then erratic current swings will be noted until the anode is "firmly" polarized. When the cell is approaching the hardware voltage limit, it may be possible to force polarization by increasing the current a bit. A polarized anode is depolarized as noted above. The time to the next polarization may be several hundred hours with PC25 or considerably less with PC60.

5.3.6. Fluorination Results

Fluorination of ethane via the CAVE proceeds smoothly and, over the range from low to complete conversion, 96% of the product comprises fluoroethanes [all possible $C_2H_xF_{(6-x)}C_2Hx$ compounds ($x = 0–5$)], 4% fluorobutanes, and traces of fluoromethanes (including CF_4) and fluoroethylenes. The production of fluoromethanes is barely above the methane level in the ethane feed; extra amounts of C_1 products can, of course, arise from ethane fragmentation of from fluorination of the anode. The products from the fluorination of ethane are those expected from the occurrence of a series of classical free-radical reactions associated with fluorination of alkanes with elemental fluorine (see Chapter 1).

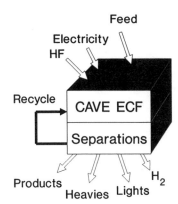

Figure 7. Stylistic representation of a generic CAVE fluorination process showing the integration of the ECF and separations steps.

Figure 7 is a sketch of a generic CAVE process. The **CAVE ECF** box takes feed, HF, and electricity and produces a mixture of products. The **separations** box takes these products and separates them into a products stream (usually perfluoro), a recycle stream (partially-fluorinated product), a heavies stream (unwanted dimers, etc.), a lights stream (fragmentation products), and a hydrogen stream. The heavies, lights, and hydrogen streams are usually waste streams. The separations box can present a significant challenge.

Substrates that fluorinate well include: 1,2-dichloroethane (80% retention of 1,2-dichloro-structure); acetyl fluoride (85% yield of fluorinated acetyl fluorides);[81] 1,1,2,2-tetrafluoro-cyclobutane (90% ring retention); and isopropyl trifluoroacetate

$$[(CH_3)_2CHOCOCF_3 \rightarrow (CF_3)_2CFOCOCF_3]$$

The production of trifluoroacetic acid from acetic acid and a little acetic anhydride (see Scheme 10 and Figure 8) is an elegant combination of electrochemistry, some simple chemistry, and separations.[82] CAVE technology also offers a route to hexafluoroacetone from isopropyl alcohol,[83] via fluorination of its trifluoroacetate and subsequent fluoride-ion initiated cleavage of the perfluoroester product.[84]

$$[(CF_3)_2CFOCOCF_3 \rightarrow (CF_3)_2CO + CF_3COF;$$

$$CF_3COF + (CH_3)_2CHOH \rightarrow (CH_3)_2CHOCOCF_3$$

for the ECF step]

Hexafluoroacetone is an important synthetic intermediate, being used in the production of polymers containing the hexafluoroisopropylidene moiety (see Chapter 19 and Ref. 85).

$$CH_3COF + HF \xrightleftharpoons{ECF} CF_3COF + H_2$$

$$CF_3COF + CH_3CO_2H \xrightleftharpoons{Metathesis} CF_3COF + CF_3CO_2H$$

$$(CH_3CO)_2O + HF \xrightleftharpoons{Makeup} CH_3COF + CH_3CO_2H$$

Scheme 10.   CAVE-based trifluoroacetic acid process.

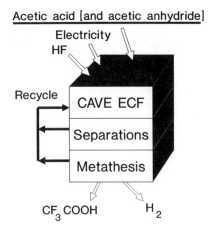

Figure 8. Integrated process for making trifluoroacetic acid.

5.3.7. Summary of the CAVE ECF Process

In many ways the CAVE/Phillips process is complementary to the Simons process. It works best with volatile substrates that do not dissolve in the electrolyte. Because of the low acidity of the system and the free radical nature of the reactions, essentially no skeletal isomerization is observed. Essential process details are summarized below.

1. The KF·2HF electrolyte does not wet the porous carbon anode.
2. Elemental fluorine is generated at the anode surface and enters the porous carbon anode.
3. Volatile organic substrates introduced into the porous carbon anode react smoothly and efficiently with fluorine within the anode; the observed products can be explained by well-known free radical reactions (substitution, disproportionation, and dimerization).
4. Polarization appears to be due to a change in the anode surface that allows the formation over much of the surface of a gas film which blocks the flow of current. A brief high-voltage treatment relieves this condition and allows normal operation to be resumed.

5.4. REFERENCES

1. J. H. Simons, *Trans. Electrochem. Soc. 95*, 47 (1949).
2. W. V. Childs, L. Christensen, F. W. Klink, and C. F. Kolpin, in: *Organic Electrochemistry*, 3rd edn. (H. Lund and M. M. Baizer, eds.), Chapter 6, p. 1103, Marcel Dekker, New York (1991).
3. H. Cheng, *Xiandai Huagong 3*, 19 (1990).
4. P. Sartori, *Bull. Electrochem. 6(4)*, 471 (1990).
5. E. Hollitzer and P. Satori, *Chem. Ing. Tech. 58*, 31 (1986).

6. I. N. Rozhkov, in: *Organic Electrochemistry*, 2nd edn. (M. M. Baizer and H. Lund, eds.), Chapter 24, p. 805, Marcel Dekker, New York (1983).
7. I. N. Rozhkov, *Russ. Chem. Rev. 45(7)*, 615 (1976).
8. T. Abe and S. Nagase, in: *Preparation, Properties and Industrial Application of Organofluorine Compounds* (R. F. Banks, ed.), p. 21, Ellis Horwood, Chichester (1982).
9. N. Watanabe, *J. Fluorine Chem. 22*, 205 (1983).
10. S. Nagase, in: *Fluorine Chemistry Reviews* (P. Tarrant, ed.), Vol. I, p. 77, Marcel Dekker, New York (1967).
11. N. L. Weinberg, in: *Technique of Electrochemical Synthesis* (N. L. Weinberg, ed.), Part II, p. 1, Wiley, New York (1975).
12. J. Burdon and J. C. Tatlow, in: *Advances in Fluorine Chemistry* (M. Stacey, J. C. Tatlow, and A. G. Sharpe, eds.), Vol. 1, p. 129, Butterworths, London (1960).
13. A. J. Rudge, in: *Industrial Electrochemical Processes* (A. T. Kuhn, ed.), p. 71, Marcel Dekker, New York (1967).
14. A. P. Tomilov, S. G. Mairanovskii, M. Ya. Fioshin, and V. A. Smirnov, *Electrochemistry of Organic Compounds* (J. Schmorak, transl.), p. 417, Wiley, New York (1972).
15. A. Dimitrov, St. Rüdiger, N. V. Ignatyev, and S. Datcenko, *J. Fluorine Chem. 50*, 197 (1990).
16. A. Dimitrov, St. Rüdiger, and M. Bartoszek, *J. Fluorine Chem. 47*, 23 (1990).
17. A. Dimitrov, H. Stewig, St. Rüdiger, and L. Kolditz, *J. Fluorine Chem. 50*, 13 (1990).
18. R. Herkelmann and P. Sartori, *J. Fluorine Chem. 44*, 299 (1989).
19. E. Hollitzer and P. Sartori, *J. Fluorine Chem. 37*, 329 (1987).
20. M. Napoli, L. Conte, G. P. Gambretto, and F. M. Carlini, *J. Fluorine Chem. 45*, 213 (1989).
21. H. W. Prokop, H-J Zhou, S-Q Xu, S. R. Chuey, and C. C. Liu, *J. Fluorine Chem. 43*, 257 (1989).
22. D. J. Wasser, P. S. Johnson, F. W. Klink, F. Kucera, and C. C. Liu, *J. Fluorine Chem. 35*, 557 (1987).
23. Y. Murata, N. Okada, Y. Hirai, and T. Tomoyasu, Japanese Patent Application 88-104160 (to Tokuyama Soda) [*CA 113*, 140887 (1990)].
24. A. J. Bard and L. R. Faulkner, in: *Electrochemical Methods: Fundamentals and Applications*, Chapter 3, p. 86, Wiley, New York (1980).
25. G. Cauquis, B. Keita, G. Pierre, and M. Jaccaud, *J. Electroanal. Chem. 100*, 205 (1979).
26. G. P. Gambaretto, M. Napoli, L. Conte, A. Scipioni, and R. Armelli, *J. Fluorine Chem. 27*, 149 (1985).
27. T. Abe, S. Nagase, and H. Baba, *Bull. Chem. Soc. Jpn. 49*, 1888 (1976).
28. M. Novak and J. Boa, *J. Electroanal. Chem. 109*, 179 (1980).
29. M. Novak, J. Jeko, and J. Boa, *Stud. Org. Chem. 30*, 67 (1987).
30. F. W. Klink, Ph.D. Thesis, Case Western Reserve University, Cleveland, Ohio, USA (1985).
31. J. H. Simons, in: *Fluorine Chemistry* (J. H. Simons, ed.), Vol. 1, Academic Press, New York (1950).
32. B. Chang, H. Yanase, K. Nakanishi, and N. Watanabe, *Electrochim. Acta 16*, 1179 (1971).
33. T. Gramstad and R. N. Haszeldine, *J. Chem. Soc.* 173 (1956).
34. R. N. Haszeldine and F. Nyman, *J. Chem. Soc.* 2684 (1956).
35. J. A. Donohue and A. Zeltz, *J. Electrochem. Soc. 115*, 1039 (1968).
36. J. A. Donohue, A. Zletz, and R. J. Flannery, *J. Electrochem. Soc. 115*, 1042 (1968).
37. J. Burdon, I. W. Parsons, and J. C. Tatlow, *Tetrahedron 28*, 43 (1972).
38. H. Meinert, Dissertation, Humboldt Universitat, Berlin (1960).
39. H. Baba, K. Kodaira, S. Nagase, and T. Abe, *Bull. Chem. Soc. Jpn. 51*, 1891 (1978).
40. G. P. Gambaretto and F. M. Conte, *J. Fluorine Chem. 19*, 427 (1982).
41. H. Meinert, R. Fackler, J. Mader, P. Reuter, and W. Rohlke, *J. Fluorine Chem. 51*, 53 (1991).
42. H. Schmidt and H. Meinert, *Angew. Chem. 72*, 109 (1960).
43. V. Plashkin, S. Mertsalov, V. Kollegov, and S. Sokolov, *Zh. Org. Khim. 6*, 1006 (1970).
44. J. Burdon and J. C. Tatlow, in: *Advances in Fluorine Chemistry* (M. Stacey, J. C. Tatlow, and A. G. Sharpe, eds.), 1, p. 160, Butterworths, London (1960).
45. T. Abe, E. Hayashi, H. Baba, and S. Nagase, *J. Fluorine Chem. 25*, 419 (1984).
46. E. A. Kauck and J. H. Simons, U.S. Patent 2 644 823 (to 3M Co.) [*CA 48*, 6469 (1954)]; British Patent 718 318 (1954).

47. M. Napoli, L. Conte, G. P. Gambaretto, and F. M. Carlini, *J. Fluorine Chem. 45*, 213 (1989).
48. W. Cao, W. Ge, and W. Huang, *Youji Huaxue*, 133 (1987) [*CA 106*, 222848 (1987)].
49. L. Conte, C. Fraccaro, M. Napoli, and M. Mistrorigo, *J. Fluorine Chem. 34*, 183 (1986).
50. G. G. I. Moore, J. C. Hansen, T. M. Barrett, J. E. Waddel, K. M. Jewell, T. A. Kestner, and R. M. Payfer, *J. Fluorine Chem. 32*, 41 (1986).
51. E. Hollitzer and P. Sartori, *J. Fluorine Chem. 35*, 329, 341 (1987).
52. St. Rüdiger, V. E. Platonov, H. Meinert, N. V. Popkova, U. Jonethal, W. Radeck, and G. Ziecinski, *Otkrytiya Izobret 34*, 275 (1990); Patent SU 1 427 780 A1, 15 Sept. 1990 [*CA 114*, 81613 (1991)].
53. T. Ono, Y. Inoue, Y. Arakawa, Y. Naito, C. Fukaya, K. Yamanouchi, and K. Yokoyama, *J. Fluorine Chem. 43*, 67 (1989).
54. Y. Inoue, Y. Arakawa, Y. Naito, T. Ono, K. Y. Fukaya, K. Yamanouchi, and K. Yokoyama, *J. Fluorine Chem. 38*, 303 (1980).
55. S. Schramm, St. Rüdiger, U. Jonethal, and M. Kupfer, Ger. (East), Patent DD 275 079, A1, 10 Jan. 1990 [*CA 113*, 180292 (1990)].
56. W. Radeck, St. Rüdiger, and U. Jonethal, Ger. (East), 5pp, Patent DD 273 290 A1, 8 Nov. 1989 [*CA 113*, 140894 (1990)].
57. E. Hayashi, T. Abe, H. Baba, and S. Nagase, *Nagoya Kogyo Gijutsu Shikensho Hokoku 37* (3), 54 (1988) [*CA 109* (22), 199847].
58. T. Abe, E. Hayashi, H. Baba, and S. Nagase, *Nippon Kagaku Kaishi 10*, 1980 (1985).
59. R. D. Chambers, R. W. Fuss, M. Jones, P. Sartori, A. P. Swales, and R. Herkelmann, *J. Fluorine Chem. 49*, 409 (1990).
60. T. Abe, K. Kodaira, H. Baba, and S. Nagase, *J. Fluorine Chem. 12*, 359 (1978).
61. T. Abe, E. Hayashi, H. Baba, and S. Nagase, *Chem. Lett.* 121 (1980).
62. T. Abe, H. Baba, E. Hayashi, and S. Nagase, *J. Fluorine. Chem. 23*, 123 (1983).
63. T. Abe, E. Hayashi, H. Fukaya, and H. Baba, *J. Fluorine Chem. 50*, 173 (1990).
64. T. Abe, E. Hayashi, H. Baba, and H. Fukaya, *J. Fluorine Chem. 48*, 257 (1990).
65. M. Napoli, L. Conte, and G. P. Gambaretto, *J. Fluorine Chem. 45*, 213 (1989).
66. M. Yonekura, S. Nagase, H. Baba, K. Kodaira, and T. Abe, *Bull. Chem. Soc. Jpn. 49*, 1113 (1976).
67. W. H. Pearlson, *J. Fluorine Chem. 32*, 29 (1986).
68. W. H. Pearlson, U.S. Patent 3 274 081 (to 3M Co.).
69. R. E. Banks (ed.), *Organofluorine Chemicals and their Industrial Applications*, Ellis Horwood, Chichester (1979); R. E. Banks (ed.), *Preparation, Properties, and Industrial Applications of Organofluorine Compounds*, Ellis Horwood, Chichester (1982).
70. M. Grayson and D. Eckroth (eds.), *Kirk-Othmer Encyclopedia of Chemical Technology*, 3rd edn., Vol. 10, p. 829 ff, Wiley, New York (1980).
71. G. V. D. Tiers, *J. Am. Chem. Soc. 77*, 4837 (1955).
72. O. Volkert, U.S. Patent 4 972 002 (to BASF) [*CA 113*, 41969 (1990)].
73. W. V. Childs, in: *Technique of Electroorganic Synthesis*, Part III, Chapter VII (N. L. Weinberg and B. V. Tilak, eds.), part of *Techniques of Chemistry*, Volume V (A. Weissberger, ed.), Wiley, New York (1982).
74. L. Bai and B. E. Conway, *Abstracts of Papers*, 177th Meeting of The Electrochemical Society, Montreal, Quebec, Canada, May 1990, Abstract 567.
75. L. Bai and B. E. Conway, *J. Appl. Electrochem. 18*, 839 (1988), and Ref. 10.
76. W. Rüdorff, U. Hoffman, G. Rüdorff, J. Endell, and G. Ruess, *Z. Anorg. Chem. 256*, 125 (1948).
77. A. J. Rudge, *The Manufacture and Use of Fluorine and Its Compounds*, Oxford University Press, Oxford (1962).
78. N. Watanabe, T. Nakajima, and H. Touhara, *Graphite Fluorides* (Studies in Inorganic Chemistry 8), pp. 1–22, Elsevier, Amsterdam (1988); T. Nakajima, T. Ogawa, and N. Watanabe, *J. Electrochem. Soc. 134*, 8 (1987); T. Tojo, Y. Chong, T. Iwasaki, K. Ikari, and N. Watanabe, *Abstracts of Papers*, 40th Meeting of the International Society of Electrochemistry, Kyoto, Japan, September 1989, Abstract a1 19-09-07-G; T. Nakajima and N. Watanabe, *Graphite Fluorides and Carbon–Fluorine Compounds*, CRC Press, Boca Raton, FL (1991).
79. M. Chemla and D. Devilliers, *J. Electrochem. Soc. 136*, 87 (1989).

80. D. Devilliers, B. Teisseyre, M. Vogler, and M. Chemla, *J. Appl. Electrochem.* *20*, 91 (1990), and Ref. 4.
81. W. V. Childs, U.S. Patent 4 022 824 (to Phillips Petroleum Co.) [*CA 87*, 22426 (1977)].
82. H. M. Fox, F. N. Ruehlen, and W. V. Childs, *J. Electrochem. Soc. 118*, 1246 (1971).
83. W. V. Childs, B. H. Ashe, Jr., and P. S. Hudson, U.S. Patent 3 900 372 (to Phillips Petroleum Co.) [*CA 85*, 192571 (1975)].
84. R. A. De Marco, D. A. Couch, and J. M. Shreeve, *J. Org. Chem. 37*, 3332 (1972).
85. P. E. Cassidy, T. M. Aminabhavi, and J. M. Farley, *Rev. Macromol. Chem. Phys. C29(2&3)*, 365 (1989).

6

Chlorofluorocarbons

ARTHUR J. ELLIOTT

6.1. INTRODUCTION

The completely halogenated chlorofluorocarbons (CFCs) have easily been the most important organic fluorine chemicals produced in the world during the past 60 years. Selected for development by Midgley and his co-workers in the late 1920s, CFC-12[*] (CF_2Cl_2) was considered to be the ideal refrigerant to replace sulfur dioxide and ammonia.[1,2] The low toxicity, nonflammability, and stability of CFC-12 quickly led to the search for other CFCs. Kinetic Chemicals, a joint venture between Du Pont and General Motors, was formed to manufacture the new materials, and later it became part of the Freon Division of Du Pont. As the list of available compounds grew, new application areas emerged such as aerosol propellants, foam blowing agents, and solvents. By the early 1970s, when they were first linked to the destruction of the ozone layer, CFCs were being produced in many countries around the world. Some major producers and their trade names are listed in Table 1.

Despite a ban on the use of CFCs as aerosol propellants in North America (effective from December 15, 1978), CFC use continued to rise until, by 1986, annual worldwide production reached some 2.5 billion pounds. In volume terms, the five compounds shown in Table 2 comprise (overwhelmingly) the bulk of CFCs produced commercially. The hydrogen-containing CFC, HCFC-22, while not strictly a CFC, is usually considered to be part of this family because of its similar use and manufacturing method. The presence of the hydrogen atom makes HCFC-22 less stable in the atmosphere; and since the ozone depletion potential[**] depends both on chlorine

---

[*]The international CFC numbering system is explained in Chapter 1.
[**]For the definition of ozone depletion potential, see Figure 1 in Chapter 7.

ARTHUR J. ELLIOT • Halocarbon Products Corporation, North Augusta, South Carolina 29841-0369.
Organofluorine Chemistry: Principles and Commercial Applications, edited by R. E. Banks *et al.* Plenum Press, New York, 1994.

Table 1.   Some Major Producers of Chlorofluorocarbons

Country	Company	Trade name
North America	Du Pont	Freon
	Allied	Genetron
	Pennwalt	Isotron
United Kingdom	ICI	Arcton
	ISC/Rhone-Poulenc	Isceon
The Netherlands	Akzo	Fcc
West Germany	Hoechst	Frigen
	Kali-Chemie	Kaltron
France	Atochem	Flugene
Italy	Montedison	Algofrene
Australia	Australian Fluorine Chemicals	Isceon
Czechoslovakia	Slovek Pro Chemickov	Ledon
Japan	Ashai Glass	Asahiflon
	Daikin Kogyo	Daiflon
USSR		Khladon

content and atmospheric lifetime, HCFC-22 was not included in the Montreal Protocol which limits the production of the CFCs shown in Table 2 (see Chapters 7 and 28).

Table 3 shows the major applications and volumes of CFCs produced in 1986. These volumes were used as annual production limits in the original Montreal Protocol (1987). These limits took effect in 1989, and a reduction by 50% was mandated by 1998. Continued pressure from environmental groups, however, has led to a revision of the Protocol (1990) which all but ensures the complete elimination of production of the CFCs listed in Table 2 by the end of the century, except for uses where they are not released into the atmosphere. The phase-out of these CFCs will cause considerable upheaval since nearly 400,000 businesses use them to generate some \$25 billion annually in goods and services in the United States alone. According to most estimates, by the year 2000 only 45% of the CFC market will have been taken over by the fluorinated alternatives now being developed (see Chapters 7 and 28). Other demand

Table 2.   Commercially Important Chlorofluorocarbons

Compounds	Formula	Boiling point (°C)	Ozone depletion potential	Atmospheric lifetime (year)
CFC-11	$CFCl_3$	23.7	1.0	75
CFC-12	CF_2Cl_2	−29.8	1.0	111
CFC-113	$CF_2ClCFCl_2$	47.7	0.8	90
CFC-114	CF_2ClCF_2Cl	3.8	1.0	185
CFC-115	CF_3CF_2Cl	−38.7	0.6	380
HCFC-22	CHF_2Cl	−40.8	0.05	20

Table 3. Global 1986 Consumption by Application of
Chlorofluorocarbons

Use	Total (10^9 pounds)	Total (%)	CFC
Aerosol	0.48	19	11,12,113,114
Refrigeration	0.75	30	11,12,113,114,115
Foams	0.73	28	11,12,113
Solvent	0.48	19	113
Other	0.10	4	

will be met by conservation, recycling, and not-in-kind replacements. Note that some of the restricted compounds, especially CFC-113, will continue to serve as chemical intermediates in processes where they are destroyed, such as the dechlorination of CFC-113 to chlorotrifluoroethylene. Some may well serve as intermediates for the alternatives presently under development (see Chapters 7 and 28).

When the phase-out is complete there will be a small but significant class of CFCs not affected by the ban, namely, *high-molecular-weight CFCs*. This class includes important inert lubricants, hydraulic fluids, damping fluids, and heat-transfer agents. Halocarbon Products Corporation (U.S.A.) is the largest producer of these materials, which are telomers of chlorotrifluoroethylene sold under the trade name Halocarbon® Oils, Greases and Waxes. Similar products are available from Hooker in the U.S. (Fluorolube®), Atochem in France (Voltalef®), and Asahi and Daikin in Japan. Plants in China and the republics of the former Soviet Union reportedly produce high-molecular-weight CFCs for internal consumption.

6.2. PRODUCTION

6.2.1. Halogen Exchange

Industrial production of CFCs is achieved by the replacement of chlorine in chlorocarbons by fluorine, using anhydrous hydrogen fluoride and a catalyst. This use consumes about 50% of the hydrogen fluoride manufactured in the world. The chlorocarbon feedstocks are prepared by direct chlorination of hydrocarbons.

Historically, most of the CFCs have been made by a liquid-phase process, employing antimony pentachloride as a catalyst. The antimony may be added either as the trichloride and converted to the pentachloride *in situ* with chlorine, or, more conveniently, as the liquid pentachloride itself. Antimony pentachloride reacts with hydrogen fluoride to give mixed chlorofluorides which are the actual catalysts, and hydrogen chloride (Eq. 1). The exchange reaction then takes place between the chlorocarbon feedstock and the mixed chlorofluoride, producing the desired CFC and regenerating antimony pentachloride (Eq. 2). Carbon tetrachloride serves as the feedstock for both CFC-11 and CFC-12 (Eqs. 3 and 4). The chlorines are exchanged stepwise, the reaction temperature determining the number of fluorines introduced. In commercial reactors, temperatures usually range from 100–150°C and pressures from 10–30 atmospheres. The hydrogen chloride is conveniently removed as a liquefied gas for other use, or absorbed in water for sale as hydrochloric acid. The crude CFC is washed with base to remove traces of hydrogen chloride and any hydrogen fluoride, dried, and purified by distillation. HCFC-22, which is used commercially to produce tetrafluoroethylene and is now growing in importance as a CFC substitute, is made similarly from chloroform (Eq. 5).

$$(1) \qquad SbCl_5 + nHF \rightarrow SbF_nCl_{5-n} + nHCl$$

$$(2) \qquad nRCl + SbF_nCl_{5-n} \rightarrow nRF + SbCl_5$$

$$(3) \qquad CCl_4 + HF \rightarrow CFCl_3 + HCl$$

$$(4) \qquad CFCl_3 + HF \rightarrow CF_2Cl_2 + HCl$$

$$(5) \qquad CHCl_3 + 2HF \rightarrow CHF_2Cl + 2HCl$$

A logical feedstock for the two-carbon CFCs is hexachloroethane. Since it is a high melting solid, it is generated *in situ* from tetrachloroethylene and chlorine in the fluorination process. The antimony pentahalides catalyze the addition of chlorine and may themselves be reduced to the trihalide before subsequent reoxidation with more chlorine. If the feed ratios are not in balance, the excess of hydrogen fluoride may add across the double bond leading to hydrogen-containing compounds. The reaction for the synthesis of CFC-113 is shown in Eq. 6. Again, the temperature of the reaction usually determines the number of chlorines exchanged. While isomers are possible, the carbon atom which is fluorinated is usually the one bearing the most chlorine atoms.

$$(6) \qquad CCl_2{=}CCl_2 + Cl_2 + 3HF \rightarrow CF_2ClCFCl_2 + 3HCl$$

Antimony-based catalysts may become deactivated with use. Simple reduction to the trihalide is easily remedied by oxidation with chlorine. Over time, however, tar formation limits the activity of the catalyst and it needs either purification or removal as waste. Owing to environmental concerns over disposal of toxic compounds, there is considerable current interest in the detoxification and recovery of antimony wastes.[3,4]

Vapor-phase exchange processes have been developed which avoid the antimony disposal problem. In these processes, vaporized mixtures of the chlorocarbon feedstock and an excess of anhydrous hydrogen fluoride vapors are passed over catalysts at temperatures ranging from 250–400°C. Once again the exchange process occurs stepwise, increasing catalyst bed temperatures giving increasing fluorination. The most common catalyst is trivalent chromium, either as the fluoride[5] or as the oxide.[6–8] The catalyst may be used on a support such as alumina.[9] Other catalysts include iron and fluorinated alumina. Recent activity related to the search for CFC substitutes has produced claims of improvements in chromia catalysts,[10,11] as well as those based on other metals such as cobalt, nickel, manganese, and lanthanum.[12,13]

Chromia (Cr_2O_3) catalysts and, to various extents, the other vapor-phase catalysts, tend to produce some rearrangement of the halogens. This scrambling may be troublesome in the two-carbon series when, for example, some CF_3CCl_3 is produced along with $CF_2ClCFCl_2$. One particular process achieves the chlorination of the hydrocarbon and the fluorination of the *in situ* generated chlorocarbon simultaneously.[14,15] The catalyst is fluorinated alumina impregnated with thorium fluoride.

Just as with the liquid-phase process, the catalyst requires regeneration from time to time. The usual problem is "coking", which requires treatment of the catalyst bed with oxygen or air at high temperatures. There have been some successful attempts to prolonging catalyst life by adding oxygen along with the feedstocks to remove the coke continually.[16] Reactivation of alumina-based catalysts with high-temperature steam has also been reported.[9]

Vapor-phase fluorinations are very attractive industrially. Reactions can be conducted continuously at modest pressures (10 atm), with appropriate recycle streams as necessary. However, since an excess of hydrogen fluoride is usually employed, downstream purification involving CFC–HF azeotropes may cause difficulties.

6.2.2. Chlorotrifluoroethylene Telomerization

While high-molecular-weight CFCs can be synthesized via a variety of methods, from a practical standpoint such materials are best prepared by free-radical telomerization of chlorotrifluoroethylene (CTFE). Either the conversion is carried out in the presence of a chain-transfer agent, or higher-molecular-weight polymers are cracked to lighter species. Carbon tetrachloride (Eq. 7)[17] and CFC-113 (Eq. 8)[18] have both been used as telogens. The telomer chains are formed predominately, but not exclusively, via radical attack at the CF_2 carbon. Trichloromethyl (CCl_3) end groups must be fluorinated, at least partially, to make useful oils or grease since these groups are too reactive for an inert oil. Whichever way the oils are made, they must be free from unsaturation, since oxygen tends to react with C=C sites leading to the formation of acid fluorides and hence unusable material. Saturation and end-group fluorination are generally accomplished using a combination of chlorine and chlorine trifluoride or cobalt trifluoride.[17] Once stabilized, the telomers are fractionated into various standard viscosities for sale as oils. Higher-molecular-weight material forms the basis of the waxes, while grease is made by the addition of suitable thickening agents.

(7) $$CCl_4 + nCF_2{=}CFCl \rightarrow CCl_3(CF_2CFCl)_nCl$$

(8) $$CF_2ClCFCl_2 + nCF_2{=}CFCl \rightarrow CF_2ClCFCl(CF_2CFCl)_nCl$$

Usually there is an overabundance of lower-molecular-weight telomers which are not as useful as the heavier products; thus, coupling processes have been developed in which two units are dimerized through the removal of end-group chlorine.[19,20] One such coupling reaction is achieved by heating telomers with a mixture of copper and zinc powder at 220 °C (Eq. 9). While coupling of CCl_3 end groups is easiest, this produces CCl_2–CCl_2 units in the middle of the chains which are difficult to fluorinate. If any remain in the product, stability is adversely affected.

(9) $$2\ CF_2ClCFClCF_2CFCl_2 \rightarrow (CF_2ClCFClCF_2CFCl)_2$$

6.3. PROPERTIES AND SPECIFICATIONS

Some properties of CFCs and related compounds are listed in Table 4. The materials are all colorless gases or liquids, and most have a slight ethereal odor. Boiling points decrease as fluorine contents increase; and isomers have very close boiling points, making separation by distillation difficult. When used as refrigerants, CFCs must meet strict purity specifications. In fact, all high-volume CFCs are manufactured to meet stringent standards, including no detectable acidity and less than 10 ppm of water.

Important physical properties of some lower-molecular-weight CTFE telomers are shown in Table 5. In practice they are not sold as single chemical entities (or mixtures of isomers), but as blends with defined properties, e.g., specific viscosity, density, or refractive index. Since they are used in such applications as flotation fluids for gyroscopes, they too must conform with specifications, just like other CFCs.

The most outstanding property of CTFE oils is their chemical inertness. They are unreactive toward many widely used industrial chemicals, including liquid oxygen. Thermal decomposition starts about 260°C and is rapid above 300°C. Thermal cracking to unsaturated species, including CTFE monomer, takes place.

As a class, CFCs are very low in toxicity. The main physiological action of volatile compounds at high inhaled vapor levels is narcosis and anesthesia; sensitization of the mammalian heart to adrenalin may also occur. Similarly, the high-molecular-weight oils display low toxicity when administered by oral or dermal routes. Of course, skin contact with liquified CFC gases may cause frostbite.

Table 4. Properties of Selected Chlorofluorocarbons

CFC No.	Formula	Molecular weight	bp (°C)	fp (°C)	Density (g cm^{-3}) [°C]	Critical temp. (°C)
10	CCl_4	153.8	76.8	−23	1.595 [20]	283
11	$CFCl_3$	137.4	23.8	−111	1.490 [20]	198
12	CF_2Cl_2	120.9	−29.8	−155	1.328 [20]	112
13	CF_3Cl	104.5	−81.4	−181	0.924 [20]	29
14	CF_4	88.0	−128	−184	1.33 [−80]	−46
22	CHF_2Cl	86.5	−40.8	−160	1.213 [20]	96
110	CCl_3CCl_3	236.8	185	186	2.091 [20]	—
111	CCl_3CFCl_2	220.3	137	100	1.74 [25]	—
112	$CFCl_2CFCl_2$	203.8	92.8	27.4	1.634 [30]	278
112a	CCl_3CF_2Cl	203.8	91.5	40.8	1.649 [20]	—
113	$CF_2ClCFCl_2$	187.4	47.7	−33	1.582 [20]	214
113a	CF_3CCl_3	187.4	45.9	14	1.579 [20]	—
114	CF_2ClCF_2Cl	170.9	3.7	−94	1.473 [20]	146
114a	CF_3CFCl_2	170.9	3.0	−57	1.454 [29]	146
115	CF_3CF_2Cl	154.5	−38	−106	1.291 [25]	80
116	CF_3CF_3	138.0	−78	−100	1.607 [−78]	24

Table 5. Properties of Chlorotrifluoroethylene Telomers $Cl(CF_2CFCl)_nCl$

$n =$	2	3	4	5
Molecular weight	304	420.5	537	653.5
Boiling point (°C)	136	205	255	300
Refractive index (n_D, 20°C)	1.383	1.397	1.403	1.407
Viscosity (100°F; 37.8°C)				
(centistokes)	0.78	1.9	5.8	25.7
(centipoise)	1.35	3.4	10.8	48.9
Density (g cm^{-3}) (100°F; 37.8°C)	1.7130	1.8079	1.8648	1.9015

6.4. APPLICATIONS

6.4.1. General

It is impossible to overstate the importance of the role that CFCs have played in making refrigeration, air-conditioning, and, to some extent, aerosols such essential facets of modern existence. Life, as we know it, would be impossible in some parts of the world without the means for cooling food and houses. Of course the very property that made the CFCs so useful, their stability, allows them to reach the stratosphere and there release chlorine atoms, which interact with the ozone layer (see Chapter 28). Over the next decade there will be a considerable decline in the use of CFCs in most of their high-volume applications. However, as mentioned earlier, there will be continuing applications of the heavier CFCs, but the volumes will be inconsequential compared to the 2.5 billion pounds of volatile CFCs produced worldwide in 1986 (see Table 3).

6.4.2. Declining Applications

6.4.2.1. Aerosols and Propellants

Since aerosols are very convenient for the consumer, worldwide demand is still growing despite the CFC ban on such use in nonessential aerosols in North America in 1978. Their use is still allowed in applications such as aerosol drug-delivery, and extensive toxicity testing will have to be performed on any alternatives used for this purpose.

Commonly used CFC propellants in other parts of the world are CFC-11 ($CFCl_3$), 12 (CF_2Cl_2), and, to some extent, 114 (CF_2ClCF_2Cl). A mixture of CFC-11 and 12 is frequently used to obtain an intermediate vapor pressure. As CFCs have begun to be replaced, there has been an increase in the use of hydrocarbons, sometimes mixed with HCFC-22 (CHF_2Cl) to lower the flammability.

Table 6. Thermodynamic Properties of Refrigerant CFCs

Property	CFCl$_3$	CF$_2$Cl$_2$	CHF$_2$Cl	CF$_2$ClCF$_2$Cl	CF$_3$CF$_2$Cl
Refrigerant no.	R-11	R-12	R-22	R-114	R-115
Boiling point (°C)	23.8	−29.8	−40.8	3.7	−38.7
Critical temperature (°C)	198	112	96	146	80
Critical pressure (atm)	43.5	40.6	49.12	32.2	30.8
Critical density (g cm^{-3})	0.554	0.558	0.525	0.582	0.596
Heat of vaporization at bp (cal g^{-1})	43.10	39.47	55.81	32.51	30.11
Specific heat of liquid at 25°C (cal g^{-1})	0.208	0.232	0.300	0.243	0.285

6.4.2.2. Refrigeration

This has been the area where CFCs have excelled in their application. They were judged to be safe alternatives to materials such as sulfur dioxide and ammonia over 60 years ago (see Chapters 1 and 28) and allowed refrigeration to become available safely in houses and automobiles. Some thermodynamic properties of the important refrigerants are shown in Table 6.

CFC-12 is commonly used in domestic and mobile air-conditioning. It will normally achieve temperatures down to −20°C, and is suitable for use with centrifugal, rotary, and reciprocating compressors. The proposed substitute for CFC-12 is HFC-134a (CF$_3$CH$_2$F), which unfortunately is not a "drop-in" replacement, thus making the future servicing of existing units a major concern. To satisfy this aftermarket of over 100 million cars worldwide, blends of HCFC-22, HCFC-124 (CF$_3$CHFCl), and HFC-152a (CHF$_2$CH$_3$) are being tested as drop-in candidates. Only relative minor changes in hoses, lubricant, and drying agent appear to be required. While the required toxicity testing is still being conducted on HCFC-124, CFC-114 may be substituted for it in the blend.

CFC-11 is used in industrial chillers and other units when a temperature of 5–10°C is adequate. The proposed substitute, HCFC-123 (CF$_3$CHCl$_2$), which still contains chlorine and will probably be subject to a phase-out later, is presently being used in some applications. HCFC-22 will cool to −40°C and may become the preferred refrigerant for domestic use. It is used where fast cooling is needed, and in frozen-food display cases. For applications where even colder temperatures are required, azeotropic blends involving CFC-115 (CF$_3$CF$_2$Cl) and CFC-12 or HCFC-22 are employed.

6.4.2.3. Foams

Polymers in a foamed or cellular state have many useful properties. Energy-efficient buildings have sheets of polystyrene foam insulation in their construction, sheets made by compounding CFC-12 under pressure with liquid polymer followed by release to atmospheric pressure via extrusion. Polyurethane foams are produced using CFC-11, which is vaporized by the heat of reaction of the isocyanate with the polyol. The foaming mass is usually heat-treated to complete the process.

When employed as a foam to fill cavities, however, CFC-12 is used to expand the polymer mass. Although some CFC, especially CFC-11, remains behind in the foam, almost all of the material is lost to the atmosphere during foaming operations. This is one application, therefore, where recycling is not feasible. It has been estimated that the use of CFCs in polyurethane foam production will decline by two-thirds by 1993, down from 461 million pounds in 1986. CFCs were completely removed from plastic foam products for food packaging, such as egg cartons and meat trays in early 1990; currently HCFC-22 mixed with carbon dioxide, pentane, or isopentane is being used. Dow Chemical has reported that it uses HCFC-142b in its polystyrene and polyethylene fabricated foams.[21]

6.4.2.4. Solvents

CFC-113 ($CF_2ClCFCl_2$) is an ideal solvent and cleaning agent for the electronics and aerospace industries.[22] It has excellent wetting properties, is noncorrosive, and is easily removed, leaving no residue. For applications where greater solvency power is needed, it may be mixed with alcohols and ketones, usually at an azeotropic composition. Solvents based on CFC-113 are used to remove flux from printed wiring boards, and as degreasers for precision and polished surfaces where no residue can be tolerated. CFC-113 is also used in the dry-cleaning of expensive, heat-sensitive fabrics. CFC-11 has also been used in dry-cleaning because of its better solvency.

A suitable nonchlorinated replacement solvent with the versatility of CFC-113 (bp 47.7°C) will be difficult to find. Solvency power increases with chlorine content, as does ozone-depletion potential. HCFC-123 (CF_3CHCl_2) is a useful solvent, but its low boiling point, 27°C, makes it unsuitable for most uses. The homolog, HCFC-225ca ($CF_3CF_2CHCl_2$), has a higher boiling point and has been recommended for toxicity testing, but it is some years away from commercialization. IBM, the world's largest electronics manufacturer, planned to eliminate its use of CFC-113 by the end of 1993. Other companies are reducing emissions by changing processes to avoid flux removal. Digital Equipment has devised a water-based process for cleaning some circuit boards. Allied-Signal has developed blends involving HCFC-141b ($CFCl_2CH_3$), HCFC-123 (CF_3CHCl_2), and methanol as a replacement solvent. Existing equipment, however, will have to be modified before the blends can be used. While one of the blends has been approved for some military applications, it remains to be seen if they will find wide market acceptance.

6.4.3. Continuing Applications

6.4.3.1. Lubricants

The CTFE telomer oils were originally developed as lubricants for the mechanical equipment needed in the separation of uranium isotopes using the extremely reactive uranium hexafluoride. These inert oils, greases, and waxes now find many applications in aggressive environments where a hydrocarbon oil will not survive. Like hydrocar-

Table 7. Typical Properties of CTFE Oil Blends

Halocarbon oil	0.8	1.8	4.2	6.3	27	56	95	700	1000N
Flash and fire points	None	—	—	—	—	—	—	—	—
Pour point (°F) (± 10°F)	−200	−135	−100	−95	−40	−30	−15	40	50
(°C) (± 5°C)	−129	−93	−73	−71	−40	−34	−26	4.5	10
Cloud point (°F) (± 10°F)	< −200	< −135	< −125	< −125	< −95	−30	−5	55	65
(°C) (< ± 5°C)	< −129	< −93	< −87	< −87	< 71	−34	−21	13	18
Viscosity (±10%)									
(centistokes)	0.79	1.8	4.2	6.3	27	56	95	700	1000
(centipoises) @ 100°F (37.8°C)	1.3	3.5	7.8	12	51	108	192	1365	1950
(centistokes)	—	0.79	1.2	1.6	3.1	4.9	6.3	17	22
(centipoises) @ 210°F (99°C)	—	1.4	2.1	2.8	5.6	8.9	12	32	41
Density (± 0.01) (g cm^{-3})									
100°F (37.8°C)	1.71	1.82	1.85	1.87	1.90	1.92	1.92	1.95	1.95
210°F (99°C)	1.60	1.71	1.75	1.77	1.81	1.82	1.82	1.86	1.86
Refractive index n_D^{20} (typical)	1.383	1.395	1.401	1.403	1.407	1.409	1.411	1.414	1.415

bon oils, they are sold as blends according to certain viscosities. Some typical commercial blends are shown in Table 7. Greases are made by mixing an appropriate oil with a thickener such as silica or higher-molecular-weight CTFE polymer.

CTFE telomer lubricants are used primarily where chemical inertness and non-flammability are required. The chemical industry and the cryogenic gas industry (primarily oxygen) are the major users of these materials. They are used to lubricate all types of process equipment, such as dryers, conveyers, pumps, valves, and compressor seals. Extreme pressure tests using the four-ball method show that they are good lubricants, there being no seizure even at an applied load of 800 kg.

In the aerospace industry, as well as finding use in the oxygen-delivery system to the space shuttle oxidizer tanks, a light CTFE oil is under development as a nonflammable hydraulic fluid for future aircraft. Since the CTFE oil is much heavier than presently used hydraulic oils, retrofitting of existing aircraft is not feasible. The CTFE oil operates in smaller lines at much higher (8000 psi) pressure. New elastomeric seals, which are compatible with the oil, have been developed. In modern aircraft, serious fires can occur during landing if a hydraulic hose ruptures and sprays fluid onto the hot braking system. Tests done with CTFE hydraulic fluid show the material is completely nonflammable during such an event.

CTFE telomers find use as vacuum pump oils in the electronics industry where reactive gases and aluminum chloride would be harmful to hydrocarbon oils. They are also used as instrument fill fluids where they substitute for glycerice or silicone fluids in pressure gauges, sensors and diaphragm seals in aggressive chemical service. Other uses include application as a cutting or drawing oil in tantalum, molybdenum, and niobium processing, as a lubricant for life-support systems in oxygen-rich atmospheres, and as a cold-temperature bath fluid in laboratory apparatus.

6.4.3.2. Gyroscopic Fluids

CTFE oils are used as flotation fluids in gyroscopes. Specific density and viscosity values are achieved by blending together different weight oils. Navigational devices containing these oils are found in many commercial and military aircraft, as well as missiles. Newer technologies such as fiber optics are now increasingly used, but the floated gyroscopes have a well-deserved reputation for accuracy and long field life.

6.4.3.3. Chemical Intermediates

CFCs subject to control under the Montreal Protocol will still function as chemical intermediates, provided they are consumed rather than released into the atmosphere. CFC-113 presently serves as the starting material for the production of CTFE monomer. Despite the many years of research into a catalytic vapor-phase process for this conversion,[23–27] the preferred current method still involves zinc dechlorination in methanol (Eq. 10).[28]

(10) $$CF_2ClCFCl_2 + Zn \rightarrow CF_2{=}CFCl + ZnCl_2$$

CTFE monomer serves as a building block for the CTFE telomer oils discussed in Section 6.4.3.1, as well as the solid higher polymer and various copolymers. CFC-113 also may be used in the production of trifluoroethylene monomer by vapor-phase reduction using hydrogen and a precious metal catalyst, usually palladium (Eq.11).[29] Copolymers of trifluoroethylene and vinylidene fluoride show interesting piezoelectric properties (see Chapter 15).[30]

(11) $$CF_2ClCFCl_2 + H_2 \rightarrow CF_2{=}CHF + 3HCl$$

CFC-113 may also serve as the starting material for some of the CFC alternatives currently under development (see Chapter 7). Conversion to CFC-113a (Eq. 12) may be accomplished using aluminum chloride as catalyst, followed by reduction with sodium sulfite under basic conditions to give HCFC-123 (Eq. 13).[31] Alternatively, CFC-113a may be reduced with isopropanol in the presence of UV light (Eq.14).[32] Both of these processes produce HCFC-123 cleanly, with no over-reduction, and essentially free from HCFC-123a ($CF_2ClCHFCl$), which is usually found in several percent yield when HCFC-123 is made by the direct fluorination of tetrachloroethylene. The presence of the less stable HCFC-123a isomer may be problematic in some projected uses of HCFC-123.

(12) $$CF_2ClCFCl_2 \rightarrow CF_3CCl_3$$

(13) $$CF_3CCl_3 + Na_2SO_3 + NaOH \rightarrow CF_3CHCl_2 + NaCl + Na_2SO_4$$

(14) $CF_3CCl_3 + (CH_3)_2CHOH \overset{UV}{\rightarrow} CF_3CHCl_2 + HCl + (CH_3)_2C{=}O$

CFC-113a can be fluorinated to CFC-114a and CFC-115 using vapor-phase techniques (Eq. 15).[33] Vapor-phase reduction of CFC-114a (Eq. 16) gives HCFC-124 and HFC-134a, both of which are under development as CFC alternatives. Similar reduction of CFC-115 gives its possible replacement HFC-125 (Eq. 17; see Chapter 7).

(15) $CF_3CCl_3 + HF \rightarrow CF_3CFCl_2 + CF_3CF_2Cl$

(16) $CF_3CFCl_2 + H_2 \rightarrow CF_3CHClF + CF_3CH_2F$

(17) $CF_3CF_2Cl + H_2 \rightarrow CF_3CHF_2 + HCl$

As mentioned earlier, HCFC-22 is usually included in a discussion of CFCs because of its similar synthesis and applications. Besides serving as a starting material for the production of tetrafluoroethylene and hexafluoropropene (Eq. 18), it acts as an *in situ* source of difluorocarbene in the synthesis of difluoromethyl 2,2,2-trifluoroethyl ether (Eq. 19), an intermediate in the production of the inhalation anesthetic isoflurane (Eq. 20; see Chapter 25).[34]

(18) $CHF_2Cl \overset{700°C}{\rightarrow} CF_2{=}CF_2 + CF_3CF{=}CF_2$

(19) $CF_3CH_2OH + CHF_2Cl + NaOH \rightarrow CF_3CH_2OCHF_2 + NaCl + H_2O$

(20) $CF_3CH_2OCHF_2 + Cl_2 \rightarrow CF_3CHClOCHF_2 + HCl$

6.5. REFERENCES

1.   T. Midgley, Jr. and A. L. Henne, *Ind. Eng. Chem.* 22, 542 (1930).
2.   T. Midgley, Jr., A. L. Henne, and R. E. McNary, U.S. Patent 1,833,847 (to Frigidare Corp.) [*CA 26*, 1047 (1932)].
3.   V. Kalcevic and J. F. McGahan, U.S. Patent 4,751,063 (to International Technology Corp.) [*CA 109*, 151923 (1988)].
4.   S. K. Lee, U.S. Patent 4,411,874 (to SCA Services, Inc.) [*CA 100*, 13276 (1984)].
5.   R. F. Ruh, R. A. Davis, and M. R. Broadworth, U.S. Patent 2,885,427 (to Dow Chemical Co.) [*CA 53*, 16962 (1969)].
6.   F. W. Swamer and B. W. Howk, U.S. Patent 3,258,500 (to Du Pont) [*CA 65*, 16861 (1966)].
7.   J. F. Knaak, U.S. Patent 3,978,145 (to Du Pont) [*CA 85*, 142607 (1976)].
8.   L. Marangoni, G. Rasia, C. Gervasutti, and L. Colombo, *Chem. Ind. (Milan) 64*, 135 (1982).
9.   A. Scipioni, G. P. Gambaretto, and D. Zanon, British Patent 1,000,485 (to Siciedison Societa) [*CA 58*, 1344 (1963)].
10. W. H. Gumprecht, L. E. Manzer, and V. N. M. Rao, European Patent 313,061 (to Du Pont) [*CA 111*, 214116 (1989)].
11. E. J. Carlson, J. N. Armor, W. J. Cunningham, and A. M. Smith, U.S. Patent 4,828,818 (to Allied Signal, Inc.) [*CA 111*, 96640 (1989)].
12. L. E. Manzer, European Patent 331,991 (to Du Pont) [*CA 112*, 98008 (1990)].
13. L. E. Manzer and V. N. M. Rao, U.S. Patent 4,766,260 (to Du Pont) [*CA 110*, 97550 (1990)].
14. M. Vecchio and L. Lodi, U.S. Patent 3,442,962 (to Montecatini Edison) [*CA 71*, 60671 (1969)].

15. L. Lodi and M. Vecchio, U.S. Patent 3,294,852 (to Montecatini Edison) [*CA 66*, 94691 (1967)].
16. Y. Oda, H. Otouma, and S. Morikawa, Japanese Patent 7,682,206 (to Asahi Glass Co.) [*CA 85*, 149662 (1976)].
17. B. F. Daniels, European Patent 140,385 (to Occidental Chemical Corp.) [*CA 103*, 124105 (1985)].
18. B. F. Daniels and D. J. Olsen, U.S. Patent 4,849,556 (to Occidental Chemical Corp.) [*CA 112*, 123789,21425 (1990)].
19. J. S. Ng, U.S. Patent 4,691,067 (to Occidental Chemical Corp.) [*CA 108*, 131011 (1988)].
20. A. J. Elliott, U.S. Patent 4,634,797 (to Halocarbon Products Corp.) [*CA 105*, 228910 (1986)].
21. *Chemical Week*, Aug. 30th, pp. 22–23 (1989).
22. K. P. Murray, *J. Test. Eval. 17*, 87 (1989).
23. L. E. Gardner, U.S. Patent 3,636,173 (to Phillips Petroleum Co.) [*CA 76*, 71992 (1972)].
24. J. T. Rucker and D. B. Stormon, U.S. Patent 2,760,997 (to Hooker Electrochemical Co.) [*CA 51*, 3653 (1957)].
25. J. W. Clark, U.S. Patent 2,704,777 (to Union Carbide and Carbon Corp.) [*CA 50*, 15574 (1956)].
26. H. R. Nychka and R. E. Eibeck, U.S. Patent 4,155,941 (to Allied Chemical Corp.) [*CA 91*, 91155 (1979)].
27. T. Morimoto, S. Morikawa, and T. Funayama, Japanese Patent 60,185,734 (to Asahi Glass Co.) [*CA 104*, 109005 (1986)].
28. E. G. Locke, W. R. Brode, and A. L. Henne, *J. Am. Chem. Soc. 56*, 1726 (1934).
29. W. J. Cunningham, R. F. Piskorz, and A. M. Smith, European Patent 53,657 (to Allied Corp.) [*CA 97*, 162332 (1987)].
30. R. B. Olsen, D. A. Bruno, J. M. Brisco, and E. W. Jacobs, *Ferroelectrics Lett. 4*, 53 (1985).
31. T. Kondo and S. Yoshikawa, German Patent 3,834,038 (to Central Glass Co.) [*CA 111*, 194095 (1989)].
32. Y. Furutaka, H. Aoyama, and T. Honda, European Patent 308,923 (to Daikin Industries) [*CA 111*, 194094 (1989)].
33. S. Morikawa, S. Samejima, and M. Yoshikawa, German Patent 3,834,038 (to Central Glass Co.) [*CA 111*, 194094 (1989)].
34. L. S. Croix and R. C. Terrell, U.S. Patent 3,637,477 (to Air Reduction Co.) [*CA 72*, 3004 (1970)].

7

Alternatives to Chlorofluorocarbons (CFCs)

V. N. M. RAO

7.1. INTRODUCTION

Volatile chlorofluorocarbons (CFCs), man-made fully halogenated compounds having remarkably long atmospheric lifetimes, are implicated in the destruction of Earth's protective stratospheric ozone layer (see Chapter 1). Studies by Lovelock in the early 1970s showed that there was a measurable concentration of trichlorofluoromethane (CFC-11)* in the atmosphere, suggesting that it can have a long atmospheric lifetime.[1] This was followed in 1974 by the publication of a paper by Molina and Rowland,[2] in which they suggested that the chlorofluoromethanes $CFCl_3$ and CF_2Cl_2 (CFC-11 and -12, respectively) might survive transport to the stratosphere, where they could be broken down photochemically releasing chlorine atoms, which would then catalyze the destruction of ozone. Bromine-containing molecules (Halons), such as CF_3Br, were suspected to be even more harmful to the ozone layer. However, in the mid-70s to mid-80s, stratospheric science had not advanced enough to confirm these postulates. Because of their long atmospheric lifetimes, CFCs were also cited as possible contributors to global warming.[3]

After the ban on most aerosol uses by certain countries, notably the USA, in the late 1970s sales of CFCs for applications such as refrigeration, cleaning, and foam insulation continued to increase until the mid-80s. Growth forecasts during this period, as well as the development of improved predictive computer modeling, led to the Alliance for Responsible CFC policy. This body called for curbs in the long-term

---

*CFC nomenclature is explained in Chapter 1.

V. N. M. RAO • DuPont Central Research and Development, Experimental Station, Wilmington, Delaware 19880-0262.

Organofluorine Chemistry: Principles and Commercial Applications, edited by R. E. Banks *et al.* Plenum Press, New York, 1994.

growth of CFCs. In September 1987, a United Nations agreement, known as the Montreal Protocol, was signed. Under the terms of this agreement, periodic reviews of developing science and a 50% cut in the production of CFCs by 1998 was proposed. Within a year, the National Aeronautics and Space Administration's (NASA) Ozone Trends Panel published new findings[4] raising fresh concern as to the adequacy of the 1998 timeframe as suggested by the Montreal Protocol. Subsequent studies resulted in a revision, in June of 1990, which called for a complete ban on production of CFCs and Halons by the year 2000 in developed countries. Subsequently some producers have announced more aggressive phase-out schedules. Carbon tetrachloride (CTC) and so-called methyl chloroform (MeClf, CH_3CCl_3) were also added to the list of controlled compounds.

   In an attempt to provide substitutes having performance properties similar to those of CFCs as well as low toxicities, industry has been focusing on the development and assessment of hydrochlorofluorocarbons (HCFCs) and hydrofluorocarbons (HFCs). HCFCs have much reduced ozone depletion potentials (ODPs), and HFCs have zero ODP compared to CFCs. Conservation and nonfluorine-containing replacements are expected to reduce consumption in the coming years, but certain applications will require the use of HCFCs and HFCs. The HCFCs would serve temporarily until suitable HFC substitutes are identified. The principal compounds involved and their numbers in the international coding system are as follows:

   CFC codes: $11 = CFCl_3$; $12 = CF_2Cl_2$; $113 = CF_2ClCFCl_2$; $114 = CF_2ClCF_2Cl$; $115 = CF_3CF_2Cl$

   HCFC and HFC codes: $22 = CHF_2Cl$; $123 = CF_3CHCl_2$; $124 = CF_3CHFCl$; $125 = CF_3CHF_2$; $134a = CF_3CH_2F$; $141b = CFCl_2CH_3$; $152a = CHF_2CH_3$.

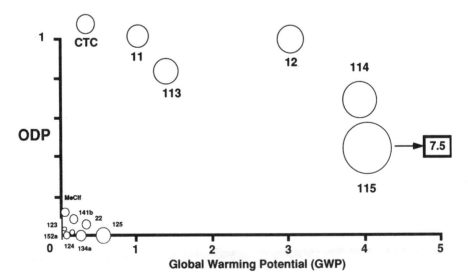

Figure 1.   Ozone depletion potential and global warming potential of selected compounds. Source: UNEP.

Other acronyms: CTC = carbon tetrachloride (CCl₄); MeClf = methyl chloroform (CH₃CCl₃).

The presence of hydrogen in the molecules of these substitutes reduces atmospheric stability (see Chapter 3). This allows for their degradation below the stratosphere, thereby reducing their ODPs as well as GWPs (Global warming potentials). The relative ODPs and GWPs of some target molecules are illustrated in Figure 1. Ozone depletion potentials (ODPs) are defined as the ratio of steady-state calculated ozone column changes for each unit mass of gas emitted into the atmosphere relative to the depletion caused by the emission of a mass unit of CFC-11 (CFCl₃). Similarly, global warming potential (GWP, or HGWP, H standing for Halocarbon) is defined as the ratio of the calculated steady-state net infrared flux change forcing at the troposphere for each unit mass of any halocarbon emitted relative to the same for CFC-11. Primarily due to vapor-pressure considerations, industry has concentrated on C_1–C_3 hydrocarbon-based molecules. Fluorinated ethers have also received some attention, and additional studies in this area have been recommended.[5] Even in the C_2 series of compounds containing only C, H, F, and Cl, eighty-eight different arrangements are possible, not counting stereoisomers! However, not all these would be satisfactory. Using the approach developed by McLinden and Didion,[6] this list can be reduced to a manageable size based on projected atmospheric lifetime, flammability, and toxicity. Potential substitutes for three CFC workhorses in industry, and their intended uses, are summarized in Table 1. The vapor-pressure curves have been determined for a number of these substitutes and some representative examples are shown in Figure 2.

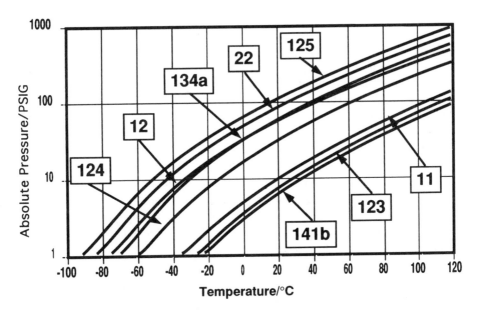

Figure 2.   Vapor pressures of fluorocarbon products. Source: DuPont.

Table 1. Applications of Some HCFCs and HFCs[a]

Application	Current CFC	Alternative
Refrigerants	CFC-12	HFC-134a
	CFC-11	HCFC-123
		Blends/azeotropes
Foaming agents	CFC-11	HFC-134a, HFC-152a
		HCFC-124, HCFC-141b
		Blends/azeotropes
Cleaning agents	CFC-113	Blends/azeotropes

[a]Source: DuPont.

7.2. SYNTHESIS

7.2.1. 1,1,1, 2-Tetrafluoroethane, CF_3CH_2F (HFC-134a)

Several approaches to this synthesis have been reported, utilizing a variety of starting materials, including hydrocarbons, halogenated hydrocarbons, olefins, and halo-olefins. The addition of HF to olefinic substrates, halogenation and halogen exchange, disproportionation, chlorofluorination, isomerization, and hydrogenolysis constitute the majority of the useful reactions employed. A multitude of catalysts has been reported or claimed for carrying out these reactions, in both liquid and vapor phase. Specific starting materials that have received considerable attention are discussed first followed by other reported approaches.

7.2.1.1. From Trichloroethylene (TCE)

The most direct route to HFC-134a comprises the reaction of TCE with HF to produce CF_3CH_2Cl (HCFC-133a) followed by replacement of the remaining chlorine with fluorine (Eqs. 1 and 2). Both liquid- and vapor-phase catalytic processes can be used for the first step,[7,8] which involves initial HF-addition across the double bond followed by a series of halogen-exchange reactions. Liquid-phase catalysts include SbX_5 (X = Cl, F), BF_3, TaF_5, NbF_5, and $MoCl_5$.[9,10] When using antimony(V) halides, it is common to employ fairly large catalyst concentrations. Improvements to such a system through the addition of various co-catalysts such as salts of Hg, Zn, Co, Ni, Fe, Zr, or Pd have been claimed.[11] Due to dissociation, reaction mixtures usually contain both Sb(III)and Sb(V) salts, and chlorine is fed to the reactor either continuously or intermittently to reoxidize most of the trivalent antimony to the pentavalent state. Because of this, some unwanted chlorination of C—H bonds occurs.

(1) $$CCl_2{=\!=}CHCl + 3\ HF \rightarrow CF_3CH_2Cl + 2HCl$$

(2) $$CF_3CH_2Cl + HF \rightarrow CF_3CH_2F + HCl$$

The vapor-phase conversion of TCE to HCFC-133a has been carried out conventionally using chromium-based catalysts. Use of unsupported chromium oxide has been reported,[12] and trivalent chromium supported on active carbon, as well as chrome salts supported on alumina, have also been advocated.[13] Normally, such catalysts are pretreated with HF at elevated temperatures prior to use.

An excess of HF is required for adequate conversion of HCFC-133a to HFC-134a (Eq. 2). This is an equilibrium limited reaction. For example, by using 6–10 moles of HF per mole of HCFC-133a and operating at 350–400°C, single-pass conversions of about 30% have been reported.[14] The need for the large excess of HF is also substantiated by earlier work on the one-step conversion of TCE to HFC-134a where only small amounts of HFC-134a were produced.[15] Again, conventional catalysts for this chemistry have been chromium(III)-based. However, such catalysts tend to deactivate rather rapidly. Various hypotheses have been presented, such as loss of HF from both starting material and product to give olefins, ultimately leading to coke formation. To overcome this difficulty small amounts of either chlorine or oxygen have been co-fed to the reactor.[16] The addition of oxygen in the ethane system leads to chlorination of both starting material and product because of the well-known Deacon chemistry for producing chlorine. This is particularly true of chrome-based systems, which have also been used for the oxidation of HCl to chlorine and water. Catalysts with much longer catalyst life, but based on using aluminum fluoride or fluorided alumina as the catalyst support, have been reported for this chemistry.[17] A variety of metal salts supported on alumina, and pretreated with HF at elevated temperature prior to use, have been disclosed.

Product streams from the reaction of HCFC-133a and HF contain small amounts of olefins, such as CF_2=CHCl. These may be toxic and have to be removed from the product. A variety of methods can be employed, one process involving treatment with aqueous permanganate.

7.2.1.2. From Tetrachloroethylene (Perchloroethylene, PCE)

A rather circuitous process based on PCE culminates in the hydrogenolysis of CF_3CFCl_2 (CFC-114a) to HFC-134a. PCE is chlorinated *in situ* to produce hexachloroethane, which then reacts with HF in the liquid phase using conventional antimony-based catalysts (usually antimony pentahalides), or in the vapor phase using chrome-based systems,[18] initially to produce $CF_2ClCFCl_2$ (CFC-113) and/or CF_2ClCF_2Cl (CFC-114) (Eqs. 3–5). Liquid-phase processes generally operate at temperatures lower than 160°C while vapor-phase systems employ temperatures ranging from 250–400°C. Several variants to the chromium-based vapor-phase systems have been proposed; these include the use of salts of iron, chromium, and nickel on aluminum fluoride.[19] Some halogen scrambling does occur, especially in vapor-phase processes, producing isomeric species (Eqs. 6 and 7). This is not detrimental, since it is these rearranged species that are needed for conversion to HFC-134a.

(3) $$CCl_2{=}CCl_2 + Cl_2 \rightarrow CCl_3CCl_3$$

(4) $$CCl_3CCl_3 + 3HF \rightarrow CF_2ClCFCl_2 + 3HCl$$

(5) $$CF_2ClCFCl_2 + HF \rightarrow CF_2ClCF_2Cl + HCl$$

(6) $$CF_2ClCFCl_2 \rightarrow CF_3CCl_3$$

(7) $$CF_2ClCF_2Cl \rightarrow CF_3CFCl_2$$

(8) $$CF_3CCl_3 + HF \rightarrow CF_3CFCl_2 + HCl$$

In one approach, the CFC-113, either by itself or mixed with its isomer CF_3CCl_3 (CFC-113a), is isomerized in the liquid phase to CFC-113a in the presence of a Friedel-Crafts catalyst such as anhydrous aluminum chloride.[20] The reaction (Eq. 6) is exothermic and very rapid, and is usually carried out under very mild conditions. The CFC-113a so produced is treated with HF either in the liquid or vapor phase to produce CFC-114a (Eq. 8). Alternatively, the CFC-114 can be isomerized to CFC-114a (Eq. 7), via either liquid- or vapor-phase processes.[21,22] Clearly, a process capable of producing CFC-114a directly from PCE without any CFC-114 formation would be desirable. Aluminum fluoride with no added co-catalyst(s) has been shown to be a good catalyst for the conversion of PCE to CFC-114a,[19] but some CFC-114 is still produced. The presence of metals salts, such as those of iron, chromium, and nickel, tend to minimize the formation of CFC-114a.

Vapor-phase hydrogenolysis of CFC-114a to HFC-134a using palladium catalysis has been reported.[23,24] In addition to HFC-134a, however, the process also produces CF_3CHFCl (HCFC-124) and CF_3CH_3 (HFC-143a) (Eq. 9). Operating temperatures of about 120°C to as high as 420°C have been reported, with various hydrogen-to-substrate ratios. High selectivities for HFC-134a have been demonstrated. Unlike CFC-114a, its isomer (CFC-114) is much less reactive at moderate temperature (about 150–250°C), and even when it does react, a product with only one chlorine replaced is obtained readily.

(9) $$CF_3CFCl_2 + H_2 \rightarrow CF_3CHFCl + CF_3CH_2F + CF_3CH_3$$

Although palladium alone has been shown to be a very effective catalyst for this hydrogenation chemistry, several improvements have been reported. For example, the presence of a Group IB element and, optionally, a lanthanide element prolongs catalyst life[25]; combinations of palladium and tungsten have also been shown to be effective.[26] Palladium is not a unique catalyst, since other metals disclosed for this chemistry include rhodium, tungsten carbide, and rhenium as well as combinations of these with other metals. In contrast, with noble-metal-catalyzed hydrogenation processes, a high-temperature process operating at 400–600°C over activated carbon has also been shown to produce good selectivities for HFC-134a production.[27]

The HCFC-124 byproduct observed in CFC-114a hydrogenolysis (Eq. 9) can be converted to HFC-134a. Similar catalysts have been used, but higher temperatures than used with CFC-114a are needed, thereby raising concerns about the sintering of palladium-based catalysts. The development of multicomponent catalysts, such as those mentioned above, is said to overcome this problem.[25] Besides conventional catalyst supports like carbon or alumina, the use of aluminum fluoride as an excellent support for this transformation has been disclosed.[28] The high-temperature process reported for the hydrogenolysis of CFC-114a over active carbon is also applicable to HCFC-124.[27]

7.2.1.3. Other Approaches

Of the alternatives to the two main approaches to HFC-134a just described, some have the potential for commercial development and some are useful methods for small-scale production.

A two-step approach from CFC-113 as outlined in Eqs. (10) and (11) is an attractive commercial possibility. However, specific catalytic systems have to be developed for the first stage (Eq. 10) since, in normal practice, CFC-113 affords chlorotrifluoroethylene (CTFE) rather than the desired trifluoroethylene (Eq. 12). Replacement of the vinylic chlorine in CTFE by hydrogen while still maintaining the olefinic linkage has been a distinct challenge, but recently catalytic systems have been reported for carrying out this transformation.[29] Alternatively, CTFE available from the well-established zinc dechlorination of CFC-113 (see Chapter 15) can be employed as the starting material. Details of the addition of HF to trifluoroethylene (Eq. 11) using chromium oxyfluoride as catalyst, especially at temperatures in the range 60–180°C, have been disclosed.[30] Methods seemingly suitable for small-scale production of HFC-134a include chlorine-for-fluorine exchange in CF_3CH_2Cl (HCFC-133a) using alkali-metal fluorides (KF, CsF) in aqueous or dipolar aprotic solvent systems at high temperatures,[31,32] or the powerful Swarts-type reagent antimony pentafluoride. Unfortunately, the reaction between HCFC-133a and SbF_5 has been reported to produce pentafluoroethane and HCFC-124 (CF_3CHFCl) in addition to the desired product[33]; operating at milder temperatures may offer some advantage.[34] Direct fluorination of the commercial monomer vinylidene fluoride ($CF_2{=}CH_2 + F_2 \rightarrow CF_3CH_2F$), while commercially unattractive by today's economic standards, offers an interesting straightforward synthesis of HFC-134a.[35]

(10) $CF_2ClCFCl_2 + 2H_2 \rightarrow CF_2{=}CHF + 3HCl$

(11) $CF_2{=}CHF + HF \rightarrow CF_3CH_2F$

(12) $CF_2ClCCl_2F + H_2 \rightarrow CF_2{=}CClF + 2HCl$

7.2.2. 2,2-Dichloro-1,1,1-Trifluoroethane, CF_3CHCl_2 (HCFC-123)

As with HCFC-134a, several routes to HFC-123 have been proposed. These are summarized in Figure 3. Although the same classes of reactions as those associated with HFC-134a production are involved, catalyst and process variables do not appear to be as extensive.

7.2.2.1. From Perchloroethylene (PCE)

The direct synthesis of HCFC-123 can be accomplished via the catalyzed reaction of PCE with HF (Eq. 13), using either liquid- or vapor-phase processes.[36,37] HF addition across the C=C bond in PCE (→ CFCl$_2$CHCl$_2$), followed by halogen exchange to give predominantly the more thermodynamically stable dichlorotri-fluoroethane (CF$_3$CHCl$_2$ versus CF$_2$ClCHFCl; Eq. 13), occurs.[37] The superiority of the liquid-phase process using tantalum salts is due to the stability of the pentavalent tantalum species; this eliminates the need for a chlorine co-feed required with anti-mony catalysts (cf. Section 7.2.1.1), thereby preserving the hydrogen in the product.[36] The success of the vapor-phase process has been attributed to the beneficial effect of the highly fluorinated alumina catalyst.[37] Some overfluorination does occur, producing CF$_3$CHFCl (HCFC-124) and CF$_3$CHF$_2$ (HFC-125). Chromium(III) oxide alone or supported on alumina are also suitable catalysts,[38] but tend to produce more overfluor-inated products due to their greater catalytic activity.

(13) $CCl_2{=}CCl_2 + 3HF \rightarrow CF_3CHCl_2 + 2HCl$

Even in conventional liquid-phase processes using PCE, Cl$_2$, and HF to produce CFC-113 (Eqs. 3 and 4), a small amount of HCFC-123 is produced as a byproduct. Small amounts of its thermodynamically less stable isomer CF$_2$ClCHFCl (HCFC-123a; Eq. 14) have also been observed. These hydrogen-containing products arise from competing HF addition versus chlorination of PCE as the initial step and subsequent halogen exchange.

(14) $CCl_2{=}CCl_2 + 3HF \rightarrow CF_2ClCHFCl + 2HCl$

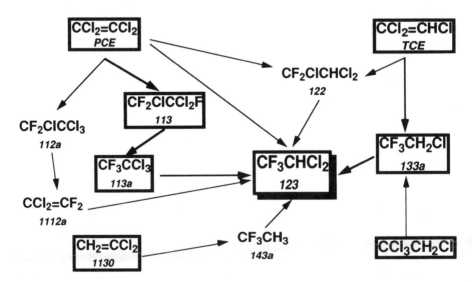

Figure 3. Potential routes to HCFC-123 (source: DuPont).

A second approach (Figure 3) from PCE follows in part the synthesis of CFC-113 and its isomerization to CFC-113a as outlined in Section 7.2.1.3. This is then subjected to hydrogenolysis along the lines reported for CFC-114a (Section 7.2.1.3). Platinum on carbon,[39] and a variety of other metals such as Rh, Ir, Pd, or Re on substrates such as carbon and alumina,[40] are reported to give good selectivities to the desired product. Generally, milder temperatures are employed, since the dichloromethyl moiety is sensitive to further hydrogenolysis. Products of further C—Cl bond hydrogenolysis, namely HCFC-133a (CF_3CH_2Cl) and HFC-143a (CF_3CH_3), are observed to varying degrees. In addition to catalytic hydrogenolysis, stoichiometric reduction techniques are also possible. For example, the reduction of CFC-113a in protic solvents such as methanol in the presence of metal salts, at room temperature, is reported[41] to afford excellent selectivities to HCFC-123. Reduction with aqueous Na_2SO_3[42] or zinc in protic solvents[43] have also been described. These methods seem attractive for small-scale synthesis. Waste disposal would be a concern in large-scale applications unless suitable recycle schemes could be developed. Also of interest are photochemical reductions.[44]

7.2.2.2. From Trichlorethylene (TCE)

Approaches based on TCE involve the initial preparation of HCFC-133a (CF_3CH_2Cl), as discussed in Section 7.2.1.2, followed by monochlorination of this intermediate. Chlorination in both the presence and absence of metal salts has been reported[45]; a variant involves *in situ* chlorine generation by co-feeding O_2, HCl, and HCFC-133a over a nickel-on-alumina Deacon catalyst at high temperatures.[46] Photochemical chlorination of HCFC-133a with high selectivities to HCFC-123 and minimum overchlorination to CFC-113a has also been reported.[47] The advantage of low temperature or photochlorination is the production of only one of the isomers, namely HCFC-123, with little chance for isomerization. The trifluoromethyl group appears to be stable under these conditions.

7.2.2.3. Miscellaneous Methods

Equilibration of different substrates can be a useful method provided high selectivities to the desired HCFC-123 can be obtained. For example, HCFC-123 is formed when mixtures of CF_3CCl_3 and CF_3CH_2Cl (HCFC-133a) are passed over chromium oxide at elevated temperatures.[48] Also, if HCFC-123a ($CF_2ClCHFCl$) is available from a certain process it can be isomerized to HCFC-123, e.g., by contact with alumina or aluminum chloride under appropriate conditions (pretreatment of the catalyst with a halogen-containing organic such as CFC-113 seems necessary for satisfactory activity).[49]

7.2.3. 2-Chloro-1,1,1,2-tetrafluoroethane, CF_3CHFCl (HCFC-124)

Methods already described for the conversion of PCE to HCFC-123 (Section 7.2.2.1) or HFC-134a (Section 7.2.1.2) involve the formation of HCFC-124 as a byproduct. Optimization of such processes to maximize HCFC-124 production and the development of new catalytic systems have been the focus of attention. For example, a cobalt salt-on-alumina catalyst precursor, fluorided prior to use, has been reported to give very little HFC-125, producing predominantly HCFC-123 and HCFC-124 from PCE.[50] Similarly, a chrome-on-alumina system, which has to be activated with HF prior to use to convert the alumina to fluorided alumina, also effects this transformation.[51] Conventional unsupported chromium(III) oxide catalysts vary in activity depending on the method of preparation.[52] Many other systems, too numerous to mention here, have been reported.

The hydrogenolysis of CFC-114a to HFC-134a produces HCFC-124 in small quantities (Eq. 9)[23,24]; if HCFC-124 is a true intermediate, perhaps its yield can be increased through catalyst modifications. A stoichiometric chemical method involving reduction of CFC-114a with an alkali-metal amalgam,[53] as well as an electrolytic method[54] for a one-step high-yield preparation of HCFC-124, have been reported.

The reaction of $CF_2{=}CFCl$ (CTFE, from dechlorination of $CF_2ClCFCl_2$) with HF in the presence of a chrome oxyfluoride catalyst can, in principle, produce HCFC-124.[30] Again, overfluorination to HFC-125 may be of some concern if this product is not desired.

7.2.4. Pentafluoroethane, CF_3CHF_2 (HFC-125)

The synthetic procedures that have been developed for HCFC-123 (Figure 3), especially those starting from PCE (Section 7.2.2.1), can also be used for the synthesis of HFC-125. Obviously, initial HF addition across PCE's double bond, followed by exchange of all the chlorines, gives HFC-125 (Eq. 15); and clearly, if HCFC-123 and HCFC-124 are available, they can be used as the starting points. If this is not feasible, PCE can be treated with Cl_2 and HF to produce chloropentafluoroethane (CFC-115) followed by hydrogenolysis of the C—Cl bond (Eqs. 16 and 17).

(15) $$CCl_2{=}CCl_2 + 5HF \rightarrow CF_3CHF_2 + 4HCl$$

(16) $$CCl_2{=}CCl_2 + Cl_2 + 5HF \rightarrow CF_3CF_2Cl + 5HCl$$

(17) $$CF_3CF_2Cl + H_2 \rightarrow CF_3CHF_2 + HCl$$

As mentioned earlier, chromium(III) oxide catalysts appear to be much more effective for both HF addition and halogen exchange. Note that the method of preparation also has a significant effect on activity: catalyst prepared by steam treatment of chromium hydroxide before calcination is more active than one prepared without this pretreatment,[55] and significant conversion of PCE to HFC-125 has been claimed at operating temperatures above 300 °C. Similarly activated, anhydrous chromium oxide as well as chromium(III) oxide gel shows high selectivity to HFC-125.[56] Such catalysts are also useful for the chlorofluorination of PCE to CFC-115. Details of the thermal hydrogenolysis of CFC-115 (Eq. 17) over platinum metals, ferrous metals, or rhenium have been disclosed.[57]

The reaction of HF with tetrafluoroethylene or chlorotrifluoroethylene over chrome oxyfluoride at elevated temperatures is an option for the synthesis of HFC-125[30]; chromium(III) oxide gel is said to be quite effective for the conversion of CTFE to HCFC-125.[56] Of interest for small-scale work is the reaction of trifluoroacetaldehyde with SF_4 or a dialkylamino trifluoride,[58] reagents well known to effect the conversion $C{=}O \rightarrow CF_2$ (see Chapter 2).

7.2.5. 1,1-Dichloro-1-fluoroethane, $CFCl_2CH_3$ (HCFC-141b)

Two direct routes are available for the synthesis of HCFC-141b, one involving halogen exchange (Eq. 18) and the other HF addition across a $C{=}C$ bond (Eq. 19). The liquid-phase reaction of HF with methyl chloroform (CH_3CCl_3) in the presence of tantalum halides has been reported to afford HCFC-141b with good selectivity.[59] Overfluorination by continuing chlorine replacement over active catalysts is a competing reaction. For example, at 100 °C in the presence of MoO_3, CH_3CClF_2 (HCFC-142b) is the major product.[60] Similar liquid-phase reactions using tin salts or organotin compounds along with oxygen-containing compounds have also been claimed to produce acceptable yields of HCFC-141b.[61] A *noncatalytic* halogen-exchange process involving CH_3CCl_3 and HF at temperatures above 100 °C has been described; both HCFC-141b and HCFC-142b can be produced, the extent of overfluorination depending on the residence time.[62] In fact, in several such processes the operational conditions can be varied to produce HCFC-142b, as well as HFC-143a by successive halogen exchange.

(18) $$CH_3CCl_3 + HF \rightarrow CH_3CFCl_2 + HCl$$

(19) $$CH_2{=}CCl_2 + HF \rightarrow CH_3CFCl_2$$

Addition of HF to vinylidene chloride according to Eq. (19) affords HCFC-141b with no HCl byproduct. Such a process using an aluminum fluoride as catalyst and operating at temperatures of 50–120 °C in the vapor phase has been reported.[63]

7.2.6. 1,1-Difluoroethane, CHF_2CH_3 (HFC-152a)

7.2.6.1. From Acetylene

The most direct route to HFC-152a is the addition of HF to acetylene (Eq. 20). Both liquid- and vapor-phase processes are available. Many catalytic systems and process improvements have been reported.

(20) $$CH{\equiv}CH + 2HF \rightarrow CH_3CHF_2$$

The liquid-phase reaction of acetylene with HF under pressure in the presence of BF_3[64] or FSO_3H[65] at low temperatures produces HFC-152a with high selectivity; the use of FSO_3H–SbF_5 to promote HF addition at 25 °C also affords excellent product selectivity.[66] To overcome catalyst losses, when BF_3 is used, KBF_4 can be employed.[67]

With this catalyst, the reaction is usually carried out in the presence of FSO_3H or a perfluoroalkanesulfonic acid; high product selectivities are claimed.

Vapor-phase processes based on acetylene operate over a wide temperature range, namely, ambient to about 300 °C. High-temperature operation tends to produce vinyl fluoride by loss of HF from the desired product. For example, the reaction of acetylene with HF over AlF_3 produces HFC-152a as the major product below 275 °C whereas vinyl fluoride is the major product at 307 °C.[68] Highly active catalysts containing pseudoboehmite, H_3BO_3, and optionally Fe_2O_3 operating at a maximum temperature of 262 °C has been reported to afford excellent yields of HFC-152a.[69]

7.2.6.2. From Vinyl Chloride

The reaction of vinyl chloride with HF to give HFC-152a (Eq. 21) has been described by several investigators. For vapor-phase processes, fluorided alumina or aluminum fluoride containing other metal salts have been widely used. Chromium(III) oxide supported on aluminum fluoride[70] or alumina[71] has been shown to be an excellent catalyst, and vanadium trichloride supported on active carbon,[72] as well as the beneficial use of steam as a co-feed, has also been advocated.[73] A liquid-phase procedure in the presence of $SnCl_4$ as catalyst and distillation of the product in the presence of anhydrous HF has been reported to produce HFC-152a free from vinyl chloride.[74]

(21) $CH_2{=}CHCl + 2HF \rightarrow CH_3CHF_2 + HCl$

7.2.6.3. Other Methods

Both dichloroethane isomers have been converted to HFC-152a with HF in liquid- and vapor-phase processes. Liquid-phase fluorination of 1,1-dichloroethane in the presence of either $SbCl_5$[75] or SbF_5[76] at −10 to 15 °C is reported to give good yields of HFC-152a. The vapor-phase reaction of 1,2-dichloroethane with HF over chromium(III) oxide also can afford good selectivities to HFC-152a, apparently via an elimination–addition–substitution sequence.[71]

7.2.7. Dichloropentafluoropropanes

Two specific positional isomers of $C_3HF_5Cl_2$ have received some attention as potential replacements for CFC-113 in solvent and cleaning applications, namely, $CF_3CF_2CHCl_2$ (HCFC-225ca) and CF_2ClCF_2CHFCl (HCFC-225cb). These two isomers can be produced in the ratio of *ca* 3:2 respectively via a Prins reaction of tetrafluoroethylene with $CHFCl_2$ (HCFC-21) at 15 °C in the presence of aluminum chloride.[77] The liquid-phase TaF_5-catalyzed halogen exchange between HF and C_3HCl_7 also is reported to give a mixture of products containing the above two isomers.[78]

7.3. COMMERCIAL ASPECTS

Production estimates for the CFC alternatives discussed here are not available. Many companies have announced plans for the commercialization of several of these alternatives, in particular HCFC-141b, HCFC-123, and HFC-134a. Developmental quantities have been available for sometime now, and major plants for the production of HFC-134a were commissioned in the U.S. (DuPont) and U.K. (ICI) in late 1991 (see Chapter 27). HFC-152a has been available commercially since 1964.

Toxicity evaluations and other environmental issues related to these alternatives have been international in scope. Programs for Alternative Fluorocarbon Toxicity Testing (PAFTT) on HFC-134a, HCFC-123, HCFC-141b, HCFC-124, HFC-123, and HCFCs-225ca and cb involve many companies from different countries. The results of initial toxicity studies performed with HCFC-141b, HCFC-124, and HFC-134a under this umbrella have been encouraging. For HCFC-123 an AEL (Acceptable Exposure Level) of 10 ppm has been suggested. Additional studies are in progress and are expected to be complete fairly soon. A similar cooperative program is the Alternative Fluorocarbon Environmental Acceptability Study (AFEAS). The aim of this study is to appraise the environmental fate of CFC alternatives, particularly their impact on the ozone layer, global warming, and acid rain.

7.4. PROPERTIES

Physical property data for some of the alternatives being developed are listed in Table 2. Studies on the thermodynamic properties of HFC-134a have been published,[79] and information on HCFC-123[80] and HFC-152a[81] is also available. Generally, these compounds are colorless liquids or gases whose boiling points decrease with increasing fluorine content, much like the CFCs. Commercial samples usually contain only

Table 2. Physical Properties of Some HFCs and HCFCs[a]

	HCFC-123	HCFC-124	HFC-125	HFC-134a	HCFC-141b	HFC-152a
Chemical formula	CF_3CHCl_2	CF_3CHFCl	CF_3CHF_2	CF_3CH_2F	$CFCl_2CH_3$	CHF_2CH_3
Molecular weight	152.9	136.5	120.0	102.0	116.95	66.0
Boiling point at 1 atm (°C)	27.9	−11.0	−48.5	−26.5	32.0	−24.7
Critical temperature (°C)	185.0	122.2	66.3	100.6	210.3	113.5
Critical pressure (atm)	37.4	35.27	34.72	40.03	45.80	44.4
Liquid density (g cm^{-3}) @25°C (77°F)	1.46	1.364	1.25[b]	1.202	1.23	0.911
Flammability limits in air (vol%)	None	None	None	None	7.3–16.0	3.9–16.9

[a]Source: DuPont.
[b]At 20°C.

traces of water (less than 10 ppm) and essentially no detectable acidity. They may form azeotropes with other organics such as alcohols, ketones, halo-olefins, CFCs, HCFCs, and HFCs.[82,83] As a class they are generally expected to be less stable than CFCs, a design feature to ensure shorter atmospheric lifetimes compared to CFCs (see Chapter 28). Because of the presence of hydrogen in the molecules, elimination reactions to produce olefins can also occur in more conventional chemical situations. Thorough toxicological testing on several of these substitutes is currently being carried out; initial results on some of them look promising (see Section 7.3).

7.5. APPLICATIONS

7.5.1. Refrigeration

Perhaps the largest area of application of these alternatives is in refrigeration and air conditioning. Home refrigerators and virtually all automobile air conditioners employ CFC-12 (CF_2Cl_2) as the refrigerant. The primary alternative is HFC-134a; HFC-125 has also been considered in applications where even lower temperatures and quick recovery are required. Although the vapor-pressure curves of CFC-12 and HFC-134a are very similar (Figure 2), there are other properties that need to be considered, such as oil solubility and permeation through hoses. Thus, while HFC-134a is an acceptable substitute, it is not a drop-in replacement for CFC-12 in automobile air conditioners. Special oils and desiccants are being developed for this purpose.

Blends of substitutes such as HCFC-22 (CHF_2Cl), HCFC-124, and HFC-152a or HCFC-22, CFC-114a, and HFC-152a are also being tested for such applications.[84] Blends such as these can be "tailored" to give the desired properties. Such an approach allows the use of compounds which, by themselves, may be marginally flammable.

CFC-11 ($CFCl_3$) has been the material of choice in industrial chillers. The alternative proposed is HCFC-123. This can serve as an interim replacement, since it contains chlorine, until other options are identified.

7.5.2. Foaming Agents

Rigid plastic foams have a cellular structure which is created by the chemical or physical action of a blowing (foaming) agent. These foams have excellent insulating properties. Polyurethane and polyisocyanurate foams are produced by using CFC-11 as a blowing agent which remains trapped in the foam. The low thermal conductivity of this blowing agent also contributes to the insulating properties of the final product. Substitutes proposed for foam blowing include HCFC-123, HCFC-141b, and HCFC-22.

Thermoplastic foams, such as those derived from polystyrene, are used in a variety of applications such as egg cartons, meat trays, disposable cups and trays. Unlike the polyurethane foams, these should contain very little, if any, of the blowing agent. CFC-12 has commonly been used as a blowing agent in this application. Possible alternates that have been suggested for this application include HCFC-124, HCFC-22, HCFC-142b, HFC-134a, and HFC-152a.

7.5.3. Solvents

The two work horses of the industry in this area have been CFC-113 ($CF_2ClCFCl_2$) and methyl chloroform (CH_3CCl_3). The largest single use has been in the area of cleaning and degreasing of metals. In addition, they are used in a variety of applications such as solder flux removal, dewatering of printed circuit boards, degreasing semiconductors, and paint and aerosol formulations, just to name a few. Finding a single alternative to meet all these needs is a monumental task and no single compound has yet been identified. The development of several compounds to suit these needs is most likely. HCFC-141b and HCFC-123, as well as their blends and azeotropes, have been proposed to satisfy some of the applications. Recently, HCFCs-225 ca and cb (Section 7.2.7) have been proposed as possible substitutes for CFC-113. Processes already reported produce these isomers as a mixture,[77] and toxicity studies are in progress. All the proposed substitutes for this application still contain chlorine and hence would serve only temporarily until satisfactory HFCs or other replacements can be found. The development of an HFC or a family of HFCs to satisfy the needs of the solvent industry is a demanding synthetic challenge.

7.6. REFERENCES

1. J. E. Lovelock, *Nature 230*, 379 (1971).
2. M. J. Molina and F. S. Rowland, *Nature 249*, 810 (1974).
3. A. Lacis, J. Hansen, P. Lee, T. Mitchell, and S. Lebedeff, *Geophys. Res. Lett. 8*, 1035 (1981).
4. R. T. Watson, M. J. Prather, and M. J. Kurylo, *NASA Ref. Publ.* 1208 (1988).
5. W. L. Kopko, *Int. J. Refrig. 13(2)*, 79 (1990).
6. M. O. McLinden and D. A. Didion, *ASHRAE J. 29*, 32 (1987).
7. A. F. Benning, U.S. Patent 2,230,925 (to Kinetic Chemicals) [*CA 35*, 3269 (1941)].
8. R. A. Firth and G. E. Foll, U.S. Patent (to ICI) 3,755,477 [*CA 74*, 63896 (1973)].
9. A. E. Feiring, U.S. Patent 4,258,225 (to DuPont) [*CA 95*, 6482 (1981)].
10. A. E. Feiring, *J. Fluorine Chem. 14*, 7 (1979).
11. K. H. Mitschke and H. Niederpruem, British Patent (to Bayer) 1,585,938 [*CA 90*, 54453 (1979)].
12. L. Marangoni, G. Rasia, C. Gervasutti, and L. Colombo, *Chim. Ind. (Milan) 64*, 135 (1982).
13. K. Maeda, M. Sano, and M. Kawagishi, Japanese Patent Application 48-072105 (Nippon Carbide Industries) [*CA 80*, 59424 (1974)].
14. S. L. Bell, U.S. Patent 4,129,603 (to ICI) [*CA 90*, 137242 (1979)].
15. M. R. Broadworth, R. A. Davis, and R. P. Ruh, U.S. Patent (to Dow Chemical) 2,885,427 [*CA 53*, 16962 (1959)].
16. Y. Ohsaka and S. Takasuki, German Patent 2,932,934 (to Daikin Kogyo Co.) [*CA 93*, 45951 (1980)].
17. L. E. Manzer, European Patent Application 328127 (DuPont) [*CA 112*, 79941 (1989)].
18. J. Knaak, U.S. Patent Application 3,978,145 (to DuPont) [*CA 85*, 142607 (1976)].
19. M. Vecchio, G. Groppelli, and J. C. Tatlow, *J. Fluorine Chem. 4*, 117 (1974).
20. W. T. Miller, *J. Am. Chem. Soc. 62*, 993 (1940).
21. F. Grozzo, N. Troiani, and P. Piccardi, U.S. Patent (to Montedison S. p. A.) 4,748,284 [*CA 99*, 139303 (1988)].
22. M. Hauptschein and A. H. Fainberg, U.S. Patent (to Atochem) 3,087,974 [*CA 59*, 9788 (1963)].

23. C. Gervasutti, L. Marangoni, and W. Marra, *J. Fluorine Chem.* *19*, 1 (1981/82).

24. J. I. Darragh, British Patent 1,578,933 (to ICI) [*CA 90*, 151557 (1980)].

25. S. Morikawa, S. Samejima, M. Yositake, and S. Tatematsu, European Patent Application 347830 (Asahi Glass Co.) [*CA 112*, 234785 (1989)].

26. S. Morikawa, M. Yoshitake, and S. Tatematsu, Japanese Patent Application 1-319443 (Asahi Glass Co.) [*CA 112*, 197613 (1989)].

27. Y. Furutaka, H. Aoyama, and Y. Homoto, Japanese Patent Application 1-093549 (Daikin Industries) [*CA 111*, 114734 (1989)].

28. C. S. Kellner and V. N. M. Rao, U.S. Patent 4,873,381 (to DuPont) [*CA 112*, 197608 (1990)].

29. L. Lerot, V. Wilmet, and J. Pirotton, European Patent Application 355907 (Solvay *et al.*) [*CA 113*, 58473 (1990)].

30. S. P. Von Halasz, German Patent 3,009,760 (to Hoechst) [*CA 95*, 186620 (1981)].

31. W. H. Gumprecht, U.S. Patent 4,311,863 (to DuPont) [*CA 96*, 180749 (1982)].

32. M. Yoshizumi and Y. Yamashita, Japanese Patent Application 1-228925 (Mitsubishi Metal Corp.) [*CA 112*, 54966 (1989)].

33. A. McCulloch, R. L. Powell, and B. R. Young, European Patent Application 300724 (ICI) [*CA 110*, 231101 (1989)].

34. W. H. Gumprecht, U.S. Patent 4,851,595 (to DuPont) [*CA 112*, 20668 (1989)].

35. Y. Furutaka and K. Sei, Japanese Patent Application 1-038034 (Daikin Industries) [*CA 111*, 133629 (1989)].

36. W. H. Gumprecht, W. G. Schindel, and V. M. Felix, PCT International Application 8,912,615 (DuPont) [*CA 112*, 234788 (1989)].

37. L. E. Manzer and V. N. M. Rao, U.S. Patent 4,766,260 (to DuPont) [*CA 110*, 97550 (1988)].

38. D. Carmello and G. Guglielmo, European Patent Application 282005 (Ausimont) [*CA 110*, 10080 (1989)].

39. Y. Furutaka, H. Aoyama, and H. Yomoto, Japanese Patent Application 1-149739 (Daikin Industries) [*CA 111*, 232067 (1989)].

40. S. Morikawa, M. Yoshitake, and S. Tatematsu, Japanese Patent Application 1-319440 (Asahi Glass Co.) [*CA 112*, 216213 (1989)].

41. S. Torii, H. Tanaka, S. Yamashita, M. Hotate, A. Suzuki, and Y. Okuma, Japanese Patent Application 2-001414 (Kanto Denka Kogyo Co., Japan) [*CA 112*, 197614 (1990)].

42. T. Kondo and S. Yoshikawa, German Patent 3,834,038 (to Central Glass Co., Japan) [*CA 111*, 194095 (1989)].

43. S. Takamatsu and S. Misaki, Japanese Patent Application 58-222036 (Daiichi Kogyo Seiyaku Japan) [*CA 100*, 156218 (1983)].

44. A. Posta and O. Paleta, Czechoslovakian Patent 136735 [*CA 75*, 19681 (1970)].

45. Y. Furutaka, Y. Homota, and T. Honda, Japanese Patent Application 1-290638 (Daikin Industries) [*CA 112*, 178061 (1989)].

46. Y. Furutaka, Y. Homoto, and T. Pponda, Japanese Patent Application 1-290639 (Daikin Industries) [*CA 112*, 178062 (1989)].

47. R. F. Sweeney and J. O. Peterson, U.S. Patent 4,060,469 (to Allied Chemical Corp.) [*CA 88*, 50257 (1978)].

48. R. F. Sweeney and B. Sukornick, U.S. Patent 4,145,368 (to Allied Chemical Corp.) [*CA 90*, 203469 (1979)].

49. H. Sonoyama and Y. Osaka, Japanese Patent Application 53-121710 (Daikin Kogyo Co.) [*CA 90*, 54456 (1978)].

50. L. E. Manzer and V. N. M. Rao, European Patent Application 349298 (DuPont) [*CA 112*, 234786 (1990)].

51. S. Hirayama, H. Kobayashi, H. Ono, and S. Tomada, PCT International Application 8,910,341 (Showa Denko K.K., Japan) [*CA 112*, 157662 (1989)].

52. W. H. Gumprecht, L. E. Manzer, and V. N. M. Rao, U.S. Patent (to DuPont) 4,843,181 [*CA 111*, 214116 (1989)].

53. D. G. H. Ballard, J. Farrar, and D. A. Laidler, European Patent Application 164954 (ICI) [*CA 104*, 185970 (1985)].
54. G. R. Davies and G. Q. Maling, U.S. Patent (to ICI) 4,938,849 [*CA 110*, 181642 (1990)].
55. R. A. Firth and G. E. Foll, U.S. Patent (to ICI) 3,755,477 [*CA 74*, 63896 (1973)].
56. F. W. Swamer, U.S. Patent (to DuPont) 3,258,500 [*CA 65*, 16861 (1966)].
57. S. Morikawa, M. Yoshitake, S. Tatematsu, S. Yoneda, and K. Ohira, Japanese Patent Application 1-258632 (Asahi Glass Co.) [*CA 112*, 118251 (1989)].
58. G. Siegemund, *Justus Liebigs Ann. Chem.* 1280 (1979).
59. V. N. M. Rao, European Patent Application 349190 (DuPont) [*CA 112*, 234787 (1990)].
60. K. Okazaki, K. Yagii, T. Manabe, and T. Komatsu, Japanese Patent Application 51-039606 (Central Glass Co., Japan) [*CA 85*, 94212 (1976)].
61. T. Ide, T. Komatsu, H. Akiyama, T. Kitamura, and S. Yamamoto, European Patent Application 187643 (Asahi Chemical Industry Co., Japan) [*CA 105*, 210771 (1986)].
62. R. Ukaji and I. Morioka, German Patent 2,137,806 (to Daikin Kogyo Co., Japan) [*CA 76*, 99093 (1972)].
63. W. H. Gumprecht, European Patent Application 353059 (DuPont) [*CA 113*, 5701 (1990)].
64. R. E. Burk, D. D. Coffman, and G. H. Kalb, U.S. Patent 2,425,991 (to DuPont) [*CA 42*, 198 (1948)].
65. J. D. Calfee and F. H. Bratton, U.S. Patent 2,462,359 (to Allied Corp.) [*CA 43*, 3834 (1949)].
66. L. Schmidhammer, German Patent 1,945,655 (to Wacher-Chemie) [*CA 74*, 111542 (1971)].
67. J. N. Meussdoerffer and H. Niederpruem, German Patent 2,139,993 (to Bayer) [*CA 78*, 123998 (1973)].
68. Y. Kanakami and W. Hircynki, French Patent 1,570,306 (to Daikin Kogyo Co., Japan) [*CA 72*, 132015 (1969)].
69. H. Pauksch, J. Massone, and H. Derleth, German Patent 2,105,748 (to Kali-Chemie) [*CA 77*, 164009 (1972)].
70. A. Akramkhodzhaev, T. S. Sirlibaev, A. A. Yul'chibaev, and Kh. U. Usmanov, *Uzb. Khim. Zh.* 17, 51 (1973) [*CA 79*, 137533 (1973)].
71. J. P. Henry, C. E. Rectenwald, and J. W. Clark, Canadian Patent 832502 (1970).
72. M. Gotfroid and G. Martens, Belgian Patent 766,395 (to Solvay *et al.*) [*CA 76*, 112656 (1971)].
73. T. R. Fiske and D. W. Baugh, Jr., U.S. Patent 4,147,733 (to Dow Chemical Co.) [*CA 91*, 91150 (1979)].
74. A. N. Golubev, A. L. Gol'dino, Yu. A. Panshin, and V. I. Kolomenskov, USSR Patent 341,788 [*CA 78*, 3663 (1972)].
75. M. Ozawa, F. Inoue, N. Koketsu, and K. Matsuoka, Japanese Patent Application 50-106904 (Central Glass Co., Japan) [*CA 83*, 205754 (1975)].
76. M. Ozawa, F. Inoue, N. Koketsu, and K. Matsuoka, Japanese Patent Application 50-106905 (Central Glass Co., Japan) [*CA 83*, 205901 (1975).
77. O. Paleta, A. Posta, and K. Tesarik, *Coll. Czech. Chem. Commun.* 36, 1867 (1971).
78. S. Morikawa, S. Samejima, M. Yashitake, S. Tatematsu, and T. Tanuma, Japanese Patent Application 2-017134 (Asahi Glass Co.) [*CA 113*, 5704 (1990)].
79. D. P. Wilson and R. S. Basu, *ASHRAE Trans.* 94, Part 2, 2095 (1988).
80. M. O. McLinden, J. S. Gallagher, L. A. Weber, G. Morrison, D. Ward, A. R. H. Goodwin, M. R. Moldover, J. W. Schmidt, H. B. Chae, T. J. Bruno, J. F. Ely, and M. L. Huber, *ASHRAE Trans.* 95, Part 2, 263 (1989).
81. A. Kamei, C. Piao, H. Sato, and K. Watanabe, *ASHRAE Trans.* 96, Part 1, 141, 1990.
82. M. Segami, T. Kamimura, and M. Fukushima, Japanese Patent Application 63-308084 (Asahi Glass Co.) [*CA 110*, 233768 (1989)].
83. E. A. E. Lund, R. S. Basu, and D. P. Wilson, U.S. Patent 4,816,175 (to Allied Signal Inc.) [*CA 111*, 32227 (1989)].
84. D. B. Bivens and H. A. Connon, U.S. Patent (to DuPont) 4,810,403 [*CA 110*, 137619 (1989)].

8

Perfluoroalkyl Bromides and Iodides

CLAUDE WAKSELMAN and ANDRÉ LANTZ

8.1. INTRODUCTION

Nowadays, perfluoroalkyl halides* of the bromo and iodo classes (R_FBr, R_FI) play an important part in organofluorine chemistry. The short-chain bromides are manufactured as fire-extinguishing agents.[1] They can be used also, as can their iodo analogs, for the introduction of fluorinated groups into organic molecules.[2,3] The compounds so formed attract interest in pharmaceutical and agrochemicals circles, owing to the lipophilic properties of aromatic and heterocyclic rings substituted by small fluorinated groups.[4,5] Long-chain iodides are important intermediates for the preparation of tensioactive agents, owing to the low free surface energy associated with such fluorocarbon groups.[6]

Initial studies on the reactivity of these perfluoroalkyl halides revealed a somewhat peculiar behavior by comparison with that of their hydrocarbon counterparts ("normal" alkyl halides),[7,8] and this prompted numerous fundamental investigations. Previously, the chemical transformations of perfluoroalkyl iodides,[9–11] and those of their bromo analogs,[2,3] have been reviewed separately. In fact their transformations are related, although iodides are understandably more reactive than the corresponding bromides. However, methods of preparation and main applications are generally different for each group.

---

*In organofluorine chemistry the general terms halogen, halogeno (or halo), and halide refer only to chlorine, bromine, and iodine, or their anions.

CLAUDE WAKSELMAN • CNRS-CERCOA, 94320 Thiais, France.   ANDRÉ LANTZ • Atochem, Centre de Recherche Rhône-Alpes, 69310 Pierre-Bénite, France.

Organofluorine Chemistry: Principles and Commercial Applications, edited by R. E. Banks *et al.* Plenum Press, New York, 1994.

8.2. PREPARATION

Bromofluoromethanes are made commercially via halogen-exchange procedures utilizing HF, followed by hydrogen replacement with molecular bromine[1] (see Section 8.4.1). Long-chain perfluoroalkyl bromides are usually prepared by the action of bromine on the corresponding iodide at high temperature.[12] A more recent method involves heating perfluoroalkanesulfonyl chlorides with gaseous hydrogen bromide at 125 °C in the presence of tetrabutylammonium bromide (Eq. 1).[13] 1,2-Dibromotetrafluoroethane and other vicinal perfluorodibromoalkanes are available by addition of bromine to fluorinated olefins.

(1) $R_FSO_2Cl + HBr \rightarrow R_FBr + SO_2 + HCl$

Trifluoromethyl iodide is made from a trifluoroacetate salt and iodine (see Chapter 1).[9] It can also be prepared from trifluoroacetyl chloride and potassium iodide,[14] from trifluoromethyl bromide and sulfur dioxide, a reductant, and iodine,[15] or from tri-fluoromethylzinc bromide and iodine chloride.[16] Perfluoroethyl iodide is made commercially by the action of iodine pentafluoride and iodine on tetrafluoroethylene; longer-chain iodides are then produced by telomerization of perfluoroethyl iodide with tetrafluoroethylene at elevated temperature (see Section 8.4.2).[6,9] This telomerization occurs at a lower temperature in the presence of copper.[17]

8.3. REACTIONS

Through their powerful inductive effect (see Chapter 3) the fluorine atoms in a perfluoroalkyl bromide and iodide induce the C—Br or the C—I bond to be polarized in the opposite direction to that found in alkyl halides,[7,18] e.g., $F_3C^{\delta-}$—$X^{\delta+}$ (X = Br, I). Also, steric effects and lone-pair repulsive forces associated with fluorine substituents shield the carbons from nucleophilic attack. Hence, perfluoroalkyl halides are notoriously resistant to displacement of chlorine, bromine, or iodine as halide ions under S_N2 or S_N1 conditions.[19] Owing to their reduction potential values,[20] they can act as electron transfer oxidants and become converted to radical {R_FX (X = Br, I) + $e^- \rightarrow R_F X^{\overline{\cdot}} \rightarrow R_F^{\bullet} + X^-$} or anionic intermediates (see later).[2] Perfluoroalkyl radicals can also be formed classically by homolytic cleavage of the C—I or (stronger[8]) C—Br bond [$D°(C_2F_5I)$ 53, $D°(C_2F_5Br)$ 68 kcal mol^{-1}].

8.3.1. "Classical" Radical Reactions

8.3.1.1. Additions to Double and Triple C—C Bonds

Radical-initiated addition of perfluoroalkyl halides across multiple C—C linkages in olefins or acetylenes (e.g., Eqs. 2 and 3) is important for synthetic purposes. Heat, UV light or a radical source (peroxides, azo initiators) are classically used to initiate the reaction.[21,22] Additions can be performed electrochemically,[23] or catalyzed with copper(I) chloride and ethanolamine (Eq. 4).[24] In fact numerous metals and their derivatives have been shown to initiate the addition of perfluoroalkyl iodides across double and triple bonds: Ru, Pt, $Ni(CO)_2(PPh_3)_2$,[25] $Ru_3(CO)_{12}$, $Fe(CO)_5$,[26] $Pd(PPh_3)_4$,[27] $SnCl_2$–AgOAc,[28] and SmI_2.[29] Triethylborane[30] and metalloenzyme-assisted[31] procedures for the additions to alkynes have been described.

$$\text{(2)} \qquad R_FI + CH_2{=}CH_2 \rightarrow R_FCH_2CH_2I$$

$$\text{(3)} \qquad CF_3I + CH{\equiv}CH \xrightarrow{\quad} CF_3CH{=}CHI \xrightarrow{KOH} CF_3C{\equiv}CH$$

$$\text{(4)} \qquad BrCF_2CFClBr + CH_2{=}CHR \rightarrow BrCF_2CFClCH_2CHBrR$$

Initiation by metals and their derivatives is generally ascribed to electron-transfer processes, as in the case of sulfinates salts,[32] and sodium dithionite.[33,34] Functional alkenes and alkynes have been used, for example, fluorinated olefins,[10,18] unsaturated esters,[35] unsaturated ketones,[36] enol ethers,[37–40] allyl acetate,[41] fluorinated alkynes,[42] or alkynylsilanes.[43] When the palladium-catalyzed reaction of perfluoroalkyl iodides with terminal alkynes or alkenes is performed in alcohols under pressurized carbon monoxide, perfluoroalkyl-substituted alkenoates (Eq. 5) or alkanoates are obtained.[44]

$$\text{(5)} \qquad R_FI + HC{\equiv}CR + CO + R'OH \rightarrow R_FCH{=}C(R)CO_2R'$$

8.3.1.2. Additions to Divalent Carbon

Additions of perfluoroalkyl iodides to divalent carbon are rather rare. It has been performed with isocyanides under thermal or copper-catalyzed conditions (Eq. 6).[45]

$$\text{(6)} \qquad R_FI + RNC \rightarrow R_FC(I){=}NR$$

8.3.1.3. Reactions with Aromatic Compounds

Direct introduction of perfluoroalkyl groups into aromatic nuclei has been accomplished through thermolysis of perfluoroalkyl iodides in the presence of benzene, halogenobenzenes, pyridine, N-alkylpyrroles, furan, and thiophene.[46–49] Isomeric mixtures are usually obtained, except with N-alkylpyrroles[48] and furan,[49] which are selectively perfluoroalkylated at the 2-position. Milder perfluoroalkylations have been performed under UV irradiation with halogenobenzenes,[50] pyridine or pyrroles,[51] and imidazoles.[52] Conversions have been performed with long-chain perfluoroalkyl iodides in the presence of copper[53] or tetrakis(triphenylphosphine)nickel.[54] Ortho and para (perfluoroalkyl)anilines have been obtained using perfluoroalkyl halides in the presence of zinc–sulfur dioxide[55] or sodium dithionite.[56] Use of a moderate pressure[55] or UV irradiation[57] enables ring trifluoromethylation of anilines, phenols, and aminophenols to be achieved with trifluoromethyl bromide (cf. p. 231) (Eqs. 7 and 8).[55]

$$\text{(7)} \qquad CF_3Br + C_6H_5NH_2 \rightarrow 2\text{- and } 4\text{-}CF_3C_6H_4NH_2$$

$$\text{(8)} \qquad CF_3Br + HOC_6H_4NH_2\text{-}2 \rightarrow 3\text{- and } 5\text{-}CF_3C_6H_3OH\text{-}1,NH_2\text{-}2$$

8.3.1.4. Reaction with Disulfides

The reaction of perfluoroalkyl iodides with dialkyl disulfides can be triggered with UV irradiation (e.g., Eq. 9).[58] Aryl trifluoromethyl sulfides have been obtained using trifluoromethyl bromide in the presence of a sulphur dioxide radical anion precursor as the source of trifluoromethyl radical.[59]

$$(9) \qquad R_FI + R_2S_2 \rightarrow R_FSR$$

8.3.1.5. Other Transformations

A host of other reactions of trifluoromethyl iodide occurring under thermal or photochemical conditions are viewed as radical in nature. Interesting ones to note are the formation of trifluoronitrosomethane from nitric oxide, of trifluoromethylated phosphines from phosphorus, bistrifluoromethyl disulfide from sulfur, and tri-fluoromethylmercuric iodide from mercury.[10,18]

8.3.2. Reactions with Nucleophiles

Pioneering studies[7,10,18] showed that a halogenophilic mechanism occurs in the reaction of strong nucleophiles with perfluoroalkyl halides (i.e., S_N2 attack on the partially positively-charged halogen[2,60]): $Nu^- + R_FX \rightarrow NuX + R_F^-$; fluorinated carbanions therefore formed abstracted proton from the media used, consequently no substitution at carbon was observed. In aprotic solvents, and at low temperature, perfluoroalkylmagnesium and lithium reagents can be formed by conventional halogen–metal exchange, as with normal alkyl halides.[61–65] These perfluorocarbanion sources are not very stable and decompose easily into carbenes or olefins. When an ionic chain mechanism can occur (Scheme 1), mixtures of two fluorinated products are often obtained. One of them seems to be the result of a formal substitution at carbon, but with the unexpected conservation of the larger halogen.[2]

More recently,[2,66] a single electron transfer (SET) process has been proposed for reactions involving soft nucleophiles because the observed conversions are inhibited by radical scavengers. In this mechanism, both charged and neutral nucleophile behave

---

$$Nu^- + XCF_2Y \rightarrow NuX + {}^-CF_2Y$$

$$^-CF_2Y \rightarrow :CF_2 + Y^-$$

$$Nu^- + :CF_2 \rightarrow NuCF_2^-$$

$$NuCF_2^- + XCF_2Y \rightarrow NuCF_2X + {}^-CF_2Y$$

$$NuCF_2^- + \text{"H" (solvent)} \rightarrow NuCF_2H$$

Scheme 1. Difluorocarbene-mediated ionic chain mechanism (X = Y = Cl or Br; X = Br, Y = F or Cl; X = I, Y = F).

$$Nu^- + XCF_2Y \rightarrow Nu^\bullet + [XCF_2Y]^{\bar{\bullet}}$$

$$[XCF_2Y]^{\bar{\bullet}} \rightarrow X^- + {}^\bullet CF_2Y$$

$$Nu^- + {}^\bullet CF_2Y \rightarrow [NuCF_2Y]^{\bar{\bullet}}$$

$$[NuCF_2Y]^{\bar{\bullet}} + XCF_2Y \rightarrow NuCF_2Y + [XCF_2Y]^{\bar{\bullet}}$$

Scheme 2. Radical-anion chain mechanism ($S_{RN}1$) for a charged nucleophile (X = Y = Cl or Br; X = Br, Y = F or Cl; X = I, Y = F).

as electron donors and initiate an $S_{RN}1$ sequence which leads only to the expected product of formal fluoroalkylation (Scheme 2).

Depending on the nature of the nucleophile and that of the perfluoroalkyl halide, therefore, condensations can involve ionic or radical intermediates.

8.3.2.1. Carbon Nucleophiles

A rare example of fluoroalkylation of organolithium reagents is the reaction of CF_2ClBr with lithium acetylides[67–69]: R—C≡C—Li + $BrCF_2Cl$ → R—C≡C—CF_2Br. Formation of a bromodifluoromethylated product conforms with the operation of an ionic chain mechanism (Scheme 1). Substituted sodiomalonates react analogously[67,70]: $RCNa(CO_2Et)_2 + CF_2Br_2 \rightarrow RC(CF_2Br)(CO_2Et)_2$.

Enamines behave like neutral nucleophiles, and their spontaneous condensation with primary perfluoroalkyl iodides follows the SET process giving, after an acidic hydrolysis, 2-perfluoroalkylketones.[48] Formation of 2-(chlorodifluoromethyl) ketones from CF_2ClBr is also in agreement with an $S_{RN}1$ mechanism.[71] During the UV irradiation of perfluoroalkyl iodides with acetylacetone in liquid ammonia, alkylation of the methylene group occurs initially.[72] A similar trend has been noted in the perfluoroalkylation of carbanions derived from malonate and acetoacetate esters.[73] These reactions too have been interpreted as radical-ion chain processes ($S_{RN}1$; Scheme 2). Evidence has been presented for the occurrence of this mechanism in the perfluoroalkylation of the 2-nitropropyl anion[32]: $R_FI + Me_2C{=}NO_2^- \rightarrow R_FC(Me)_2NO_2$.

8.3.2.2. Phosphorus Nucleophiles

The bromodifluoromethylphosphonium salt $Ph_3PCF_2Br^+Br^-$ can be prepared by treating CF_2Br_2 with triphenylphosphine.[74] Dehalogenation of this salt with the same phosphine or with a metal generates the Wittig reagent $Ph_3P{=}CF_2$, which can be used to prepare difluoroolefins.[75,76]

In situ capture by trimethylsilyl chloride of the trifluoromethyl anion formed in the reaction between CF_3Br and an aminophosphine gives a trifluoromethylsilane[77]: $(Et_2N)_3P + CF_3Br \rightarrow (Et_2N)_3PBr^+CF_3^- \rightarrow Me_3SiCF_3$. Desilylation of this fluorinated silane with fluoride ion regenerates trifluoromethyl anion, which can be trapped with carbonyl compounds.[78,79] Similar preparations of bromo- and chloro-difluoromethyl-silanes from CF_2Br_2 and CF_2ClBr have been reported recently.[80]

The Michaelis–Becker reaction between diethylsodiophosphite and CF_3Br gives tetraethyldifluoromethylene-bis-phosphonate.[81]

8.3.2.3. Sulfur Nucleophiles

Several groups have reported the perfluoroalkylation of thiols: $RS^- + R_FX \rightarrow R_FSR$ (X = I, Br). Reactions involving perfluoroalkyl iodides were initially performed in DMSO with methanethiolate,[82] or in liquid ammonia under UV irradiation with various thiols.[83] The condensation can be carried out under phase-transfer catalyzed conditions, photochemically[84] or without irradiation.[85] Even the poorly reactive CF_3Br has been transformed to thioethers in DMF under moderate pressure,[86,87] or under UV irradiation in liquid ammonia.[88] Perfluoroalkyl radicals are intermediates in these condensations, as demonstrated by inhibition with electron scavengers and trapping experiments.[85–87] The phase-transfer catalyzed conditions were previously used for the condensation of thiophenoxides with CF_2Br_2 and CF_2ClBr[89,90]: $ArS^- + BrCF_2Y \rightarrow ArSCF_2Br + ArSCF_2H$ (Y = Br, Cl). The formation of bromo derivatives from CF_2ClBr and the presence of byproducts containing C—H bonds support the operation of the ionic mechanism (Scheme 1) under these conditions. Another ionic process, involving tetrafluoroethylene as an intermediate, occurs in the condensation of 1,2-dibromotetrafluoroethane with thiophenoxides[91]: $ArS^- + BrCF_2CF_2Br \rightarrow ArSCF_2CF_2Br + ArSCF_2CF_2H$.

The UV-induced perfluoroalkylation of aromatic sulfinic acids occurs in liquid ammonia[92]: $ArSO_2H + R_FI \rightarrow R_FSO_2Ar$.

Dehalosulfination of long-chain perfluoroalkyl halides with sodium dithionite in the presence of sodium bicarbonate in aqueous acetonitrile has been described[33]: $^-O_2SSO_2^- + R_FX \rightarrow R_FSO_2^- + SO_2 + X^-$ (X = I, Br). No condensation was observed with CF_3Br under these conditions, but the reaction did occur under moderate pressure in aqueous DMF in the presence of disodium hydrogenophosphate,[93] initiation being ascribed to an SET process involving sulfur dioxide radical-anion. The triflinate salt, which was transformed into triflic acid, was also obtained with sodium hy-droxymethanesulfinate in anhydrous DMF ($HOCH_2SO_2^- + CF_3Br \rightarrow CF_3SO_2^-$).[93] This reaction was applied to various R_FI species[94]; however, others reported that this condensation in aqueous DMF gives sodium perfluorocarboxylates ($HOCH_2SO_2^- + R_FCF_2I \rightarrow R_FCO_2^-$)[95]. Transformation of the sulfinate to the carboxylate can result via photochemical electron transfer from the sulfinate to oxygen.[96]

8.3.2.4. Selenium Nucleophiles

Relatively little attention has been paid to selenium-based nucleophiles. Mono-substitution occurs with CF_2Br_2,[97] and disubstitution with $BrCF_2CF_2Br$[98]: $PhSe^- + BrCF_2Br \rightarrow PhSeCF_2Br$; $MeSe^- + BrCF_2CF_2Br \rightarrow MeSeCF_2CF_2SeMe$.

8.3.2.5. Oxygen Nucleophiles

Reactions of phenoxides with bromofluoroalkanes do not always proceed spontaneously; if not, good nucleophiles, like thiolates, can be used for their initiation[2,60,89,99]: $ArO^- + BrCF_2Br \rightarrow ArOCF_2Br + ArOCF_2H$; $ArO^- + BrCF_2CF_2Br \rightarrow ArOCF_2CF_2Br + ArOCF_2CF_2H$.

8.3.2.6. Nitrogen Nucleophiles

The reaction of sodium azide with vicinal dihalogenopolyfluoroalkanes has been studied.[100] Product structures are in agreement with an ionic mechanism: $N_3^- + BrCF_2CFClBr \rightarrow N_3CF_2CFClBr + N_3CF_2CFClH$

On the other hand, an SET mechanism has been proposed to account for alkylation at the 4-position of sodium imidazolide by long-chain perfluoroalkyl halides.[101]

8.3.3. Reactions with Metals

Perfluoroalkylorganometallics are often prepared by ligand exchange[61,102,103] (see Section 8.3.2). Direct attack of perfluoroalkyl halides on metals can also occur, mainly under Barbier conditions. Recent developments have focused on derivatives of zinc, cadmium, and copper.[104]

8.3.3.1. Zinc and Cadmium

Perfluoroalkylzinc iodides, prepared from R_FI and zinc in ether solvents, were reported long ago to be very poorly reactive.[105,106] For example, they were found not to react with carbon dioxide.[106] However, treatment of long chain perfluoroalkyl iodides with carbon dioxide, or sulfur dioxide in DMSO in the presence of a zinc–copper couple, led to their corresponding carboxylic or sulfonic derivatives.[107–109] These reactions, which can be performed with zinc alone in DMF,[110] were thought to occur at the metallic surface, where the organometallic is adsorbed and consequently activated.[108,109] More recently, it has been shown that trifluoromethyl bromide can be transformed readily into its corresponding sulfinate, provided that the reaction is performed under moderate pressure.[111] This peculiar reaction is believed to occur mainly in solution by an SET process involving $SO_2^{\bullet -}$,[55,111] like the dithionite-mediated reaction[93] (see Section 8.3.2.3).

Barbier conditions have been used for condensations with other electrophiles. Aldehydes and ketones can be transformed into trifluoromethylcarbinols with trifluoromethyl bromide and zinc in DMF in the presence of $(Ph_3P)NiCl_2$, or Cp_2TiCl_2 under ultrasonic irradiation.[112] Similar transformations can be achieved in pyridine under slight pressure,[113] and the same conditions have been used for condensations involving activated esters, for example, the transformation of ethyl oxalate to ethyl trifluoropyruvate[114]: $CF_3Br + Zn + EtO_2CCO_2Et \rightarrow CF_3COCO_2Et$. Use of alkyl thiocyanates as substrates gives fluorinated sulfides,[115] attack on the sulfur atom occurring as in the case of perfluoroalkylmagnesium halides.[116]

Studies on the reactivity of perfluoroalkyl iodides with zinc–copper couples have been extended to other aprotic solvents: alkyl carbonates and alkyl phosphates,[117] e.g., $R_FI + Zn—Cu + PO(OR)_3 \rightarrow (R_F)_2PO(OR)$. When reactions in DMF are performed in the presence of a radical initiator, perfluoroalkylamides and aldehydes are obtained.[118] In an acid solvent, like propionic acid, perfluoroalkylation of olefins occurs,[119] e.g., $R_FI + Zn/H^+ + CH_2{=}CHCN \rightarrow R_FCH_2CH_2CN + R_FCH_2CH(CN)CH(CN)CH_2R_F$.

A remarkably simple preparation of trifluoromethyl-zinc and -cadmium reagents directly from difluorodihalogenomethanes is available (Eq. 10).[120] Transformation of the difluoromethyl unit into a trifluoromethyl group is ascribed to the intermediacy of difluorocarbene, which in part is trapped by the solvent (DMF): CF_2X_2 (X = Br, Cl) $+ M$ (Cd, Zn) $\rightarrow :CF_2 \rightarrow$ (with DMF) $CO + Me_2N^+ = CHF\ F^-$; $F^- + :CF_2 \rightarrow CF_3^-$. Cadmium reagents can readily be prepared via the direct reaction of perfluoroalkyl iodides with cadmium powder in DMF at room temperature.[121]

(10) $2\ CF_2X_2 + 2\ Cd + HCONMe_2 \rightarrow CF_3CdX + CdX_2 + Me_2N^+{=}CHFX^-$

$\qquad\qquad$ (X = Br,Cl)

8.3.3.2.   Reactions with Copper

Long-chain perfluoroalkylcopper compounds can be generated by metallation of perfluoroalkyl iodides in DMSO[122] $(R_FI + 2\ Cu \rightarrow R_FCu + CuI)$ and used *in situ* for the perfluoroalkylation of halogenoaromatic compounds[122–125]: $R_FI + 2\ Cu + ArX \rightarrow R_FAr + CuI$. Coupling occurs also with vinyl halides,[126,127] and with iodoalkynes.[128] A solution of a trifluoromethyl copper complex prepared in HMPA has been used for the trifluoromethylation of vinyl, allyl, and benzyl halides.[129] The coupling reaction can also be applied to halogenoheterocyclic compounds (pyrimidines,[130] benzofuran,[131] thiophenes,[132] pyridines, and furans[133]); formation of positional isomers is sometimes observed and a carbenoid intermediate may be involved with 3-halogenofurans[131] and thiophenes.[132] A striking example of cine-substitution occurs with certain bromoalkenes[134]: $R_FI + Cu + R_F'\ C(Br){=}CH_2 \rightarrow R_F'\ CH{=}CHR_F$. Other unexpected products can be observed in trifluoromethylcopper condensations because difluorocarbene formed by decomposition can react further to give homologated products[131,135]: $CF_3Cu \rightarrow :CF_2 \rightarrow CF_3CF_2Cu \rightarrow$ etc.

Another route to trifluoromethylcopper involves metal exchange between trifluoromethyl-cadmium, or -zinc reagents prepared from CF_2X_2 with copper salts at low temperature in the presence of HMPA,[135] e.g., $CF_3CdX + CuI \rightarrow CF_3Cu$; the solutions can be used to convert iodobenzenes to the corresponding trifluoromethylbenzenes. Trifluoromethylcopper prepared directly from CF_2Br_2 and copper in dimethylacetamide at 100 °C can be condensed with activated chlorobenzenes.[136]

8.3.4.   Reactions with Acidic Reagents

Treatment of *n*-perfluoroalkyl iodides with oleum gives carboxylic acids (see later).[137] Treatment of perfluorinated α,ω-dibromoalkanes with chlorine fluorosulfate in fluorosulfonic acid medium containing antimony pentafluoride leads to substitution of the bromine atoms by fluorosulfate groups; methanol in the presence of potassium fluoride converts these products to esters of perfluorodicarboxylic acids: $Br(CF_2)_{n+2}Br + 2\ ClOSO_2F \rightarrow FO_2SO(CF_2)_{n+2}OSO_2F \rightarrow MeOCO(CF_2)_nCO_2Me$ ($n = 1, 2$).[138]

Long-chain perfluoroalkyl iodides can be oxidized with hydrogen peroxide in trifluoroacetic anhydride to iodoso bistrifluoroacetates,[139] which can be transformed into arylperfluoroalkyliodonium salts with trifluoroacetic acid or triflic acid in the presence of benzene or toluene[139–141]: $R_FI \rightarrow R_FI(OCOCF_3)_2 \rightarrow R_FI(X)Ar$ ($X = Cl$, CF_3SO_3). These iodonium triflates have been used for the electrophilic perfluoroalkylation of aromatic compounds, alkenes, carbanions, and thiols. Unfortunately, the trifluoromethyliodonium salt could not be prepared, because the oxidation of trifluoromethyl iodide did not produce bis(trifluoroacetoxy)iodotrifluoromethane but tris(trifluoroacetoxy)iodine.[142]

8.3.5. Electrochemical Transformations

Electrochemical reduction of perfluoroalkyl iodides at a mercury electrode in DMF in the presence of sulfur dioxide or carbon dioxide produces perfluoroalkanesulfonic and carboxylic acids.[143] With SO_2 and a carbon fiber cathode, sulfonic acids are formed when the DMF is wet and carboxylic acids when it is dry.[144] Electroreduction of trifluoromethyl bromide in DMF containing carbonyl compounds affords the corresponding trifluoromethylated alcohols.[145–147] Use of CO_2 and SO_2 as traps gives access to trifluoroacetic acid[145] and trifluoromethanesulfinic acid derivatives,[148,149] respectively. In the latter case, interpretation of the condensation was based on an inner-sphere SET from $SO_2^{\cdot}$ to CF_3Br.[149]

Primary perfluoroalkyl bromides are converted to their corresponding fluorosulfates by electrooxidation in fluorosulfuric acid.[150] Perfluorinated sulfonic esters are obtained by anodic oxidation of perfluoroiodoalkanes in perfluoroalkanesulfonic acids[151]: $R_FI + R_F'SO_3H \rightarrow R_F'SO_3R_F$.

8.4. APPLICATIONS

The main industrial products are Halons (fire extinguishing agents) and long-chain perfluoroalkyl iodides used as intermediates for the manufacture of fluorinated tensioactives and surface-treatment agents (textiles, leather, papers, and so on). Applications of long-chain perfluoroalkyl bromides as radiopaques[12] or as blood substitutes[152] are under study; perfluoro-octyl bromide (PFOB) may thus soon have a commercial use.

8.4.1. Halons

Halons are volatile bromine-containing halogenofluorocarbons (Table 1) related to CFCs and manufactured from chloroform (Scheme 3). The coding is different from

Table 1. Properties of Halons

FE code number	1211	1301	2402
Chemical formula	CF$_2$ClBr	CF$_3$Br	CF$_2$BrCF$_2$Br
Molecular weight	165.4	148.9	259.9
Boiling point at 1013 mbar[a] (°C)	–3.9	–57.8	47.3
Melting point (°C)	–161	–168	–110.5
Density[b] (kg dm^{-3}) (at °C)	1.83 (20 °C)	1.539 (25 °C)	2.163 (25 °C)
Critical temperature (°C)	154	67	214.5
Ozone depletion potential	3	10	6
Toxicity[c] (% in air)	32	80	12.6

[a]1 atmosphere.
[b]Of liquid.
[c]"Approximate lethal concentration" for 15 min exposure.

that of CFCs, however; thus the first digit of the Halon numbering system represents the number of carbon atoms in the molecule, the second the number of fluorine atoms, the third the number of chlorine atoms, and the fourth the number of bromine atoms. Total worldwide Halon production during the Montreal Protocol base year (1986; see Chapter 28) was estimated by UNEP (United Nations Environmental Program) to be 25000 tonnes, made up as follows: Halon 2402, 1000 tonnes; Halon 1211, 14,000 tonnes; Halon 1301, 10,000 tonnes (approximately 10% of which was used as a low-temperature refrigerant). Halon 2402, which is more toxic than the other two, has been used mainly in eastern Europe and in Italy. The main producers of Halons are: ICI (1211), Atochem (1211 and 1301), and Kali-Chemie (1211 and 1301) in Europe; ICI (1211), Du Pont (1301), and Great Lakes (1211 and 1301) in the U.S.; and Daikin, Asahi Glass, and Nippon Halon in Japan.

Halons came on the fire-fighting scene in the 1960s, their development arising from a search for a new fire-extinguishing agents to replace the more toxic extinguishants methyl bromide and carbon tetrachloride. They exhibit exceptional fire-fighting effectiveness, are electrically nonconductive, dissipate quickly, leave no residue, and are remarkably safe for human exposure. The two most suitable compounds from the viewpoints of efficiency, low toxicity, and physical properties are 1301 and 1211 (Table 1). In total-flooding applications, the extinguishing agent is discharged into a closed room to provide the concentration needed to suppress any

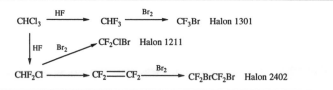

Scheme 3. Commercial routes to Halons.

fire. This type of release, which is usually performed with fixed, often automated systems, is particularly effective if the extinguishing agent used is sufficiently volatile to vaporize rapidly on discharge, thus mixing rapidly with air to create an extinguishing concentration. Halon 1301 is very suitable for this purpose. Its concentration in air necessary to extinguish most fires is 5% by volume or less, a level at which human exposure for up to ten minutes is generally acceptable. Hence, there is no absolute need to evacuate personnel before releasing the fire-fighting agent. Its great efficiency, low toxicity, and lack of physical effects on the most delicate of equipment explain the rapid growth of Halon 1301 for the protection of computer installations, telephone exchanges, art galleries, libraries, shipboard machinery spaces, pipeline pumping stations, oil rigs, aircraft engines, offshore platforms, and so on. Owing to their higher boiling points, Halons 1211 and 2402 cannot be used in total flooding because, on discharge, they produce a cloud of vapor and droplets of liquid, and it is difficult to obtain the required homogeneous concentration of extinguishing agent throughout the room or space to be protected. Thus, they are used in both indoor and outdoor "streaming" applications, steams of liquid Halon being discharged onto a fire, generally from portable or other types of mobile extinguishers.

Halons can have two modes of action against fire: physical (dilution, cooling, etc.) and chemical.[153] It is believed that the action is mainly chemical, i.e., inhibition of the complex chain reactions which occur between fuel and oxygen. The species which support the chain process are free radicals in type, notably $H^\bullet$ and $HO^\bullet$; when a Halon is released into the flames (in which the temperature is above 500 °C) it suffers decomposition, principally via homolysis of the C—Br bond ($CF_3Br \rightarrow CF_3^\bullet + Br^\bullet$) and the bromine atoms scavenge radicals resulting from the combustion: $Br^\bullet + H^\bullet \rightarrow HBr$; $HBr + HO^\bullet \rightarrow Br^\bullet + H_2O$. In these reactions, the two propagating species $H^\bullet$ and $HO^\bullet$ are destroyed and bromine atoms regenerated to participate in the removal of other propagating radicals. Thus, the chain reaction is broken causing extinction of the fire in a very short time. The recycling of the bromine atoms ($Br^\bullet$) explains why relatively small quantities of Halon are required for extinguishing a fire.

Unfortunately, Halons are implicated in the depletion of stratospheric ozone and global warming. The participants of the 1992 Montreal Protocol Meeting in Copenhagen agreed to phase out Halon production by the year 1994, except for some essential fire-fighting uses. Research on alternative agents has been initiated in order to find new products with low or zero ODP (Ozone Depletion Potential) and GWP (Global Warming Potential). As in the case of CFCs (see Chapter 7), the most obvious way to lower ODP and GWP values of a Halon is to introduce a hydrogen atom into the molecule, which can be abstracted by atmospheric hydroxyl radicals ($HO^\bullet$), consequently lowering its atmospheric lifetime. Great Lakes Chemical announced in 1990 the development of CHF_2Br (Halon 1201, bp –15 °C), which has about the same fire-fighting performance as Halon 1301 or 1211. Nevertheless, this compound is slightly more toxic than Halon 1211 (LC_{50}: 10.8% in air) and still has an ODP not equal to zero (0.5); it is difficult to predict the future of such a compound. Bromine-free analogs of Halons do possess some fire-fighting ability, and Du Pont has announced the development of HFC 125 (CF_3CHF_2) and HCFC 123 (CF_3CHCl_2) as substitutes for Halon 1301 and 1211. The HFC has a zero ODP, and the HCFC a very low value (0.02), but both are significantly less efficient as fire extinguishants than Halons [extinguishing concentration (heptane)/vol% in air: Halon 1301, 3.5; Halon 1211, 3.8; HCFC-123, 7.1; HFC-125, 10.1; Halon 1201, 4.0].

8.4.2. Perfluoroalkyl Iodides

Manufacturing processes for these iodides involve two steps:

$$5\ C_2F_4 + IF_5\ (\text{from } I_2 + F_2) + 2\ I_2 \overset{\text{catalyst}}{\longrightarrow} 5\ C_2F_5I$$

$$C_2F_5I + nC_2F_4 \rightarrow C_2F_5(C_2F_4)_nI$$

The properties of the starting materials [highly aggressive (I_2, IF_5), potentially explosive (C_2F_4), and expensive] explain why there are only a few producers of R_FI compounds and why prices are exceptionally high. It is believed that the first step is carried out in the liquid phase under pressure, while both liquid- (radical initiation) and vapor-phase (thermal initiation) reactions are used for the telomerization step. The latter delivers a mixture of telomers ($n = 1$ to 7), but for most applications the preferred chain lengths are C_6F_{13}, C_8F_{17}, and $C_{10}F_{21}$, ($n = 2$ to 4); to obtain a distribution corresponding to request, lower iodides (C_4F_9I and part of the $C_6F_{13}I$) are recycled, but the problem is then to minimize the formation of higher iodides ($n > 5$).

These straight-chain even carbon-numbered perfluoroalkyl iodides are colorless, odorless, and seemingly nontoxic liquids (Table 2). By exposure to daylight they turn (rapidly in sunlight) pink, then red, due to decomposition with the formation of iodine. The main producers are Du Pont, Atochem, Hoechst, and Asahi Glass [other Japanese

Table 2.   Properties of Some Perfluoro-n-alkyl Iodides and Their Derivatives

	R_F	Melting point (°C)	Boiling point (°C)	Density (at °C)
R_FI	C_2F_5	−92	11	2.16 (0)
	C_4F_9	−88	68	2.07 (15)
	C_6F_{13}	−45	116	2.06 (20)
	C_8F_{17}	18	158	2.04 (25)
$R_FC_2H_4I$	C_4F_9	−15	140	1.94 (20)
	C_6F_{13}	21	177	1.84 (32)
	C_8F_{17}	55	204	1.84 (56)
$R_FCH{=}CH_2$	C_4F_9	<−100	59	1.46 (20)
	C_6F_{13}	<−60	105	1.58 (20)
	C_8F_{17}	−48	152	1.68 (20)
$R_FC_2H_4OH$	C_6F_{13}	<−20	170	1.68 (25)
	C_8F_{17}	42	196	1.67 (50)
$R_FC_2H_4SH$	C_6F_{13}		162	1.62 (24)
	C_8F_{17}	21	194	1.64 (50)

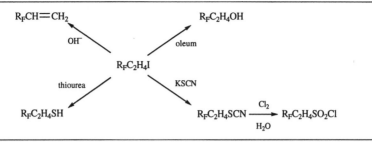

Scheme 4.   Reactions of β-(perfluoroalkyl)ethyl iodides.

companies (Daikin, Nippon Mektron) are active in this area, too]. Worldwide production is estimated to be about 2000 tonnes per annum (1992).

Several companies have developed routes to iodides containing perfluoroisopropyl (Pennwalt) or perfluoroisopropoxy groups (Allied), and branched perfluoroalkyl chains stemming from anionic polymerization of tetrafluoroethylene (ICI),[6] but the manufacture of all these compounds has now ceased. Nowadays, all the products containing a long perfluoroalkyl chain are prepared via the C_2F_5I-based telomerization process or by electrochemical fluorination (production of $C_7F_{15}COF$ and $C_8F_{17}SO_2F$, mainly by 3M; see Chapter 5).

When a long perfluoroalkyl chain is incorporated into an organic compound, it induces the following properties: thermal and chemical stability, water resistance, oil resistance, and surface tension reduction in both aqueous and solvent systems.

Owing to the strong electronegativity of the perfluoroalkyl chain, the iodine atom in R_FI is difficult to substitute, as discussed earlier. Hence, only two important industrial compounds are prepared *directly* from C_2F_5I-derived long-chain perfluororo-*n*-alkyl iodides, namely, perfluorocarboxylic acids [$R_FCF_2I \rightarrow$ (with oleum) R_FCO_2H],[137] and perfluoroalkanesulfonyl chlorides [$R_FCF_2I \rightarrow$ (first with SO_2/Zn then Cl_2) $R_FCF_2SO_2Cl$].[110] However, preparation of β-(perfluoroalkyl)ethyl iodides $R_FC_2H_4I$ by ethylenation of R_FI is easy, and members of this class of iodo compounds can be converted to many different intermediates via more or less conventional iodide conversions (Scheme 4).

In short, telomeric perfluoroalkyl iodides of type $C_2F_5(CF_2CF_2)_nI$ (Table 2) are the main industrial intermediates for the manufacture of finished products, surfactants, and surface-treatment agents. For most of the applications (surfactants, textile finishes), a mixture of telomers gives often a better result than a pure telomer. An example of such a mixture is C_6F_{13} (50 ± 5%), C_8F_{17} (28 ± 5%), $C_{10}F_{21}$(12 ± 3%), ≥ $C_{12}F_{25}$ (<10%). Applications of the finished products have been reviewed in a survey still providing up-to-date information,[6] so revision appears to be unnecessary. The main applications can be summarized as follows.

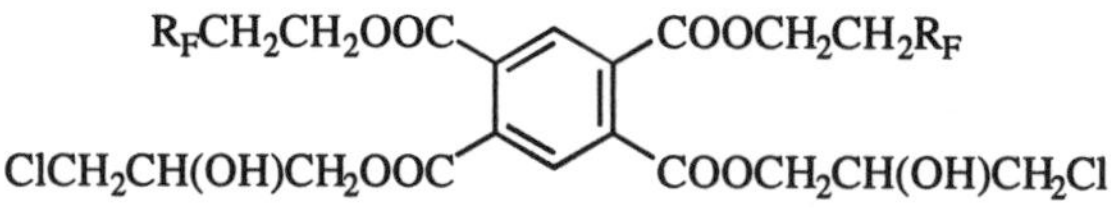

$$R_FCH_2CH_2OOC \diagdown \diagup COOCH_2CH_2R_F$$
$$ClCH_2CH(OH)CH_2OOC \diagup \diagdown COOCH_2CH(OH)CH_2Cl$$

1

8.4.2.1. Surfactants

Fluorochemical surfactants are used for a wide variety of applications: fire-fighting foams; emulsifiers in the polymerization of fluorinated monomers (C_2F_4, $CF_2{=}CH_2$, etc.); additives in chromium-plating baths; wetting agents for scourers and scalers; spreading agents for paints, inks, waxes, polishes; and so forth.

Fire-fighting foams provide still the most important outlet for these surfactants. Amphoteric surfactants, like $R_FC_2H_4SO_2NHCH_2CH_2CH_2{}^+N(CH_3)_2CH_2CO_2{}^-$, have in particular been developed for Aqueous Film Forming Foams, and are very efficient.[154] Compounds of this type can be used also for the preparation of Fluoro-protein Foams. However, more specific surfactants, such as acrylamide telomers $\{R_FC_2H_4S[CH_2CH(CONH_2)]_nH^{155}$ and $R_F[CH_2CH(CONH_2)]_nI^{156}\}$, have been designed for that application. These foams are used mainly for extinguishing hydrocarbon fuel fires, because polar solvents tend to destroy the foam too rapidly. Special foams for polar solvent fires have been developed; generally they are made from a fluorinated surfactant and a thixotropic polysaccharide, which allows the formation of a polymeric film on the surface solvent.[157]

8.4.2.2. Textile, Leather, and Paper Finishes

In the textile field, an aqueous latex formed by copolymerization of a monomer [acrylate or methacrylate of a fluorinated alcohol: $R_FCH_2CH_2OCOC(R){=}CH_2$; R = H, CH_3] with one or several nonfluorinated monomers, is usually applied in finishing. In the case of carpets the treatment can be similar to that of textiles. The result is better when the product is incorporated as spin finish during the manufacture of the fiber. Esters of type 1 are particularly efficient.[158]

The market for leather treatment is much less important. Copolymers used in aqueous media during the leather manufacture are similar to those found in the textile industry. Otherwise, copolymers in organic solvents are applied by spraying as finish and cleaning materials.

In the paper field, the fluorinated products, which are often phosphoric esters such as $(R_FCH_2CH_2O)_2P(O)OH,NH(CH_2CH_2OH)_2$, are used mainly as grease-resistance additives.

8.5. REFERENCES

1. A. K. Barbour, in: *Organofluorine Chemicals and Their Industrial Applications* (R.E. Banks, ed.), p. 44, Ellis Horwood, Chichester (1979).
2. C. Wakselman and C. Kaziz, *J. Fluorine Chem.* 33, 347 (1986).

3. T. S. Everett, *J. Chem. Educ. 64*, 143 (1987).
4. R. Filler, in Ref. 1, p. 123.
5. G. T. Newbold, in Ref. 1, p. 169.
6. H. C. Fielding, in Ref. 1, p. 214.
7. J. Banus, H. J. Emeléus, and R. N. Haszeldine, *J. Chem. Soc.* 60 (1951).
8. B. E. Smart, in: *The Chemistry of Functional Groups, Supplement D, The Chemistry of Halides, Pseudo-Halides and Azides* (S. Patai and Z. Rappoport, ed.), p. 603, Wiley, New York (1983).
9. P. Tarrant, *J.Fluorine Chem. 25*, 69 (1984).
10. R. N. Haszeldine, *J. Fluorine Chem. 33*, 307 (1986).
11. L. E. Deev, T. I. Nazarenko, K. I. Pashkevich, and V. G. Ponomarev, *Russ. Chem. Rev. 61*, 40 (1992).
12. D. M. Long, C. B. Higgins, R. F. Mattrey, R. M. Mitten, F. K. Multer, C. M. Sharts, and D. F. Shellhamer, in: *Preparation, Properties and Industrial Applications of Organofluorine Compounds* (R. E. Banks, ed.), p. 139, Ellis Horwood, Chichester (1982).
13. G. Drivon, P. Durual, B. Gurtner, and A. Lantz, European Patent 298 870 (to Atochem) [*CA 111*, 173590 (1989)].
14. C. G. Krespan, *J. Org. Chem. 23*, 2016 (1958).
15. M. Tordeux and C. Wakselman, European Patent 266 281 (to Rhône-Poulenc) [*CA 109*, 230 258 (1988)].
16. D. Naumann, W. Tyrra, B. Kock, W. Rudolph, and B. Wilkes, European Patent 291 860 (to Kali-Chemie) [*CA 110*, 156 515 (1989)].
17. Q. Y. Chen, D. B. Su, Z. Y. Yang, and R. X. Zhu, *J. Fluorine Chem. 36*, 483 (1987).
18. R. E. Banks, *Fluorocarbons and their Derivatives*, p. 82, MacDonald, London (1970).
19. R. D. Chambers, in: *Fluorine in Organic Chemistry*, p. 98, Wiley, New York (1973).
20. I. Rozhkov, G. Becker, S. Igumnov, S. Pletnev, G. Rempel, and Yu. Borisov, *J. Fluorine Chem. 45*, 114 (1989).
21. W. A. Sheppard and C. M. Sharts, *Organic Fluorine Chemistry*, p. 185, W. A. Benjamin Inc., New York (1969).
22. N. O. Brace, *J. Org. Chem. 31*, 2879 (1966).
23. P. Calas, P. Moreau, and A. Commeyras, *J. Chem. Soc., Chem. Commun.* 433 (1982).
24. D. J. Burton and L. J. Kehoe, *J. Org. Chem. 36*, 2586 (1971).
25. K. von Werner, *J. Fluorine Chem. 28*, 229 (1985).
26. T. Fuchikami and I. Ojima, *Tetrahedron Lett. 25*, 303 (1984).
27. Q. Y. Chen, Z. Y. Yang, C. X. Zhao, and Z. M. Qiu, *J. Chem. Soc., Perkin Trans. 1*, 563 (1988).
28. T. Ishihara and M. Kuroboshi, *Synth. Commun. 19*, 1611 (1989).
29. S. Ma and X. Lu, *Tetrahedron 46*, 357 (1990).
30. Y. Takeyama, Y. Ichinose, K. Oshima, and K. Utimoto, *Tetrahedron Lett. 30*, 3159 (1989).
31. T. Kitazume and T. Ikeya, *J. Org. Chem. 53*, 2350 (1988).
32. A. E. Feiring, *J. Org. Chem. 50*, 3269 (1985).
33. W. Y. Huang, *J. Fluorine Chem. 32*, 179 (1986).
34. W. Y. Huang, L. Q. Hu, and W. Z. Ge, *J. Fluorine Chem. 43*, 305 (1989).
35. N. O. Brace, *J. Org. Chem. 27*, 4491 (1962).
36. G. H. Rasmusson, R. D. Brown, and G. E. Arth, *J. Org. Chem. 40*, 672 (1975).
37. W. O. Godtfredsen and S. Vangedal, *Acta Chem. Scand. 15*, 1786 (1961).
38. P. Tarrant and E. C. Stump Jr., *J. Org. Chem. 29*, 1198 (1964).
39. H. Molines, M. Tordeux, and C. Wakselman, *Bull. Soc. Chim. II*, 367 (1982).
40. J. Leroy, H. Molines, and C. Wakseman, *J. Org. Chem. 52*, 290 (1987).
41. N. O. Brace, *J. Fluorine Chem. 20*, 313 (1982).
42. F. Jeanneaux, G. Santini, M. Leblanc, A. Cambon, and J. G. Riess, *Tetrahedron 30*, 4197 (1974).
43. M. G. Voronkov, O. G. Yarosh, and L. N. Il'icheva, *Bull. Acad. Sci. USSR. Div. Chem. Sci. 31*, 1271 (1982).
44. H. Urata, H. Yugari, and T. Fuchikami, *Chem. Lett.* 833 (1987).
45. M. Tordeux and C. Wakselman, *Tetrahedron 37*, 315 (1981).
46. G. V. D. Tiers, *J. Am. Chem. Soc. 82*, 5513 (1960).

47. L. M. Yagupolskii, A. G. Galushko, and V. A. Troitskaya, *Zh. Obshch. Khim.* **38**, 1736 (1968).
48. D. Cantacuzène, C. Wakselman, and R. Dorme, *J. Chem. Soc., Perkin Trans. 1*, 1365 (1977).
49. A. B. Cowell and C. Tamborski, *J. Fluorine Chem.* **17**, 345 (1981).
50. J. M. Birchall, G. P. Irvin, and R. A. Boyson, *J. Chem. Soc., Perkin Trans. 2*, 435 (1975).
51. Y. Kobayashi, I. Kumadaki, A. Ohsawa, and S. Murakami, *Chem. Pharm. Bull.* **26**, 1247 (1978).
52. H. Kimoto, S. Fujii, and L. A. Cohen, *J. Org. Chem.* **47**, 2867 (1982).
53. T. Fuchikami and I. Ojima, *J. Fluorine Chem.* **22**, 54 (1983).
54. Q. L. Zhou and Y. Z. Huang, *J. Fluorine Chem.* **43**, 385 (1989).
55. C. Wakselman and M. Tordeux, *J. Chem. Soc., Chem. Commun.* 1701 (1987); M. Tordeux, B. Langlois, and C. Wakselman, *J. Chem. Soc., Perkin Trans. 1*, 2293 (1990).
56. M. Tordeux, C. Wakselman, and B. Langlois, European Patent 298 803 (to Rhône-Poulenc) [*CA 111*, 173 730 (1989)].
57. T. Akiyama, K. Kato, M. Kajitani, Y. Sakaguchi, J. Nakamura, H. Hayashi, and A. Sugimori, *Bull. Chem. Soc. Jpn.* **61**, 3531 (1988).
58. R. N. Haszeldine, B. Hewitson, and A. E. Tipping, *J. Chem. Soc., Perkin Trans. 1*, 1706 (1974).
59. C. Wakselman, M. Tordeux, J. L. Clavel, and B. Langlois, *J. Chem. Soc., Chem. Commun.* 993 (1991); J. L. Clavel, B. Langlois, R. Nantermet, M. Tordeux, and C. Wakselman, *J. Chem. Soc., Perkin Trans. 1*, 3371 (1992).
60. X. Y. Li; X. K. Jiang, H. Q. Pan, J. S. Hu, and W. M. Fu, *Pure Appl. Chem.* **59**, 1015 (1987).
61. H. Gilman, *J. Organomet. Chem.* **100**, 83 (1975).
62. C. F. Smith, E. J. Sokolski, and C. Tamborski, *J. Fluorine Chem.* **4**, 35 (1974).
63. T. Nguyen, *J. Fluorine Chem.* **5**, 115 (1975).
64. R. Albadri, P. Moreau, and A. Commeyras, *Nouv. J. Chim.* **6**, 581 (1982).
65. P. G. Gassman and N. J. O'Reilly, *J. Org. Chem.* **52**, 2481 (1987).
66. C. Wakselman, *J. Fluorine Chem.* **59**, 367 (1992).
67. I. Rico, D. Cantacuzène, and C. Wakselman, *J. Chem. Soc., Perkin Trans. 1*, 1063 (1982).
68. P. Y. Kwok, F. W. Muellner, C. K. Chen, and J. Fried, *J. Am. Chem. Soc.* **109**, 3684 (1987).
69. Y. Hanzawa, K. Inazawa, A. Kon, H. Aoki, and Y. Kobayashi, *Tetrahedron Lett.* **28**, 659 (1987).
70. T. S. Everett, S. T. Purrington, and C. L. Bumgardner, *J. Org. Chem.* **49**, 3702 (1984).
71. I. Rico, D. Cantacuzène, and C. Wakselman, *Tetrahedron Lett.* **22**, 3405 (1981).
72. L. M. Yagupolskii, G. I. Matyushecheva, N. V. Pavlenko, and V. N. Boiko, *Zh. Org. Khim.* **18**, 14 (1982).
73. Q. Y. Chen and Z. M. Qiu, *J. Fluorine Chem.* **35**, 343 (1987).
74. D. J. Burton and D. G. Naae, *J. Am. Chem. Soc.* **95**, 8467 (1973).
75. I. Hayashi, T. Nakai, N. Ishikawa, D. J. Burton, D. G. Naae, and H. S. Kesling, *Chem. Lett.* 983 (1979).
76. W. B. Motherwell, B. C. Ross, and M. J. Tozer, *J. Chem. Soc., Chem. Commun.* 1437 (1989).
77. I. Ruppert, K. Schlich, and W. Volbach, *Tetrahedron Lett.* **25**, 2195 (1984).
78. G. P. Stahly and D. R. Bell, *J. Org. Chem.* **54**, 2873 (1989).
79. G. K. S. Prakash, R. Krishnamurti, and G. A. Olah, *J. Am. Chem. Soc.* **111**, 393 (1989); for a recent review, see: G. K. S. Prakash, in *Synthetic Fluorine Chemistry* (G. A. Olah, R. D. Chambers, and G. K. S. Prakash, eds.), p. 227, Wiley, New York (1992).
80. V. Broicher and D. Geffken, *J. Organomet. Chem.* **381**, 315 (1990).
81. G. M. Blackburn and G. E. Taylor, *J. Organomet. Chem.* **348**, 55 (1988).
82. B. Haley, R. N. Haszeldine, B. Hewitson, and A. E. Tipping, *J. Chem. Soc., Perkin Trans. 1*, 525 (1976).
83. V. N. Boiko, G. M. Schchupack, and L. M. Yagupolskii, *J. Org. Chem. USSR* **13**, 972 (1977).
84. V. I. Popov, V. N. Boiko, and L. M. Yagupolskii, *J. Fluorine Chem.* **21**, 365 (1982).
85. A. E. Feiring, *J. Fluorine Chem.* **24**, 191 (1984).
86. C. Wakselman and M. Tordeux, *J. Chem. Soc., Chem. Commun.* 793 (1984).
87. C. Wakselman and M. Tordeux, *J. Org. Chem.* **50**, 4047 (1985).
88. N. V. Ignatev, V. N. Boiko, and L. M. Yagupolskii, *J. Fluorine Chem.* **29**, 210 (1985).
89. I. Rico and C. Wakselman, *Tetrahedron* **37**, 4209 (1981).

90. I. Rico, D. Cantacuzène, and C. Wakselman, *J. Org. Chem.* *48*, 1979 (1983).
91. I. Rico and C. Wakselman, *J. Fluorine Chem.* *20*, 759 (1982).
92. N. V. Kondratenko, V. I. Popov, V. N. Boiko, and L. M. Yagupolskii, *J. Org. Chem. USSR 13*, 2086 (1977).
93. M. Tordeux, B. Langlois, and C. Wakselman, *J. Org. Chem.* *54*, 2452 (1989).
94. M. Tordeux, B. Langlois, and C. Wakselman, European Patent 278 822 (to Rhône Poulenc) [*CA 110*, 94 514 (1989)].
95. B. N. Huang, A. Haas, and M. Lieb, *J. Fluorine Chem.* *36*, 49 (1987).
96. C. M. Hu. Z. Q. Xu, and F. L. Qing, *Tetrahedron Lett.* *30*, 6717 (1989).
97. C. Wakselman, I. Rico, and M. Tordeux, *J. Fluorine Chem.* *23*, 486 (1983).
98. K. K. Bhasin, R. J. Cross, D. S. Rycroft, and D. W. A. Sharp, *J. Fluorine Chem.* *14*, 171 (1979).
99. I. Rico and C. Wakselman, *J. Fluorine Chem.* *20*, 759 (1982).
100. S. A. Postovoi, Yu. V. Zeifman, and I. L. Knunyants, *Bull. Acad. Sci. USSR. Div. Chem. Sci.* 1183 (1986).
101. Q. Y. Chen and Z. M. Qiu, *J. Chem. Soc., Chem. Commun.* 1241 (1987).
102. J. A. Morrison, *Adv. Inorg. Chem. Radiochem.* *27*, 293 (1983).
103. H. Lange and D. Naumann, *J. Fluorine Chem.* *26*, 435 (1984).
104. D. J. Burton and Z. Y. Yang, *Tetrahedron 48*, 189 (1992); D. J. Burton, in *Synthetic Fluorine Chemistry* (G. A. Olah, R. D. Chambers, and G. K. S. Prakash, eds.), p. 205, Wiley, New York (1992).
105. R. N. Haszeldine and E. G. Walaschewski, *J. Chem. Soc.* 3606 (1953).
106. W. T. Miller, E. Bergman, and H. Fainberg, *J. Am. Chem. Soc.* *79*, 4159 (1957).
107. H. Blancou, P. Moreau, and A. Commeyras, *J. Chem. Soc., Chem. Commun.* 885 (1976).
108. A. Commeyras, *Ann. Chim. Fr. 9*, 673 (1984).
109. J. A. Grondin, P. J. A. Vottero, H. Blancou, and A. Commeyras, *l'Actualité Chimique* 57 (1987).
110. A. Commeyras, H. Blancou, and A. Lantz, French Patent 2 374 287 (to Ugine Kuhlman) [*CA 89* 108 161 (1978)].
111. C. Wakselman and M. Tordeux, *Bull. Soc. Chim.* 868 (1986).
112. T. Kitazume and N. Ishikawa, *J. Am. Chem. Soc.* *107*, 5186 (1985).
113. C. Francèse, M. Tordeux, and C. Wakselman, *J. Chem. Soc., Chem. Commun.* 642 (1987).
114. C. Francèse, M. Tordeux, and C. Wakselman, *Tetrahedron Lett.* *29*, 1029 (1988).
115. M. Tordeux, C. Francèse, and C. Wakselman, *J. Fluorine Chem.* *43*, 27 (1989).
116. T. Nguyen, M. Rubinstein, and C. Wakselman, *J. Org. Chem.* *46*, 1938 (1981).
117. S. Bénéfice-Malouet, H. Blancou, and A. Commeyras, *J. Fluorine Chem.* *35*, 80 (1987).
118. S. Bénéfice-Malouet, H. Blancou, and A. Commeyras, *J. Fluorine Chem.* *45*, 87 (1989).
119. H. Blancou, R. Teissedre, and A. Commeyras, *J. Fluorine Chem.* *35*, 19 (1987).
120. D. J. Burton and D. M. Wiemers, *J. Am. Chem. Soc.* *107*, 5014 (1985).
121. P. L. Heinze and D. J. Burton, *J. Fluorine Chem.* *29*, 359 (1985).
122. V. C. R. McLoughlin and J. Thrower, *Tetrahedron 25*, 5921 (1969).
123. P. L. Coe and N. E. Milner, *J. Fluorine Chem.* *2*, 167 (1972–73).
124. J. Leroy, C. Wakselman, P. Lacroix, and O. Kahn, *J. Fluorine Chem.* *40*, 23 (1988).
125. G. J. Chen and C. Tamborski, *J. Fluorine Chem.* *43*, 207 (1989).
126. J. Burdon, P. L. Coe, C. R. Marsh, and J. C. Tatlow, *J. Chem. Soc., Chem. Commun.* 1259 (1967).
127. J. Burdon, P. L. Coe, C. R. Marsh, and J. C. Tatlow, *J. Chem. Soc., Perkin Trans. 1*, 639 (1972).
128. P. L. Coe and N. E. Milner, *J. Organomet. Chem.* *70*, 147 (1974).
129. Y. Kobayashi, K. Yamamoto, and I. Kumadaki, *Tetrahedron Lett.* 4071 (1979).
130. Y. Kobayashi, I. Kumadaki, and K. Yamamoto, *J. Chem. Soc., Chem. Commun.* 536 (1977).
131. Y. Kobayashi and I. Kumadaki, *J. Chem. Soc., Perkin Trans. 1*, 661 (1980).
132. J. Leroy, M. Rubinstein, and C. Wakselman, *J. Fluorine Chem.* *27*, 291 (1985).
133. G. J. Chen and C. Tamborski, *J. Fluorine Chem.* *46*, 137 (1990).
134. G. Santini, M. Le Blanc, and J. G. Riess, *Tetrahedron 29*, 2411 (1973).
135. D. M. Wiemers and D. J. Burton, *J. Am. Chem. Soc.* *108*, 832 (1986).
136. J. H. Clark, M. A. McClinton, and R. J. Blade, *J. Chem. Soc., Chem. Commun.* 638 (1988).
137. M. Hauptschein, *J. Am. Chem. Soc. 83*, 2500 (1961).

194   Claude Wakselman and André Lantz

138. A. V. Fokin, I. V. Martynov, A. N. Chekhlov, A. I. Rapkin, and A. S. Tatarinov, *Bull. Acad. Sci. USSR, Div. Chem. Sci. 32* (2, part 2), 329 (1989).
139. L. M. Yagupolskii, I. I. Maletina, N. V. Kondratenko, and V. V. Orda, *Synthesis* 835 (1978).
140. T. Umemoto, Y. Kuriu, H. Shuyama, O. Miyano, and S. I. Nakayama, *J. Fluorine Chem. 20*, 695 (1982).
141. T. Umemoto, Y. Kuriu, H. Shuyama, O. Miyano, and L. M. Nakayama, *J. Fluorine Chem. 31*, 37 (1986).
142. V. V. Lyalin, V. V. Orda, L. A. Alekseeva, and L. M. Yagupolskii, *J. Org. Chem. USSR 7*, 1524 (1971).
143. P. Calas and A. Commeyras, *J. Electroanal. Chem. 89*, 363 (1978).
144. S. Bénéfice-Malouet, H. Blancou, P. Calas, and A. Commeyras, *J. Fluorine Chem. 39*, 125 (1988).
145. F. Leroux and M. Jaccaud, European Patent 203 851 (to Atochem) [*CA 107*, 96 317 (1987)].
146. S. Sibille, E. d'Incan, L. Leport, and J. Perichon, *Tetrahedron Lett. 27*, 3129 (1986).
147. S. Sibille, S. Mcharek, and J. Perichon, *Tetrahedron 45*, 1423 (1989).
148. J. C. Folest, J. Y. Nedelec, and J. Perichon, *Synth. Commun. 18*, 1491 (1988).
149. C. P. Andrieux, L. Gelis, and J. M. Savéant, *J. Am. Chem. Soc. 112*, 786 (1990).
150. A. Germain, D. Brunel, and P. Moreau, *J. Fluorine Chem. 43*, 249 (1989).
151. A. Germain and A. Commeyras, *Tetrahedron 37*, 487 (1981).
152. D. M. Long, European Patent 231 070 [*CA 108*, 26 949 (1988)].
153. R. S. Sheinson, J. E. Penner-Hahn, and D. Indritz, *Fire Saf. J. 15*, 437 (1989) [*CA 113*, 43 467 (1990)].
154. R. Bertocchio and L. Foulletier, French Patent 2 088 941 (to Ugine Kuhlman) [*CA 76*, 33 810 (1972)].
155. E. K. Kleiner, T. W. Cooke, and R. A. Falk, European Patent 19 584 (to Ciba-Geigy) [*CA 94*, 104115 (1981)].
156. B. Boutevin, Y. Pietrasanta, M. Taha, and A. Lantz, European Patent 189 698 (to Atochem) [*CA 106*, 69 170 (1987)].
157. P. J. Chiesa, French Patent 2 206 958 (to National Foam System Inc.) [*CA 81*, 155 367 (1974)].
158. R. U. Thomas, W. R. Hammond, M. P. Friedberger, and A. W. Archie, U.S. Patent 4 605 587 (to Allied Corp.) [*CA 99*, 159 927 (1983)].

9

Industrial Routes to Ring-Fluorinated Aromatic Compounds

J. S. MOILLIET

9.1. INTRODUCTION

This chapter concerns compounds in which fluorine is bonded to carbon forming part of an aromatic ring (ring-fluorinated aromatics, the simplest being fluorobenzene C_6H_5F) as distinct from those carrying fluorine at side chain positions (e.g., benzotrifluoride, $C_6H_5CF_3$; see Chapter 10). Naturally, both types of C—F bond can occur in the same molecule, and all three types of compound are referred to as fluoroaromatics. Only placement of fluorine at the ring sites is considered here, and only methods developed for doing so on an industrial scale are discussed.

From fluorine's position in the Periodic Table it can be appreciated that much of its synthetic chemistry involves the use of fluoride ion in one form or another (F^-). Indeed, unlike the well known chlorination or bromination of hydrocarbon aromatics, in which a ring hydrogen is replaced electrophilically and therefore lost as a proton, the introduction of fluorine is normally carried out via nucleophilic attack by an F^- source. Fluoronium ion (F^+) has never been shown to exist in normal chemical systems, although reagents are known which will fluorinate electron-rich organic substrates seemingly via an electrophilic mechanism (e.g., CH_3COOF, CF_3OF, XeF_2; see Chapter 2); but these have found very little application outside the laboratory.

Except under rather special conditions (see Chapter 2), the reaction of fluorine itself with an aromatic hydrocarbon proceeds via fluorine atom ($F^•$) attack, with little selectivity; invariably considerable breakdown of the ring and tar formation occurs. Fluorine does feature in the "cobalt fluoride route" to perfluoroaromatics (see Section

J. S. MOILLIET • Zeneca Agrochemicals, Huddersfield HD2 1FF, England. *Present address:* BNFL Fluorochemicals, Springfields, Salwick, Preston PR4 OXJ, England.

Organofluorine Chemistry: Principles and Commercial Applications, edited by R. E. Banks *et al.* Plenum Press, New York, 1994.

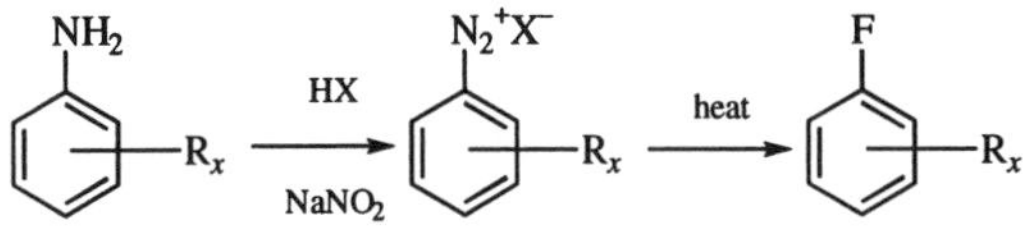

1 2 3

$X^- = F^-, BF_4^-$

R_x = H or one or more (x) substituents

Scheme 1. Ring fluorination by the diazo-route.

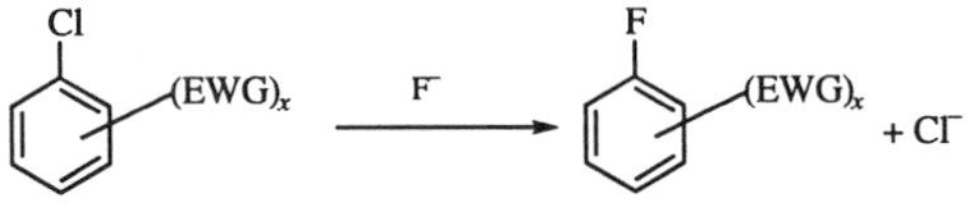

4 5

$(EWG)_x$ = electron-withdrawing (activating) group or groups (x)

Scheme 2. Ring fluorination by the Halex route.

9.4.1), and in the production of 5-fluorouracil (see Section 9.4.3). In general, however, methods based on the use of elemental fluorine at any stage are of limited commercial interest owing to the cost involved.

There are two principal commercial methods for the selective introduction of fluorine into an aromatic ring. The first (Scheme 1) is based on the diazotization of an aromatic amine (**1**) to produce a diazonium fluoride or tetrafluoroborate (**2**), which is then decomposed thermally to give nitrogen (plus boron trifluoride in the case of the tetrafluoroborate) and a fluoroaromatic (**3**). The second step is called a fluoro-dediazoniation in modern literature.

In the second method (Scheme 2) the source of fluoride ion is an alkali metal fluoride (normally KF), which is usually suspended in a polar aprotic solvent. Fluoride ion is used to displace chlorine as chloride in an "activated" situation (**4**) to give the corresponding fluoroaromatic (**5**). This nucleophilic *hal*ogen *ex*change is usually referred to as *Halex fluorination*, and proceeds via an addition–elimination mechanism (S_NAr).

9.2. "HALEX" FLUORINATIONS

9.2.1. General Considerations

Halex (halogen-exchange) methodology involves the displacement of a heavier halogen from a ring carbon by a fluorine in the form of fluoride ion.[1] The reaction does not proceed at an acceptable rate unless the leaving group is "activated." Chlorobenzene itself, for example, will not react with fluoride ion except under extremely forcing

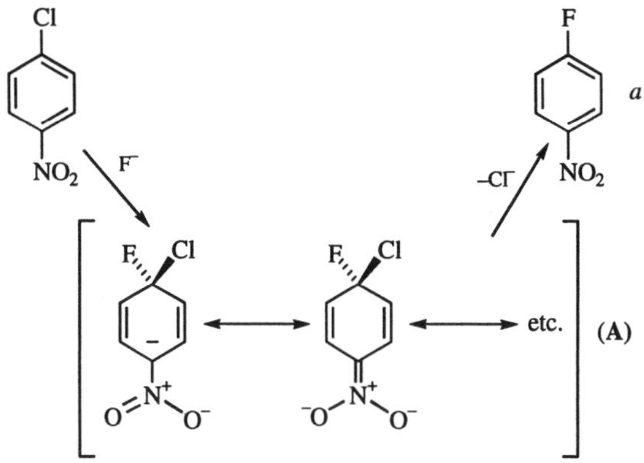

Scheme 3. Halex fluorination involving formation of a "Meisenheimer complex." [a]Produced in *ca.*70% yield.[2]

conditions. Activation is achieved with electron-withdrawing groups (EWG) situated *ortho* or *para* to the halogen being exchanged, and the reaction proceeds via the formation (rate determining) of an anionic intermediate ("Meisenheimer complex"; **A**, Scheme 3). The most efficient activating groups withdraw electron density via both inductive (-I) and mesomeric (-M) effects (NO_2, CN, COF, CHO, SO_2F, etc.); more drastic reaction conditions are needed if activation is provided by substituents with only -I character (e.g., CF_3; see Scheme 4).

Halogen exchange at 2- (6-) and 4-positions in halogenopyridines is readily performed because the ring nitrogen acts to stabilize the anionic Meisenheimer complexes. Note also that nitro groups can be displaced (as nitrite anion) from activated sites by fluoride[2]; this presents the complication that the nucleofuge can then act to produce phenolic nitrite esters (e.g., Scheme 5). The order of ease of nucleophilic displacement of halogen from activated aromatic systems (EWG-Ar-X) is X = F> Cl > Br > I, since a C—X bond is not broken in the rate-determining step. This allows the use of the cheaper chloro-aromatics as feedstocks. Relatively cheap potassium fluoride normally suffices as the source of fluoride ion.

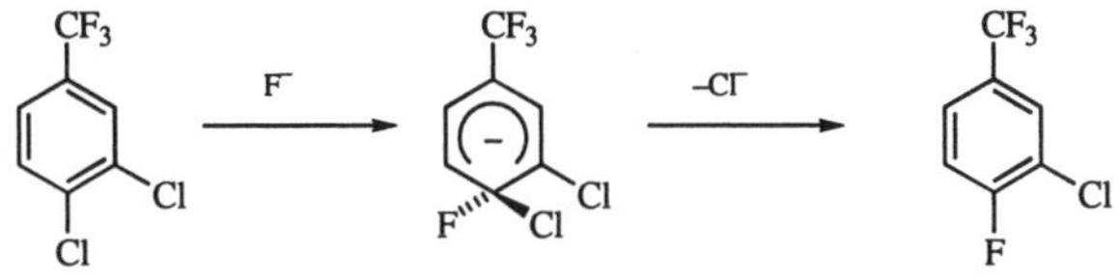

Scheme 4. Halex fluorination with activation by CF_3 and Cl substitutents.

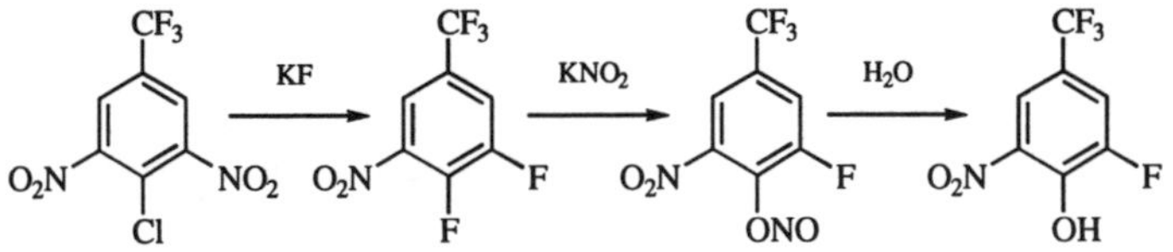

Scheme 5. Secondary reactions in Halex fluorination.

9.2.2. Examples

Some examples of the preparation of commercially significant fluorobenzenes via Halex fluorination of chloroaromatics are shown in Schemes 6–9. Since each chlorine in 2,4-dichloronitrobenzene occupies a position either *ortho* or *para* to the powerfully activating nitro group, (-I, -M), both can be replaced (Scheme 6), while in 3,4-dichloronitrobenzene the *para* chlorine is preferentially replaced (Scheme 7). Examples of *ortho/para* activation by the cyano group are shown in Schemes 8 and 9. Like the nitro group, CN is capable of being chemically manipulated once

Scheme 6. Halex replacement of two activated chlorine atoms.[3]

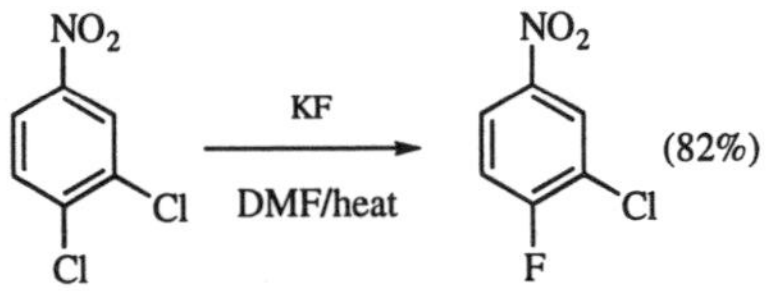

Scheme 7. Preferential replacement of a more highly activated chlorine atom.

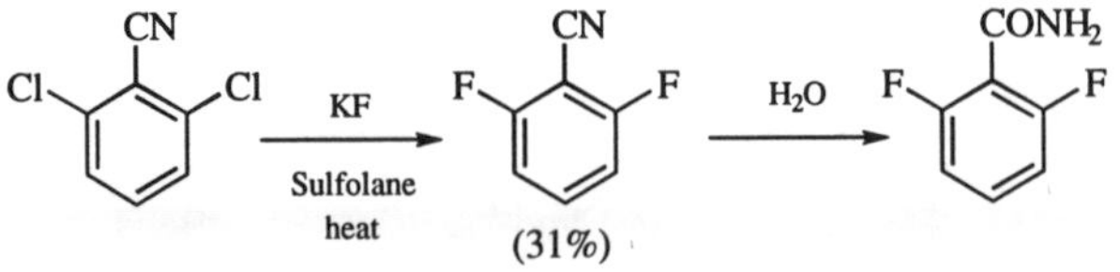

Scheme 8. Synthesis of 2,6-difluorobenzamide.[4]

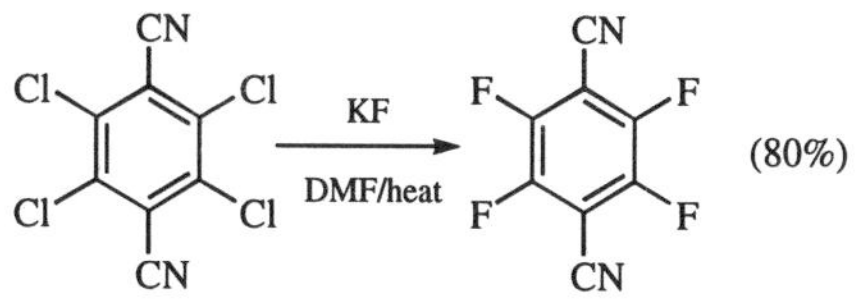

Scheme 9. Halex route to the dinitrile of tetrafluoroterephthalic acid.[5]

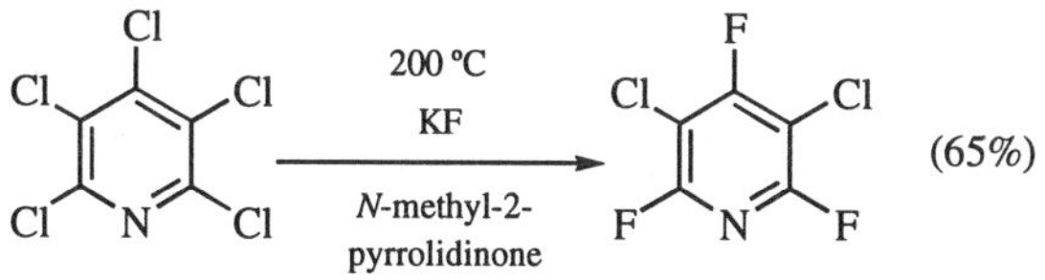

Scheme 10. Synthesis of 3,5-dichloro-2,4,6-trifluoropyridine.[6]

the fluorines are in position. For example, 2,6-difluorobenzonitrile can be converted to 2,6-difluorobenzamide (Scheme 8) and thence, after coupling with 4-chlorophenyl isocyanate, to the urea-type insecticide "Diflubenzuron." Activation by ring nitrogen enables the 2-, 4-, and 6-chlorines in pentachloropyridine to be replaced (Scheme 10). Under more drastic conditions (no solvent with KF at 500°C) pentafluoropyridine is produced in >80% yield.[6]

9.2.3. Fluoride Sources

Alkali-metal fluorides are the best source of fluoride ion for Halex fluorinations in the aromatic field (alkaline-earth fluorides are unsuitable), and the order of reactivity is CsF > RbF > KF >> NaF >> LiF as expected from consideration of the crystal lattice energies. Cesium fluoride, the most reactive reagent, is not appropriate for use on an industrial scale on grounds of both cost and availability, and the same would apply even more so in the case of rubidium fluoride.

Potassium fluoride provides the best compromise between expense and reactivity, and is the most widely used reagent in aromatic Halex processes. Sodium fluoride, the cheapest alkali metal fluoride, could be used with certain highly activated nitrogen heterocycles, but its reactivity is clearly much lower than that of the potassium salt.[2] Mono- and di-fluorination of polychlorinated pyrimidines with NaF in Sulfolane has been studied. At 180°C tetrachloropyrimidine can be converted to 6-fluoro-2,4,5-trichloropyrimidine[7] (cf Chapter 13). Catalytic amounts of cesium fluoride, either neat or deposited on silica, have been used to improve the production of 2-fluoronitrobenzene from 2-chloronitrobenzene and KF.[8]

Unfortunately it has not been possible to utilize the cheapest metal fluoride, CaF_2 (fluorite or fluorspar), for Halex reactions of chloroaromatics owing to its high lattice energy and insolubility. Hydrogen fluoride, which is used to manufacture the alkali metal fluorides, also cannot be used to effect nucleophilic displacements of aromatic chlorines, except in special circumstances (see Section 9.4.2).

9.2.4. Solvents

Aromatic Halex reactions may be carried out in a melt,[3] but inevitably this leads to thermal decomposition of both substrates and products. Initially, few solvents were known that would dissolve sufficient of both the polar inorganic fluoride and the organic substrate; furthermore, since the reaction must be carried out in a dry, aprotic system, this limited the choice of solvents available. With the development of dipolar aprotic solvents this problem was largely overcome. The ones most widely used on a large scale are shown in Table 1.

The best solvent for a particular reaction has to be determined by experimentation, since there are no hard and fast rules. A major factor is the temperature required for the reaction to take place. Acetonitrile is generally too low boiling, and although benzonitrile has the possible work-up advantage of being immiscible with water, it is not widely used. The most useful are DMF, DMAc, and Sulfolane. The first two are reasonably stable up to their boiling points, although both decompose to some extent with the formation of dimethylamine, a nucleophile capable of attacking the substrate and product, with consequent loss of yield. Sulfolane[10] is usable to a higher temperature, but at about 220°C it starts to decompose slowly with the evolution of sulfur dioxide and forms dark tarry material. Other effective solvents which have found use in the laboratory, but are not widely employed on a large scale, are hexamethylphosphoric triamide, which is suspected of being a carcinogen, and N-methyl-2-pyrrolidinone. Dimethyl sulfoxide (DMSO) is also a good solvent, but its use is plagued by the fact that it is almost impossible to prevent its decomposition to dimethyl sulfide, which

Table 1. Dipolar Aprotic Solvents Used for Halex Reactions[9]

Name	Structure	bp (°C)	Dipole moment (debye)
Acetonitrile	CH_3CN	82	4.10
Benzonitrile	C_6H_5CN	188	4.37
N,N-Dimethylformamide (DMF)	$HCON(CH_3)_2$	152	3.90
N,N-Dimethylacetamide (DMAc)	$CH_3CON(CH_3)_2$	166	3.72
Dimethyl sulfone	$(CH_3)_2SO_2$	238	4.26
Tetrahydrothiophene-1,1-dioxide (Sulfolane)	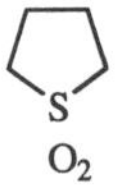	285 (dec.)	4.71
N-Methyl-2-pyrrolidinone (NM2P)	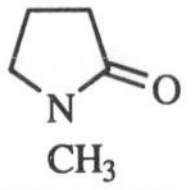	202	4.08

at even ppm levels has a foul smell. These dipolar aprotic organic solvents dissolve the substrate and also sufficient of the inorganic fluoride to enable the reactions to proceed in solution. Their mode of action as electron pair donors is to solvate selectively the metal cation (order: $Li^+ > Na^+ > K^+ > Rb^+ > Cs^+$) leaving the fluoride ion free to react.

The solubility of potassium fluoride in DMF is low even at reaction temperatures. It is necessary therefore to stir the fluoride as a slurry and rely on continuous entry of part of it into solution for the reaction to take place. The potassium chloride formed is likewise only partially soluble, so this creates the problem that it deposits onto the KF particles. Clearly the Halex reaction would cease if undissolved fluoride became completely coated with KCl. To reduce this effect to a minimum the surface area of the potassium fluoride should be as high as possible, hence the use of spray-dried material (surface area 1.5 m^2/g).[5] This is very much more effective than standard calcined material (0.1 m^2/g). Since KCl always coats the potassium fluoride to some extent, it is necessary always to work with an excess of the fluoride, the exact amount varying from reaction to reaction.

The efficiency and rate of the reaction may be improved by the addition of catalysts which increase the solubility and rate of solution of the fluoride ion. The most widely used are quaternary ammonium and phosphonium salts. These act as phase transfer catalysts and provide increased mobility of the fluoride ion from the solid phase into solution.[3] A popular example, cetyltrimethylammonium bromide [$(CH_3)_3C_{16}H_{33}N^+$ Br^-] is effective up to its decomposition temperature of 150°C, above which it produces amines which may react with halogenoaromatics to form undesirable byproducts. For operation up to 220°C, such as in Sulfolane, a phosphonium salt such as tetraphenylphosphonium bromide (TPPB) [$(C_6H_5)_4P^+$ Br^-] is more effective.

Alternative methods of increasing the concentration of fluoride ion in solution involve the addition of small amounts of cesium fluoride or a crown ether.[11,12] Both of these have been employed with success in the laboratory, but are too expensive to be used on a large scale except perhaps for high value products.

9.2.5. Procedure for the Halex Process

A generalized layout of a typical batch plant for the Halex reaction is outlined in Figure 1. The reactor is a stirred heated vessel which is made out of a fluoride-resistant, nickel-based alloy such as Inconel or Monel. Once charged with the solvent of choice and the potassium fluoride, heat is applied to dry the system by distilling out any water with some of the solvent. This can be helped by the addition of a recoverable compound, such as toluene, which will form an azeotrope with the water. The toluene may then be separated and, if necessary, recycled. The substrate and, if required, phase transfer catalyst are then added, and the reaction is carried out at the predetermined temperature for the time required.

Isolation of the product presents a number of options:

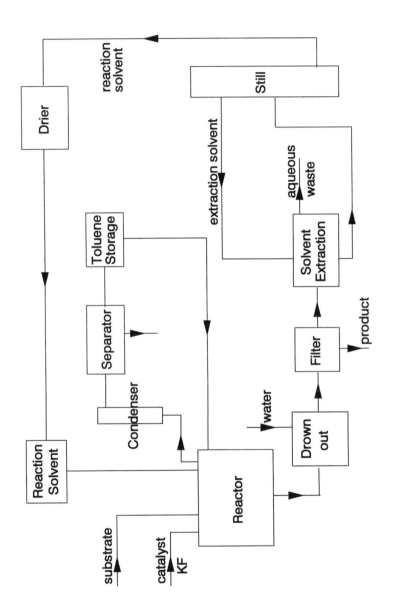

Figure 1.   Halex plant.

1. If the boiling point of the product is lower than that of the solvent, it can be distilled out as it is formed in the reaction. Solvent may then be recovered by distillation for recycling, leaving the potassium salts (KF and KCl) as a residue with any involatile organic byproducts.

2. The reaction mass can be "drowned out" into water and the organic product layer removed and washed with water to remove any traces of solvent. Solvent can then be recovered from the aqueous material (which contains the KF and KCl) by either extraction or distillation, and dried before storage for recycle. This variant is shown in Figure 1.

3. If the solvent is more volatile than the product, it can be removed by distillation, and water added to the residue containing the product and potassium halides. The product is filtered off, leaving an aqueous solution of KF and KCl.

4. Since the potassium salts are largely insoluble in the reaction solvent, the completed batch can be filtered while still warm to remove them before work-up to recover product and the solvent.

From both a cost and an environmental point of view, no matter which procedure is adopted, the solvent should be recovered. As may be deduced from the four options above, the capital cost of solvent recovery adds considerably to the overall expense, so if possible a "drown out" stage is to be avoided. Other factors which affect the cost, and these have to be optimized for each reaction, are as follows:

a. Determination of the reactant ratios. Since reaction mixtures are heterogeneous and potassium chloride deposits on the potassium fluoride particles, it is necessary to work with an excess of the latter. The actual amount of KF employed varies with temperature and solvent, but excesses of 25% over stoichiometric are quite usual. It is not economic to separate the unreacted potassium fluoride from the chloride, so the mixed salts have to be disposed of after completion of the reaction.

b. Choice and amount of solvent. The best solvent for the conversion in hand needs to be determined. Regarding the amount, this varies from reaction to reaction. The substrate needs to be completely dissolved and the reaction mixture properly fluidized. This may lead to a low reactor inventory; a ratio in the region of 6 moles of starting material to 80 moles of solvent is typical.

c. Catalyst. A phase transfer catalyst is not always necessary, particularly with the more reactive substrates, but the advantage is that its presence will speed up the replacement of a less activated chlorine, or lower the reaction temperature, therefore reducing the corrosion of the vessel.

9.3. DIAZOTIZATION METHODS

9.3.1. The Balz–Schiemann Reaction

The first widely used method[13,14] for the introduction of a fluorine atom onto an aromatic ring dates from 1927.[15] Discovered by Balz and developed by his co-worker Schiemann[11] it involves a three stage process starting from an amino arene: $ArNH_2 \rightarrow ArN_2^+ Cl^- \rightarrow ArN_2^+ BF_4^- \rightarrow ArF + N_2 + BF_3$. The key to the method is the insolubility of the arenediazonium tetrafluoroborate in an aqueous medium. The reaction was not used commercially until fairly recently (mid-1980s).[16]

The most usual way of carrying out the first stage is to diazotize the aminoarene with sodium nitrite and hydrochloric acid. The diazonium chloride formed is soluble, but on the addition of either an aqueous solution of sodium tetrafluoroborate or an approximately 50% solution of fluoroboric acid, the diazonium tetrafluoroborate precipitates (**2**, $X^- = BF_4^-$ in Scheme 1). This salt is then isolated and washed with a cold 5% solution of fluoroboric acid to remove as much as possible of the occluded sodium chloride from the first stage. (If this is not done, then chloroaromatics may be formed in the final stage.[14]) The solid is dried by first washing it with ethanol and then gently warming in a stream of warm air. This is necessary, since arenediazonium tetrafluoroborates are usually less stable when wet than dry. Also, if any water is present during the decomposition, then phenolic byproducts are formed.

In the dediazoniation step, the solid tetrafluoroborate is heated until it decomposes with the evolution of nitrogen and boron trifluoride (temperatures of up to 200°C are employed). There are a number of techniques for carrying out this stage, but in all cases care has to be taken since the reaction is exothermic. At no time should a large amount of the solid be heated, and it is best carried out by feeding the salt through a heated reactor in a continuous manner, at no time having a large inventory. Alternatively, the heat may be dispersed by means of an inert diluent, mixed with the tetrafluoroborate. Inert media such as sand, high boiling liquid paraffin, and polychlorobenzenes have all been tried with success.[14]

In order for the process to be economically viable, the evolved boron trifluoride must be recovered as fluoroboric acid and hence recycled via the second stage for recycle. In effect, therefore, the HF in the scrubbing tower is the source of fluorine in the fluoroaromatic product, and the boron trifluoride simply plays the role of fluoride carrier.

9.3.1.1. Scope and Mechanism

If the correct conditions for diazotization are applied, and the aminoarene is pure, no preparative difficulties are encountered in the first two stages of the Balz–Schiemann process, and yields of isolated arenediazonium tetrafluoroborates are generally high, often in excess of 90%. Certain ring substituents can markedly affect yields by increasing the solubility of the salts in water. Hydroxyl and carboxylic acid groups are notorious in this respect and cause yields to drop to less than 40% overall. By carrying out the reaction at as low a pH value as possible, this effect may be reduced, but in general these groups are to be avoided in the Balz–Schiemann method.

Formation of aromatic C—F bonds when solid arenediazonium tetrafluoroborates mixed with sand, or slurried with inert liquids, are heated is thought to proceed via an ion-pair mechanism involving nucleophilic transfer of fluorine from tetrafluoroborate anion to an incipient aryl cation[17–19]:

$$ArN_2^+BF_4^- \rightarrow [Ar^+ \, N_2] \, BF_4^- \rightarrow ArF + N_2 + BF_3$$

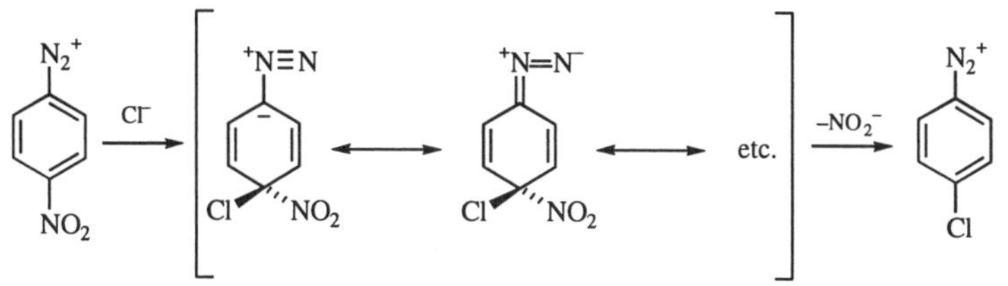

Scheme 11. Interference with the Balz–Schiemann sequence due to chloride ion impurity.[20]

Even in solution, for example, 4-Me$_3$CC$_6$H$_4$N$_2^+$BF$_4^-$ in CH$_2$Cl$_2$, the reaction does not require dissociation of BF$_4^-$ to F$^-$ and BF$_3$.[17] Regarding the effect of ring substituents, arenediazonium tetrafluoroborates containing electron-donating substituents generally decompose at lower temperatures than those carrying electron-withdrawing groups, in keeping with an S$_N$1 (aromatic) mechanism. This and other effects are discussed in more detail later when thermal dediazoniation of arenediazonium fluorides is addressed.

Product yields in the dediazoniation step are invariably lower than in the other two stages. Inevitably tar formation arising from free-radical side reactions is hard to suppress, particularly when nitro groups are involved. Nucleophilic impurities left in the diazonium salts, which are not purified other than by a washing procedure, can create losses (e.g., NaCl gives rise to chloroaromatics, and water to phenols). Note that the diazonium group acts as a powerful activator for the displacement of nucleofugal groups in *ortho* and *para* positions (see Scheme 11).[20,21]

Other side reactions may be caused by the BF$_3$ released, which is a strong Lewis acid. For example, yields of fluoroaromatics are lower when alkoxy groups are present on the ring.[14] This problem may be alleviated by operating under reduced pressure to remove the boron trifluoride as quickly as possible, or alternatively pyrolyzing the diazonium hexafluorophosphate [ArN$_2^+$PF$_6^- \rightarrow$ ArF + N$_2$ + PF$_5$] since PF$_5$ is a weaker Lewis acid. Such measures have not yet been used on a plant scale.[22]

9.3.1.2. Plant Procedure

A generalized flow diagram for a Balz–Schiemann plant is shown in Figure 2. The first stage of the process is carried out in a diazotization vessel which must be cooled and fitted with an efficient agitator. The aromatic amine is dissolved in an approximately 2.25 molar excess of hydrochloric acid, then diazotized at less than 10°C by adding a solution of sodium nitrite (about 30%). When the reaction is complete, an excess of 50% fluoroboric acid is added, while the cooling and agitation are maintained. The arenediazonium tetrafluoroborate is recovered by filtration, washed with cold 5% fluoroboric acid then ethanol, and dried in a stream of warm air to remove the last traces of water and alcohol.

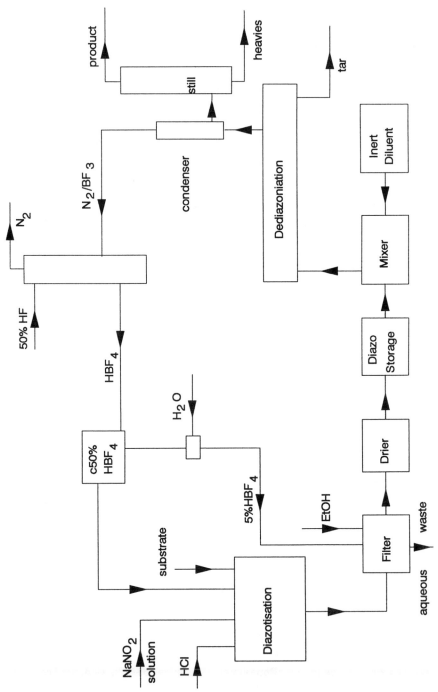

Figure 2. Balz–Schiemann plant.

The diazonium tetrafluoroborate may be fed directly to the dediazoniation reactor where it is heated to a predetermined temperature, or first it can be mixed with an inert medium such as sand or paraffin. Rotating drum or screw feeds can be used, but arrangements must be made so that at no time is a large amount of the salt being heated. The temperature must be closely controlled to prevent a run-away reaction. The volatile reaction products are fractionated to separate the fluoroaromatic from any byproducts. Nitrogen and boron trifluoride are scrubbed with aqueous hydrofluoric acid in a tower, constructed of a suitable plastic material, to convert the latter to fluoroboric acid. This is then stored for recycle, while the nitrogen is freed from acidic impurities in a caustic wash and vented to atmosphere.

There are a number of hazards associated with the Balz–Schiemann process which require strict attention.[23] Although, in general, dry arenediazonium tetrafluoroborates are safer than the wet pastes, proper precautions still need to be taken over their storage in whatever form. Some salts, such as those with electron-donating groups on the ring, should not be stored at all because of their enhanced instability. Also, arenediazonium tetrafluoroborates are suspect skin sensitizers and severe irritants, so their handling in dry powder form should be undertaken with suitable precautions. Clearly, detailed thermal stability tests should be carried out at an early stage on any arenediazonium tetrafluoroborates of commercial interest to ascertain whether its decomposition is controllable, and to determine precisely the temperature at which it starts to decompose.

9.3.2. HF-Diazotization/Dediazoniation

9.3.2.1. General Remarks

As an extension of the Balz–Schiemann reaction, a number of alternative counterions to the tetrafluoroborate have been tried, namely, F^-, PF_6^-, SiF_6^{2-}, AsF_6^-, and SbF_6^-.[14] Of these only fluoride (Scheme 1) has featured in research developed to full plant scale.[24] In this case an aromatic amine is diazotized with nitrosyl fluoride, generated $in\ situ$ from hydrogen fluoride and sodium nitrite: $NaNO_2 + HF \rightarrow NOF + H_2O + NaF$; $ArNH_2 + NOF \rightarrow ArN_2^+F^- + H_2O$. In contrast to the Balz–Schiemann procedure, no attempt is made to isolate the fluoride prior to its thermal decomposition, and no recycling of a gaseous product is required: $ArN_2^+F^- \rightarrow ArF + N_2$. The anhydrous HF acts as both solvent and reagent, in the region of 15 moles of the acid being necessary for each mole of arylamine. Furthermore, since both water and sodium fluoride are produced in the diazotization reaction, the boiling point of the medium at these concentrations is about 40°C. For many arenediazonium fluorides this value is too low to effect their decomposition to fluoroarenes within a reasonable time, hence it is necessary to raise the boiling point of the reaction solution. The easiest and cheapest way to do this is to add water to the initial charge of HF. This, however, leads to the formation of phenolic byproducts, so a compromise must be reached between decomposition rate and yield. For example, in the conversion of anthranilic acid to 2-fluorobenzoic, salicylic acid is also formed[25] (Scheme 12); a temperature of at least

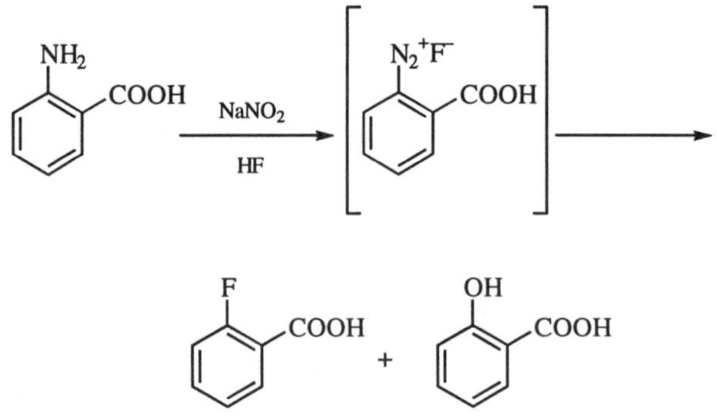

Scheme 12. HF diazotization/dediazoniation of anthranilic acid.

58°C must be reached to achieve any fluoro-dediazoniation of the diazonium salt, and this requires the addition of 10 wt% of water. However, in order to ensure that the reaction proceeds to completion within 5 hours, three times that amount of water must be added, and this causes significant amounts of salicylic acid (19% yield) to be formed at the expense of 2-fluorobenzoic acid (71% yield).[26] The addition of water is usually done by recycling crude HF containing NaF. There are other ways of elevating the boiling point of a batch of arenediazonium fluoride solution, such as the addition of a miscible organic solvent, such as glyme, when a yield of 86% of 2-fluorobenzoic acid from anthranilic acid can be achieved,[27] or by operating at elevated pressures,[36] but these techniques add significantly to the capital cost. In general, if decomposition requires the addition of enough water to achieve a temperature above 85°C, only phenolic products and tars are formed.

The temperature at which the decomposition of an arenediazonium fluoride takes place is largely determined by the nature and position of other substituents on the ring. Electron-donating (+I) groups promote reaction, i.e., fluoro-dediazoniation takes place at a lower temperature than with benzenediazonium fluoride; conversely, electron-withdrawing (–I) groups create a demand for higher temperatures. For example, diazonium fluorides derived from 2- and 3-methyl anilines will decompose within 3 to 5 hours at ambient temperatures, whereas the *para* isomer must be heated to 50°C; and while diazonium fluorides derived from 3- and 4-nitroaniline require heating to 80°C, that produced from 2,4,6-trimethylaniline decomposes rapidly at subzero temperatures. The presence of a +I substituent offsets to some extent the effect of a –I group. For example, the HF-diazotization of 3-nitroaniline and 2-methylaniline, followed by *in situ* dediazoniation of the arenediazonium fluorides at 83 and 25°C, respectively, gives 3-fluoronitrobenzene and 2-fluorotoluene in 42% and 73% yields, respectively. Using the "mixed" substrate 2-methyl-5-nitroaniline, dediazoniation can be achieved at 76°C to give the corresponding fluorobenzene in 57% yield.

An important structural effect is illustrated by the three nitroanilines. When they are converted to the fluoronitrobenzenes, the 3- and 4-isomers give 42% and 34%

yield, respectively, but the 2-isomer gives no fluoroaromatic product at all. In the last case, hindrance of the formation of an aryl cation is not only maximized by the $-I$ effect of the adjacent nitro group, but possibly the diazonium ion may be stabilized via interaction between the N_2^+ group and an electron pair on the oxygen in the nitro function.[28] Analogous arguments may be invoked to explain why certain other groups, such as alkoxy, hydroxy, and halogeno, cause arenediazonium fluorides to need higher decomposition temperatures leading to little or no fluoroaromatic product and much tar, when in the *ortho* position.

9.3.2.2. Plant Procedure

A schematic diagram of a plant using this reaction is shown in Figure 3. Two HF-resistant reactors with heating/cooling coils and a separating vessel are involved. Anhydrous HF is fed into the first reactor and a measured amount of recycled HF containing water and NaF is added so that the acid strength will give the correct reflux temperature for the dediazoniation. The reactor is cooled while the aromatic amine and then solid $NaNO_2$ (in 2% excess) are added, care being taken to maintain the temperature below the decomposition point of the diazonium salt. After the excess NOF has been destroyed with urea or sulfamic acid, the reaction mixture is pumped into the second vessel and heated under reflux. The evolved nitrogen passes through the condenser, which is cooled to retain as much HF as possible, and is then scrubbed with NaOH before being vented to atmosphere. The decomposition may be carried out in the presence of an immiscible organic solvent to remove the product as it is formed in order to prevent its involvement in side reactions.

When nitrogen evolution ceases, the batch is transferred to the separating vessel and the HF layer run off to a storage tank. From there it is either pumped back to the diazotization reactor or fed to a still charged with sulfuric acid, and the HF recovered by distillation. The fluoroaromatic product in the separator is washed with water, followed by dilute NaOH, then fractionated. Any solvent used in the dediazoniation reactor is recovered and recycled. Tars remain as a still-bottoms residue.

An essential feature of this reaction is the recycle of the HF, which is used in considerable excess, since it acts as both solvent and reagent. It is not economical to recover the sodium fluoride formed in the generation of the diazotizing agent, NOF.

Because HF diazotization/dediazoniation involves aqueous HF, potentially serious corrosion problems are encountered.[29] In the 1930s, when this method was developed,[24] no metals were known which would resist attack by the reaction mixture, so the reactors had to be replaced regularly. Improved alloy technology, and the ability to produce HF-resistant plastic-lined vessels means that reactors no longer have to be seen as sacrificial. Corrosion does, however, remain a serious hurdle to be overcome by companies investing in this process. As far as possible, plastics such as PTFE [poly(tetrafluoroethylene)], FEP [poly(tetrafluoroethylene-*co*-hexafluoropropylene)], PFA {poly[tetrafluoroethylene-*co*-perfluoro(propylvinyl ether)]}, and PVDF [poly(vinylidene fluoride)] are used. Where both heating and cooling are required, e.g.,

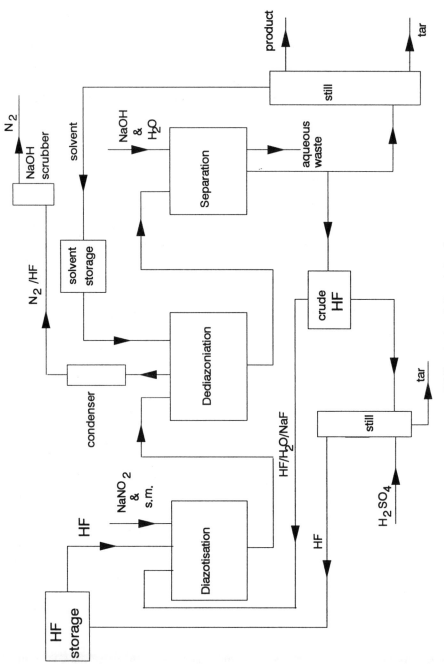

Figure 3.   HF diazotization–dediazoniation plant.

in heat-exchange coils, a more exotic, and therefore expensive, nickel-based alloy needs to be used.

Hazards associated with HF diazotization/dediazoniation arise owing to factors listed below.

1. HF is toxic and causes unique burns to skin which require prompt medical attention.
2. HF is highly corrosive, particularly when it contains up to 30% water and sodium fluoride.
3. Exothermic reactions are encountered at all stages of the process (mixing fresh AHF with crude recycled HF; protonation of the amine feedstock; diazotization; dediazoniation; and washing to remove traces of HF from products).
4. The volatility of HF allows its easy loss in the stream of nitrogen generated.
5. Tarry residues containing fluorobiphenyls and phenolic products of questionable toxicity are encountered, and need to be disposed of safely by incineration. The sulfuric acid soluble residues from the HF recycle are neutralized with lime prior to disposal.

9.4. OTHER FLUORINATION METHODS

9.4.1. Cobalt Fluoride Fluorination

The methods covered so far enable selective fluorination to be achieved. Per- or polyfluorination of carbocyclic aromatic compounds can be achieved by Halex fluorination of highly chlorinated analogs, but an alternative method is to employ a high-valency metal fluoride. This is particularly true of polynuclear perfluorinated aromatics of the fused ring variety. Cobalt trifluoride has been found the most useful reagent on a commercial scale (see Chapter 4). Produced by oxidizing cobalt(II)

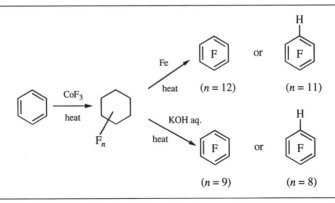

Scheme 13.   Syntheses of hexa- and penta-fluorodenzene (see Chapter 1).

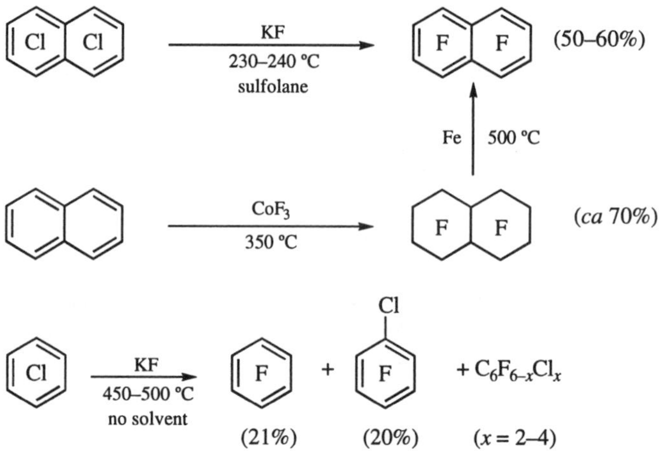

Scheme 14. Syntheses of hexafluorobenzene and of octafluoronaphthalene.

fluoride with fluorine, it can be used to convert benzene, naphthalene, etc. to mixtures of saturated per- and polyfluoroalicyclic compounds, which can subsequently be aromatized.[30] For example, benzene may be converted to either hexafluoro- or pentafluorobenzene by this process (Scheme 13) and naphthalene to perfluoronaphthalene (Scheme 14). Note, however, that both perfluorobenzene and perfluoronaphthalene can be obtained via Halex fluorination of the analogous perchlorocompounds under vigorous conditions. (Scheme 14).[31,32]

The disadvantage of this process is that it involves the use of the highly reactive gas fluorine, which is an expensive item at approximately ten times the cost of HF. Furthermore, much of the fluorine is lost as HF in the fluorination stage and as FeF_3 or KF in the aromatization stage. It remains, however, an acceptable way of producing highly fluorinated carbocyclic aromatics, particularly perfluorinated polynuclear compounds, where the high cost may be tolerated.

9.4.2. Using Hydrogen Fluoride

Although HF is widely used in the manufacture of aliphatic fluorine containing compounds by halogen exchange of the Swarts type, it is not much use for Halex

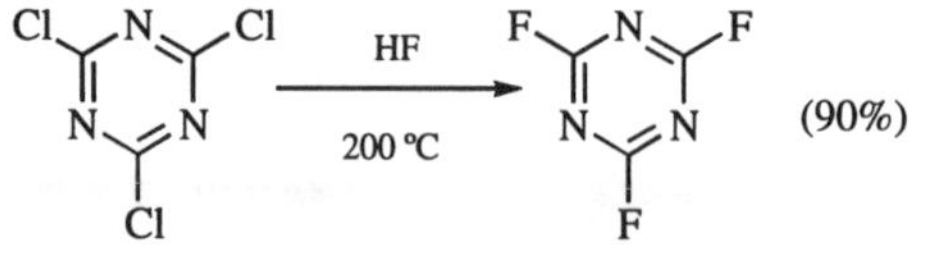

Scheme 15. Synthesis of trifluoro-s-triazine.[33]

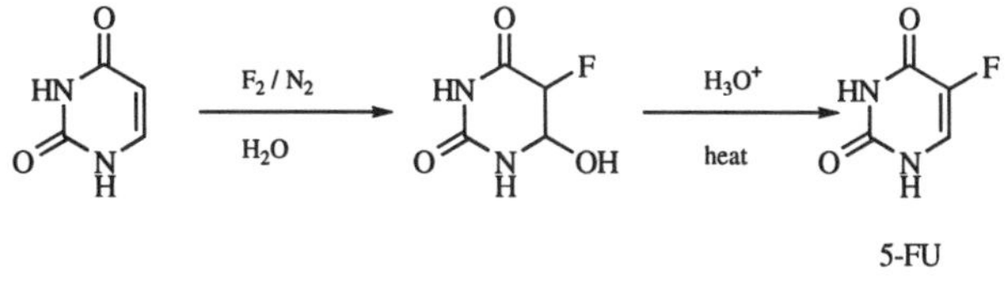

Scheme 16. 5-Fluorouracil from direct fluorination.

production of fluoroaromatics because of its high degree of association. It is therefore a poor source of F⁻, but can be used to replace chlorine in a very highly activated substrate (e.g., see Scheme 15 and Chapter 13).[7,33]

9.4.3. Direct Fluorination

The only example of the direct use of fluorine to produce a fluoroaromatic compound on a large scale is found with the anti-cancer agent 5-fluorouracil[34] (Scheme 16).

9.5. COMPARISON OF THE THREE PRINCIPAL METHODS

Halex fluorination and the two diazotization methods for producing fluoroaromatics on an industrial scale are in many ways complementary to one another. Halex methodology is used with feedstocks activated toward nucleophilic attack, and is limited to chloro or bromo substrates in which –I/–M groups (e.g., NO₂, CN) are located *ortho* and/or *para* to the halogen (usually chlorine) being replaced by F⁻ (normally supplied by KF). Two or more chlorines can be replaced (see Schemes 6, 8–10 and 14) and the method is particularly useful for the conversion of chlorinated N-heterocycles to fluorinated analogs. A problem which can arise is that other nucleofugal groups (such as NO₂) may also be displaced (e.g., Scheme 5).

In contrast to the Halex method, the diazotization methods are suitable only for the introduction of one or, in the case of the Balz–Schiemann method, two fluorines into a single benzene nucleus (e.g., see Scheme 17). The Balz–Schiemann method has been used in the laboratory for the tedious stepwise introduction of three and even four fluorines into a benzene ring in low overall yields. As with most diazotization-based

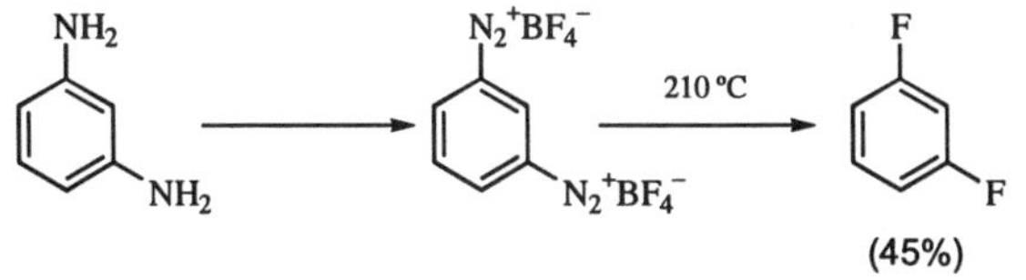

Scheme 17. Balz–Schiemann route to 1,3-difluorobenzene.[35]

syntheses, tar formation always occurs and lowers yields. A convenient practical way of reducing this problem in the HF diazotization case is to use an immiscible solvent during the decomposition stage. Ring substituents need to be stable under acidic conditions, particularly when the HF route is employed (for example, methoxy groups tend to demethylate). Suitable ring substituents for the Balz–Schiemann method include halogen, alkyl, ester, and alkoxy; low yields are encountered when OH, COOH, and NO_2 groups are present, especially when these occupy *ortho* positions. Yields in the HF diazotization method are often badly affected by *ortho* groups carrying lone-pair electrons.

In both the Halex process and the dediazoniation stage of the Balz–Schiemann process, the reaction mixture must be free of water to ensure efficient conversions and to prevent vessel corrosion by fluoride ion. The HF diazotization method involves aqueous acid, so serious corrosion problems arise from the outset. All three are exothermic, but the fluorodediazoniation steps in both diazotization methods must be controlled very carefully. Not only are serious chemical hazards associated with substrates and products, there is a more serious problem for the diazotization routes associated with the inorganic components of the systems since both HF and BF_3 are considerably more toxic than KF. The capital costs of operating the Halex process are relatively lower than those found with the diazotization routes, mainly because the reactants are less corrosive. However, the raw material cost can be higher, inflated by the need to use an excess of KF in the Halex process. Economic factors demand that the BF_3 be recovered and recycled as HBF_4 in the Balz–Schiemann method, and the same consideration applies to HF in the other diazotization method.

9.6. THE INDUSTRIAL SCENE

All three principal production processes outlined in this chapter are operated on a large scale by a number of companies in both Europe and the USA. Since Halex and the diazotization methods are essentially complementary to each other, major fluoroaromatics producers have tended to build plants for both types of process.

9.6.1. HF-Diazotization/Dediazoniation

One of the largest producers of fluoroaromatics is Riedel de Haën (Seelze, Hanover) owned by Hoechst. Using the HF diazotization method first developed by the old I.G. Farben during the Second World War for the production of fluorobenzene from aniline (see Chapter 1), it has expanded to a total capacity of about 1600 tonnes per annum (tpa), using both batch and continuous technology with the flexibility to produce a wide range of compounds. Zeneca (once part of the ICI group), which first manufactured 2-fluorobenzoic acid, has recently expanded its total output of fluoroaromatics by commissioning a new unit for the manufacture of fluorobenzene at Grangemouth (UK). Also in Europe both Rhône-Poulenc (Salindres) and MitEni

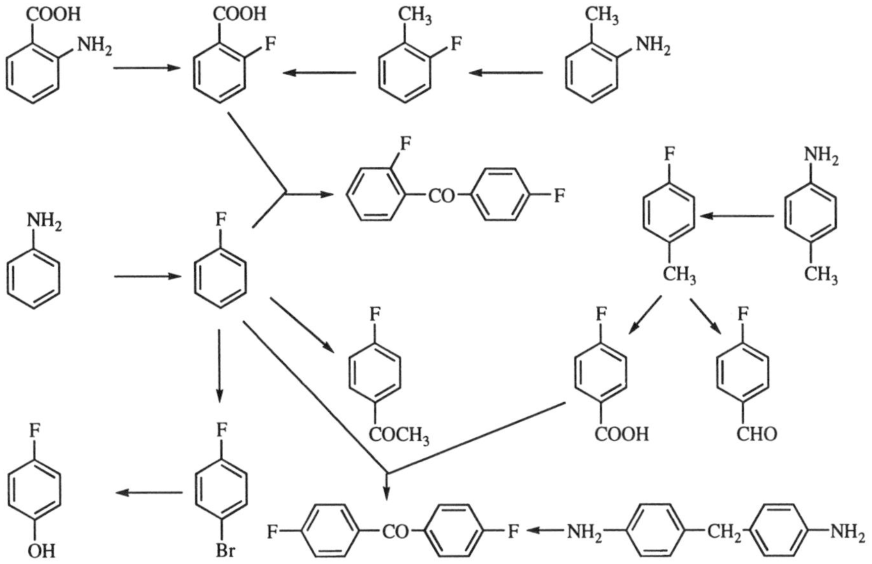

Scheme 18. Some important synthesis relationships based on HF diazotization-dediazoniation of aminoarenes.

(Trissino), a joint venture between Enimont and Mitsubishi, possess plants of about 1000 tpa. In the USA, which until recently did not have any indigenous large-scale production, both Mallinckrodt (1200 tpa) and Du Pont (1400 tpa) are active. The former owns a flexible unit for making a number of different fluorobenzenes and toluenes, while Du Pont mainly manufactures fluorobenzene.

The principal product made by this method is fluorobenzene, which is an important starting point for the synthesis of a variety of fluoroaromatic materials (Scheme 18). Electrophilic bromination, for example, provides 4-bromofluorobenzene and thence 4-fluorophenol (Riedel de Haën, ZENECA). Friedel-Crafts acylation produces 4-fluoroacetophenone (ZENECA), oxidation of which provides 4-fluorobenzoic acid for use in the production of the important monomer 4,4′-difluorobenzophenone which may also be manufactured by the HF diazotization of 4,4′-diaminodiphenylmethane, followed by oxidation of the methylene bridge. Note that electrophilic monofunctionalization of fluorobenzene is highly *para* selective.[30]

All compounds shown in Scheme 18 are used as building blocks for the manufacture of more complex molecules. For example, 2,4′-difluorobenzophenone is used to make the ZENECA fungicide Flutriafol and 4-bromofluorobenzene is used in the manufacture of Du Pont's silicon-containing fungicide Flusilazole. These and other examples are in Chapters 11 and 12.

Table 2.   Some Intermediates Made via Halex Reactions

Starting material	Fluoroaromatic intermediate	End product	Manufacturer
NO_2 / Cl (ortho benzene)	NO_2 / F (ortho benzene)	Flubiprofen	Boots
NO_2 / Cl (para benzene)	NO_2 / F (para benzene)	Fluorimide	Mitsubishi
NO_2, Cl, Cl (benzene)	NO_2, F, F (benzene)	Diflufenican	Rhône-Poulenc
		Diflunisal	Merck
NO_2, Cl, Cl (benzene)	NO_2, Cl, F (benzene)	Flampropisopropyl	Shell
		Norfloxacin	Kyorin
NO_2, Cl, F, Cl (benzene)	NO_2, F, F, F (benzene)	Perfloxacin	Rhône Poulenc
		Ofloxacin	Daiichi
CN / Cl (ortho benzene)	CN / F (ortho benzene)	Flurazepam	Roche
Cl, CN, Cl (benzene)	F, CN, Cl (benzene)	Floxacillin	Shell, Beecham
		Diflubenzuron	Duphar
N, Cl (pyridine)	N, F (pyridine)	275	Dow
Cl (pyridine)	Cl, F, Cl (pyridine)	Starane	Dow

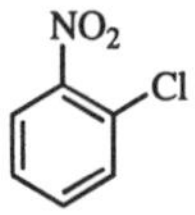

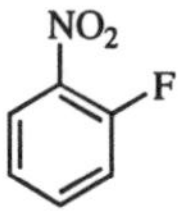

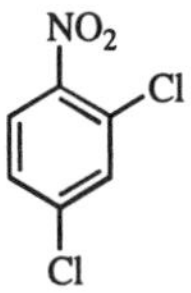

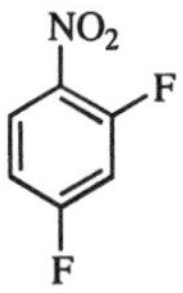

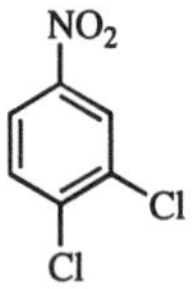

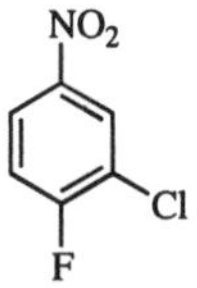

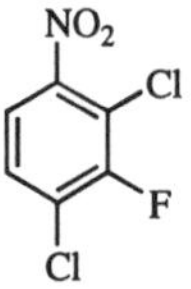

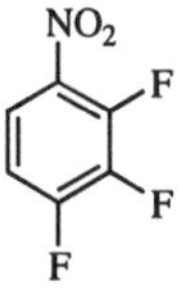

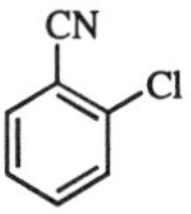

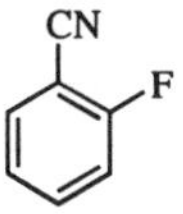

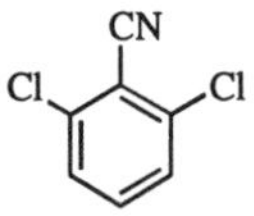

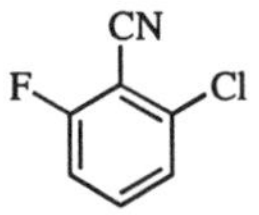

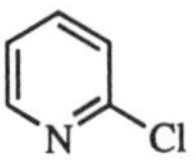

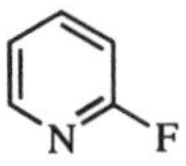

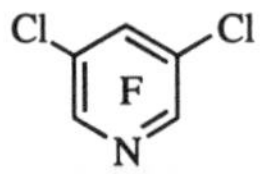

9.6.2. Halex Fluorination

Halex conversions are carried out commercially on quite a range of substrates and, unlike HF diazotization, the scene is not dominated by the production of a single compound (fluorobenzene)(see also Chapter 13). The activating groups in the chloroarenes used tend to be NO_2 or CN (*ortho* and/or *para*).The former gives a convenient entry to fluoroanilines by reduction, while the latter may be readily converted to COOH or $CONH_2$ by hydrolysis, or to CH_2NH_2 by reduction. The derivatives thus obtained can be used to synthesize a selection of more complex molecules.

A number of companies operate Halex plants. Shell has production facilities at Pernis, Netherlands to make 3-chloro-4-fluoroaniline for their herbicide Flampropiso-propyl, and Duphar (Netherlands) operates a plant solely to manufacture the insecticide Diflubenzuron, for which 2,6-difluorobenzamide (Scheme 8) is needed. Halex exchange is also used by ZENECA (en route to its insecticide Tefluthrin), by Rhône-Poulenc at Avonmouth (U.K.) and Salindres (France), and Hoechst (Germany). Some of the Halex fluoroaromatics used to make pharmaceuticals and agrochemicals are listed in Table 2.

9.6.3. Balz–Schiemann Methodology

Although the Balz–Schiemann reaction was the first means developed for the introduction of a fluorine atom onto an aromatic ring, it has been the last of the three principle methods to be used for large-scale production. This situation arose through the inherent dangers associated with the manipulation of arenediazonium tetrafluoroborates, as discussed earlier. However, Wendstone in the U.K., which later became part of Laporte, developed the technique in the 1980s to make 4,4′-difluorodiphenyl methane from the diamino compound on a 400-tpa scale, and in doing so overcame considerable technical problems. Other companies have looked again at the process with the result that both Bayer and Riedel de Haën in Germany now have plants with Balz–Schiemann capacities in the region of 200 tpa.

As stated earlier, the method can be used to introduce two fluorine atoms onto one aromatic ring without the need for an activating group as in the halex procedure. Clearly there is room for further development of the Balz–Schiemann process, which in many ways has a wider scope than the HF diazotization method provided the scale-up problems can be mastered.

9.7. CONCLUDING REMARKS

During the 1970s and 1980s considerable interest was shown in the incorporation of fluoroaromatic rings into pharmaceuticals and agrochemicals. Many compounds were patented and some, usually containing one or two fluorine atoms per molecule, have found a niche in the marketplace.

As a result of this ongoing interest and progress, a spate of plant building took place which seems to have led to an overcapacity. Since the mid-1980s, when the total world capacity for fluoroaromatic production was approximately 6000 tpa (principally Riedel de Haën, Rhône-Poulenc, Shell, and ISC), several new plants have been built (Zeneca, Du Pont, Mallinckrodt, MitEni). These have more than doubled the world capacity. Recent estimates of the expected rate of growth vary from about 6% (Riedel de Haën) to an optimistic 20%; but even though reality may lie somewhere between the two, there is likely to be too much plant on the ground for some years to come. This situation is aggravated by the fact that, although more new compounds are coming forward, invariably they are more active than those they are designed to replace. Furthermore, because increasingly they have a narrower spectrum of applications, particularly in the case of agrochemicals, more compounds will be required in smaller tonnages. From a technical viewpoint, because a greater number of quite different chemicals are being made, the reactions being used and the methodology will become more varied. This leads not only to greater interest, but also to greater scope for inventiveness in developing new processes.

9.8. REFERENCES

1. For reviews, see A. K. Barber, L. J. Belf, and M. W. Buxton, in: *Advances in Fluorine Chemistry* (M. Stacey, J. C. Tatlow, and A. G. Sharpe, eds.), Vol. 3, p. 181, Butterworths, London (1963); and L. Dolby-Glover, *Chem. Ind. (London)*, 518 (1986).
2. G. C. Finger and C. W. Kruse, *J. Am. Chem. Soc.* 78, 6034 (1956).
3. R. J. Tull, L. M. Weinstock, and I. Shinkai, U.S. Patent 4,140,719 (to Merck) [*CA 90*, 186558 (1979)]; R. A. North, U.K. Patent 2,042,507 (to Boots Co.) [*CA 93*, 185932 (1980)].
4. G. Fuller, German Patent 2,902,887 (to ISC Chemicals) [*CA 91*, 157474 (1979)].
5. S. Fujii and K. Inukai, U.S. Patent 3,975,424 (to Ind. Sci. Tech.) [*CA 86*, 5194 (1977)].
6. R. E. Banks, R. N. Haszeldine, J. V. Latham, and I. M. Young, *J. Chem. Soc.* 594 (1965).
7. E. Klauke, L. Oehlmann, and B. Baasner, *J. Fluorine Chem.* 21, 495 (1982).
8. Belgian Patent 867,463 (to BASF) [*CA 90*, 5631 (1979)].
9. C. Reichardt, *Solvents and Solvent Effects in Organic Chemistry*, 2nd edn., VCH, Weinheim (1988).
10. "The Properties of Shell Sulfolane", Shell Industrial Chemicals Bulletin, ICS:65:21.
11. H. G. Oeser, K. H. Koenig, and D. Mangold, German Patent 2,803,259 (to BASF) [*CA 91*, 157418 (1979)].
12. R. Markezich, O. S. Zamek, P. E. Donahue, and J. F. Williams, *J. Org. Chem.* 42, 3435 (1977).
13. For a historical account, see R. E. Banks and J. C. Tatlow, *J. Fluorine Chem.* 33, 71 (1986).
14. For a detailed review, see H. Suschitzky, in: *Advances in Fluorine Chemistry* (M. Stacey, J. C. Tatlow, and A. G. Sharpe, eds.), Vol. 4, p. 1, Butterworths, London (1965).
15. G. Balz and G. Schiemann, *Chem. Ber.* 60, 1186 (1927).
16. *Speciality Chemicals* 5 (2), 16 and 18 (1985).
17. C. G. Swain and R. J. Rogers, *J. Am. Chem. Soc.* 97, 799 (1975).
18. I. Szele and H. Zollinger, *J. Am. Chem. Soc.* 100, 2811 (1978).
19. H. G. O. Becker and G. Israel, *J. Prakt. Chem.* 321, 579 (1979).
20. G. C. Finger, F. H. Reed, D. M. Burness, D. M. Fort, and R. R. Blough, *J. Am. Chem. Soc.* 73, 145 (1951).
21. G. C. Finger and R. Oesterling, *J. Am. Chem. Soc.* 78, 2593 (1956).
22. K. G. Rutherford, W. Redmond, and J. Rigamonti, *J. Org. Chem.* 26, 5149 (1961).
23. C. F. Coates and W. Riddell, *Inst. Chem. Eng. Symp.* 68 4/Y:1 (1981); P. D. Storey, *Inst. Chem. Eng. Symp.* 68 3/P:1 (1981).

24. P. Osswald and O. Scherer, German Patent 600,706 (to I. G. Farben) [*CA 28*, 7260 (1934)].
25. Anon, *Research Disclosures 247*, 004 (1984) [*CA 102*, 95343 (1985)].
26. J. S. Moilliet, *J. Fluorine Chem. 35*, abstract 031 (1986).
27. J. S. Moilliet, British Patent 2,173,188 (to ICI) [*CA 107*, 23064 (1987)].
28. R. L. Ferm and C. A. VanderWerf, *J. Am. Chem. Soc. 72*, 4809 (1950).
29. G. A. Nelson, *Corrosion Data Survey*, National Association of Corrosion Engineers, Houston (Texas) (1967).
30. R. E. Banks, *Fluorocarbons and their Derivatives*, 2nd edn., Macdonald, London (1970); R. E. Banks and J. C. Tatlow, Ref. 13, p. 227.
31. N. N. Vorozhtsov, V. E. Platonov, and G. G. Yakobson, *Bull. Acad. Sci. USSR* 1380 (1963).
32. G. Fuller, *J. Chem. Soc.* 6264 (1965).
33. G. Seifert and S. Staeubli, German Patent 2,814,450 (to Ciba-Geigy AG) [*CA 90*, 23123 (1979)]; E. Kysela, E. Klauke, and H. Schwarz, German Patent 2,729,762 (to Bayer AG) [*CA 90*, 168649 (1979)].
34. R. Filler, *J. Fluorine Chem. 33*, 361 (1986) and references cited therein.
35. German Patent 1,096,889 (to Deutsch Gold- und Silber-Scheidenanstalt) [*CA 56*, 3408g (1962)].
36. M. S. Howarth and D. Tomkinson, European Patent 258,985 (to ICI) [*CA 109*, 229723 (1988)].

10

Side-Chain Fluorinated Aromatic Compounds
Routes to Benzotrifluorides

BERNARD LANGLOIS

10.1. INTRODUCTION

The market for benzotrifluoride [(trifluoromethyl) benzene] derivatives is much larger than that associated with fluorobenzenes, i.e., compounds with fluorine at ring sites (see Chapter 9), and three compounds are of major importance:

- 4-chlorobenzotrifluoride (1), which is the starting material for trifluralin-like herbicides [in 1986 trifluralin itself (2) was manufactured on a 50,000 tonne-per-year world scale, with 12,000 tonnes originating from the American industry; see Chapter 11];
- 3,4-dichlorobenzotrifluoride (3), on which a large family of agrochemicals of the diphenyl ether type is based (see Chapter 11);
- 3-aminobenzotrifluoride (4), from which a wide range of products are manufactured for pharmaceutical and agrochemical purposes.

Benzotrifluoride is produced in large quantities, but chiefly for captive transformation to 3-aminobenzotrifluoride (4) which is mainly transformed into:

- 3-isocyanatobenzotrifluoride by phosgenation;[1]
- 3-cyanobenzotrifluoride by dehydration of the corresponding formanilide;[2]
- its diazo derivative, which is then reduced to hydrazino compounds[3] or subjected to copper-mediated reactions with water, sulfur dioxide, cyanides;

BERNARD LANGLOIS • Laboratoire de Chimie Organique III, Université Claude Bernard-Lyon I, 69622 Villeurbanne Cedex, France.

Organofluorine Chemistry: Principles and Commercial Applications, edited by R. E. Banks *et al.* Plenum Press, New York, 1994.

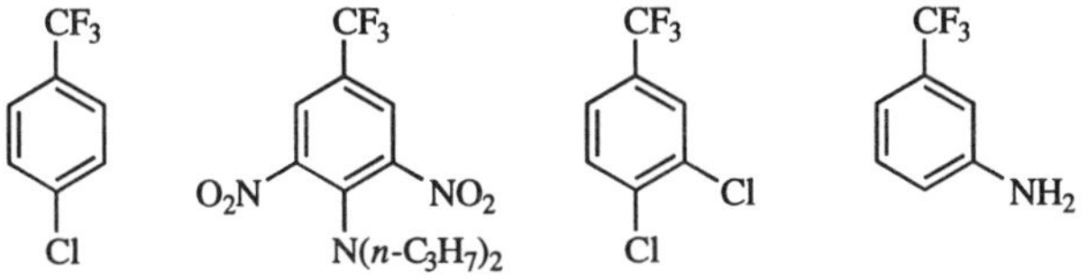

or vinylidene chloride to achieve, overall, substitution of the amino group by hydroxyl,[4] chlorosulfonyl,[5] cyano,[6] and 2,2,2-trichloroethyl[7] moieties, respectively.

10.2. SYNTHESIS OF BENZOTRIFLUORIDES FROM THE CORRESPONDING TOLUENES: LIQUID-PHASE METHODS

10.2.1. General Considerations

Though one-step syntheses of trifluoromethylated aromatics from benzoic acids and sulfur tetrafluoride,[10] and from benzoyl chlorides and molybdenum hexafluoride,[11] have been described, inadequate availabilities and high toxicities of these fluorinating agents have prevented industrial development. At present, therefore, the only process used on an industrial scale is a two-step one: first, radical chlorination of toluenes, to give benzotrichlorides, which, in the second stage, are fluorinated by means of anhydrous hydrogen fluoride (AHF). It will be recalled that CCl_3 groups carried on arene rings are exchanged more readily than those attached to saturated aliphatic systems (Chapter 1).

Synthesis strategy is outlined in Scheme 1, where Ar represents an aromatic ring optionally substituted by moieties resistant to strong anhydrous acids, oxidants, and radicals.

The trifluralin intermediate 4-chlorobenzotrifluoride (1) is produced by the route:

$$C_6H_5-CH_3 \rightarrow 4\text{-}Cl\text{-}C_6H_4\text{-}CH_3 \rightarrow 4\text{-}Cl\text{-}C_6H_4\text{-}CCl_3 \rightarrow 4\text{-}Cl\text{-}C_6H_4\text{-}CF_3$$

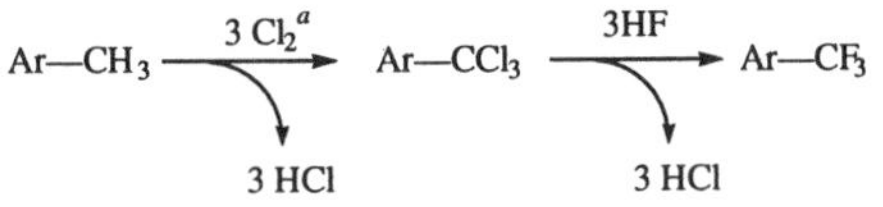

Scheme 1. General synthesis of benzotrifluoride derivatives. [a]Free-radical chlorination is commonly initiated by light as well as via thermolysis of benzoyl peroxide or azobisisobutyronitrile (AIBN). Direct thermal homolysis of molecular chlorine at temperatures above 150°C has been used for methylpyridines.[8,9] Nuclear chlorination competes with side-chain chlorination when 2- and 3-methylpyridine are chlorinated at 400°C, giving 2-Cl-6-CCl₃-C₅H₃N and 2-Cl-5-CCl₃-C₅H₃N, respectively.[9]

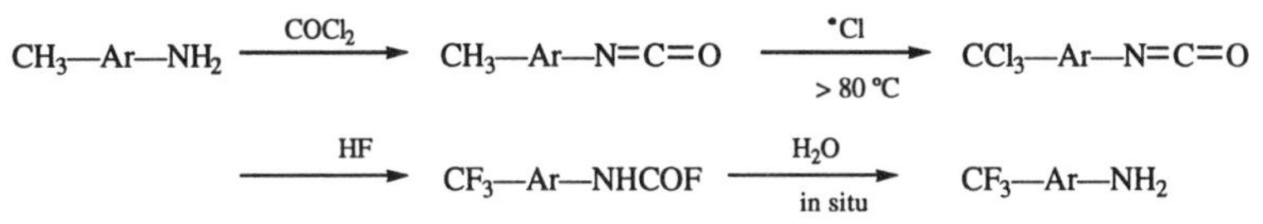

$$CH_3\!-\!Ar\!-\!NH_2 \xrightarrow{\text{COCl}_2} CH_3\!-\!Ar\!-\!N\!=\!C\!=\!O \xrightarrow[> 80\ ^\circ C]{\cdot Cl} CCl_3\!-\!Ar\!-\!N\!=\!C\!=\!O$$

$$\xrightarrow{\text{HF}} CF_3\!-\!Ar\!-\!NHCOF \xrightarrow[\text{in situ}]{H_2O} CF_3\!-\!Ar\!-\!NH_2$$

Scheme 2. Synthesis of 3- and 4-aminobenzotrifluorides.

An amino substituent must be protected by conversion to an isocyanate, a subsequent chlorination performed at a temperature which precludes formation of a carbamoyl chloride (Scheme 2); the carbamoyl fluoride[12a] can be hydrolyzed *in situ,*[12b] if necessary.

Bromo and iodo substituents are excluded as ring substituents because of their *ipso* displacement by chlorine atoms. Since the chlorination step is a radical chain process, radical scavengers like nitro or hydroxyl groups are not suitable either; hydroxyl functions can be masked and recovered after fluorination, as shown in Scheme 3.

All methyl and certain methyl-containing groups attached to the aromatic nucleus are transformed during the radical chlorination. For instance, *meta* and *para* xylenes give the corresponding bis(trichloromethyl)benzenes, and *meta* and *para* methoxy- or methylthio-toluenes afford (trichloromethoxy)- or (trichloromethylthio)-substituted benzotrichlorides. These hexachloro compounds are easily transformed with AHF into their hexafluoro analogs. *Ortho*-xylene cannot be hexachlorinated because of steric hindrance: $\alpha,\alpha,\alpha,\alpha',\alpha'$-pentachloro-*o*-xylene is the ultimate chlorination product. With AHF this gives $1\text{-}(CF_3)\text{-}2\text{-}(CHCl_2)C_6H_4$, and thence $1\text{-}(CF_3)\text{-}2\text{-}(CHF_2)C_6H_4$; both of these give 2-formylbenzotrifluoride, $1\text{-}(CF_3)\text{-}2\text{-}(CHO)C_6H_4$, when hydrolyzed (H_2SO_4),[13] and both afford 1,2-bis(trifluoromethyl)benzene by further chlorination followed by halogen exchange using AHF.[13]

Methoxycarbonyl moieties are also chlorinated but, provided the chlorination temperature is sufficient high (120–150°C), phosgene is extruded and the resultant chlorocarbonyl function undergoes halogen exchange during the fluorination step (Scheme 4).[14]

Because of the above-mentioned limitations, the chlorination–fluorination technique is usually applied to toluenes substituted only by CH_3, F, Cl, NCO, COCl, OCOCl, or CO_2CH_3 groups.

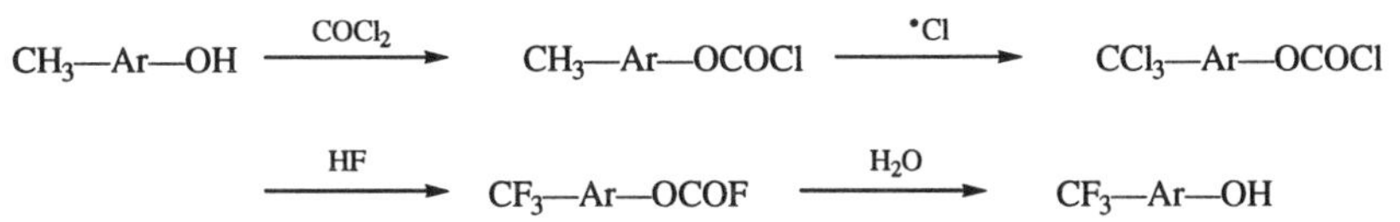

$$CH_3\!-\!Ar\!-\!OH \xrightarrow{\text{COCl}_2} CH_3\!-\!Ar\!-\!OCOCl \xrightarrow{\cdot Cl} CCl_3\!-\!Ar\!-\!OCOCl$$

$$\xrightarrow{\text{HF}} CF_3\!-\!Ar\!-\!OCOF \xrightarrow{H_2O} CF_3\!-\!Ar\!-\!OH$$

Scheme 3. Synthesis of 3- and 4-hydroxybenzotrifluorides.

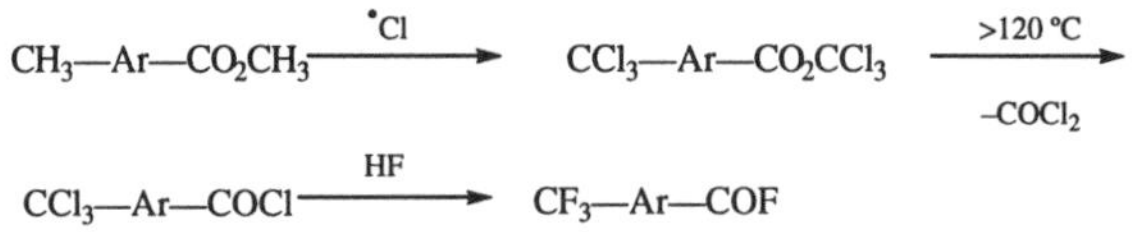
•Cl
CH3—Ar—CO2CH3
CCl3—Ar—CO2CCl3
>120 °C
-COCl2
HF
CCl3—Ar—COCl
CF3—Ar—COF

Scheme 4. Synthesis of trifluoromethylbenzoyl fluorides.

10.2.2. Fluorination of Benzotrichlorides

On an industrial scale, benzotrichlorides are usually converted to benzotrifluorides using anhydrous hydrogen fluoride (Scheme 1). Reactions can be performed under ambient conditions with gaseous HF provided that a metal-based catalyst ($MoCl_4$,[15] $MoCl_5$,[16] TaX_5, NbX_5, RhX_5[17]) or fluorosulfuric acid[18] is present. Though mild, this method excludes the presence of nucleophilic sites, like ether or carbonyl groups, which would complex with and deactivate the catalyst. The most popular technique on an industrial scale involves use of liquid AHF under autogenous pressure in steel equipment, the HF acting as both solvent and halogen-exchange reagent, and presenting no recycling problems. During batchwise operations, the two first chlorine atoms are readily exchanged around room temperature but, for the last one, temperatures of 100 to 130°C are required, so that a final pressure of about 10 to 13 bars is needed to keep the hydrogen fluoride in the liquid state. Fluorination can be also performed in a continuous mode by feeding, countercurrently, liquid benzotrichlorides and gaseous hydrogen fluoride to a column heated at 90 to 120°C under a pressure of 8 to 15 bars.[19]

As mentioned previously, when treated with AHF, (trichloromethyl)phenyl isocyanates and chloroformates provide the expected trifluoromethylated compounds (Schemes 2 and 3). However, when the trichloromethyl moiety is *ortho* to an isocyanato[20,21] or chloroformyloxy[21,22] group, rearrangements occur via intramolecular cyclization: trifluoromethoxy- or (trifluoromethylamino)-benzoyl fluorides are finally formed (Scheme 5). 2-(Trifluoromethyl)phenyl isocyanate can be obtained by fluorination of the trichloromethyl analog with antimony trifluoride.[23] This technique is

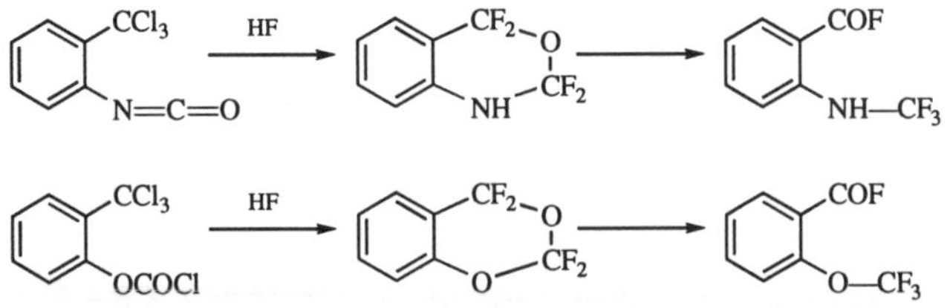
CCl3
HF
N=C=O
CF2
O
CF2
NH
COF
NH—CF3
CCl3
HF
OCOCl
CF2
O
CF2
O
COF
O—CF3

Scheme 5. Rearrangements in the reactions of AHF with *ortho*-substituted trichloromethyl compounds.

unsatisfactory on an industrial scale owing to the cost of the fluorinating reagent and problems connected to recovery and disposal of the antimony byproducts, but conditions have been found recently under which 2-(trifluoromethylamino)benzoyl fluoride isomerizes to the required 2-(trifluoromethyl)phenyl carbamoyl fluoride in AHF.[24]

10.3. VAPOR-PHASE ROUTES TO BENZOTRIFLUORIDES AND (TRIFLUOROMETHYL)PYRIDINES

Fluorination in the liquid phase is not suitable for (trichloromethyl)pyridines, which are deactivated by protonation. So vapor-phase conditions, under which pyridinium fluorides are less stable and dissociate, have been proposed for these types of compounds.[8] The method has also been applied successfully to certain benzotrichlorides, using AHF containing catalytic amounts of chlorine at 500°C under atmospheric pressure.[25] The use of solid catalysts, like aluminum fluoride or fluorinated chromium(III) oxide, allows lower reaction temperatures (270–350°C) to be employed.[26]

A one-step gas-phase synthesis of benzotrifluorides has been developed, e.g., toluene, chlorine, and hydrogen fluoride at 475°C give a mixture of benzotrifluoride and its ring-chlorinated derivatives.[27] Nuclear chlorination tends to occur extensively, and advantage has been taken of this in the direct manufacture of 2-chloro-5-(trifluoromethyl)pyridine, a useful pesticide precursor (see Chapter 11), from 3-methylpyridine.[9]

10.4. SYNTHETIC MANIPULATION OF BENZOTRIFLUORIDES

10.4.1. Electrophilic Substitution

Benzotrifluoride undergoes electrophilic attack preferentially at the *meta* position.[28] Information is available concerning chloro- and fluoro-sulfonylation,[29–31] chloromethylation,[32,33] benzoylation,[34] benzenesulfonylation,[35] benzophenone and sulfone formation,[36–39] amidomethylation,[40,41] nitration, and halogenation. Only the last two types of functionalization are dealt with here owing to constraints on space.

10.4.1.1. Nitration

Treatment of benzotrifluoride with mixed concentrated nitric and sulfuric acids at 0°C gives a mixture of mono-nitro derivatives in which the *meta* isomer greatly predominates (*o: m: p* = 6: 91: 3).[42–44] Separation of the different isomers by distillation is rather tedious and is more easily performed on the mixture of corresponding anilines produced by catalytic hydrogenation, presumably owing to hydrogen-bonding phenomena. The most volatile component, 2-aminobenzotrifluoride, is easily separated but purification of the major product, 3-aminobenzotrifluoride (**4**), requires precise distillation. In order to avoid hydrolysis of the trifluoromethyl moiety (see Section 10.4.3) during the nitration step, NO_2^+ sources producing less water than HNO_3–

H₂SO₄ have been claimed: $KNO_3–H_2SO_4$,[45a] $KNO_3–BF_3,H_2O$,[45b] $NO_2F–BF_3$,[46a] $N_2O_5–BF_3$,[46b] or $HNO_3–(CF_3SO_2)_2O$.[47]

Dinitration of 2-chlorobenzotrifluoride [→ 2-chloro-3,5-dinitrobenzotrifluoride, an important pesticide precursor] and of 4-chlorobenzotrifluoride (1) [→ 4-chloro-3,5-dinitrobenzotrifluoride (5), the largest tonnage product in the fluoroaromatics industry] requires much higher reaction temperatures (110°C) *and* more acidic and dehydrating media (oleum containing 20 to 65% by wt of sulfur trioxide). Such drastic conditions give unsatisfactory results from a chemoselectivity viewpoint, hence a two-step procedure has been developed in which the mononitrated product is isolated.[48] Traces of carcinogenic nitrosoamines contaminate products but can be eliminated physically by recrystallization under very strict temperature control,[49] or chemically using phosphorus oxychloride or thionyl chloride.[50]

Nitration of 1,3-bis(trifluoromethyl)benzene leads to 3,5-bis(trifluoromethyl)nitrobenzene only, but requires such severe conditions that hydrolysis of the trifluoromethyl groups cannot be completely avoided. The nitration kinetics have been studied in detail.[51]

10.4.1.2. Halogenation

a. Chlorination. The action of chlorine on liquid benzotrifluoride has been studied in the presence of Lewis acids, which have been classified with respect to the yield of *meta* chlorination they provide ($SbCl_5$ > Fe + $FeCl_3$ > $FeCl_3$).[52] The *meta* regioselectivity achieved with $FeCl_3$ can also be improved by adding small amounts of sulfur monochoride.[53] The isomeric distribution resulting from $FeCl_3$-catalyzed overchlorination of benzotrifluoride has been reported.[54] 3-Chlorobenzotrifluoride can also result directly and selectively from chlorine gas and benzotrichloride in liquid AHF.[55]

Electrophilic chlorination of benzotrifluoride provides a means of blocking the *meta* position during appropriate functionalization. For example, 2-aminobenzotrifluoride, which cannot be prepared satisfactorily either via direct nitration or by chlorination–fluorination of 2-tolyl isocyanate, is obtainable as shown in Scheme 6.[56]

Lack of total regioselectivity during the chlorination and nitration steps prevents crude 2-aminobenzotrifluoride from being obtained isomerically pure, but, provided that the first step is taken to about 15–20% overchlorination, a final aminobenzotrifluorides mixture rich in the *ortho* isomer (*o:m: p* = 75: 23: 2) is obtained, from which the desired product is easily separated.

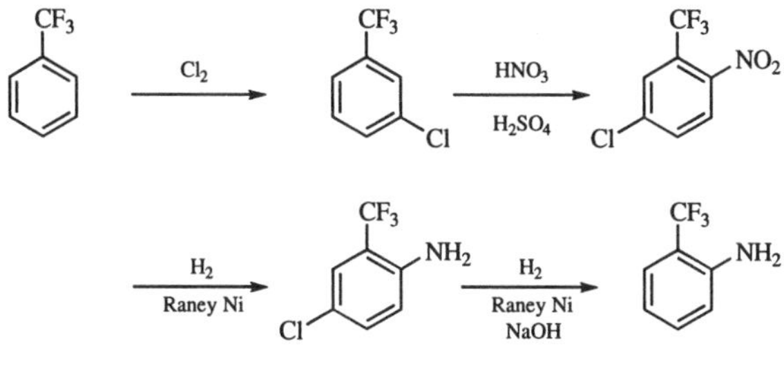

Scheme 6. Synthesis of 2-aminobenzotrifluoride.

b. Bromination. Benzotrifluoride reacts with bromine in the presence of ferric chloride,[57–60] iron,[60] aluminum chloride,[60] or antimony pentachloride,[60] but, under normal conditions, yields and regioselectivities are not completely satisfying. For example, with $SbCl_5$, 67% of 3-bromobenzotrifluoride is obtained along with 5–7% of 4-bromobenzotrifluoride[60] and, with ferric chloride as catalyst, some HF is produced, causing corrosion; this can be prevented by adding silica powder to the reactants[61]; ferric sulfide, claimed as an efficient catalyst for the bromination of 4-chlorobenzotrifluoride,[62] can play the same role. Addition of chlorine enhances yields and regioselectivities,[59] presumably owing to the production of some bromine monochloride, which is known to be a highly efficient brominator of benzotrifluoride.[63]

10.4.2. Nucleophilic Displacement of Nuclear Halogen from 4-Halogenobenzotrifluorides

Owing to the electron-withdrawing character of the trifluoromethyl group, benzotrifluorides bearing good leaving groups in *ortho* or *para* positions react with nucleophiles through S_NAr processes. However, as CF_3 exerts only a -*I* effect, halobenzotrifluorides require more severe reaction conditions than do halonitrobenzenes. As expected, 2- or 4-fluorobenzotrifluorides are more reactive than the corresponding chlorobenzotrifluorides.

10.4.2.1. Nitrogen-Centered Nucleophiles

The best known example is the manufacture of the important herbicide trifluralin (**2**) from di-*n*-propylamine and 4-chloro-3,5-dinitrobenzotrifluoride. Several analogs have been developed from 2- or 4-chloro-3,5-dinitrobenzotrifluorides and different secondary amines or ammonia.

Nucleophilic substitution is much more difficult to achieve when no nitro group is present. For example, 4-aminobenzotrifluoride can be obtained from 4-chlorobenzotrifluoride and anhydrous ammonia only under pressure at 150–240°C and in the presence of potassium fluoride and cuprous chloride.[64] From an economics viewpoint, this synthesis can compete with the milder chlorination–fluorination–hydrolysis sequence from 4-tolyl isocyanate (Scheme 2), but against this must be balanced numerous technical parameters like hazards, stability of the products (4-aminobenzotrifluoride begins to decompose thermally at 155°C[21]), corrosion, work-up ease, recycle of reagents and wastes, and energy consumption.

In connection with the manufacture of insecticides based on 1-[2,6-dichloro-4-(trifluoromethyl)phenyl]pyrazoles (Chapter 11), the reactions of ammonia and hydrazine on 3,4,5-trichloro- and 3,5-dichloro-4-fluoro-benzotrifluoride have been studied.[65] The latter substrate (6) is obtained via Halex fluorination (with KF; Chapter 9) of the trifluralin precursor (5), followed by thermal chlorinolysis to remove the NO_2 groups.

10.4.2.2.   Oxygen-Centered Nucleophiles

Replacement of chlorine at the 4-position in 4-chloro-, 4-chloro-3-nitro-, and 3,4-dichloro-benzotrifluoride using phenols, hydroquinones, or 4-hydroxyphenoxypropionic esters has been widely used for the manufacture of herbicides. The reaction is not totally regioselective with 3,4-dichlorobenzotrifluoride, and about 10% of the 3-substituted compound must be separated.

All three (trifluoromethyl)phenols can be prepared via independent hydrogenolysis of the corresponding benzyl ethers obtained from chlorobenzotrifluorides and benzyl alcohol[66]:

$$CF_3C_6H_4\text{-}Cl \quad \xrightarrow[\text{DMAc reflux}]{\text{BnOH–NaH}} \quad CF_3C_6H_4\text{–}OCH_2C_6H_5 \quad \xrightarrow[\text{4–7 bar}]{\text{H}_2\text{–Pd/C}} \quad CF_3C_6H_4OH$$

Although of little interest commercially for the production of 3-hydroxybenzotrifluoride (best obtained via diazotization of the corresponding amino compound), this route is suitable for the preparation of the two other isomers (note the neutral conditions of step 2), which lose fluorine very easily in basic media and also to some extent under strongly acidic conditions (see below). Moreover, 2-hydroxybenzotrifluoride cannot be obtained from 2-tolyl chloroformate (Scheme 5).

10.4.3.   Hydrolysis of Trifluoromethyl Groups

The trifluoromethyl group in benzotrifluoride resists hydrolysis, but can be transformed into the carboxylic acid moiety with hot concentrated sulfuric acid. A difluoromethyl group is less stable, and advantage has been taken of this fact in the preparation of 2-(trifluoromethyl)benzaldehyde from 2-(difluoromethyl)benzotrifluoride and aqueous sulfuric acid at 80°C.[13] However, when a heteroatom bearing p-electron pairs is located *ortho* or *para* to the trifluoromethyl group, hydrolysis with strong (12 M) aqueous sodium hydroxide becomes possible. 4-(Trifluoromethyl)anthranilic acid has been manufactured this way, starting from 2-amino-1,4-bis(trifluoromethyl)benzene.[67]

10.5. NEWER COMMERCIALLY INTERESTING METHODS FOR THE TRIFLUOROMETHYLATION OF AROMATIC COMPOUNDS

The $ArCH_3 \rightarrow ArCCl_3 \rightarrow ArCF_3$ route to trifluoromethylated arenes suffers from important restrictions, as discussed in Section 2. Briefly:

- the chlorination–fluorination sequence is compatible with only a few substituents (notably, phenols and anilines must be previously protected; and methyl groups are excluded);
- unexpected rearrangements can occur with toluenes substituted in the *ortho* position by groups carrying carbonylic sites (isocyanates, chloroformates; see Scheme 5);
- the method is by no means ideal for the synthesis of trifluoromethylated heterocycles.

In addition, post-synthesis functionalization of benzotrifluorides is also of limited use because the CF_3 group is powerfully *meta*-directing in electrophilic substitution situations, and benzotrifluorides substituted in *ortho* positions by electron-donating groups need to be obtained via tedious S_NAr reactions on 2-chlorobenzotrifluorides.

These problems associated with gaining ready access to benzotrifluorides bearing methyl groups or electron-donating moieties like NH_2 and OH (mainly in *ortho* positions) have prompted on-going investigations into new aromatic trifluoromethylation methodology which is capable of placing CF_3 groups *ortho* or *para* to synthetically useful electron-donating aromatic substituents. Significant progress in this important endeavor from the viewpoint of the development of viable industrial processes is summarized briefly below (see also Chapter 8).

10.5.1. Electrophilic Trifluoromethylation

An old patent (1942) from Du Pont de Nemours describing the trifluoromethylation of naphthalene with carbon tetrachloride, hydrogen fluoride, and copper[68] was seemingly neglected for a long time. The technique was reinvestigated in the early 1980s and the copper found unnecessary, simple thermal treatment of naphthalene (NapH) and a number of its derivatives (NapX, X = 1- or 2-Cl, 1- or 2-COCl, 1- or 2-SO$_2$Cl, 1-NO$_2$) with AHF–CCl$_4$ in a stainless steel autoclave leading to trifluoromethylated derivatives with structures consistent with a Friedel-Crafts type mechanism.[69] Similarly, benzene, toluene, xylenes, mesitylene, biphenyl, and their monohalogeno derivatives were found to participate in the reaction (e.g., see Scheme 7).[69]

Unfortunately, the scope of the method has notable limitations: diaryldifluoro- and triarylfluoro-methanes are produced extensively from excessively electron-rich substrates, like *ortho*-xylene, while somewhat deactivated compounds, like

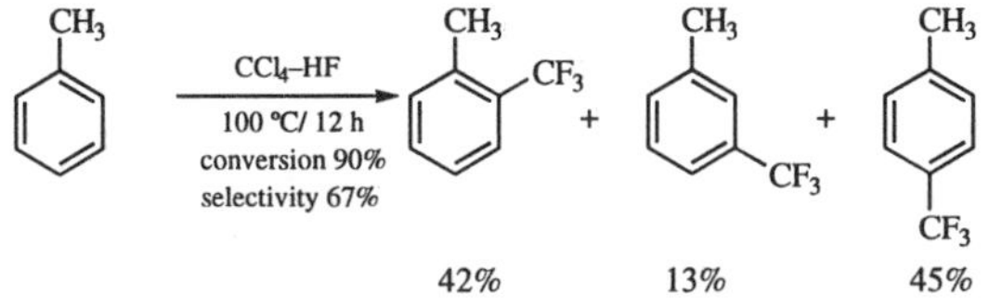

Scheme 7. Trifluoromethylation of toluene by CCl₄-HF.

ortho-dichlorobenzene, remain unreactive. Trifluoromethylation of tetra- and pentachlorobenzene can be achieved, it is claimed,[70] if antimony pentafluoride or titanium tetrafluoride is added to the reagents. However, these powerful systems produce polynuclear products from dichlorobenzenes, and HF–BF₃ mixtures are preferred for trifluoromethylation of such substrates.[71]

More recently, a variation on this method has been reported whereby aromatics can be trifluoromethylated by fluorotrichloromethane in the presence of aluminum chloride.[72] Owing to methathetic redistributions, reaction time is an important parameter: short contact times favor trifluoromethylated products, while trichloromethylated compounds predominate if equilibrium is established: ArH + CFCl₃-AlCl₃ → ArCF₃ → (with AlCl₃) ArCCl₃.

10.5.2. "Nucleophilic" Trifluoromethylation

Mixtures of sodium trifluoroacetate and copper(I) iodide have been used to substitute halogens by a trifluoromethyl group in iodo- or bromo-benzenes and pyridines.[73–76] A true (possibly [CF₃CuI]⁻) or incipient trifluoromethyl copper species, stabilized by the polar aprotic solvent used (*N*-methylpyrrolidin-2-one or *N,N*-dimethylacetamide), is probably involved: ArX + CF₃CO₂Na–CuI → (heat) ArCF₃ + CO₂ + NaX (order of reactivity = ArI > ArBr >> ArCl). The main drawback lies in the large excesses of reagents required over stoichiometric (typically required 1 ArX: ≥ 4CF₃CO₂Na: ≥ 2CuI); this method is restricted to the synthesis of compounds representing high added values. When polymethylhalobenzenes are reacted, smaller excesses of sodium trifluoroacetate and copper iodide can be employed, but the method is then limited to iodoaromatics and extensive hydrodeiodination occurs.[77]

Very recently, related techniques, using chlorodifluoroacetic[78,79] or trifluoroacetic[80] esters, copper iodide, and alkali metal fluorides, has been reported for the substitution of halogenoaromatics.

10.5.3. Radical Trifluoromethylation

Reduction of the commercial fire-extinguishing agent trifluoromethyl bromide (Halon 1301) by the sulfur dioxide radical-anion in dimethylformamide (DMF) or

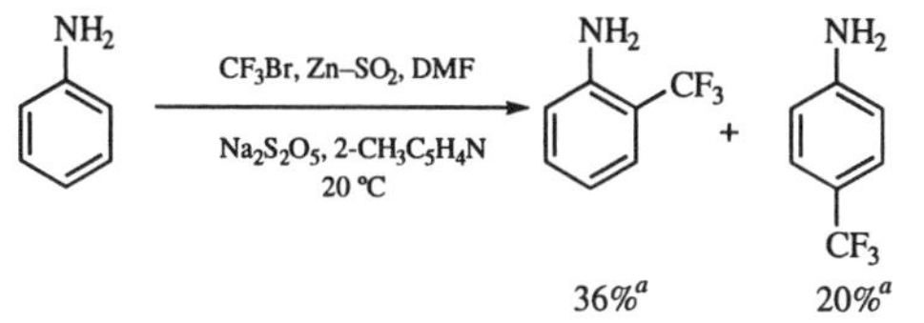

Scheme 8. Radical-type trifluoromethylation of aniline. [a]Isolated material.

DMF–water mixtures generates trifluoromethyl radicals under mild conditions: CF_3Br + $SO_2^{\bullet-}$ → SO_2 + $[CF_3Br]^{\bullet-}$ → $CF_3^{\bullet}$ + Br^-. Provided that gaseous trifluoromethyl bromide is introduced under a slight pressure (≥ 3 bar), the reaction proceeds smoothly at temperatures between 20 and 60°C, the required value being dependent on the nature of the $SO_2^{\bullet-}$ source; with the zinc–sulfur dioxide system, room temperature suffices, but with sodium dithionite ($Na_2S_2O_4$) or metallic hydroxymethanesulfinates [Rongalite ($HOCH_2SO_2Na$) or the corresponding zinc salt, Decroline],[81] temperatures in the range 40–50°C are necessary. The trifluoromethyl radicals generated can be trapped with modest-to-good efficiencies by electron-rich aromatics like phenols, anisoles, anilines, aminophenols, toluenes, or pyrroles; a base (B) such as $Na_2S_2O_5$, Na_2HPO_4, or 2-methylpyridine is included in the recipe to neutralize the HBr formed: $ArH + CF_3Br + Zn–SO_2 + B → ArCF_3 + BH^+ Br^-$.[82,83] Strongly electron-donating substituents, such as NH_2, are necessary in the Ar group in order to obtain reasonable yields (cf. p. 179).

The electrophilic character of the trifluoromethyl radical is revealed by the product isomer distributions observed (e.g., see Scheme 8). The *ortho:para* trifluoromethylation ratio is around 2 for substrates bearing mesomeric electron-donating groups, like phenols or anilines; and the π-electron excessive heterocycles pyrrole

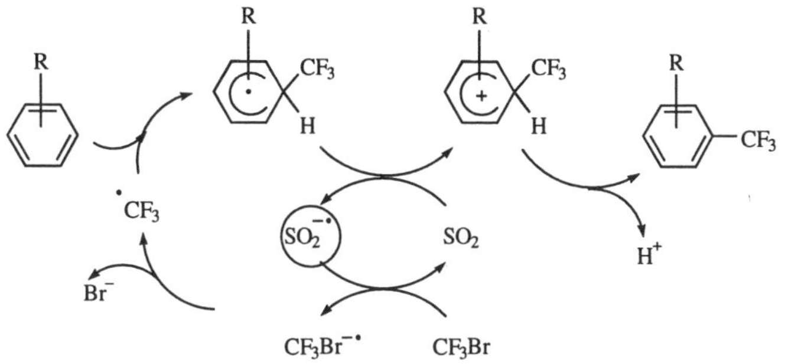

Scheme 9. Pathway for trifluoromethylation via radical-ions.

and N-methylpyrrole are regiospecifically trifluoromethylated in the 2-position in fair yield. Trifluoromethylation of pyridine (π-electron deficient) gives poor yields of all three possible isomers, as expected (e.g., 4%, 1%, and 5% of 2-, 3-, and 4-$CF_3C_5H_4N$ with CF_3Br-Na_2HPO_4-$Na_2S_2O_4$ aq.).

The sulfur dioxide radical-anion source can be employed in lower quantities than stoichiometric (0.1 eq. for Zn–SO_2 and 0.5 eq. for the less stable $Na_2S_2O_4$) since a catalytic cycle involving the redox system $SO_2/SO_2^{\bullet}$ can be maintained (Scheme 9).

Perfluoroalkyl iodides, $F(CF_2)_xI$ (x = 1, 2, 4, 6, 8), have also been used to perfluoroalkylate electron-rich aromatics in this fashion.[83b]

10.6. CONCLUDING REMARKS

The new trifluoromethylation methods (Section 5) are not yet fully developed for industrial-scale use, but without doubt some of them will be utilized in the future, especially for the production of certain trifluoromethylheterocycles and of trifluoromethyltoluenes; easy access to the latter will provide new opportunities for trifluoromethylated benzyl derivatives.

Taking these recent methods in conjunction with the classical routes to trifluoromethylated aromatics (including functionalizations of benzotrifluorides), it is clear that a rich palette of synthetic tools is available in the field of trifluoromethylated aromatics. Owing to the enormous structural diversity these tools allow, the importance of benzotrifluoride derivatives and their analogs will probably grow in the future at a higher rate than that of ring-fluorinated (fluorobenzenic) compounds.

REFERENCES

1. J. H. Werntz, U.S. Patent 2,625,561 (to Du Pont) [*CA 47*, 11243 (1953)].
2. R. Jacques, M. Reppelin, and L. Seigneurin, European Patent 65,447 (to Rhône-Poulenc Spécialités Chimiques) [*CA 98*, 160448d (1983)]; French Patent 2,505,830 (European Patent 66,482) [*CA 99*, 5367d (1983)]; French Patent 2,505,831 (European Patent 66,483) [*CA 99*, 38239s (1983)].
3. F. C. Copp, A. G. Caldwell, and D. Collard, European Patent 56,465 (to Wellcome Foundation) [*CA 97*, 216171g (1982)].
4. United States Rubber Co., British Patent 934,577 [*CA 60*, 456b (1964)]; D. N. Gray, *J. Chem. Soc.* 2243 (1960).
5. M. Delarge, *Ann. Pharm. Fr.* 119 (1977).
6. P. M. Maginnity, *J. Am. Chem. Soc.* 74, 6119 (1952).
7. A. Marhold, German Patent 3,314,249 (European Patent 123,187) (to Bayer) [*CA 102*, 113053n (1985)].
8. E. T. McBee and E. M. Hodnett, U.S. Patent 2,516,402 (to Purdue Research Foundation) [*CA 45*, 670 (1951)]; H. Johnston, M. S. Tomita, F. H. Norton, and W. H. Taplin, Belgian Patent 624,800 (to Dow Chemical) [*CA 61*, 1841a (1964)].
9. G. Whittaker, European Patent 14,033 (to I.C.I.) [*CA 94*, 83959j (1981)].
10. W. R. Hasek, W. C. Smtih, and V. A. Engelhardt, *J. Am. Chem. Soc. 82*, 543 (1960); L. M. Yagupolskii, A. I. Burmakov, and L. A. Alekseeva, *J. Gen. Chem. USSR 39*, 2007 (1969); L. M. Yagupolskii, A. I. Burmakov, and L. A. Alekseeva, *Zh. Org. Khim. 6*, 144, 2498 (1970); W. Dmowski, *J. Fluorine Chem. 32*, 255 (1986) (review).
11. F. Mathey and J. Bensoam, French Patent 2,214,674 (to I.R.CH.A.) [*CA 82*, 155757 (1975)].

12. (a) E. Klauke, E. Kuehle, H. Hack, and I. Eue, German Patent 1,768,634 (South African Patent 69 03,471) (to Bayer) [*CA 73*, 25153x (1970)]; (b) H. C. Lin and B. R. Cotter, U.S. Patent 4,481,370 (to Occidental Chem.) [*CA 102*, 95380k (1985)]; M. Desbois, French Patent 2,558,464 (to Rhône-Poulenc Spécialités Chimiques) [*CA 104*, 168096a (1986)].

13. O. Scherer, W. Schumacher, and F. Muller, German Patent 668,033 (to I.G. Farben) [*CA 33*, 2149 (1939)]; R. Belcher, A. Sykes, and J. C. Tatlow, *Anal. Chim Acta 10*, 34 (1954); J. Fernandez-Bolaños, W. G. Overend, A. Sykes, J. C. Tatlow, and E. H. Wiseman, *J. Chem. Soc.* 4003 (1960).

14. E. Klauke, F. Doering, and H. Schwarz, German Patent 2,325,089 (to Bayer) [*CA 83*, 9539d (1975)].

15. L. P. Sendlak, U.S. Patent 4,130,594 (to Hooker Chemicals and Plastics) [*CA 90*, 121165z (1979)]; U.S. Patent 4,129,602 (to Hooker Chemicals and Plastics) [*CA 90*, 121166a (1979)].

16. Y. A. Baxasuma and S. Robota, German Patent 2,739,218 (to Hooker Chemicals and Plastics) [*CA 89*, 42740j (1978)] and U.S. Patent 4,183,873 [*CA 92*, 215034z (1980)].

17. L. P. Sedlak, U.S. Patent 4,129,602 (to Hooker Chemicals and Plastics) [*CA 90*, 121166a (1979)].

18. J. J. Maul and V. A. Pattison, U.S. Patent 4,061,688 (to Hooker Chemicals and Plastics) [*CA 88*, 50453n (1978)].

19. Ramanadin and L. Seigneurin, European Patent 36,352 (to Rhone Poulenc Industries) [*CA 96*, 34809t (1982)].

20. E. Kuehle and E. Klauke, *Kem. Kenn. 1*, 596 (1974).

21. E. Kuehle and E. Klauke, *Angew. Chem., Int. Ed. Engl. 16*, 735 (1977).

22. K. U. Alles, E. Klauke, and D. Dauerer, *Justus Liebigs Ann. Chem. 730*, 16 (1966).

23. G. Buettner and E. Klauke, German Patent 2,133,467 (to Bayer) [*CA 79*, 115295y (1973)].

24. H. C. Lin and B. R Cotter, U.S. Patent 4,466,927 (to Occidental Chem.) [*CA 101*, 191396v (1984)].

25. T. Nakagawa, U. Hiramatsu, and T. Honda, German Patent 2,756,235 (to Daikin Kogyo) [*CA 89*, 108599d (1978)]; Y. Ohsaka, European Patent 4,636 (to Daikin Kogyo) [*CA 92*, 58413c (1980)].

26. Y. Aisaka and H. Sonoyama, Japanese Patent 87 12,758 (to Daikin Kogyo) [*CA 106*, 176182g (1987)].

27. U. Hiramatsu, T. Honda, and Y. Ohsaka, German Patent 2,758,164 (to Daikin Kogyo) [*CA 89*, 129231t (1978)].

28. W. A. Sheppard, *J. Am. Chem. Soc. 87*, 2410 (1965); W. A. Sheppard and C. M. Sharts, *Organic Fluorine Chemistry*, W. A. Benjamin Inc., New York (1969); G. Schiemann and B. Cornils, *Chemie und Technologie Cyclischer Fluorverbindungen*, Enke Verlag, Stuttgart (1969).

29. M. J. Miller, Belgian Patent 900,519 (to Monsanto) [*CA 103*, 22261h (1985)].

30. Sagami Chemican Research Center, Japanese Patent 82,108,060 [*CA 97*, 215751j (1982)].

31. Anonymous, French Patent 2,066,754 (to Bayer) [*CA 76*, 140207 (1972)].

32. W. Dowd and T. H. Fisher, U.S. Patent 4,144,265 (to Dow Chem.) [*CA 91*, 20137g (1979)].

33. H. J. Treiber, German Patent 1,568,938 (to Knoll) [*CA 81*, 37371 (1974)].

34. M. Desbois, French Patent 2,519,975 (to Rhône-Poulenc Spécialités Chimiques) [*CA 100*, 6086n (1984)]; European Patent 84,742 [*CA 100*, 6085m (1984)].

35. M. Desbois, European Patent 85,265 (to Rhône-Poulenc Spécialités Chimiques) [*CA 100*, 6086n (1984)].

36. M. Desbois, European Patent 147,299 (to Rhône-Poulenc Spécialités Chimiques) [*CA 103*, 214986v (1985)] .

37. M. Desbois, European Patent 147,298 (to Rhône-Poulenc Spécialités Chimiques) [*CA 103*, 141625q (1985)].

38. M. Desbois, European Patent 92,627 (to Rhône-Poulenc Spécialités Chimiques) [*CA 100*, 102945n (1984)].

39. P. Kovacic, *J. Org. Chem. 26*, 2541 (1961).

40. M. Desbois, European Patent 67,080 (to Rhone Poulenc Industries) [*CA 99*, 38172q (1983)].

41. M. Desbois, European Patent 66,484 (to Rhone Poulenc Industries) [*CA 99*, 38356c (1983)].

42. R. J. Alberts and E. C. Kooymann, *Recl. Trav. Chim. Pays-Bas 83*, 930 (1964).

43. N. Ishikawa, *Senryo to Yakuhin 26*, 106 (1981).

44. C. L. Coon and M. E. Hill, U.S. Patent 3,714,272 (to Stanford Research Inst.) [*CA 78*, 97303x (1973)].

45. (a) R. A. Smith, K. G. Salisbury, and M. A. Leaffer, German Patent 2,131,561 (to U.S. Borax and Chem.) [*CA 77*, 34130x (1972)]; (b) G. A. Olah, Q. Wang, X. Y. Li, and I. Bucsi, *Synthesis* 1085 (1992).

46. (a) G. A. Olah, *J. Am. Chem. Soc. 80*, 6541 (1958); (b) G. Bryant Bachman, *J. Am. Chem. Soc. 80*, 5871 (1958).

47. G. A. Olah, V. Prakash Reddy, and G. K. Surya Prakash, *Synthesis* 1087 (1992).

48. L. Schneider and D. E. Graham, U.S. Patent 4,096,195 (to GAF) [*CA 89*, 215053 (1978)].

49. K. Habig and K. Baessler, German Patent 2,926,947 (to Hoechst) [*CA 94*, 191896 (1981)].

50. W. N. Cannon and R. F. Eizember, Belgian Patent 870,613 (to Eli-Lilly) [*CA 91*, 74327 (1979)].

51. R. C. Miller, D. S. Noyce, and T. Vermeulen, *Ind. Eng. Chem. 56*, 43 (1964).

52. I. Kageyama, K. Maruo, and T. Shimoike, *Kogyo Kagaku Zasshi 65*, 1203 (to Osaka Kinzoku Kogyo) (1962) [*CA 58*, 5543b (1963)].

53. S. Robota and E. A. Bellmore, U.S. Patent 3,234,292 (to Hooker Chem.) [*CA 64*, 11125e (1966)].

54. A. A. Ushakov, P. P. Alikhanov, G. V. Motsarev, V. R. Kalinachenko, G. N. Kuznetsova, and V. I. Kolbasov, *Zh. Org. Khim.* 20, 2187 (1984) [*CA 102*, 112944s (1985)]; P. P. Alikhanov, A. A. Ushakov, G. N. Kusnetsova, V. R. Kalinachenko, and G. V. Motsarev, *Zh. Org. Khim. 21*, 809 (1985) [*CA 103*, 104204f (1985)].

55. M. Desbois and C. Disdier, European Patent 130,876 (to Rhône-Poulenc Spécialités Chimiques) [*CA 102*, 148864e (1985)].

56. P. J. Martinaud, C. Disdier, and J. Sullivan, European Patent 54,464 (to Rhone Poulenc Industries) [*CA 97*, 181935e (1982)].

57. E. T. McBee, R. A. Sanford, and P. J. Graham, *J. Am. Chem. Soc. 72*, 1651 (1950).

58. M. Markarian, *J. Am. Chem. Soc. 74*, 1858 (1952).

59. M. Ponomarev, USSR Patent 160,173 [*CA 61*, 5563 (1964)].

60. K. Inukai and T. Ueda, *Kogyo Kagaku Zasshi 64*, 2156 [*CA 57*, 2105 (1962)].

61. S. Misaki, Y. Furutake, and T. Shimoike, Japanese Patent 75 76,029 (to Daikin Kogyo) [*CA 83*, 178535 (1975)].

62. A. Marhold and E. Klauke, German Patent 2,708,190 (to Bayer) [*CA 89*, 215044r (1978)].

63. J. F. Mills and J. A. Schneider, *Ind. Eng. Chem., Prod. Res. Dev. 12*, 160 (1973).

64. L. P. Seiwell, U.S. Patent 4,096,185 (to du Pont de Nemours) [*CA 89*, 163243h (1978)].

65. R. Braden, A. Marhold, and L. Oelhmann, *IXth European Symposium on Fluorine Chemistry*, commun. 0-60, Leicester (U.K.), September 4–8, 1989; *J. Fluorine Chem. 45*, 142 (1989).

66. E. R. Lavagnino,, B. B. Molloy, and P. Pranc, U.S. Patent 4,168,388 (to Eli-Lilly) [*CA 90*, 186513 (1979)]; *Org. Prep. Proced. Int. 11*, 23 (1979).

67. M. Hudlicky, *Symposium on Fluorine Chemistry for Organic Chemists: Principles and Applications*, Winston-Salem (North Carolina, U.S.A.), October 10, 1989.

68. A. F. Benning and H. B. Gottlieb, U.S.Patent 2,273,922 (to Du Pont) [*CA 36*, 3812 (1942)].

69. A. Marhold and E. Klauke, German Patent 2,837,499 (European Patent 8,453) (to Bayer) [*CA 93*, 71248t (1980)]; German Patent 2,928,745 (European Patent 23,584) (to Bayer) [*CA 94*, 192005r (1981)]; *J. Fluorine Chem. 18*, 281 (1981).

70. T. R. Opie, U.S. Patent 4,207,266 (to Rohm and Haas) [*CA 93*, 167863r (1980)].

71. M. Desbois, French Patent 2,538,380 (to Rhône-Poulenc Spécialités Chimiques) [*CA 102*, 61914x (1985)].

72. J. Riera, J. Castañer, A. Robert, J. Carilla, E. Molins, and C. Miravitlles, *7th IUPAC Conference on Organic Synthesis*, Commun 4-R41, Nancy (France), July 4–7, 1988.

73. K. Matsui, E. Tobita, M. Ando, and K. Kondo, *Chem. Lett.* 1719 (1981); G. E. Carr, R. D. Chambers, T. F. Holmes, and D. G. Parker, *J. Chem. Soc., Perkin Trans. 1*, 921 (1988).

74. O. Miyano, H. Shuyama, and Y. Tsutsumi, Japanese Patent 85 204,759 (to Toyo Soda) [*CA 104*, 109470v (1986)].

75. O. Miyano, H. Shumaya, and Y. Tsutsumi, Japanese Patent 86 72,736 (to Toyo Soda) [*CA 105*, 133503z (1986)].

76. V. Ramachandran, R. I. Davidson, and J. R. Maloney, U.S. Patent 4,590,010 (to Ethyl) [*CA 105*, 78690f (1986)]; R. L. Davidson, U.S. Patent 4,814,480 (to Ethyl) [*CA 111*, 41807x (1989)].

77.   H. Suzuki, Y. Yoshida, and A. Osuka, *Chem. Lett.* 135 (1982).
78.   D. B. Su, J. X. Duan, and Q. Y. Chen, *Tetrahedron Lett.* 32, 7689 (1991).
79.   J. G. MacNeil Jr. and D. J. Burton, *J. Fluorine Chem.* 55, 225 (1991).
80.   B. Langlois and N. Roques, French Demande 92/ 08582 (to Rhône-Poulenc Chimie) (7/10/1992).
81.   M. Tordeux, B. Langlois, and C. Wakselman, *J. Org. Chem.* 54, 2453 (1989).
82.   C. Wakselman and M. Tordeux, European Patent 206,951 (to Rhône-Poulenc Spécialités Chimiques) [*CA 106*, 101853t (1987)]; *J. Chem. Soc., Chem. Commun.* 1701 (1987).
83.   (a) M. Tordeux, C. Wakselman, and B. Langlois, European Patent 298,803 (to Rhône-Poulenc Chimie) [*CA 111*, 173730b (1989)]; (b) *J. Chem. Soc., Perkin Trans. 1*, 2293 (1990).

11

Recent Developments in Fluorine-Containing Agrochemicals

DAVID CARTWRIGHT

11.1. INTRODUCTION

The past 15 years has been a very exciting time for research into agrochemicals, with many significant advances being achieved. Highly active and effective new series of compounds have been discovered in each of the agrochemical disciplines (herbicides, insecticides, fungicides, and plant growth regulators), and compounds have been introduced to the market which are often an order of magnitude more active than earlier products. This has all been achieved against a background of change in the agrochemicals industry. Finney has presented figures to show that while there has been a progressive decline in the profitability of the industry, expenditure on research and development has increased very substantially.[1] This reflects the substantial increase in development costs required to satisfy increasing regulatory demands. Graham-Bryce has argued that these demands need to be considered in their proper perspective.[2] Fluorine as a substituent has played a significant and increasingly important part in the development of new agrochemicals and is likely to continue to do so in the future.

This review concentrates on compounds introduced to the market during the last twelve years, i.e., since Newbold's review.[3] Examples have been selected to demonstrate the variety of fluorine substituents and substitution patterns used in agrochemicals, and to show how they are assembled from key building blocks. Since fluorine-containing molecules are generally more expensive than nonfluorinated analogs, a clear advantage needs to be gained to justify the inclusion of fluorine. The required fluorine-containing starting materials are obtained wherever possible from

DAVID CARTWRIGHT • Zeneca Agrochemicals, Jealott's Hill Research Station, Bracknell, Berkshire RG12 6EY, England.

Organofluorine Chemistry: Principles and Commercial Applications, edited by R. E. Banks *et al.* Plenum Press, New York, 1994.

Table 1.   Common and Trade Names

Common name	Trade name	Company (approx year of introduction)
Acifluorfen	Blazer	Rohm & Haas (1977)
	Tackle	Mobil (Rhône Poulenc)
Bifenox	Modown	Mobil (Rhône Poulenc) (1975)
Bifenthrin	Brigade, Capture, Talstar	FMC
Bioallethrin	Bioallethrine	Roussel Uclaf
	Esbiol, Esbiothrin	
Bioresmethrin	Resbuthrin,	Wellcome,
	Combat White Fly Insecticide,	Fisons (Schering), Sumitomo
	Chryson, Forte	
Bromethalin	Vengeance	Eli Lilly (Dow Elanco)
Chlorazifop	—	Ishihara
Chlorfluazuron	Atabron, Aim	Ishihara, Ciba Geigy,
	Jupitor	ICI[a]
Cyhalothrin	Grenade, Karate, Icon	ICI
Cyfluthrin	Baythroid, Eulan, Responsar, Solfac	Bayer
Cypermethrin	Polytrin	Ciba Geigy
	Cymbush, Imperator	ICI
	Ripcord, Stockade	Shell
Diclofop-methyl	Hoegrass Hoelon, Illoxan	Hoechst
Diflubenzuron	Dimilin, Astonex Larvakil	Duphar (1980)
Diflufenican	Jaguar, Cougar Ardent	Rhône-Poulenc (1987)
Fenvalerate	Sumicidin,	Sumitomo
	Bellmark, Pydrin	Shell
Flocoumafen	Storm, Stratagem	Shell
Fluazifop-butyl	Fusilade,	ICI (1982)
	Hache UnoSuper Onecide	Ishihara
Flucythrinate	Cythrin, Pay-off, Cybolt	American Cyanamid
Flufenoxuron	Cascade	Shell
Flumetralin	Prime	Ciba Geigy
Fluoroglycofen ethyl	Super Blazer Complete	Rohm & Haas (BASF)
Fluotrimazole	Persulon	Bayer
Fluridone	Brake, Pride	Eli Lilly (Dow Elanco) (1978)
Flurochloridone	Racer, Winner	Stauffer (ICI) (1983)
Fluroxypyr 1-methylheptyl	Starane	Dow (Dow Elanco)
Flurprimidol	Cutlass	Eli Lilly (Dow Elanco)
Flusilazole	Nustar, Olymp, Punch	Du Pont
Flutolanil	Moncut	Nihon Nohyaku
Flutriafol	Impact	ICI
Fluvalinate	Mavrik, Aquaflow, Spur	Zoecon (Sandoz)
Fomesafen	Flex, Reflex	ICI
Haloxyfop ethoxyethyl	Gallant, Verdict	Dow (Dow Elanco)
Hydramethylnon	Amdro, Combat, Matox, Maxforce, Wipeout	American Cyanamid
Imazalil	Florasan, Fungaflor	Janssen
Lactofen	Cobra	PPG
Mefluidide	Embark, Vistar	3M

Table 1.   (Continued)

Common name	Trade name	Company (approx year of introduction)
Nitrofen	Tok, Tokurn	Rohm & Haas
Norflurazon	Zorial, Evital, Solicam, Telok	Sandoz (1975)
Nuarimol	Trimidal, Triminol, Gauntlet, Murox	Eli Lilly (Dow Elanco)
Oxyfluorfen	Goal, Koltar	Rohm & Haas
Perfluidone	Destun	3M
Permethrin	Ambush, Kafil, Picket Pounce, Pramex, Talcord, Outflank, Eksmin, Coopex	ICI, Penick, Shell Sumitomo, Wellcome
Primisulfuron methyl	Beacon, Rifle, Tell	Ciba Geigy
Sulfometuron methyl	Oust	Du Pont
Teflubenzuron	Nomolt, Dart, Diarect	Celamerck (Shell)
Tefluthrin	Force, Forca, Forza	ICI
Triadimefon	Bayleton, Baylon	Bayer
Triarimol	—	Eli Lilly (Dow Elanco)
Triflumizole	Duo Top, Procure, Trifmine	Nippon Soda
Trifluralin	Treflan	Eli Lilly (Dow Elanco)

[a]The relationship between ICI and Zeneca is outlined in Section 27B.5 (p. 607).

commercial sources. If unavailable, they are synthesized using the techniques devised by dedicated fluorine chemists (see Chapters 2, 9, and 10). Examples of both these approaches are given in the subsequent sections. The introduction of fluorine into a molecule does not necessarily lead to an improvement in activity. Indeed, fluorine substitution can often lead to loss of activity or selectivity. The compounds described below are those which have found their way to the marketplace. They have done so because they each offer some saleable advantage over their nonfluorinated counterparts. Compounds are selected for development once a thorough program of synthesis and screening has taken place to identify the preferred molecular structure. The fact that a higher proportion of today's agrochemicals contain fluorine than 13 years ago reflects the increasingly important part that fluorine is playing in agrochemicals research.

Throughout this review, agrochemicals are referred to by approved common names, which are listed in Table 1 together with the most frequently used trade names.

11.2.   OCCURRENCE OF FLUORINE IN AGROCHEMICALS

Newbold[3] found that only about 4% of the pesticides listed in the 1977 edition of the *Pesticide Manual*[4] contain fluorine, and noted that the largest subgroup of these contained a CF_3 substituent bonded to an aromatic ring. A computer search of the 1990 edition of this manual has revealed that the proportion subsequently rose to 9.3%.[5] A parallel search of the *Agrochemicals Handbook* database provided a value of 8.7%.[6]

This rise indicates that the proportion of new agrochemicals containing fluorine introduced to the market since 1977 has grown at a faster rate than for nonfluorine-containing compounds.

The following conclusions can be drawn from the data in the *Agrochemicals Handbook.*[6]

1. The distribution of fluorine-containing compounds between the various agrochemical types is: herbicides, 48%; insecticides, 23%; fungicides, 18%; plant growth regulators, 5.5%; and rodenticides, 5.5%. These values roughly equate to the relative market size of the sectors,[7] indicating that the introduction of fluorine into molecules has been equally relevant to each of the disciplines.

2. The relative frequency of occurrence of different types of fluorinated substituent is as follows: aromatic-type (sidechain) CF_3 groups, 54.5%; fluoroaromatics (nuclear F), 26%; aliphatic-type fluoroalkyl groups (14%); fluoroalkoxy-aromatics (3%); and trifluoromethanesulfonyl derivatives (2.5%).

11.3. FLUORINE-CONTAINING HERBICIDES

11.3.1. Herbicides Containing an Aromatic-Type CF_3 Group

11.3.1.1. Dinitroanilines

This group of compounds has been known for many years, trifluralin (**1**) being the first major fluorine-containing agrochemical to be marketed. About 18,000 tons were sold in the United States during 1982.[7] The compound is used for pre-emergence weed control in cotton and a variety of other crops.[8,9] Trifluralin is synthesized by reaction of di-*n*-propylamine with 4-chloro-3,5-dinitrobenzotrifluoride (**2**, X = Y = NO_2) which is obtained by dinitration of 4-chlorobenzotrifluoride (**2**; X = Y = H). Benzotrifluorides of this type can be synthesized from the corresponding toluene by chlorination of the methyl group followed by fluorine/chlorine exchange (see Chapter 10). Because of the ready availability of these compounds, they have become a popular and widely used building block in agrochemical synthesis.

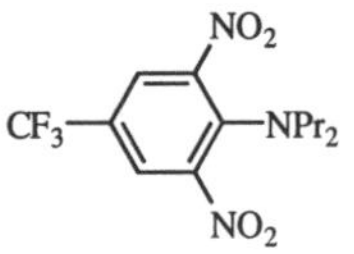

1
Trifluralin

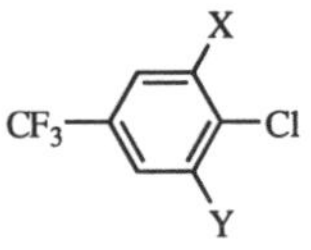

2

11.3.1.2. Diphenyl Ethers

The herbicidal properties of diphenyl ethers have been known for some time,[10] and early examples such as nitrofen[11] (**3**; R = H) and bifenox[12] (**3**; R = CO_2Me) were derived from 2,4-dichlorophenol. Interest in the area increased when it was discovered

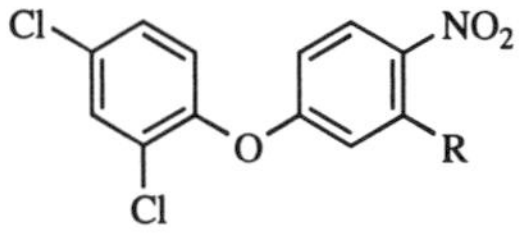

3

Nitrofen [R=H]
Bifenox [R=CO$_2$Me]

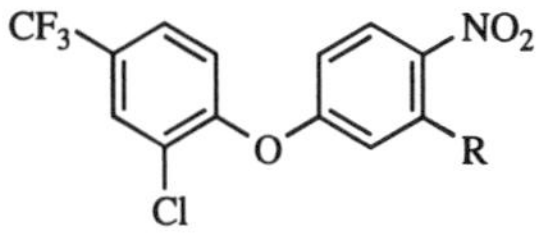

4

Oxyfluorfen [R=OEt]
Acifluorfen [R=CO$_2$H]
Fomesafen [R=CONHSO$_2$Me]
Fluoroglycofen-ethyl [R=CO$_2$CH$_2$CO$_2$Et]
Lactofen [R=CO$_2$CHMeCO$_2$Et]

that replacement of the 4-chloro substituent by a trifluoromethyl group causes a substantial increase in activity and a broadening of the weed spectrum. Following the introduction of oxyfluorfen (**4**; R = OEt),[13] acifluorfen (**4**; R = CO$_2$H)[14] was introduced for post-emergence weed control in soyabean. This has been followed by the introduction of fomesafen (**4**; R = CONHSO$_2$Me),[15,16] fluoroglycofen-ethyl (**4**: R = CO$_2$CH$_2$CO$_2$Et),[17] and lactofen (**4**; R = CO$_2$CHMeCO$_2$Et).[18] Recent biochemical studies have shown that these compounds act by inhibiting protoporphyrinogen IX oxidase.[19,20]

Two main synthetic approaches have been used to synthesize these compounds. The key process is the formation of the diphenyl ether bond via the reaction of a phenol with a halobenzene. To achieve a facile coupling reaction, the phenolic moiety needs to be electron-rich and the halobenzene electron-deficient. Reaction of 2-chloro-4-(trifluoromethyl)phenol (**5**; Y = OH) with the 5-chloro-2-nitrobenzoate (**6**; X = Cl) does not readily occur; however, if the more reactive 5-fluoro-2-nitrobenzoate (**6**; X = F) is used, the transformation proceeds smoothly to give the desired product.[21,22] The second approach utilizes the reaction of *meta*-substituted phenol (**7**) with 1,2-dichloro-4-(trifluoromethyl)benzene (**5**; Y = Cl), followed by nitration. The phenols used in this approach are the *meta*-hydroxy benzoate (**7**; R = CO$_2$R′)[23] and *meta*-cresol (**7**; R = Me), the methyl group of which is oxidized following coupling.[24,25] The acid thus obtained (**8**; R = CO$_2$H) can then be derivatized or nitrated, as necessary, to give **4**.[15] 2-Chloro-4-(trifluoromethyl)phenol (**5**; Y = OH) and 1,2-dichloro-4-(trifluoromethyl) benzene (**5**; Y = Cl) are readily available from chlorotoluene.

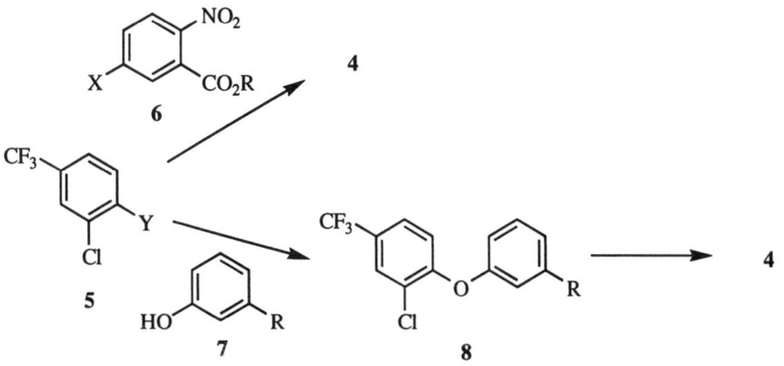

11.3.1.3. Carotenoid Biosynthesis Inhibitors

This group of compounds acts by inhibition of carotenoid biosynthesis. Carotenoids help to prevent overoxidation of chlorophyll in plants; in their absence plants become bleached and unable to photosynthesize.[26] Two examples of this class are norfluazon (9)[3] and fluridone (10).[27] Both compounds contain a *meta* disposed CF_3 group, which seems to be important for activity. Two new compounds, flurochloridone (11)[28] and diflufenican (12),[29–31] have been introduced. Flurochloridone (11) is sold for pre-emergence weed control in sunflowers. Diflufenican (12) is used in cereal crops.

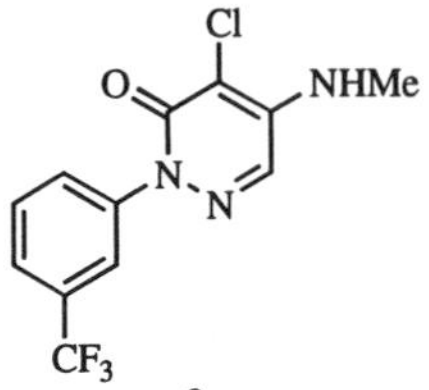

9	**10**
Norflurazon	Fluridone
11	**12**
Flurochloridone	Diflufenican

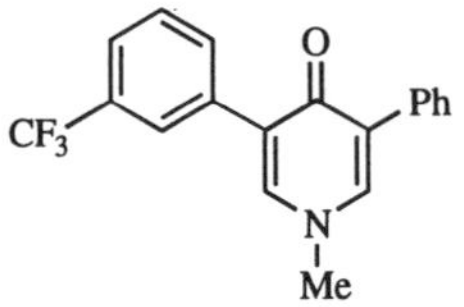

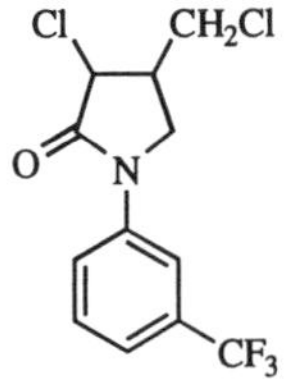

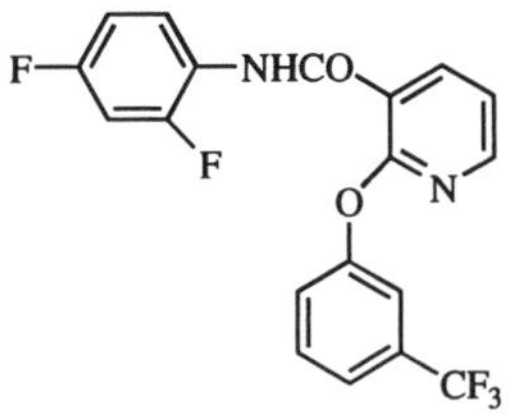

Flurochloridone (11) is synthesized from 3-trifluoromethyaniline (13; X = NH_2), which is obtained by nitration of trifluoromethylbenzene followed by reduction. Monoalkylation of the amine (13; X = NH_2) is achieved by converting it to the corresponding trifluoroacetamide using trifluoroacetic anhydride, followed by alkylation with allyl bromide and deprotection to give 14. Acetylation with dichloroacetyl chloride gives the amide 15, which is cyclized using ferric chloride.[32] The product 11 is obtained as a mixture of *cis* and *trans* isomers.

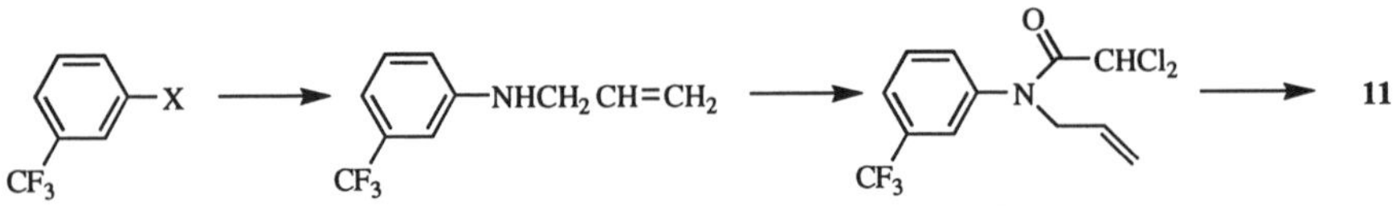

13	**14**	**15**	**11**
(X=NH_2)			

Diflufenican (12) is prepared via condensation of 3-trifluoromethylphenol (13; X = OH) with a 2-chloropyridine-3-carboxylic acid derivative 16. Conversion of the product **17** to the acid chloride is followed by amidation with 2,4-difluoroaniline to give diflufenican (**12**).[33] 2,4-Difluoroaniline can be prepared from 2,4-dichloro-1-nitrobenzene via a halex reaction with potassium fluoride, followed by reduction (see Chapter 9).

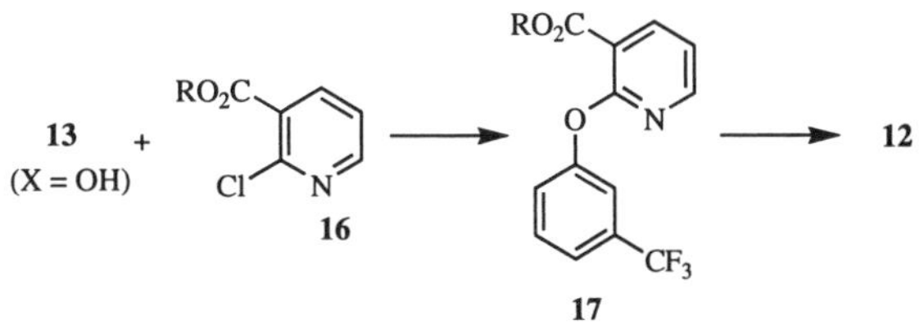

$$13 + (X = OH) \quad \textbf{16} \longrightarrow \textbf{17} \longrightarrow \textbf{12}$$

11.3.1.4. Aryloxyphenoxypropionates

These new and important herbicides show selective activity against grass species, and are widely used for post-emergence grass control in a wide range of broad leaf crops.[34] The compounds interfere with fatty acid biosynthesis via inhibition of acetyl-CoA carboxylase.[35] The initial compound discovered in this area was diclofop-methyl (**18**; Z = CH; R = Me), which is derived from 2,4-dichlorophenol. Further research led first to chlorazifop (**18**; Z = N; R = H),[36] and then to fluazifop-butyl (**19**; X = CF$_3$; Y = H; R = Bu^n) and haloxyfop-ethoxyethyl [**19**; X = CF$_3$; Y = Cl; R = EtO(CH$_2$)$_2$], which have found widespread use.[37–41] Introduction of the trifluoromethyl group increased activity and produced molecules which are active against a wide range of annual and perennial grasses. Because no suitable fluorinated starting materials were commercially available, synthetic approaches had to be developed—the work was being carried out in several companies' laboratories.

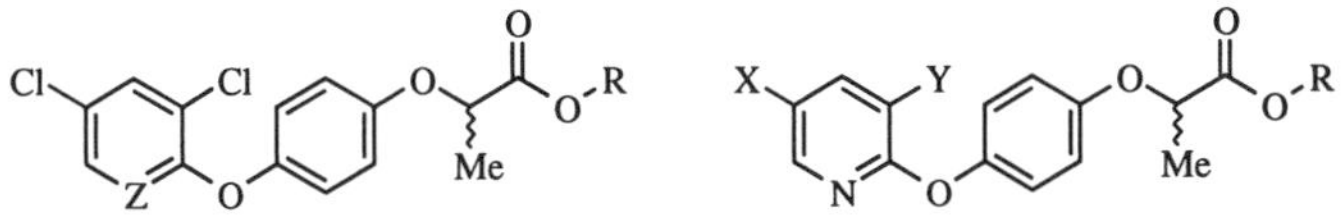

18
Diclofop-methyl [Z=CH, R=Me]
Chlorazifop [Z=N, R=H]

19
Fluazifop-butyl [X=CF$_3$, Y=H, R=n-Bu]
Haloxyfop-ethoxyethyl [X=CF$_3$, Y=Cl, R=EtO(CH$_2$)$_2$-]

Research in Japan[37] showed that the trifluoromethyl group could be introduced at the final stage of the synthesis using an approach developed by Kobayashi *et al.*[42] Coupling of the 5-bromo or 5-iodopyridyloxyphenoxypropionic acid derivative (**19**; X = Br or I) with trifluoroiodomethane in the presence of copper gave the desired product (**19**; X = CF$_3$) directly.

In other laboratories the trifluoromethyl group was introduced at an earlier stage. The key 2-halo-5-(trifluoromethyl)pyridine intermediates **22** can be produced either by treatment of the corresponding 2-chloropyridine-5-carboxylic acid (**20**) with sulfur tetrafluoride in the presence of anhydrous hydrogen fluoride,[39] or by the action of antimony trifluoride and chlorine on the appropriate trichloromethylpyridine (**21**).[37–39]

Various methods for the synthesis of 2- and 2,3-dihalo-5-(trichloromethyl)pyridines have been elucidated.[43–47] Reaction of the 2-halopyridines **22** with a 2-(4-hydroxyphenoxy)propionic acid derivative **23** gives **19** ($X = CF_3$).

20 21 22 23

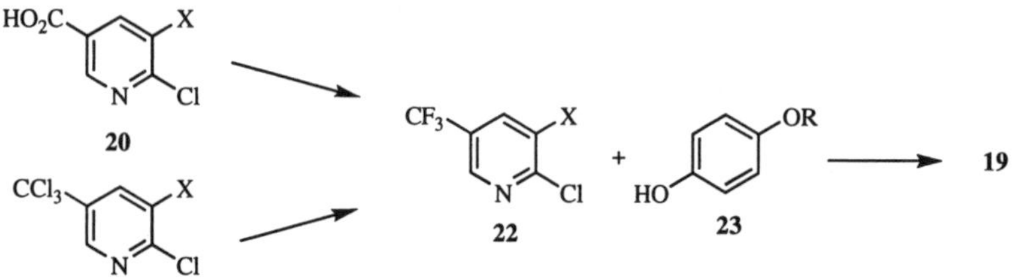

11.3.2. Herbicides Containing a Fluoroaromatic Group: Pyridyloxyacetic Acids

Fluroxypyr (**24**; R = 1-methylheptyl) is a translocated herbicide which controls a large range of broad leaf weeds in cereals. It enjoys a similar mode of action to 2,4-D (**25**),[48] but has the advantage of controlling resistant weed species.

24
Fluroxypyr [R= —$CHMe(CH_2)_5Me$]

25
2,4-D

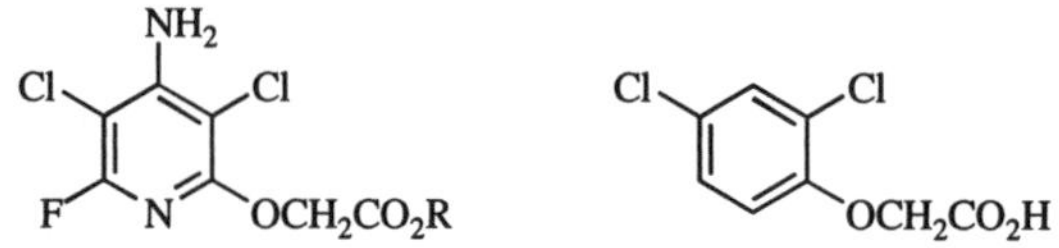

The compound is synthesized in several stages from pyridine.[49] Gas-phase chlorination gives pentachloropyridine, which readily undergoes halogen exchange with potassium fluoride at the activated 2-, 4-, and 6-positions to give 3,5-dichloro-2,4,6-trifluoropyridine.[50] On treatment with ammonia this gives the 4-aminopyridine **26**, which can then be hydrolyzed to give the corresponding 2-hydroxy compound **27**. Carboxymethylation of **27** with chloroacetic acid gives fluroxypyr acid (**24**, R = H),[51] which is converted to the methylheptyl ester by standard means.

26 **27**

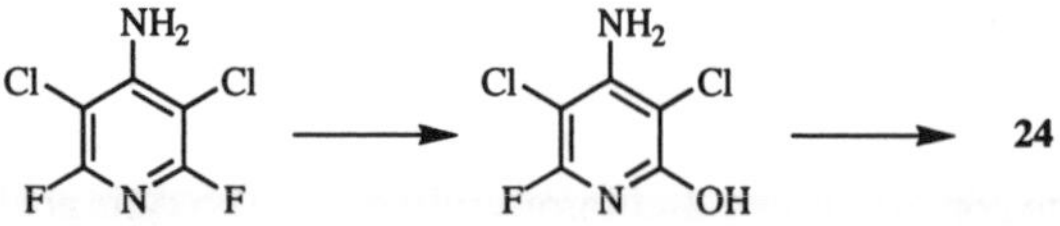

11.3.3. Herbicides Containing a Fluoroalkoxy Group: Sulfonylureas

The most important herbicide discovery of the past decade has been a group of compounds which act by the inhibition of acetolactate synthetase (ALS).[52] These compounds are very highly active, with the sulfonyl ureas being applied at rates of a few grams per hectare. Several different families have been discovered, including sulfonyl ureas (the most important group), imidazolidinones, and sulfonanilides. So far, only one commercial product containing fluorine has been introduced, namely, primsulfuron methyl (**28**; R = OCHF$_2$).[53,54] This resembles sulfometuron methyl (**28**; R = Me) but contains haloalkoxy substituents on the pyrimidine ring: it shows selectivity in maize.

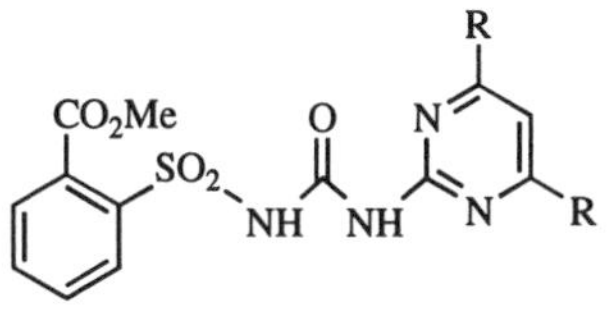

28
Primsulfuron methyl [R=OCHF$_2$]
Sulfometuron methyl [R=Me]

Synthesis of **28** (R = OCHF$_2$) involves treatment of the pyrimidine **30** with a sulfonyl isocyanate (**31**) derived from saccharin. The required pyrimidine is obtained via capture by dihydroxypyrimidine (**29**; X = SMe) of difluorocarbene, generated *in situ* from chlorodifluoromethane and sodium hydroxide.[55–57] The methylthio group is oxidized to methanesulfonyl, which is readily displaced with ammonia. Addition of the amine (**30**) thus procured to the isocyanate (**31**) gives primsulfuron methyl (**28**; R = OCHF$_2$).[58]

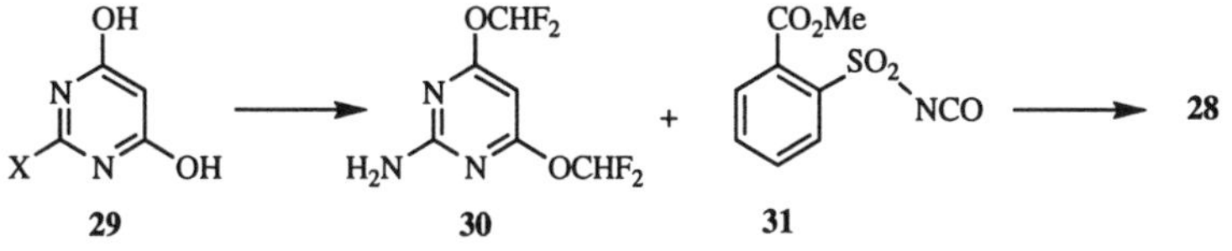

29   **30**   **31**   **28**

11.3.4. Herbicides Containing a Trifluoromethanesulfonyl Group: Trifluoromethanesulfonanilides

Although mefluidide (**32**; X = Me; Y = NHAc) and perfluidone (**32**; X = SO$_2$Ph; Y = H) were introduced in the mid-seventies,[59,60] there have been no subsequent developments in this area. The compounds are worthy of mention because, within the agrochemical area, they are rare examples which contain a fluorinated substituent arising from an electrochemical process: they are prepared via treatment of appropriate anilines with trifluoromethanesulfonyl fluoride, which is produced by Simons electrochemical fluorination of methanesulfonyl fluoride (see Chapter 5).[61]

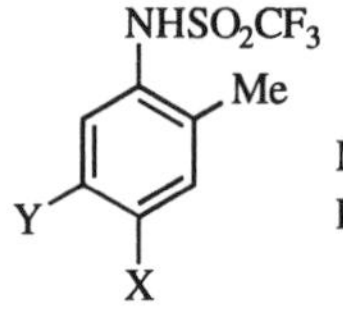

Mefluidide [X=Me, Y=NHAc]
Perfluidone [X=SO₂Ph, Y=H]

32

11.4. FLUORINE-CONTAINING INSECTICIDES

11.4.1. Compounds Affecting Insect Growth

The insect growth hormones were discovered in the mid-1970s and act by inhibiting chitin synthesis in insects. Hassall[62] and Maas *et al.*[63] have reviewed work on their chemistry and mode of action. Diflubenzuron (**33**) was the first compound introduced to the market[3]; subsequent research has led to the discovery of chlorfluazuron (**34**),[64] teflubenzuron (**35**),[65] and flufenoxuron (**36**).[66]

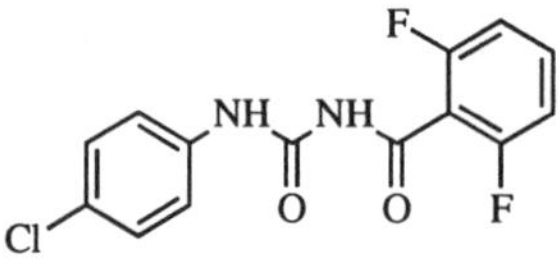

33
Diflubenzuron

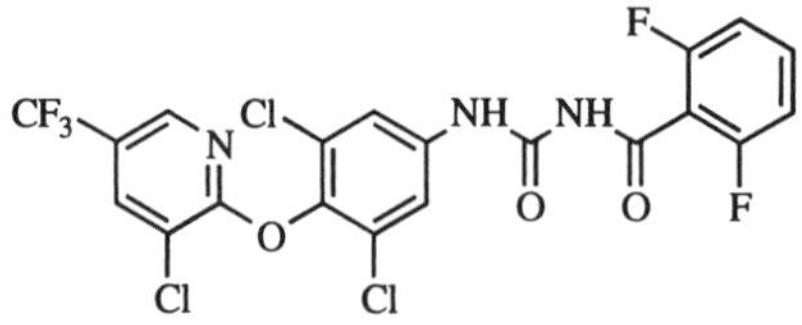

34
Chlorfluazuron

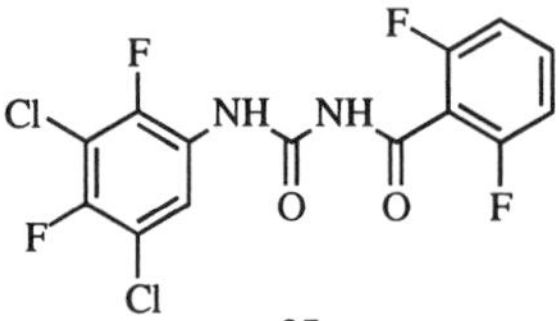

35
Teflubenzuron

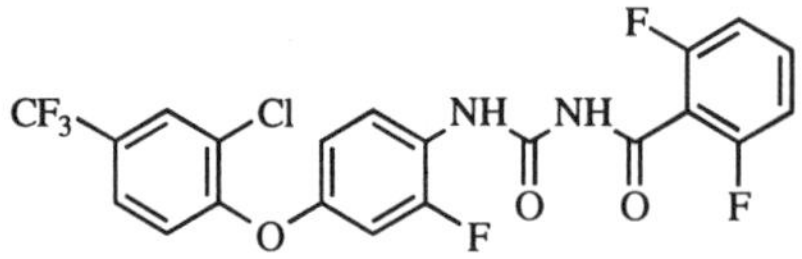

36
Flufenoxuron

These compounds can be synthesized from the appropriate isocyanate (**37**) and 2,6-difluorobenzamide (**38**). The benzamide **38** is obtained from 2,6-dichlorobenzonitrile by halogen exchange using potassium fluoride, followed by hydrolysis of the nitrile group to amide. The aryl isocyanates are prepared from the corresponding aniline and phosgene.

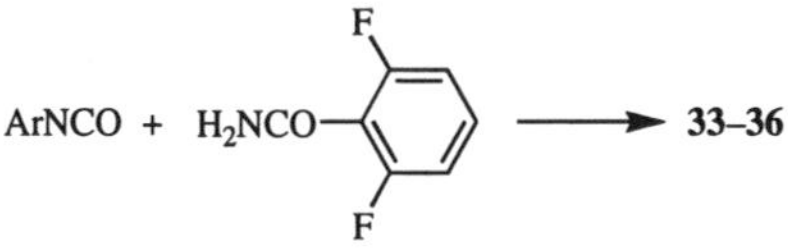

$$ArNCO + H_2NCO\text{—}\quad\quad \longrightarrow \quad 33\text{–}36$$

37   38

The anilines required for chlorflurazon (34) and flufenoxuron (36) are synthesized by conventional approaches. Reaction of 2,3-dichloro-5-(trifluoromethyl)pyridine (22; X = Cl) (see Section 11.3.1.4) with 2,6-dichlorophenol gives the phenoxypyridine (39; Z = N; R^1 = 2,6-diCl; R^2 = H), which is converted to the corresponding benzenamine by nitration followed by reduction.[64] Similarly, reaction of 1,2-dichloro-4-(trifluoromethyl)benzene (5: Y = Cl) (see Section 11.3.1.2) with 3-fluoro-4-nitrophenol gives the diphenyl ether (39; Z = CH; R^1 = 3-F; R^2 = 4-NO$_2$) which is reduced to the required aniline.[67,68]

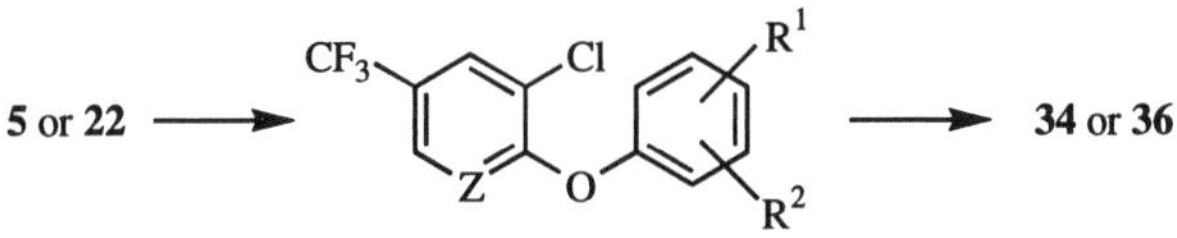

$$5 \text{ or } 22 \quad \longrightarrow \quad \text{[structure 39]} \quad \longrightarrow \quad 34 \text{ or } 36$$

39

The aniline required for teflubenzuron (35) is obtained via mononitration of 1,2,3,4-tetrachlorobenzene to give 40. The nitro group in 40 activates the *ortho* and *para* chlorine substituents to nucleophilic attack, hence they can be replaced by fluorine using potassium fluoride.[69,70] The resulting 1,3-dichloro-2,4-difluoro-5-nitrobenzene (41) is converted to the benzenamine intermediate by reduction.

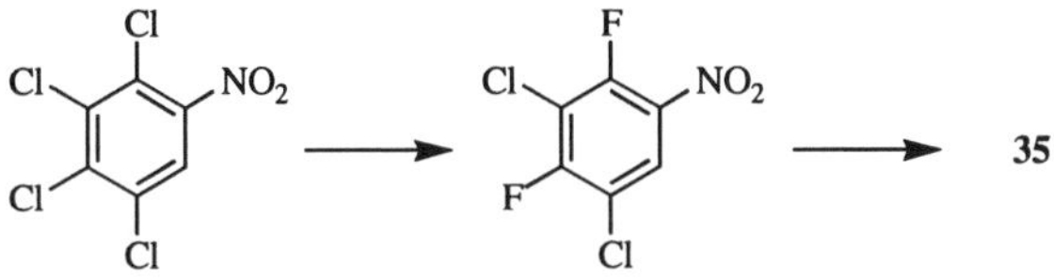

40   41   35

11.4.2. Pyrethroid Insecticides

11.4.2.1. Introduction

The pyrethrins are potent insecticides derived from the plant species *Chrysanthemum cinerariafolium*. They have extremely good knock-down properties but are insufficiently stable for commercial use. Bioallethrin (42) and bioresmethrin (43), which contain modified "alcohol portions," were introduced to the domestic (household) market but were still insufficiently stable for more widespread use.[71] In the early 1970s Elliott made a major breakthrough with the discovery of permethrin (44; X = Y = Cl; R = W = V = H; Z = OPh) and cypermethrin (44; X = Y = Cl; R = CN; W = V =

H; Z = OPh).[72,73] These are considerably more photolytically and metabolically stable than **42** or **43** and suitable for widespread use in agriculture. The increased stability of these molecules was achieved by replacing the dimethylvinyl group with dichlorovinyl, and by using 3-phenoxybenzyl alcohol as the "alcohol." Subsequently, more active compounds containing fluorine have been discovered, as discussed below. Other research led to the discovery of a new type of pyrethroid, derived from 2-(4-chlorophenyl)-3-methylbutanoic acid.[74] The parent compound of this series is fenvalerate (**45**, X = Cl),[75] and recently two fluorinated analogs have been introduced.

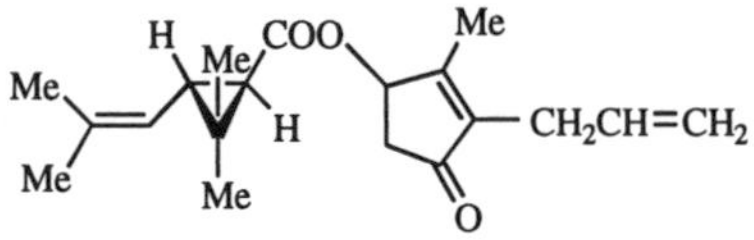

42
Bioallethrin

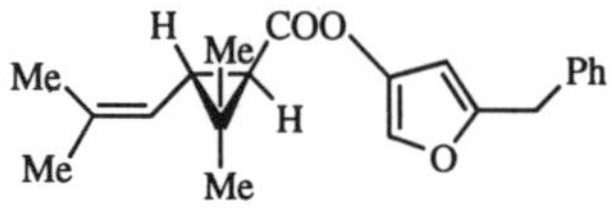

43
Bioresmethrin

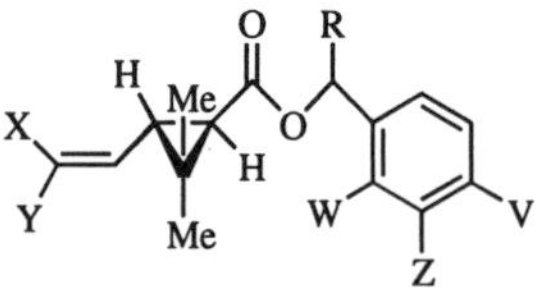

44
Permethrin [X=Cl, Y=Cl, R=H, W=H, V=H, Z=OPh]
Cypermethrin [X=Cl, Y=Cl, R=CN, W=H, V=H, Z=OPh]
Cyhalothrin [X=CF₃, Y=Cl, R=CN, W=H, V=H, Z=OPh]
Bifenthrin [X=CF₃, Y=Cl, R=H, W=Me, V=H, Z=Ph]
Cyfluthrin [X=Cl, Y=Cl, R=CN, W=H, V=F, Z=OPh]

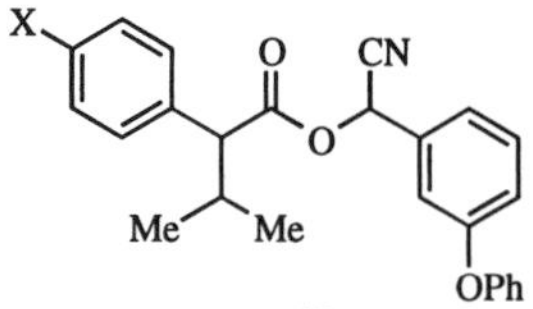

45
Fenvalerate [X=Cl]
Flucythrinate [X=OCHF₂]

11.4.2.2. Acid Variants

Modification studies on the "acid portion" of permethrin led to the discovery that replacement of one of the vinylic chlorines by a trifluoromethyl group enhanced activity considerably. Three compounds, cyhalothrin (**44**; X = CF₃; Y = Cl; R = CN; Z = OPh; W = V = H),[76–78] bifenthrin (**44**; X = CF₃; Y = Cl; R = V = H; Z = Ph; W = Me),[79,80] and tefluthrin (**46**),[81] based on this structural change, are marketed, each one being synthesized by esterification of acid **49** (R = H) with the appropriate alcohol or cyanohydrin.

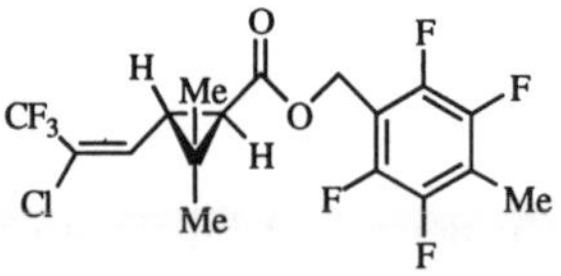

46
Tefluthrin

The alcohols required for cyhalothrin and bifenthrin are readily available; the synthesis of the tetrafluorobenzyl alcohol required for tefluthrin (46) is described later (Section 11.4.2.4). The acid portion (49; R = H) stems from olefin 47 and 1,1,1-trichloro-2,2,2-trifluoroethane, which combine under radical conditions [using copper(I) chloride and 2-aminoethanol] to give adduct 48.[82-86] This can be cyclized and dehydrohalogenated in one step using sodium butoxide in 1,2-dimethoxyethane. Hydrolysis of the ester obtained (49; R = Et) (largely the *trans* isomer) gives the desired acid (49; R = H). Bentley *et al.* deal with the various approaches available for the synthesis of 49 (R = H) and related acids.[77]

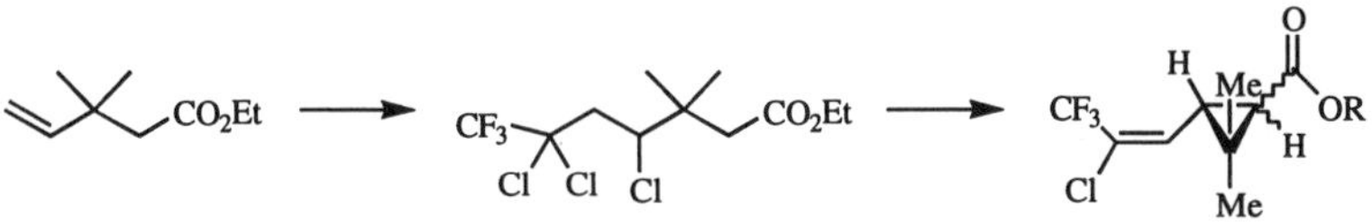

11.4.2.3. Fenvalerate Types

The introduction of fenvalerate (45; X = Cl) to the market was followed by the discovery of flucythrinate (45; X = OCHF$_2$)[87] and fluvalinate (50).[88-90] Flucythrinate (45; X = OCHF$_2$) can be synthesized from 4-hydroxytoluene via a multistage process[91] in which the OH group is first converted to OCHF$_2$ by reaction with difluoromethylene (CHF$_2$Cl/NaOH),[92,93] before the nitrile group (see 52) is introduced by controlled radical chlorination followed by treatment of the benzyl chloride thus produced with sodium cyanide in dimethyl sulfoxide. Subsequently, the nitrile (52) is alkylated with isopropyl bromide under basic conditions and the product (53; Z = CN) hydrolyzed to the corresponding racemic acid (53; Z = CO$_2$H), which can be resolved via treatment with an optically active base at this stage, if desired.[94] Finally, treatment of the acid chloride of 53 (Z = COCl) with 3-phenoxybenzaldehyde in the presence of sodium cyanide gives flucythrinate (45; X = OCHF$_2$).

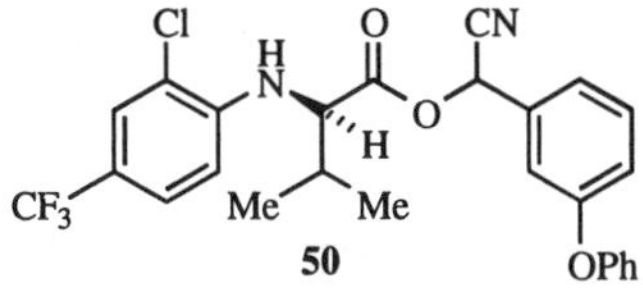

50
Fluvalinate

Alternatively, the difluoromethoxy substituent may be introduced at a later stage in the sequence. In this case the starting material is 4-methoxytoluene. A parallel reaction sequence is carried out, the OH protecting group being removed at the nitrile hydrolysis stage and the resulting phenol treated with the difluorocarbene source.

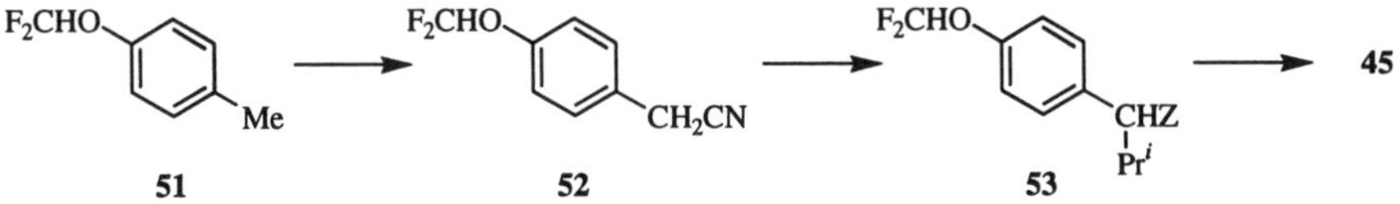

Two similar approaches have been suggested for the synthesis of fluvalinate (50).[95,96] In one, aniline **54** (X = H) derived from 1-chloro-4-(trifluoromethyl)benzene is condensed with bromoacid **55**, the reaction proceeding with inversion of stereochemistry to give **56** (X = H). Chlorination with *N*-chlorosuccinamide gives the desired acid (**56**; X = Cl). Alternatively, 2-chloro-4-(trifluoromethyl)aniline (**54**; X = Cl) is converted to the aminonitrile **57** with isobutyraldehyde in the presence of cyanide. Hydrolysis of the CN group gives acid **56** (X = Cl), which can be resolved to give the desired stereoisomer; acid chloride formation followed by reaction with the cyanohydrin of 3-phenoxybenzaldehyde gives fluvalinate (**50**).

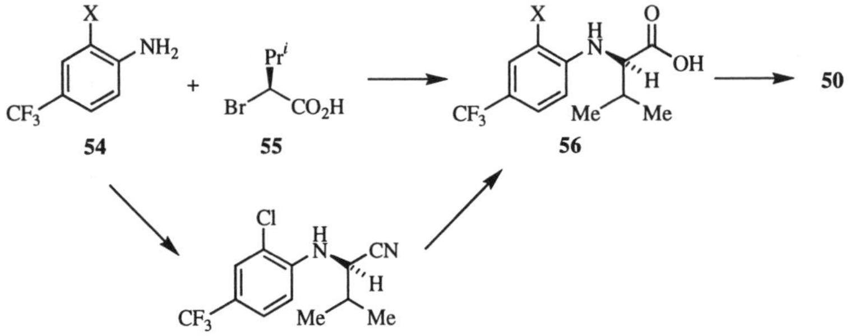

11.4.2.4.   Alcohol Variants

Research on the introduction of fluorine into the "alcohol portion" of pyrethroids has led to the development of cyfluthrin (**44**; X = Cl; Y = Cl; R = CN; W = H; Z = OPh; V = F)[97–99] and tefluthrin (**46**).[81] While cyfluthrin services traditional pyrethroid markets, tefluthrin (**46**) is the first pyrethroid to be introduced for the control of soil borne pests, effective control of which is usually only achieved with volatile materials able to permeate through the soil. Traditional pyrethroids are insufficiently volatile to do this, and tefluthrin (**46**) was designed to combine high activity with increased volatility. The compound is used to control corn root worms.[100]

Several approaches have been used to synthesize the alcohol portion of cyfluthrin. All rely on the synthesis of a compound related to 5-methyl or 5-carboxy-2-fluorobromobenzene (**58**; E = Me or CO_2R) which reacts with phenoxide in the presence of copper or copper(I) oxide to give a diphenyl ether (**59**).[101,102] The 5-substituent is then converted to aldehyde **60** by either oxidation or reduction as appropriate and cyfluthrin is made from the cyanohydrin of **60** and the appropriate acid chloride.

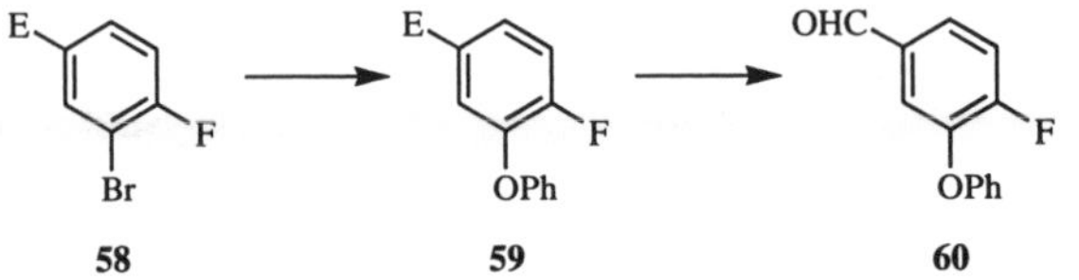

Bromination of 1-chloro-4-(trichloromethyl)benzene gives 1-bromo-2-chloro-5-(trichloromethyl)benzene (**61**) which can be partially hydrolyzed to give the acid chloride (**62**; R = X = Cl).[103] Reaction of the acid chloride with potassium fluoride allows selective exchange of the activated 4-chloro substituent to give the bromofluorobenzoyl fluoride (**63**), which can then be converted to the diphenyl ether (**60**).

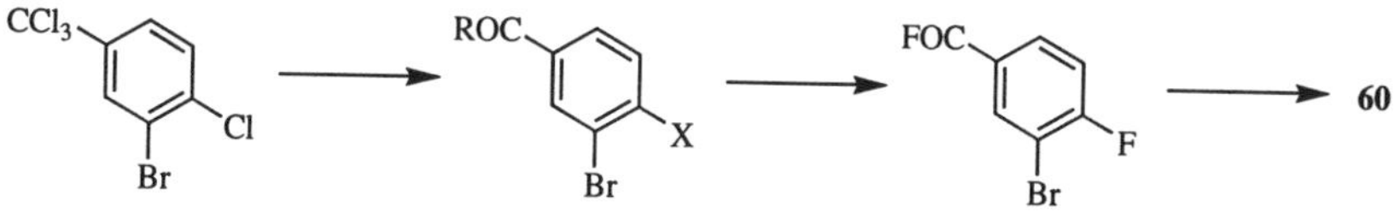

CCl₃ structure with Cl and Br (**61**) → ROC structure with X and Br (**62**) → FOC structure with F and Br (**63**) → **60**

Alternatively, acylation of fluorobenzene with acetyl chloride and aluminum trichloride under Friedel-Crafts conditions gives the 4-acetyl derivative **64**. Electrophilic bromination occurs *ortho* to the fluorine substituent to give the 4-bromophenylethanone (**65**), which is converted to the acid (**62**; R = OH; X = F) by reaction with sodium hypochlorite.[104] In a different approach, diazotization of 2-bromo-4-methylbenzenamine in aqueous hydrofluoroboric acid gives the diazonium fluoroborate, which can be decomposed to give the bromofluorotoluene (**66**). The intermediate is then converted to the diphenyl ether as described above.[105]

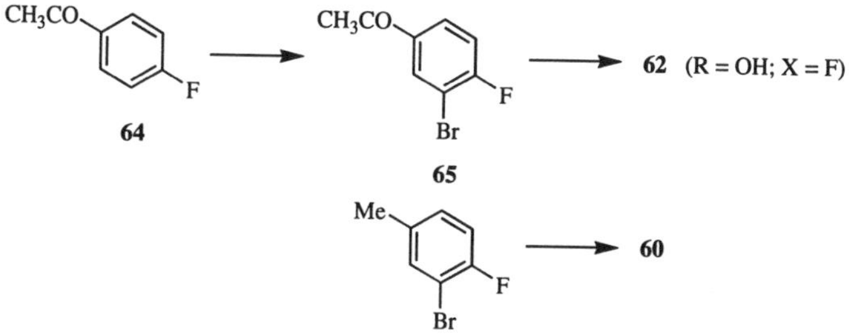

CH₃CO structure with F (**64**) → CH₃CO structure with Br and F (**65**) → **62** (R = OH; X = F)

Me structure with Br and F (**66**) → **60**

The alcohol portion of tefluthrin (**46**) is derived from 1,2,4,5-tetrafluoro-3,6-dimethylbenzene (**67a**), the monobromination product of which (**67**; A = Me; B = CH₂Br) is treated with the acid **49** (R = H; see Section 11.4.2.2) to give tefluthrin.[106] The synthesis of the key xylene has been approached in a number of ways. Lithiation of 1,2,4,5-tetrafluorobenzene (**67**; A = B = H) gives a dilithium derivative, which can be methylated to give the desired product.[107] Alternatively, tetrafluoro-terephthalonitrile (**67**; A = B = CN), which is readily obtained by gas-phase chlorination of terephthalonitrile followed by chlorine for fluorine exchange with potassium fluoride, is converted to **68** with a methylmagnesium halide and thence via further alkylation to **67a**.[108] The Grignard reagent presumably attacks the ring carbon rather than the nitrile because of the presence of the electronegative ring fluorine atoms. Reduction of tetrafluoroterephthalonitrile followed by exhaustive methylation, then hydrogenation of the resultant bis(trimethylammonium)derivative **69** (R = Me), provides a third route to **67a**.[109]

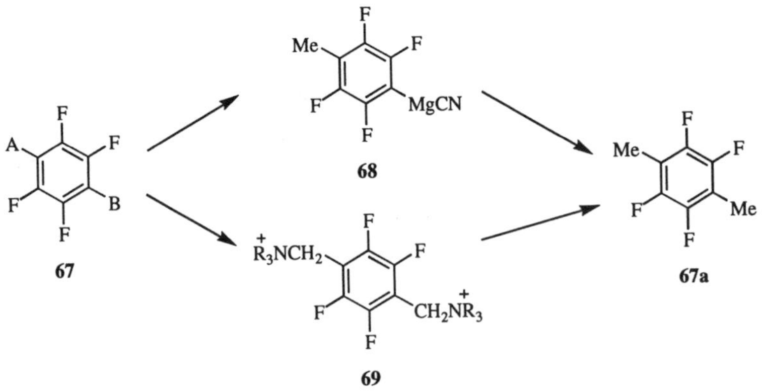

11.4.3. Others

Hydramethylnon (72) is used to control ants.[110,111] It is synthesized by the condensation of two moles of 4-(trifluoromethyl)benzaldehyde with acetone to give 70, which reacts with the hydrazine 71 to give 72.[112]

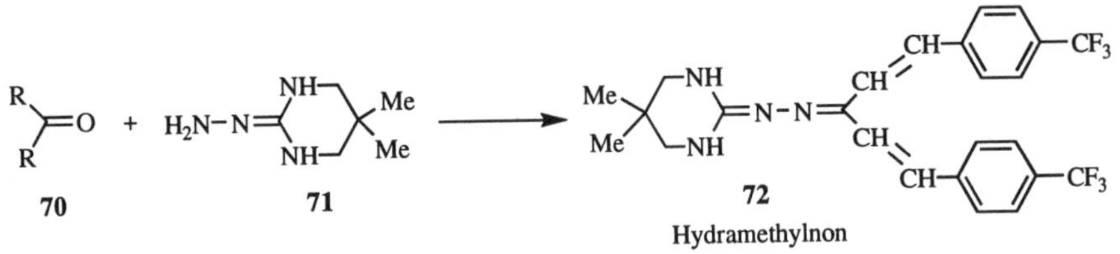

11.5. FLUORINE-CONTAINING FUNGICIDES

11.5.1. Sterol Biosynthesis Inhibitors

The major focus of attention in fungicide chemistry for the past 15 years has been the exploration of sterol biosynthesis inhibitors. These compounds are extremely effective fungicides which act by blocking the C14 alpha demethylation step in ergosterol biosynthesis.[113] The first examples of the main classes of compound are triadimefon (73), imazalil (74), and triarimol (75). These compounds bear a close structural resemblance to certain plant growth regulators which have a similar mode of action (see Section 11.6.1).

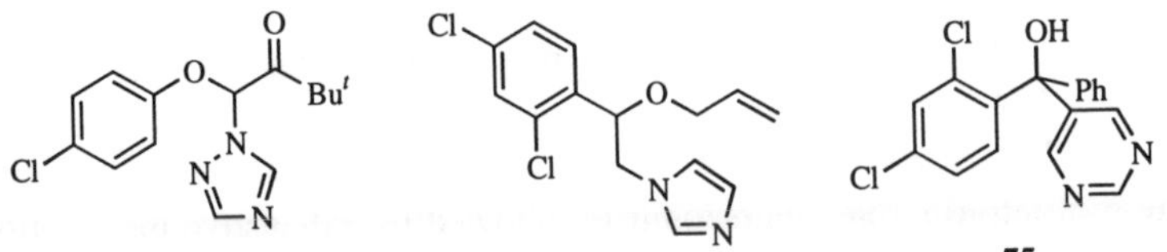

73	74	75
Triadimefon	Imazalil	Triarimol

Extensive research in the area has uncovered a wealth of interesting fungicides, and many of those introduced to the market contain fluorine. The work has been described in detail[114–118]; representative fluorine-containing products are flutriafol (**76**)[119,120] flusilazole (**77**),[121,122] triflumizole (**78**),[123] fluotrimazole (**79**),[124–126] and nuarimol (**80**; X = F; R = 2-ClC$_6$H$_4$).[127,128]

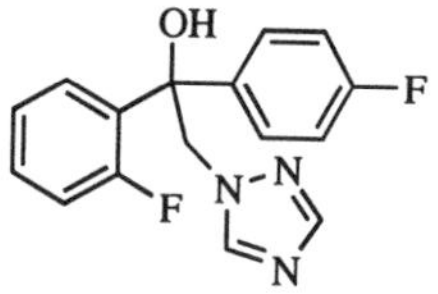

76
Flutriafol

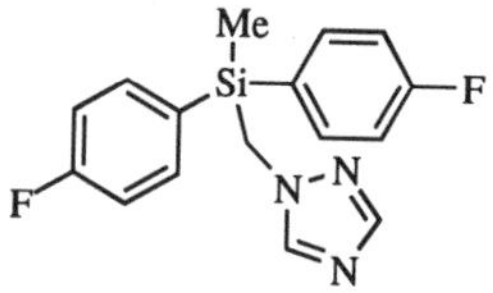

77
Flusilazole

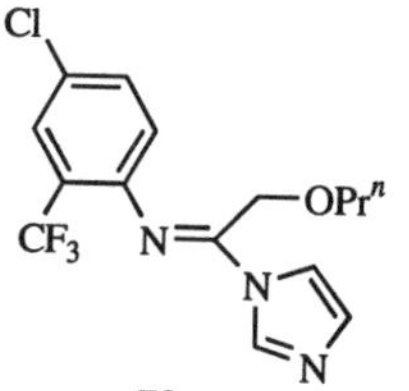

78
Triflumizole

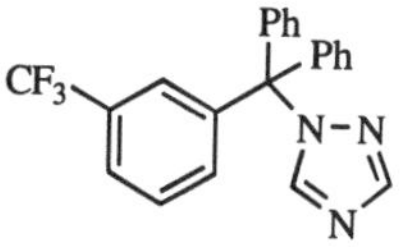

79
Fluotrimazole

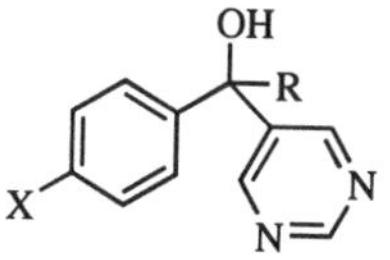

80
Nuarimol [X = F, R = 2-ClC$_6$H$_4$-]
Flurprimidol [X = OCF$_3$, R = i-Pr]

11.5.1.1. 1,2,4-Triazoles

The synthesis of flutriafol (**76**) starts from the Friedel-Crafts acylation of fluorobenzene with chloroacetyl chloride to give **81**, which reacts with the Grignard reagent **82** to give the chlorohydrin **83**. This is condensed with triazole under basic conditions to yield flutriafol.[129] Alternatively, syntheses start with (2-fluorophenyl)(4-fluorophenyl)methanone, which is converted to either the epoxide or the corresponding halohydrin (**83**), and thence to flutriafol (**76**).

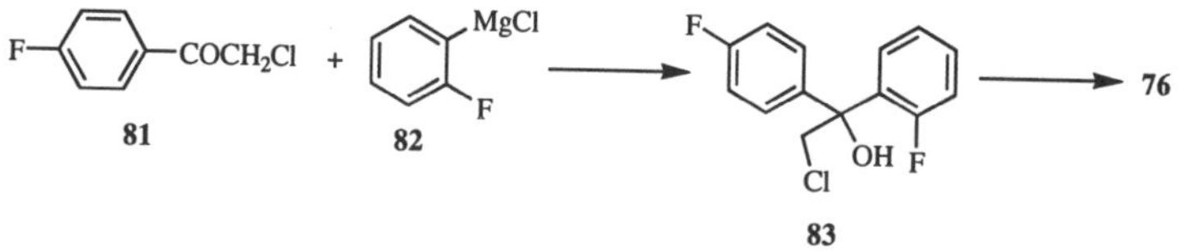

81   **82**   **83**   **76**

A recently introduced triazole fungicide, flusilazole (**77**), contains both fluorine and silicon. Chlorination of dichlorodimethylsilane gives dichloro(chloromethyl)methylsilane (**84**), which is treated with 4-fluorophenyl-lithium to cause replacement of the two chlorine atoms bonded to silicon to give the (chloromethyl)silane **85**. Substitution of the carbon-bonded chlorine in **85** using the sodium salt of 1,2,4-triazole gives flusilazole (**77**).[130]

ArLi + **84** ───→ **85** ───→ **77**

Fluotrimazole (**79**) is synthesized by an interesting sequence of reactions. Radical chlorination of *meta*-xylene gives 1,3-bis(trichloromethyl)benzene (**86**; X = Y = Cl), which can be converted to the corresponding bis(trifluoromethyl)benzene (**86**; X = Y = F) by halogen exchange. When treated with aluminum trichloride and hydrochloric acid, a mixture of these two intermediates gives 1-(trichloromethyl)-3-(trifluoromethyl)benzene (**86**; X = F; Y = Cl). This benzotrichloride derivative combines with benzene in the presence of aluminum trichloride to provide 1-(chlorodiphenylmethyl)-3-(trifluoromethyl)benzene (**87**), which is converted to fluotrimazole (**79**) with 1,2,4-triazole.[124]

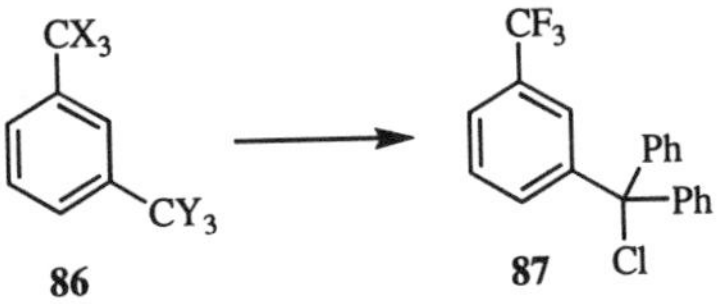

86 ───→ **87**

11.5.1.2. Imidazoles

Only one fluorine-containing imidazole derivative, triflumizole (**78**), has reached the market so far. This is synthesized via treatment of 4-chloro-2-trifluoromethylaniline (**88**) with propyloxyacetyl chloride; the resultant amide (**89**) is then converted to triflumizole (**78**) via the imidoyl chloride **90**.[131]

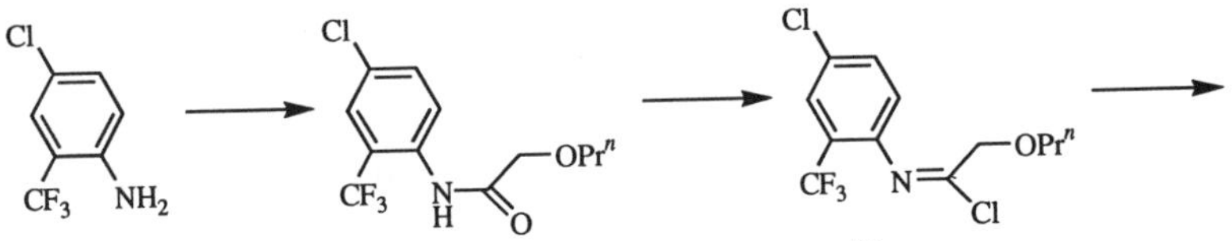

88 ───→ **89** ───→ **90** ───→ **78**

11.5.1.3. Pyrimidines

Interestingly, while only one fluorine-containing pyrimidine, nuarimol (**80**; X = F; R = 2-ClC$_6$H$_4$), is sold as a fungicide, a close analog, flurprimidol (**80**, X = OCF$_3$; R = Pr^i), is marketed as a plant growth regulator. Both compounds are synthesized via addition of pyrimidin-5-yl-lithium (**92**; from metalation of 5-bromopyrimidine) to an appropriate ketone (**91**).[132]

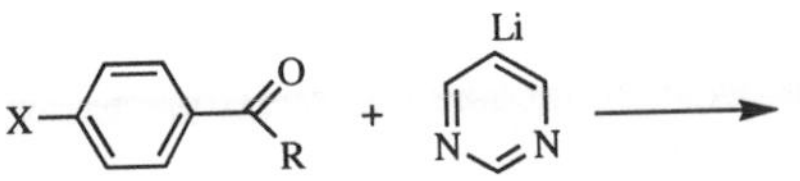

91 + **92** ───→ **80**

11.5.2. Amide Fungicides

Flutolanil (**93**), a fungicide active against *Basidomycetes*, is used in rice, cereals, and vegetables.[133] It can be prepared by the reaction of a 2-trifluoromethylbenzoic acid derivative (**96**) with 3-isopropoxyaniline.[134]

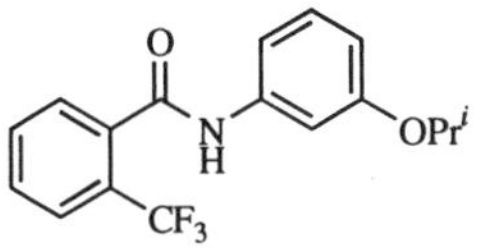

93
Flutolanil

Several procedures are reported for the synthesis of the benzoic acid. Radical chlorination of 2-methylbenzoyl chloride gives its trichloromethyl analog **94**, which can be converted to the trifluoromethyl derivative **95** via a halogen-exchange reaction with hydrofluoric acid[135] that proceeds in two steps, giving **95** via a tetrafluoro ether intermediate.[136] Acid **95** can also be synthesized by fluorination of the half ester of phthalic acid (**97**) with sulfur tetrafluoride; this converts the acid group to a trifluoromethyl group without affecting the ester. Hydrolysis then liberates the desired acid.[137] Conversion of the acid to the acid chloride (**96**, R = Cl) and amidation with 3-aminophenol gives an intermediate phenol, which can be alkylated under basic conditions to provide **93**.[134]

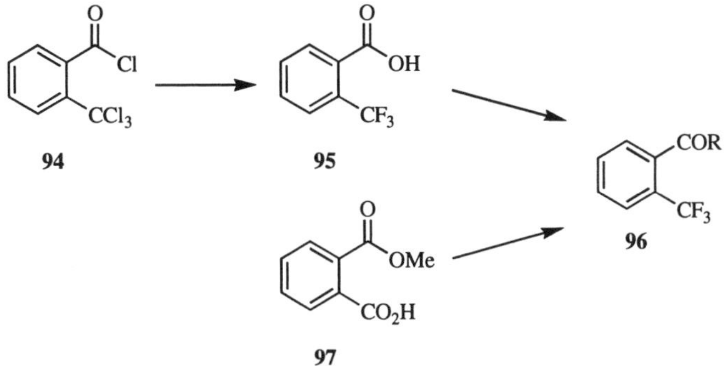

11.6. FLUORINE-CONTAINING PLANT GROWTH REGULATORS

11.6.1. Compounds That Interfere with Gibberellin Biosynthesis

Members of this group of chemicals are structurally related to the fungicidal inhibitors of sterol biosynthesis discussed in Section 11.5.1, but act by interfering with Gibberellin biosynthesis. The only compound containing fluorine which has progressed to the market from this group is fluprimidol (**80**; X = OCF$_3$; R = Pr^i).[138] It is synthesized by the approach described for its analog nuarimol (**80**; X = F; R = 2-ClC$_6$H$_4$) in Section 11.5.1.3. In this case, the ketonic intermediate **91** (X = OCF$_3$; R = Pr^i) required for use with lithio-pyrimidine **92** can be prepared from 4-methoxybenzoyl chloride (**98**). This is first chlorinated (radical conditions) to give the trichloromethoxy derivative **99** which is converted by halogen exchange to **100**.[139] The nitrile **101** is then prepared from the acid fluoride by standard means, and is converted by treatment with a Grignard agent to the ketonic precursor (**91**; X = OCF$_3$; R = Pr^i) of fluprimidol. Alternatively, a 4-hydroxybenzoate (**102**) is converted to the fluoroformate (**103**) by reaction with carbonyl fluoride; reaction with sulfur tetrafluoride then gives the trifluoromethoxy intermediate (**100**).[140]

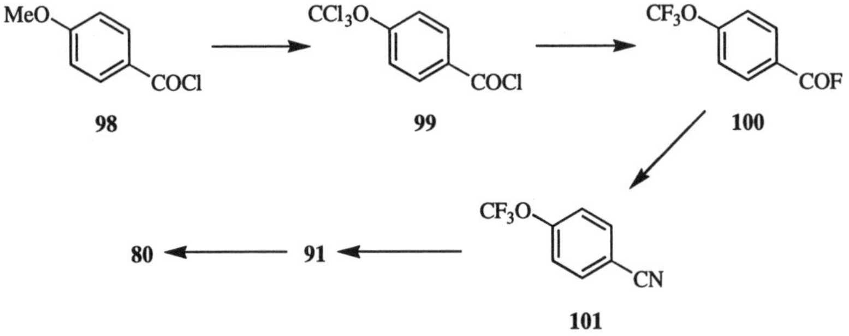

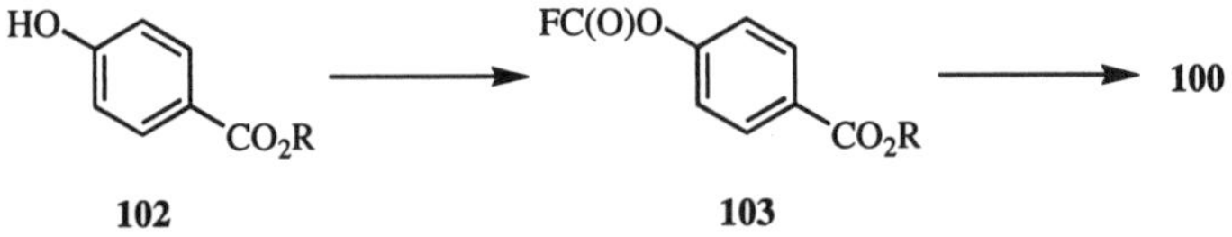

11.6.2. Benzylamine Derivatives

Flumetralin (**106**) is a plant growth regulator used to control suckers on tobacco.[141] It is prepared via an S$_N$Ar reaction between the trifluralin precursor 2-chloro-1,3-dinitro-5-(trifluoromethyl)benzene (**2**; X = Y = NO$_2$; see Section 11.3.1.1) and a benzylamine (**105**) made from 2-chloro-6-fluorobenzonitrile.[142] Conversion of this nitrile to aldehyde **104** is followed by a reaction with ethylamine and reduction of the resultant imine to give **105**. Reaction of **105** with the electron-deficient trifluoromethylbenzene **2** proceeds smoothly to give flumetralin (**106**).[143]

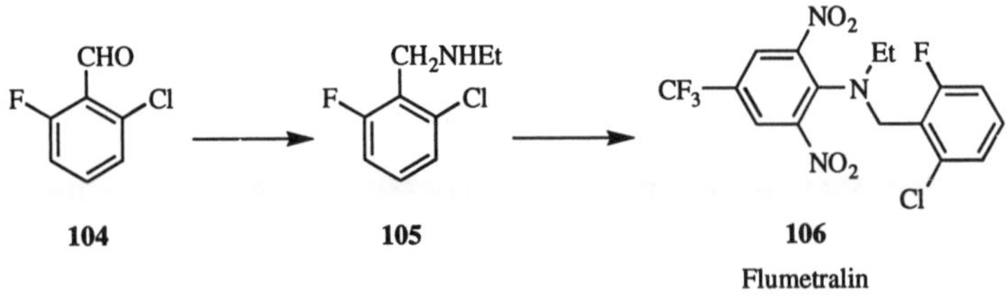

11.7. RODENTICIDES

Flocoumafen (**107**) is an anticoagulant which is active against rats resistant to other agents. The fluorine-containing portion is derived from 1-methyl-4-(trifluoromethyl)benzene and can be introduced into the molecule at a late stage of the synthesis.[144] Bromethalin (**108**) has a different mode of action[145] and, although appearing straightforward to synthesize, poses several constructional problems owing to the difficulty encountered in producing an N-methyl diphenylamine which has four bulky substituents *ortho* to the amino group.[146] One successful route employs the reaction of N-methylaniline (**110**) with 2-fluoro-1,5-dinitro-3-(trifluoromethyl)benzene (**109**) to give diphenylamine **111** (the ring-fluoro derivative is used rather than the chloro to enhance reactivity). Bromination of **111** proceeds smoothly to give **108**.[146] In an alternative synthesis, 2,4,6-tribromoaniline (**113**) is coupled with 1-chloro-4-nitro-2-(trifluoromethyl)benzene (**112**) to give a diphenylamine (**114**, R = H) which is methylated under basic conditions to give **114** (R = Me) which is then nitrated to give **108**.[147]

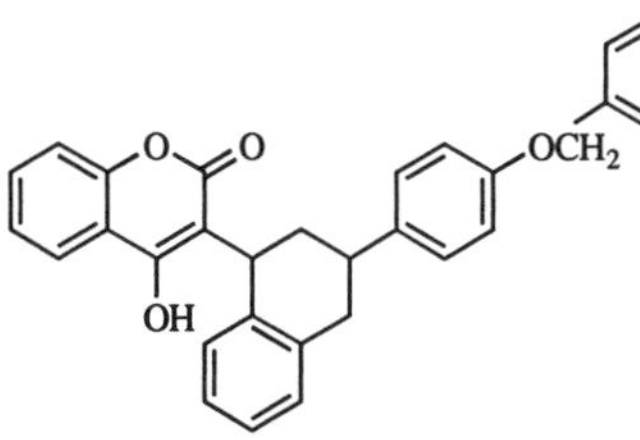

| 107 | 108 |
| Flocoumafen | Bromethalin |

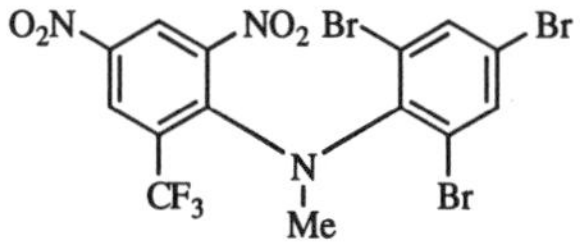

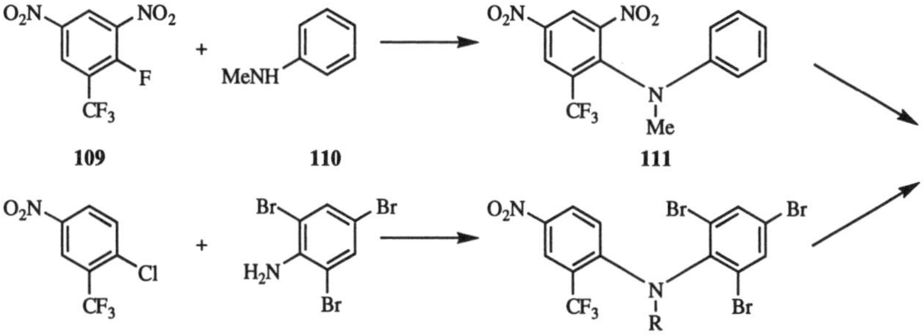

109 + 110 → 111 → 108

112 + 113 → 114 →

11.8. REFERENCES

1. J. R. Finney, *Proc. Brighton Crop Protection Conf.—Pests & Diseases 1*, 3 (1988).
2. I. J. Graham-Bryce, *Proc. Brighton Crop Protection Conf.—Weeds 1*, 3 (1989).
3. G. T. Newbold, in: *Organofluorine Chemicals and their Industrial Applications* (R. E. Banks, ed.), p. 169, Ellis Horwood, London (1979).

4. H. Martin and C. R. Worthing (eds.), *Pesticide Manual*, 5th edn., British Crop Protection Council, London (1977).
5. C. R. Worthing (ed.), *Pesticide Manual*, 9th edn., British Crop Protection Council, London (1990).
6. *Agrochemicals Handbook*, 2nd edn., The Royal Society of Chemistry (1988).
7. M. B. Green, G. S. Hartley, and T. F. West, *Chemicals for Crop Protection and Pest Control*, 3rd edn., Pergamon Press, Oxford (1987).
8. G. W. Probst and W. L. Wright, in: *Herbicides, Chemistry, Degradation and Mode of Action* (P. C. Kearney and D. D. Kaufman, eds.), Vol.1, p. 453, Marcel Dekker, New York and Basel (1975).
9. E. F. Alder, W. L. Wright, and Q. F. Soper, *Proc. N. Central Weed Conf.* 23 (1960).
10. S. Matsunaka, in: *Herbicides, Chemistry, Degradation and Mode of Action* (P. C. Kearney and D. D. Kaufman eds.), Vol. 2, p. 709, Marcel Dekker, New York and Basel (1976).
11. H. F. Wilson and D. H. McRae, U.S. Patent 3,080,225 (to Rohm & Haas) [*CA 59*, 2114 (1963)].
12. R. J. Theissen, U.S. Patent 3,652,645 (to Mobil, now Rhône-Poulenc) [*CA 74*, 30981 (1971)].
13. R. Y. Yih and C. Swithenbank, *J. Agric. Food Chem.* 23, 592 (1975).
14. W. O. Johnson, G. E. Kollman, G. Swithenbank, and R. Y. Yih, *J. Agric. Food Chem,* 26, 285 (1978).
15. H. O. Bayer and R. Y. Yih, U.S. Patent 3,928, 416 and 3,798, 276 (to Rohm & Haas) [*CA 80*, 3253 (1974)].
16. D. Cartwright and D. J. Collins, European Patent 3,416 (to ICI, now part of Zeneca) [*CA 92*, 6284 (1980)].
17. S. R. Colby, J. W. Barnes, T. A. Sampson, J. L. Shoham, and D. J. Osborn, *Proc. 10th Int. Cong. of Plant Protection 1*, 295 (1983).
18. P. H. Maigrot, A. Perrot, L. Hede-Hauy, and A. Murray, *Brighton Crop Prot. Weeds 1,* 47 (1989).
19. *5th Crop Protection Chemicals Reference*, p. 2205, Chemical and Pharmaceutical Press, New York and Paris (1989).
20. M. Matringe, J. M. Carnadro, R. Labbe, and R. Scalla, *Biochem. J. 260*, 231 (1989).
21. D. Cartwright and D. J. Collins, U.S. Patent 4,285,725 (to ICI, now part of Zeneca) [*CA 92*, 62843 (1980)].
22. W. O. Johnson, European Patent 19,388 (to Rohm & Haas) [*CA 94*, 156538 (1981)].
23. W. O. Johnson, U.S. Patent 4,031,131 (to Rohm & Haas) [*CA 87*, 84689 (1977)].
24. A. E. Kaye, J. O. Morley, and A. C. Tucker, British Patent 2,068,945 (to ICI, now part of Zeneca) [*CA 96*, 34878 (1982)].
25. R. W. Etherington and R. J. Thiessen, European Patent 22,610 (to Mobil, now Rhône-Poulenc) [*CA 94*, 208557 (1981)].
26. S. M. Ridley, in: *Carotenoid Chemistry & Biochemistry* (C. B. Britton and T. W. Goodwin, eds.), p. 353, Pergamon, Oxford (1982).
27. T. Waldrep and H. M. Taylor, *J. Agric. Food Chem.* 24, 1250 (1976).
28. F. Pereiro and J. C. Ballaux, *Proc. Brit. Crop Prot. Conf. Weed 1*, 225 (1982).
29. M. C. Cramp, J. Gilmour, L. R. Hatton, R. H. Hewett, C. J. Nolan, and E. W. Parnell, *Proc. Brit. Crop Prot. Conf. Weeds 1*, 23 (1985).
30. M. C. Cramp, J. Gilmour, L. R. Hatton, R. H. Hewitt, C. J. Nolan, and E. W. Parnell, *Pestic. Sci. 18*, 15 (1987).
31. C. F. A. Kyndt and M. T. F. Turner, *Proc. Brit. Crop. Conf. Weeds 1*, 29 (1985).
32. E. G. Teach, German Patent 2,612,731 (to Stauffer, now ICI, now part of Zeneca) [*CA 86*, 5308 (1977)].
33. M. C. Cramp, J. Gilmour, and E. W. Parnell, British Patent 2,087,887 (to Rhône-Poulenc) [*CA 97*, 144785 (1982)].
34. H. J. Nestler, in: *Chemie der Pflanzenschutz und Schadlinsbekampfungsmittel* (R. Wegler, ed.), Band 8, p. 2, Springer-Verlag, Berlin–Heidelberg (1982).
35. J. D. Burton, J. W. Greenwald, D. A. Somers, J. A. Connelly, B. G. Gegenbach, and D. L. Wyse, *Biochem. Biophys. Res. Commun. 148*, 1039 (1987).
36. R. Takashita, K. Fujikawa, I. Yokomichi, and N. Sakashita, U.S. Patent 4,046,553 (to I.S.K.) [*CA 85*, 78015 (1976)].

37. R. Nishiyama, T. Haga, and N. Sakashita, British Patent 1,599,121 (to I.S.K.) [*CA 90*, 152017 (1979)].
38. H. Johnson and L. Troxell, European Patent 483 (to Dow, now Dow Elanco) [*CA 90*, 203882 (1979)].
39. D. Cartwright, European Patent 1,473 (to ICI, now part of Zeneca) [*CA 92*, 58618 (1980)].
40. R. E. Plowman, W. C. Stonebridge, and J. N. Hawtree, *Proc. Brit. Crop Prot. Conf. Weeds 1*, 29 (1980).
41. H. Rempfler, W. Foery, and R. Schurter, British Patent 1,550,574 (to Ciba Geigy) [*CA 90*, 103846 (1979)].
42. Y. Kobayashi, I. Kumadaki, S. Sato, N. Hara, and E. Chikami, *Chem. Pharm. Bull. 18*, 2334 (1970).
43. R. Nishiyama, K. Fujikawa, and T. Haga, British Patent 1,599,123 and 1,599,124 (to I.S.K) [*CA 90*, 72070 (1979)].
44. R. D. Bowden, U.S. Patent 4,205,175 (to ICI, now part of Zeneca) [*CA 92*, 76311 (1980)].
45. R. Nishiyama, K. Fujikawa, I. Yokomichi, R. Nasu, A. Awazu, and J. Kawashima, European Patent 9,212 (to I.S.K.) (*CA 93*, 186180 (1980)].
46. J. Werner, C. A. Wilson, and C. E. Mixan, U.S. Patent 4,331,811 (to Dow, now Dow Elanco) [*CA 97*, 92160 (1982)].
47. C. A. Wilson and J. A. Werner, U.S. Patent 4,309,548 (to Dow, now Dow Elanco) [*CA 96*, 181152 (1982)].
48. M. A. Loos, in: *Herbicides, Chemistry, Degradation and Mode of Action* (P. C. Kearney and D. D. Kaufman, eds.), Vol. 1, p. 1, Marcel Dekker, New York and Basel (1975).
49. L. H. McKendry, U.S. Patent 4,108,629 (to Dow, now Dow Elanco) [*CA 90*, 98563 (1979)].
50. R. D. Chambers, J. Hutchinson, and W. K. R. Musgrave, *Proc. Chem. Soc.* 83 (1964).
51. T. J. Adaway, U.S. Patent 4,701,531 (to Dow, now Dow Elanco) [*CA 108*, 131594 (1988)].
52. E. M. Beyer Jr., M. J. Duffy, J. V. Hay, and D. D. Schlueter, in: *Herbicides, Chemistry, Degradation and Mode of Action* (P. C. Kearney and D. D. Kauffman, eds.), Vol. 3, p. 117, Marcel Dekker, New York and Basel (1988).
53. W. Maurer, H. R. Gerber, and J. Rufener, *Proc. Brit. Crop. Prot. Conf. Weeds 1*, 43 (1987).
54. F. Werner, K. Gass, W. Meyer, and R. Schurter, European Patent 70,802 to (Ciba Geigy) [*CA 99*, 22504 (1983)].
55. R. W. Pfluger, U.S. Patent 4,542,216 (to Ciba Geigy) [*CA 104*, 88586 (1986)].
56. R. Hassig, U.S. Patent 4,692,524 (to Ciba Geigy) [*CA 108*, 6041 (1988)].
57. W. Meyer, European Patent 70, 804 (to Ciba Geigy) [*CA 99*, 22486 (1983)].
58. W. Meyer, K. Gass, W. Topil, R. Schurter, and G. Pissiotas, European Patent 84,020 (to Ciba Geigy) [*CA 99*, 158457 (1983)].
59. R. D. Trepka, J. K. Harrington, J. E. Robertson, and J. T. Waddington, *J. Agric. Food Chem. 18*, 1176 (1970).
60. R. D. Trepka, J. K. Harrington, J. W. McConville, and K. T. McGurran, *J. Agric. Food Chem. 22*, 1111 (1974).
61. T. L. Fridinger, D. R. Pauly, R. D. Trepka, and G. G. I. Moore, *Proc. 166th A.C.S. National Mtg*, Abs. No. PEST 063 (Aug 1974).
62. K. A. Hassall, *The Chemistry of Pesticides*, p. 164, Macmillan, London (1982).
63. W. Mass, R. van Hes, A. C. Grosscurt, and D. H. Deul, in: *Chemie der Pflanzenschutz und Schadlingsbekampfungsmittel* (R. Wegler, ed.), Band 6, p. 424, Springer-Verlag, Berlin (1981).
64. R. Nishiyama, K. Fujikawa, R. Nasu, and T. Toki, South African Patent 7,802,440 (to I.S.K.) [*CA 91*, 175204 (1979)].
65. H. M. Becher, P. Becker, R. Prokic-Immel, and W. Wurtz, *Proc. Int. Conf. Plant Prot. 1*, 408 (1983).
66. M. Anderson, J. P. Fisher, and J. Robinson, *Proc. Brit. Crop Prot. Conf. Pests and Diseases 1*, 89 (1986).
67. R. S. Twyfell and J. M. Radcliffe, European Patent 275,132 (to Shell) [*CA 109*, 165741 (1988)].
68. M. Anderson, Australian Patent 8,540,924 (to Shell) [*CA 104*, 148560 (1986)].
69. H. M. Becher, R. Prokic-Immel, and W. Wurtz, European Patent 52,833 (to Celamerck, now Shell) [*CA 97*, 182003 (1982)].
70. H. M. Becher, German Patent 3,435,889 (to Celamerck, now Shell) [*CA 104*, 168101 (1986)].

71. K. A. Hassall, *The Chemistry of Pesticides*, p. 235, Macmillan, London (1982).
72. M. Elliott, A. W. Farnham, N. F. James, P. H. Needham, D. A. Pulman, and J. H. Stevenson, *Proc. 7th Brit. Insect & Fung. Conf.* 2, 721 (1973).
73. M. Elliott, A. W. Farnham, N. F. James, P. H. Needham, D. A. P. Pulman, and J. H. Stevenson, *Nature (London) 246*, 169 (1973).
74. N. Ohno, K. Fujimoto, Y. Okuno, T. Mizutani, M. Hirano, N. Itaya, T. Honda, and H. Yaskioka, *Agric. Biol. Chem. 38*, 881 (1974).
75. M. D. Mowlam, D. P. Highwood, R. J. Dowson, and J. Hattori, *Proc. Brit. Crop Conf. Pests & Diseases* 2, 649 (1977).
76. A. R. Jutsum, M. D. Collins, R. M. Perrin, D. D. Evans, R. A. H. Davies, and C. N. E. Ruscoe, *Proc. Brit. Crop Prot. Conf. Pests and Diseases* 2, 421 (1984).
77. P. D. Bentley, R. Cheetham, and R. K. Huff, *Pestic. Sci. 11*, 156 (1980).
78. V. K. Stubbs, *Aust. Vet. J. 59*, 152 (1982).
79. H. J. H. Doel, A. R. Crossman, and L. A. Bourdouxhe, *Meded. Fac. Landbouwwet., Rijksuniv. Gent 49*, 929 (1984).
80. A. R. Crossman, L. A. Bourdouxhe, and H. J. H. Doel, *Meded. Fac. Landbouwwet., Rijksuniv. Gent 49*, 939 (1984).
81. A. R. Jutsum, R. F. S. Gordon, and C. N. E. Ruscoe, *Proc. Brit. Crop Prot. Conf. Pests and Diseases 1*, 97 (1986).
82. R. K. Huff, British Patent 2,000,764 (to ICI, now part of Zeneca) [*CA 92*, 22137 (1980)].
83. E. McDonald, N. Punja, and A. R. Jutsum, *Proc. Brit. Crop Conf. Pests and Diseases* 199 (1986).
84. J. F. Engel, European Patent 3,336 (to F.M.C.) [*CA 92*, 93950 (1980)].
85. J. F. Engel, U.S. Patent 4,332,815 (to F.M.C.) [*CA 98*, 72491 (1983)].
86. W. G. Scharp and M. S. Glenn, World Patent 8,300,485 (to F.M.C.) [*CA 99*, 22243 (1983)].
87. W. K. Whitney and K. Wettstein, *Proc. Brit. Crop Prot. Pests and Diseases* 2, 1979 (1979).
88. C. A. Hendrick, B. A. Garcia, G. B. Staal, D. C. Gerf, R. J. Anderson, K. Gill, H. R. Chinn, J. N. Labovitz, M. M. Leippe, S. L. Woo, R. L. Carney, D. C. Gordon, and G. K. Kohn, *Pestic. Sci. 11*, 224 (1980).
89. R. J. Anderson. K. G. Adams, and C. A. Hendrick, *J. Agric. Food Chem. 33*, 508 (1985).
90. C. A. Hendrick, R. J. Anderson, R. L. Carney, B. A. Garcia, and G. B. Staal, in: *Recent Advances in the Chemistry of Insect Control* (N. F. James, ed.), Special Publ. No. 53, p. 2, Royal Society of Chemistry, London (1984).
91. G. Berkelhammer and V. Kameswaran, U.S. Patent 4,199,595 (to American Cyanamid) [*CA 95*, 80502 (1981)].
92. J. K. Siddens and R. Swaraman, U.S. Patent 4,405,529 (to American Cyanamid) [*CA 100*, 22410 (1984)].
93. J. K. Siddens, U.S. Patent 4,407,760 (to American Cyanamid) [*CA 100*, 22412 (1984)].
94. V. Kameswaran, U.S. Patent 4,454,344 (to American Cyanamid) [*CA 100*, 6094 (1984)].
95. R. J. Anderson, K. G. Adams, and C. A. Hendrick, U.S. Patent 4,260,633 (to Zoecon now Sandoz) [*CA 95*, 81538 (1981)].
96. R. J. Anderson and T. A. Baer, U.S. Patent 4,226,802 (to Zoecon, now Sandoz) [*CA 94*, 66075 (1981)].
97. I. Hamman and R. Fuchs, *Pflanzenschutz-Nachr. (Engl. Ed.) 34*, 121 (1981).
98. W. Behrenz, A. Elbert, and R. Fuchs, *Pflanzenschutz-Nachr. (Engl. Ed.) 36*, 127 (1983).
99. R. Fuchs and I. Hamman, German Patent 2,730,515 (to Bayer) [*CA 90*, 151658 (1979)].
100. E. McDonald, N. Punja, and A. R. Jutsum, *Proc. Brit. Crop Prot. Conf. Pests and Diseases* 199 (1986).
101. R. Colln, U. Priesnitz, and E. Klauke, European Patent 17,882 (to Bayer) [*CA 94*, 156550 (1981)].
102. R. Fuchs, F. Maurer, U. Priesnitz, and H. H. Riebel, European Patent 24,019 (to Bayer) [*CA 95*, 61745 (1981)].
103. A. Marhold, E. Kysela, and E. Klauke, German Patent 2,928,987 (to Bayer) [*CA 95*, 42681 (1981)].
104. F. Maurer, European Patent 48,375 (to Bayer) [*CA 97*, 55484 (1982)].
105. R. Hagemann, B. Baasner, and E. Klauke, European Patent 22,942 (to Bayer) [*CA 95*, 42606 (1981)].

106. R. A. Hann, European Patent 113,185 (to ICI, now part of Zeneca) [*CA 101*, 191309 (1984)].
107. N. Punja, European Patent 31,199 (to ICI, now part of Zeneca) [*CA 95*, 186837 (1981)].
108. D. J. Milner, British Patent 2,135,306 (to ICI, now part of Zeneca) [*CA 102*, 45787 (1985)].
109. P. J. Richardson and G. R. Davies, British Patent 2,134,109 (to ICI, now part of Zeneca) [*CA 102*, 45594 (1985)].
110. J. B. Lovell, *Proc. Brit. Crop Prot. Conf. Pests and Diseases 2*, 575 (1979).
111. J. B. Lovell, U.S. Patent 4,163,102 (to American Cyanamid) [*CA 91*, 211437 (1979)].
112. N. A. Cortese Jr. and W. H. Gastrock, U.S. Patent 4,521,629 (to American Cyanamid) [*CA 103*, 87633 (1985)].
113. P. A. Worthington, in: *Synthesis and Chemistry of Agrochemicals* (D. R. Baker, J. G. Fenyes, W. K. Moberg, and B. Cross, eds.), p. 302, ACS Symposium Series No. 355, Washington (1987).
114. P. A. Worthington, in: *Sterol Biosynthesis Inhibitors* (D. Berg and M. Plempel, eds.), p. 19, Ellis Horwood, London (1988).
115. K. H. Buckel, in: *Fungicide Chemistry, Advances and Practical Applications* (M. B. Green and D. A. Spilker, eds.), p. 1, ACS Symposium Series No. 304, Washington (1986).
116. D. Berg, in: *Fungicide Chemistry, Advances and Practical Applications* (M. B. Green and D. A. Spilker, eds.), p. 25, ACS Symposium Series No. 304, Washington (1986).
117. H. Scheinpflug and K. H. Kuck, in: *Modern Selective Fungicides* (D. H. Lyr, ed.), p. 173, Longman, Harlow (1987).
118. H. Buckenauer, in: *Modern Selective Fugicides* (D. H. Lyr, ed.), p. 205, Longman, Harlow (1987).
119. A. M. Skidmore, P. N. French, and W. G. Rathmell, *Proc. 10th Int. Cong. Plant Prot. 1*, 368 (1983).
120. P. J. Northwood, A. Horellou, and K. H. Heckele, *Proc. 10th Int. Cong. Plant Prot. 13*, 930 (1983).
121. W. K. Moberg, in: *Synthesis and Chemistry of Agrochemicals* (D. R. Baker, J. G. Fenyes, W. K. Moberg, and B. Cross, eds.), p. 288, ACS Symposium Series No. 355, Washington (1987).
122. T. M. Fort and W. K. Moberg, *Proc. Brit. Crop Prot. Pests & Diseases 2*, 413 (1984).
123. A. Nakata, *Proc. 5th Cong. Pest. Chem.*, Abs. No. 116-9, Kyoto (1982).
124. K. H. Buchel, F. Grewe, and H. Kaspers, British Patent 1,237,509 (to Bayer) [*CA 74*, 100062 (1971)].
125. F. Grewe and K. H. Buckel, *Mitt. Biol. Bundesanst. Land-Forstwirtsch., Berlin-Dahlem 151*, 208 (1973).
126. K. H. Kuck, and H. Schiempflug, in: *Chemistry of Plant Protection* (G. Haug and H. Hoffman, eds.), p. 65, Springer, Berlin (1986).
127. I. F. Brown Jr, H. M. Taylor, and R. E. Hackler, in: *Pesticide Synthesis Through Rational Approaches* (P. S. Magee, G. K. Kohn, and J. J. Mehn, eds.), p. 65, ACS Symposium Series No. 255, Washington, (1969).
128. I. F. Brown Jr, H. M. Taylor, and H. R. Hall, *Proc. Am. Phytopathol. Soc. 2*, 31 (1975).
129. K. P. Parry, P. A. Worthington, and W. G. Rathmell, European Patent 15,756 (to ICI now part of Zeneca) [*CA 94*, 103388 (1981)].
130. W. K. Moberg, U.S. Patent 4,510,136 (to Du Pont) [*CA 104*, 207438 (1986)].
131. K. Ikura, K. Katsuura, M. Kataoka, A. Nakata, and M. Mizuno, Belgian Patent 865,569 (to Nippon Soda) [*CA 90*, 38917 (1979)].
132. J. D. Davenport, R. E. Hackler, and H. M Taylor, British Patent 1,218,623 (to Eli Lilly now Dow Elanco) [*CA 72*, 100745 (1970)].
133. F. Araki and K. Vabutani, *Proc. Brit. Crop Prot. Conf. Pests & Diseases 1*, 3 (1981).
134. K. I. Yabutani, K. T. Ikeda, S. S. Hatta, and T. K. Harada, German Patent 2,731,522 (to Nihon Nohyaku) [*CA 89*, 6122 (1978)].
135. T. Kodaira, Y. Kobayashi, and H. Kurona, European Patent 15,557 (to Nihon Nohyaku) [*CA 94*, 4907 (1981)].
136. T. Kodaira, Y. Kobayashi, and H. Kurona, Japanese Patent 55,059,135 (to Nihon Nohyaku) [*CA 93*, 185962 (1980)].
137. T. Kodaira, Y. Kobayashi, and H. Kurona, Japanese Patent 55,129,242 (to Nihon Nohyaku) [*CA 94*, 139472 (1981)].
138. R. L. B. Manilla and E. V. Krumkalns, U.S. Patent 4,002,628 (to Eli Lilly, now Dow Elanco) [*CA 86*, 155686 (1977)].

139. R. L. Robey and J. C. Smirz, European Patent 39,157 (to Eli Lilly, now Dow Elanco) [*CA 96*, 85264 (1982)].
140. W. A. Sheppard, *J. Org. Chem. 29*, 1 (1964).
141. M. Wilcox, I. Y. Chen, P. C. Kennedy, Y. Y. Li, L. R. Kincaid, and N. T. Helseth, *Proc. Plant Growth Regul. Working Group 4*, 194 (1977).
142. O. Ward Hopkins, British Patent 2,128,603 (to Ciba Geigy) [*CA 101*, 191314 (1984)].
143. M. Wilcox, U.S. Patent 4,169,721 (to Ciba Geigy) [*CA 92*, 22231 (1980)].
144. D. J. Bowler, I. D Entwistle, and A. J. Porter, *Proc. Brit. Crop Prot. Conf. Pests and Diseases 2*, 397 (1984).
145. B. A. Driekhorn, G. O. P. Docherty, A. J. Clinton, and K. E. Kramer, *Proc. Brit. Crop Prot. Conf. Pests & Diseases 2*, 491 (1979).
146. B. A. Driekhorn, U.S. Patent 4,187,318 (to Eli Lilly, now Dow Elanco) [*CA 92*, 215040 (1980)].
147. W. S. Briggs, European Patent 197,706 (to Eli Lilly, now Dow Elanco) [*CA 106*, 18078 (1987)].

12

Fluorinated Liquid Crystals

TAKESHI INOI

12.1. INTRODUCTION

The liquid crystalline state, located between the crystalline and the isotropic liquid states of matter, can be observed in certain types of organic compound, including polymers. A liquid crystal can flow like an ordinary liquid; however, other properties, such as birefringence, are remains of the crystalline phase. Through their anisotropy, liquid crystals are very sensitive to external changes such as electric field. Thus liquid crystals have been used as a practical means for monitoring, imaging, and other purposes.

As with other specialty organic chemicals, interest in the practical applications of liquid crystalline compounds containing carbon–fluorine bonds has increased substantially during the past decade. This chapter provides a brief background to the subject of liquid crystals (basic features of liquid crystals, liquid crystal displays, types of molecule leading to liquid crystalline phases) before dealing with the molecular design and synthesis of fluorinated materials of this class. No attempt has been made to give a complete description of all the fluorinated liquid crystals known at this time, especially for display or imaging; rather, the objective has been to focus attention on current interest in the molecular design of low-molecular-weight liquid crystals.

12.2. PROPERTIES AND STRUCTURAL CLASSIFICATION OF LIQUID CRYSTALS

Liquid crystals have long been known as the anisotropic fluids that may exist between the boundaries of the solid crystalline phase and the isotropic liquid phase. The liquid crystal phase is mainly a result of long-range orientational ordering of

TAKESHI INOI • Technical Division, Chisso Corporation, Tokyo, Japan.   *Present address:* 8-8-6 Sugita, Isogo-ku, Yokohama-shi, Kanagawa-ken 236, Japan.

Organofluorine Chemistry: Principles and Commercial Applications, edited by R. E. Banks *et al.* Plenum Press, New York, 1994.

"linear" molecules that occurs within certain ranges of temperature during the molten stage and in the clear liquid of many organic compounds. Although this ordering is sufficient to impart some solid crystalline properties on the fluid, typically birefringence, for example, the intermolecular attractive forces usually are not strong enough to suppress flow. This unique combination of characteristics accounts for the term *liquid crystal.* Liquid crystallinity as a phenomenon is referred to as *mesomorphism*, hence the terms *mesophase* (i.e., intermediate between fully ordered crystal and isotropic liquid) and *mesogenic compound* (i.e., capable of forming liquid crystal phases or mesophases).

Many thousands of organic substances,[1,2] including polymers,[3] exhibit liquid crystallinity. The general, common molecular feature is an elongated, narrow framework, which usually is depicted as a rod-shaped entity. Some disk-like shaped molecules such as hexa(alkanoyl)benzenes also form mesophases (discotic crystals).[4] The geometric anisotropy of individual molecules translates uniformly throughout the fluid medium. Due to the delicacy of the intermolecular forces involved, liquid crystals are extremely sensitive to external changes, e.g., temperature, pressure, electric or magnetic fields, or the presence of foreign vapors. Thus liquid crystals have been used as an important part of practical devices to monitor ambient changes, or to transform an environmental alteration into a useful output.

Liquid crystals have been divided into two categories, i.e., *thermotropic* and *lyotropic* liquid crystals. Sole-component systems that show mesomorphic behavior in a definite temperature range are called thermotropic, while lyotropic liquid crystalline phases show mesomorphic behavior in solution. Characterization of these phases has been thoroughly discussed.[5–8] The main subdivision of thermotropic liquid crystals is into *nematic* (Greek: thread-like) and *smectic* (Greek: soap-like) mesophases (Figure 1). In the nematic phase the centers of mass of the molecules have three translational degrees of freedom, and are thus distributed at random. Smectic phases are characterized in addition by a positional order in at least one dimension. The centers of the molecules are arranged in an equidistant plane (or sheet). Smectic phases are further classified into ten or more subdivided phases, which are indicated as S_A, S_B, ..., based on the orientation of the preferred directions of molecules with respect to the layer, and on the organization of the centers of the molecules within the layers. A unique situation may occur when a smectic liquid crystal is derived from optically active compounds or is "doped" with smaller quantities of a chiral molecule. In these cases, each layer is turned through a small angle with respect to the next layer, and the cumulated twist results in a strongly optical material. So-called S_C^* compounds belong to this optically active series of materials. Similarly, ordinal nematic liquid crystals can also be converted into optically active compounds by adding other optically active compounds. These phenomena are important for practical liquid crystal display. The third type of liquid crystal phase is the classical cholesteric phase, so named because this type of mesophase is associated with cholesterol derivatives [the phenomenon of liquid crystallinity was discovered more than a hundred years ago (F. Reinitzer, 1888) for cholesteryl benzoate]. This phase is essentially a kind of nematic phase, however,

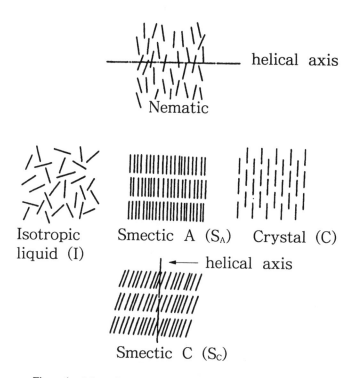

Figure 1. Schematic representation of nematic and smectic phases.

and is distinguished by its extremely optically active nature and by having a "blue phase."

In the case of nematic liquid crystals, the average alignment (or internal order) of the molecules with their long axes parallel to each other leads to a preferred direction in space. This local direction of alignment is usually described by a unit vector $\mathbf{n}$, called a director, which gives the direction of the preferred axis at each point in a phase. The order parameter S is a measure of the alignment of the long axes of the molecules; and $S = \frac{1}{2} \langle 3 \cos^2 \theta \rangle$, where θ is the angle between the long axis of the representative molecule and some preferred direction, the director.

Liquid crystals possess various useful anisotropic physical properties. When a light beam is incident along the director, the ordinary refractive index n_o will be measured. If an incident light beam is perpendicular to the director, and is polarized along $\mathbf{n}$, the other situation is found, giving the extraordinary refractive index n_e. Indicating the directions parallel and perpendicular to $\mathbf{n}$ by the subscript $\parallel$ and $\perp$, respectively, it is usual to define the birefringence as $\Delta n = n_e - n_o = n_{\parallel} - n_{\perp}$. Liquid crystals can be strongly birefringent, with Δn reaching values of up to 0.5. Similarly, depending on the sign of the dielectric anisotropy (defined as $\Delta \varepsilon = \varepsilon_{\parallel} - \varepsilon_{\perp}$) of a mesogenic material, nematic directors align parallel ($\Delta \varepsilon > 0$; called positive) or perpendicular ($\Delta \varepsilon < 0$; called negative) to the applied field direction. All distortions of

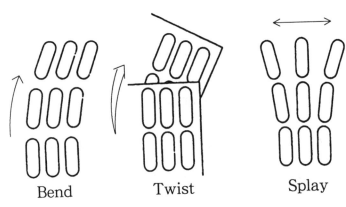

Figure 2. The three basic curvatures of the moleuclar director.

the nematic phase may be changed into three basic curvatures of the molecular director: splay, twist, and bend (Figure 2). The force constants opposing these strains are denoted k_{11} (splay), k_{22} (twist), and k_{33} (bend).

A liquid crystalline compound can adopt more than one type of mesomorphic structure as the conditions of temperature (or solvent) are changed. In thermotropic liquid crystals, transitions between various structural subclasses occur at definite temperatures, usually accompanied by definite changes in the latent heat (Figure 3). When the transitions are reversible upon heating and cooling, such multiphase transitions are called *enantiotropic*, while a given liquid crystal phase, that can only be approached in one direction in the thermal cycle, is called *monotropic*.

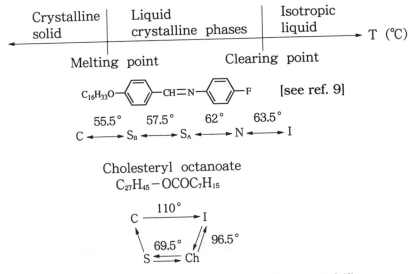

Figure 3. Transitions between structural subclasses. (see Ref. 10)

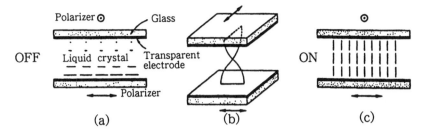

Figure 4. Twisted nematic display. Between crossed polarizers oriented with the parallel director axes on the two glasses the sample would appear transparent. If the nematic/chiral host has a positive dielectric anisotropy ($\varepsilon_\parallel > \varepsilon_\perp$), the application of voltage above the threshold tilts the director parallel to the field. This eliminates the twist in the sample and the display appears dark.

12.3. APPLICATIONS

The most important application of liquid crystals lies in electrooptical display.[11] In the late 1960s, several important phenomena capable of producing images were explored. Owing to their potential technological applicability, the dynamic scattering effect,[12] guest–host effect,[13] and field-induced cholesteric–nematic phase change effect[14] provided the impetus for a wider interest in liquid crystals.

A very useful development of electrically controlled birefringence is the twisted nematic effect,[15,16] which is predominantly used in today's liquid crystal displays (Figure 4) [for a review, see Ref. 17]. The twisted nematic effect has now been developed into the supertwisted nematic effect.[18] Other important effects and modes, such as ferroelectric crystal displays, are under development.[19,20]

12.4. MOLECULAR DESIGN

The particular mesomorphic structure that occurs not only depends on linear molecular shape but is intimately connected with the strength and position of polar groups within the molecule, the molecule's overall polarizability, and the presence of chiral centers. These factors have been reviewed.[6,21,22] Molecular interactions that lead to attraction are dipole–dipole interactions, dispersion forces, and hydrogen bonding. In order for dipole–dipole interactions to be effective for liquid crystallinity, the molecule must contain polar groups and polarizable groups, as mentioned above. In this case, electronic distortion is favored by the presence of cyclic groups and conjugated double or triple bonds.

Many mesogenic compounds are of the type:

$$y - \boxed{A} - \boxed{x} - \boxed{B} - z$$

In this specified structure, there are three basic moieties: core units (A, B), a linking or bridging group (x), and terminal groups (y, z). The *core* (A and B) is the rigid part,

Table 1. Various Types of Ring Systems in Mesomorphic Compounds

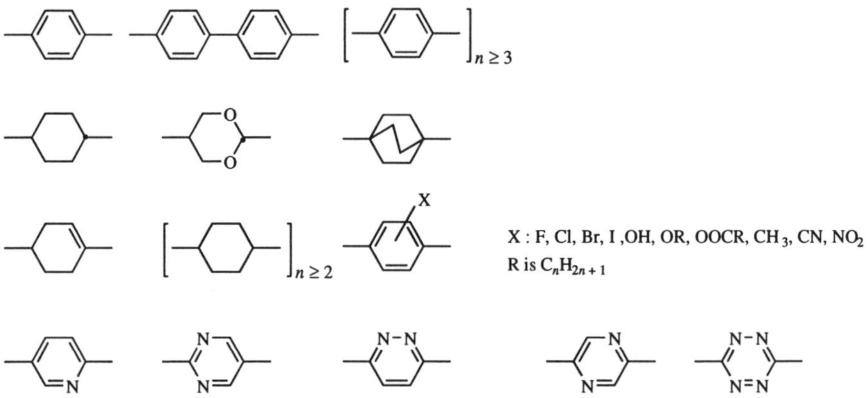

$n \geq 3$
$n \geq 2$
X : F, Cl, Br, I ,OH, OR, OOCR, CH_3, CN, NO_2
R is C_nH_{2n+1}

comprised of cyclic structures, often benzene or cyclohexane rings or their heterocyclic analogs being used (see Table 1). Each molecule must contain at least two ring systems. Examples of linking groups (x) used to connect the ring systems are given in Table 2.

Terminal or end groups (y and z) can be either low-polarity aliphatic chains (*n*-alkyl and *n*-alkoxy groups are commonly used) or rigid polar groups (CN, CF_3, OCF_3). Terminal cyano compounds occupy a special position in the history of liquid crystals because the synthesis of 4-*n*-alkyl- and 4-*n*-alkoxy-4′-cyanobiphenyls in the early 1970s promoted the commercial exploitation of liquid crystals for electrooptic display devices. An asymmetric center is necessary for the cholesteric, i.e., spontaneously twisted nematic, and chiral smectic mesophases. Examples are given in Table 3.

Table 2. Various Types of Bridging Groups in Mesomorphic

—C≡C—
—N=C=N—
$—(CH_2)_n—$
$—NH(CH_2)_n—$
$—O(CH_2)_nO—$
$—COO(CH_2)_nOOC—$
$—(CH_2)_nCOO—$

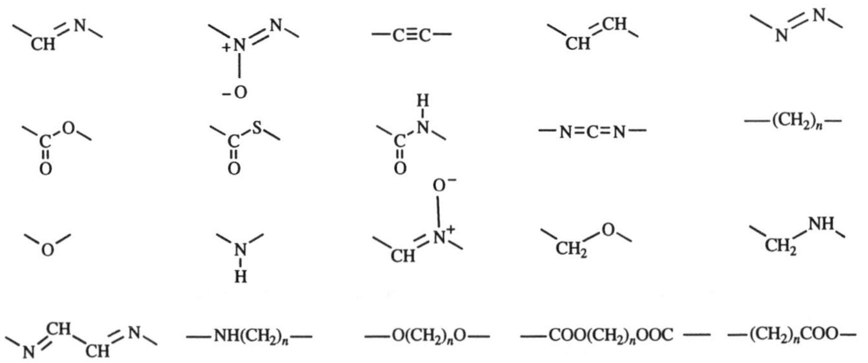

Table 3.   Various Types of Terminal Groups in Mesomorphic Compounds

–OR	–R	–COOR	–COR
–OOCR	–OOCOR	$\diagdown_{CH}\diagup^{CH}\diagdown_{COOR}$	
–CN	–Cl	–NO$_2$	
–H	–F	–Br	
–I	–R′	–N=C=O	
–NH$_2$	–COOR′	–CR=CR–COOR	
–SR	–NHR	–NR$_2$	
	–NHCOR	– N=C=S	
	–O(CH$_2$)$_n$OR	–CF$_3$	–OCF$_3$

R = *n*-alkyl.
R′ = branched or unsaturated alkyl.

The fundamental target in liquid crystal molecular design is to produce a wide range of room temperature thermotropics with appropriate physical properties and with enhanced chemical and photochemical stabilities for use in devices. Table 4 contains details of typical series of nematic liquid crystals.

Several types (Table 4) of liquid crystal are currently employed for commercial liquid crystal displays. Schiff base materials possess a sufficiently wide nematic range, low operating voltages, and easy alignment characteristics; however, their sensitivity to moisture limits their use. Azoxy materials with structures similar to the Schiff bases can be used, but require extra UV protection because of their photochemical sensitivity. Esters, which have been extensively studied, usually have lower dielectric anisotropy and higher viscosity. Owing to their relatively flat threshold voltage versus temperature characteristics, cyanophenyl benzoates have been found useful; however, slower response times caused by higher viscosity values are experienced. The biphenyls are stable compounds and have large optical anisotropy and other balanced physical properties. Cyclohexyl carboxylates, however, have proved to be much more compatible. These compounds generally have a wider nematic phase and lower viscosity, leading to shorter response times than found with the corresponding benzoates; also they are applicable as mixtures for practical use. The phenylcyclohexanes have better chemical and photochemical stability, and have been applied in some cases owing to their lower viscosities and wider mesomorphic temperature ranges.

Phenylpyrimidines, which have relatively higher birefringence (Table 4), have contributed to increasing the multiplexable lines of twisted nematic liquid crystal displays. Their phenyldioxan analogs have larger dielectric anisotropy, but otherwise are similar. The cyclohexylcyclohexanes, which are well recognized as materials with low birefringence, prove the point that nonaromatic rigid cores can lead to nematic phases; and this low birefringence made possible the operation of twisted nematic displays with wide viewing angles. The increasing sophistication of twisted and super-twisted liquid crystal display technology has placed greater demands on the mesogen

Table 4.   Nematic Liquid Crystal Types and Their Averaged Physical Properties[a]

Structural type	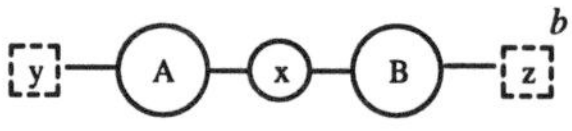 $[y]$—(A)—(x)—(B)—$[z]$[b]	Birefringence anisotropy (Δn)	Dielectric anisotropy ($\Delta \varepsilon$)	Viscosity (η_{20}) / cSt
Schiff base	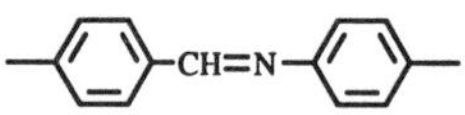 —◯—CH=N—◯—	0.24 to 0.28	−0.7 to +20	30 to 90
Benzoate	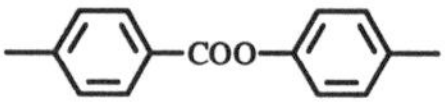 —◯—COO—◯—	0.14 to 0.20	−0.3 to +20	20 to 100
Biphenyl	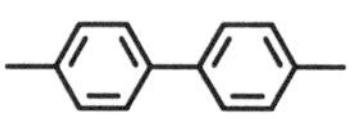 —◯—◯—	0.18 to 0.23	+10 to +14	30 to 80
Cyclohexyane-carboxylates	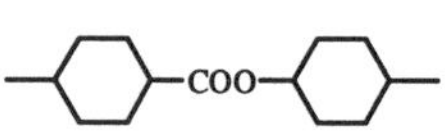 —◯—COO—◯—	0.08 to 0.12	−0.6 to +8	to 10
Phenylcyclohexane	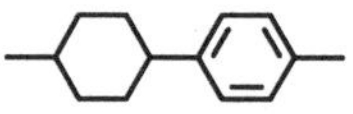 —◯—◯—	0.08 to 0.14	−0.5 to +14	5 to 30
Phenylpyrimidine	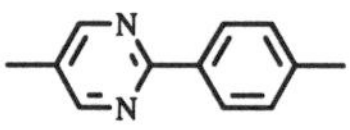 —◯(N,N)—◯—	0.16 to 0.25	+1 to +25	80 to 110
Phenyldioxane	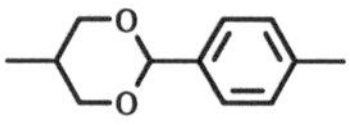 —◯(O,O)—◯—	0.10 to 0.15	+13 to +22	40 to 70
Cyclohexylcyclohexane	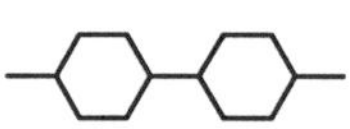 —◯—◯—	0.02 to 0.06	to +5	60 to 80

[a]Source: S. Matsumoto, *Kagaku to Kogyo* (Chemistry and Chemical Industry) *41*, 1014 (1988).
[b]y, z: alkyl, alkoxy, or cyano group.

material requirements; thus, hybrid molecular structures containing multicore systems and ferroelectric liquid crystals are the current targets.

12.5. FLUORINATED LIQUID CRYSTALS

Since the mid-1950s, progress in organic fluorine chemistry has been translated into useful applications in a number of specialty chemical areas, particularly medicine. Not unexpectedly, the effects and advantages of fluoro substitution in low-molecular-weight liquid crystals have been studied. The argument for introducing fluorine into the molecules centered on its high electronegativity, or strong electron-withdrawing inductive effect, in relation to its small size, fairly close to that of hydrogen (see Chapter 3); therefore its presence could be expected to exert a large influence on permittivity, while causing minimum change in molecular shape.

In theory, there exist several different sites of substitution in mesogenic molecules. Usually, fluorine is introduced into a terminal group and/or the rigid cyclic core. A review based on literature published before 1984 is available.[23]

12.5.1. Semifluorinated Alkanes

Simple elongated molecules, such as n-alkanes, will not display any liquid crystalline properties, however, the semifluorinated alkanes, $F(CF_2)_{10}(CH_2)_9H$, $F(CF_2)_{10}(CH_2)_{10}H$, and $F(CF_2)_{10}(CH_2)_{11}H$, undergo a reversible transition between tilted smectic phases above room temperature.[24] These compounds were prepared via addition of appropriate perfluoroalkyl iodides to α-olefins and subsequent reductive de-iodination of the adducts.

Cholesteryl Esters

Cholesteryl n-alkanoates are typical liquid crystal-forming compounds, however, cholesteryl n-ω-monohydropolyfluoroalkanoates [e.g., $C_{27}H_{45}OCO(CF_2)_nCHF_2$] and cholesteryl n-perfluoroalkanoates generally show an attenuated liquid crystalline phase, probably due to a hindrance factor caused by the bulk of the fluoroalkanoate moieties.[25]

12.5.2. Schiff Bases (Azomethines)

Several terminally substituted (trifluoromethyl)benzylideneanilines have been synthesized.[26] The following transition temperatures (°C) were measured, and smectic B phases were produced:

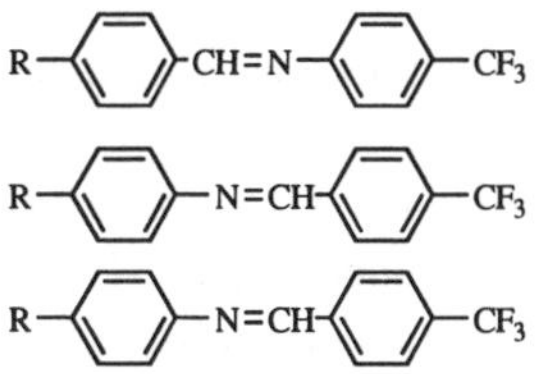

	R	C—S	S—I
$R-\langle\rangle-CH=N-\langle\rangle-CF_3$	$C_{10}H_{21}O-$	67.8	77.8
$R-\langle\rangle-N=CH-\langle\rangle-CF_3$	$C_{10}H_{21}O-$	95.2	96.6
$R-\langle\rangle-N=CH-\langle\rangle-CF_3$	$C_{10}H_{21}-$	42.5	42.9

Both 2- and 2′-halogeno-substituted 4-p-n-alkoxybenzylideneaminobiphenyls have been synthesized and compared with their nonhalogenated analogs.[27] The data listed in Table 5 show that for each 2′-halogeno substituent, (i) the azomethines (right column) have only a slightly higher nematic thermal stability than their analogs (left column), and (ii) reversal of the CH=N linkage leaves the nematic thermal stability virtually unaffected. The 2′-halogenobiphenyl-4-carboxaldehydes required for the preparation of Schiff bases via conventional condensation with $para$-substituted anilines were synthesized as shown in Scheme 1.

12.5.3. Benzoates

The effects of fluoro substitution in simpler esters have been extensively studied. Benzoate esters, 4-$RC_6H_4C(O)OC_6H_4R'$-4, where R, R′ = n-alkyl or n-alkoxy, have found an important use as host nematogens for mixtures used in twisted nematic display devices. A large number of laterally fluorinated 4-n-alkyl- and 4-n-alkoxyphenyl, 4′-n-alkyl-, and 4′-n-alkoxybenzoates and related cyclo-

Table 5. Average Transition Temperatures (°C) Representing Nematic (N) and Smectic (S$_A$) Thermal Stabilities of 2- and 2′-Halogeno-Substituted 4-*p*-*n*-Alkoxybenzylideneaminobiphenyls and 4-*n*-Alkoxy-*N*-(biphenyl-4-ylmethylene)anilines

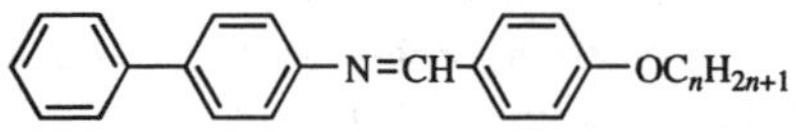

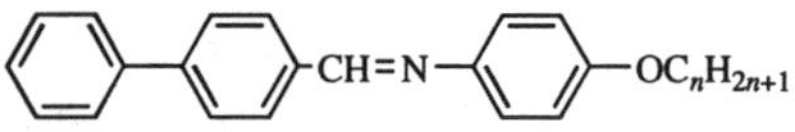

Substituent	C	N^a	S$_A{}^b$	C	N^a	S$_A{}^b$
2′-F	110.1		89.8	113.3		—
2′-Cl	59^c		—	61.6		—
2′-Br	44.1^c		—	46.1		—
2′-I	29.5^{c,d}		—	32.5^e		—
2-F	113.3^c		92.2	121.5		108.2

^{a}Obtained from average N—I transition temperature, $n = 7$–10.
^{b}Obtained from average S$_A$—N transition temperatures; $n = 8$–10.
^{c}Obtained from data given by D. J. Byron *et al.*, *J. Chem. Soc.* 2246 (1963).
^{d}Value for $n = 10$ only.
^{e}This is a "virtual," extrapolated, value for $n = 10$ only.

hexane-1-carboxylates have been synthesized.[28] The 2-fluorobenzoate esters have lower melting points than their nonfluorinated analogs. The melting points for the 3-fluorobenzoate esters are much higher than those of either the 2-fluorobenzoate esters or their nonfluorinated analogs.

    Diverse series of fluorine-containing phenyl benzoates have been synthesized in a search for much larger values of positive dielectric anisotropy. The 4-cyanophenyl 4-*n*-alkyl- and 4-*n*-alkoxy-benzoates are well known as practical liquid crystals. Misaki *et al.* have produced a series of 4-(trifluoromethyl)phenyl 4-*n*-alkoxybenzoates (synthesized as shown in Scheme 2) and compared them with 4-cyano analogs.[29] Table 6 shows the effect on the mesomorphic properties. 4-Trifluoromethyl 4-*n*-alkoxybenzoates show smectic liquid crystalline properties when the alkoxy chain contains six carbon atoms or more, and apparently the trifluoromethyl group gives a strongly positive dielectric anisotropy. The benzoate 4-*n*-C$_5$H$_{11}$C$_6$H$_4$C(O)OC$_6$H$_4$F-4 has a lower viscosity, resulting in a shorter response time[30]; also, use of a trifluoromethoxy group in the 4-position of the phenol component has been claimed to be preferable.[31]

$$\text{ArH} \xrightarrow{1} \text{ArCOCH}_3 \xrightarrow{2} \text{ArCO}_2\text{H} \xrightarrow{3} \text{ArCO}_2\text{Me} \xrightarrow{4} \text{ArCH}_2\text{OH} \xrightarrow{5} \text{ArCHO}$$

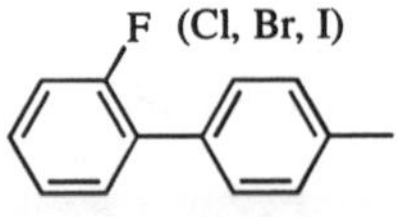

Scheme 1. Synthesis of 2′-halogenobiphenyl-4-carboxaldehydes: 1. Friedel-Crafts acylation; 2. Hypobromite oxidation; 3. Esterification, CH$_3$OH—BF$_3$·CH$_3$OH; 4. LiAlH$_4$; 5. Borane-dimethyl sulfide.

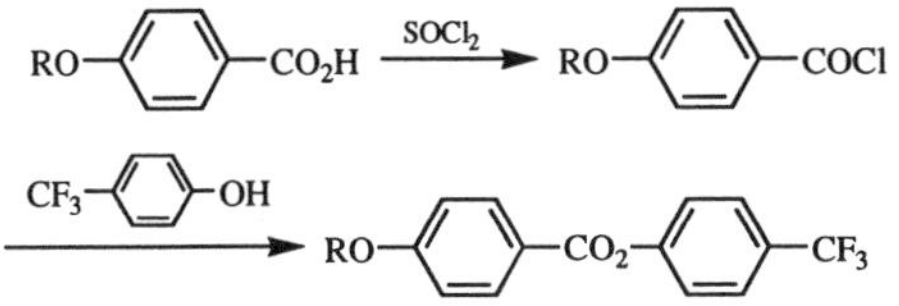

Scheme 2.   Synthesis of 4-(trifluoromethyl)phenyl alkoxybenzoates.

Table 6.   Comparison of Transition Temperatures (°C) of Fluorinated Benzoates

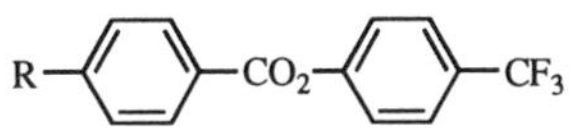

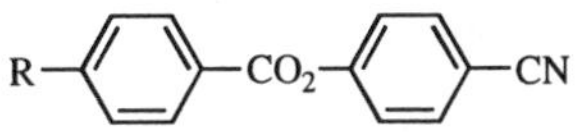

R	C—I	C—S	S—I	C—I	C—N	N—I
C_4H_9O	93.4			110.0		
$C_5H_{11}O$	78.5				87.0	96.0
$C_6H_{13}O$		60.0	74.7		70.0	81.0
$C_7H_{15}O$		62.9	74.3		71.5	80.0
$C_8H_{17}O$		62.5	72.5		75.0	83.0

Extremely high positive values of dielectric anisotropy in simpler terminal 4-cyanophenyl systems has been achieved by Schad and Kelly.[32–34] In this series, the ester 4-cyano-3-fluorophenyl 4-heptylbenzoate exhibits a narrow-range enantiotropic nematic phase (C—N: 28°C; N—I: 28.5°C; supercoolable to below 0°C) of very high positive dielectric anisotropy, estimated to be of the order of 30. The key intermediates of this series, monofluorinated 4-hydroxybenzonitriles, were prepared as outlined in Scheme 3.

Further detailed study was carried out by Gray *et al.* from the viewpoint of molecular design.[35] Transition temperatures of fluoro-substituted 4-cyanophenyl 4-alkyl and 4-alkoxy-benzoates are listed in Table 7. The effect of fluoro substitution on the melting points varies irregularly. However, a pattern emerges with transition temperature, $T_{N—I}$ values. Monofluoro substitution in the phenol ring has a greater effect on the $T_{N—I}$ values of esters than monofluoro substitution in the acid ring; a second fluoro substituent in the phenol causes a further depression in $T_{N—I}$; also, a second fluoro substituent in the acid causes a larger depression in $T_{N—I}$ than that caused by a single fluoro substituent in the acid. Part of the synthetic route is shown in Scheme 4.

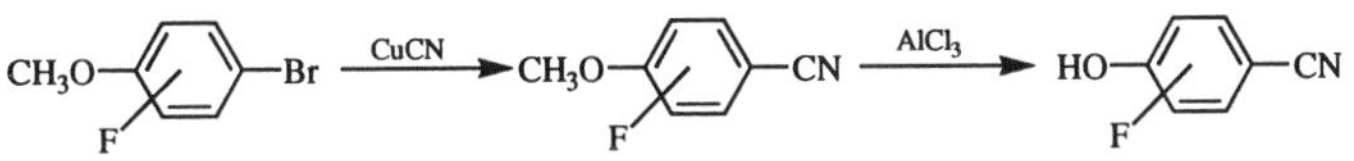

Scheme 3.   Synthesis of fluorinated 4-hydroxybenzonitriles.

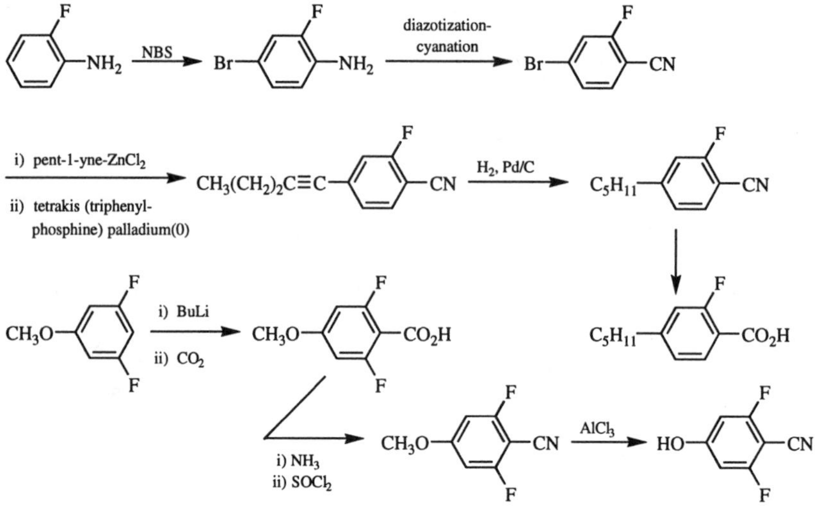

Scheme 4. Synthetic routes to 2-fluoro-4-pentylbenzoic acid and 2,6-difluoro-4-hydroxybenzonitrile.

Table 7. Transition Temperatures (°C) of Fluoro-Substituted 4-Cyanophenyl 4-Pentyl- and 4-Butoxy-benzoates[a]

Acid	Phenol		
	HO—⟨⟩—CN	HO—⟨F⟩—CN	HO—⟨F,F⟩—CN
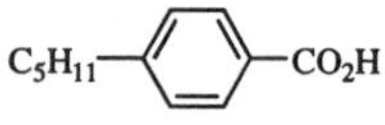 C$_5$H$_{11}$—⟨⟩—CO$_2$H	C—I 64.5 N—I (55.5)	C—I 30.5 N—I (24.5)	C—I 29.5 N—I (−8.0)
C$_4$H$_9$O—⟨⟩—CO$_2$H	C—N 92.0 N—I (104.0)	C—I 72.0 N—I (48.5)	C—I 71.5 N—I [7.5]
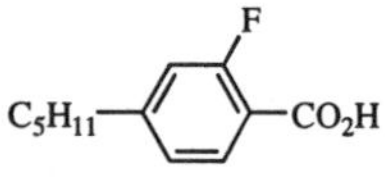 C$_5$H$_{11}$—⟨F⟩—CO$_2$H	C—I 65.5 N—I (32.0)	C—I 39.5 N—I (−3.0)	C—I 55.0 N—I (−20.5)
C$_4$H$_9$O—⟨F⟩—CO$_2$H	C—I 90.0 N—I (61.0)	C—I 55.5 N—I (21.5)	C—I 68.0 N—I [6.0]
C$_5$H$_{11}$—⟨F,F⟩—CO$_2$H	C—I 74.5 N—I [−18.0]	C—I 32.0 N—I [−45.0]	C—I 36.5 N—I [−67.0]
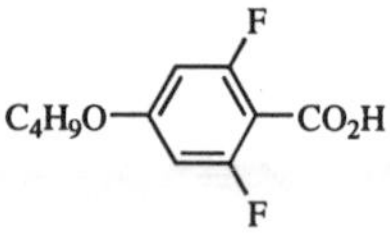 C$_4$H$_9$O—⟨F,F⟩—CO$_2$H	C—I 101.0 N—I [25.0]	C—I 56.0 N—I (−1.5)	C—I 63.0 N—I [−31.5]

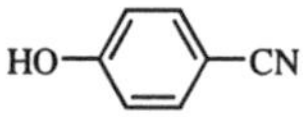

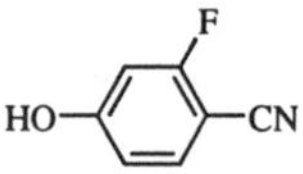

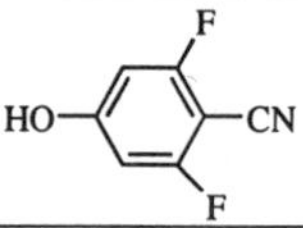

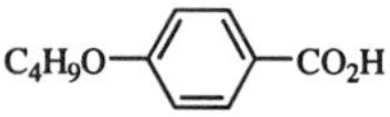

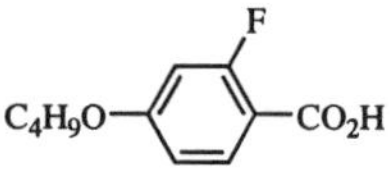

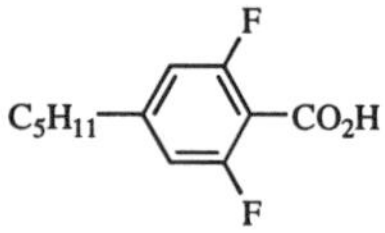

[a]( ) = Monotropic transition; [ ] = virtual transition.

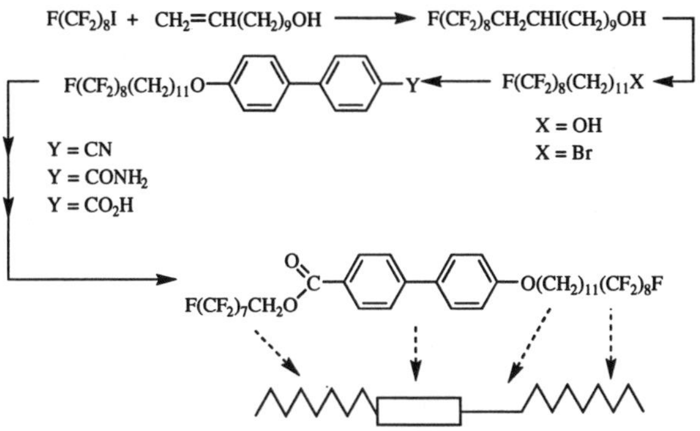

Scheme 5. Route to biaryl derivatives.

12.5.4. Biphenyls

The synthesis of a biaryl derivative substituted with two perfluoroalkylated terminal chains has been reported by Tournilhac et al.[36] In this "four block mesogen", a smectic C structure was found between 92°C and 111°C. The synthetic pathway used is outlined in Scheme 5.

Table 8. Transition Temperatures and Enthalpies (ΔH) for the 3,4-Difluorophenyl *trans*-4'-Substituted Cyclohexane-1'-carboxylates

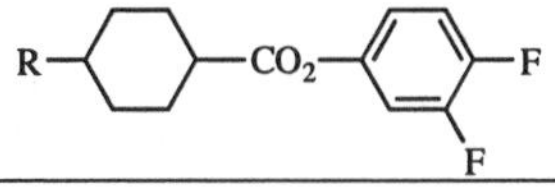

R	Transition temp. (°C)[a]		ΔH (kcal mol^{-1})
	C—I	N—I	C—I
C_2H_5	19	—	—
C_3H_7	32	—	7.40
C_4H_9	25	(−26)	6.95
C_5H_{11}	37	(−15)	7.78
C_6H_{13}	38	—	8.25
C_7H_{15}	49	(4)	9.58

	C—N	N—I	C—N	N—I
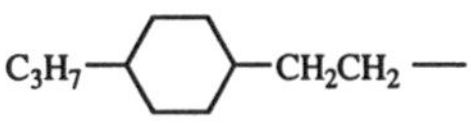 C_3H_7—⬡—CH_2CH_2—	56	132	7.46	0.197
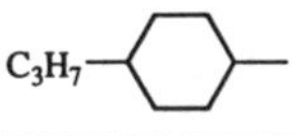 C_3H_7—⬡—	57	153	8.29	0.190

[a]( ) = Monotropic transition.

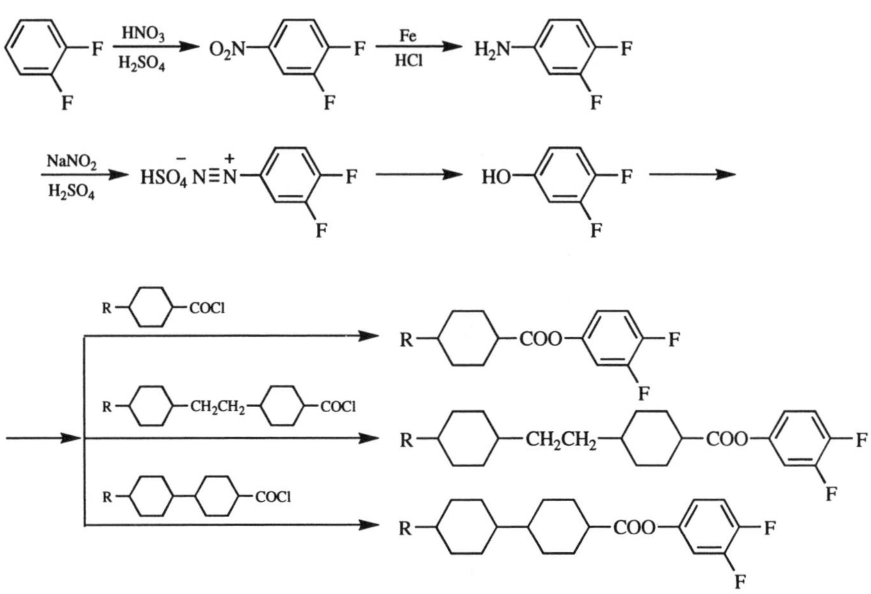

Scheme 6. Synthetic pathway for 3,4-difluorophenyl esters.

12.5.5. Cyclohexanecarboxylates

Takatsu *et al.* have synthesized a series of 3,4-difluorophenyl *trans*-4′-substituted cyclohexane-1′-carboxylates in order to achieve new nematic compounds of positive dielectric anisotropy and especially low viscosity.[37] Thermal data are listed in Table 8 and the synthetic pathway is shown in Scheme 6. These compounds are good as components for liquid crystal mixtures.

12.5.6. Liquid Crystals with Hybridized Structures and Multiring Systems

For improving liquid crystal display, liquid crystals of greater variation in physical properties are needed today; and to meet this requirement compounds with hybrid structures have been studied. Also, molecules characterized by the presence of three or more ring systems are of practical interest.

12.5.6.1. Cyclohexyl(4-phenylcyclohexyl)ethanes

A series of 1-cyclohexyl-2-(4-phenylcyclohexyl)ethane derivatives (synthesized as shown in Scheme 7) have a nematic phase in a wide temperature range with low viscosity and low birefringence; melting point, clearing point, and other physical properties are listed in Table 9.[38]

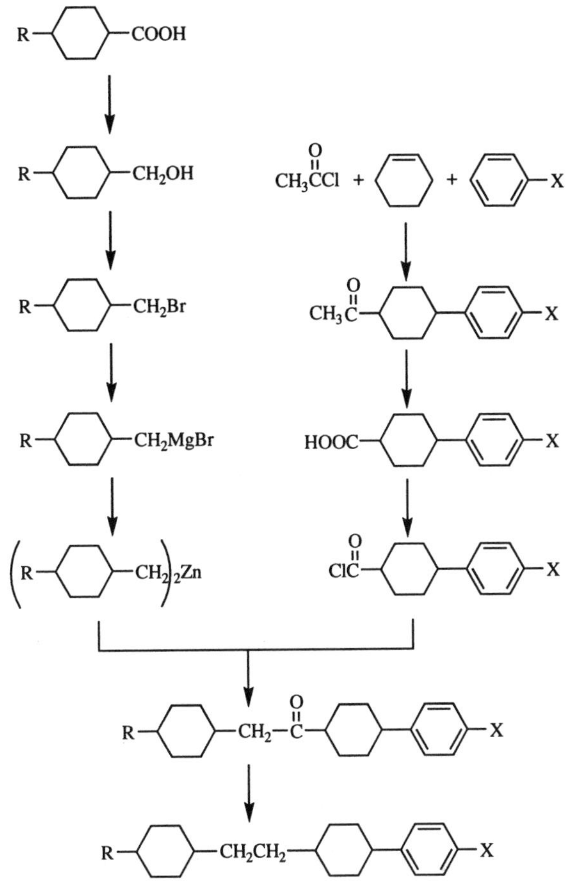

Scheme 7. Synthesis of 1-cyclohexyl-2-(4-phenylcyclohexyl)ethane derivatives.

Table 9. Physical Properties of 1-Cyclohexyl-2-(4-Phenylcyclohexyl)ethane Derivatives

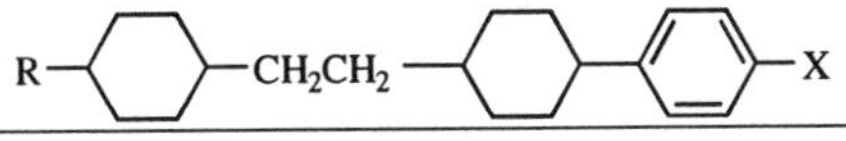

R	X	Tm (°C)[a]	Tc (°C)[a]	Δn	$\Delta \varepsilon$	ν (cSt)[b]
n-C$_3$H$_7$	F	64.3	130.2	0.062	4.7	—
n-C$_4$H$_9$	F	72.3	126.5	0.063	4.3	46.6
n-C$_3$H$_7$	Cl	77.0	155.0	0.092	5.9	—
n-C$_4$H$_9$	Cl	54.3	149.2	0.085	5.4	48.4
n-C$_3$H$_7$	CN	86.3	197.1	0.106	16.4	—
n-C$_4$H$_9$	CN	64.4	191.6	—	20.9	50.4
n-C$_3$H$_7$	n-C$_3$H$_7$	141.8[c]	144.0	—	0	46.7

[a]Tm: Melting point; Tc: clearing point.
[b]Viscosity of 10 wt% mixture in a parent liquid crystal mixture, [25 wt% *trans*-1-n-propyl-4-(4-ethoxyphenyl)cyclohexane, 20 wt% *trans*-1-n-propyl-4-(4-n-butoxyphenyl)cyclohexane, and 55 wt% ZLI-1083 (Merck & Co., Inc.)].
[c]Smectic-to-nematic transition.

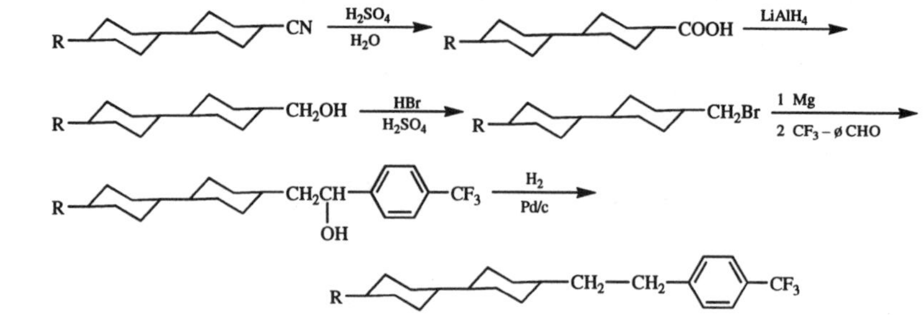

Scheme 8.   Synthetic pathway for bicyclohexyl[(trifluoromethyl)phenyl]ethanes.

12.5.6.2.   Bicyclohexyl[(trifluoromethyl)phenyl]ethanes

In addition to alkylcyclohexyl benzotrifluorides, compounds of the general structure R—CY—CY—CH_2CH_2—$C_6H_4CF_3$ have been synthesized (Scheme 8)[39] where
CY is a *trans*-1,4-disubstituted cyclohexyl ring and R is a normal alkyl group. In the
case where R = *n*-propyl, the compound has nematic and smectic phases, $T_{N—I}$:
110.3°C, $T_{N—S}$: 104.6°C, and $T_{S—C}$: 51.7°C. This compound will be useful for a fast
switching liquid crystal mixture.

12.5.6.3.   Biphenyls

In this series, Le Barny *et al.* have synthesized two families of 4-alkoxytetra-
fluorobenzoates and 4-trifluoromethyl-4'-substituted derivatives.[40] The introduction
of a perfluorinated 1,4-phenylene ring leads in general to compounds having lower
melting points and decreased mesomorphic stability. These compounds are additives
of interest for display mixtures (Table 10). The 4-alkoxytetrafluorobenzoic acids

Table 10.   Thermodynamic Data (°C) for a Typical Mixture of
Cyanobiphenyls

Compounds	Molar fraction
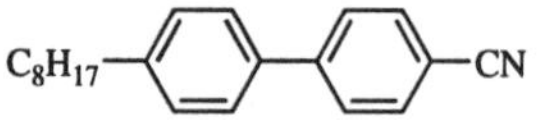 C_8H_{17}—⟨⟩—⟨⟩—CN	66.50
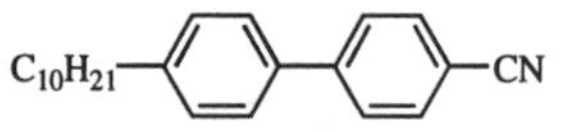 $C_{10}H_{21}$—⟨⟩—⟨⟩—CN	22.40
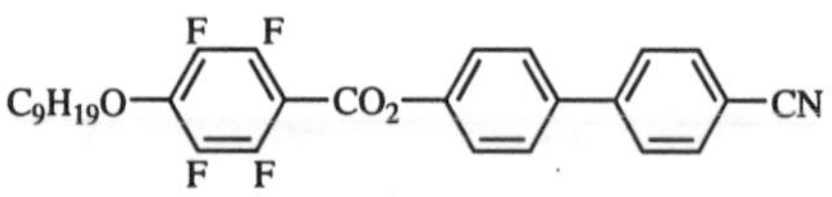 $C_9H_{19}O$—⟨F F/F F⟩—CO_2—⟨⟩—⟨⟩—CN	11.10

Range C 4.5 S_A 45   N 53   I

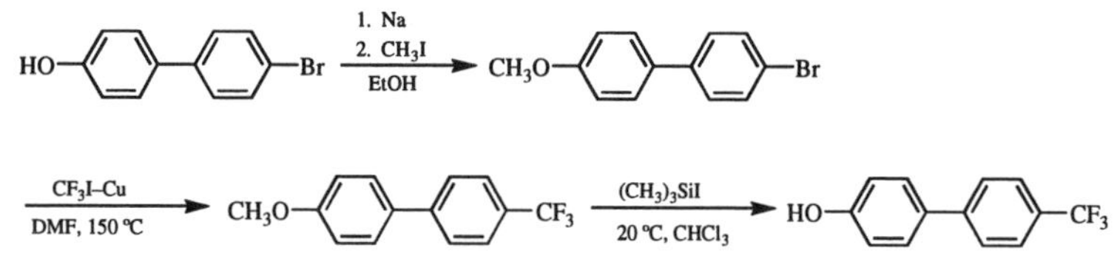

Scheme 9. Synthesis of 4-hydroxy-4'-(trifluoromethyl)biphenyl.

required as starting materials were obtained by treating pentafluorobenzoic acid with appropriate sodium alkanoates. 4-Hydroxy-4'-(trifluoromethyl)biphenyl was procured as indicated in Scheme 9.

12.5.6.4. Pyrimidines

Chiral esters derived from 2-(4-n-hexylphenyl)- and 2-(4-n-hexyloxyphenyl)-5-(4-hydroxyphenyl)pyrimidine and α-fluorocarboxylic acids have been synthesized by Bömelburg et al.[41] All compounds exhibit wide-range smectic phases with higher values of spontaneous polarization (ca 4 × 10^{-7} C/cm^2) (Table 11). Spontaneous polarization is directly concerned with ferroelectricity of liquid crystals (see Section 12.6).

Chiral α-fluorocarboxylic acids and esters were synthesized as indicated in Scheme 10.

Table 11. Transition Temperatures for Alkyl- and Alkoxy-Pyrimidine Derivatives[a]

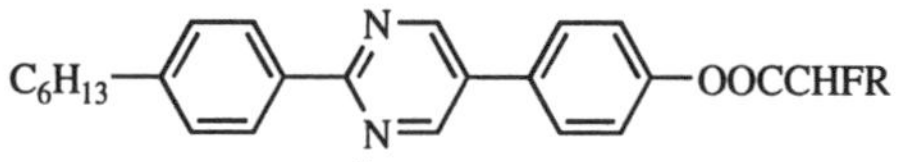

R	C	S$_3$	S$_C^*$	I
−CH(CH$_3$)$_2$	165.1	156.5	172.7	202.6
−CH$_2$CH(CH$_3$)$_2$	150.7	138.8	175.2	190.5
−C*H(CH$_3$) (C$_2$H$_5$)	152.4	143.8	162.5	172.5
−C$_3$H$_7$	150.3	135.1	171.8	208.5

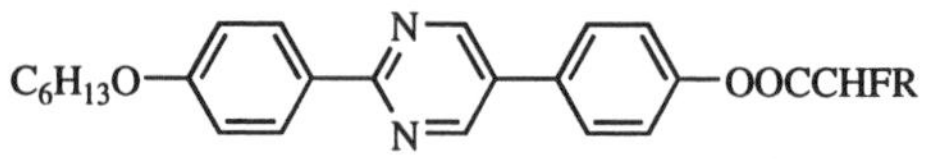

R	C	S$_5$	S$_4$	S$_3$	S$_C^*$	S$_A$	Ch	I
−CH(CH$_3$)$_2$	146.1	—	146.9	—	192.7	215.0		—
−CH$_2$CH(CH$_3$)$_2$	135.2	—	128.0	128.3	190.6	200.6		—
−C*H(CH$_3$)(C$_2$H$_5$)	140.8	124.1	132.3	—	179.5	185.4		185.8
−C$_3$H$_7$	134.0	—	—	—	193.0	218.1		—

[a]S$_3$, S$_4$, and S$_5$ are unidentified smectic phases.

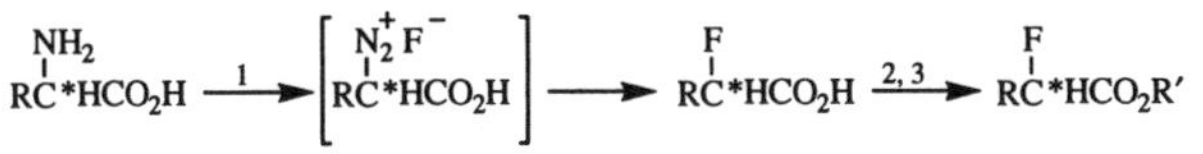

$$\underset{\underset{\text{RC*HCO}_2\text{H}}{\overset{\text{NH}_2}{|}}}{} \xrightarrow{\ 1\ } \left[\underset{\underset{\text{RC*HCO}_2\text{H}}{\overset{\text{N}_2^+\ \text{F}^-}{|}}}{} \right] \longrightarrow \underset{\underset{\text{RC*HCO}_2\text{H}}{\overset{\text{F}}{|}}}{} \xrightarrow{\ 2,3\ } \underset{\underset{\text{RC*HCO}_2\text{R'}}{\overset{\text{F}}{|}}}{}$$

1. (HF)-Pyridine/NaNO$_2$; 2. SOCl$_2$; 3. R'OH/pyridine

Scheme 10. Synthesis of chiral α-fluorocarboxylic acids and esters.

12.5.6.5. Tolans

To improve the response time for twisted nematic liquid crystal displays, a series of tolans have been used as components of liquid crystal mixtures with low viscosity (Table 12). The tolans were prepared via coupling of *n*-alkylphenylacetylenes and the corresponding iodides.[42]

12.5.6.6. Terphenyls

Homologs of several compounds containing the 2,3-difluoro-1,4-phenylene moiety have been prepared[43–46] to provide higher negative dielectric anisotropies (Table 13). Some of the key steps in the synthetic routes are shown in Scheme 11. A series of laterally fluorinated phenyl carboxylates was synthesized.[44] Generally, the number and position of the fluoro groups have an effect on the parent liquid crystal transition temperatures. Compounds containing a 2,3-difluorophenyl group provide useful materials which have a large negative dielectric anisotropy (Table 14).

Table 12. Transition Temperatures for 2- or 3-Substituted Tolans[42]

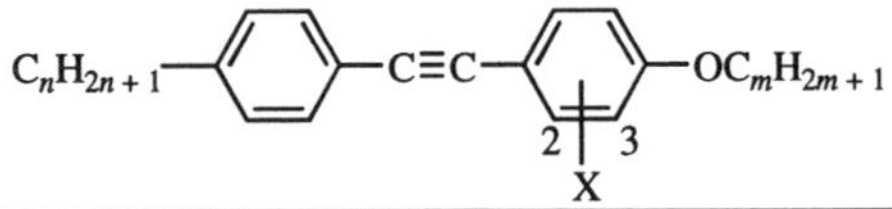

$$\text{C}_n\text{H}_{2n+1}\text{—}\langle\!\bigcirc\!\rangle\text{—C}\!\equiv\!\text{C—}\underset{\underset{\text{X}}{2\ |\ 3}}{\langle\!\bigcirc\!\rangle}\text{—OC}_m\text{H}_{2m+1}$$

				Transition temperature (°C)[a]		
n	*m*	X	Position (2 or 3)	C	N	I
3	2	F	2	70	(70)	
4	2	F	2	45	51	
4	2	F	3	56	—	
5	2	F	2	61	66	
5	5	F	3	40	(35)	
4	2	CH$_3$	2	58	(42)	
4	2	CH$_3$	3	55	—	
5	2	CH$_3$	2	42	54	
5	2	CH$_3$	3	70	(45)	

[a]( ) = Monotropic transition.

Table 13. Physical Properties of Dialkylterphenyls

Chemical structure	Transition temp. (°C)				
	C	S/N	I	$\Delta\varepsilon$	Δn
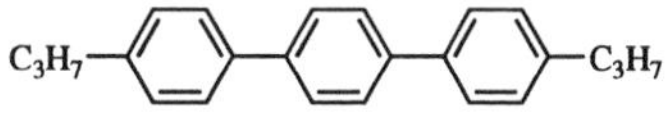 C_3H_7—⬡—⬡—⬡—C_3H_7	221	S	228		
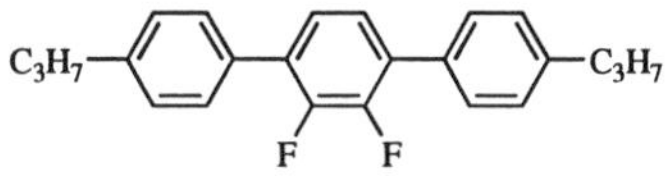 C_3H_7—⬡—⬡(F,F)—⬡—C_3H_7	96	N	132	−1.7	0.25
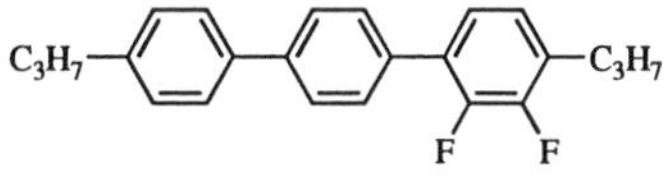 C_3H_7—⬡—⬡—⬡(F,F)—C_3H_7	132	N	148	−1.9	0.25

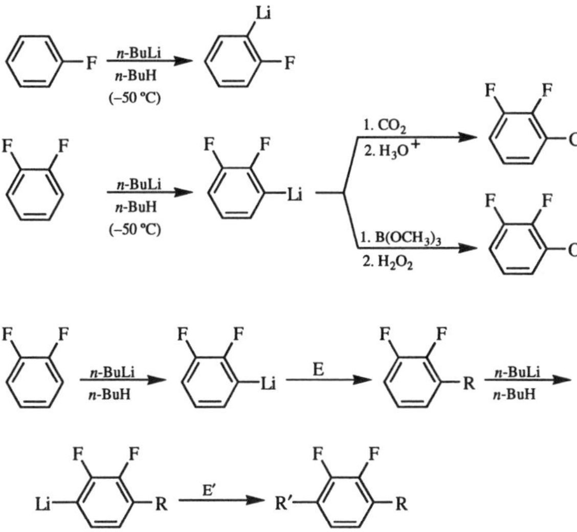

Scheme 11. General route to terphenyls: E and E′ represent appropriate electrophiles.

Table 14. Transition Temperatures (°C) for 2,3-Difluoro Derivatives of Some Phenyl Biphenylcarboxylates

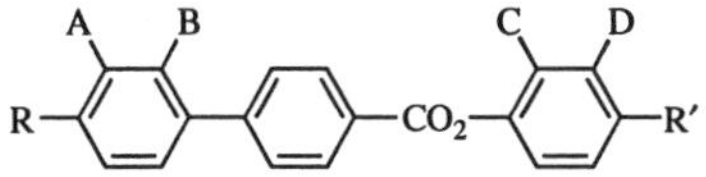

R	R′	A	B	C	D	C	S_C	S_A	N	I
$C_8H_{17}O$	$C_8H_{17}O$	F	F	H	H	83.8	151.7	154.9		165.4
$C_8H_{17}O$	C_8H_{17}	F	F	H	H	83.6	121.0	139.0		144.6
$C_8H_{17}O$	$C_7H_{15}O$	H	H	F	F	94.0	156.9	166.2		174.7
C_8H_{17}	$C_7H_{15}O$	H	H	F	F	86.0	125.2	131.2		145.5

Table 15. Transition Temperatures (°C) for Several Difluoro-Substituted
4,4″-Dialkylterphenyls

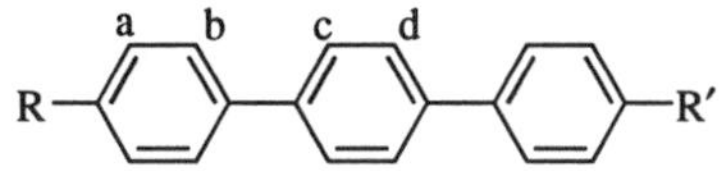

R	R'	a	b	c	d	C	S_I	S_C	S_A	N	I
C_5H_{11}	C_5H_{11}	H	H	F	F	60.0	—	—	—	120.0
C_5H_{11}	C_5H_{11}	F	F	H	H	81.0	—	115.5	131.5	142.0
C_5H_{11}	C_7H_{15}	H	H	F	F	36.5	—	$(24.0)^a$	—	111.5
C_5H_{11}	C_7H_{15}	F	F	H	H	65.5	74.5	118.5	135.0	137.0
C_7H_{15}	C_5H_{11}	F	F	H	H	56.0	—	105.5	131.0	136.0
C_5H_{11}	C_9H_{19}	H	H	F	F	42.5	—	66.0	—	110.0
C_7H_{15}	C_9H_{19}	H	H	F	F	49.0	—	77.0	93.0	108.5

^{a}( ) = Monotropic transition.

A lateral monofluoro substituent in terphenyls causes a reduction in the melting point of the parent system and suppression of the more ordered smectic phases. Furthermore, it generates compounds with a tendency to form tilted smectic phases such as S_C, S_I, and S_F; however, these compounds still retain ordered smectic phases, and with other factors it is then somewhat difficult to use as the host material for a practical S_C mixture. The new series of 2,3-difluoro substituted terphenyls synthesized by Gray et al.[45] are low-melting liquid crystals with wide-range S_C phases and no underlying smectic phase, thus these compounds will be good hosts for ferroelectric systems (Table 15). What facilitates the synthesis of these compounds is the ability to lithiate a position *ortho* to an aryl fluorine atom (see Scheme 11). A successive coupling reaction is generally useful for the synthesis of this series (Scheme 12).

Note that a useful route to 2-trifluoromethyl-biphenyls and -terphenyls was reported by Hiyama and Sato.[46] This involves Diels-Alder cyclization of 1-aryl-3,3,3-trifluoropropynes with 1,3-dienes, followed by aromatization, and has provided compounds which may be applicable to liquid crystal preparation (see Table 16).

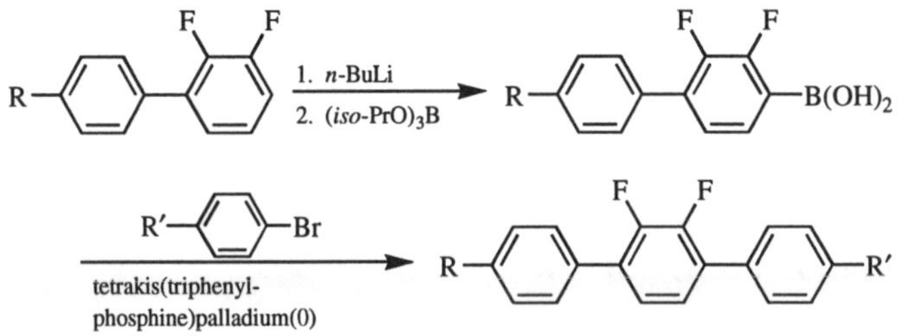

Scheme 12. Synthesis of difluoro-substituted 4,4″-dialkylterphenyls.

Table 16.   Synthesis of 2-(Trifluoromethyl)-biphenyls and -terphenyls[46]

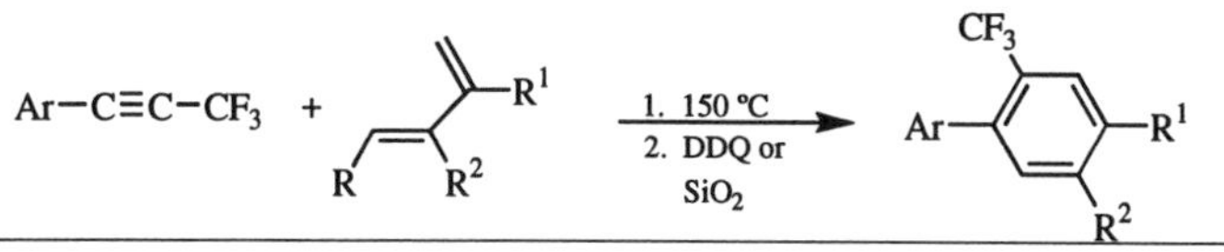

Ar	Aromatization step	Product	Yield (%)
4-PhC6H4–[a]	DDQ[c]	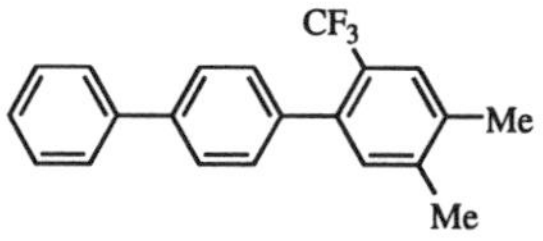	37
4-PhC6H4–[b]	SiO2		73
4-ClC6H4–[b]	SiO2	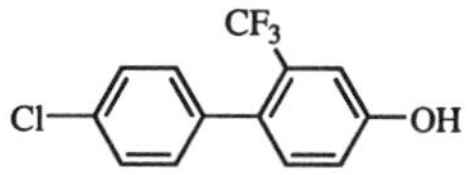	82
4-n-C8H17OC6H4–[b]	SiO2	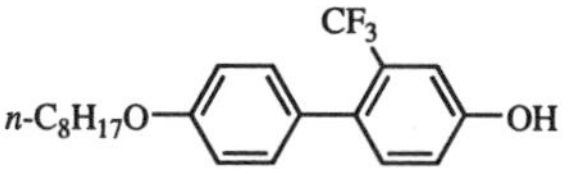	75
4-n-C3H7C6H4–[b]	SiO2	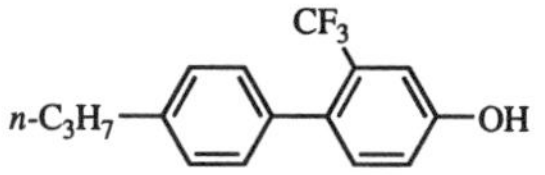	90

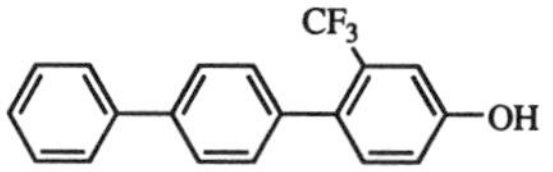

[a] $R^1=R^2=Me$, R=H.
[b] $R^1=OSiMe_3$, $R^2=H$, R=MeO.
[c] DDQ = 2,3-dichloro-5,6-dicyano-1,4-benzoquinone.

12.6. FERROELECTRIC LIQUID CRYSTALS

The discovery of ferroelectricity, which is significant for the spontaneous polarization in the chiral smectic C phase (S_C^*) of (S)-2-methylbutyl-p-[(p-n-decyloxybenzylidene)amino]cinnamate (DOBAMBC, Figure 5), by Meyer *et al.*[47,48] has stimulated the study of the fundamental properties of ferroelectric liquid crystals and their practical applications. On the applications side, significant advances in the "surface stabilized ferroelectric liquid displays" of the Clark–Lagerwall type[49] have been under development. Liquid crystals which display ferroelectricity must possess at least a chiral carbon and a dipole moment, and also must show a smectic phase with some tilt angle. In most ferroelectric liquid crystals, bonds such as C=O, C—Cl, and C—Br have been utilized as the source of dipole moment. Many ferroelectric compounds in the Schiff base, azoxy compound, benzoate, etc., series have been reported; the C—F bond will have a range of useful applications in a number of compound types.

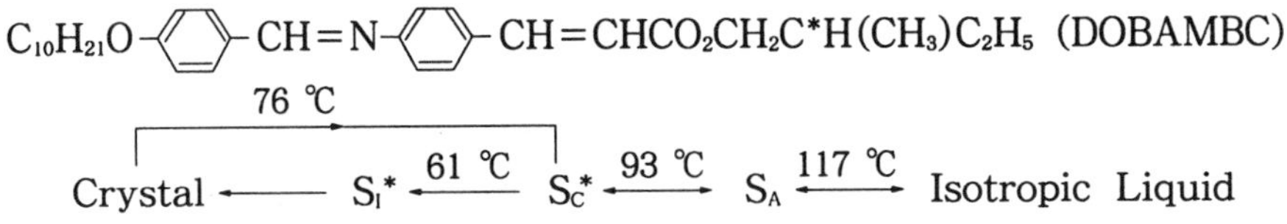

$$C_{10}H_{21}O-\langle\rangle-CH=N-\langle\rangle-CH=CHCO_2CH_2C^*H(CH_3)C_2H_5 \quad (DOBAMBC)$$

$$\text{Crystal} \xleftarrow{\hspace{1cm}} S_I^* \xleftarrow{61\ ^\circ C} S_C^* \xrightarrow[\hphantom{x}]{93\ ^\circ C} S_A \xleftrightarrow{117\ ^\circ C} \text{Isotropic Liquid}$$

(with 76 °C marked above the Crystal → S_I* transition)

Figure 5. Formula and phases of DOBAMBC.

Table 17. Characteristics of Biphenylcarboxylates[a]

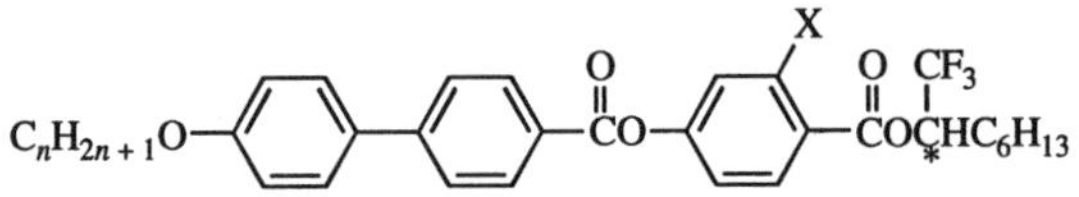

$$C_nH_{2n+1}O-\langle\rangle-\langle\rangle-\overset{\overset{O}{\|}}{C}O-\langle\rangle\overset{X}{}-\overset{\overset{O}{\|}}{C}O\overset{CF_3}{\underset{*}{C}}HC_6H_{13}$$

X	n	C	$S_?^{*b}$	S_Y^{*c}	S_C^*	S_A^*	I	Ps[d]
H	8	68.0	65.5	118.5	122.0	149.8		+ 70
F	7	66.5	—	120.4	121.2	145.4		+ 53
F	8	36.3	—	123.4	124.2	142.1		+ 60
F	10	30.5	—	118.4	133.4	133.4		+ 62
Cl	8	29.5	—	94.9	116.3	116.3		+ 78

[a]Spontaneous polarization was measured at $T = T_c - 10^\circ C$.
[b]$S_?^*$ = not identified.
[c]S_Y^* = tentatively named (a kind of S_C^*).
[d]Ps (nC cm^{-2}).

(–)-4'-(3-Ethoxycarbonyl-1-trifluoromethyl-2-propoxycarbonyl)phenyl 4-[4-(octyl-oxy)phenyl]benzoate and related compounds have been synthesized[50,51]; spontaneous polarization of this S_C^* compound is of the order of 10^{-7} C/cm^2, a value much larger than that of DOBAMBC ($ca\ 10^{-9}$ C/cm^2). A "third stable state" was found with the chiral biphenylcarboxylate

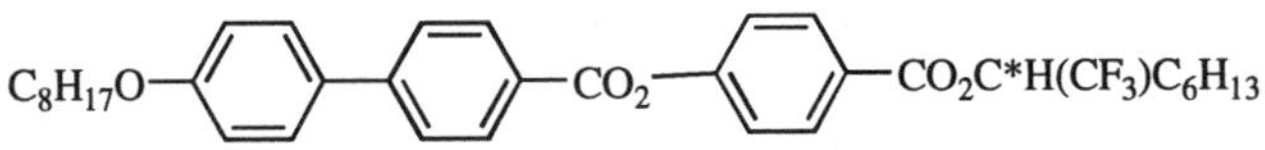

$$C_8H_{17}O-\langle\rangle-\langle\rangle-CO_2-\langle\rangle-CO_2C^*H(CF_3)C_6H_{13}$$

which is expected to be useful for display devices.[52] The laterally fluorine-substituted multiring homologs of this compound were synthesized in order to obtain large spontaneous polarization (Table 17).[53] In these compounds, a new type of smectic phase (tentatively named as S_Y^*) was observed, and this is now characterized as the new "antiferroelectric phase."[54]

12.7. CONCLUSIONS

In general, terminal fluoro-, trifluoromethyl-, and trifluoromethoxy-substituted mesogenic molecules have been shown to be useful by themselves, or as additives for formulated practical mixtures with other more conventional cyano-terminated mesogens, because they are chemically stable and possess moderate positive dielectric anisotropy with lower viscosity. Lateral substitution with fluorine on the core side, especially with simpler 4-cyanophenyl benzoates, exerts two opposing effects on the thermal stability of mesophases: the presence of the halogen increases the polarity of the molecule and results in mesophase stability; on the contrary, the size of the lateral fluorine group deforms the three-dimensional molecular shape, i.e., the length-to-breadth ratio, and this tends to decrease the stability of mesophases.

Since the beginning of the 1980s, the effect and advantages of fluoro substitution in lower-molecular-weight liquid crystals have become well known, however, things become less straightforward when this kind of trade-off is applied to multicore systems. The availability of theoretical studies has provided the necessary insights into the relationships between molecular structure and physical properties. Thus, computational chemistry will be one of the ways to design molecules that more closely approximate the actual system of liquid crystals.

ACKNOWLEDGMENTS

The author is indebted to Dr. T. Inukai, Dr. S. Sugimori, Mr. K. Furukawa and his research group (Chisso Corp.) for useful discussion and for helpful information; and to the late Prof. Dr. N. Ishikawa for his advice and for the provision of necessary information.

12.8. REFERENCES

1. D. Demus, H. Demus, and H. Zaschke, *Flüssige Kritalle in Tabellen*, Deutscher Verlag für Grundstoffindustrie, Leipzig (1974).
2. D. Demus and H. Zaschke, *Flüssige Kritalle in Tabellen II*, Deutscher Verlag für Grundstosffindustrie, Leipzig (1984).
3. E. T. Samulski and D. B. Dupré, in *Advances in Liquid Crystals*, Vol. 4 (G. H. Brown, ed.), Academic Press (1979).
4. S. Chandrasekhar, B. K. Sadashiva, and K. A. Suresh, *Pramana 9*, 471 (1977).
5. D. Demus and L. Richter, *Texture of Liquid Crystals*, Verlag Chemie, Berlin (1978).
6. G. R. Luckhurst and G. W. Gray (ed.), *The Molecular Physics of Liquid Crystals*, Academic Press, New York (1979).
7. G. W. Gray and J. W. Goodby, *Smectic Liquid Crystals—Textures and Structures*, Leonard Hill (1984).
8. G. Vertogen and W. H. de Jeu, *Thermotropic Liquid Crystals, Fundamentals*, Springer-Verlag, Berlin (1988); for a concise review of liquid crystals, see Kirk-Othmer, *Encyclopedia of Chemical Technology*, 3rd edn., Vol. 15, p. 395, Wiley, New York (1981).
9. C. Destrade, F. Vinet, P. Maelstaf, and H. Gasparoux, *Mol. Cryst. Liq. Cryst. 68*, 175 (1981).
10. G. W. Gray, *J. Chem. Soc.* 3733 (1956).
11. M. Schadt, *Liq. Cryst. 5*, 57 (1989).
12. G. H. Heilmeier, L. A. Zanoni, and L. A. Barton, *Proc IEEE 56*, 1162 (1968).
13. G. H. Heilmeier and L. A. Zanoni, *Appl. Phys. Lett. 13*, 91 (1968).
14. J. J. Wysocki, J. Adams, and W. Hass, *Phys. Rev. Lett. 20*, 1024 (1968).
15. M. Schadt and W. Helfrich, *Appl. Phys Lett. 18*, 127 (1971).

16. J. L. Fergason, *Appl. Opt.* 7, 1729 (1968).
17. A. Miyaji, M. Yamaguchi, A. Toda, H. Mada, and S. Kobayashi, *IEEE Trans. Electron Devices 24*, 811 (1977).
18. T. J. Scheffer and J. Nehring, *Appl. Phys. Lett.* 45, 1021 (1984).
19. R. B. Meyer, *Mol. Cryst. Liq. Cryst. 40*, 33 (1977).
20. N. A. Clark and S. T. Lagerwall, *Appl. Phys. Lett. 36*, 899 (1980).
21. G. W. Gray, *Mol. Cryst. Liq. Cryst. 21*, 161 (1973).
22. D. Demus, *Liq. Cryst. 5*, 75 (1989).
23. P. Balkwill, D. Bishop, A. Pearson, and I. Sage, *Mol. Cryst. Liq. Cryst. 123*, 1 (1985).
24. C. Viney, R. J. Twieg, T. P. Russel, and L. E. Depero, *Liq. Cryst. 5*, 1783 (1989).
25. S. Yano, T. Adachi, H. Oyaidz, M. Kato, and K. Morita, *Liq. Cryst. 2*, 429 (1987).
26. N. A. Vaz, S. L. Arora, J. W. Doane, and A. DeVries, *Mol. Cryst. Liq. Cryst. 128*, 23 (1985).
27. J. W. Brown, J. L. Butcher, D. J. Byron, E. S. Gunn, M. Rees, and R. C. Wilson, *Mol. Cryst. Liq. Cryst. 159*, 255 (1988).
28. G. W. Gray, C. Hogg, and D. Lacey, *Mol. Cryst. Liq. Cryst. 67*, 1 (1981).
29. S. Misaki, S. Takamatsu, and M. Suefuji, *Mol. Cryst. Liq. Cryst. 66*, 9123 (1981).
30. S. Sugimori, Japanese Patent 56-104884 (to Chisso) [*CA 96* 199323t (1982)].
31. B. S. Scheuble, Inst. Television Eng. (Jpn.) IV Symp. (1989), preprint, p. 557.
32. S. M. Kelly, *Helv. Chim. Acta 67*, 1572 (1984).
33. S. M. Kelly and Hp Schad, *Helv. Chim. Acta 67*, 1580 (1984).
34. Hp. Schad and S. M. Kelly, *J. Chem. Phys. 81*, 1514 (1984).
35. G. W. Gray, M. Hird, D. Lacey, and K. J. Toyne, *Mol. Cryst. Liq. Cryst. 172*, 165 (1989).
36. F. Tournilhac, L. Bosio, J. F. Nicoud, and J. Simon, *Chem. Phys. Lett. 145*, 452 (1988).
37. H. Takatsu, K. Takeuchi, and H. Sato, *Mol. Cryst. Liq. Cryst. 112*, 165 (1984).
38. T. Gunjima and R. Takei, Reports Res. Lab. Asahi Glass Co. Ltd. *36*, 275 (1986).
39. J. C. Liang, J. O. Cross, and L. Chen, *Mol. Cryst. Liq. Cryst. 167*, 199 (1989).
40. P. Le Barny, G. Ravaux, and J. C. Dubois, *Mol. Cryst. Liq. Cryst. 127*, 413 (1985).
41. J. Bömelburg, G. Heppke, and A. Ranft, *Z. Naturforsch. 44b*, 1127 (1989).
42. H. Takatsu, K. Takeuchi, Y. Tanaka, and M. Sasaki, *Mol. Cryst. Liq. Cryst. 141*, 279 (1986).
43. V. Reiffenrath, J. Krause, H. J. Plach, and G. Weber, *Liq. Cryst. 5*, 159 (1989).
44. M. Chambers, R. Clemitson, D. Coates, S. Greenfield, J. A. Jenner, and J. C. Sage, *Liq. Cryst. 5*, 153 (1987).
45. G. W. Gray, M. Hird, D. Lacey, and K. J. Toyne, *J. Chem. Soc., Perkin Trans. 2*, 2041 (1989).
46. T. Hiyama and K. Sato, *Synth. Lett.* 53 (1990).
47. R. B. Meyer, *Mol. Cryst. Liq. Cryst. 40*, 33 (1977).
48. R. B. Meyer, L. Liébert, L. Strzelecki, and P. Keller, *J. Phys. (Paris) 36*, L69 (1975).
49. N. A. Clark, M. A. Handschy, and S. T. Lagerwall, *Appl. Phys. Lett. 36*, 899 (1980).
50. K. Yoshino, M. Ozaki, H. Taniguchi, M. Itoh, K. Satoh, N. Yamasaki, and T. Kitazume, *Jpn. J. Appl. Phys. 26*, L77 (1987).
51. Y. Suzuki, T. Hagiwara, I. Kamura, N. Okamura, T. Kitazume, M. Kakimoto, Y. Imai, Y. Ouchi, H. Takezoe, and A. Fukuda, *Liq. Cryst. 6*, 167 (1989).
52. A. D. L. Chandani, T. Hagiwara, Y. Suzuki, Y. Ouchi, H. Takezoe, and A. Fukuda, *Jpn. J. Appl. Phys. 27*, L729 (1988).
53. K. Furukawa, K. Terashima, M. Ichihashi, S. Saitoh, K. Miyazawa, and T. Inukai, *Ferroelectrics 85*, 451 (1988).
54. A. D. Chandani, Y. Ouchi, H. Takezoe, H. Fukuda, A. Terashima, K. Furukawa, and A. Kishi, *Jpn. J. Appl. Phys. 27*, L1261 (1989).

13

Fluorine-Containing Dyes

A. Reactive Dyes

K. J. HERD

13A.1. INTRODUCTION[*]

The economic importance of reactive dyes for cotton, and in particular of fluorine-containing reactive dyes, has increased over the last twenty years.[1–10] In 1990 an estimated 50,000 tonnes of reactive dyes were used for dyeing and printing cellulosic fibers. This represents 15–20% of total consumption of dyes for cotton. Only sulfur and direct dyes achieve higher tonnages. In industrialized countries, such as the USA, Japan, and the Western Europe group, reactive dyes already account for about 30–40% of the market for cotton dyes. World sales of fluorine-containing reactive dyes in 1990 is estimated at 10,000 tonnes.[2,9–13]

This development started in 1956 with the launch of chlorotriazine dyes by ICI and vinylsulfone dyes by Hoechst. These were the first textile dyes that were able to form a covalent chemical bond with the cotton fiber during dyeing due to the presence of a reactive group in the dyestuff molecule. The result was dyeings with excellent wet-fastness. These new dyes were easy and reliable to use. They produced clear, brilliant dyeings and full tones, which had previously been unobtainable on cotton, and were available in a wide range of shades.

Ten years later, in 1966, Bayer and Sandoz launched the first fluorine-containing pyrimidine reactive system (5-chloro-2,6-difluoro-4-pyrimidinyl). Ciba-Geigy and ICI followed with monofluorotriazine dyes. This development led to a flood of new

---

[*]This review complements and expands an article by W. Harms and the literature cited there.[1]

K. J. HERD • Bayer AG, Research and Development—Dyes and Pigments, D-51368 Leverkusen 1, Germany.

Organofluorine Chemistry: Principles and Commercial Applications, edited by R. E. Banks *et al.* Plenum Press, New York, 1994.

patent applications and is still going on. Recently, economic and ecological demands and environmental legislation have given rise to a need for improved reactive systems which save energy and water and reduce effluent load.

In 1981 Sumitomo brought the first bifunctional reactive dyes onto the market. This new type of dyestuff had a monochlorotriazine and a vinylsulfone group in the dyestuff molecule.[11,12,14–19] A few years later Bayer and Ciba-Geigy developed the first high-fixation bifunctional dyes, comprising a combination of a fluorine-containing heterocyclic reactive radical and a vinylsulfonyl radical. The higher fixation probability of two reactive groups gives far higher fixation yields (*ca.* 90%) than conventional monofunctional dyes, which lose, on average, a quarter of the dyestuff through hydrolysis during dyeing.[10,13,20–22] Dyestuff manufacturers are thus concentrating their research activities on this promising new reactive dyestuff type.

13A.2.   DYESTUFFS WITH ONE REACTIVE SYSTEM

13A.2.1.   General Information on Reactive Dyes

Reactive dyes are colored, water-soluble substances with structures comprising two parts, a sulfonated chromophore system, e.g., with an azo or anthraquinone unit, and a fragment carrying a reactive center (leaving group).

In reactive dyes with halogen as the leaving group, these two partial structures are nearly always linked by an amino function in the chromophore, which makes an

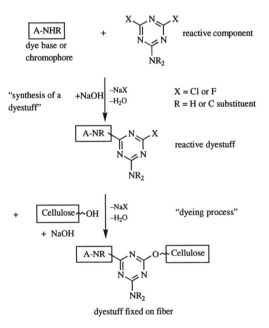

Figure 1.   Functioning of a reactive component.

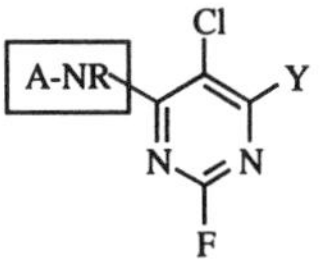

R = H or C substituent

1 (Y = F)	LEVAFIX® EA dyestuffs (Bayer)	Drimaren® K dyestuffs (Sandoz)
2 ($Y = CH_3$)	LEVAFIX® PN dyestuffs (Bayer)	

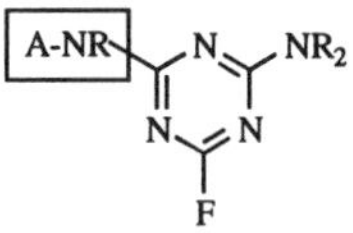

3	Cibacron® F and C dyestuffs (Ciba-Geigy)	LEVAFIX® EN dyestuffs (Bayer)

Figure 2. Fluorine-containing reactive dyestuff ranges.

important contribution to the chromophore conjugated π-electron system. In practical terms, this means that the reactive component must have at least two reactive groups, one that reacts with the dye base and one that reacts with the cellulosic fiber. Acylation of the dye base at the N atom generally involves a hypsochromic change in shade. The principle behind this reaction and dyeing system can be illustrated by monohalogen triazine dyestuffs (X = Cl or F; Figure 1).

The dyestuff can only be fixed in basic conditions (pH 11–13). One drawback is the competitive reaction between the dyestuff and hydroxide anions, generating unusable dyestuff hydrolysate.[23–26]

If they are to be suitable for dyeing and printing, reactive dyes have to satisfy a number of technical requirements and fastness standards. Properties such as solubility in water and substantivity,[27] washing-off of unfixed dyestuff hydrolysate, and brilliance, lightfastness, and fastness to chlorine are mainly determined by the chromophore component of the dyestuff. Wet-fastness properties, fastness to acids and alkalis (stability of the dyestuff–fiber bond), rate of fixation,* and optimum dyeing temperature are generally determined by the properties of the reactive component.[24–32]

The introduction of fluorine-containing reactive components in place of chlorine-containing ones has substantially improved these dyeing parameters.[6,9,28] In particular, fixation yields have been increased by *ca.* 10–15% while the temperature needed for exhaust dyeing has been reduced from 75–90°C to 40–60°C, thus saving energy.[14]

13A.2.2. Heterocyclic Carrier Systems with Fluorine as the Leaving Group

Of all the patented and published reactive components containing fluorine, only three have attained any notable technical and economic significance: 5-chloro-2,4,6-trifluoropyrimidine,[33,34] 5-chloro-2,4-difluoro-6-methylpyrimidine,[35] and 2,4,6-trifluoro-*s*-triazine.[36,37] Figure 2 shows the relevant dyestuff ranges **1–3**, respectively, and their manufacturers.

*Rate of fixation is defined as the percent of dyestuff fixed on the fiber, taking the initial quantity as 100%.[25]

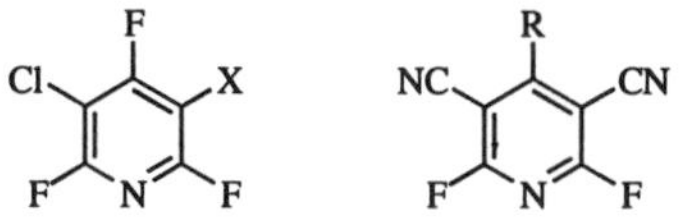

4, X = Cl, CN   5, R = phenyl, 2-furanyl

Figure 3. Fluorinated pyridines.

This shows that six-membered heterocycles are the key to these systems. The other known reactive components, for example, sulfonic acid fluorides,[38] mono- and di-fluoroalkanes,[39] activated di-,[40] tri-,[41] and tetra-fluorobenzenes,[42,43] and five-membered heterocycles,[44] are of no commercial relevance.

Unlike these species, six-membered heterocycles have adequate reactivity and guarantee optimum dyeings with stable fixation on the fiber. On the basis of charge density calculations and reactivity studies, the following reactivity sequence for nucleophilic substitution can be established for perfluorinated heterocyclic systems[45–47]:

pyridine < pyrazine << pyridazine < pyrimidine < sym.-triazine

By introducing an amino substituent (cf 3), the reactivity of the trifluoro-*s*-triazine can be adjusted to suit that of the pyrimidine. Under certain conditions, the pyrimidine system forms a more stable dyestuff–fiber bond than this substituted triazine system.[48] Another advantage of these two six-membered heterocycles is that they are relatively easy to synthesize.

13A.2.2.1. Pyridines

Only a few fluorinated pyridines[49–51] (Figure 3) and quinolines,[52] which are additionally activated with chlorine or cyano substituents, are used as reactive groups in the synthesis of dyestuffs. It is not yet clear which F atom is substituted in the 3-chloro-5-cyano-2,4,6-trifluoropyridine (4) during reaction with an amino-containing chromophore.[49]

Perfluorinated pyridine undergoes nucleophilic substitution in the 4 position.[45,47] Activation by the cyano group (4, 5) is obviously comparable with the effect of a hetero-N atom in this position. The reactivity of dyes containing these pyridinyl radicals thus corresponds to that of dyes of structure 1.[23]

As far as we know, fluorinated pyrazines[45,46] have not yet been used as reactive components in dyestuff chemistry.

13A.2.2.2. Pyridazines and Pyridazinones

Pyridazines with chlorine as the leaving group (Figure 4) are used in the Solidazol® dyestuffs range (Cassella) while corresponding pyridazinones (6, X = Cl) are used in the Primazin® P range (BASF).[6,53,54] Fluorine-containing dyestuffs (6, X = F) and dyestuffs based on pyridazinone (7), e.g., as a diazo component, are more reactive and

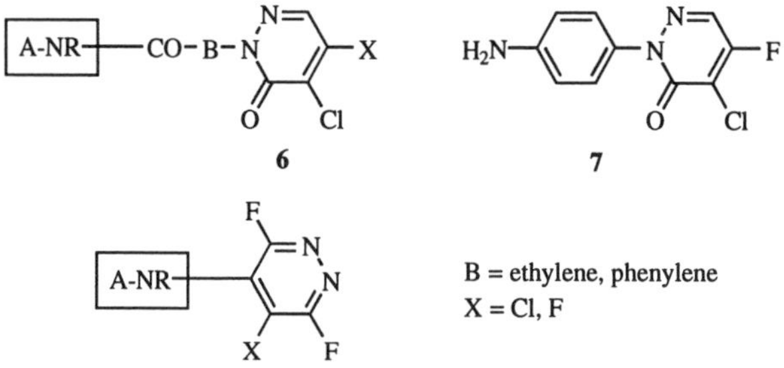

Figure 4. Pyridazinone and pyridazine dyestuffs.

should produce dyeings with comparable fastness properties.[55,56] Substitution by hydroxy groups from the cellulose always occurs in the 4 position in **7**.[57]

The use of 4-chloro-3,5,6-trifluoro- or tetrafluoro-pyridazine produces the reactive dyestuffs **8**, which are suitable for dyeing cotton from long liquors at 60°C.[58] The fluorine-containing dyestuffs **6** and **8** have not been marketed. The reactive components needed for these dyes can be produced from the corresponding dichloro and tetrachloro precursors by halogen exchange with potassium fluoride under various reaction conditions.[59,60]

13A.2.2.3. Pyrimidines

The first examples of fluorine-containing pyrimidines in the literature were linked to the reactive dyestuff **9** (Figure 5) via a carbonyl bridge similar to **6**, rather than by a direct link between the hetero-C atom and the amino function of the chromophore. Since Reaktofil® dyestuffs (**9**, X = Cl) (Ciba-Geigy) already have extremely high reactivity and a low dyeing temperature, corresponding dyes with X = F are too reactive for fixation on the fiber and extremely susceptible to hydrolysis, because they have three strong acceptor substituents.[61-63]

Direct linking of 5-chloro-2,4,6-trifluoropyrimidine (FCP) with dye bases to give the commercial dyestuffs **1** was both technically and economically successful. Analagous trichloropyrimidine dyes for dyeing at 80°C have been marketed under the brand names Drimaren® (Sandoz) and Reacton® (Geigy) since 1960.[64-66] The wide range of LEVAFIX® EA[67] and Drimaren® K dyestuffs[24] is illustrated by Figure 6, showing typical azo, anthraquinone, formazan, and phthalocyanine products.

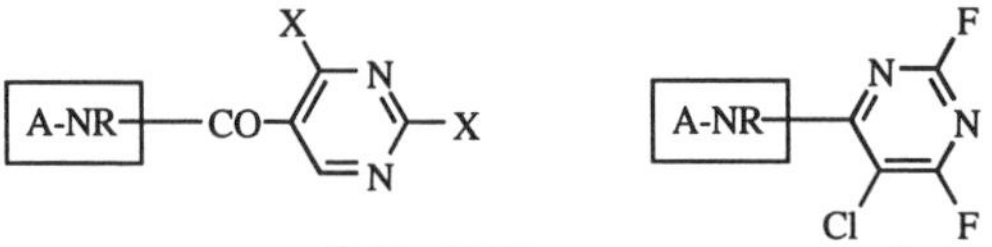

Figure 5. Pyrimidine dyestuffs.

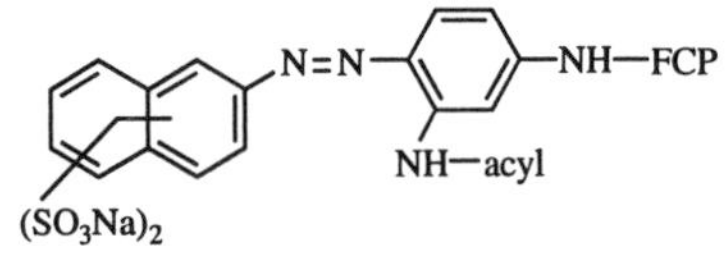

10 Yellow[34,68]

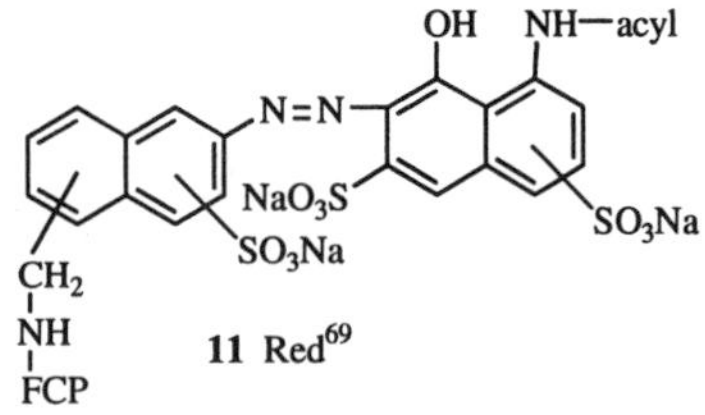

11 Red[69]

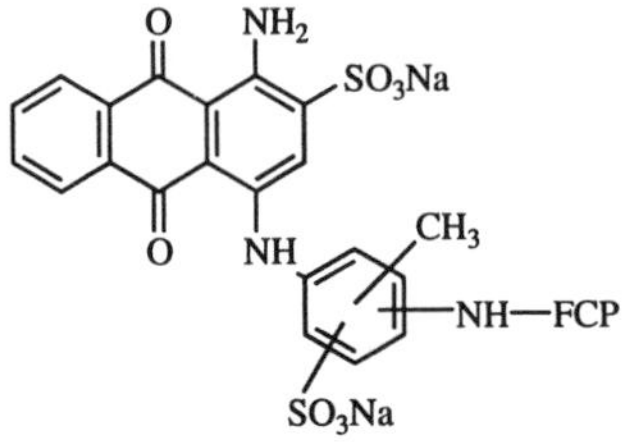

12 Blue[34,68]

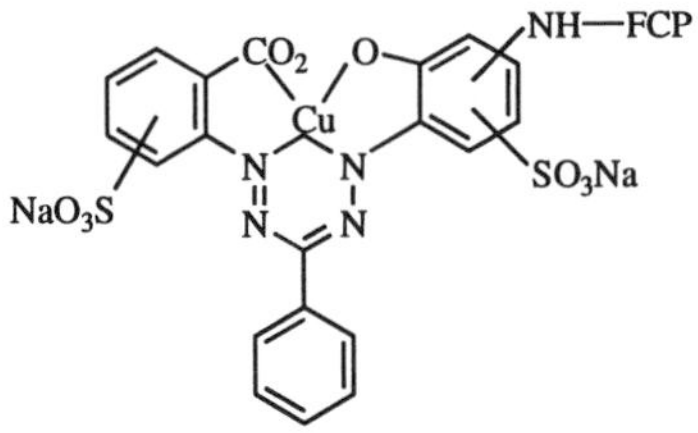

13 Blue[70]

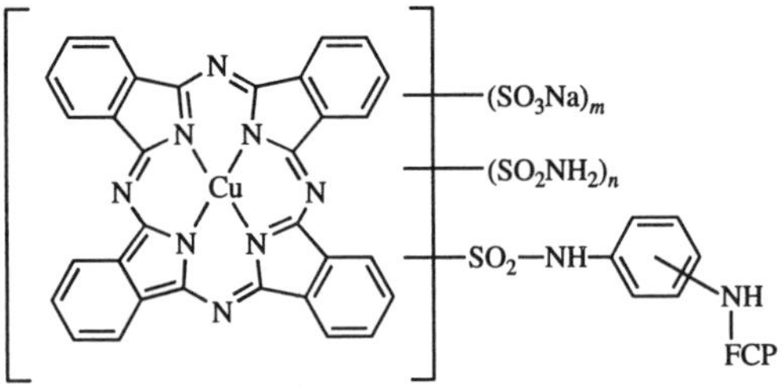

14 $m + n = 3$
Turquoise[34, 71]

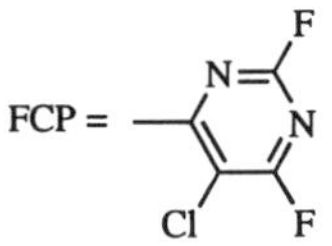

Figure 6. FCP dyestuffs.

FCP dyestuffs with less well-known chromophores have also been described, for example, diarylaminoxanthene,[72] red and blue triphenedioxazine,[73–75] and novel complex dyestuffs.[76,77]

The most favorable temperature for application of these dyes **1** by the exhaust method is 40–50°C. In order to achieve optimum rates of fixation and fixation yields (*ca.* 80%) and to improve reproducibility of the dyeings, automated alkali and salt metering should be used, e.g., the Levametering process.[27,78,79] Dyeings with **1** have excellent fastness properties, but the stability of the dyestuff–fiber bond in peroxide or perborate-containing scouring agents leaves something to be desired.[80,81]

The stability of the dyestuff–fiber bond in reactive dyeings in relation to the acceptor function of various substituted pyrimidines is a subject for quantum-mechanical calculations.[29] These show that true bireactive dyes,[4,20] which have two reactive fluorine atoms, tend to react with the cellulosic fiber in the 6-pyrimidinyl position.[29]

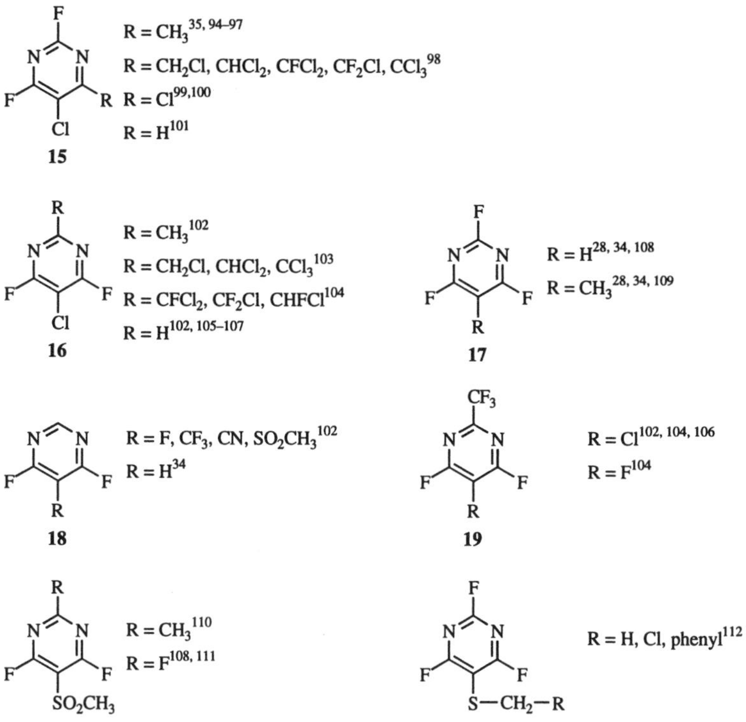

15
R = CH$_3$ [35,94-97]
R = CH$_2$Cl, CHCl$_2$, CFCl$_2$, CF$_2$Cl, CCl$_3$ [98]
R = Cl [99,100]
R = H [101]

16
R = CH$_3$ [102]
R = CH$_2$Cl, CHCl$_2$, CCl$_3$ [103]
R = CFCl$_2$, CF$_2$Cl, CHFCl [104]
R = H [102,105-107]

17
R = H [28,34,108]
R = CH$_3$ [28,34,109]

18
R = F, CF$_3$, CN, SO$_2$CH$_3$ [102]
R = H [34]

19
R = Cl [102,104,106]
R = F [104]

20
R = CH$_3$ [110]
R = F [108,111]

21
R = H, Cl, phenyl [112]

Figure 7. Fluorine-containing pyrimidines for dyestuff synthesis (the superscript numbers refer to the relevant dyestuff patents).

FCP dyestuffs (**1**) include the LEVAFIX® PA and Drimaren® ranges for reactive printing and continuous dyeing,[82] and the Verofix®/Drimalan® ranges for reactive dyeing of wool.[83-87] If dyeing is performed in weakly acidic conditions, a covalent bond is formed between the pyrimidinyl system and the free amino and mercapto functions in the polypeptide chain in the wool. Wool dyed in this way has extremely good wet-fastness properties.[83,88] Azo red (**11**) and anthraquinone blue (**12**) are examples of this type of reactive dye for wool.[87]

Other applications for **1** are the dyeing of polyester/cotton blends,[89-91] wool/cotton blends,[92] and silk.[93]

Following successful introduction of the FCP reactive system, the patent literature now includes a large number of similar fluorine-containing pyrimidine derivatives, some of which are shown in Figure 7.

Very little has been published about the technical significance of these systems. Only a few data on reaction kinetics have been compiled.[47,113] The reactivity sequence for nucleophilic substitution of pyrimidines substituted with fluorine is as follows:

2,4-difluoro- < 4,6-difluoro- << 2,4,6-trifluoro-
< 2,4,5-trifluoro- << 2,4,5,6-tetrafluoropyrimidine

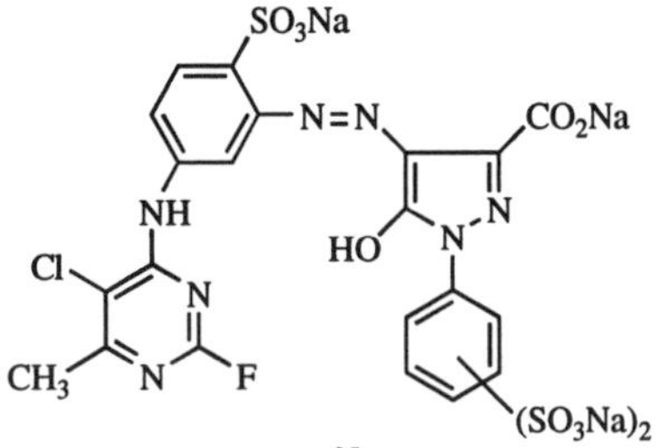

Figure 8. MFCP dyestuffs.

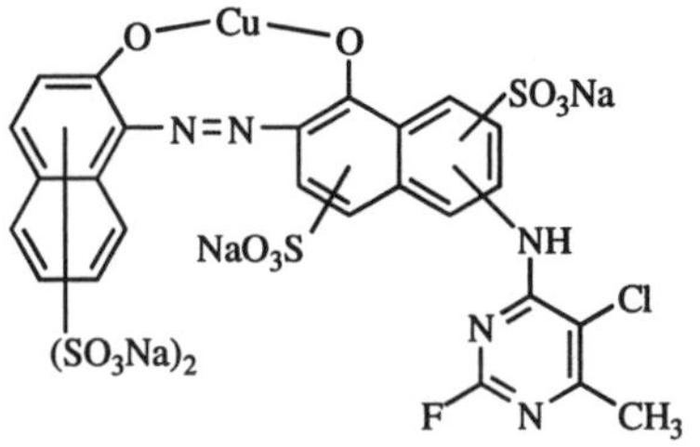

As far as the individual dyestuffs are concerned, this means that for each additional F atom introduced into the reactive system, the optimum dyeing temperature decreases by 10–20°C. The fixation behavior and fastness to acids and alkalis of dyestuffs containing the components **17** have also been described. The only reactive system that seems suitable for practical application is where R = H.[28]

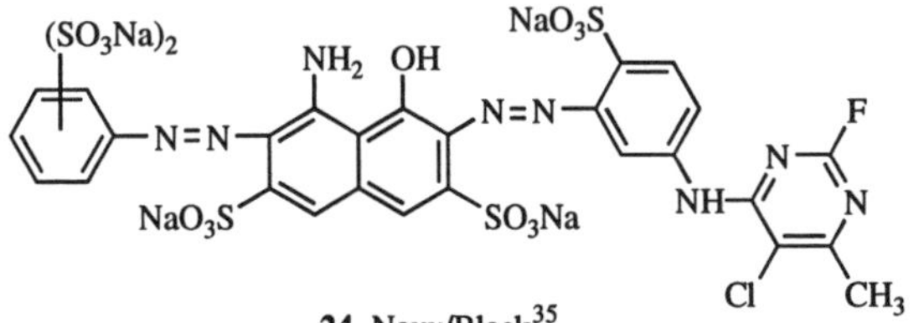

The only pyrimidine illustrated in Figure 7 that is produced on an industrial scale is 5-chloro-2,4-difluoro-6-methylpyrimidine (MFCP) (**15**, R = CH₃), which is used as the reactive component in LEVAFIX® PN dyestuffs (**2**). These dyestuffs, a selection of which are illustrated in Figure 8, were developed specially for reactive printing of cotton.[2,35,114]

The methyl substituent in MFCP reduces reactivity and thus increases the dyeing temperature by 40°C compared with FCP. MFCP dyestuffs are therefore extremely suitable for the rapid steam processes (100–180°C) used in printing and continuous dyeing. Fixation time is about 30% shorter than for monochlorotriazine dyestuffs for reactive printing. Under alkaline fixation conditions, the methylpyrimidinyl system is extremely resistant to hydrolysis. Fixation yields of 80% are generally obtained with most of the dyestuffs illustrated here. These dyestuffs have more sulfo groups in the dyestuff molecule than FCP dyestuffs. This improves solubility in water, guarantees sufficient diffusion in the fiber during the short reaction time, and reduces the amount of unfixed dyestuff washed out of the fiber during after-treatment.[114]

Dyes of type **25** (Figure 9), with donor substituents in the reactive system which can be obtained by substitution of FCP dyestuffs with alcoholates, phenols, pyridine-carboxylic acids, and amines in pH-regulated conditions, have been developed for similar applications.[115]

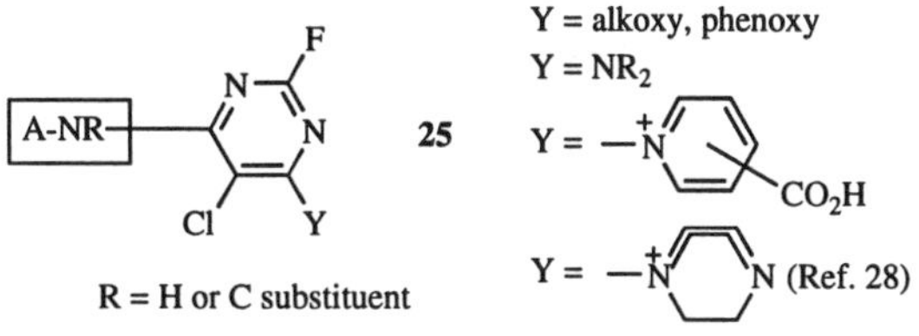

Figure 9. Variation of the substituent on FCP dyestuffs.

Unlike phenol-substituted dyes,[116] aniline and alkylamine-substituted dyes[117] are so unreactive that only moderate fixation yields can normally be obtained in printing. Dyes with nicotinic acid as the pyrimidine substituent have been proposed for one-bath dyeing of polyester/cellulosic fiber blends.[118,119] Dyestuff types with extremely poor resistance to hydrolysis, where the FCP radical is linked to an oxygen atom in the chromophore system, are of no practical significance.[120]

13A.2.2.4. Triazines

The development of fluorine-containing triazine dyes was a logical step from the introduction of monochlorotriazine and FCP dyestuffs. At first this was hampered by the extremely high reactivity of 2,4,6-trifluoro-s-triazine. Dyestuff synthesis in an aqueous medium seemed to be an impossibility.[4] Ciba-Geigy and later Bayer were only able to build up complete dyestuff ranges (Cibacron® F and LEVAFIX® EN) once suitable methods of synthesis had been developed (see Section 13A.2.4.2).[121] The choice of the second substituent for the trifluorotriazine was of decisive importance.

Sulfonated difluorotriazine dyestuffs (26, Y = F) for dyeing from aqueous liquors have been patented, as have alkoxy- and phenoxy-substituted types (26),[66,122,123] but their instability and susceptibility to hydrolysis makes them unsuitable for industrial application (see Figure 10).

By contrast, amino substituents stabilize the triazine system because of their strong donor character, providing the right reactivity for application and fixation on the fiber. Model studies and kinetic measurements of the hydrolysis of fluorotriazine (FT) dyestuffs (3) confirm this.[25,26,124]

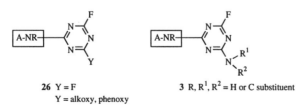

26 Y = F   3 R, R¹, R² = H or C substituent
    Y = alkoxy, phenoxy

Figure 10.   Fluorine-containing triazine dyestuffs.

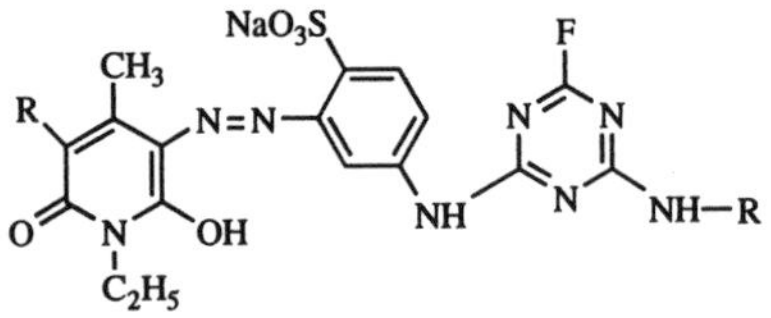

27 Yellow[132]

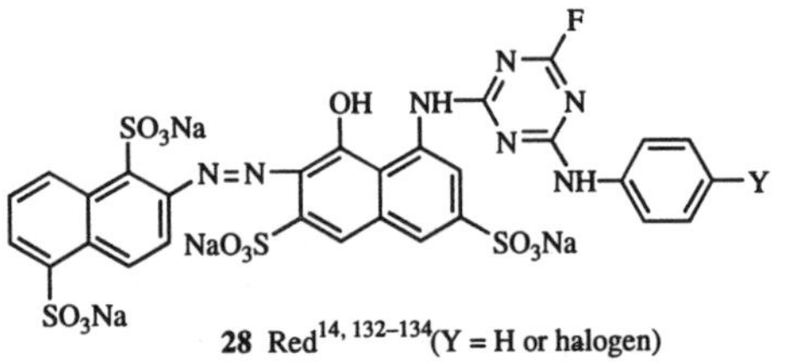

28 Red[14, 132-134](Y = H or halogen)

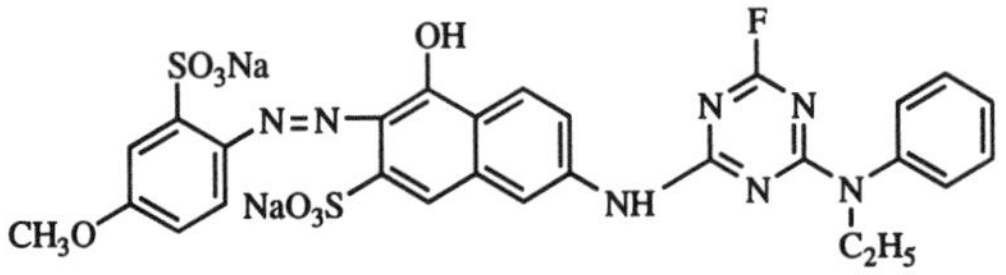

29 Red[25]

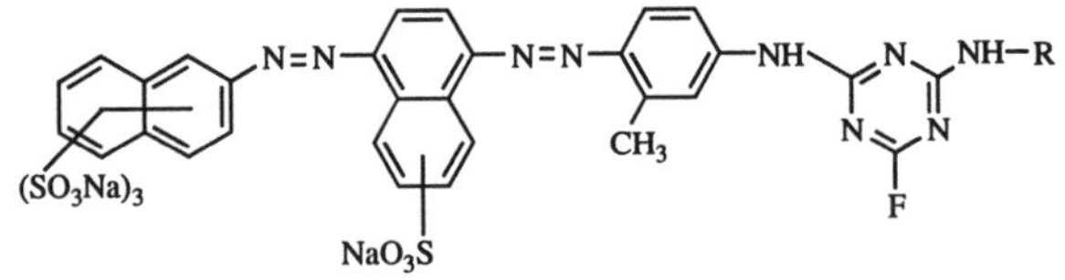

30 Brown[135]

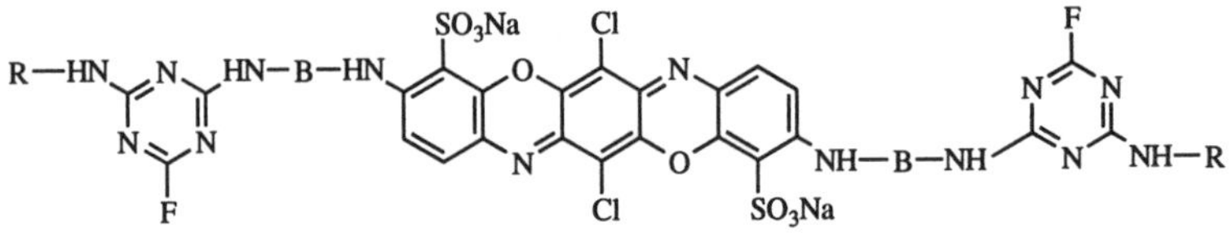

31 Blue[136] (B = alkylene)

Figure 11. Monofluorotriazine dyestuffs (R = C substituent).

Synthesis with trifluorotriazines offers one small advantage over synthesis with pyrimidine derivatives: by varying the second substituent, the amine, solubility, substantivity, diffusion, and washing-off of the hydrolysate can be optimized.[125,126]

Accordingly, aniline, for example, increases the substantivity and build-up of **3** (R^1 = phenyl, R^2 = H), but considerably reduces its solubility. Taurine (R^1 = $CH_2CH_2SO_3H$, R^2 = H) and metanilic acid (R^1 = 3-sulfophenyl, R^2 = H) have virtually the opposite effect, and also reduce the fixation yield. The N-ethylaniline radical (R^1 = phenyl, R^2 = ethyl) seems to have optimum application properties. Variation of radical R on the chromophore can also alter substantivity and the properties obtained.[125]

In principle, all chromophore dye bases can be converted to FT reactive dyestuffs in all imaginable shades using trifluorotriazine.[73,74,127–131] Figure 11 shows some examples. FT dyes are mainly used in exhaust dyeing at 40–50°C, but they are also used in the cold pad-batch process and, to a lesser extent, in textile printing.[82,114,125,126] Fixation yields vary between 60% and 85%, depending on the class of dyestuff. They can also be used to dye polyester/cotton blends[91,137] and silk.[138]

Two interesting new developments are the yellow pyridone dyestuff **27** and the blue triphenedioxazine dyestuff **31**. These chromophore systems have been taken over from disperse and direct dyestuffs and have very high color strength. They are thus far superior to the older pyrazolone and anthraquinone types, such as **22** and **12**. Replacement of some brilliant anthraquinone dyestuffs by comparable triphenedioxazine and formazan dyes is on the horizon.[139–141]

Apparently, none of the efforts to develop unsulfonated reactive disperse dyes for one-step printing and continuous dyeing of polyester/cellulose blends has been successful so far.[142–144] The aim is to produce one dye that is capable of dyeing both types of fiber. A number of reactive systems have been tested. The best fixation rates on cotton have been obtained with FT disperse dyestuffs (**26**) with a methoxy radical as the second substituent Y.[145,146]

We do not know of any dyestuffs with a 2,5,6-[147,148] or a 4,5,6-trifluoro-triazine[149–151] as the reactive component.

13A.2.3. Preparation of the Reactive Compounds

13A.2.3.1. Fluoropyrimidines

Because of the economic importance of fluorinated pyrimidines, a wide variety of production processes has been developed for this class of substances. All of these processes include selective Cl/F exchange starting from the relevant chloropyrimidines.[152] Hydrofluoric acid, sodium, potassium or cesium fluoride, or KF/CaF blends[112] can be used as the fluorine donor (cf Chapter 9).

The following high-temperature/high-pressure methods of fluorinating chloropyrimidines, especially tetrachloropyrimidine, have been described:

- conversion with fluorides without solvents[153–156] or with antimony trioxide as the catalyst[155];
- conversion with fluorides in benzonitrile,[157] sulfolane,[103,112,152,158] nitrobenzene,[94] DMF or NM2P (N-methyl-2-pyrrolidone),[159,160] toluene or xylene[110,161] or in 18-crown-6 as the solvent and catalyst[162];
- conversion with anhydrous hydrofluoric acid[152,163,164] or with trialkylamine trishydrofluorides such as $Et_3N \cdot 3HF$[165,166] in the liquid phase;
- conversion with anhydrous hydrofluoric acid in the gaseous phase.[167–169]

The chlorine/fluorine exchange reaction is an equilibrium reaction and can be influenced by altering the pressure, time, and temperature, and the ratio of the reactants. The best results are obtained with sodium fluoride/sulfolane.[152] 5-Chloro-2,4-

difluoro-6-methylpyrimidine (15, R = CH₃) has been used to investigate the kinetics and solvent-dependence of fluorination. The results are compared with a CNDO/2 calculation. The result was a pseudo-first-order consecutive reaction.[170]

Less drastic fluorination conditions can be used, e.g., for methylsulfonyl-substituted pyrimidines, by working with $FSO_2^-K^+$ in toluene.[110] In the case of pyrimidines with a chlorinated side chain, such as 2,4,5-trichloro-6-(trichloromethyl)pyrimidine, alkali fluoride exchange only takes place in the nucleus, while with hydrofluoric acid exchange takes place in both the nucleus and the side chain.[152,171] Using antimony trifluoride and a catalytic amount of antimony pentachloride, the chlorine/fluorine exchange reaction can be performed selectively on the side chain only.[152]

Processes of scientific interest only are electrophilic fluorination with chlorine pentafluoride (e.g., the synthesis of tetrafluoropyrimidine from 2,4,6-trifluoropyrimidine)[172] and reductive dehalogenation (e.g., catalytic hydrogenation of 4,6-dichloro-2,5-difluoropyrimidine to 4-chloro-2,5-difluoropyrimidine using hydrogen/triethylamine and Pd/C as the catalyst).[173,174]

Tetrachloropyrimidine (32), the intermediate required for industrial production of 5-chloro-2,4,6-trifluoropyrimidine (FCP), can be obtained by a variety of routes. Apart from the standard procedure starting from barbituric acid (33), the most significant processes are those shown in Figure 12, i.e., the chlorination of dichlorobarbituric acid (34),[175] 1-methyl-5,6-dihydrouracil (35),[176] and (dimethylamino)propionitrile (36)[153,177] and its derivatives.[178,179]

The reaction starting from 36 has been studied in detail and the reaction mechanism discussed.[177,180,181] Depending on the reaction conditions, in addition to 32, 4,5,6-trichloropyrimidine[153,177] or 4,5,6-trichloro-2-trichloromethyl-pyrimidine[181,182] can be produced by this method. The trichloromethyl derivative can be converted to tetrachloropyrimidine (32) and carbon tetrachloride by high-temperature chlorinolysis.[183]

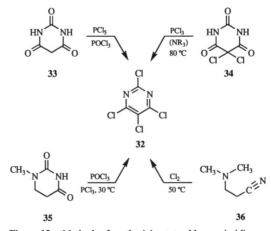

Figure 12. Methods of synthesizing tetrachloropyrimidine.

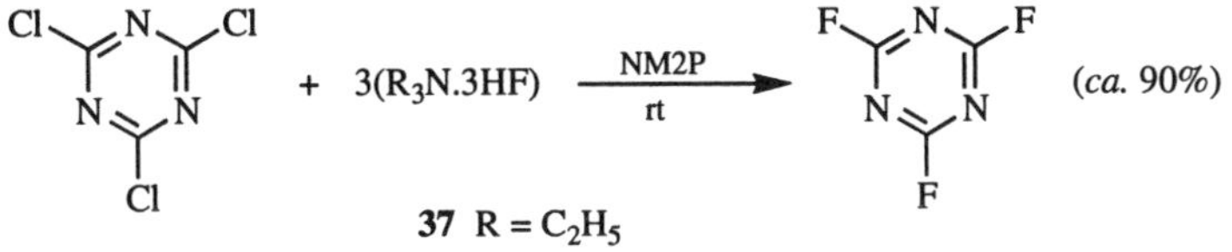

Figure 13.   One method of synthesizing TFT (see also Scheme 15, Chapter 9).

13A.2.3.2.   2,4,6-Trifluoro-s-triazine (TFT)

There are various methods of producing TFT, all of which are based on Cl/F substitution of cyanuric chloride. One of the most important of these is conversion with sodium fluoride in sulfolane, which was described as early as 1960.[184] Optimum yields are obtained in a one-hour reaction at 190°C.[185] Liquid TFT (bp 71–72°C) is usually isolated by distillation. The sulfolane can be recycled if the sulfolane and sodium chloride residues from distillation of the TFT are neutralized with calcium hydroxide and the sulfolane is then distilled again.[186] For reactions with potassium fluoride, it is advisable to use 18-crown-6[187] and perfluoroacyl fluorides[188] as catalysts.

Cyanuric chloride can also be converted with anhydrous hydrofluoric acid under various reaction conditions. The reaction can take place without addition of solvents at 200°C[189] or 150°C/5 bar.[190] An approximately 10 molar excess of HF is used, and the excess HF is distilled off at the end of the reaction without cooling the mixture by gradually reducing the pressure in the reactor.[168] The TFT is then distilled. The small percent of 2-chloro-4,6-difluorotriazine or 2,4-dichloro-6-fluorotriazine generated can be recycled for use in subsequent fluorination reactions.[185,190] By adding a catalyst such as antimony pentafluoride, the reaction temperature can be lowered and the reaction time reduced.[191]

Cyanuric chloride can also be reacted with HF at moderate temperatures (20–30°C) using TFT as an inert solvent.[192] Another alternative is to convert cyanuric chloride into TFT in the presence of TFT and antimony pentafluoride.[191,193] Partly fluorinated s-triazines can be produced by this method.[193] An interesting, mild, and low-corrosive process is conversion of cyanuric chloride with the relatively stable liquid triethylamine trishydrofluoride complex (37) (Figure 13; cf Section 13A.2.3.1).[166,194]

A weakly basic solvent such as N-methyl-2-pyrrolidone (NM2P) or ε-caprolactam binds any HF released, which can be recovered as 37 by adding triethylamine after the TFT has been distilled out.

13A.2.4.   Preparation of the Reactive Dyes

13A.2.4.1.   Fluoropyrimidine Dyes

Reaction of the liquid fluoropyrimidines, which are highly susceptible to hydrolysis, with amino-containing color bases in water is relatively free from problems,

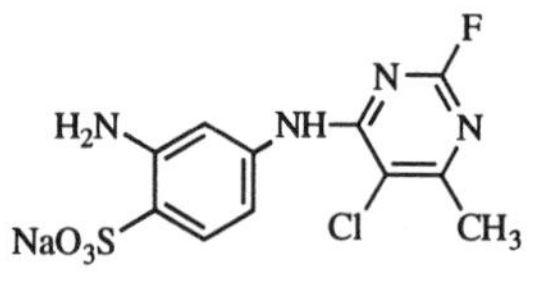

Figure 14.   Fluoropyrimidine-containing intermediates.

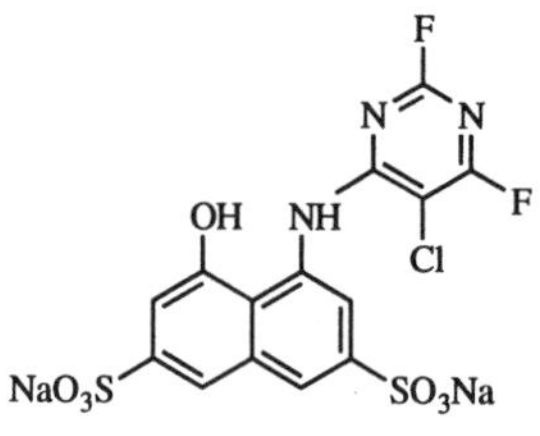

provided suitable safety precautions are taken and the conversion reaction takes place in the neutral pH range between 5 and 9. Alkali hydroxide or sodium carbonate solution is added to neutralize the HF released. Alternatively, the reactive components can be condensed with nonchromophore dyestuff intermediates. Subsequent synthesis of the chromophore system must then be carried out under conditions which do not allow reaction with the second reactive group in the pyrimidine. For example, to produce the yellow azo reactive dyestuff (**22**), 2,4-diaminobenzenesulfonic acid is first condensed with MFCP (**15**, R = CH₃) and the condensation product **38** is then diazotized and coupled with the pyrazolone derivative.[35]

Conversely, coupling components such as 1-amino-8-hydroxynaphthalene-3,6-disulfonic acid can be condensed with FCP (**15**, R = F) to form the intermediate **39** and then coupled with diazonium salts in a second step, forming azo dyes. In such cases, the condensation reaction is best carried out at a constant pH of 10.[195] The most suitable condensation temperature is determined by the reactivity, i.e., the acceptor substitution pattern, of the pyrimidine in question. For FCP, the most suitable condensation temperature is 5–25°C,[34] while for MFCP and comparable derivatives, **16** (R = H)[35,102] or **17**,[28] it is ca. 40–60°C. Reaction times vary from one to six hours, depending on the reactivity of the pyrimidines and the nucleophilic properties of the amino compounds. The reaction can be accelerated by adding heterogeneous catalysts such as zeolites, silicic acid, and basic ion exchange resins.[196,197]

Pyrimidines such as FCP and MFCP, which have fluorine substituents in the 2 and 4 (or 6) positions, tend to react with nucleophilic substances in the 4 position (see **38**, and **39**).[59,152,155,156,160] One possible explanation for this is that the acceptor function of the halogen atom in the 5 position activates the *ortho* 4 position more strongly than the *para* 2 position. Kinetic results and comparisons confirm the importance of ion–dipole interactions for this activation by *ortho*-halogen.[198] In the conditions used for dyestuff synthesis and dyeing, the halogen atom in the 5 position is completely inactive.[152,155]

MNDO studies on chloropyrimidines provide further evidence of the selectivity of the substitution reaction. These studies show that the transition state leading to the formation of the Meisenheimer-type complex determines the reaction rate. Because of its geometry, the nucleophile apparently finds it easier to approach the 4 position and requires less activation energy for this than for substitution in the 2 position. The influence of the type of substituent in the 6 position should not be underestimated.[199]

13A.2.4.2. Fluorotriazine Dyes

A wide range of reaction sequences can be used to produce FT dyes (**3**). Figure 15 illustrates two possible routes.

In many cases, either the amine/aniline derivative[36] or the dye base A-NHR[200] can be condensed with TFT at the start of the process. Alternatively, the reaction can be carried out with a colorless intermediate instead of a dye base, subsequently producing the reactive dye (**3**) by azo coupling (cf Section 13A.2.4.1).[201,202]

However, the critical, indeed decisive, step in the reaction sequence is the initial condensation with TFT, generating relatively unstable difluorotriazine intermediates such as **40** and **41**. In this reaction, there is no need to use inert organic solvents.[203] Although TFT can react very quickly and violently with water, it can be selectively substituted with nucleophilic amino compounds in water. Various processes for the initial condensation of TFT have been patented. All of these processes use reaction temperatures of between –5°C and +5°C and the initial substitution is immediately followed by a second substitution of the reaction products **40** and **41** which have not been isolated.

Optimum yields from initial condensation with sulfonated anilines or aminohy-droxynaphthalenesulfonic acids are obtained at pH 3.5–5.0.[3,204–206] The HF released can be neutralized by adding alkali hydroxide, sodium carbonate solution, or solid sodium hydrogen carbonate. The pH and the reaction can be regulated more easily if buffer substances like sodium fluoride[207] or boric acid and their salts[208] are added before the start of the reaction.

Continuous feed of both reactant streams into the reaction vessel at the same time prevents formation of undesired dicondensation products.[200] The best results are obtained by mixing the reactants thoroughly in a tubular reactor with dwell and reaction times of a few seconds.[209] This continuous method can also be used to condense dye bases with TFT.[210]

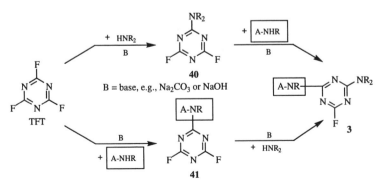

Figure 15.   Synthesis of FT reactive dyes.

TFT can be converted with ammonia or alkyl amines at pH 7–9.[211] In such cases, FT dyes (3) can be produced in a one-stage reaction process. A solution of the corresponding ammonium salt and dye base is placed in the reaction vessel and condensed directly with TFT at pH 8.5 and 0°C.[212]

13A.2.4.3. Dye Formulations

The dyestuffs synthesized by these methods generally undergo further processing to the required supply form. In order to ensure that the commercial grades are as stable and as soluble as possible, and that they have a standard color strength, small amounts of buffer substances, solubilizing agents, and dedusting agents and larger quantities of diluents such as naphthalenesulfonic acid formaldehyde condensation products are added. The mixtures are then homogenized in powder mills. Alongside these powder brands, new commercial grades, such as liquid formulations and low-dusting granular brands, are gaining in importance.[2,4,213–215]

"Liquid dyestuffs" are stable, *ca.* 30–40% aqueous dyestuff solutions. They allow dust-free processing and easy and rapid formulation of dye liquors, save time and energy because they do not have to be dissolved and are particularly suitable for accurate metering in modern automated dye stations, especially for continuous processes. Liquid formulations must have sufficient resistance to hydrolysis, adequate storage stability, unlimited combinability, and good rheological properties if they are to succeed commercially. A special inverse osmosis process on semipermeable microporous membranes[216–219] has been developed to prepare highly concentrated, low-salt dyes with good storage stability. The stability of fluorinated reactive dyestuffs in solution is determined by selecting the correct pH, lithium salt preparations, and the addition of stabilizers, functional additives, and solubilizing agents such as ε-caprolactam, dimethylsulfone, and urea derivatives.[214,220–222]

Although the hydrolysis of monofluorotriazine dyestuffs (3) has been studied in detail,[25,26] so far no-one has succeeded in producing a liquid dyestuff of this type with adequate storage stability.[2,222] Liquid FCP dyestuffs (1) are available under the tradenames LEVAFIX® EA liq. 40% and Drimaren® K liq. 40%.[213,215] Reactive dyestuffs in granular or bead form also have ecological and economic advantages. LEVAFIX® EA/EN Macrolat dyestuffs and Drimaren® CDG and Cibacron® F/C granular dyestuffs, which are produced in fluid spray driers (FSD)[217] or special spray driers,[215] have high abrasion resistance, high bulk density, and excellent cold solubility, ensuring reliable and cost-effective metering.[4,214,215,223,224]

13A.3. DYES WITH TWO OR MORE REACTIVE SYSTEMS

Bi- and poly-functional reactive dyes can be divided into four groups, as shown in Figure 16. If the two reactive groups in the dyestuff molecule are the same, they are known as homobifunctional dyes, and if they are different, they are known as heterobifunctional types.[20]

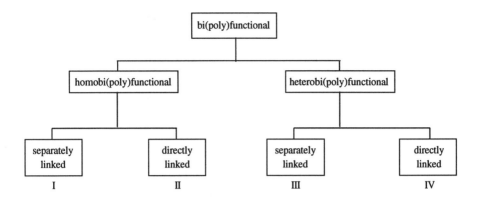

Figure 16. General classification of bifunctional reactive dyestuffs.

13A.3.1. Homobifunctional Reactive Dyes

Dyestuffs with this structure have been known since the start of reactive dyestuff chemistry. Hoechst's established Remazol Black B (C.I. Reactive Black 5)[14,15] and two recent developments, LEVAFIX Royal Blue E-FR[20,140,141] (Bayer) and Cibacron Navy C-G[13] (Ciba-Geigy), all have two sulfatoethylsulfonyl (= vinylsufonyl) groups as their reactive groups and accordingly manifest good fixation. There are many commercial homobifunctional monochlorotriazine dyestuffs, most of which have symmetrical or double chromophore structures.[8] This has the effect of enlarging the molecule, but often impairs the fixation behavior of such dyestuffs.[15] Fluorinated homobifunctional reactive dyes can be subdivided into two groups (Figures 16 and 17). In examples 42 and 43 in Group I, the two heterocyclic reactive radicals are separated by the chromophore system. This is only possible if the bifunctional dye base contains two amino functions which can be condensed. One commercial product of the patented dyestuffs with structures 42[225] and 43[226–231] is the triphenedioxazine blue (31),[136,140] which contains two fluorotriazine groups due to its symmetrical structure (cf 43, Figure 17).

In the examples in Group II, the heterocyclic reactive components are linked by a bivalent bridge, such as a diaminoalkylene or a diaminophenylene radical.[232–236] Examples 44 can only be produced by a sequence of four condensation steps while the formula indicates that 45 can be produced by doubling the known monofunctional FT dyestuffs. Unlike comparable hot-dyeing monochlorotriazine types, bifunctional warm-dyeing dyestuffs (40–60°C) in Group II have not yet been used on an industrial scale.

13A.3.2. Heterobifunctional Reactive Dyes

In the last ten years, high-fixation reactive dyes with two different reactive groups have become of increasing importance.[8,13,14,16,20] A variety of complex dyestuff

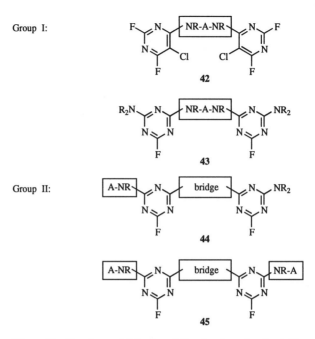

Figure 17.   Fluorine-containing homobifunctional reactive dyestuffs.

structures can be obtained by combining the well-known fluoropyrimidines and trifluorotriazines with all other reactive systems. No new processes are needed to synthesize these dyes as the methods discussed in Section 13A.2.4 are basically suitable for these new bifunctional dyestuffs. They can be subdivided into two further groups according to whether the reactive groups are linked to one another (IV) or to separate parts of the dye bases (III).

Like all homobifunctional reactive systems, examples 46[237] and 49[238–240] (Figure 18) comprise two fluorinated reactive systems. Consequently, there is less scope for variation than in the other four examples. The following combinations have been described for **47, 48, 50,** and **51**:

47,   Z =   vinylsulfonyl (or 2-sulfatoethylsulfonyl),[241–247] bromoacryloyl[248]

48,   Z =   vinylsulfonyl,[245–247,249–252] bromoacryloyl[248] 4-alkoxy-6-chloro-2-triazinyl,[237]

50,   Z =   vinylsulfonyl[253–269] bromoacryloyl,[270,271] 4-alkoxy-6-chloro-2-triazinyl,[272] chloropyridazinone,[273]

51, T/P =   monofluorotriazinyl,[14,274] FCP,[14,239,274–278] MFCP,[238] fluoropyridazinone.[55]

This list can be lengthened to include MFCP dyes, which are additionally substituted with a vinylsulfonyl radical, analogous to **47,**[279] and dyes with activated fluoropyrimidine, which are linked to a further reactive system Z as in **50.**[280–282]

Group III:

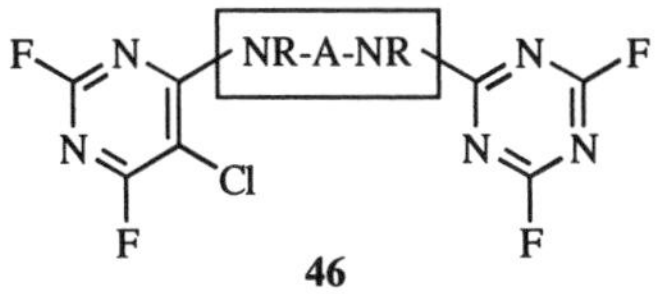

46

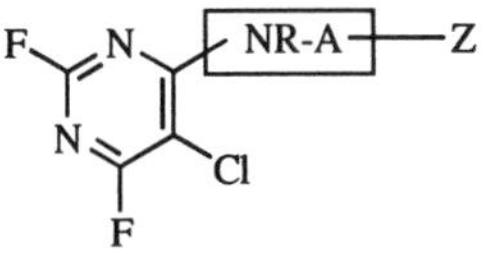

47 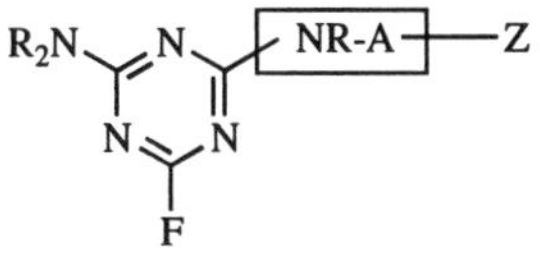   **48**

Z = fiber-reactive residue free of fluorine

Group IV:

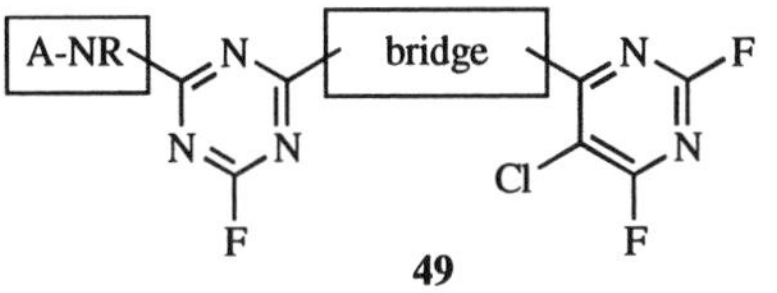

49

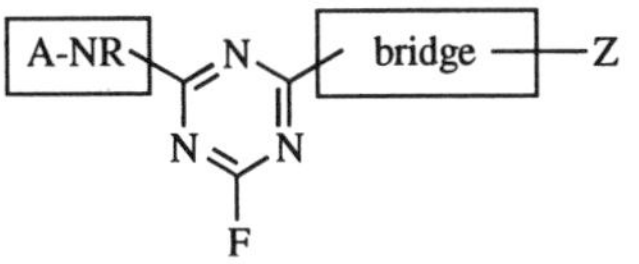

50 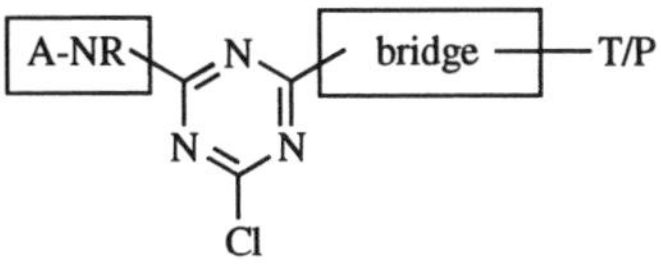   **51**

T/P = triazinyl/pyrimidinyl residue containing fluorine

Figure 18. Fluorine-containing heterobifunctional reactive dyes.

LEVAFIX Navy Blue E-BNA, a vinylsulfonyl-containing FCP dyestuff with a 92% fixation yield, belongs in category **47**.[20,241]

However, a combination of different reactive systems is usually unsatisfactory. The introduction of a second reactive group alters important physical and dyeing properties such as solubility, diffusion capacity, affinity for the fiber, and often also the reactivity of the first reactive group.[10]

The optimum temperature range for dyeing is widened, so dyeing is safer and less sensitive to differences in temperature.[14] Hence, the vinylsulfonyl group in Sumifix® Supra dyestuffs **52**[14–16] is more reactive than the monochlorotriazine radical (Figure 19). Optimum dyeing conditions are thus 50–80°C. The wet-fastness properties and hydrolysis behavior of the dyeing are thus predominantly determined by the vinylsulfonyl system, which fixes on the fiber first but is sensitive to alkali.[13,15]

It seems to be advantageous if both reactive systems have similar reactivity as can be found in some of the dyestuffs in the new Cibacron® C range (**53**). By changing the chlorine atom to a fluorine atom, the vinylsulfonyl group becomes slightly less reactive than the triazine group. At the moment, the commercial range with this bifunctional

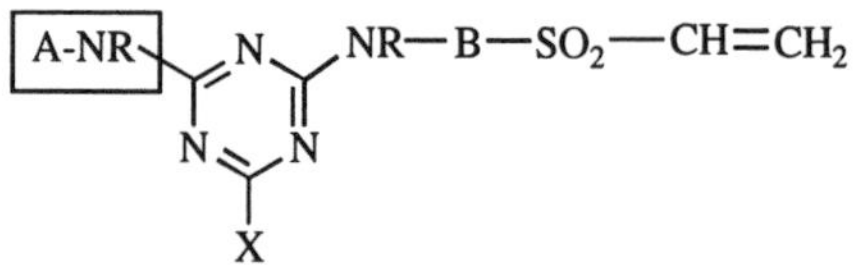

A-NR—⬡(triazine ring with N's)—NR—B—SO$_2$—CH=CH$_2$, X

52 (X = Cl; B = phenylene)
Sumifix supra dyestuffs (Sumitomo Chem.)

53 (X = F; B = bivalent aliphatic ether)
Cibacron C dyestuffs (Ciba-Geigy)

Figure 19. Heterobifunctional reactive dyestuff ranges.[283]

link comprises four dyestuffs: Cibacron Yellow C-5G, Cibacron Yellow C-RN, Cibacron Orange C-G, and Cibacron Blue C-R. The properties of the bridging element B in **53** were selected to give reactive dyestuffs with excellent solubility and good washing-off properties in continuous processes (e.g., pad-batch, pad-steam, and pad-thermofix) and in the cold pad-batch process.[13]

The main advantage of these new heterobifunctional fluorinated dyestuffs is their greatly improved fixation yield. When dyeing from long liquors, the fixation yield of Group III dyes is *ca.* 90–92% and of Group IV dyes *ca.* 80–82%.[20,21] In the pad-batch process, Cibacron® C dyes are said to have a fixation yield of 90–98%.[13] Note that every additional reactive group reduces the molar color strength of the dye and raises production costs.[20] The success of these relatively expensive dyes on the market is primarily due to their ecological benefits. By raising the fixation yield from 80% to 90%, the effluent load from dyehouses is reduced from 20% to 10%, i.e., both effluent load and COD are halved.

13A.3.3. Polyfunctional Reactive Dyes

Cibacron Red C-2G[13] and Remazol Red SBB[4] are commercial trifunctional reactive dyestuffs containing one monochlorotriazinyl and two vinylsulfonyl groups. Analogous fluorine-containing dyestuffs are only known in patent literature.[240,284–296] No new reactive components are described, but established bifunctional dyes are extended to include a further reactive group, most commonly vinylsulfonyl, usually by substitution or by doubling the molecule. One example is doubling of vinylsulfonyl-containing dye bases via a monofluorotriazinyl radical.[284,285] Another variation comprises additional substitution of the chromophore A-NR in **48**,[290,291] **49**,[296] **51**,[288,289,296] or **43**[285] with a vinylsulfonyl radical. The same third reactive group can be introduced into the bridge in the dyestuffs **44**[292] and **53**[287,293] In **53**, B is then a bisvinylsulfone-containing bridge. Only the future will tell whether these new polyfunctional reactive dyes will give fixation yields of around 98–99%.

13A.4. CONCLUSION

Reactive dyes have better fastness properties than direct dyes, are more economical than vat dyes, and are ecologically and toxicologically superior to naphthol developing dyes. Their main disadvantage, unsatisfactory bath exhaustion, in other words, an unsatisfactory fixation yield and highly colored residual liquors, is being overcome by the introduction of fluorinated reactive systems. A third generation of bifunctional dyes is helping to improve the productivity and cost-effectiveness of dyehouses and solving their ecological problems, as well as encouraging the development of new dyeing technology. And for the man in the street, the wide range of reactive dyestuffs now available brings a touch of color to the often gray reality of everyday life.

13A.5. REFERENCES

1. W. Harms, in: *Organofluorine Chemicals and Their Industrial Applications* (R. E. Banks, ed.), Chapter 9, p. 188, Ellis Horwood, Chichester (1979).
2. E. Brunnschweiler and G. Siegrist, *Textilveredlung 19*, 305 (1984).
3. K. H. Schündehütte, in: *Ullmann's Encyklopaedie der technischen Chemie*, 4th edn., Vol. 20, p. 113, Verlag Chemie, Weinheim (1981).
4. H. Zollinger, *Melliand Textilber. 68*, 644 (1988).
5. P. Rys and H. Zollinger, in: *The Theory of Coloration of Textiles* (C. L. Bird and W. S. Boston, eds.), p. 345, Dyers Company Publ. Trust, Bradford (1975).
6. I. D. Rattee, *Rev. Prog Coloration 14*, 50 (1984).
7. K. Aeberhard, *Textilveredlung 16*, 442 (1981).
8. A. H. M. Renfrew and J. A. Taylor, *Rev. Prog. Coloration 20*, 1 (1990).
9. D. W. Ramsay, *J. Soc. Dyers Colourists 97*, 102 (1981).
10. H. Bürgin, "Cibacron-C-Farbstoffe," paper read at the Meeting of the Society of Northern Textile Engineers, Boras, Sweden (1989).
11. S. Abeta and T. Omura, *Nikkakyo Geppo*, 8-15 (1987) [*CA 107*, 200367 (1987)].
12. S. Abeta and K. Imada, *Rev. Prog Coloration 20*, 19 (1990).
13. J. P. Luttringer and A. Tzikas, *Textilveredlung 25*, 311 (1990).
14. S. Fuijoka and S. Abeta, *Dyes and Pigments 3*, 281 (1982).
15. M. Matsui, U. Meyer, and H. Zollinger, *J. Soc. Dyers Colourists 104*, 425 (1988).
16. S. Abeta, K. Imada, N. Harada, and T. Yoshida, *JTN—Int. Textile Magazine 79* (1987).
17. K. Imada, S. Abeta, and T. Yoshida, *JTN—Int. Textile Magazine 64*, 72 (1983).
18. H. Liqi, Z. Zhenghua, C. Kongchang, and Z. Faxiang, *Dyes and Pigments 10*, 195 (1989).
19. N. Yamauchi, K. Imada, and S. Ikeou, European Patent 234573 (to Sumitomo) [*CA 107*, 238518 (1987)].
20. J. Wolff and H. Henk, *Textilveredlung 25*, 213 (1990).
21. D. Hildebrand and J. Wolff, "Modern Aspects of the Dye–Fibre Reaction in Reactive Dyestuff," paper read at 1st International Conference on the Chemistry and Application of Reactive Dyes, Bodington Hall, Leeds (1989).
22. K. Imada, S. Kikkawa, N. Harada, and T. Yoshida, "Dyeing Properties of Reactive Dyes with a Mixed Bifunctional Reactive System," paper read at 1st International Conference on the Chemistry and Application of Reactive Dyes, Bodington Hall, Leeds (1989).
23. C. V. Stead, *Dyes and Pigments 3*, 161 (1982).
24. U. Ruf and W. B. Egger, *Textilveredlung 13*, 304 (1978).
25. J. P. Luttringer and P. Dussy, *Melliand Textilber. 62*, 84 (1981).
26. L. Xiao-Tu, Z. Zhenghua, and C. Kongchang, *Dyes and Pigments 11*, 123 (1989).

27. D. Hildebrand, *Bayer Farbenrevue 37*, 30 (1985).
28. F. Lehr, *Dyes and Pigments 14*, 239 (1990).
29. M. Dörr and M. Gisler, "Quantum Mechanical Calculations on the Stability of Reactive Dyeings," paper read at 1st International Conference on the Chemistry and Application of Reactive Dyes, Bodington Hall, Leeds (1989).
30. C. J. Bent, P. A. Davies, and D. A. S. Phillips, *J. Soc. Dyers Colourists 98*, 327 (1982).
31. D. Hildebrand and W. Beckmann, *Melliand Textilber. 45*, 1138 (1964).
32. D. Hildebrand, *Chemtech. 8*, 224 (1978).
33. H. Schroeder, E. Kober, H. Ulrich, R. Rätz, H. Agahigian, and C. Grundmann, *J. Org. Chem. 27*, 2580 (1962).
34. H. S. Bien and E. Klauke, British Patent 1169254; Ger. Offen. 1644204 (to Bayer) [*CA 72*, 122901 (1970)].
35. K. H. Schündehütte and E. Klauke, Ger. Offen. 2817780 (to Bayer) [*CA 92*, 78095 (1980)].
36. H. S. Bien, E. Klauke, and K. Wunderlich, Ger. Offen. 1644208; French Patent 1561704 (to Bayer) [*CA 73*, 36551 (1970)].
37. H. Seiler and G. Hegar, Ger. Offen. 2556640 (to Ciba-Geigy) [*CA 85*, 194076 (1976)].
38. References 16–20 cited in Ref. 1.
39. References 23–28 cited in Ref. 1.
40. References 33–42 cited in Ref. 1.
41. B. Anderson, Ger. Offen. 2916715 (to I.C.I.) [*CA 92*, 112247 (1980)].
42. A. Takaoka, O. Yokokohji, Y. Yamayuchi, T. Isono, M. Motoyoshi, and N. Ishikawa, *Nippon Kagaku Kaishi* 2155 (1985) [*CA 105*, 152649 (1986)].
43. European Patent 120575 (to Nippon Shokubai Kagaku) [*CA 102*, 45595 (1985)].
44. References 43–46 cited in Ref. 1.
45. R. D. Chambers, *Dyes and Pigments 3*, 183 (1982).
46. R. D. Chambers, J. A. H. MacBride, and W. K. R. Musgrave, *Chem. Ind. (London)* 1721 (1966).
47. R. D. Chambers, M. J. Seabury, D. L. H. Williams, and N. Hughes, *J. Chem. Soc., Perkin Trans. 1*, 255 (1988).
48. D. Hildebrand, *Textil-Praxis 26*, 564 (1971).
49. D. A. S. Phillips, B. Anderson, C. V. Stead, and A. T. Costello, Ger. Offen. 2814206 (to I.C.I.) [*CA 91*, 22409 (1979)].
50. C. A. Wilson and A. P. Fung, U.S. Patent 4746744 (to Dow Chemical) [*CA 109*, 57061 (1988)].
51. A. Hackenberger and M. Patsch, Ger. Offen. 3528459 (to BASF) [*CA 106*, 178118 (1987)].
52. A. Hackenberger and M. Patsch, Ger. Offen. 3620856 (to BASF) [*CA 108*, 169185 (1988)].
53. D. M. Lewis, *J. Soc. Dyers Colourists 93*, 165 (1982).
54. H. R. Hensel and G. Lützel, *Angew. Chem., Int. Ed. Engl. 4*, 312 (1965).
55. D. Augart, H. Eilingsfeld, H. Krüger, H. Lardon, O. Schaffer, and G. Seybold, Ger. Offen. 3305881 (to BASF) [*CA 102*, 26353 (1985)].
56. H. Lardon and G. Seybold, Ger. Offen. 3229325 (to BASF) [*CA 100*, 193541 (1984)].
57. R. D. Chambers, J. A. H. MacBride, and W. K. R. Musgrave, *J. Chem. Soc. (C)* 2116 (1968).
58. H. S. Bien, E. Klauke, and K. Wunderlich, Ger. Offen. 1644206; French Patent 1568321 (to Bayer) [*CA 73*, 46654 (1970)].
59. R. D. Chambers, J. A. H. MacBride, and W. K. R. Musgrave, *Chem. Ind. (London)* 904 (1966).
60. E. Klauke, L. Oehlmann, and B. Baasner, *J. Fluorine Chem. 23*, 301 (1983).
61. H. Ackermann, H. Frei, and H. Meindl, Belgian Patent 644495 (to Geigy) [*CA 63*, 11743 (1965)].
62. H. Ackermann, H. Frei, and H. Meindl, U.S. Patent 3433781; Ger. Offen. 1252824 (to Geigy).
63. D. Günther and D. Bosse, Ger. Offen. 2929594; European Patent 23033 (to Hoechst) [*CA 94*, 175159 (1981)].
64. P. Dussy, J. Ammann, and W. Bossard, Ger. Offen. 1109807 (to Geigy) [*CA 56*, 2536 (1962)].
65. J. Benz and A. Schweizer, Ger. Offen. 1228013; British Patent 916094 (to Sandoz) [*CA 59*, 8914 (1963)].
66. K. Kabitzke and H. Nickel, French Patent 1573149; British Patent 1176898 (to Bayer) [*CA 73*, 57166 (1970); CA 73, 57159 (1970)].

67. P. Elzer, *Bayer Farbenrevue 26*, 40 (1975).
68. H. S. Bien and E. Klauke, Ger. Offen. 1644171; British Patent 1169254 (to Bayer) [*CA 72*, 122901 (1970)].
69. H. Jäger, Ger. Offen. 2232541 (to Bayer) [*CA 80*, 109835 (1974)].
70. P. Dussy and W. Bossard, French Patent 2000892 (to Geigy) [*CA 73*, 100038 (1970)].
71. F. Müller and F. Bachmann, Ger. Offen. 1959342 (to Sandoz) [*CA 73*, 57099 (1970)].
72. R. Raue and K. Wunderlich, Ger. Offen. 3347246 (to Bayer) [*CA 103*, 19737 (1985)].
73. W. Harms, Ger. Offen. 3635312 (to Bayer) [*CA 109*, 232747 (1988)].
74. W. Harms and K. J. Herd, European Patent 299328 (to Bayer) [*CA 111*, 8891 (1989)].
75. K. Wunderlich, W. Harms, K. J. Herd, and H. Jäger, Ger. Offen. 3510612 (to Bayer) [*CA 106*, 19995 (1987)].
76. H. Henk, Ger. Offen. 3516667 (to Bayer) [*CA 106*, 68733 (1987)].
77. H. Henk, European Patent 240839; Ger. Offen. 3612016 (to Bayer) [*CA 108*, 39644 (1988)].
78. D. Hildebrand, B. Renziehausen, and D. Heilmann, *Melliand Textilber 69*, 895 (1988).
79. D. Hildebrand, Ger. Offen. 3715545 (to Bayer) [*CA 110*, 136915 (1989)].
80. I. D. Rattee and J. I. N. Rocha Gomes, *J. Soc. Dyers Colourists 101*, 319 (1985).
81. I. D. Rattee and K. F. So, *Dyes and Pigments 1*, 121 (1980).
82. D. Hildebrand, K. Rompf, H. Koch, and E. Becker, *Textil Prax. Int. 42*, 1242, 1359 and 1480 (1987).
83. D. Hildebrand and G. Meier, *Textil Prax. Int. 26*, 499 and 557 (1971).
84. D. M. Lewis, *Rev. Prog. Coloration 8*, 10 (1977).
85. D. M. Lewis, *J. Soc. Dyers Colourists 93*, 165 (1982).
86. D. M. Lewis, *Melliand Textilber. 67*, 717 (1986).
87. K. Neufang, E. R. Fritze, and R. Kuth, Ger. Offen. 3335957 (to Bayer) [*CA 103*, 55389 (1985)].
88. A. Datyner, E. Finnimore, and U. Meyer, *J. Soc. Dyers Colourists 93*, 278 (1977).
89. D. Hildebrand and J. Fiegel, *Melliand Textilber. 64*, 290, and 357 (1983).
90. D. Hildebrand and H. Koch, European Patent 43984 (to Bayer) [*CA 96*, 144414 (1982)].
91. D. Hildebrand and J. Fiegel, *Bayer Farbenrevue 34*, 31 (1983); *38*, 24 (1986).
92. D. Hildebrand, *Tinctoria 83*, 68 (1986).
93. U. Meyer, J. Wang, Y. Xia, J. Yang, and H. Zollinger, *J. Soc. Dyers Colourists 102*, 6 (1986).
94. X. Minren, C. Zhusheng, and Z. Zenghua, *Huadong Huagong Xueyuan Xuebao 12*, 413 (1986) [*CA 106*, 178094 (1987)].
95. K. H. Schündehütte and H. Henk, Ger. Offen. 3010161 (to Bayer) [*CA 96*, 8138 (1982)].
96. H. Henk and K. H. Schündehütte, Ger. Offen. 3122425 (to Bayer) [*CA 98*, 91027 (1983)].
97. R. Weitz, R. Neeff, and R. Schwaebel, Ger. Offen. 3406232 (to Bayer) [*CA 104*, 111356 (1986)].
98. O. Schallner, K. H. Schündehütte, and E. Klauke, Ger. Offen. 3118699; European Patent 65479 (to Bayer) [*CA 98*, 55567 (1983)].
99. H. Jäger, E. Klauke, and E. Kysela, Ger. Offen. 3426008; European Patent 168703 (to Bayer) [*CA 107*, 60644 (1987)].
100. H. Jäger and E. Klauke, Ger. Offen. 2906191 (to Bayer) [*CA 93*, 206150 (1980)].
101. K. H. Schündehütte and E. Klauke, Ger. Offen. 2920949 (to Bayer) [*CA 94*, 141188 (1981)].
102. H. S. Bien and E. Klauke, Ger. Offen. 1644203; British Patent 1165661 (to Bayer) [*CA 72*, 80336 (1970)].
103. K. H. Schündehütte and E. Klauke, Ger. Offen. 3317651 (to Bayer) [*CA 102*, 80272 (1985)].
104. K. H. Schündehütte and E. Klauke, Ger. Offen. 3335987 (to Bayer) [*CA 103*, 161865 (1985)].
105. K. H. Schündehütte, M. Groll, and J. W. Stawitz, European Patent 371332 (to Bayer) [*CA 114*, 124554 (1991)].
106. K. H. Schündehütte and F. M Stöhr, European Patent 377189; Ger. Offen. 3900182 (to Bayer) [*CA 113*, 233337 (1990)].
107. K. H. Schündehütte and K. J. Herd, European Patent 377902; Ger. Offen. 3900535 (to Bayer) [*CA 114*, 25811 (1991)].
108. K. Hoegerle, Ger. Offen. 2819837 (to Ciba-Geigy) [*CA 90*, 103997 (1979)].
109. L. Schneider, Swiss Patent 501715 (to Sandoz) [*CA 75*, 7453 (1971)].
110. K. Hoegerle and K. Ohnemus, European Patent 55214 (to Ciba-Geigy) [*CA 97*, 182452 (1982)].

111.   K. Seitz and K. Hoegerle, European Patent 69703 (to Ciba-Geigy) [CA 98, 181117 (1983)].
112.   K. Hoegerle and U. Lehmann, European Patent 356394 (to Ciba-Geigy) [CA 113, 193448 (1990)].
113.   Ref. 1, pp. 198 and 199.
114.   R. Schwaebel and H. Nordmeyer, Bayer Farbenrevue 33, 28 (1983).
115.   H. S. Bien and E. Klauke, Ger. Offen. 2132765 (to Bayer) [CA 78, 137946 (1973)].
116.   H. Jäger, K. Neufang, D. Hildebrand, K. Langheinrich, and M. Söll, Ger. Offen. 3407934 (to Bayer)
       [CA 104, 90480 (1986)].
117.   K. Neufang, R. Kuth, E. R. Fritze, and H. Jäger, Ger. Offen. 3335956 (to Bayer) [CA 103, 161873
       (1985)].
118.   T. Miyamota, T. Omura, Y. Kaneya, A. Takeshita, and N. Harada, Japanese Patent 60, 208367 (to
       Sumitomo) [CA 104, 226326 (1986)].
119.   T. Shirasaki and M. Kojima, Japanese Patent 63, 06181 (to Nippon Kayaku) [CA 109, 56514 (1988)].
120.   W. Frey, P. Grandjeau, L. Schneider, and A. Schweizer, Ger. Offen. 2000753; French Patent 2028214
       (to Sandoz) [CA 73, 110860 (1970)].
121.   References 78, 83–88, 95–109 cited in Ref. 1.
122.   W. Weissauer and M. Ruske, Belgian Patent 614405 (to BASF) [CA 59, 4077 (1963)].
123.   T. Niwa, T. Hihara, and K. Sato, Japanese Patents 54, 61230 and 54, 68835 (to Mitsubishi) [CA 91,
       124863 and 159028 (1979)].
124.   A. H. M. Renfrew and J. A. Taylor, J. Soc. Dyers Colourists 105, 441 (1989).
125.   M. Haelters, Melliand Textilber. 61, 1016 (1980).
126.   G. Siegrist and M. Haelters, Melliand Textilber. 60, 590 (1979).
127.   H. Henk and H. Jäger, Ger. Offen. 2840380 and 2842687 (to Bayer) [CA 93, 73754 (1980)].
128.   W. Härms, K. Wunderlich, and G. von Oertzen, Ger. Offen. 2817733 (to Bayer) [CA 92, 78104
       (1980)].
129.   K. Wunderlich and W. Harms, Ger. Offen. 2729240 (to Bayer) [CA 90, 123058 (1979)].
130.   K. Seitz and G. Hegar, Ger. Offen. 2924228 (to Ciba-Geigy) [CA 92, 165202 (1980)].
131.   H. Jäger, Ger. Offen. 3423581 (to Bayer) [CA 105, 192862 (1986)].
132.   K. Franke and E. Ruhlmann, European Patent 172790 (to Ciba-Geigy) [CA 105, 228492 (1986)].
133.   P. Scheibli, A. Känzig, and A. Schaub, European Patent 144093 (to Ciba-Geigy) [CA 103, 124947
       (1985)].
134.   H. Jäger, Ger. Offen. 3545462 (to Bayer) [CA 107, 200387 (1987)].
135.   H. Henk, Ger. Offen. 2804248 and 2733109 (to Bayer) [CA 90, 170152 (1979)].
136.   K. Seitz, European Patent 101665 (to Ciba-Geigy) [CA 100, 211658 (1984)].
137.   G. Siegrist, Ger. Offen. 3616532 (to Ciba-Geigy) [CA 106, 68685 (1987)].
138.   R. Rohrer, Textilveredlung 20, 85 (1985).
139.   A. H. M. Renfrew, Rev. Prog. Coloration 15, 15 (1985).
140.   B. Parton, "Triphendioxazine Reactive Dyes," paper read at 1st International Conference on the
       Chemistry and Application of Reactive Dyes, Bodington Hall, Leeds (1989).
141.   H. Jäger, W. Harms, and K. J. Herd, "Neue Reaktivfarbstoffe aus der Reihe der Triphendioxazine,"
       paper read at 10th Int. Farbensymposium, Trier (1988).
142.   See the chapter entitled "Reactive Disperse Dyes" in Ref. 1.
143.   W. Harms, R. Kuth, R. Neef, and K. Wunderlich, Ger. Offen. 2918881 (to Bayer) [CA 94, 104896
       (1981)].
144.   K. Himeno, T. Hihara, and Y. Shimizu, Japanese Patent 61,75885 (to Mitsubishi) [CA 106, 19890
       (1986)].
145.   T. Niwa, K. Himeno, and T. Hihara, Senshoku Kogyo 30, 496 (1982) [CA 98, 55387 (1983)].
146.   T. Hihara, Book Pap.—Int. Conf. Exhib., AATCC 135 (1987) [CA 108, 133262 (1988)].
147.   M. G. Barlow, R. N. Haszeldine, C. Simon, D. J. Simpkin, and G. Ziervogel, J. Chem. Soc., Perkin
       Trans. 1 1251 (1982).
148.   M. G. Barlow, R. N. Haszeldine, and D. J. Simpkin, J. Chem. Soc., Perkin Trans. 1 1245 (1982).
149.   R. D. Chambers, T. Shepherd, and M. Tamura, J. Chem. Soc., Perkin Trans. 1 975 (1990).
150.   R. D. Chambers, T. Tamura, J. A. K. Howard, and O. Johnson, J. Chem. Soc., Chem. Commun. 1697
       (1987).

151. R. D. Chambers, T. Shepherd, and M. Tamura, *Tetrahedron 44*, 2583 (1988).
152. E. Klauke, L. Oehlmann, and B. Baasner, *J. Fluorine Chem. 21*, 495 (1982).
153. French Patent 1545313 (to Bayer) [*CA 72*, 3497 (1969)].
154. M. M. Boudakian, E. H. Kober, and E. R. Shipkowski, U.S. Patent 3280124 (to Olin Mathieson) [*CA 66*, 2582 (1967)].
155. R. E. Banks, D. S. Field, and R. N. Haszeldine, *J. Chem. Soc. (C)* 1822 (1967).
156. R. E. Banks, D. S. Field, and R. N. Haszeldine, *J. Chem. Soc. (C)* 1280 (1970).
157. O. Kaieda, K. Hirota, N. Tominaga, and T. Nakamura, Japanese Patent 61, 047465 (to Nippon Shokubai Kagaku) [*CA 105*, 60633 (1986)].
158. M. M. Boudakian and C. W. Kaufmann, U.S. Patent 3314955 (to Olin Mathieson) [*CA 68*, 59604 (1968)].
159. O. P. Shkurko, S. G. Baram, and V. P. Mamaev, *Izv. Sib. Otd. Akad. Nauk SSSR, Ser. Khim. Nauk* 81 (1973) [*CA 80*, 59913 (1974)].
160. T. Okano, S. Goya, and H. Matsumoto, *Yakugaku Zasshi 87*, 1315 (1967) [*CA 68*, 114540 (1968)].
161. K. Seitz and K. Hoegerle, "Reaktivfarbstoffe aus neuen Pyrimidinderivaten," paper read at 8th Int. Farbensymposium, Baden-Baden (1982).
162. Japanese Patent 58, 219163 (to Dainippon Ink and Chem.) [*CA 100*, 174855 (1984)].
163. E. Klauke and H. S. Bien, British Patent 1158300; French Patent 1545174 (to Bayer) [*CA 71*, 124481 (1969)].
164. E. Klauke and H. S. Bien, Ger. Offen. 1670780 (to Bayer).
165. M. Bimmler, M. von Janta-Lipinski, P. Langen, and H. Plaul, Ger. (East) Patent 221736 (to Akademie der Wissenschaften der DDR) [*CA 104*, 68880 (1986)).
166. H. Muffler and R. Franz, Ger. Offen. 2823969 (to Hoechst) [*CA 92*, 197322 (1980)].
167. H. U. Alles, E. Klauke, and H. S. Bien, Ger. Offen. 1931640 (to Bayer) [*CA 74*, 76439 (1971)].
168. A. Günther, M. Lenthe, and G. Dankert, Ger. Offen. 3131735 (to Bayer) [*CA 99*, 38485 (1983)].
169. E. Klauke and K. H. Schündehütte, European Patent 04945 (to Bayer) [*CA 92*, 94429 (1980)].
170. W. Tang, Y. X. Zhao, Z. H. Zhu, and K. C. Chen, *J. Fluorine Chem. 35*, 373 (1987).
171. E. Klauke, O. Schallner, and K. H. Schündehütte, Ger. Offen. 3118700 (to Bayer) [*CA 98*, 73842 (1983)].
172. M. M. Boudakian and G. A. Hyde, *J. Fluorine Chem. 25*, 435 (1984).
173. B. Baasner and E. Klauke, *J. Fluorine Chem. 45*, 417 (1989).
174. B. Baasner, E. Klauke, and K. H. Schündehütte, Ger. Offen. 3402194 (to Bayer) [*CA 104*, 5890 (1986)].
175. G. Steffan and G. Zahl, Ger. Offen. 3228712 (to Bayer) [*CA 100*, 174674 (1984)].
176. G. Beck, Ger. Offen. 2745497 (to Bayer) [*CA 91*, 39521 (1979)].
177. G. Beck, H. Heitzer, and H. Holtschmidt, *Angew. Chem. 86*, 134 (1974); *Angew. Chem., Int. Ed. Engl. 13*, 210 (1974).
178. G. Beck, F. Doering, H. Holtschmidt, and K. Ley, Ger. Offen. 2307863 (to Bayer) [*CA 81*, 152261 (1974)].
179. G. Beck, G. Dankert, and F. Döring, Ger. Offen. 2753204 (to Bayer) [*CA 91*, 91664 (1979)].
180. L. Lin, J. Lu, and C. Chen, *Tianjin Daxue Xuebao*, 12 (1985) [*CA 105*, 42045 (1986)].
181. G. Beck, H. Holtschmidt, and H. Heitzer, *Justus Liebigs Ann. Chem. 731*, 45 (1970).
182. G. Beck and H. Heitzer, European Patent 127827; Ger. Offen. 3319957 (to Bayer) [*CA 102*, 45978 (1985)].
183. R. Fauss, K. Findeisen, K. H. Schündehütte, and B. Thelen, Ger. Offen. 3609801 (to Bayer) [*CA 110*, 135255 (1989)].
184. C. W. Tullock and D. D. Coffman, *J. Org. Chem. 25*, 2016 (1960).
185. E. Klauke, E. Kysela, A. Stüwe, and A. Dorlars, Ger. Offen. 3008923 (to Bayer) [*CA 95*, 204009 (1981)].
186. K. Hungerbühler, Ger. Offen. 3727973 (to Ciba-Geigy) [*CA 109*, 8413 (1988)].
187. J. Vencl, Z. Vidner, and V. Chmatal, Czech. Patent 247969 [*CA 108*, 188926 (1988)].
188. W. X. Liang and Q. Y. Chen, *Hua Hsueh Hsueh Pao 38*, 269 (1980) [*CA 94*, 65630 (1981)].
189. G. Seifert and S. Stäubli, Ger. Offen. 2814450 (to Ciba-Geigy) [*CA 90*, 23123 (1979)].

190. E. Kysela, E. Klauke, and H. Schwarz, Ger. Offen. 2729762 (to Bayer) [*CA 90*, 168649 (1979)].
191. E. Klauke and A. Dorlars, Ger. Offen. 2643335 (to Bayer) [*CA 88*, 190911 (1978)].
192. E. Klauke and A. Dorlars, Ger. Offen. 2643251 (to Bayer) [*CA 88*, 190914 (1978)].
193. E. Klauke, E. Kysela, and A. Dorlars, Ger. Offen. 2702625 (to Bayer) [*CA 89*, 163605 (1978)].
194. R. Franz, *J. Fluorine Chem. 15*, 423 (1980).
195. H. Seiler and R. Deitz, European Patent 41922 (to Ciba-Geigy) [*CA 96*, 164154 (1982)].
196. J. Wolff, K. Wolf, and P. Wegner, Ger. Offen. 3400411 (to Bayer) [*CA 104*, 90488 (1986)].
197. J. Wolff, K. Wolf, R. M. Klipper, and P. M. Lange, Ger. Offen. 3625693 (to Bayer) [*CA 108*, 206264 (1988)].
198. R. D. Chambers, P. A. Martin, J. S. Waterhouse, D. L. H. Williams, and B. Anderson, *J. Fluorine Chem. 20*, 507 (1982).
199. M. Yukawa, T. Niiya, Y. Goto, T. Sakamoto, H. Yoshizawa, A. Watanabe, and H. Yamanaka, *Chem. Pharm. Bull. 37*, 2892 (1989).
200. G. Altorfer, S. Gati, and G. Hegar, Ger. Offen. 2746109 (to Ciba-Geigy) [*CA 89*, 112417 (1978)].
201. K. Wunderlich and W. Harms, European Patent 14844, Ger. Offen. 2903594 (to Bayer) [*CA 94*, 4942 (1981)].
202. H. Jäger and K. Wunderlich, Ger. Offen. 2711150 (to Bayer) [*CA 90*, 40195 (1979)].
203. A. Dorlars, Ger. Offen. 1076696 (to Bayer) [*CA 55*, 16579 (1961)].
204. J. Otten and F. Meininger, European Patent 36133; Ger. Offen. 3010502 (to Hoechst) [*CA 96*, 36883 (1982)].
205. E. Bonometti and H. Seiler, Ger. Offen. 2747011 (to Ciba-Geigy) [*CA 89*, 59903 (1978)].
206. H. Seiler, Ger. Offen. 2838540 (to Ciba-Geigy) [*CA 90*, 205774 (1979)].
207. E. Bonometti and H. Seiler, European Patent 228348 (to Ciba-Geigy) [*CA 107*, 200388 (1987)].
208. H. Jäger, European Patent 227983; Ger. Offen. 3545460 (to Bayer) [*CA 107*, 154352 (1987)].
209. K. H. Franke and E. Ruhlmann, European Patent 172790 (to Ciba-Geigy) [*CA 105*, 228492 (1986)].
210. G. Altorfer, S. Gati, and G. Hegar, Swiss Patent 630107 (to Ciba-Geigy) [*CA 97*, 129107 (1982)].
211. G. Hegar and H. Riat, Ger. Offen. 2901498 (to Ciba-Geigy) [*CA 91*, 194627 (1979)].
212. R. Deitz and H. Seiler, European Patent 41919 (to Ciba-Geigy) [*CA 96*, 124510 (1982)].
213. K. Lesche, *Bayer Farbenrevue 37*, 47 (1985).
214. B. Bruttel, *Textilveredlung 25*, 194 (1990).
215. W. Beckmann, F. Hoffmann, and J. Wolff, *Textilveredlung 24*, 81 (1989).
216. P. Hugelshofer, B. Bruttel, H. Pfenninger, and R. Lacroix, European Patent 41240 (to Ciba-Geigy) [*CA 96*, 144405 (1982)].
217. B. Bruttel and H. Pfenninger, European Patent 37382 (to Ciba-Geigy) [*CA 96*, 53751 (1982)].
218. J. Wolff, K. Wolf, R. Hörnle, R. Ditzer, and K. Falkenberg, European Patent 87703; Ger. Offen. 3207534 (to Bayer) [*CA 99*, 196640 (1983)].
219. J. Koll, H. H. Mölls, R. Ditzer, G. Martiny, and K. H. Steiner, Ger. Offen. 3301870 (to Bayer) [*CA 101*, 172 955 (1984)].
220. J. Wolff, K. Wolf, and R. Hörnle, European Patent 87705; Ger. Offen. 3207533 (to Bayer) [*CA 100*, 8427 (1984)].
221. J. Wolff, K. Wolf, and P. Wegner, European Patent 148496; Ger. Offen. 3400412 (to Bayer) [*CA 103*, 197306 (1985)].
222. J. Wolff and K. Wolf, European Patent 217217; Ger Offen 3534729 (to Bayer) [*CA 107*, 79420 (1987)].
223. D. Link and J. Moreau, *Textilveredlung 24*, 87 (1989).
224. W. Zysset, *Textilveredlung 24*, 91 (1989).
225. K. Seitz, European Patent 131543 (to Ciba-Geigy) [*CA 102*, 150920 (1985)].
226. K. Seitz and H. Riat, Ger. Offen. 2730581 (to Ciba-Geigy) [*CA 88*, 137890 (1978)].
227. K. Seitz, Ger. Offen. 2738823 (to Ciba-Geigy) [*CA 89*, 7580 (1978)].
228. W. Scholl, Ger. Offen. 2749647 (to Bayer) [*CA 91*, 75702 (1979)].
229. A. Tzikas and A. Käser, European Patent 131542 (to Ciba-Geigy) [*CA 102*, 150918 (1985)].
230. J. L. Leng and D. W. Shaw, Ger. Offen. 2600490 (to I.C.I.) [*CA 85*, 110105 (1976)].
231. H. Jäger, European Patent 189081; Ger. Offen. 3502104 (to Bayer) [*CA 106*, 68725 (1987)].

232. H. Seiler and G. Hegar, Ger. Offen. 2611550 (to Ciba-Geigy) [*CA 85*, 194089 (1976)].

233. H. Henk, W. Harms, K. H. Schündehütte, and K. Wunderlich, Ger. Offen. 2752224 (to Bayer) [*CA 91*, 109000 (1979)].

234. W. Koch and K. Brenneisen, Ger. Offen. 3138019; French Patent 2491486 (to Sandoz) [*CA 97*, 93948 (1982)].

235. I. K. Barben and C. V. Stead, British Patent 1461125 (to I.C.I.) [*CA 87*, 69745 (1977)].

236. K. Brenneisen and W. Koch, Ger. Offen. 3043915 (to Sandoz) [*CA 95*, 117059 (1981)].

237. K. Seitz, P. Scheibli, and H. Seiler, European Patent 132223 (to Ciba-Geigy) [*CA 102*, 150922 (1985)].

238. K. H. Schündehütte, R. Moll, H. Nickel, and W. Scholl, Ger. Offen. 2847938 (to Bayer) [*CA 93*, 96792 (1980)].

239. H. S. Bien, D. Hildebrand, and W. Harms, Ger. Offen. 2603670 (to Bayer) [*CA 87*, 153417 (1977)].

240. J. L. Jessen, K. Pandl, H. Löffler, B. Siegel, and M. Patsch, European Patent 391264 (to BASF) [*CA 114*, 83890 (1991)].

241. H. Jäger, European Patents 84314 and 203505; Ger. Offen. 3201114 and 3519551 (to Bayer) [*CA 99*, 141530 (1983) and *CA 106*, 103809 (1987)].

242. H. Jäger, Ger. Offen. 3318146 and 3503747 (to Bayer) [*CA 102*, 80279 (1985) and *CA 106*, 68747 (1987)].

243. A. Tzikas, European Patents 134193, 133843 and 221013 (to Ciba-Geigy) [*CA 103*, 7733 (1985) and *CA 107*, 98203 (1987)].

244. H. Jäger and K. J. Herd, Ger. Offen. 3503567; European Patent 192067 (to Bayer) [*CA 107*, 41682 (1987)].

245. T. Hihara, K. Shimizu, and Y. Shimizu, European Patent 286113 (to Mitsubishi) [*CA 110*, 156103 (1989)].

246. F. M. Stöhr, H. Henk, and K. J. Herd, European Patent 318785 (to Bayer) [*CA 111*, 176189 (1989)].

247. H. Eilingsfeld and R. Iden, European Patent 286021 (to BASF) [*CA 110*, 116663 (1989)].

248. P. Scheibli, K. Seitz, H. Seiler, J. Markert, and A. Käser, European Patent 131545 (to Ciba-Geigy) [*CA 102*, 150919(1985)].

249. F. Meininger and J. Otten, Ger. Offen. 3019936 (to Hoechst) [*CA 96*, 164125 (1982)].

250. F. Meininger, E. Hoyer, and R. Fass, Ger. Offen. 3202688 (to Hoechst) [*CA 99*, 177496 (1983)].

251. F. M. Stöhr, H. Henk, and K. J. Herd, European Patent 308787 (to Bayer) [*CA 111*, 116818 (1989)].

252. H. Jäger, Ger. Offen. 3513261 and 3513260 (to Bayer) [*CA 106*, 34634 and 34636 (1987)].

253. H. Seiler, Ger. Offen. 2927102 (to Ciba-Geigy) [*CA 92*, 199745 (1980)].

254. H. Henk, W. Harms, and E. Siegel, Ger. Offen. 2842640 (to Bayer) [*CA 93*, 133818 (1980)].

255. Japanese Patent 56, 100861 (to Sumitomo) [*CA 96*, 8128 (1982)].

256. J. Otten and M. Hähnke, Ger. Offen. 3011447 (to Hoechst) [*CA 96*, 8142 (1982)].

257. H. Seiler, European Patents 70806 and 70807 (to Ciba-Geigy) [*CA 98*, 198832 and 98, 217156 (1983)].

258. P. Scheibli and H. Seiler, European Patent 74928 (to Ciba-Geigy) [*CA 99*, 72172 (1983)].

259. J. Markert and H. Seiler, European Patents 85025 and 85654 (to Ciba-Geigy) [*CA 99*, 177499 and 196635 (1983)].

260. F. Meininger, R. Hoyer, and R. Fass, Ger. Offen. 3204259 (to Hoechst) [*CA 100*, 8502 (1984)].

261. A. Tzikas, P. Scheibli, and H. Seiler, European Patent 144766 (to Ciba-Geigy) [*CA 103*, 143355 (1985)].

262. A. Tzikas, European Patents 159292 and 315585 (to Ciba-Geigy) [*CA 104*, 170101 (1986) and *111*, 116823 (1989)].

263. A. Tzikas, P. Äschlimann, and P. Herzig, WO Patent 87/01123; European Patent 214093 (to Ciba-Geigy) [*CA 107*, 79475 (1987)].

264. R. Iden, Ger. Offen. 3628090 (to BASF) [*CA 109*, 18870 (1988)].

265. A. Tzikas and P. Herzig, European Patent 278904 (to Ciba-Geigy) [*CA 111*, 24958 (1989)].

266. H. Springer, W. Helmling, L. Schläfer, and W. H. Russ, European Patent 279351; Ger. Offen. 3704660 (to Hoechst) [*CA 110*, 40490 (1989)].

314   K. J. Herd

267. M. Patsch, U. Nahr, J. L. Jessen, K. Pandl, and F. Wirsing, European Patent 307817 (to BASF) [*CA 111*, 79892 (1989)].
268. H. M. Büch and H. Springer, European Patent 361440 (to Hoechst) [*CA 113*, 193449 (1990)].
269. H. Springer, K. Hussong, R. Hähnle, and W. Russ, Ger. Offen. 3825658 (to Hoechst) [*CA 113*, 61299 (1990)].
270. P. Scheibli, H. Seiler, K. Seitz, and A. Tzikas, European Patent 179019 (to Ciba-Geigy) [*CA 106*, 68720 (1987)].
271. U. Lehmann, European Patent 352222 (to Ciba-Geigy) [*CA 112*, 236898 (1990)].
272. P. Scheibli, European Patent 89923 (to Ciba-Geigy) [*CA 100*, 53191 (1984)].
273. U. Bergmann, A. Hackenberger, and M. Patsch, Ger. Offen. 3622080 (to BASF) [*CA 106*, 157965 (1987)].
274. H. Seiler and G. Hegar, Ger. Offen. 2653199 (to Ciba-Geigy) [*CA 87*, 203057 (1977)].
275. H. S. Bien and D. Hildebrand, Ger. Offen. 2607028 and 2603670 (to Bayer) [*CA 87*, 169244 and 153417 (1977)].
276. K. Brenneisen, Ger. Offen. 2706417 (to Sandoz) [*CA 87*, 186058 (1977)].
277. W. Scholl and H. Nickel, Ger. Offen. 2749597 (to Bayer) [*CA 91*, 40898 (1979)].
278. P. Doswald and W. Koch, Ger. Offen. 3700846 and 3800093 (to Sandoz) [*CA 109*, 212370 (1988)].
279. K. J. Herd, H. Henk, H. Jäger, K. H. Schündehütte, and F. M. Stöhr, Ger. Offen. 3800261 (to Bayer) [*CA 112*, 79431 (1990)].
280. K. Seitz, European Patent 141776 (to Ciba-Geigy) [*CA 103*, 106293 (1985)].
281. K. Seitz and K. Hoegerle, European Patent 285571 (to Ciba-Geigy) [*CA 111*, 8882 (1989)].
282. K. Hoegerle, European Patent 304924 (to Ciba-Geigy) [*CA 111*, 59548 (1989)].
283. W. Schindler, E. Tannwaldt, and R. Krüger, *Melliand Textilber. 71*, 388 (1990).
284. E. Hoyer, F. Meininger, and R. Fass, Ger. Offen. 2748929 and 2748965 (to Hoechst) [*CA 91*, 75700 and 75690 (1979)].
285. E. Hoyer, F. Meininger, W. Noll, and R. Fass, Ger Offen. 2748966 (to Hoechst) [*CA 91*, 75691 (1979)].
286. E. Hoyer, F. Meininger, and R. Fass, European Patent 65732; Ger. Offen. 3120187 (to Hoechst) [*CA 98*, 91020 (1983)].
287. H. Seiler, European Patent 70808 (to Ciba-Geigy) [*CA 98*, 199833)].
288. H. J. Bredereck, E. Hoyer, and F. Meininger, European Patent 94055; Ger. Offen. 3217812 (to Hoechst) [*CA 100*, 69858 (1984)].
289. P. Scheibli, European Patent 141367 (to Ciba-Geigy) [*CA 103*, 106294 (1985)].
290. H. Jäger, K. Langheinrich, and K. J. Herd, Ger. Offen. 3439755 (to Bayer) [*CA 105*, 62209 (1986)].
291. H. Springer, G. Schwaiger, and W. Helmling, Ger. Offen. 3544982 (to Hoechst) [*CA 108*, 39649 (1988)].
292. A. Tzikas, European Patents 297044 and 302006 (to Ciba-Geigy) [*CA 111*, 24960 and 8884 (1989)].
293. R. Hähnle, European Patent 385426; Ger. Offen. 3906778 (to Hoechst) [*CA 114*, 187556 (1991)].
294. H. M. Büch, Ger. Offen. 4009066; European Patent 388864 (to Hoechst) [*CA 115*, 94741 (1991)].
295. H. Löffler, K. Pandl, M. Patsch, and B. Siegel, European Patents 387579 and 387589 (to BASF) [*CA 114*, 83885 and 83886 (1991)].
296. K. J. Herd, H. Henk, and F. M. Stöhr, European Patent 395951 (to Bayer) [*CA 115*, 31103 (1991)].

13

Fluorine-Containing Dyes

B. Other Fluorinated Dyestuffs

A. ENGEL

13B.1. INTRODUCTION*

From the industrial standpoint, the reason for the synthesis of fluorine-containing dyestuffs is the continuing search for improved properties for certain applications. Emphasis has been placed on the incorporation of the CF₃ group into azo dyes, though examples of all classes of organic dyes with fluorine-containing substituents are known. Whether the practical utilization of these dyes leads to commercial success, depends mainly on the synthesis costs, i.e., the price–performance ratio.

13B.2. PROPERTIES OF FLUORINE-CONTAINING DYES

Aminobenzenes normally undergo N-alkylation with 1,1,1-trifluorochloroethane, but in the presence of iron, nickel, cobalt, or copper salts, ring substitution occurs to give 2-(2,2,2-trifluoroethyl)anilines, which can be used as diazo components in the production of disperse dyes.[1]

As fluorine-containing substituents lower the energy of the intermolecular interactions and the heat of evaporation of disperse azo dyes by 10–40 kJ mol⁻¹ compared with the corresponding fluorine-free dyes,[2] they often are useful in the heat-transfer printing of polyester or polyamide fabrics. Because of the low fastness to sublimation, the pyrazolone dye (1) containing the SO₂F group,[3] as well as the CF₃-substituted dyes 2,[4] 3,[5] 4,[6] 5,[7] and 6[8] are useful in the heat-transfer printing of polyester fabrics.

---

*This review updates that by G. Wolfrum (see Ref. 13).

A. ENGEL • Bayer AG, Research and Development—Dyes and Pigments, D-51368 Leverkusen 1, Germany.

Organofluorine Chemistry: Principles and Commercial Applications, edited by R. E. Banks *et al.* Plenum Press, New York, 1994.

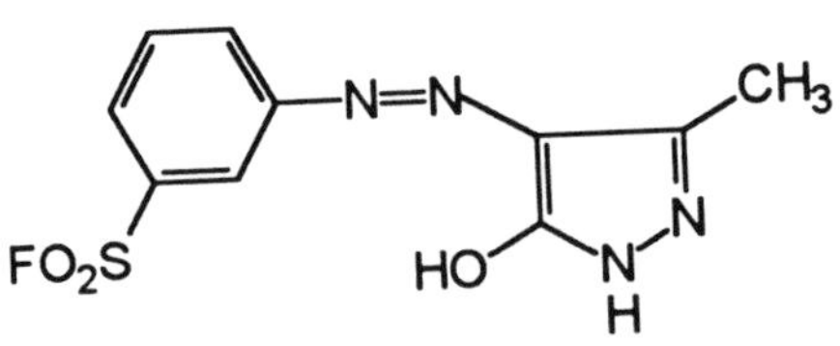

1: yellow

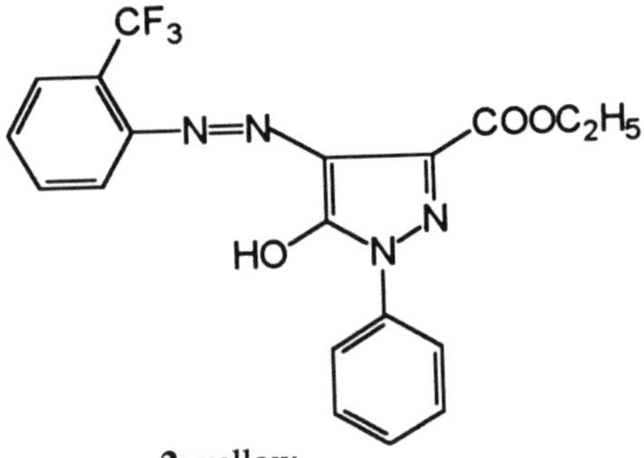

2: yellow

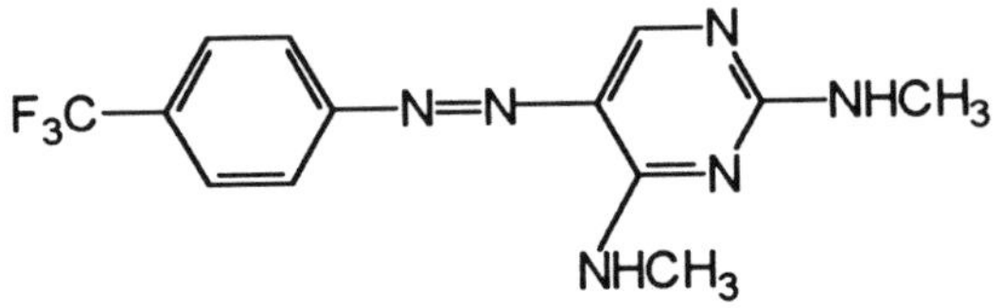

3: greenish yellow

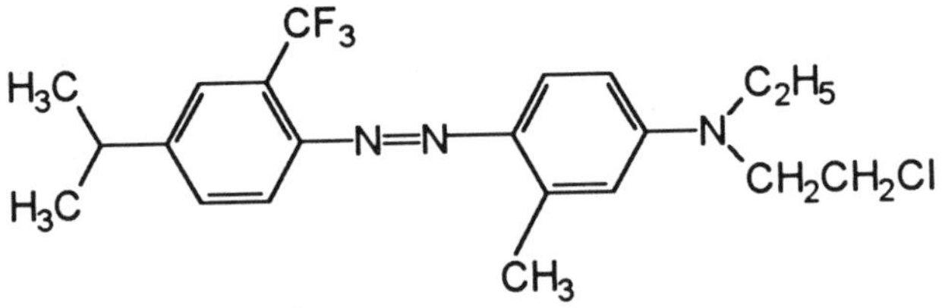

4: red

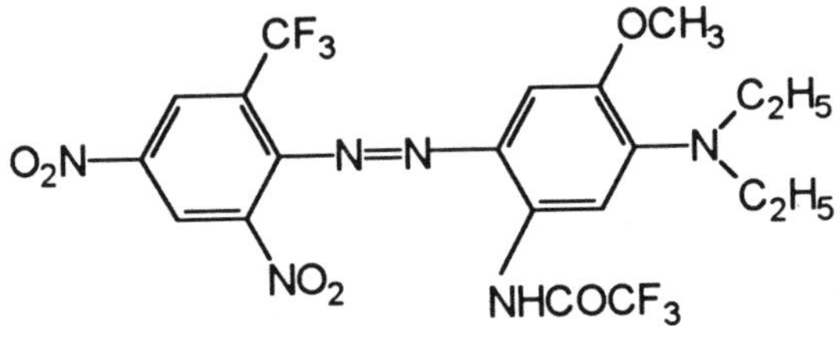

5: blue

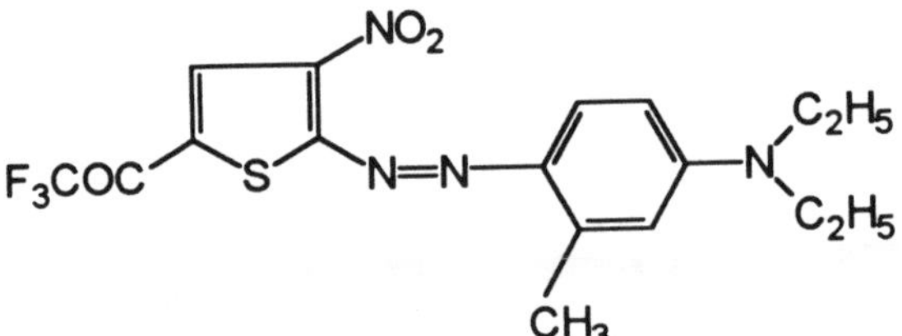

6: blue

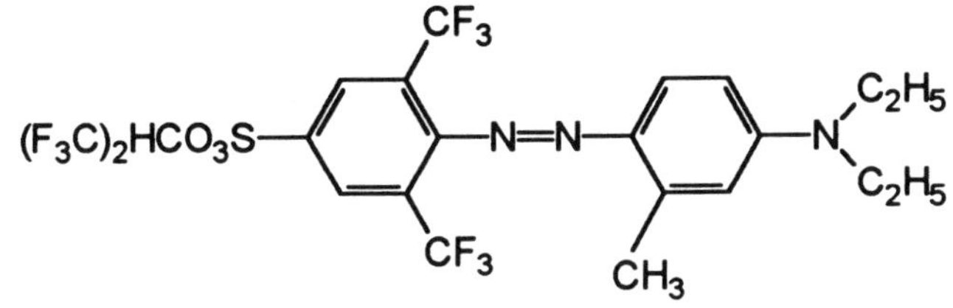

7: blue

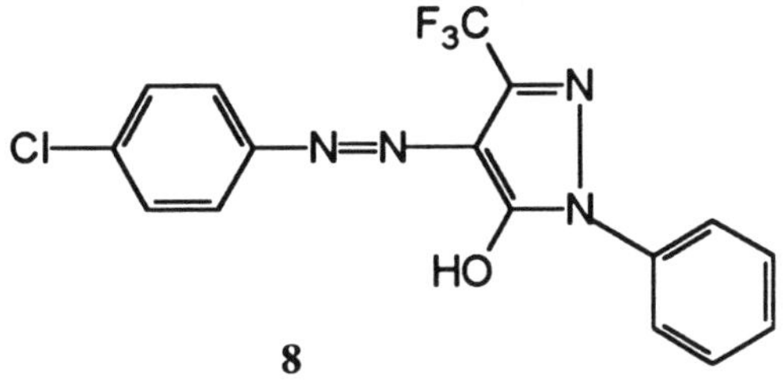

8

Dyes with fluorinated sulfonic ester groups, which are stable with respect to the hydrolytic conditions used in the exhaust dyeing process, can be applied for both dyeing and heat-transfer printing of textiles, such as **7**.[9]

The CF$_3$ group present instead of the CH$_3$ group in the yellow polyester dye (**8**) leads to higher fiber affinity and better lightfastness.[10]

1,2,4-Thiadiazolylazo dyes with a CF$_3$ group in the 5 position of the heterocyclic ring are recommended for dyeing and printing of polyester and cellulose fibers.[11] Clear blue shades on polyester are obtained with disperse azo dyes based on 2-aminothiophene as diazo compound with SO$_2$F groups in the 3 and 5 position, such as **9**.[12]

Because of the outstanding lightfastness on polyamides, the red azo dyes from 2-trifluoromethylanilines and 2-amino-8-hydroxy-6-naphthalenesulfonic acid[13] remain commercially interesting, e.g., **10**,[6] **11**,[14] and **12**.[15]

The cationic dyestuff **13** for polyacrylonitrile is tinctorially stronger than that in which fluorine is substituted by chlorine or bromine, and also possesses advantages in lightfastness and brightness.[16]

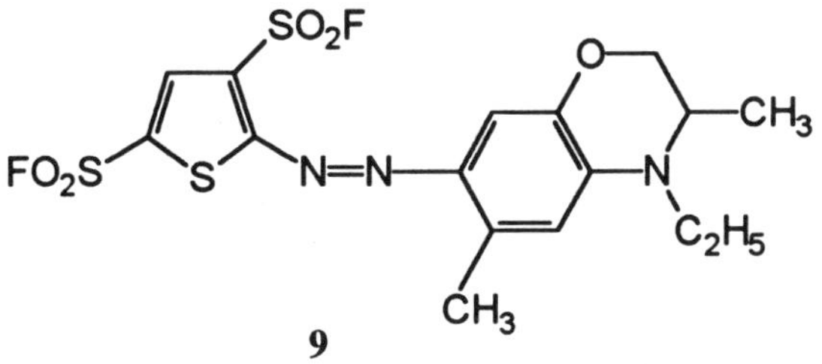

9

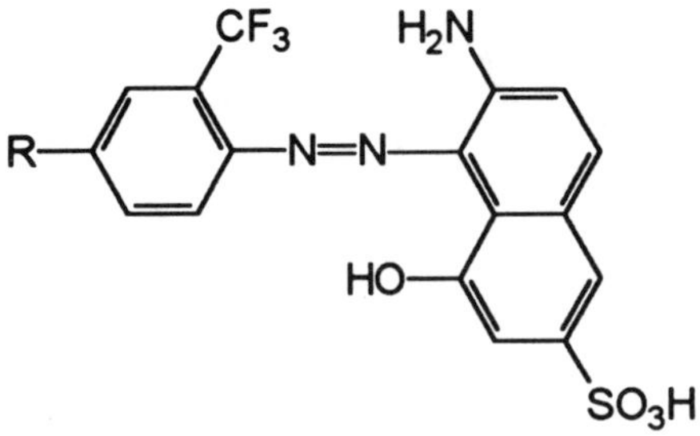

10: R = —CH(CH₃)₂
11: R = —OCH₃
12: R = —NHCOCH₃

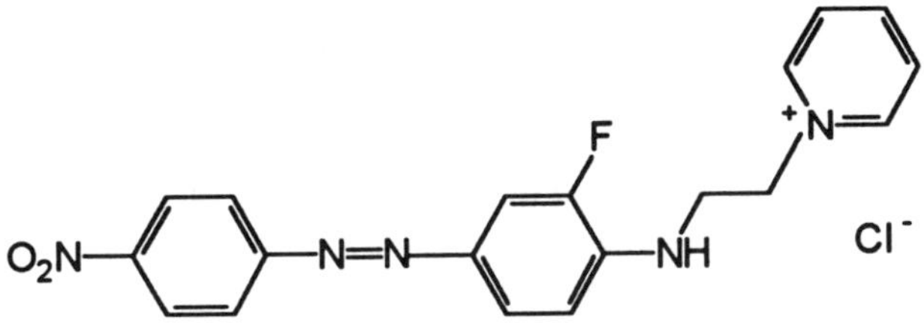

13: reddish yellow

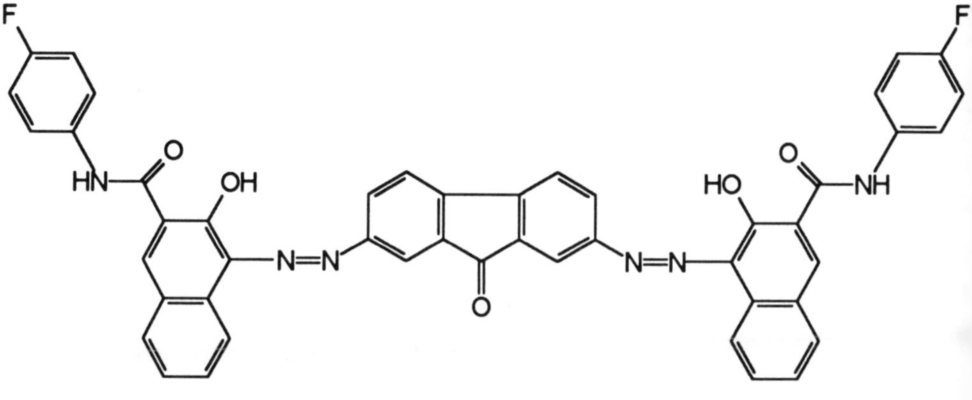

14

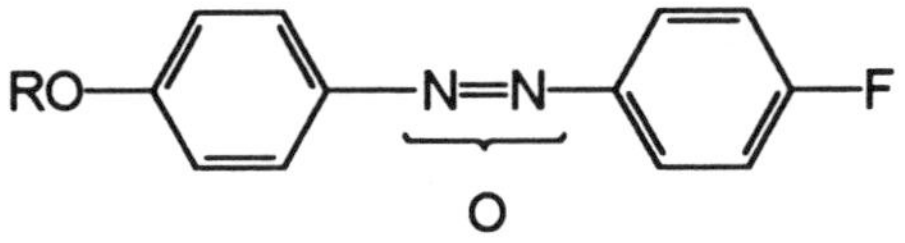

15: R = C_1–C_{15}-alkyl

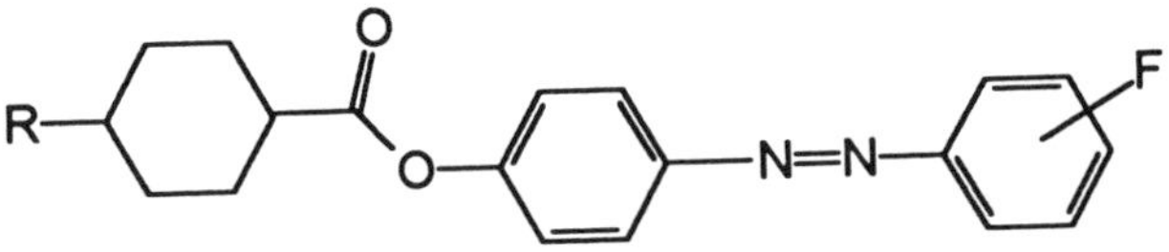

16: R = C_1–C_{15}-alkyl
F in 3 or 4 position

The fluorine-containing pigment **14** is patented as a charge-generating component in electrophotography.[17]

Azo pigments with CF_3 groups are claimed.[18–20] They are faster to light than the corresponding pigments with chlorine atoms instead of CF_3 groups.[19]

A mixture of the isomeric azoxy compounds **15** improve the properties of nematic liquid crystals by enhancing the resistance to moisture and electricity or by reducing the viscosity.[21]

The orange liquid crystalline compounds **16** show liquid crystalline behavior over a wide temperature range with positive dielectric anisotropy, high transparency, and low viscosity.[22]

Noteworthy, though not of commercial importance, is the fact that 4-aminotetrafluoropyridine can be diazotized with $NaNO_2$ in a mixture of sulfuric acid, acetic acid, and propionic acid at 0 °C and coupled with mesitylene in 77% yield,[23] while the same azo compound is isolated only in 30% yield after diazotizing the amine in 80% aqueous fluoric acid.[24] Pentafluoroaniline is advantageously diazotized in a two-phase system (water/dichloromethane) under phase-transfer conditions and coupled with anisole (80% yield), 1-methoxynaphthalene (75% yield), and mesitylene (44% yield).[25]

Fluorine-containing fluorans as color formers[26] in heat- and pressure-sensitive copying papers and fluorophthalocyanine compounds,[27,28] useful for optical recording medium, have been patented.

The influence of the fluorine atom on the color of polymethine dyes has been investigated extensively.[29] The fluorine atom behaves as an electron-donor substituent and causes a bathochromic shift of the absorption maxium in even positions of the chain and a hypsochromic shift in odd positions; in sterically hindered merocyanines, however, it effects a hypsochromic shift, even in even-numbered positions of the polymethine chain.[30]

13B.3. SUMMARY AND OUTLOOK

In spite of the numerous patents dealing with fluorine-containing dyes and their applicability for dyeing and printing, only a few products, principally based on compounds containing CF$_3$ groups, have achieved any degree of commercial success to date.

13B.4. REFERENCES

1. L. Foulletier, J. P. E. Pechmeze, and R. F. M. Sureau, U.S. Patent 4059626; Ger. Offen. 2513802 (to Ugine Kuhlmann) [*CA 84*, 16935 (1976)].
2. V. V. Karpov, G. N. Rodionova, J. V. Krutovskaya, L. Z. Gandel'sman, L. A. Khomenko, and L. M. Yagupol'skii, *Dyes and Pigments 5*, 285 (1984).
3. D. R. Waring and R. M. Bellas, British Patent 1503907 (to Kodak).
4. R. I. Steiner, U.S. Patent 4234481 (to Crompton and Knowles) [*CA 94*, 104890 (1981)].
5. J. Gray and D. R. Waring, British Patent 1558403 (to Kodak) [*CA 92*, 165209 (1980)].
6. A. Marhold and G. Wolfrum, European Patent 84320; DE 3201112 (to Bayer) [*CA 99*, 177492 (1983)].
7. S. Imahori, Y. Murata, K. Abe, and S. Suzuki, Japanese Patent 53058083 (to Mitsubishi) [*CA 89*, 131060 (1978)].
8. D. Mullen, British Patent 2041961 (to Kodak) [*CA 94*, 210209 (1981)].
9. R. N. Gourley, U.S. Patent 4283332; British Patent 2065693 (to Eastman Kodak) [*CA 96*, 36857 (1982)].
10. J. Bernardin and J. Pechmeze, Ger. Offen. 2627110 (to Ugine Kuhlmann) [*CA 86*, 107238 (1977)].
11. Japanese Patent 57095381 (to Sumitomo) [*CA 98*, 36075 (1983)].
12. R. N. Gourley, British Patent 2108993 (to Kodak) [*CA 99*, 124073 (1983)].
13. G. Wolfrum, in: *Organofluorine Chemicals and Their Industrial Applications* (R. E. Banks, ed.), Chapter 10, p. 208, Ellis Horwood, London (1979).
14. G. Wolfrum, E. Klauke, and H.-G. Otten, Ger. Offen. 2728073 (to Bayer) [*CA 90*, 123067 (1979)].
15. G. Wolfrum, E. Klauke, and H.-G. Otten, Ger. Offen. 2712170 (to Bayer) [*CA 90*, 24775 (1979)].
16. P. Gregory, British Patent 1551613 (to I.C.I.) [*CA 92*, 165208 (1980)].
17. Japanese Patent 57102665 (to Ricoh) [*CA 98*, 188988 (1983)].
18. A. Rouèche, European Patent 31798 (to Ciba-Geigy) [*CA 96*, 53806 (1982)].
19. G. Cseh and A. Rouèche, Ger. Offen. 2734659 (to Ciba-Geigy) [*CA 88*, 154323 (1978)].
20. K. Ronco, European Patent 90777 (to Ciba-Geigy) [*CA 100*, 23555 (1984)].
21. Japanese Patent 56123959 (to Chisso Corp.) [*CA 96*, 208466 (1982)].
22. Japanese Patent 59170141 (to Chisso Corp.) [*CA 102*, 63583 (1985)].
23. A. C. Alty, R. E. Banks, A. R. Thompson, and B. R. Fishwick, *J. Fluorine Chem. 26*, 263 (1984).
24. R. E. Banks, A. R. Thompson, and H. S. Vellis, *J. Fluorine Chem. 22*, 499 (1983).
25. H. Iwamoto, T. Sonoda, and H. Kobayashi, *J. Fluorine Chem. 24*, 535 (1984).
26. Japanese Patent 58065754 (to Nippon Kayaku) [*CA 99*, 196644 (1983)].
27. O. Kaieda, I. Okitaka, and H. Ito, Japanese Patent 63030566 (to Nippon Shokubai Kagaku) [*CA 109*, 75212 (1988)].
28. O. Kaieda, I. Okitaka, H. Ito, K. Yoshitoshi, and N. Takatani, Japanese Patent 01045474 (to Nippon Shokubai Kagaku) [*CA 111*, 196752 (1989)].
29. L. M. Yagupol'skii, A. Ya. Il'chenko, and L. Z. Gandel'sman, *Usp. Khim. 52*, 1732 (1983); Engl. Transl. *Russ Chem. Rev. 52*, 993 (1983).
30. Zh. A. Krasnaya, T. S. Stytsenko, V. S. Bogdanov, N. V. Monich, M. M. Kul'chitskii, S. V. Pazenok, and L. M. Yagupol'skii, *Izv. Akad. Nauk SSSR Ser. Khim.* 636 (1989); Engl. Transl. *Bull. Acad. Sci. USSR, Div. Chem. Sci. 38*, 562 (1989).

14

Textile Finishes and Fluorosurfactants

NANDAKUMAR S. RAO and BRUCE E. BAKER

14.1. TEXTILE REPELLENT FINISHES

Fluorochemicals carrying perfluoroalkyl groups (R_F) are widely used to modify the surfaces of textiles and carpets, and hence impart resistance to water, oils, soils, and staining. In fact, fluorinated materials have essentially replaced silicones and hydrocarbon-based finishes as textile repellents, and function by imparting a condition of limited wettability to the treated substrate. The efficiency of the surface modification depends on the intrinsic repellency of the active fluorochemical, the extent of coverage of the textile by the fluorochemical, the orientation of the perfluoroalkyl segments, and the amount and location of the fluorochemical on the textile. Typically, 0.05 to 0.50 percent of the fluorochemical by weight of the textile is used to deliver durable repellency. The repellents are applied to textiles and carpets in mills as aqueous dispersions, and in some after-market applications, as solutions in halogenated solvents.

14.1.1. Intrinsic Repellency and Fluorocarbon Structure

The empirical wetting relationship developed by Zisman is a useful way to present differences in intrinsic repellency for hydrophobic surfaces, and has guided the development of textile repellent finishes.[1] Figure 1 shows Zisman "straight line" plots for condensed perfluoroalkanecarboxylic acid monolayers, $F(CF_2)_nCO_2H$, on glass,[2] contact angles (θ) having been measured for a series of normal alkanes, and cos θ plotted as a function of the surface tensions of the alkane liquids. The resulting relationships are linear, and extrapolation of any line to an intercept where cos θ = 1

NANDAKUMAR S. RAO and BRUCE E. BAKER • DuPont Specialty Chemicals, Jackson Laboratory, Chambers Works, Deepwater, New Jersey 08023.

Organofluorine Chemistry: Principles and Commercial Applications, edited by R. E. Banks *et al.* Plenum Press, New York, 1994.

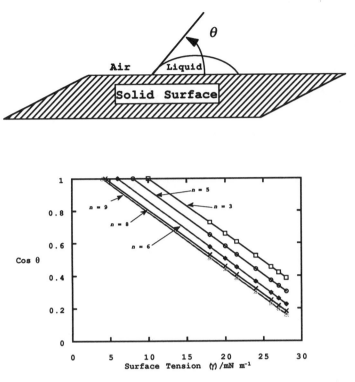

Figure 1. Zisman plots for perfluoroalkanoic acid [F(CF₂)ₙCO₂H] monolayers in contact with *n*-alkanes.[2]

defines the critical surface tensions (γ_c) of wetting for the given surface (i.e., the surface tension of a liquid which will just completely wet that surface). Zisman's group determined γ_c values for a number of "low energy" monolayers and polymers (Tables 1 and 2),[1] and proposed that liquids spread on a surface only when their surface tensions lie below γ_c; closely packed trifluoromethyl surfaces were found to be the most difficult to wet, i.e., they exhibit the lowest γ_c (Table 1). Substitution of just one of the fluorines of each trifluoromethyl group by a hydrogen atom more than doubles the γ_c value!

The length of the perfluoroalkyl segment and the nature of the polymer backbone impact the intrinsic repellencies of fluorochemicals. Critical surface tensions of films of fluoroacrylate homopolymers as a function of perfluoroalkyl side-chain length are shown in Figure 2; similar plots for monolayers of perfluoroalkanoic acids are included for comparison. For a given fluorochemical surface, a normal perfluoroalkyl chain length of ten carbons or more results in minimum γ_c values (maximum repellency), though decreases achieved by increasing chain lengths from seven to ten carbons are relatively small.

Monolayers of normal (straight-chain) perfluoroalkanoic acids have lower critical surface tensions than their isomeric perfluoroisoalkanoic acids,[3] as illustrated in Figure

Table 1.   Critical Surface Tensions of Polymeric
Fluorocarbon and Hydrocarbon Surfaces[a]

Surface assembly (monolayer)	γ_c (mN m^{-1})
CF_3 CF_3 CF_3 $\mid$ $\mid$ $\mid$ $(CF_2)_n$ $(CF_2)_n$ $(CF_2)_n$ $\mid$ $\mid$ $\mid$	6
CHF_2 CHF_2 CHF_2 $\mid$ $\mid$ $\mid$ $(CF_2)_n$ $(CF_2)_n$ $(CF_2)_n$ $\mid$ $\mid$ $\mid$	15
CH_3 CH_3 CH_3 $\mid$ $\mid$ $\mid$ $(CH_2)_n$ $(CH_2)_n$ $(CH_2)_n$ $\mid$ $\mid$ $\mid$	24

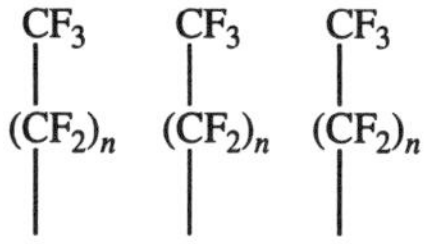

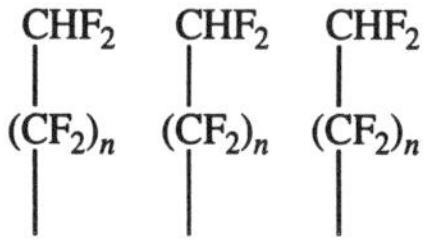

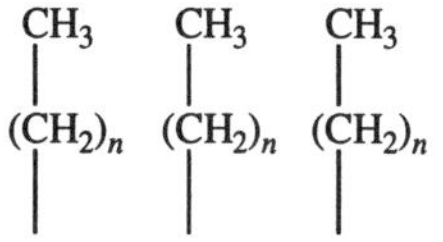

[a]See Ref. 1.

3. Similarly, films of a polymer with branched perfluorocarbon side chains should have higher γ_c values, presumably because they would not achieve the necessary close-packing density to inhibit molecular penetration of alkanes.[4]

Thus, polymers which manifest the highest intrinsic repellencies contain C_7 (or greater) *normal* perfluoroalkyl side chains attached to backbones which do not inhibit their close-packing. The low γ_c values of surfaces treated with such fluoropolymers provide repellency to oils as well as to water, a property not achievable with silicone or paraffinic finishes.

Table 2.   Critical Surface Tensions (γ_c) of Polymer Surfaces[a]

Polymer	Surface	γ_c (mN m^{-1})
Poly(hexamethylene adipamide)	$\sim\sim\sim$NHCO(CH$_2$)$_6$CONH$\sim\sim\sim$	46
Poly(ethylene terephthalate)	$\sim\sim\sim$CH$_2$CH$_2$O$_2$C–C$_6$H$_4$–CO$_2$CH$_2$CH$_2$$\sim\sim\sim$	43
Poly(ethylene)	$\sim\sim\sim$CH$_2$CH$_2$CH$_2$CH$_2$CH$_2$CH$_2$$\sim\sim\sim$	31
Poly(dimethylsiloxane)	CH$_3$ CH$_3$ CH$_3$ CH$_3$ $\mid$ $\mid$ $\mid$ $\mid$ $\sim\sim\sim$Si–O–Si–O–Si–O–Si–O$\sim\sim\sim$ $\mid$ $\mid$ $\mid$ $\mid$ CH$_3$ CH$_3$ CH$_3$ CH$_3$	24
Poly(tetrafluoroethylene)	$\sim\sim\sim$CF$_2$CF$_2$CF$_2$CF$_2$CF$_2$CF$_2$$\sim\sim\sim$	18.5

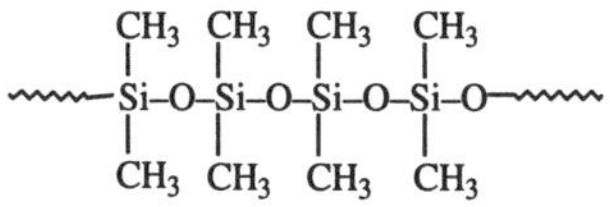

[a]See Ref. 1.

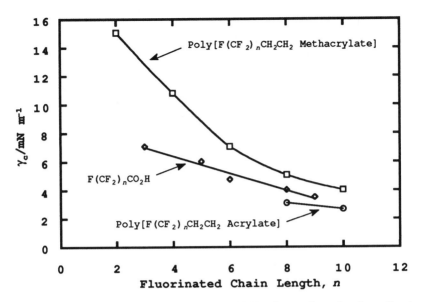

Figure 2. Effect of fluorinated chain length on the critical surface tensions of perfluoroalkanoic acid monolayers and films of acrylic fluoropolymers.[2] Note that the values for the perfluoroalkanecarboxylic acids differ from those reported earlier by Bernett and Zisman (see Figure 3).[3]

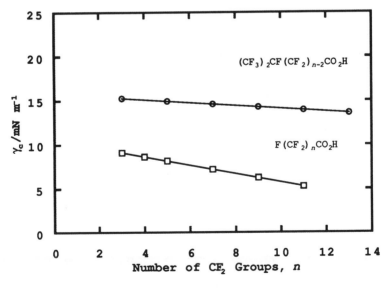

Figure 3. Effect of terminal branching on critical surface tensions of perfluoroalkanoic acid monolayers.[3]

14.1.2. Synthesis of Fluoroalkyl Intermediates

The most important commercial process for producing functionalized fluoroalkyl intermediates are based on electrochemical fluorination of hydrocarbon structures (see Chapter 5),[5] and telomerization of tetrafluoroethylene (TFE) with iodopentafluoroethane (see Chapter 8).[6] Another approach involves treating a difluorodihalogenomethane with copper (Chapter 8) followed by iodination,[7] but this is not yet significant commercially.

14.1.2.1. Electrochemical Fluorination

Simons electrochemical fluorination (ECF)[5] of alkanoic acid derivatives and alkanesulfonyl chlorides (or fluorides) in anhydrous hydrogen fluoride was commercialized by the 3M Company in 1958 as the first route to perfluoroalkyl intermediates for textile repellent-finishes (see Chapter 5). The process generates perfluorinated carboxylic (Eq. 1) or sulfonic acid fluorides (Eq. 2), which are subsequently converted to amides or other intermediates required for the production of repellents.

(1) $$C_7H_{15}\,COOH + 16\,HF \rightarrow C_7F_{15}COF + 16\,H_2$$

(2) $$C_8H_{17}SO_2F + 17\,HF \rightarrow C_8F_{17}SO_2F + 17\,H_2$$

14.1.2.2. Fluorotelomers

The term *fluorotelomer* arises from the most common synthetic route to commercial perfluoroalkyl compounds: telomerization of TFE with pentafluoroiodoethane (see Chapter 8). The chemistry and manufacturing process were developed by the DuPont Company[6] by analogy with the telomerization of TFE with trifluoroiodomethane reported by Haszeldine.[8]

Industrial preparation involves two steps: fluoroiodination of TFE to give pentafluoroiodoethane (Eq. 3) and subsequent reaction with TFE to give perfluoroalkyl iodide telomers (Eq. 4) of even carbon number (R_FI), sometimes referred to as *Telomer A*. Both steps are catalyzed by strong Lewis acids such as antimony or tantalum pentafluoride. The average telomer chain length is controlled via the TFE:C_2F_5I molar ratio, and conditions have been optimized in commercial processes to yield average carbon numbers ($2n + 2$) of 8 to 9, consistent with structural requirements for maximum intrinsic repellency (see Section 14.1.1).

(3) $$CF_2{=}CF_2 + I_2 + IF_5 \rightarrow C_2F_5I$$

(4) $$CF_3CF_2I + n\,CF_2{=}CF_2 \rightarrow CF_3CF_2(CF_2CF_2)_nI$$

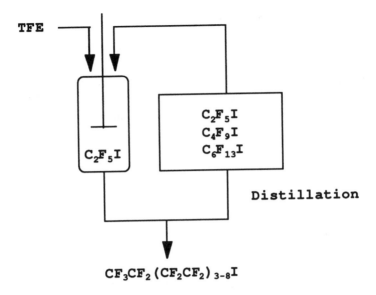

Figure 4.   Schematic process for the manufacture of perfluoroalkyl iodides.

14.1.2.3.   Perfluoroalkyl Iodide Process

TFE Telomerization with pentafluoroiodoethane (C-2 Iodide) is highly exother-
mic ($\Delta H = 192$ kcal mol^{-1}) and requires a large excess of an inert solvent or the starting
iodide as a heat-transfer medium.[9] The reaction is typically carried out under a positive
pressure, and the excess of C-2 iodide plus the C-4 and some of the C-6 iodide are
removed by distillation and recycled (Figure 4). Typical distributions of the remaining
homologous mixtures of normal perfluoroalkyl iodides (average C-8 to C-9) are shown
in Table 3. Two types of undesired side products are encountered in this route to
long-chain perfluoro-n-alkyl iodides (Eqs. 5 and 6) which fortunately are easily
removed by distillation.

Table 3.   Compositions of Mixtures Produced by C_2F_5I–TFE
Telomerization (wt %)

	Three typical telomer distributions[a]		
Perfluoro-n-alkyl iodide	Mix 1	Mix 2	Mix 3
C_4F_9I	5	4	—
$C_6F_{13}I$	50	35	6
$C_8F_{17}I$	30	30	50
$C_{10}F_{21}I$	10	17	30
$C_{12}F_{25}I$	4	8	10
$\geq C_{14}F_{29}I$	1	6	4

[a]Obtained by varying the relative amounts of C_2F_5I, C_4F_9I, and $C_6F_{13}I$ used as
starting material in the telomerization (see Figure 4).

(5)
$$R_FI + IF_5 \rightarrow R_FF + (I_2F_4)$$

(6)
$$2\,R_FI \rightarrow R_FR_F + I_2$$

14.1.2.4. Reactions of Telomer Iodides

Perfluoroalkyl iodides (R_FI)) do not by themselves have significant direct commercial applications; rather, they are converted to other intermediates, notably through the characteristic ease with which homolytic cleavage of the CF_2–I bond occurs and allows the thermal insertion of ethylene (Eq. 7) to yield $1H,1H,2H,2H$-1-iodopolyfluoroalkanes (perfluoroalkylethyl iodides referred to as *Telomer B*) (see also Chapters 8 and 27).[10] This is a crucial commercial step toward the introduction of a host of various nonfluorinated structural features. Under the conditions typically employed, a single molecule of ethylene is inserted per C–I bond; multiple ethylenations are possible, but require more forcing conditions.[10,11]

(7)
$$C_2F_5(CF_2CF_2)_nI + CH_2{=}CH_2 \rightarrow C_2F_5(CF_2CF_2)_nCH_2CH_2I$$

Additions of perfluoroalkyl iodides to substituted olefins[10–12] (Eqs. 8 and 9), aromatics[13] (Eq. 10), and to alkynes[14] (Eq. 11) are possible, but have so far found limited commercial use. They do not react with nucleophiles such as OH^- or NH_3, but will participate in organometallic-mediated transformations, including halogen exchange[15] (Eq. 12), carboxylation[15,16] (Eq. 13), and coupling[17] (Eq. 14). For further information, see Chapter 8.

(8)
$$R_F{-}I + H_2C{=}CHR \rightarrow \underset{\text{major}}{R_FCH_2CHRI} + \underset{\text{minor}}{R_FCHR{-}CH_2I}$$

(9)
$$R_F{-}I + CH_2{=}CHR_F \rightarrow R_FCH_2CHIR_F$$

(10)
$$R_FI + PhI \xrightarrow{Cu} PhR_F + CuI$$

(11)
$$R_F{-}I + HC{\equiv}CH \rightarrow \underset{cis+trans}{R_FCH{=}CHI}$$

(12)
$$C_8F_{17}I + Br_2 \xrightarrow{300°C} C_8F_{17}Br + IBr$$

(13)
$$R_FI + CO_2 + Zn\text{-}Cu \xrightarrow{\text{(aq.)HCl}} R_FCO_2H$$

(14)
$$2\,R_FI + Zn \xrightarrow{\text{acetic anhydride}} R_FR_F + ZnI_2$$

Dehydroiodination of Telomer B iodides is easily achieved, producing synthetically versatile perfluoroalkylethylenes (*Telomer B olefins*); for example, telomer nitriles,[18] silanes,[19] and epoxides,[20] respectively, can be prepared by hydrocyanation, hydrosilylation, and epoxidation of these olefins (Scheme 1). Displacement of iodine from Telomer B iodides with soft nucleophiles occurs with sulfides[21] and thiocyanates,[21] for example (Scheme 1). Perfluoroalkylethylmercaptans are thereby manufactured commercially for use as intermediates in repellent products and fluorosurfactants.

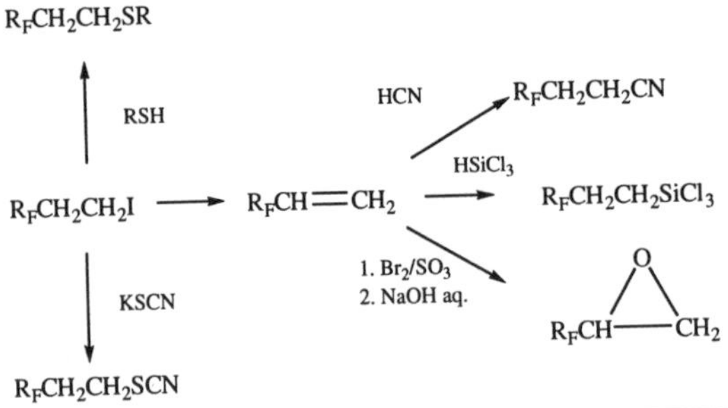

$$R_FCH_2CH_2SR$$

Scheme 1. Some conversions of Telomer B [(β-perfluoroalkyl)ethyl iodides] and Telomer B olefins.

14.1.2.5. Telomer Alcohols and Their Derivatives

Fluorotelomer alcohols and their esters are the most versatile and widely utilized types of fluorotelomer intermediate in use today. Perfluoroalkylethyl alcohols are prepared commercially from perfluoroalkylethyl iodides. The specific reagents employed and intermediates involved vary widely,[22] and are chosen to minimize elimination (loss of HI) as a side reaction. A commercial oleum process involves initial formation of sulfate anhydrides, which are subsequently hydrolyzed to produce the alcohol (Eq. 15). Similarly, hydrolysis in the presence of dimethylformamide or N-methylpyrrolidone involves the intermediacy of solvent adducts (Eq. 16).[23]

(15) $$R_FCH_2CH_2I \xrightarrow{SO_3} [(R_FCH_2CH_2OSO_2)_2O] \xrightarrow{H_2O} R_FCH_2CH_2OH$$

(16) $$R_FCH_2CH_2I \xrightarrow[H_2O]{DMF} R_FCH_2CH_2OC(O)H \rightarrow R_FCH_2CH_2OH$$

Once formed, these fluorotelomer alcohols with at least 2 carbons separating the perfluoroalkyl segment and the OH function, undergo reactions similar to those of their hydrocarbon analogs. Thus, they are readily esterified with carboxylic acids or acid chlorides, or transesterified with lower alcohol esters of carboxylic acids, and also react readily with isocyanates to give fluorocarbamates. Formation of esters directly from perfluoroalkylethyl iodides has also been reported.[24] Perfluoroalkyl carbinols (R_FCH_2OH) and other perfluoroalkylalkanols are known, but have far less commercial significance than perfluoroalkylethanols.[25] Iodohydrins are made by addition of perfluoroalkyl iodides to allyl alcohol or its derivatives (Eq. 17), and are used as intermediates in the formation of telomer epoxides.[26]

(17) $$R_FI + CH_2{=}CHCH_2OH \rightarrow R_FCH_2CH(I)CH_2OH \rightarrow R_FCH_2\overset{O}{\overset{\triangle}{CH\text{-}CH_2}}$$

14.1.2.6. Physical Properties of Telomer Derivatives

Melting and/or boiling points of selected homologs of commercial telomer derivatives are presented in Table 4. As the intermediates are not perfluorinated, they

Table 4. Some Physical Constants of Telomer Derivatives

$F(CF_2)_nX$	Boiling point (°C) (mm Hg)			
X	$n = 4$	$n = 6$	$n = 8$	$n = 10$
I	67 (760)	118 (760)^a	163 (760)^b	203 (760)^c
CH_2CH_2I	139 (760)	183 (760)	124 (50)	281 (760)
$CH{=}CH_2$	56 (760)	105 (760)	149 (760)	193 (760)
CH_2CH_2OH	155 (760)	182 (760)	213 (760)^d	144 (50)^e
$CH_2CH_2OC(O)CH{=}CH_2$	—	65 (20)	65 (0.7)	Solid
$CH_2CH_2OC(O)C(CH_3){=}CH_2$	70 (9)	96 (10)	90 (2)	104 (1)
$\overset{\text{O}}{\overbrace{CH_2CH{-}CH_2}}$	86 (150)	87 (50)	113 (45)	—
CO_2H	—	104 (50)^f	112 (28)^f	109 (5)^g

^{a}mp = –45; ^{b}mp = 21; ^{c}mp = 66; ^{d}mp = 44; ^{e}mp = 94; ^{f}solid at room temperature; ^{g}mp = 102°C.

exhibit limited thermal stability, typically to about 200°C. At high temperature, thermal elimination of HF and oxidation occurs, though the intermediates are nonflammable. Densities of the fluorochemical intermediates typically fall in the range of 1.5–1.8 g cm^{-3}.

14.1.3. Synthesis of Fluorochemical Repellents

Until the mid-50s, paraffinic and hydrocarbon siloxane products dominated much of the post-war market for repellent outerwear. However, their soiling tendencies toward oily material and their poor durability to dry cleaning were found to be objectionable. Reviews of the technological progress of repellent technology have been written by Harding and Kissa.[27,28]

14.1.3.1. Monomeric Fluorochemical Repellents for Textiles

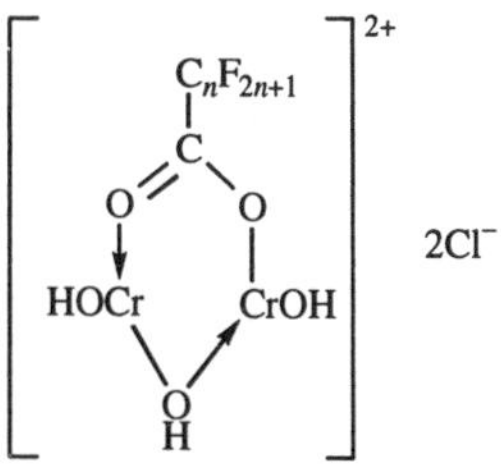

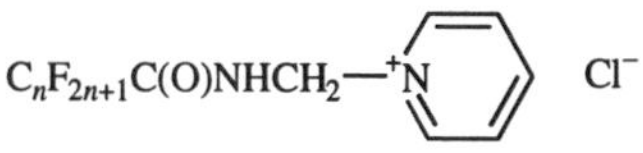

By analogy with the monomeric structures once characteristic of hydrocarbon repellent finishes,[27] the first commercial fluoroalkylated textile repellents were monomeric and self-dispersing in water. Werner-type chromium complexes of perfluoroalkanoic acids (1),[29] zirconium perfluoroalkanecarboxylates (2),[30] and pyridinium (3) or other ammonium derivatives[31] exhibited good oil and water repellencies, but lacked durability to laundering and dry cleaning. This deficiency prompted the investigation of polymeric fluoroalkylated materials.

14.1.3.2. Polymeric Fluorochemical Repellents for Textiles

Polymeric fluorochemical repellent finishes have set performance standards in repellency and durability. Those in wide commercial use today are solution or emulsion copolymers of perfluoroalkyl-containing acrylate or methacrylate esters. The specific structures of the fluoromonomers vary widely, depending upon the fluorochemical intermediate source. The most prevalent fluoromonomers for textile repellent finishes are perfluoroalkylethyl acrylates [$R_FCH_2CH_2OC(O)CH{=}CH_2$], methacrylates [$R_FCH_2CH_2OC(O)C(CH_3){=}CH_2$], and sulfonamides of the type $R_FSO_2N(R)CH_2CH_2OC(O)CH{=}CH_2$ (R = CH_3, CH_3CH_2). Nonfluorinated co-monomers serve the functions of extending fluorine efficiency, providing functionality to enhance durability, and/or modifying the hand of the treated fabric. The monomers are typically co-polymerized by standard emulsion polymerization processes, using cationic or nonionic surfactants.[32] In some cases, solution polymerizations are conducted in organic solvents such as 1,1,1-trichloroethane.

Fluorinated polyurethanes are also coming into use, particularly for the treatment of 100% cotton and cotton-blend textiles, and are based on the reaction of fluorinated diol intermediates with diisocyanates.[33]

Representative examples of fluoropolymer repellents are provided in a previous review of patents on the subject by Kissa.[28]

Fluorinated textile finishes are usually applied in combination with hydrocarbon water-repellent adjuvants,[34] resins, and/or monomeric fluorochemicals that demonstrate some synergistic effect.[35] Application baths and test methods used by the industry (US and abroad) have also been reviewed by Kissa.[28]

14.1.3.3. Carpet Soil-Resistant Finishes

Fluorochemical finishes are used on carpets to reduce the accumulation of particulate soil transferred from shoe soles, and to resist staining by water- or oil-borne staining components. Soiling studies have established that adhesion of soil particles to the fiber is the dominant soiling mechanism (by van der Waals forces).[36] Residual lubricant on carpet fiber is known to contribute heavily to soil adhesion.[37] Thus, fluorochemicals impart repellency by lowering the critical surface tensions of fiber surfaces, hence reducing adhesion and limiting wicking of water-borne or oily substances into the yarn bundle. The area soiled/stained is thereby kept to a minimum. Fluorochemicals further retard the reappearance of soiled spots after cleaning. However, repellency alone is not a sufficient feature in providing resistance to particulate soil. The coating hardness, which impacts the area of contact with the soil particle, is also critical. A soft, deformable coating has a higher contact area with soil particles than a hard brittle one. Commercial soil-resistant finishes have been developed with these critical features in mind. Polymer hardness is carefully controlled by cross-linking, or by the use of organic or inorganic hardeners.

Fluorinated intermediates with hydroxyl functions are the most useful intermediates for carpet soil-resistant chemicals. Polyesters or polyurethanes are prepared from fluorotelomer alcohols (Section 14.1.2.5), and subsequently dispersed in water using anionic, cationic, or nonionic surfactants. A survey of typical compositions of soil-resistant agents is available.[28]

14.1.3.4. Stain-Resistant Nylon Carpet

The use of nylon fibers in carpets has been preferred because of their resilience, wearing characteristics, dyeability, and esthetics. The drawback has been their lack of soil and stain resistance. Fluorochemical finishes improve resistance to staining by reducing wetting of the yarn by oil- or water-borne stains. However, on carpets with fluorochemical treatments alone, staining agents must be removed rapidly to avoid chemisorption or "dyeing" of fibers.

In 1986, DuPont commercialized technology for nylon carpets that essentially eliminates chemisorption, in particular of acid dyes, and greatly facilitates removal of stains, even on long standing. Acid dyes are the most frequently encountered, and the more-difficult-to-remove colorants present in food, beverages, cosmetics, and medicines. The new technology, generically referred to as stain blocking, requires, in addition to fluorochemical finishes, the co-application or post-dye finishing of nylon carpets with an anionic hydrocarbon polymer designed to adsorb and slowly diffuse into the nylon filaments to block dye sites. Low-molecular-weight phenol-formaldehyde polymers (syntans) were the first stain-blocking agents to be commercialized.[38] Styrene-maleic acid copolymers,[39] and also polymethacrylic acid copolymers,[40] are reported to be effective stain blockers. Structures of products and postulated mechanisms for stain blocking are presented in several recent reviews.[41] Fluorochemical finishes continue to be used on stain-resistant nylon carpet to provide the necessary resistance to soiling, and to promote resistance to staining by dispersed or oil-based colorants.

14.1.4. Soil-Release Finishes

In contrast with repellent finishes that function by imparting a barrier to wetting of textile substrates, soil-release finishes are hydrophilic and facilitate removal of fatty or oily soils containing solid matter from fabric. An excellent review of the soil-release mechanism and structural requirements is available.[42] Though the vast majority of commercial textile soil-release finishes are hydrocarbons, fluorinated soil-release finishes have proven more effective for cotton and synthetic-cotton blended fabric.

Smith and Sherman pointed out that a soil-release finish should: (a) reduce the surface energy of fibers and prevent spontaneous spreading; (b) increase the hydrophilicity of fibers; and (c) form a uniform thin film on the fibers.[43] The opposing effects of (a) and (b) have been resolved by designing hybrid block co-polymers containing hydrophilic and hydrophobic segments. It is proposed that hybrid fluoropolymers assume different orientations in water or at the air interface. In air, the perfluoroalkyl segments dominate the treated fabric surface, while in water the hydrophilic segments dominate the surface.[44]

A review of patents dealing with fluorinated soil-release polymers is included in Kissa's account on soil-release finishes.[42] The technology spans polyacrylic, polyurethane, and polyester block co-polymers containing perfluoroalkyl and poly(ethylene glycol) segments.

14.1.5. Future Developments

A great deal of industrial research continues to be devoted to the development of improved fluorochemical finishes for textiles. At the heart of the effort is the search for new fluorinated intermediates that impart unique properties to the final finish. Perfluoroalkyl-containing organic amines, silanes, epoxides, and difunctional intermediates offer numerous new synthetic possibilities. Also, many new applications apart from textile finishes are being developed for fluorinated intermediates. Though water-repellent finishing is a mature technology, research is continuing to improve the repellents and application processes to ease pressure on their cost. A soft, supple hand while maintaining oil and water repellency is desirable for the apparel market segments. Fluorochemicals will remain the most effective and durable repellents. Silicones and hydrocarbons will share the market where cost is the overriding factor.

Finishing of carpets with fluorochemicals for dry-soil protection will continue, but strong pressures on cost reduction are likely. As polypropylene carpets gain market share at the expense of nylon, fluorochemicals added during the extrusion process will become more and more prevalent. Combinations of fluorochemical finishes and hydrocarbon stain-resistant finishes applied simultaneously may become the norm.

Foam finishing with fluorochemicals is gaining acceptance over spray and exhaust methods because of lower energy demand and environmental impact.

Finally, fluorinated soil-release finishes are trending toward increased durability, efficiency, and improved anti-redeposition effects. As new durable press resins replace the current formaldehyde based resins, a shift in soil-release technology is likely.

14.1.6. Major Manufacturers

Fluorochemical intermediates and industrial products derived from them are sold under the tradenames of Zonyl®, Zepel®, and Teflon® by DuPont, Fluorad®, Scotchgard®, and Scotchrelease® by 3M, Asahigard® by Asahi Glass, Milease® by ICI, Lodyne® and Oleophobal® by Ciba Geigy, Forafac® by Atochem, and Fluowet® and Nuva® by Hoechst.

14.2. FLUOROSURFACTANTS

The hydrophobicity of linear or branched alkanes has long been used as a basis for surfactants. Perfluoroalkanes have the unique property of being both hydrophobic and oleophobic. As one would therefore expect, molecules with discrete perfluoroalkyl and hydrophilic segments reduce aqueous surface tensions. In fact, surfactants containing fluoroalkyl segments are much more surface active than their hydrocarbon analogs (see Chapter 3).[45–48] Surface tensions of aqueous solutions in the range of 15–25 dyn cm^{-1} are common at concentrations of 0.005 to 0.10% by weight. The lipophobicity of the perfluoroalkyl segment also enables surface-tension lowering of organic solvents.[49]

Fluorosurfactants can reach extremes in surfactancy behavior (detergency, wetting, foaming, emulsifying, dispersing) unattainable with nonfluorinated surfactants. In addition, their stability against acidic, alkaline, oxidative and reductive reagents, and to elevated temperatures permit special industrial applications.

14.2.1. Fluorosurfactant Synthesis

Fluoroalkyl intermediates, available from electrochemical fluorination or telomer chemistries (Section 14.1.2), serve as precursors to commercially prepared fluorosurfactants.[50] The hydrophilic portion is altered to attain specific properties. Commercial synthetic pathways to cationic, anionic, amphoteric, and nonionic fluorosurfactants from telomer intermediates are shown in Scheme 2. Related surfactant compositions can be prepared from the perfluoroalkanecarboxylic acid and perfluoroalkanesulfonyl fluoride intermediates acquired via electrochemical fluorination (Scheme 3).

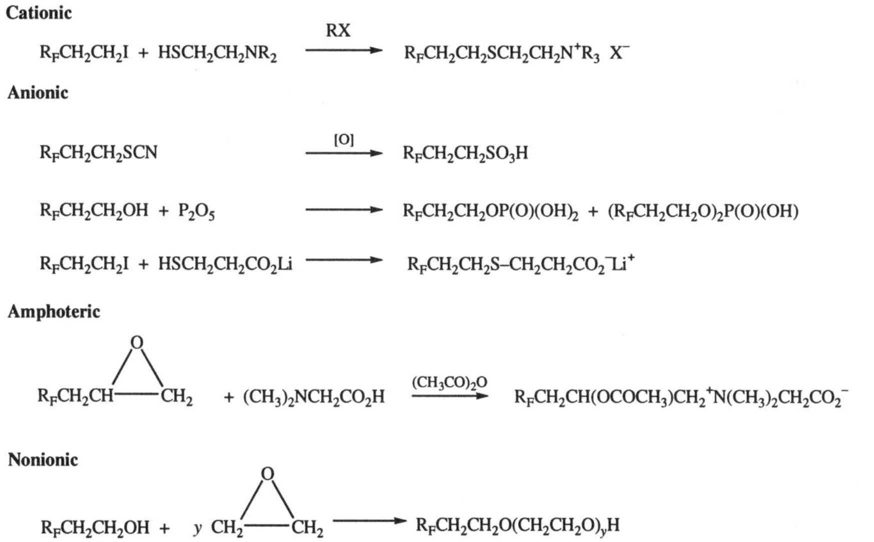

Cationic

$$R_FCH_2CH_2I + HSCH_2CH_2NR_2 \xrightarrow{RX} R_FCH_2CH_2SCH_2CH_2N^+R_3\ X^-$$

Anionic

$$R_FCH_2CH_2SCN \xrightarrow{[O]} R_FCH_2CH_2SO_3H$$

$$R_FCH_2CH_2OH + P_2O_5 \longrightarrow R_FCH_2CH_2OP(O)(OH)_2 + (R_FCH_2CH_2O)_2P(O)(OH)$$

$$R_FCH_2CH_2I + HSCH_2CH_2CO_2Li \longrightarrow R_FCH_2CH_2S-CH_2CH_2CO_2^-Li^+$$

Amphoteric

$$R_FCH_2CH\overset{O}{\overbrace{}}CH_2 + (CH_3)_2NCH_2CO_2H \xrightarrow{(CH_3CO)_2O} R_FCH_2CH(OCOCH_3)CH_2^+N(CH_3)_2CH_2CO_2^-$$

Nonionic

$$R_FCH_2CH_2OH + y\ CH_2\overset{O}{\overbrace{}}CH_2 \longrightarrow R_FCH_2CH_2O(CH_2CH_2O)_yH$$

Scheme 2. Routes to fluorinated surfactants from fluorotelomer intermediates.

Cationic

$$R_FSO_2F + H_2N(CH_2)_nNR_2 \rightarrow R_FSO_2NH(CH_2)_nNR_2$$

$$R_FSO_2NH(CH_2)_nNR_2 + R'X \rightarrow R_FSO_2NH(CH_2)_n{}^+NR_2R'X^-$$

Anionic

$$R_FC(O)F + H_2O \rightarrow R_FCO_2H$$

$$R_FCO_2H + MOH \rightarrow R_FCO_2^-M^+$$

$$R_FSO_2F + MOH \rightarrow R_FSO_3^-M^+$$

(M = Na, K, NH$_4$, etc.)

Amphoteric

$$R_FC(O)NH(CH_2)_nNR_2 + \overline{CH_2C(O)OCH_2} \rightarrow$$
$$R_FC(O)NH(CH_2)_n{}^+NR_2CH_2CH_2CO_2^-$$

$$R_FSO_2NH(CH_2)_nNR_2 + ClCH_2CH_2OH \rightarrow R_FSO_2N(CH_2CH_2OH)(CH_2)_nNR_2$$

$$R_FSO_2N(CH_2CH_2OH)(CH_2)_nNR_2 + \overline{CH_2C(O)OCH_2} \rightarrow$$
$$R_FSO_2N(CH_2CH_2OH)(CH_2)_n{}^+NR_2CH_2CH_2CO_2^-$$

Nonionic

$$R_FSO_2NHR + ClCH_2CH_2OH \rightarrow R_FSO_2N(R)CH_2CH_2OH$$

$$R_FSO_2N(R)CH_2CH_2OH + m\ CH_2\overset{O}{-}CH_2 \rightarrow$$
$$R_FSO_2N(R)CH_2CH_2O(CH_2CH_2O)_mH$$

Scheme 3. Routes to fluorosurfactants based on acid fluorides produced by the Simons ECF Process.

14.2.2. Aqueous Solutions of Fluorosurfactants

The fluoroalkyl portions of fluorosurfactants give them an extreme tendency to orient at air interfaces because of their low affinity for water, and the low interaction between adjacent fluorocarbon chains. Consequently, fluorosurfactants lower the surface tensions of aqueous solutions to a greater extent than do hydrocarbon surfactants. Surface tension versus concentration curves for a number of pure surfactants have been published.[45–47]

Commercial surfactants derived from telomer intermediates typically contain homologous mixtures. Since the critical micelle concentration (CMC) of each component of a homologous mixture is different (a function of chain length), a homologous mixture does not manifest a distinct CMC when solution surface tension is plotted as a function of the logarithm of the concentration. Surface tension plots of representative

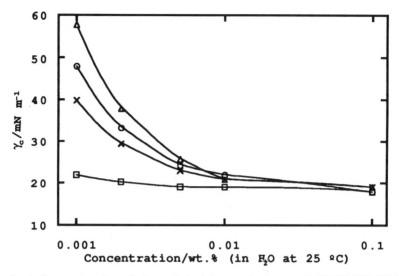

Figure 5. Surface-tension plots of telomer-derived fluorosurfactants: △, $R_FCH_2CH_2SCH_2CH_2N^+$-$(CH_3)_3 CH_3SO_4^-$; ○, $R_FCH_2CH_2SCH_2CH_2CO_2Li$; ×, $R_FCH_2CH(OCOCH_3)CH_2N^+(CH_3)_2CH_2CO_2^-$; □, $R_FCH_2CH_2O(CH_2CH_2O)_yH$.

telomer derived fluorosurfactants are shown in Figure 5. For pure homologs, data listed in Table 5 demonstrate general trends of decreasing surface tensions with increasing chain length and the effects of gegenions.[48]

The temperature above which a hydrated solid surfactant melts and dissolves as micelles in water is defined as the Krafft point of the surfactant.[51] The Krafft points of fluorinated surfactants are generally high, so those most efficient at room temperature are mainly limited to chain lengths C_7–C_8. Above the Krafft point the solubility in water increases abruptly.

Table 5.   Surface Tension Minima in Aqueous
Solutions at 25°C

Fluorosurfactant	Minimum aqueous surface tension (mN m^{-1})	Ref.
n-$C_6F_{13}CO_2Li$	27.8	47
n-$C_8F_{17}CO_2Li$	24.6	47
n-$C_{10}F_{21}CO_2Li$	20.5	47
n-$C_7F_{15}CO_2Na$	24.6	46
n-$C_8F_{17}CO_2Na$	21.5	47
n-$C_8F_{17}CO_2K$	20.7	47
n-$C_8F_{17}SO_3Li$	29.8	46
n-$C_8F_{17}SO_3Na$	40.5	46
n-$C_8F_{17}SO_3K$	34.5	46
$(CF_3)_2CF(CF_2)_4CO_2K$	19.5	46

14.2.3. Properties and Uses

Fluorosurfactants are available in solid, liquid, or paste form, as concentrates or as aqueous (and, more rarely, organic) solutions. In pure form the liquids have densities characteristic of fluorochemicals (1.4 to 1.8 g cm^{-3}). Some have limited solubility in polar organic solvents, such as acetone, methyl isobutyl ketone, ethyl acetate, and low-molecular-weight alcohols.

Their effectiveness at low concentrations is the key to their industrial utility, as they are considerably more expensive on a per-pound basis than nonfluorinated surfactants. Nonetheless, on a per-application basis they can be considerably less expensive, especially when performance extremes demand their capabilities, or when the harshness of the environment obviates the use of nonfluorinated alternatives.

Fluorosurfactants are useful as emulsifying and dispersing agents where the dispersed phase is a fluorocarbon. For example, perfluoro-octanoic acid is used as surfactant in the emulsion polymerization of tetrafluoroethylene (see Chapter 15).[52]

Leveling is a special case of surface tension reduction and a particularly important property for aqueous-based coating compositions. Fluorosurfactants are used in floor polishes and latex paints at concentrations of 50 to 150 ppm to minimize cratering and peeling. They are used in some water-based adhesive compositions to improve leveling, and to impart semirelease characteristics.[53,54]

Many ink formulations contain fluorosurfactants to enhance ink flow and leveling, to improve cylinder life, and to eliminate snowflaking or nonuniform printing.[55,56]

When surfactants are required in harsh environments, fluorosurfactants are extremely useful. They are added, for example, to zinc bromide battery analyte to prevent zinc dendrites formation, for scale removal during the pickling process of metals, and in flotation processes in mining operations for uranium recovery.

Selected fluorosurfactants added to hair-conditioning formulations enhance wet combing and render hair oleophobic.[57]

There are several manufacturers of fluorosurfactants: Zonyl® (DuPont), Fluorad® (3M), Surflon® (Asahi Glass), Atsurf-F® (ICI), Lodyne® (Ciba-Geigy), Forafac® (Atochem), and Fluowet® (Hoechst).

ACKNOWLEDGMENT

The authors thank Drs. Robert H. Dettre and K. Spencer Prowse for valuable discussions and expertise.

14.3. REFERENCES

1. W. A. Zisman, *Advances in Chemistry*, ACS Symposium Series No. 43 (F. M. Fowker and W. A. Zisman, eds.), p. 1, American Chemical Society, Washington, D.C. (1963); E. G. Shofrin and W. A. Zisman, in: *ibid.*, p. 145: A. G. Pitmann, in: *Fluoropolymers*, High Polymers Vol. XXV (L. A. Wall, ed.), p. 381, Wiley, New York (1972).
2. R. E. Johnson, Jr. and R. H. Dettre, *Polym. Prepr., Am. Chem. Soc., Div. Polym. Chem. 28*, 48 (1987).
3. M. K. Bernett and W. A. Zisman, *J. Phys. Chem. 71*, 2075 (1967).

4. C. O. Timmons and W. A. Zisman, *J. Colloid Interface Sci.* 22, 165 (1966).

5. I. N. Rozhov, in: *Organic Electrochemistry*, 2nd edn. (M. M. Baizer and H. Lund, eds.), p. 805, Marcel Dekker, New York (1983); W. V. Childs, in: *Techniques of Electroorganic Synthesis*, Part III, p. 341, Wiley, New York (1982); A. J. Rudge, in: *Industrial Electrical Processes* (A. T. Kuhn, ed.), p. 71, Elsevier, Amsterdam (1971).

6. R. E. Parsons, U.S. Patent 3,234,294 (to DuPont) [*CA 62*, 13045 (1965)]; R. E. Parsons, U.S. Patent 3,132,185 (to DuPont) [*CA 61*, 1755 (1964)]; W. A. Blanchard and J. C. Rhode, U.S. Patent 3,226,449 (to DuPont) [*CA 64*, 8031 (1966)].

7. D. J. Burton, D. M. Wiemers, and J. C. Easden, U.S. Patent 4,749,802 (to Univ. of Iowa Research Foundation) [*CA 105*, 97696 (1986)].

8. R. N. Haszeldine, *J. Chem. Soc.* 2856 (1949).

9. S. Mizaki, T. Kamibukigoshi, and M. Suefuji, Japanese Patent 58-192837 (to Daikin Kogyo Co.) [*CA 100*, 102733 (1984)].

10. N. O. Brace, U.S. Patent 3,145,222 (to DuPont) [*CA 61*, 10589 (1964)]; N. O. Brace, U.S. Patent 3,016,406 (to DuPont) [*CA 57*, 2078 (1962)]; A. N. Alekseenko, V. P. Nazaretyan, A. Ya. Il'Chenko, and L. M. Yagupol'skii, *J. Org. Chem. USSR, (Engl. Transl.)* 24, 2130 (1988).

11. N. O. Brace, *J. Org. Chem.* 27, 4491 (1962); *J. Org. Chem.* 27, 3033 (1962); *J. Org. Chem.* 49, 2361 (1964); G. V. D. Tiers, U.S. Patent 2,951,051 (to 3M Co.) [*CA 55*, 5347 (1961)]; K. Von Werner, *J. Fluorine Chem.* 28, 229 (1985).

12. F. Jeanneaux, M. le Blanc, A. Cambon, and J. Guion, *J. Fluorine Chem.* 4, 261 (1974); K. Von Werner, German Patent DE 3,338,299 (to Hoechst) [*CA 103*, 214858 (1985)].

13. V. C. R. McLoughlin and J. Thrower, *Tetrahedron* 25, 5921 (1969); Y. Kobayashi and I. Kumadaki, *Tetrahedron Lett.* 4095 (1969); P. L. Coe and N. E. Milner, *J. Organomet. Chem.* 39, 395 (1972).

14. Q. Chen and Z. Yong, *J. Chem. Soc., Chem. Commun.* 498 (1956); Y. Takeyama, Y. Ichinose, K. Oshima, and K. Utimoto, *Tetrahedron Lett.* 30, 3159 (1989).

15. M. Hudlicky, *Chemistry of Organic Fluorine Compounds*, 2nd edn., Ellis Horwood, Chichester (1976); R. Chambers, *Fluorine in Organic Chemistry*, Wiley, New York (1973).

16. J. J. Blancou, P. Moreau, and A. A. Commeyras, *J. Chem. Soc., Chem. Commun.* 885 (1976).

17. J. F. Harris, U.S. Patent 3,048,569 (to DuPont) [*CA 57*, 16887 (1962)].

18. S. D. Ittel, A. E. Feiring, C. C. Cumbo, R. J. McKinney, and D. R. Anton, U.S. Patent Appl. CR-8414 (by DuPont).

19. T. D. Talcott, U.S. Patent 3,006,878 (to Dow Chemical Corp.) [*CA 56*, 10374 (1962)]; German Patent 1,445,324 (to Dow Corning Corp.) (1966).

20. C. G. Krespan, U.S. Patent Appl. CR-8816 (by DuPont).

21. J. T. Barr and F. E. Lawlor, U.S. Patent 2,894,991 (to Atochem North America) [*CA 55*, 2572 (1961)].

22. W. V. Cohen, U.S. Patent 3,017,421 (to DuPont) [*CA 56*, 15366 (1962)]; A. H. Ahlbrecht, U.S. Patent 3,171,861 (to 3M) [*CA 64*, 11089 (1966)]; L. H. Beck, U.S. Patent 4,618,731 (to DuPont) [*CA 110*, 212120 (1989)].

23. N. O. Brace, *J. Fluorine Chem.* 31, 151 (1986) and references cited therein.

24. K. C. Smeltz, U.S. Patent 3,504,016 (to DuPont) [*CA 72*, 122867 (1970)]; H. Ukihashi, T. Uchino, T. Hayashi, and R. Aihara, *Chem. Engineering (Japan)* 42, 330 (1978).

25. A. H. Ahlbrecht and D. R. Husted, U.S. Patent 2,666,797 (to 3M Co.) [*CA 49*, 1772 (1955)]; A. H. Ahlbrecht, D. R Husted, and T. S. Reid, U.S. Patent 2,642,416 (to 3M Co.) [*CA 48*, 5880 (1954)].

26. S. E. Krahler, U.S. Patent 4,489,006 (to DuPont) [*CA 102*, 221011 (1985)].

27. T. R. Harding, *J. Text. Inst.* 42, 691 (1951).

28. E. Kissa, in: *Handbook of Fiber Science and Technology: Vol. II, Chemical Processing of Fibers and Fabrics Functional Finishes, Part B* (M. Lewin and S. B. Sello, eds.), Chapter 2, p. 143, Marcel Dekker, New York (1984).

29. R. J. Pavlin, *Tappi* 36, 107A (1953); R. K. Iler, *Ind. Eng. Chem.* 46, 766 (1954).

30. E. B. Higgins, in: *Waterproofing and Water Repellency* (J. C. Moilliet, ed.), p. 188, Elsevier, Amsterdam (1963); E. B. Higgins, *Text. Inst. Ind.* 4, 255 (1966); W. B. Blumenthal, *Ind. Eng. Chem.* 42, 640 (1950).

31. R. J. W. Reynolds, E. E. Walker, and C. S. Woolvin, British Patent 466,817 [*CA 31*, 8195 (1937)]; F. V. Davis, *J. Soc. Dyers Colour. 63*, 260 (1947).

32. A. H. Ahlbrecht and H. A. Brown, U.S. Patent 2,841,573 (to 3M Co.) [*CA 52*, 19945 (1958)]; R. W. Fasik and S. R. Raynolds, U.S. Patent 3,282,905 (to DuPont) [*CA 66*, 85490 (1967)]; E. K. Kleiner and P. L. Pacini, U.S. Patent 3,497,575 (to Geigy Chemical Corp.) [*CA 72*, 80281 (1970)].

33. R. A. Falk, K. P. Kirkland, A. Karydas, and M. Jacobson, U.S. Patent 4,898,981 (to Ciba-Geigy) [*CA 112*, 217687 (1990)]; R. E. A. Dear and R. A. Falk, U.S. Patent 3,935,277 (to Ciba-Geigy) [*CA 83*, 192559 (1975)].

34. H. B. Goldstein, *Text. Res. J. 31*, 377 (1961).

35. C. G. DeMario, A. J. McQuade, and S. J. Kennedy, *Modern Textile Magazine 41*, 50 (1960).

36. V. Kling and H. Mahl, *Melliand Textilber 35*, 640 (1954); E. Kissa, *Text. Res. J. 43*, 86 (1973).

37. J. Compton and W. J. Hart, *Ind. Eng. Chem. 43*, 1564 (1951).

38. C. C. Cook, *Rev. Prog. Coloration 12*, 73 (1982); R. C. Blyth and P. A. Ucci, U.S. Patents 4,592,940 [*CA 108*, 152107 (1988)] and 4,680,212 [*CA 108*, 114161 (1988)] (to Monsanto Co.); I. Greschleri, C. P. Malone, and A. P. Zinnato, U.S. Patent 4,780,099 (to DuPont) [*CA 108*, 7466 (1988)].

39. P. H. Fitzgerald, N. S. Rao, Y. V. Vinod, and J. R. Alender, U.S. Patent 4,948,650 (to DuPont) [*CA 114*, 104307 (1991)].

40. T. H. Moss, III, R. R. Sargent, and M. S. Williams, U.S. Patent 4,940,757 (to Peach State Labs) [*CA 113*, 154244 (1990)]; J. C. Chang, M. H. Olson, and I. A. Muggli, U.S. Patent 4,937,123 (to 3M Co.) [*CA 112*, 120560 (1990)].

41. T. F. Cooke, *TRI Report No. 31* (August 1989); P. W. Harris and D. A. Hangey, *Textile Chemist and Colorist 21*, 25 (1989).

42. E. Kissa, in: *Handbook of Fiber Science and Technology: Vol. II,. Chemical Processing of Fibers and Fabrics, Functional Finishes, Part B* (M. Lewin and S. B. Sello, eds.), Chapter 3, p. 211, Marcel Dekker, New York (1984).

43. S. Smith and P. O. Sherman, *Text. Res. J. 39*, 441 (1969).

44. P. O. Sherman and S. Smith, U.S. Patent 3,574,791 (to DuPont) [*CA 71*, 92605 (1969)].

45. H. G. Bryce, in: *Fluorine Chemistry* (J. H. Simmons, ed.), Vol. 5, pp. 370–390, Academic Press, New York (1964).

46. K. Shinoda and K. Mashio, *J. Phys. Chem. 64*, 54, (1960).

47. K. Shinoda, M. Hato, and T. Mayashi, *J. Phys. Chem. 76*, 909 (1972).

48. H. Kunieda and K. Shinoda, *J. Phys. Chem. 80*, 2468 (1976).

49. A. H. Ellison and W. A. Zisman, *J. Phys. Chem. 63*, 1121 (1959); N. L. Jarvis and W. A. Zisman, *J. Phys. Chem. 64*, 150 (1960); M. K. Bernett, N. L. Harvis, and W. A. Zisman, *J. Phys. Chem. 66*, 328 (1962); A. R. Katritzky, G. W. Rewcastle, T. L. Davis, G. O. Rubel, and M. T. Pike, *Langmuir 4*, 732 (1988); S. Raynolds and L. B. Fournier, U.S. Patent 4,432,882 (to DuPont) [*CA 100*, 158192 (1984)].

50. H. C. Fielding, in: *Organofluorine Chemicals and Their Industrial Applications* (R. E. Banks, ed.), Chapter 11, pp. 218–222, Ellis Horwood, Chichester (1979).

51. K. Shinoda, T. Nakagawa, B. Tamamushi, and T. Isemura, *Colloidal Surfactants*, pp. 7–9, Academic Press, New York (1963).

52. C. A. Sperati, in: *High Performance Polymers: Their Origin and Development* (R. B. Seymour and G. S. Kirshenbaum, eds.), pp. 267–278, Elsevier Applied Science, London (1986); S. G. Bankoff, U.S. Patent 2,559,752 (to DuPont) [*CA 46*, 3064 (1952)].

53. H. Aoyama and Y. Amimoto, Eur. Patent 230,115 (to Daikin Industries) [*CA 110*, 40688 (1989)].

54. O. Sata and K. Sumiya, German Patent DE 3,614,439 (to Hitachi Maxell) [*CA 107*, 3062 (1987)].

55. Japanese Patent 58-213,057 (to Dainippon Ink and Chemicals, Inc.) [*CA 100*, 176610 (1984)].

56. J. R. Larry, German Patent DE 2,062,051 (to DuPont) [*CA 77*, 11239 (1972)].

57. M. Waki and E. Tsuruta, Japanese Patent 62-250,074 (to Daito Kasei Kogyo Co. and Fuji Kasei K. K.) [*CA 108*, 192601 (1988)].

15

Fluoroplastics

ANDREW E. FEIRING

15.1. INTRODUCTION

Fluorinated polymers have achieved commercial importance as thermoplastics, elastomers, membranes, and coatings due to their unique combination of properties. The highly fluorinated plastics, in particular, have high thermal stability, low dielectric constant, low moisture absorption, excellent weatherability, low flammability, low surface energy, and outstanding resistance to most chemicals. The presence of only strong[1] C—F and C—C bonds in perfluoropolymers imparts a high degree of oxidative and hydrolytic stability which extends to a remarkable degree to less highly fluorinated analogs.

Although many fluorinated plastics have been prepared, the major commercial products are homopolymers of only four monomers: tetrafluoroethylene (TFE) (see Table 1 for abbreviations), chlorotrifluoroethylene (CTFE), vinyl fluoride (VF), and vinylidene fluoride (VDF), and selected copolymers of these olefins with hexafluoropropylene (HFP), perfluoropropyl vinyl ether (PPVE), or ethylene. All are prepared by free-radical polymerization in water or a fluorinated medium under relatively mild conditions. In general, the polymerization processes are not unusual; fluoroplastics are relatively expensive due largely to the cost of producing and purifying monomers and the relatively small scale on which the polymers are made.

Fluoropolymer history began in the late 1930s with the synthesis of low-molecular-weight polychlorotrifluoroethylene (PCTFE)[2] and the accidental discovery of high-molecular-weight polytetrafluoroethylene (PTFE).[3] Although the remarkable properties of PTFE were immediately apparent, so were the difficulties of converting it into shaped objects since it could be neither solution nor melt processed by traditional techniques. However, the need for materials

ANDREW E. FEIRING • Dupont Central Research & Development, Experimental Station, Wilmington, Delaware 19880-0328.

Organofluorine Chemistry: Principles and Commercial Applications, edited by R. E. Banks *et al.* Plenum Press, New York, 1994.

Table 1.   Abbreviations

CTFE	chlorotrifluoroethylene
ECTFE	ethylene–chlorotrifluoroethylene copolymer
ETFE	ethylene–tetrafluoroethylene copolymer
FEP	fluorinated ethylene–propylene; tetrafluoroethylene–hexafluoropropylene copolymer
HFP	hexafluoropropylene
HFPO	hexafluoropropylene epoxide
PCTFE	poly(chlorotrifluoroethylene)
PFA	perfluoroalkoxy copolymer; tetrafluoroethylene–perfluoro(propyl vinyl ether) copolymer
PPVE	perfluoro(propyl vinyl ether)
PTFE	poly(tetrafluoroethylene)
PVDF	poly(vinylidene fluoride)
PVF	poly(vinyl fluoride)
TFE	tetrafluoroethylene
VDF	vinylidene fluoride
VF	vinyl fluoride

of construction for handling corrosive uranium hexafluoride in the Manhattan project during World War II stimulated development of unique processing techniques for PTFE and the preparation of a high-molecular-weight form of the more readily processed PCTFE.[4] Both polymers were introduced commercially after the war as Teflon® (by Du Pont) and Kel-F® (by M. W. Kellog Co.), respectively. PTFE remains, by far, the largest volume fluoroplastic. The need for a more easily processed form of PTFE led to the discovery in the 1950s and 60s of the melt processible perfluorinated copolymers Teflon® FEP and PFA and, most recently, to the development of amorphous perfluoroplastics. The last materials are unique among the perfluoropolymers in having solubility in certain fluorinated solvents, thus permitting the casting of thin films. It was also found that TFE and CTFE will copolymerize with many nonfluorinated olefins, often in an alternating one-to-one fashion, which led to commercial introduction of the ethylene/CTFE and ethylene/TFE copolymers; these generally possess better mechanical properties than the perfluoropolymers. The homopolymers of vinyl fluoride and vinylidene fluoride were also developed as commercial products with their own unique properties.

This chapter focuses on the preparation, properties, and applications of the commercially important fluoroplastics. A brief description will also be given of some fluorine-containing plastics which have not yet achieved commercial significance to further illustrate the effects of fluorination on polymer properties. Other applications of fluoropolymers as elastomers, coatings, and membranes are treated elsewhere in this book. Recent reviews have appeared on fluoropolymers,[5] tetrafluoroethylene polymers,[6] chlorotrifluoroethylene homopolymer[7] and copolymers,[8] polyvinyl fluoride[9] and poly(vinylidene fluoride).[10,11]

15.2. FLUORINATED VINYL MONOMERS

TFE is produced by a noncatalytic gas-phase pyrolysis of chlorodifluoromethane at 600–900°C and atmospheric or subatmospheric pressures (Eq. 1).[12,13] Steam or other inert gases may be added as diluents. The reaction, probably occurring through the formation and dimerization of difluorocarbene, gives yields as high as 95%. Major byproducts, including HFP, perfluoroisobutylene, octafluorocyclobutane, and saturated chlorofluoroalkanes, are removed by a complex distillation process. HFP, obtained as a co-product in the TFE synthesis,[14] is also prepared by heating TFE to 700–900°C under reduced pressure (Eq. 2).[15] The relative amounts of TFE and HFP from the pyrolysis processes can be controlled by reaction conditions. Other syntheses of TFE and HFP have been described,[12,16] but are not practiced commercially.

(1) $$2\ CHClF_2 \rightarrow CF_2{=}CF_2 + 2\ HCl$$

(2) $$3\ CF_2{=}CF_2 \rightarrow 2\ CF_3CF{=}CF_2$$

PPVE is prepared from HFP in three steps (Eqs. 3–5). HFP is converted to hexafluoropropylene epoxide (HFPO)[17] by reaction with oxygen,[18] by electrochemical oxidation,[19] or by reaction with hypochlorites[20] or hydrogen peroxide[21] in alkaline media. The fluorinated epoxide undergoes a rearrangement and dimerization in the presence of a catalyst, such as an alkali metal fluoride, to afford the acid fluoride **1**.[22] PPVE is formed by pyrolysis of **1** over an alkali metal carbonate or phosphate at 75–300°C.[23]

(3) $$CF_3CF{=}CF_2 \xrightarrow{[O]} CF_3\overset{O}{\overset{\diagup\diagdown}{CF{-}CF_2}}$$

(4) $$2\ CF_3\overset{O}{\overset{\diagup\diagdown}{CF{-}CF_2}} \rightarrow CF_3CF_2CF_2OCF(CF_3)COF$$
                                                                   1

(5) $$\mathbf{1} \rightarrow CF_3CF_2CF_2OCF{=}CF_2$$

CTFE is prepared from 1,1,2-trichloro-1,2,2-trifluoroethane by dehalogenation using zinc in methanol (Eq. 6),[24] or in the vapor phase using aluminum fluoride-nickel phosphate[25] or metal oxide catalysts in the presence of hydrogen.[26] VF is prepared from acetylene by addition of two moles of HF to form 1,1-difluoroethane, followed by elimination of HF,[27] or in a one-step process in the presence of mercury catalyst (Eq. 7).[28] VF can also be obtained by elimination of HCl from 1-chloro-1-fluoroethane or 1-chloro-2-fluoroethane in the vapor phase.[29] Other routes have been reported.[9]

(6) $$CF_2ClCFCl_2 + Zn \rightarrow CF_2{=}CFCl + ZnCl_2$$

(7) $$HC{\equiv}CH + HF \xrightarrow{Hg} \begin{array}{c} CH_2{=}CHF \\ \diagup \\ CH_3CHF_2 \end{array}$$

VDF is most typically prepared by elimination of HCl from 1-chloro-1,1-difluoroethane at temperatures of 700–900°C in the gas phase (Eq. 8).[13,30] The

Table 2. Properties of Fluorinated Monomers

Monomer	CAS Registry no.	Molecular weight	Boiling point (°C)	Melting point (°C)	Critical temp (°C)	Critical pressure (MPa)
TFE	[116-14-3]	100.02	−75.6	−142.5	33.3	3.82
HFP	[116-15-4]	150.02	−29.4	−156.2	86.2	2.75
PPVE	[1623-05-8]	266.04	36.0		423.6	1.9
CTFE	[359-29-5]	116.47	−28.4	−158.2	105.8	3.93
VF	[75-02-5]	46.04	−72.2	−160.5	54.7	5.43
VDF	[75-38-7]	64.03	−82.0	−144.0	30.1	4.29

elimination can be run at higher temperatures,[31] in the presence of copper catalysts,[32] or at lower temperatures in the presence of steam.[33] VDF can also be obtained by elimination of HF from 1,1,1-trifluoroethane,[34] by dechlorination of 1,2-dichloro-1,1-difluoroethane,[35] and from other intermediates.[10]

(8) $CH_3CF_2Cl \rightarrow CH_2{=}CF_2 + HCl$

Physical properties of the fluoromonomers are collected in Table 2. They are colorless gases, except for PPVE which is a liquid at room temperature. The fluoromonomers are relatively nontoxic, with approximate lethal concentrations ranging from about 1000 ppm for CTFE to over 800,000 ppm for VF and VDF, but the possible presence of the exceptionally toxic perfluoroisobutylene or other contaminants must be considered.[36] With the exception of HFP, the fluoromonomers are flammable and TFE, in particular, must be handled with great care. It can homopolymerize exothermically, undergo a violent disproportionation to CF_4 and carbon, and form explosive peroxides in the presence of oxygen.[37] It is normally stored with an inhibitor which is removed immediately before polymerization. CTFE is less reactive than TFE, but can form explosive peroxides and other oxygenated compounds with oxygen. It is stable when kept free of oxygen. VF and VDF can be handled without an inhibitor, although one is often used to avoid undesired polymerization. In contrast, HFP and PPVE homopolymerize only under extreme conditions.[38]

15.3. CRYSTALLINE PERFLUOROPLASTICS

The crystalline perfluoroplastics are the homopolymer of TFE and its copolymers with HFP and PPVE.

15.3.1. Poly(tetrafluoroethylene) (PTFE)

PTFE is the linear homopolymer of TFE with a repeat unit —$(CF_2CF_2)_n$— and is made by polymerizing TFE in water using free-radical initiators. The monomer is polymerized to extremely high molecular weight to limit crystallinity in the final product and give the desired mechanical properties. The high molecular weight, however, results in a very high melt viscosity and the need for unusual processing techniques.

15.3.1.1. Polymerization

Polymerization of TFE is typically conducted at 70–120°C and 1–6 MPa pressure using a water-soluble initiator, such as ammonium persulfate,[39] an alkali metal permanganate,[40] or disuccinic acid peroxide.[41] Radiation-induced polymerization has also been reported.[42] Two distinctly different processes are used, depending on the presence or absence of a dispersing agent and the amount of agitation. Although the chemical structure of the polymer from both processes is identical, the particle size and subsequent processing steps vary greatly. A suspension polymerization process uses little or no dispersing agent and vigorous agitation to give a coagulated granular resin.[43] In the dispersion polymerization process,[44] gentle agitation and an emulsifying agent, such as ammonium perfluorooctanoate,[45] are used to give a stable aqueous dispersion of small (0.2 μm) polymer particles. A hydrocarbon wax can be added to prevent coagulation during polymerization.[45] Since the amount of emulsifying agent in the dispersion process is usually less than its critical micelle concentration, it is not a typical emulsion polymerization, although it shares some characteristics. The rate of polymerization and the shape of the polymer particles are strong functions of the amount of emulsifying agent, with long rod-like single crystals observed at very high concentrations.[46] Small amounts of a comonomer, such as HFP or PPVE, may be added in either the suspension[47] or dispersion[48] process to control processing and cold-flow characteristics of the polymer. Particles with a core-shell structure may also be formed.[49]

15.3.1.2. Processing of PTFE

The initially prepared, high-molecular-weight PTFE has crystallinity of over 90% with a melting point of about 341°C. This initial melting temperature is irreversible and subsequent melting transition occurs at 327°C. Although the polymer is stable at its melting point, the extremely high viscosity of the melt (*ca* 10^{11} poise at 380°C) prevents application of the usual melt processing techniques which involve a large displacement.[50] Small amounts of PTFE can be dissolved above its melting point in high-boiling perfluoroalkanes,[51] but its lack of solubility in all solvents under practical conditions also prevents solution processing. Consequently, three distinct techniques were developed for preparing shaped objects: sintering for granular resins from suspension polymerization, and dispersion coating or paste extrusion for the fine powder produced by dispersion polymerization.

a. Granular Resin. The stringy, porous and irregularly shaped particles from a suspension polymerization are dried and ground to different average particle size, depending on the product requirements. Course cut product has better flow properties, while finer cut material gives products with fewer voids. A typical resin may have a particle size of 400–800 μm as a balance for these requirements.[52] The ground resin is compressed in a mold at moderate temperatures to produce a preform, which can be handled without breaking. The preform is sintered in an oven at 380°C to coalesce the particles and cooled at a controlled rate to produce the desired crystallinity. The resulting molded article may be used as formed, skived as a billet on a lathe to produce sheets, or machined into precision parts.

b. Fine Powder Resin.   A dispersion polymerized PTFE is gently coagulated and dried to give small particles with few voids ("fine powder") for paste extrusion. In the paste extrusion process the fine powder is mixed with 15–25 wt% of a lubricant (e.g., kerosene), shaped into a preform at low pressures, and pushed through a die mounted on an extruder at ambient temperatures. The shear stress encountered during extrusion fibrillates the powder for longitudinal strength. The lubricant is evaporated and the product sintered at about 380°C. This process is particularly applicable for producing tubing and wire insulation.

c. Dispersions.   The aqueous dispersions used for making fine powder can also be used for coatings and for making PTFE fibers. The raw dispersions are stabilized by adding additional surfactant, concentrated to 50–60% solids, and applied to various substrates by spraying, flow-coating, dipping, coagulating, or electrodepositing. In the fiber process,[53] the dispersion is mixed with a matrix-forming resin such as viscose and spun into fibers. The coagulated fibers are heated to remove the matrix polymer, sintered, and drawn.

All three types of PTFE resin may be compounded with fillers such as glass fibers, graphite, or powdered metals for improved mechanical performance. Detailed processing procedures are available in technical bulletins from the various manufacturers (see below).

15.3.2. Perfluorinated Copolymers (FEP and PFA)

The perfluorinated copolymers combine most of the desirable properties of PTFE with a melt viscosity low enough for conventional melt processing. The lower melt viscosity results from the use of comonomers, rather than extremely high molecular weight, to control the degree of crystallinity for desirable mechanical properties. In contrast to PTFE, where the molecular weight can be over 10^7, the copolymer molecular weights of 10^5–10^6 are more typical for thermoplastics. Melt processibility permits the injection molding of complex parts and extrusion of longer continuous lengths than can be achieved by the batchwise paste extrusion of PTFE.

FEP (*f*luorinated *e*thylene *p*ropylene) and PFA (*per*fluoro*a*lkoxy) resins are random copolymers of TFE with HFP or PPVE, respectively. The comonomer content in FEP is about 10–12 wt%, while only 2–4 wt% of the comonomer is required in PFA to achieve the desired level of properties because the larger perfluoropropoxy group in PPVE is more efficient at reducing crystallinity. PPVE also copolymerizes more readily than HFP, so a smaller excess is needed during polymerization. In addition, the high-temperature properties of PFA are better than FEP. On the other hand, PPVE is

$$\text{wwv}(CF_2-CF_2)_x-(CF_2-CF)\text{wwv} \qquad \text{wwv}(CF_2-CF_2)_x-(CF_2-CF)\text{wwv}$$

$$\underset{\text{FEP}}{}\underset{CF_3}{|} \qquad\qquad \underset{\text{PFA}}{}\underset{OC_3F_7}{|}$$

far more expensive than HFP, so PFA is the most costly of the crystalline perfluoro-plastics.

15.3.2.1. Polymerization

Copolymerizations of HFP[54] or PPVE[55] with TFE can take place in aqueous or nonaqueous media. Aqueous processes are similar to the dispersion polymerization of TFE using a dispersing agent, such as ammonium perfluorooctanoate or other fluorinated surfactants,[56] and a water-soluble free-radical initiator, such as ammonium persulfate. Nonaqueous polymerizations tend to be run at slightly lower temperatures (30–60°C) in a fluorinated solvent using a soluble organic initiator such as perfluoropropionyl peroxide. Polymerizations in mixed aqueous/nonaqueous media using higher-molecular-weight perhalogenated peroxide initiators have been reported,[57] as have polymerizations in the gas phase.[58] FEP is generally produced commercially by an aqueous process while both aqueous and nonaqueous processes are used for PFA. Molecular-weight control is achieved by the use of chain-transfer agents, such as methanol in the nonaqueous PFA process and hydrogen or an alkane in aqueous polymerizations.[55,59]

Reactive end groups, such as acyl fluoride or carboxylic acid, resulting from initiator fragments in aqueous polymerizations and from a chain-transfer process involving the PPVE comonomer in PFA, must be removed from the raw polymer to avoid formation of bubbles or dark colors during high-temperature processing. Stabilization processes include heating the polymer in the presence of steam, or reaction with methanol, ammonia, or amines to convert the end groups to more stable ester, amide, or difluoromethyl groups.[60] Treatment of PFA with elemental fluorine provides a finished polymer with stable CF_3 end groups and a very low level of ionic contamination,[61] which is of special importance to the semiconductor industry.[62]

15.3.2.2. Processing

Both FEP and PFA can be processed by conventional methods, such as extrusion or injection molding.[63] Various melt viscosity grades are provided to optimize performance for individual processing techniques. Due to the potential for formation of HF or other corrosive gases during high-temperature processing, corrosion-resistant parts must be used. The copolymers, especially FEP, tend to be more difficult to process than conventional thermoplastics due to a high melt viscosity and a lower critical shear rate, above which melt fracture occurs. Fortunately, melt strengths are high, so high draw down ratios can be used. PFA resins have higher critical shear rates than FEP and can be handled at higher temperatures due to a higher inherent thermal stability; hence they can be processed faster and provide superior molding behavior for intricate parts. Dispersion coating processes can also be used with both resins, but their insolubility in all solvents under practical conditions precludes solution processing.

15.3.3. Properties of the Perfluoroplastics

The characteristic properties of the perfluoroplastics derive from a combination of their strong *intra*molecular C—F and C—C bonds with relatively weak *inter*molecular associations (cf. Chapter 3). The great strength of the C—F bond also prevents chain transfer processes during polymerization, so PTFE possesses a strictly linear chain,[64] in contrast to many polymers prepared by free-radical methods. The larger size of the fluorine atom relative to hydrogen tends to stiffen[65] the PTFE chain and forces it to adopt a helical conformation, in contrast to the planar zigzag structure of poly(ethylene).[66] The copolymers have deliberately introduced short-chain branches, but are similarly free of long-chain branches.

The insolubility of the perfluoroplastics usually prevents direct determination of molecular weight. Initially, molecular weights were estimated by use of radioactively labeled end groups.[67] In practice, however, they are measured using empirically derived correlations of molecular weight with heat of fusion or specific gravity of specially prepared PTFE samples, or by melt flow measurements on the copolymers.[68] Recently, the weight-average molecular weight of a commercial PTFE sample dissolved in a perfluorokerosene at high temperatures has been measured as 2.1×10^6 by laser light scattering.[51] Commercial PTFE samples have been shown by viscoelastic spectroscopy to possess a bimodal molecular weight distribution with M_w values of 5×10^6 and 5×10^7, while FEP samples show a single molecular weight peak.[69]

The as-produced polymers are very highly crystalline, ranging from 50–75% for the copolymers to > 90% for PTFE. Crystallinity is reduced to 40–70% during processing for improved mechanical performance, and depends on processing conditions, rate of cooling from the melt, and molecular weight of the resin.

The mechanical, electrical, chemical, and surface properties of the perfluoroplastics are similar. Several property compilations are available.[5,6,70] Perhaps, the most significant differences reside in their thermal behavior and, of course, the melt processibility of the copolymers. PTFE shows, in addition to its melting point of 327 °C, other first-order (crystalline) transitions at 19, 30, and about 90°C.[71] The 19°C transition is especially important, since it occurs around ambient temperature and affects polymer properties. Second-order transitions are observed at 130°C, which is often quoted as the glass transition temperature, and at subambient temperatures.[50,70] Recently, yet another order–disorder transition at 370°C has been reported in high-molecular-weight sheared samples.[72] In contrast, FEP and PFA show only single first-order transitions at their melting points of 260 and 305°C, respectively, plus a number of second-order transitions.[73]

PTFE and PFA possess excellent thermal stability with rapid decomposition occurring only above 450°C, and mechanical properties which are retained for long periods at the upper-use temperature of 260°C. Thermal decomposition of PTFE in

Table 3. Properties of Perfluorinated Plastics

	PTFE	FEP	PFA
Melting point (°C)	327	260	305
Upper use temperature (°C)	260	200	260
Relative density	2.18	2.14–2.17	2.13–2.17
Tensile strength (MPa)	7–28	20–31	28–31
Elongation (%)	100–600	300	300
Flexural modulus (MPa)	280–630	580–660	650–690
Impact strength (J m^{-1})	160	no break	no break
Dielectric constant (1 MHz)	2.1	2.02	2.06
Dissipation factor (1MHz)	0.003	0.008	0.0008
Refractive index	1.376	1.345	1.350
Critical surface tension (mN m^{-1})	18.5	17.8	17.8

vacuum gives mostly monomer.[74] The thermal stability of FEP is lower, and physical strength decreases significantly when samples are held above 200°C.

While not especially strong or stiff compared to some modern engineering thermoplastics,[75] the perfluoroplastics do show excellent toughness and flexibility and high elongation. Properties are well maintained from cryogenic to their upper-use temperatures. The copolymers do have an advantage, since they lack the marked change in volume which occurs at room temperature in PTFE due to its 19°C thermal transition. Processing variables can significantly affect the mechanical properties, especially with PTFE, and the wear, stiffness, and creep can be improved with appropriate fillers. Typical properties are shown in Table 3.

Of greater significance to their applications are the outstanding electrical properties, chemical resistance, and low surface energy of the perfluoroplastics. They are excellent insulators with high dielectric strengths and with dielectric constants (2.0–2.1) among the lowest known for solid materials. The dielectric constant does not change significantly over wide temperature or frequency ranges, and their low moisture absorption results in electrical properties which are little affected by humidity. The perfluoroplastics are inert to nearly all chemicals and solvents, even at elevated temperatures and pressures, reacting only with strong reducing agents such as molten alkali metals, elemental fluorine, strong fluorinating agents, and sodium hydroxide above 300°C. With the exception of perfluorinated liquids, few chemicals are absorbed into or swell perfluoroplastics, and their high crystallinity leads to relatively low gas permeability, so they act as excellent barrier resins. Their low surface energy and coefficients of friction against other materials lead to the well known anti-stick applications.

The low surface energy of the perfluoroplastics makes it difficult to bond these materials with other polymers or metals, so a variety of methods have had to be developed to improve adhesion.[76] Reported techniques include: wet-chemical etching[77] with alkali metals in liquid ammonia, alkali metal amalgams or sodium naphthalide; electrochemical reductions; treatment with corona discharge, ion or electron beams; vacuum deposition of metals; and plasma treatment.[78] A relatively mild reducing agent, the benzoin dianion in DMSO, also corrodes the surface of PTFE[79] and the copolymers,[80] and has been used to introduce functional groups, such as hydroxyl or carboxylate, on the polymer surface.[81] The reduced surface has been characterized as polyacetylenic.[82] The perfluoropolymers are relatively sensitive to ionizing radiation, tending to undergo chain cleavage rather than crosslinking, especially in the presence of air.[83] Irradiation-induced grafting of olefinic monomers, such as acrylic acid,[84] onto perfluoropolymers may also be used to create a more hydrophilic surface.

15.3.4. Applications and Commercial Aspects

The largest application for the perfluoroplastics is wiring insulation, including electrical, coaxial, and computer cables, where advantage is taken of the outstanding electrical properties and low flammability of these resins. Tapes from granular PTFE, paste extruded PTFE fine powder, melt extruded FEP and PFA, and a foamed FEP[85] are used. Other important uses are for gaskets, seals, bearing, hoses, vessel liners, and laboratory equipment where chemical resistance and surface properties are important. Fluoropolymer release coatings, e.g., for cookware surfaces, are well known; the most recent versions employ a three-layer coating for improved adhesion to the metal substrate.[86] In construction, architectural fabrics are prepared by coating glass fabrics with PTFE dispersions, and FEP film is used to cover solar collectors, because of the excellent weatherability of the perfluoroplastics. The unique fibrillation capability of PTFE fine powder is used to prepare porous fabrics (Gore-Tex[87]) which have high permeability to water vapor, but resist passage of liquid water. PTFE fibers find major application as pump packing. PTFE micropowders, prepared by gamma or electron beam irradiation to decrease their molecular weight,[88] are used as additives in plastics, inks, finishes and lubricants, to provide nonstick and sliding properties. Complex molded articles from PFA (e.g., wafer baskets) are used to handle corrosive liquids and gases in the semiconductor industry, where requirements for very pure materials are paramount. Printed circuit boards, which are laminates of copper on a fiber-reinforced fluoropolymer layer, take advantage of the low dielectric constant provided by the fluoropolymer.[89]

In medicine, the inert and nonadhesive character of the perfluoroplastics makes them suitable materials for implants, and enzyme immobilization and bioaffinity separation systems using perfluorocarbon polymer beads have recently been described.[90]

The major producers of the perfluoroplastics in the USA, Western Europe, and Japan are listed with their product trade names in Table 4. PTFE is also produced in Russia and the People's Republic of China. PTFE is the largest volume commercial fluoropolymer, with an estimated worldwide production of about 33×10^6 kg in 1988. FEP and PFA are far smaller at about 6×10^6 and 1.4×10^6 kg per year, respectively. Selling prices depend on resin type, but range from 13–20 \$ kg^{-1} for PTFE, 24–26 \$ kg^{-1} for FEP, and 42–46 \$ kg^{-1} for PFA in the USA.

Table 4. Perfluoroplastics Manufacturers

Product	Company	Tradename
PTFE	Du Pont	Teflon PTFE
	ICI	Fluon
	Hoechst	Hostaflon TFE
	Montefluos	Algoflon
	Asahi-ICI Fluoropolymers	Fluon
	Daikin Industries	Polyflon
FEP	Du Pont	Teflon FEP
	Hoechst	Hostaflon
	Montefluos	Algoflon
	Daikin Industries	Neoflon
PFA	Du Pont	Teflon PFA
	Hoechst	Hostaflon PFA
	Daikin Industries	Neoflon

15.4. AMORPHOUS PERFLUOROPLASTICS

Amorphous perfluoroplastics, which combine the outstanding electrical properties, chemical resistance, and thermal stability of the crystalline perfluoropolymers, while adding high optical clarity, improved mechanical properties, and limited solubility in perfluoroether solvents, have recently been commercialized by two companies. The products are Teflon® AF (Du Pont) and Cytop® (Asahi Glass Company). While both have cyclic, oxygen-containing units in the polymer backbone, the approaches to obtaining the cyclic structures are quite different.

Teflon® AF is the copolymer of TFE and a cyclic monomer, perfluoro-(2,2-dimethyl-1,3-dioxole) (PDD).[91] PDD is surprisingly reactive, and will homopolymerize or copolymerize with TFE in essentially any ratio. The glass transition temperature of the copolymers is a strong function of the comonomer content and can range from <100°C with low PDD content to >300°C for PDD homopolymer. Initial commercial products contain about 65 and 80 mol% PDD giving resins with T_g values of 160 and 240°C, respectively. The cyclic monomer is prepared in four steps from hexafluoroacetone.

In contrast, Cytop® is prepared by cyclopolymerization of linear perfluorodienes of general structure $CF_2=CFO(CF_2)_nCF=CF_2$. Although the exact structure of the commercial product has not been described, comparisons of the glass transition temperature described in a technical bulletin[92] with patent descriptions[93] suggest that it is the homopolymer of the above diene with $n = 2$. Copolymers with TFE and other fluoro-olefins have also been reported.

Polymerization processes for Teflon® AF and Cytop® appear similar to those used for the crystalline perfluoroplastics, and can be run in either aqueous or nonaqueous media.

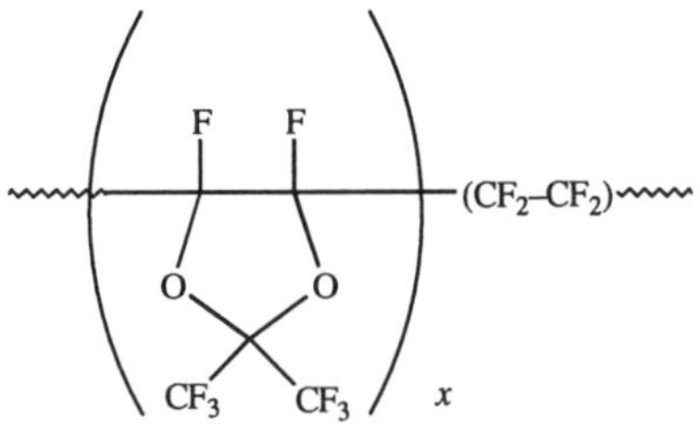

Telfon® AF

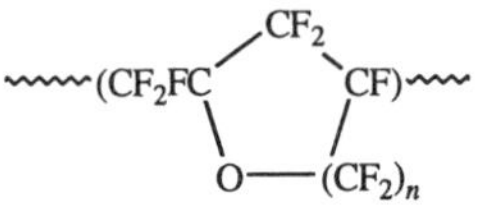

and/or

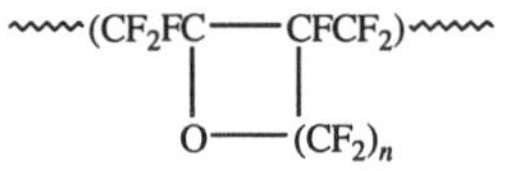

Cytop®

Table 5.   Properties of Perfluoroplastics

	Teflon AF	Cytop	PTFE
Glass transition temp (°C)	160, 240	108	(130)
Melting point (°C)	—	—	327
Tensile modulus (MPa)	950–2150	1200	300–500
Elongation (%)	3–40	50	100–600
Tensile strength (MPa)	25–27	32	7–28
Dielectric constant	1.89–1.93	2.1–2.2	2.1
Refractive index	1.29–1.31	1.35	1.38
Optical transmittance (%)	>95	95	opaque

Key physical properties for the amorphous polymers are compared with PTFE in Table 5. Noteworthy are the increase in room temperature modulus and the high degree of optical clarity for the new materials. The dielectric constant for Teflon® AF is the lowest for any known polymer. Their solubility in perfluoroethers, such as perfluoro-(2-butyltetrahydrofuran), permits solution coating, spin casting, or spray coating of clear film less than one micron in thickness. Samples may also be extruded or injection molded above their glass transition temperatures.

Major applications are likely to be in optics or electronics for cladding of optical fibers, antireflection coatings, and dielectric interlayers. Current production is likely to be very small, with selling prices in the range of hundreds of dollars per kg.

15.5. POLY(CHLOROTRIFLUOROETHYLENE)

PCTFE is a linear homopolymer, $—[CF_2CFCl]_n—$ made by the free-radical polymerization of CTFE. Small amounts (< 5%) of vinylidene fluoride may be added to improve processibility with little effect on polymer properties. In comparison with

Table 6. Properties of PCTFE

Melting point (°C)	210
Upper use temperature (°C)	180
Relative density	2.08–2.19
Tensile strength (MPa)	30–40
Elongation (%)	100–200
Tensile modulus (MPa)	1200–1500
Flexural modulus (MPa)	1000–2000
Izod impact strength (J m^{-1})	133–187
Refractive index	1.425
Dielectric constant (1 MHz)	2.3–2.5
Dissipation factor (1 MHz)	0.01
Critical surface tension (mN m^{-1})	31

PTFE, PCTFE is melt processible and has improved mechanical properties, better optical clarity and lower permeability to gases, but inferior thermal stability, chemical resistance and electrical properties. It also has a somewhat higher critical surface tension and coefficient of friction.

15.5.1. Production

PCTFE is prepared by free-radical-initiated polymerization of CTFE in bulk, aqueous suspension or emulsion using a variety of organic or water-soluble initiators. For bulk polymerization,[94] halogenated peroxides, such as trichloroacetyl or dichlorotrifluoropropionyl peroxide, are used. Halogenated diluents may be added,[95] although the polymer is insoluble in its monomer and in all other solvents under normal polymerization conditions.[96] Aqueous suspension polymerizations generally employ redox initiators, such as alkali metal persulfates and bisulfites.[97] Iron, silver, or copper salts may be used as accelerators and buffers may be added to control pH. Perhalogenated surfactants, such as perfluoro-octanoic acid salts, are used in emulsion polymerization.[98] Polymerization conditions are similar to those used for PTFE or the perfluorinated copolymers.

PCTFE resins can be processed by conventional thermoplastic techniques, such as compression and injection molding and extrusion. Typical processing temperatures lie in the range of 275–300°C. Thermal degradation occurs at these temperatures so corrosion-resistant parts must be used. Coatings may also be applied using finely divided powder or dispersions in a liquid carrier, followed by baking.

15.5.2. Properties

Like PTFE, PCTFE has a strong tendency to crystallize, and molecular weights of commercial products must be kept sufficiently high to maintain the degree of crystallinity at 40–80% for optimum mechanical properties. Unlike PTFE, this can be achieved by proper processing conditions with molecular weights in the range of 1–5 $\times 10^5$ where melt processing is still possible. The degree of crystallinity can be determined by density or infrared measurements.[99] In addition to the melting point at 210°C, thermal transitions have been observed at 150, 90, and –37°C in highly crystalline samples.[100] The polymer chain is helical with a mostly syndiotactic arrangement of the monomer units.[101]

Selected properties of PCTFE are shown in Table 6. Noteworthy are its significantly better mechanical properties, but inferior thermal and electrical properties as compared to PTFE. Thermal decomposition gives a mixture of monomer and low-molecular-weight oils.[102] It is resistant to most chemicals, being attacked only by strong reducing agents, such as alkali metals, some organic amines, and powerful nucleophiles. Reactions with sulfur, selenium, and phosphorus nucleophiles,[103] organolithium reagents,[104] and reducing agents[105] have been used to functionalize the polymer surface. It swells in some highly chlorinated or fluorinated solvents and can be dissolved in a few, e.g., benzotrifluoride,[96] at elevated temperatures. PCTFE is somewhat less sensitive to radiation than PTFE[106] and has lower permeability to gases.

15.5.3. Applications and Commercial Aspects

Due to its mechanical properties, chemical resistance, and low permeability, the major uses for CTFE are in molded parts for chemical process equipment and cryogenic applications, including seals, gaskets, pump parts, tubes, linings, and electrical insulators. In addition, films are used for packaging air and moisture-sensitive materials, such as pharmaceuticals and electronic equipment.

PCTFE and its lightly modified VDF copolymers are produced by 3M ("Kel-F"), Allied-Signal ("Aclon"), Atochem ("Voltalef"), and Daikin Industries ("Daiflon") with worldwide annual production of about 1×10^6 kg in 1988 and a selling price of 60–150 $ kg^{-1} in the USA.

15.6. PARTIALLY FLUORINATED PLASTICS

The commercially important, partially fluorinated, plastics are the copolymers of TFE or CTFE with ethylene and the homopolymers of VDF or VF. Key properties of all four materials are collected in Table 7.

15.6.1. Ethylene–Tetrafluoroethylene Copolymer

Poly(ethylene-*co*-tetrafluoroethylene) (ETFE) is a linear copolymer of TFE and ethylene in a nearly one-to-one molar ratio. Since the monomers have a strong tendency to alternate during polymerization,[107] the polymer structure is mainly — [CF$_2$CF$_2$CH$_2$CH$_2$]$_n$. ETFE retains most of the desirable thermal, electrical, and chemical properties of the perfluoroplastics, while possessing significantly better mechanical properties and radiation resistance.

Table 7. Properties of the Partially Fluorinated Plastics

	ETFE	ECTFE	PVDF	PVF
Melting point (°C)	275	240	155–192	185–210
Upper use temperature (°C)	150–200	150	150	110
Relative density	1.70	1.70	1.76	1.38–1.57
Tensile strength (MPa)	45	50–65	40–60	48–120
Elongation (%)	100–300	150–250	30–200	115–250
Tensile modulus (MPa)	800–900	1600–1700	1000–2300	1700–2600
Flexural modulus (MPa)	1400	1800–2500	1400–3000	
Izod impact strength (23°C)	no break	no break	75–235	
Dielectric constant (1 MHz)	2.6	2.6	8–9	6.2–7.7
Dissipation factor (1 MHz)	0.005	0.016	0.18	0.17–0.28
Refractive index	1.403		1.42	1.46
Critical surface tension (mN m^{-1})	22.1		25	28

15.6.1.1. Production

ETFE is produced by free-radical, suspension polymerization of the monomers in aqueous, nonaqueous, or mixed systems, under conditions where TFE, but not ethylene, will homopolymerize. Polymerization conditions are similar to those described above for the perfluoroplastics.[108] Emulsion polymerization, using fluorinated surfactants, may also be employed.[109] The nonaqueous or mixed systems generally use an inert solvent such as 1,1,2-trichloro-1,2,2-trichloroethane and peroxide initiators, while redox initiators are used in the aqueous or emulsion systems. Commercial ETFE resins contain small amounts (0.1–10%) of a third monomer which improves their resistance to stress cracking at elevated temperatures; typical termonomers are perfluoroalkoxyvinyl compounds or perfluoroalkylethylenes.[110]

ETFE resins can be processed by the usual melt fabrication techniques, including injection molding, blow molding, compression molding and extrusion. Typical process temperatures are 300–340°C. Various melt viscosity grades are available.

15.6.1.2. Properties

ETFE is a semicrystalline polymer with the degree of crystallinity ranging from 40–60%, depending on processing conditions. It exists in an extended zigzag molecular conformation with the molecular packing in the orthorhombic configuration.[111] Its melting point is in the range of 200–300°C, depending on the TFE/ethylene ratio and degree of alternation; commercial resins have a melting temperature of about 275°C, corresponding to about 92% alternation. Three second-order transitions have been observed at 110, –25, and –120°C by dynamic mechanical analysis.[112] The polymer is insoluble in all solvents at room temperature, but will dissolve in certain high-boiling esters (e.g., di-isobutyl adipate) at temperatures above 230°C,[113] permitting determination of a weight-average molecular weight (5.6 to 1.13×10^6) by light scattering. Solution viscosity data under similar conditions suggest that the polymer exists as a slightly expanded coil.[114]

ETFE resins are stronger, tougher, stiffer, and more creep-resistant than the crystalline perfluoroplastics (Table 7). Significant improvements in mechanical properties can be achieved by glass-fiber reinforcement, and properties are well maintained from $-100°C$ to an upper-use temperature in the range of $150–200°C$. They have excellent electrical-insulating properties. Although ETFE's dielectric constant (2.6) is greater than that of PTFE (2.1),[112] it is still among the lowest of the organic polymers. They are less sensitive to high-energy irradiation than PTFE, tending to undergo crosslinking rather than chain cleavage under moderate conditions. Radiation crosslinking, in fact, is used to improve the high-temperature mechanical properties of ETFE wire coatings.[115]

ETFE resins are resistant to most chemicals and solvents, except for strong oxidizing agents, organic bases, and sulfonic acids. They will dissolve, as noted above, in certain esters at high temperatures.

15.6.1.3. Applications and Commercial Aspects

The most important uses for ETFE resins are for the jacketing and insulation of electrical cable, including power, communication, and control wiring in aircraft and other transport systems. Higher viscosity grades or radiation crosslinked materials find particular application for insulation in hot environments, or where the insulation may be subjected to mechanical stress. Foamed resins provide insulation with an even lower dielectric constant.[116] Other applications include seal glands, pipe plugs, fasteners, and pump parts for the chemical process industry, and laboratory wear. ETFE films are also used as tough and flexible windows in greenhouses and conservatories due to their high transparency to both UV and visible light and excellent resistance to weathering.[117]

ETFE resins are made in the USA, Europe, and Japan. Suppliers and their trade names include Du Pont ("Tefzel"), Hoechst ("Hostaflon ET"), Asahi Glass Co. ("Aflon COP"), and Daikin Industries ("Neoflon"). Worldwide annual production (1988) can be estimated at 2.4×10^6 kg. Resin selling prices in the USA are about $30–35$ \$ kg^{-1}.

15.6.2. Ethylene–Chlorotrifluoroethylene Copolymer

Poly(ethylene-co-chlorotrifluoroethylene) (ECTFE) is a copolymer with the two monomers in a nearly alternating, one-to-one molar ratio. The polymer structure is mainly —$[CF_2CFClCH_2CH_2]_n$—. ECTFE has many similarities to ETFE in its preparation, processing, and properties. As might be expected, substitution of CTFE for TFE in the copolymers results in improvement in mechanical and barrier properties at the cost of a modest decrease in thermal and chemical stability.

15.6.2.1.   Production

ECTFE is prepared by the free-radical copolymerization of the monomers in aqueous, organic, or mixed media.[108,118–121] Peroxides, such as bis(trichloroacetyl) peroxide, or the combination of a trialkylboron and oxygen may be used as initiators. The latter system has the advantage of permitting polymerization at very low temperatures (<10°C) which improves the degree of alternation of the comonomers. Chain transfer agents (e.g., chlorinated alkanes) are used to control molecular weight. The commercial polymers contain small amounts of a termonomer, such as a perfluoroalkylethylene[122] or 2-trifluoromethyl-3,3,3-trifluoropropene,[123] to improve the polymer's resistance to catastrophic mechanical failure when stressed at temperatures above 150°C.

ECTFE resins are processed by the usual thermoplastic techniques, such as extrusion, injection molding, rotomolding, or powder coating. Typical processing temperatures are in the range of 260–300°C. Thermal stability of the polymer at processing temperatures is improved by adding stabilizers, including polyhydric phenol phosphites, zinc alkanoates, and other commercially available materials.[124]

15.6.2.2.   Properties

ECTFE is a semicrystalline polymer with an extended zigzag structure and a degree of crystallinity of 50–60%, depending on processing conditions. The commercial polymer has a melting point of 240°C and second-order transitions at about 140, 90, and –65°C. The melting point is a function of the comonomer ratio and degree of alternation with the perfectly alternating material melting at 264°C.[125] The interesting feature that the copolymer has a higher melting point than either homopolymer has been explained as arising from intermolecular interactions between fluorine and hydrogen.[121] Molecular weights of the commercial polymers are in the range of 1 to 5×10^5. The polymer is insoluble in all solvents at ambient temperatures, but will dissolve in certain chlorofluorinated hydrocarbons (such as 2,5-dichlorobenzotrifluoride) at temperatures above 120°C and can be attacked by amines, esters, and ketones when warm.

Properties of ECTFE and ETFE resins are compared in Table 7. The major differences are in the slightly better mechanical properties of ECTFE, while ETFE has the edge in thermal stability and chemical resistance. ECTFE has excellent barrier properties to water vapor and other gases, excellent weatherability, good electrical properties, and resistance to high-energy irradiation. Mechanical properties can be improved by appropriate fillers and by irradiation crosslinking.[126]

15.6.2.3.   Applications and Commercial Aspects

As with ETFE, the major applications of ECTFE are as wire and cable insulation. Its lack of flammability, toughness, and abrasion resistance permit its use in office buildings as insulation for low-voltage plenum wiring without the need for metal conduit. A foamed version provides an even lower dielectric constant.[127] Other applications include coating and linings for chemical equipment and containers for corrosive chemicals.

ECTFE resins are produced commercially by Ausimont USA ("Halar") with annual production in 1988 estimated at about 1.6×10^6 kg and a selling price of 23–70 $ kg^{-1}.

15.6.3. Poly(Vinylidene Fluoride)

Poly(vinylidene fluoride) (PVDF, often abbreviated in the literature as PVF2) is a linear, semicrystalline polymer with a repeat structure consisting of mostly head-to-tail monomer units —$[CH_2CF_2]_n$—. Second to PTFE in commercial volume among the fluoroplastics, it is in many respects the most unusual. Although isomeric with ETFE, its properties, especially electrical properties, are quite different. It is one of the most studied polymers due to its remarkable piezoelectric and pyroelectric properties and complicated polymorphism. Many copolymers of vinylidene fluoride are known, including the commercially important VDF/HFP elastomers (see Chapter 16), but this section will focus on the VDF homopolymer and its crystalline copolymers containing minor amounts of HFP or trifluoroethylene.

15.6.3.1. Production

Vinylidene fluoride undergoes free-radical initiated polymerization under a variety of conditions. Generally, aqueous suspension or emulsion processes and organic percarbonates, diacyl peroxides, other organic peroxides or persulfate initiators are used. Although initially prepared at high pressures (30 MPa),[128] current processes employ more moderate pressures (1–10 MPa) and temperatures of 10 to 150°C depending on the initiator.[129] Fluorinated surfactants, such as ammonium perfluorooctanoate, are used as stabilizers in emulsion polymerizations.[130] Molecular-weight control is achieved by the use of chain-transfer agents, such as acetone, chloroform, or trichlorofluoromethane.[131,132] Nonaqueous polymerizations employing liquid chlorofluoroalkane solvents and fluorinated acyl peroxide initiators are also known.[133]

The semicrystalline copolymers of VDF, such as the material containing 1–13 mole percent of HFP,[134] are prepared under similar aqueous polymerization conditions.[132]

Conventional melt-processing techniques, such as extrusion or injection, compression or transfer molding, are used to prepare shaped objects from PVDF. Typical processing temperatures are in the range of 200–290°C. The polymer is generally stable up to 300°C, so special precautions or added stabilizers are usually not required. Organosols (dispersions of very small particles of the polymer in an organic solvent) are used to apply coatings to metals. In addition, and unlike most commercial fluoroplastics, PVDF has sufficient solubility in polar solvents, such as dimethylacetamide, so that thin films for membranes can be cast from solution.[135]

15.6.3.2. Properties

Among the most significant features of PDVF is its crystalline polymorphism, which has been studied in great detail.[11,136–141] At least four crystal structures have been detected, the most important being the α (phase II) and the β (phase I). The α form is the thermodynamically most stable phase and is produced by melt or solution crystallization under normal pressures. The chain conformation is *trans-gauche* which minimizes steric interaction among the fluorines and adjacent hydrogens. The β-phase is produced by mechanical deformation (stretching or drawing[142]) of the melt crystallized polymer, or by crystallization from the melt under high pressures.[138,140] The chain conformation is all-*trans* planar zigzag, with fluorines on one side and hydrogens on the other. It is this highly polar phase which gives rise to the piezo- and pyroelectric properties of PVDF. Copolymers of VDF with trifluoroethylene or TFE also show different crystalline phases but, unlike the homopolymer, can crystallize directly into the polar β-phase under normal conditions.[143,144]

A second unusual feature of PVDF is its relatively high solubility in polar solvent, including *N,N*-dimethylacetamide, dimethyl sulfoxide, acetone, and ethyl acetate. It is also miscible with certain carbonyl-containing polymers, such as poly(methyl methacrylate), poly(ethyl acrylate), poly(vinyl acetate), and poly(methyl vinyl ketone). Some of these blends have been of commercial interest.[145] The solubility behavior presumably arises from interactions between the polar PVDF segments and carbonyl oxygens.

Although composed of mostly head-to-tail linkages, 3–6% of the monomer units in PDVF are reversed, the number increasing as the polymerization temperature increases.[146] These chain defects have been measured by fluorine NMR spectroscopy,[147] and their effects on crystal structure and polymorphism have been studied.[148] Light-scattering measurements on fractionated commercial samples showed weight-average molecular weights in the range of 4 to 9.5×10^6.[149] Commercial resins are about 35–70% crystalline, depending on processing conditions. Due to polymorphism, the polymer melting point is not precise, but is generally in the range of 155 to 192°C. Additional thermal transitions have been observed at about 90, 50, –35, and –70°C, with the highest temperature peak being assigned to a crystalline transition and the others to transitions in the amorphous regions.[138]

Properties of a typical PVDF resin are summarized in Table 7. As with ETFE and ECTFE, mechanical properties of PVDF are superior to those of the crystalline perfluoroplastics. The tensile strength of PVDF is a bit higher and elongation lower than those of ETFE, while moduli are similar. Modification of the polymer with a small amount of HFP leads to lower modulus, increased elongation, and higher impact strength. The largest differences between the isomeric PVDF and ETFE are the much lower melting point (*ca* 170°C versus 275°C) and larger dielectric constant (8–9 versus 2.6) of the former.

PVDF has good resistance to inorganic acids and bases, alcohols, hydrocarbons, and aromatics but, as already noted, is soluble in some polar solvents. It will react with organic or inorganic bases[150] or strong Lewis acids[151] in solution, undergoing dehydrofluorination. Loss of HF also occurs at high temperatures, resulting in formation of significant amounts of char. The high char yield gives superior performance in severe flammability tests. It also has a high tolerance to ionizing radiation[152] and can be radiation crosslinked to improve mechanical properties. Like the other fluoroplastics, PVDF shows excellent weatherability and resistance to UV irradiation.

15.6.3.3. Applications and Commercial Aspects

A major use of PVDF is as a weather-resistant coating for metals in architectural applications. Both dispersion coatings and thin films are used. Solid PVDF is also used to make pipes, fittings, and liners for fluid-handling applications. Although its high dielectric constant and dissipation factor limit its usefulness as an insulator for high-frequency circuits, PVDF does find application as insulation for low-frequency plenum cable due to its mechanical properties and high char yield on exposure to heat.

Small in volume, but of increasing importance, are applications which depend on the piezoelectric and pyroelectric properties of PVDF films.[144,153] These properties appear when PVDF films are poled in their β form, and provide an electrical signal in response to mechanical or thermal signals or, inversely, mechanical motion or a change in heat content in response to an applied electrical field. Copolymers with trifluoroethylene are of particular interest due to their greater activity and spontaneous crystallization in the β phase. Applications include speakers, transducers, hydrophones, and electromagnetic radiation detectors. Another small volume, but high value, application for PVDF is as ultrafiltration and microporous membranes in biotechnology.[135,154]

Manufacturers of PVDF resins are Atochem USA ("Kynar"), Atochem ("Foraflon"), Solvay ("Solef"), Daikin Industries ("Neoflon"), and Kureha Chemical Industry ("KF"). Worldwide annual production in 1988 was about 7.4×10^6 kg with a selling price in the USA of about 15.00 \$ kg^{-1}.

15.6.4. Poly(Vinyl Fluoride)

Poly(vinyl fluoride) (PVF) is a semicrystalline polymer, prepared by free-radical polymerization of vinyl fluoride (VF). It is the least highly fluorinated of the commercial fluoroplastics, but the presence of fluorine still has a significant effect on its properties. Although the polymer repeat structure is generally written as —[CH$_2$CHF]$_n$—, chain branching and monomer reversals result in a somewhat irregular structure.

15.6.4.1. Production

Preparation of PVF is generally similar to the processes used for the other fluoroplastics. VF is less reactive toward free-radical polymerization than other vinyl monomers, such as TFE, VDF, or vinyl chloride,[155] so polymerization conditions tend to be more severe. Initially PVF was prepared in an organic solvent at pressures as high as 600 MPa,[156] but current processes generally employ an aqueous medium at temperatures of 50–165°C and pressures of up to 100 MPa.[157] Typical free-radical initiators, including acyl peroxides and water-soluble azo compounds,[158] are employed. Aqueous dispersion polymerizations using a variety of suspending agents[159] and continuous, one-[160] and two-stage[161] aqueous suspension polymerizations at temperatures of 80–100°C and pressures of 50–60 MPa have been described. Hydrocarbon olefins containing at least three carbon atoms (e.g., propylene or butylene) may be added as chain-transfer agents. Emulsion polymerizations can be run under somewhat milder conditions at temperatures of 40–50°C and pressures of around 4 MPa.[162]

VF has also been polymerized using trialkylboron/oxygen initiators or modified vanadium/aluminum Ziegler-Natta catalysts at temperatures below 30°C.[163,164] The latter system was considered to initiate a free-radical rather than coordination polymerization, and gave no improvement in polymer stereoregularity.[165] The lower polymerization temperatures, however, gave polymer with a higher melting point and degree of crystallinity, presumably due to a reduction in the number of monomer reversals and/or chain branches. A wide variety of copolymers, including graft copolymers, has also been reported.[164–166]

Although considered to be a thermoplastic, PVF is not processed by melt techniques, such as injection molding, since the polymer is thermally unstable above its melting point. Free-standing films are prepared using organosol technology in which the polymer is mixed with a "latent" solvent, pigments, and stabilizers, heated to a temperature sufficient to convert the mixture into a fluid, single-phase composition, and extruded.[167] The solvent, which is an organic liquid with a boiling point of above 100°C, is then evaporated and recovered. Typical solvents include γ-butyrolactone, propylene carbonate, N-acetylmorpholine, and tetramethylenesulfone. Various methods, such as flame or corona treatment or wet chemical etching, may be employed to increase adhesion of the film to substrates. PVF may also be applied by dispersion or powder coating methods.

15.6.4.2. Properties

PVF is a semicrystalline polymer with an atactic, planar zigzag structure and a degree of crystallinity of 20–60%, depending on polymerization and processing conditions.[146,163] The melting point varies with polymerization conditions and is generally in the range of 185–210°C. Second-order transitions have been detected at −15 to −20°C and at 40 to 70°C. Data from a number of solution techniques in DMF gave a number-average molecular weight of $7.6–23.4 \times 10^4$ and a molecular weight distribution of 2.5–5.6.[169] This relatively wide distribution was ascribed to chain branching. Recent NMR data indicate that 10–13% of the monomer units are reversed (i.e., head-to-head) in commercial samples,[170,171] although higher values were initially reported.[172] Monomer reversals have also been identified in low-molecular-weight oligomers produced by telomerization of VF with trifluoromethyl iodide.[173] The presence of tertiary fluorines was detected by high-resolution fluorine NMR, suggesting the presence of chain branching which increases as the polymerization temperature increases.[171]

Properties of PVF films are summarized in Table 7. Electrical properties, such as dielectric constant, fall between those of PVDF and the more highly fluorinated polymers. The polymer is insoluble in organic solvents and resistant to acids and bases below 100°C, but will dissolve in polar solvents, such as amides, ketones, esters, and sulfones at higher temperatures. It has excellent resistance to weathering and UV light, and films with high transmission of both UV and visible light can be prepared. PVF show good resistance to high-energy irradiation. Thermal decomposition in air or nitrogen generates hydrogen fluoride and an unsaturated residue, which results in fairly high char yield.

15.6.4.3. Applications and Commercial Aspects

The major application for PVF is in the form of decorative or protective films for indoor or outdoor applications. Both clear and pigmented films are used. PVF-clad metal or plastic laminates are used as exterior siding in industrial and residential buildings, affording excellent resistance to weathering and soiling. A significant indoor application is in the coating of wall and ceiling panels in passenger aircraft. PVF films are also used as release sheeting in the molding of thermoset plastics, and for covering solar cells.

PVF films are manufactured by Du Pont ("Tedlar"). Production in 1988 is estimated to be 1.6×10^6 kg, the selling price ranging from 24 to 57 \$ kg^{-1}.

15.7. OTHER FLUORINE-CONTAINING PLASTICS

In addition to the major commercial products already discussed, many other fluoroplastics have been prepared. The motivation for introducing fluorine is generally to impart some fluoropolymer-like features to the substrate, such as lower surface energy, improved optical or electrical properties, or increased thermal stability and solvent resistance, although increased solubility in organic solvents is often observed in partially fluorinated polymers. Considering that some or nearly all hydrogens in organic polymers may be replaced with fluorines, the number of theoretically possible fluorinated polymers is nearly limitless. Space does not permit a complete or detailed treatment of this subject, but a few examples, including both "addition" (chain-growth) and "condensation" (step-growth) polymers, will be cited to illustrate trends and approaches. Some may be available commercially, although on a very limited scale, while others are laboratory curiosities at present.

15.7.1. Addition Polymers

Fluorinated olefins, including TFE, CTFE, VDF, and VF, will copolymerize with a wide variety of olefins besides ethylene. The copolymers with greatest commercial significance, at present, are elastomers (e.g., the HFP/VDF and TFE/propylene co-polymers; see Chapter 16), or are used mostly as coatings (e.g., copolymers of TFE or CTFE with nonfluorinated vinyl ethers; see Chapter 17) or membranes (e.g., the perfluorinated ionomers; see Chapter 18). Representative thermoplastics include the copolymers of TFE with isobutylene,[174] of VDF with hexafluoroisobutylene (2-trifluoromethyl-3,3,3-trifluoropropene),[175] and of hexafluoroisobutylene with vinyl acetate,[176] all largely alternating structures. The VDF/hexafluoroisobutylene copolymer is said to be a limited commercial product.[177] It is melt processible and, compared with PTFE, has the same melting point (327°C), a higher continuous use temperature (280°C), and better mechanical properties, but inferior solvent resistance.

Other widely explored classes of addition polymers are the poly(fluoroalkyl (meth)acrylates), which are of particular interest for contact lenses, as cladding materials for optical fibers, and as electron beam or X-ray resists for use in microelectronics.[178] Introduction of fluorine, usually in the alcohol portion of these esters, can provide a lower refractive index, higher transparency, greater chemical stability, and better low-temperature flexibility as compared with nonfluorinated analogs. In one recent example, a polymer of 2,2-bis(trifluoromethyl)propyl methacrylate was reported to be useful as an optical fiber sheath, having an index of refraction of 1.395 and a glass transition temperature of 150°C.[179] Acrylates and methacrylates of fluorinated alcohols, typically $CF_3(CF_2)_n(CH_2)_mOH$ with $n = 0$–10 and $m = 1$ or 2, give polymers which can have lower surface energies than PTFE, due to the high percent of trifluoromethyl groups on the polymer surface,[180] suggesting the use of these materials as leveling agents, dispersants, and oil and water repellants. The fluorinated side-chains also tend to crystallize more readily than their straight-chain hydrocarbon analogs.[181] Polymers of methyl (trifluoromethyl)acrylate,[182] 2,2,2-trifluoroethyl-α-chloroacrylate,[183] and related compounds[184] are useful as positive electron-beam resists due to the high yield of chain cleavage products formed upon irradiation.

Poly(styrenes) and poly(α-methylstyrenes) with fluoroalkyl or fluoroalkenyl side-chains are known,[185] some of which are useful as resists, coatings, and mold release agents with low surface tensions. The homopolymer of perfluoroallyl vinyl ether is reported to provide optical lenses with low refractive index and high transparency.[186] Cationic ring-opening polymerization of 2-perfluoroalkyloxazolines provides poly(N-perfluoroacyl ethyleneimines) with strictly linear backbones and low surface energies[187]; their block copolymers with 2-alkyloxazolines behave as surfactants with low critical micelle concentrations.[188]

Fluorine-containing poly(acetylenes) show a variety of physical properties. Hexafluoro-2-butyne undergoes both radical and anionic polymerization; the products are insoluble and infusible white solids with thermal stabilities greater than PTFE.[189] An insoluble, thermally stable polymer is also produced by anionic polymerization of perfluoro-1,3-butadiene.[190] The (perfluoroalkyl)acetylene $C_6F_{13}C\equiv CH$ gives a soluble, low-molecular-weight polymer using a tungsten hexachloride/tetraphenyl tin catalyst.[191] This and related transition metal catalysts will also polymerize o-(trifluoromethyl)phenylacetylene to a high-molecular-weight polymer which is soluble in organic solvents and more thermally stable than poly(phenylacetylene).[192] Stable, glassy resins have been produced by thermal polymerization of diacetylenes of the type HC≡C(CF$_2$)$_n$C≡CH (n = 6, 8, 10).[193] The fluorine-containing polyacetylenes show, at best, semiconductor properties when doped.

Fluorinated carbonyl compounds have unusual properties as compared with their hydrocarbon analogs, and several are known to undergo unique addition homopolymerizations. Trifluoroacetaldehyde and chlorofluoroacetaldehydes polymerize under cationic, anionic, or free-radical conditions to give either crystalline or amorphous homopolymers.[194] Thiocarbonyl fluoride will polymerize under anionic or free-radical conditions to give a tough elastomer with a very low glass transition temperature (−118°C) and outstanding resiliency, but it slowly crystallizes to a stiff, low melting (35°C) plastic.[195] Copolymers of thiocarbonyl fluoride with vinyl monomers, and polymers of other thiocarbonyl compounds have also been prepared. Tetrafluorothiirane homopolymerizes under radical conditions to a tough resin with a melting point of about 175°C.[196] These sulfur-containing fluoropolymers have relatively poor thermal stability as compared to other highly fluorinated polymers, generally decomposing at temperatures of 300°C or less, and have received little attention in recent years.

15.7.2. Condensation Polymers

Fluorinated groups have been introduced into a variety of condensation polymers, generally in an effort to improve their electrical properties, moisture resistance, thermal stability, or solubility in organic solvents. The fluorinated group has to be located with care to insure stability in the resulting product. For example, alcohols and amines with α-fluorines and their corresponding esters and amides have, at best, fleeting stability, and amides and esters of α-fluorinated acids hydrolyze readily, so fluoroalkyl groups are generally insulated from the functional groups by spacers.

Perhaps the most common moiety in fluorinated condensation polymers is the 2,2-diaryl-1,1,1,3,3,3-hexafluoropropyl unit 2, which has been incorporated into poly-amides, -esters, -carbonates, -imides, -urethanes, -formals, and -benzoxazoles (see Chapter 19).[197] Monomers based on 2, including diamines, dianhydrides, diphenols, dicarboxylates, and bis(aminophenols), are generally prepared by condensation of substituted arenes with hexafluoroacetone, followed by normal functional group manipulation. In one recent example, polyformals were prepared from the 4,4′-diphenol based on structure 2 ("Bisphenol AF") and its nonfluorinated analog ("Bisphenol A"), and mixtures of the two.[198] Increased amounts of the fluorinated diphenol resulted in modest decreases in glass transition temperature and thermal stability, and increases in contact angle and solubility in organic solvents.

Fluorine-containing polyimides are of particular interest for applications in electronics. In recent examples, polyimides from diamines or dianhydrides based on 2 have lower dielectric constants (<3) and moisture absorption (<1%), but also reduced

2

mechanical properties, glass transition temperatures, and solvent resistance as compared with nonfluorinated analogs.[199] Little effect was observed on thermal stability and thermal expansion coefficients. Polyimides based on aromatic diamines with fluoroalkoxy side chains have also shown significantly improved dielectric properties, lower indices of refraction, and decreases in glass transition temperature.[200] Many variations in the fluorinated diamines or dianhdrides have been reported in an effort to achieve an optimum combination of thermal, mechanical and electrical properties.[201]

In other aromatic polymers, placement of trifluoromethyl groups on 4,4'-diamino- or dihydroxy-biphenyl units has resulted in a significant increase in solubility and optical clarity in polyamides and thermotropic polyesters,[202] addition of a perfluoroalkenyloxy group to the diester portion of an aromatic polyester has produced improved chemical resistance, stain resistance, and weatherability,[203] and incorporation of a fluorinated side chain in a poly(phenylene oxide) has led to improved heat resistance, solubility in organic solvents, and electrical properties.[204] Phenolic and epoxy resins have been modified by the incorporation of perfluoroalkylene groups for enhanced oil and water repellency.[205] Incorporation of perfluoroalkyl segments into the backbones of poly(ether sulfones)[206] and aromatic polyethers[207] decreased dielectric constants and improved optical clarity, but lowered glass transition temperatures. A variety of polymers with fluorine atoms bonded directly to the aromatic rings is also known.[208]

Aliphatic poly-esters,[209] -amides,[210] and -carbonates[211] with fluorinated units in the backbone or on side chains have also been investigated. Small amounts of fluorine can have dramatic effects on surface properties (e.g., water repellency) when incorporated as side chains, while incorporation in the polymer backbone tends to have greater impact on melting behavior and solubility. Fluorine-containing segmented polyurethanes have superior blood compatibility and durability for use in an artificial heart pump.[212] Fluorine-containing metathesis polymers, generally based on substituted bicyclo[2.2.1]-heptenes and -heptadienes, have been studied, with an emphasis on effects of substituents and polymerization catalysts on the stereo- and regiochemistry of the resultant polymers.[213]

15.7.3. Surface-Fluorinated Plastics

In addition to the use of fluorinated monomers, fluorine may also be incorporated via postpolymerization reactions on polymers. Since many of the desired effects of fluorine (decreased surface energy and moisture absorption, improved solvent resistance and biocompatibility) are largely surface phenomena, this approach could offer a more efficient use of this relatively expensive element and retain the often desirable bulk properties (e.g., mechanical performance and processibility) of the polymeric substrate. Thus, poly(ethylene) surfaces can be fluorinated by careful treatment with elemental fluorine,[214] a process which is used commercially to improve the solvent resistance of automobile gasoline tanks (see Chapter 22). Fluorine has also been used to modify the surfaces of polysulfone and polystyrene membranes to increase selectivity in gas separations,[215] and to treat aromatic ladder polymers to develop conductive patterns.[216] Other approaches to surface modification include grafting of fluorinated monomers onto polymers such as poly(ethylene) or poly(propylene),[217] or perfluoroacylation of hydroxy-containing polymers such as cellulose diacetate.[218]

15.8. OUTLOOK AND CONCLUSIONS

The estimated total annual production of fluorinated plastics was at 55–60 *million* kg in 1988. In comparison, over 20 *billion* kg of polyethylene were produced in the same period. The comparison becomes less dramatic, however, when value-in-use is considered, since the selling prices of the fluoroplastics are from one to several orders of magnitude greater than that of poly(ethylene).

For several reasons, the fluorinated plastics seem well positioned for continued growth: their major applications are in fields such as electronics, optics, and solar energy, which are likely to experience significant growth over the next decades; and new fluorine-containing polymers are being developed with properties tailored for specific applications. The long service lifetimes provided by fluorinated plastics are increasingly attractive features in a world which is less tolerant of breakdowns and environmental disasters Long lifetimes also reduce concern about disposal.

Research interest in fluoroplastics continues at a high level, as evidenced by the volume of recent publications dealing with new structures, applications, and studies of physical properties. It is interesting to note the recent results on the molecular weight and molecular-weight distribution of PTFE fifty years after its discovery, and the continuing fascination with the structural properties of PVDF and its copolymers.

One major challenge facing the fluoroplastics industry stems from the concern over the destruction of Earth's ozone layer by the volatile chlorofluorocarbons (CFCs, see Chapter 28). CFCs and related hydrochlorofluorocarbons (HCFCs) are precursors of the major commercial fluoromonomers and some are also widely used as polymerization solvents. As these compounds come under increasing regulation, the economics, at least, of the fluoroplastics industry are certain to change. On the other hand, creative chemists should view this challenge as an opportunity, since whole new families of fluorine-containing intermediates will become available commercially as CFC replacements over the next decade.[219] They could well lead to a variety of interesting new monomers and polymers.

15.9. REFERENCES

1. B. E. Smart, in: *Supplement D, The Chemistry of Functional Groups*, (S. Patai and Z. Rappoport, eds.), Chapter 14, Wiley, New York (1983).
2. Brit. Pat. 465,520 (1937) to I. G. Farbenindustrie [*CA 31*, 7145 (1937)].
3. R. J. Plunkett, U.S. Pat 2,230,654 (1941) to Kinetic Chemicals, Inc. [*CA 35*, 3365 (1941)].
4. W. T. Miller, in: *Preparation, Properties and Technology of Fluorine and Organofluorine Compounds* (C. Slesser and S. R. Schram, eds.), McGraw-Hill, New York (1951).

5. D. P. Carlson and W. Schmiegel, in: *Ullmann's Encyclopedia of Industrial Chemistry*, 5th edn., Vol. A11, pp. 393–429; S. Kawachi and M. Nanba, *J. Synth. Org. Chem. (Japan) 42*, 829 (1984); R. F. Brady, *Chemistry in Britain,* 427 (1990).

6. S. V. Gangal, in: *Encyclopedia of Polymer Science and Engineering* (H. F. Mark, N. M. Bikales, C. G. Overberger and G. Menges, eds.), Vol. 16, pp. 577–642, Wiley, New York (1989).

7. S. Chandrasekaran, in: *Encyclopedia of Polymer Science and Engineering* (H. F. Mark, N. M. Bikales, C. G. Overberger, and G. Menges, eds.), Vol. 3, pp. 463–480, Wiley, New York (1985).

8. W. A. Miller, in: *Encyclopedia of Polymer Science and Engineering* (H. F. Mark, N. M. Bikales, C. G. Overberger, and G. Menges, eds.) Vol. 3, pp. 480–491, Wiley, New York (1985).

9. D. Brasure and S. Ebnesajjad, in: *Encyclopedia of Polymer Science and Engineering* (H. F. Mark, N. M. Bikales, C. G. Overberger, and G. Menges, eds.), Vol. 17, pp. 468–491, Wiley, New York (1989).

10. J. E. Dohany and J. S. Humphrey, in: *Encyclopedia of Polymer Science and Engineering* (H. F. Mark, N. M. Bikales, C. G. Overberger, and G. Menges, eds.), Vol. 17, pp. 532–548, Wiley, New York (1989).

11. A. J. Lovinger, in: *Developments in Crystalline Polymers* (D. C. Bassett, ed.), Chapter 5, Applied Science Publishers, London (1982).

12. J. M. Hamilton, Jr., in: *Advances in Fluorine Chemistry*, Vol. 3 (M. Stacey, J. C. Tatlow, and A. G. Sharpe, eds.), Butterworth, London (1963).

13. F. B. Downing, A. F. Benning, and R. C. McHarness, U.S. Pat. 2,551,573 (1951) to Du Pont [*CA 45*, 9072 (1956)].

14. H. Ukihashi and M. Hisasue, U.S. Pat. 3,459,818 (1966) to Asahi Glass Co., Ltd [*CA 71*, 123506 (1969)].

15. D. A. Nelson, U.S. Pat. 2,758,138 (1956) to Du Pont [*CA 51*, 3654 (1957)]; E. H. Ten Eyck and G. P. Larson, U.S. Pat. 2,970,176 (1961) to Du Pont [*CA 55*, 17498 (1961)]; H. Niimiya, Brit. Pat. 1,016,016 (1963) to Daikin Industries [*CA 64*, 14089 (1966)]; N. E. West, U.S. Pat. 3,873,630 (1972) to Du Pont [*CA 80*, 82053 (1974)].

16. G. Siegemund, W. Schwertfeger, A. Feiring, B. Smart, F. Behr, H. Vogel and B. McKusick, in: *Ullmann's Encyclopedia of Industrial Chemistry*, 5th edn., Vol. A11, pp. 361–362, VCH, Weinheim (1988).

17. H. Millauer, W. Schwertfeger, and G. Siegemund, *Angew. Chem., Int. Ed. Engl. 24*, 161 (1985).

18. H. H. Biggs and J. L. Warnell, Fr. Pat. 1,322,597 (1963) to Du Pont [*CA 59*, 11423 (1963)]; D. P. Carlson, U.S. Pat. 3,536,733 (1970) to Du Pont [*CA 74*, 22681 (1971)].

19. M. Millauer and W. Lindner, Ger. Pat. 2,658,382 (1978) to Hoechst [*CA 89*, 75641 (1978)].

20. M. Ikeda, M. Miura, and A. Aoshima, EP-OS 64,293 (1982) to Asahi Chemical [*CA 98*, 126783 (1983)]; Daikin Industries, Ltd, Jpn. Kokai Tokkyo Koho JP 58,131,976 (1983) [*CA 100*, 6309 (1984)].

21. H. S. Eleuterio and R. W. Meschke, U.S. Pat. 3,358,003 (1967) to Du Pont [*CA 68*, 29573 (1968)]; R. Sulzbach and F. Heller, Ger. Pat. 2,557,655 (1977) to Hoechst [*CA 87*, 151997 (1977)].

22. E. P. Moore, A. S. Milian, and H. S. Eleuterio, U.S. Pat 3,250,808 (1966) to Du Pont [*CA 65*, 13554 (1966)]; E. P. Moore, U.S. Pat. 3,322,826 (1967) to Du Pont [*CA 67*, 44292 (1967)]; T. Martini, U.S. Pat. 4,118,421 (1978) to Hoechst [*CA 88*, 104700 (1978)].

23. C. G. Fritz and S. Selman, U.S. Pat. 3,291,843 (1966) to Du Pont [*CA 66*, 37427 (1967)]; C. G. Fritz, E. P. Moore, and S. Selman, U.S. Pat. 3,114,778 (1963) to Du Pont [*CA 60*, 6750 (1964)].

24. O. A. Blum, U.S. Pat. 2,590,433 (1949) to Kellog [*CA 47*, 1180 (1953)]; J. W. Jewell, U.S. Pat. 3,014,015 (1961) to 3M [*CA 58*, 6945 (1963)].

25. L. E. Gardner, U.S. Pat. 3,789,016 (1974) to Phillips Petroleum [*CA 80*, 82052 (1974)].

26. K. Ohira, S. Yoneda, and I. Goto, Jpn. Kokai Tokkyo Koho JP 01 29,328 (1989) to Asahi Glass Co. [*CA 111*, 173591 (1989)].

27. J. Harmon, U.S. Pat. 2,599,631 (1952) to Du Pont [*CA 47*, 1725 (1953)].

28. D. D. Coffman and R. D. Cramer, U.S. Pat. 2,461,523 (1949) to Du Pont [*CA 43*, 3437 (1949)]; D. D. Coffman and T. A. Ford, U.S. Pat. 2,419,010 (1947) to Du Pont [*CA 41*, 4964 (1947)]; A.

Bluemcke, P. Fischer, and W. Krings, Ger. Pat. 1,259,329 (1968) to Dynamit Nobel AG [*CA 68*, 68452 (1968)].

29.   J. W. Hamersma, U.S. Pat 3,621,067 (1971) to Atlantic Richfield [*CA 76*, 86375 (1972)]; J. W. Hamersma, U.S. Pat. 3,642,917 (1972) to Atlantic Richfield [*CA 76*, 154387 (1972)]; D. Sianesi and G. Nelli, U.S. Pat. 3,200,160 (1965) to Montecatini [*CA 61*, 1755 (1964)].

30.   C. B. Miller, U.S. Pat. 2,628,989 (1953) to Allied Chemical [*CA 48*, 1406 (1954)]; O. Scherer, A. Steinmetz, H. Kuhn, W. Wetzel, and K. Grafen, U.S. Pat. 3,183,277 (1965) to Hoechst [*CA 55*, 12295 (1961)].

31.   C. F. Feasley and W. H. Stover, U.S. Pat. 2,627,529 (1953) to Mobil [*CA 48*, 1406 (1954)].

32.   W. S. Bernhart and R. M. Mantell, U.S. Pat. 2,774,799 (1956) to Kellog [*CA 51*, 12955 (1957)].

33.   J. N. Meussdoerffer and H. Niederpruem, Ger. Offen. 2,044,370 (1972) to Farbenfab Bayer AG [*CA 77*, 20345 (1972)].

34.   M. Hauptschein and A. H. Fainberg, U.S. Pat. 3,188,356 (1965) to Pennsalt [*CA 63*, 6859 (1965)].

35.   J. D. Calfee and C. B. Miller, U.S. Pat. 2,734,090 (1956) to Allied Chemical [*CA 50*, 9441 (1956)]; M. W. Farlow and E. L. Muetterties, U.S. Pat. 2,894,996 (1959) to Du Pont [*CA 55*, 2481 (1961)]; M. E. Miville and J. J. Earley, U.S. Pat. 3,246,041 (1966) to Pennsalt [*CA 64*, 19410 (1966)].

36.   J. W. Clayton, in: *Fluorine Chemistry Reviews*, Vol. 1 (P. Tarrant, ed.), Marcel Dekker, New York (1967).

37.   A. Pajaczkowski and J. W. Spoors, *Chem. Ind. (London)* 695 (1964).

38.   H. S. Eleuterio, U.S. Pat. 2,958,685 (1960) to Du Pont [*CA 55*, 6041 (1961)].

39.   S. V. Gangal, U.S. Pat. 4,186,121 (1980) to Du Pont [*CA 92*, 129920 (1980)].

40.   S. Malhotra, U.S. Pat. 4,725,644 (1988) to Du Pont [*CA 108*, 168554 (1988)].

41.   A. E. Kroll, U.S. Pat. 2,750,350 (1956) to Du Pont [*CA 50*, 13507 (1956)].

42.   T. Suwa, M. Takehisa, and S. Machi, *J. Appl. Polym. Sci.* 18, 2249 (1974).

43.   M. M. Brubaker, U.S. Pat. 2,393,967 (1946) to Du Pont [*CA 40*, 3648 (1946)].

44.   K. L. Berry, U.S. Pat. 2,559,752 (1951) to Du Pont [*CA 46*, 3064 (1952)].

45.   S. G. Bankoff, U.S. Pat. 2,612,484 (1952) to Du Pont [*CA 47*, 3618 (1953)].

46.   T. Folda, H. Hoffmann, H. Chanzy, and P. Smith, *Nature 333*, 55 (1988); B. Luhmann and A. E. Feiring, *Polymer 30*, 1723 (1989); S. Yamaguchi, M. Tatemoto, and M. Tsuji, *Kobunshi Ronbunshu 47*, 105 (1990).

47.   M. B. Mueller, P. O. Salatiello, and H. S. Kaufman, U.S. Pat. 3,759,883 (1973) to Allied Chemical [*CA 80*, 4241 (1974)]; T. R. Doughty, C. A. Sperati, and H. Un, U.S. Pat. 3,855,191 (1974) to Du Pont [*CA 82*, 73855 (1975)].

48.   A. J. Cardinal, W. L. Edens, and J. W. VanDyk, U.S. Pat. 3,142,665 (1964) to Du Pont [*CA 61*, 16259 (1964)].

49.   J. M. Downer, W. G. Rodway, and L. S. J. Shipp, Ger. Offen 2,521,738 (1975) to ICI [*CA 84*, 60465 (1976)].

50.   C. A. Sperati and H. W. Starkweather, *Adv. Polym. Sci.* 2, 465 (1961).

51.   B. Chu, C. Wu, and W. Buck, *Macromolecules 22*, 831 (1989).

52.   R. Roberts and R. F. Anderson, U.S. Pat. 3,766,133 (1973) to Du Pont [*CA 65*, 13899 (1966)].

53.   P. E. Frankenburg, in: *Ullmann's Encyclopedia of Industrial Chemistry*, 5th edn., Vol. A10, pp. 649–650, VCH, Weinheim, (1987).

54.   M. I. Bro and B. W. Sand, U.S. Pat. 2,946,763 (1960) to Du Pont [*CA 54*, 26015 (1960)]; M. J. Couture, D. L. Schindler, and R. B.Weiser, U.S. Pat. 3,132,124 (1964) to Du Pont [*CA 61*, 1970 (1964)].

55.   W. F. Gresham and A. F. Vogelpohl, U.S. Pat. 3,635,926 (1972) to Du Pont [*CA 75*, 37034 (1971)]; J. F. Harris and D. I. McCane, U.S. Pat. 3,132,123 (1964) to Du Pont [*CA 61*, 1968 (1964)]; D. P. Carlson, U.S. Pat. 3,536,733 (1970) to Du Pont [*CA 74*, 54360 (1971)].

56.   A. A. Khan and R. A. Morgan, U.S. Pat. 4,380,618 (1983) to Du Pont [*CA 99*, 6231 (1983)]; J. Blaise and J.-L. Herisson, U.S. Pat. 4,384,092 (1983) to Produits Chimiques Ugine Kuhlmann [*CA 96*, 69988 (1982)].

57. H. Wachi, M. Kaya, and S. Shintani, Jpn. Kokai Tokkyo Koho JP 62,223,210 (1987) to Asahi Glass Company [*CA 108*, 132795 (1988)]; Y. T. Kometani, M. Tatemoto, and S. Sakata, Ger. Offen. 2,104, 077 (1971) to Daikin Industries [*CA 75*, 118787 (1971)].

58. E. W. Slocum, A. C. Sobrero, and R. C. Wheland, U.S. Pat. 4,861,845 (1989) to Du Pont [*CA 112*, 99506 (1990)].

59. D. P. Carlson, U.S. Pat. 3,642,742 (1972) to Du Pont [*CA 74*, 54360 (1971)].

60. R. C. Schreyer, U.S. Pat. 3,085,083 (1963) to Du Pont [*CA 59*, 1806 (1963)]; D. P. Carlson, D. L. Kerbow, T. J. Leck, and A. H. Olson, U.S. Pat. 4,599,386 (1986) to Du Pont [*CA 106*, 177121 (1987)]; D. P. Carlson, U.S. Pat. 3,674,758 (1972) to Du Pont [*CA 77*, 89390 (1972)]; M. D. Buckmaster, PCT Int. Appl WO 89 11,495 (1989) to Du Pont [*CA 112*, 159259 (1990)].

61. J. F. Imbalzano and D. L. Kerbow, U.S. Pat. 4,943,658 (1988) to Du Pont [*CA 107*, 116143 (1987)].

62. J. Goodman and S. Andrews, *Solid State Technol. 33*, 65 (1990).

63. R. S. Altland, *Mod. Plast. 62*, 200 (1985).

64. W. M. D. Bryant, *J. Polym. Sci. 56*, 277 (1962); A. D. English and O. T. Garza, *Macromolecules 12*, 352 (1979).

65. D. F. Eaton and B. E. Smart, *J. Am. Chem. Soc. 112*, 2821 (1990).

66. C. W. Bunn and E. R. Howells, *Nature 174*, 549 (1954); B. L. Farmer and R. K. Eby, *Polymer 22*, 1487 (1981); R. I. Beecroft and C. A. Swenson, *J. Appl. Phys. 30*, 1793 (1959); M. Springborg and M. Lev, *Phys. Rev. B 40*, 3333 (1989).

67. K. L. Berry and J. H. Peterson, *J. Am. Chem. Soc. 73*, 5195 (1951).

68. T. Suwa, M. Takehisa, and S. Machi, *J. Appl. Polym. Sci. 17*, 3253 (1973); ASTM D 1457-83.

69. H. Starkweather and S. Wu, *Polymer 30*, 1669 (1989); W. H. Tuminello, *Polym. Eng. Sci. 29*, 645 (1989).

70. C. A. Sperati, in: *Polymer Handbook*, 2nd edn. (J. Brandrup and E. H. Immergut, eds.), Wiley-Interscience, New York (1975).

71. Y. Araki, *J. App. Polym. Sci. 9*, 3585 (1965); *9*, 3575 (1965).

72. Y. P. Khanna, G. Chomyn, R. Kumar, N. S. Murthy, K. P. O'Brien, and A. C. Reimschuessel, *Macromolecules 23*, 2488 (1990).

73. R. K. Eby and F. C. Wilson, *J. Appl. Phys. 33*, 2951 (1962).

74. J. C. Siegle, L. T. Muus, T.-P. Lin, and H. A. Larsen, *J. Polym. Sci., Part A 2*, 391 (1964).

75. R. B. Seymour and C. E. Carraher, *Structure–Property Relationships in Polymers*, Plenum Press, New York (1984).

76. R. Dahm, in: *Surface Analysis and Pretreatment of Plastics and Metals* (D. Brewis, ed.), Applied Science Publishers, New York (1982); L. M. Siperko and R. R. Thomas, *J. Adhesion Sci. Technol. 3*, 157 (1989).

77. R. R. Rye and G. W. Arnold, *Langmuir 5*, 1331 (1989).

78. D. T. Clark and D. R. Hutton, *J. Polym. Sci., Polym. Chem. Ed. 25*, 2643 (1987).

79. C. A. Costello and T. J. McCarthy, *Macromolecules 17*, 2940 (1984).

80. R. C. Bening and T. J. McCarthy, *Polym. Prepr. 29*, 336 (1988).

81. C. A. Costello and T. J. McCarthy, *Macromolecules 20*, 2819 (1987).

82. Z. Iqbal, D. M. Ivory, J. S. Szobota, R. L. Elsenbaumer, and R. H. Baughman, *Macromolecules 19*, 2992 (1986).

83. R. Clough, in: *Encyclopedia of Polymer Science and Engineering*, Vol. 13, Wiley, New York (1988); J. A. Kelber and J. W. Rogers, in: *Encyclopedia of Materials Sciences and Engineering*, Supplementary Vol. 1 (R. W. Cahn, ed.), Pergamon Press, Oxford (1988).

84. E. A. Hegazy, I. Ishigaki, A. Rabie, A. M. Dessouki, and J. Okamoto, *J. Appl. Polym. Sci. 26*, 3871 (1981); B. D. Gupta and A. Chapiro, *Eur. Polym. J. 25*, 1145 (1989).

85. Y. Ando, I. Seki, H. Yagiyu, K. Endo, F. Nakahigashi, and T. Endo, Jpn. Kokai Tokkyo Koho JP 63,250,027 (1988) to Hitachi Cable [*CA 110*, 77716 (1989)].

86. K. Batzar, U.S. Pat. 4,818,350 (1989) to Du Pont [*CA 111*, 59693 (1989)].

87. R. W. Gore, U.S. Pat. 3,962,153 (1976) to Gore Associates [*CA 83*, 11632 (1975)].

88. K. Lunkwitz, H. J. Brink, D. Handte, and A. Ferse, *Radiat. Phys. Chem. 33*, 523 (1989).

89. C. S. McEwen, Eur. Pat. Appl. EP 320,901 (1989) to Du Pont [*CA 112*, 47203 (1990)].

90.   J. Eveleigh, W. DeLouche, and R. K. Kobos, Eur. Pat. Appl. EP 281,368 (1988) to Du Pont [*CA 110*, 227752 (1989)].
91.   E. N. Squire, U.S. Pat. 4,754,009 (1988) to Du Pont [*CA 110*, 77691 (1989)]; P. R. Resnick, *Polym. Prepr. 31*, 312 (1990).
92.   "Cytop" Technical Bulletin, Asahi Glass Company, Ltd. (1988).
93.   H. Matsuo, I. Kaneko, M. Kanba, and H. Nakamura, Jpn. Kokai Tokkyo Koho JP 63,238,115 (1988) to Asahi Glass Co. [*CA 110*, 135936 (1989)]; H. Nakamura, S. Samejima, M. Kanba, and G. Kojima, Jpn. Kokai Tokkyo Koho JP 63,238,111 (1988) to Asahi Glass Co. [*CA 110*, 115551 (1989)]; M. Nakamura, I. Kaneko, K. Oharu, G. Kojima, M. Matsuo, S. Samejima, and M. Kamba, U.S. Pat. 4,897,457 (1990) to Asahi Glass Co. [*CA 111*, 135000 (1989)].
94.   W. T. Miller and J. T. Maynard, U.S. Pat. 2,626,254 (1953) to US Atomic Energy Commission [*CA 47*, 4652 (1953)]; A. L. Dittman and J. M. Wrightson, U.S. Pat. 2,705,706 (1955) to M. W. Kellogg Co. [*CA 49*, 13695 (1955)]; A. L. Dittman and J. M. Wrightson, U.S. Pat. 2,784,176 (1957) to M. W. Kellogg [*CA 51*, 7758 (1957)]; Brit Pat. 838,651 (1960) to Hoechst [*CA 54*, 23437 (1960)]; J. W. Jewell, U.S. Pat. 3,014,015 (1961) to 3M [*CA 58*, 6945 (1963)].
95   W. E. Hanford, U.S. Pat. 2,820,027 (1958) to 3M [*CA 52*, 5884 (1958)].
96.   W. M. Thomas and M. T. O'Shaughnessy, *J. Polym. Sci. 11*, 455 (1953).
97.   J. M. Hamilton, U.S. Pat. 2,569,524 (1951) to Du Pont [*CA 46*, 1299 1952)]; A. L. Dittman, H. J. Passino, and J. M. Wrightson, U.S. Pat. 2,689,241 (1954) to M. W. Kellogg [*CA 49*, 11681 (1955)]; H. J. Passino, A. L. Dittman, and J. M. Wrightson, U.S. Pat. 2,783,219 (1957) to M. W. Kellogg [*CA 51*, 7758 (1957)]; H. J. Passino, A. L. Dittman, and J. M. Wrightson, U.S. Pat. 2,820,026 (1958) to 3M [*CA 52*, 5884 (1958)]; R. L. Herbst and B. F. Landrum, U.S. Pat 2,842,528 (1958) to 3M [*CA 52*, 21245 (1958)].
98.   Brit. Pat. 805,103 (1958) to 3M [*CA 53*, 11890 (1959)].
99.   H. Matsuo, *J. Polym. Sci. 25*, 234 (1957).
100. A. H. Scott, D. J. Scheiber, A. J. Curtis, J. T. Lauritzen, and J. D. Hoffman, *J. Res. Natl. Bur. Stand. 66A*, 269 (1962).
101. G. V. D. Tiers and F. A. Bovey, *J. Polym. Sci., Part A-1, 1*, 833 (1963).
102. S. L. Madorsky and S. Straus, *J. Res. Natl. Bur. Stand. 55*, 223 (1955).
103. M. W. Pelter and R. T. Taylor, *J. Polym. Sci., Polym. Chem. Ed. 26*, 2651 (1988).
104. B. U. Kolb, P. A. Patton, and T. J. McCarthy, *Macromolecules 23*, 366 (1990); A. J. Diaz and T. J. McCarthy, *Macromolecules 18*, 1826 (1985).
105. Z. Plzak, F. P. Dousek, and J. Jansta, *J. Chromatogr. 147*, 137 (1978); A. Oku, S. Nakagawa, H. Kato, and H. Taguchi, *Nippon Kagaku Kaishi* 2021 (1988) [*CA 110*, 76244 (1989)].
106. A. Monnet and R. Bensa, *Energ. Nucl. 13*, 123 (1971) [*CA 75*, 77784 (1971)].
107. M. Modena, C. Garbuglio, and M. Ragazzini, *Polym. Lett. 10*, 153 (1972).
108. W. E. Hanford and J. R. Roland, Jr. U.S. Pat. 2,468,664 to Du Pont (1949) [*CA 43*, 5410 (1949)].
109. H. Sulzbach, U.S. Pat. 4,338,237 (1982) to Hoechst A.G. [*CA 96*, 123539 (1982)].
110. D. P. Carlson, U.S. Pat. 3,624,250 (1971) to Du Pont [*CA 73*, 67223 (1970)]; M. M. Mueller and S. Chandrasekaran, U.S. Pat. 3,847,881 (1974) to Allied Chemical [*CA 82*, 112677 (1975)]; H. Ukihashi and M. Yamabe, U.S. Pat. 4,123,602 (1978) to Asahi Glass Co. [*CA 90*, 88051 (1979)].
111. F. C. Wilson and H. W. Starkweather, *J. Polym. Sci., Polym. Phys. Ed. 11*, 919 (1973).
112. H. W. Starkweather, *J. Polym. Sci., Polym. Phys. Ed. 11*, 587 (1973).
113. B. Chu and C. Wu, *Macromolecules 20*, 93 (1987).
114. B. Chu, C. Wu, and W. Buck, *Macromolecules 22*, 371 (1989); Z. Wang, A. Tontisakis, W. H. Tuminello, W. Buck, and B. Chu, *Macromolecules 23*, 1444 (1990).
115. D. P. Carlson and N. E. West, U.S. Pat. 3,738,923 (1973) to Du Pont [*CA 77*, 49164 (1972)]; E. Aronoff and K. S. Dhami, Ger. Offen. 2,445,795 (1975) to International Standard Electric Corp. [*CA 83*, 29225 (1975)].
116. S. K. Randa, C. R. Frywald, and D. P. Reifschneider, *Proc. Int. Wire Cable Symp.* 14–22 (1987) [*CA 110*, 58965 (1989)].
117. J. Emsley, *New Scientist*, p. 46 (April 22, 1989).
118. W. E. Hanford, U.S. Pat. 2,392,378 (1946) to Du Pont [*CA 40*, 5959 (1948)].

119. D. Carcano and M. Ragazzini, U.S. Pat. 3,501,446 (1970) to Montedison SPA [*CA 57*, 12741 (1962)].
120. M. Ragazzini, C. Garbuglio, D. Carcano, B. Minasso, and G. Cevidalli, *Eur. Polym. J. 3*, 129 (1967).
121. J. P. Sibilia, L. G. Roldin, and S. Chandrasekaran, *J. Polym. Sci.. Part A-2, 10*, 549 (1972).
122. H. K. Reimschuessel and F. J. Rahl, *J. Polym. Sci., Polym. Chem. Ed. 25*, 1871 (1987).
123. N. S. Muthy, S. Chandrasekaran, and H. K. Reimschuessel, *Polymer 29*, 829 (1988).
124. G. Khattab and A. Stoloff, U.S. Pat. 3,745,145 (1973) to Allied Chemical [*CA 79*, 127042 (1973)].
125. C. Garbuglio, M. Razazzini, O. Pilati, D. Carcano, and G. Cevidalli, *Eur. Polym. J. 3*, 137 (1967).
126. E. J. Aronoff, K. S. Dhami, and T.-C. Shieh, U.S. Pat. 3,894,118 (1975) to International Telephone & Telegraph Company [*CA 83*, 165204 (1975)].
127. D. Chung and H. F. Finelli, *Research & Development 79* (March 1986).
128. T. A. Ford and W. E. Hanford, U.S. Pat. 2,435,537 (1948) to Du Pont [*CA 42*, 3215 (1948)].
129. M. Hauptschein, U.S. Pat. 3,193,539 (1965) to Pennsalt Chemical [*CA 63*, 13443 (1965)]; Kureha Chemical Industry, Fr. Pat. 1,419,741 (1965) [*CA 65*, 9049 (1966)]; Daikin Kogyo Co., Brit. Pat. 1,178,227 (1970) [*CA 72*, 67510 (1970)]; J. P. Stallings. Ger. Pat. 2,063,248 (1971) to Diamond Shamrock [*CA 75*, 110713 (1971)].
130. G. H. McCain, J. R. Semancik, and J. J. Dietrich, Fr. Pat. 1,530,119 (1968) to Diamond Shamrock [*CA 71*, 3923 (1969)]; Daikin Kogyo Co., Brit. Pat. 1,179,078 (1970) [*CA 72*, 56095 (1970)].
131. Kureha Chemical Industry, Brit. Pat. 1,094,558 (1967) [*CA 68*, 40292 (1968)]; Kureha Chemical Industry, Neth. Pat. 6,613,154 (1967) [*CA 67*, 65368 (1967)]; J. E. Dohany, U.S. Pat. 3,857,827 (1974) to Pennwalt [*CA 82*, 99161 (1975)]; J. E.Dohany, U.S. Pat. 3,781,265 (1973) to Pennwalt [*CA 80*, 121515 (1974)].
132. L. A. Barber, Eur. Pat. Appl. EP 169,328 (1986) to Pennwalt [*CA 104*, 187079 (1986)].
133. D. P. Carlson, Ger. Pat. 1,806,426 (1969) to Du Pont [*CA 71*, 13533 (1969)].
134. E. S. Lo, U.S. Pat. 3,178,399 (1965) to 3M [*CA 63*, 1900 (1965)].
135. W. D. Benzinger and D. N. Robinson, U.S. Pat. 4,383,047 (1983) to Pennwalt [*CA 96*, 86613 (1982)].
136. R. Gerhard-Multhaupt, *Ferroelectrics 75*, 385 (1987).
137. K. Loufakis, K. J. Miller, and B. Wunderlich, *Macromolecules 19*, 1271 (1986).
138. C. C. Hsu and P. H. Geil, *J. Appl. Phys. 56*, 2404 (1984).
139. M. G. Broadhurst, G. T. Davis, and J. E. McKinney, *J. Appl. Phys. 49*, 4992 (1978).
140. R. Hasegawa, Y. Tanabe, M. Kobayashi, H. Tadokoro, A. Sawaoka, and N. Kawai, *J. Polym. Sci., Polym. Phys. Ed. 8*, 1073 (1970).
141. J. B. Lando and W. W. Doll, *J. Macromol Sci. Phys. B2* 219 (1968); *B4*, 896 (1970).
142. A. Richardson, P. S. Hope, and I. M. Ward, *J. Polym. Sci., Polym. Phys. Ed. 21*, 2525 (1983).
143. W. Doll and J. Lando, *J. Macromol. Sci. Phys. B4*, 897 (1970).
144. N. A. Suttle, *Mater. Des. 9*, 318 (1988).
145. D. R. Paul and J. W. Barlow, *J. Macromol. Sci., Rev. Macromol. Chem. C18*, 109 (1980); A Tanaka, H. Sawada, and Y. Kojima, *Polym. J. (Japan) 22*, 463 (1990).
146. M. Gorlitz, R. Minke, W. Trautvetter, and G. Weisgerber, *Angew. Makromol. Chem. 29/30*, 137 (1973).
147. R. C. Ferguson and E. G. Baume, *J. Phys. Chem. 83*, 1397 (1979).
148. A. J. Lovinger, D. D. Davis, R. E. Cais, and J. M. Kometani, *Polymer 28*, 617 (1987).
149. S. Ali and A. K. Raina, *Makromol. Chem. 179*, 2925 (1978).
150. H. Kise, H. Ogata, and M. Nakata, *Angew. Makromol. Chem. 168*, 205 (1989).
151. R. D. Chambers, M. Salisbury, G. Apsey, T. F. Holmes, and S. Modena, *J.Chem. Soc., Chem. Commun.* 679 (1988).
152. R. Timmerman and W. Greyson, *J. Appl. Polym. Sci. 6*, 456 (1962).
153. T. Furukawa, *Kobunshi 36*, 868 (1987).
154. S. Steinberg, U.S. Pat. 4,340,482 (1982) to Millipore [*CA 97*, 164098 (1982)].
155. A. E. Newkirk, *J. Am. Chem. Soc. 68*, 2467 (1946).
156. H. W. Starkweather, *J. Am. Chem. Soc. 56*, 1870 (1934).
157. D. D. Coffman and T. A. Ford, U.S. Pat. 2,419,008 (1947) to Du Pont [*CA 41*, 4963 (1947)]; F. J. Johnson and D. C. Pease, U.S. Pat. 2,510,783 (1950) to Du Pont [*CA 46*, 1299 (1952)]; G. H. Kalb, D. D. Coffman, T. A. Ford, and F. L. Johnston, *J. Appl. Polym. Sci. 4*, 55 (1960).

158. V. E. James, U.S. Pat. 3,129,207 (1964) to Du Pont [*CA 58*, 9253 (1963)].
159. R. A. Bonsall and B. Hopkins, U.S. Pat. 3,627,744 (1971) to Monsanto [*CA 73*, 121079 (1970)].
160. J. L. Hecht, U.S. Pat. 3,265,678 (1966) to Du Pont [*CA 58*, 6947 (1963)].
161. J. L. Hecht and T. Hughes, Brit. Pat. 1,077,728 (1967) to Du Pont [*CA 65*, 3994 (1966)].
162. J. G. Frielink, Brit. Pat. 1,161,958 (1969) to Deutsche Solvay Werke G.m.b.H. [*CA 71*, 81896 (1969)].
163. D. Sianesi, G. Caporiccio, and E. Strepparola, Brit. Pat 1,029,635 (1966) to Montecatini [*CA 62*, 663 (1965)]; R. N. Haszeldine, T. G. Hyde, and P. J. T. Tait, *Polymer 14*, 221 (1973).
164. G. Natta, G. Allegra, l. W. Bassi, D. Sianesi, G. Caporiccio, and E. Torti, *J. Polym. Sci., Part A 3*, 4263 (1965).
165. D. Sianesi and G. Caporiccio, *J. Polym. Sci. Part A 1, 6*, 335 (1968).
166. K. U. Usmanov, *Russ. Chem. Rev. 46*, 462 (1977).
167. L. R. Bartron, U.S. Pat. 2,953,818 (1960) to Du Pont [*CA 55*, 4052 (1961)].
168. J. L. Koenig and J. J. Mannion, *J. Polym. Sci. Part A-2, 4*, 401 (1966).
169. M. L. Wallach and M. A. Kabayama, *J. Polym. Sci. Part A-1, 4*, 2667 (1966).
170. A. E. Tonelli, F. C. Schilling, and R. E. Cais, *Macromolecules 14*, 560 (1981); *15*, 849 (1982).
171. D. W. Ovenall and R. E. Uschold, *Macromolecules 24*, 3235 (1991).
172. C. W. Wilson and E. R. Santee, *J. Polym. Sci. C 8*, 97 (1965).
173. T. J. Dougherty, *J. Am. Chem. Soc. 86*, 460 (1964).
174. J. N. Coker, *J. Polym. Sci., Polym. Chem. Ed. 13*, 2473 (1975).
175. P. S. Minhas and F. Petruccelli, *Plastics Engineering 60* (March 1977).
176. C. Wu, R. Brambilla, and J. T. Yardley, *Macromolecules 23*, 997 (1990).
177. H. G. Elias, *New Commercial Polymers 2*, p. 154, Gordon and Breach Science Publishers, New York (1986).
178. O. Paleta, V. Delek, and H -J Timpe, *Chem. Tech. (Leipzig) 40*, 459 (1988); J. A. Delaire, M. Lagarde, D. Broussoux, and J. C. Dubois, *J. Vac. Sci. Technol. B 8*, 33 (1990); H. Inukai and T. Kitahara, Eur. Pat. Appl. EP 331,056 (1989) to Daikin [*CA 112*, 100780 (1990)].
179. H. Inukai, T. Yasuhara, and T. Kitahara, PCT Int. Appl. WO 89 05,287 (1989) [*CA 111*, 195615 (1989)].
180. S. Ohtoshi, *J. Plastics (Japan) 30*, 117 (1984); B. Boutevin, G. Rigal, A. Rousseau, and D. Bose, *J. Fluorine Chem. 38*, 47 (1988); R. Ramharack, *Polym. Prepr. 29*, 146 (1988); Y. X. Zhang, A.-H. Da, and T. E. Hogen-Esch, *J. Polym. Sci., Polym. Lett. 28*, 213 (1990).
181. K. Yokota and T. Hirabayashi, *Polymer J. (Tokyo) 17*, 991 (1985).
182. H. Ito and R. Schwalm, *Polym. Prepr. 27*, 196 (1986).
183. M. Kataoka, Jpn. Kokai Tokkyo Koho JP 61,170,735 (1986) [*CA 106*, 147112 (1987)].
184. T. Narita, T. Hagiwara, H. Hamana, and S. Maesaka, *Polym. J. (Tokyo) 20*, 51 (1988).
185. B. Bomer and H. Hagemann, *Angew. Makromol. Chem. 109–110*, 285 (1982); K. Ishihara, R. Kogure, and K. Matsui, *Kobunshi Ronbunshu 45*, 653 (1988) [*CA 110*, 24348 (1989)]; T. Oishi and S. Togami, Jpn. Kokai Tokkyo Koho JP 62,158,234 (1987) to Mitsui Toatsu Chemicals [*CA 109*, 111043 (1988)].
186. M. Kanba, T. Sugiyama, and G. Kojima, Jpn. Kokai Tokkyo Koho JP 01,147,501 (1989) to Asahi Glass Company [*CA 112*, 62712 (1990)].
187. M. Miyamoto, K. Aoi, and T. Saegusa, *Macromolecules 21*, 1880 (1988).
188. M. Miyamoto, K. Aoi, and T. Saegusa, *Macromolecules 22*, 3540 (1989).
189. J. F. Harris, U.S. Pat. 3,037,010 (1962) to Du Pont [*CA 57*, 7465 (1962)]; J. A. Jackson, *J. Polym. Sci., Polym. Chem. Ed. 10*, 2935 (1972); R. D. Chambers, D. T. Clark, D. Kilcast, and S. Partington, *J. Polym. Sci., Polym. Chem. Ed. 12*, 1647 (1974).
190. T. Narita, T. Higiwara, H. Hamana, M. Sezaki, A. Nagai, S. Nishimura, and A. Takahashi, *Macromolecules 22*, 3183 (1989).
191. K. Tsuchihara, T. Masuda, and T. Higashimura, *Polym. Bull. (Berlin) 20*, 343 (1988).
192. T. Masuda, T. Hamano, T. Higashimura, T. Ueda, and H. Muramatsu, *Macromolecules 21*, 281 (1988).
193. K. Baum, P. G. Cheng, R. J. Hunadi, and C. D. Bedford, *J. Polym. Sci., Polym. Chem. Ed. 26*, 3229 (1988).

194. S. Temple and R. L. Thornton, *J. Polym. Sci., Part A-1, 10*, 709 (1972); B. Yamada, R. W. Campbell, and O. Vogl, *Polym. J. 9*, 23 (1977).

195. W. J. Middleton, H. W. Jacobson, R. E. Putnam, H. C. Walter, D. G. Pye, and W. H. Sharkey, *J. Polym. Sci., Part A, 3*, 4115 (1965); A. L. Barney, J. M. Bruce, J. N. Coker, H. W. Jacobson, and W. H. Sharkey, *J. Polym. Sci., Part A-1, 4*, 2617 (1966).

196. F. C. McGrew, U.S. Pat. 3,136,744 (1964) to Du Pont [*CA 61*, 4312 (1964)].

197. P. E. Cassidy, T. M. Aminabhavi, and J. M. Farley, *J. Macromol. Sci., Rev. Macromol. Chem. Phys. C29*, 365 (1989).

198. S. Nakamura, Y. Saegusa, M. Kuriki, A. Kawai, and S.-G. Shan, *Polym. Prepr. 31*, 352 (1990).

199. D. L. Goff and E. L. Yuan, *Polym. Mater. Sci. Eng. 59*, 186 (1988); A. K. St. Clair, T. L. St. Clair, and W. P. Winfree, *Polym. Mater. Sci. Eng. 59*, 28 (1988); R. A. Hayes, Eur. Pat. Appl. EP 336,998 (1989) to Du Pont [*CA 112*, 159207 (1990)].

200. T. Ichino, S. Sasaki, T. Matsura, and S. Hishi, *J. Polym. Sci., Polym. Chem. Ed. 28*, 323 (1990).

201. F. W. Harris and L. H. Lanier, in: *Structure Solubility Relationships in Polymers* (F. W. Harris and R. B. Seymour, eds.), Academic Press, New York (1977); T. Kobayashi and M. Kubo, *Polym. Preprints Japan 37*, 2728 (1988); G. Hougham, G. Tesoro, and J. Shaw, *Polym. Mater. Sci. Eng. 61*, 369 (1989); M. K. Gerber, J. R. Pratt, A. K. St. Clair, and T. L. St. Clair, *Polym. Prepr. 31*, 340 (1990); D. A. Scola, U.S. Pat. 4,742,152 (1988) [*CA 109*, 111124 (1988)] and 4,801,682 (1989) to United Technologies Corporation [*CA 110*, 232318 (1989)]; M. Fryd, U.S. Pat. 4,588,804 (1986) to Du Pont [*CA 105*, 154083 (1986)].

202. H. G. Rogers, R. A. Gaudiana, W. C. Hollinsed, P. S. Kalyanaraman, J. S. Manello, C. McGowan, R. A. Minns, and S. Sahatjian, *Macromolecules 18*, 1058 (1985); R. Sinta, R. A. Gaudiana, R. A. Minns, and H. G. Rogers, *Macromolecules 20*, 2374 (1987).

203. H. Tomota, Jpn. Kokai Tokkyo Koho JP 01 16,828 (1989) to Neos Co. [*CA 111*, 58581 (1989)]; O. Shinonome, T. Kitahara, and S. Murakami, Jpn. Kokai Tokkyo Koho JP 62,197,419 (1987) and 62,206,019 (1987) to Unitika [*CA 108*, 113197 (1988) and *108*, 206205 (1988)].

204. K. Matsui, Y. Nagase, and T. Ueda, Jpn Kokai Tokkyo Koho JP 61,247,729 (1986) [*CA 106*, 138944 (1987)].

205. H. Tomota, H. Fujii, H. Hase, and K. Ito, Jpn. Kokai Tokkyo Koho JP 01 45,423 (1989) to Neos Co. [*CA 111*, 41024 (1989)]; S. Tamaru, M. Kubo, and M. Kashiwagi, Eur. Pat. Appl. EP 284,993 (1988) to Daikin [*CA 110*, 136377 (1989)]; A. Nishikawa and T. Koyama, Jpn. Kokai Tokkyo Koho JP 63 57,632 (1988) to Hitachi [*CA 109*, 150956 (1988)]; A. Nishikawa, T. Sugawara, and A. Nagai, Jpn. Kokai Tokkyo Koho JP 01 108,221 (1989) to Hitachi [*CA 111*, 196327 (1989)]; S. Sasaki, *J. Polym. Sci., Polym. Lett. 24*, 249 (1986); S. Sasaki and K. Nakamura, *J. Polym. Sci., Polym. Chem. Ed. 22*, 831 (1984).

206. A. E. Feiring, E. R. Wonchoba, and S. D. Arthur, *J. Polym. Sci., Polym. Chem. Ed. 28*, 2809 (1990).

207. J. W. Labadie and J. L. Hedrick, *Polym. Prepr. 31*, 344 (1990).

208. M. A. Shimizu, M.-A. Kikimoto, and Y. Imai, *J. Polym. Sci., Polym. Chem. Ed. 25*, 2385 (1987); M.-A. Kakimoto, S. Harada, Y. Oishi, and Y. Imai, *J. Polym. Sci., Polym. Chem. Ed. 25*, 2747 (1987); Y. Oishi, S. Harada, M. Kamimoto, and Y. Imai, *J. Polym. Sci., Polym. Chem. Ed. 27*, 3393 (1989).

209. K. Matsuo, W. H. Stockmayer, and G. F. Needham, *J. Polym. Sci., Polym. Symp. 71*, 95 (1984); P. Johncock, S. P. Barnett, and P. A. Rickard, *J. Polym. Sci., Polym. Chem. Ed. 24*, 2033 (1986); K. Narita and S. Nakanishi, Eur. Pat. Appl. EP 217,641 (1987) to Sanyo Electric Co., Ltd. [*CA 107*, 31357 (1987)].

210. T. Kiyotsukuri, N. Tsutsumi, and T. Sandan, *J. Polym. Sci., Polym. Chem. Ed. 28*, 315 (1990); Y. Chujo, A. Hiraiwa, H. Kobayashi, and Y. Yamashita, *J. Polym. Sci., Polym. Chem. Ed. 26*, 2991 (1988).

211. W. J. Feast and P. J. Tweedale, *Br. Polym. J. 16*, 314 (1984).

212. T. Takakura and M. Kato, *J. Syn. Org. Chem. Japan 42*, 822 (1984).

213. P. M. Blackmore and W. J. Feast, *J. Mol. Catal. 36*, 145 (1986); E. Perez, J. P. Laval, M. Bon, I. Rico, and A. Lattes, *J. Fluorine Chem. 39*, 173 (1988).

214. C. Bliefert, H.-M. Boldhaus, F. Erdt, and M. Hoffmann, *Kunstoffe 76*, 235 (1986).

215. C. C. Chiao, U.S. Pat. 4,828,585 (1989) to Dow [*CA 111*, 176891 (1989)].

216. I. Belaish, D. Davidov, H. Selig, M. R. McLean, and L. Dalton, *Angew. Chem., Int. Ed. Engl. 28*, 1569 (1989).
217. B. Boutevin, J. Mouanda, Y. Pietrasanta, and M. Taha, *Eur. Polym. J. 21*, 181 (1985); T. Momose, I. Ishigaki, and J. Okamota, *J. Appl. Polym. Sci. 36*, 55, 63 (1988).
218. N. Ishikawa and H. Matsuhisa, *Nippon Kagaku Kaishi* 1247 (1985).
219. L. E. Manzer, *Science 249*, 31 (1990).

16

Fluoroelastomers

ANESTIS L. LOGOTHETIS

16.1. INTRODUCTION

Although discovered more than 50 years ago, fluoropolymers are still growing in terms of production volume and of introduction of new compositions to meet industrial needs. Also, they still represent an active field of research holding great promise for new discoveries in the form of macromolecules with unusual properties. As a class fluoroelastomers account for less than ten percent of total fluoropolymer production, which is dominated by PTFE (mainly) and related crystalline fluoroplastics (see Chapter 15).

Fluoroelastomers, lacking crystallinity, derive their strength by being crosslinked (cured or vulcanized) to produce permanent three-dimensional networks. Once vulcanized to maintain rubber-like elasticity at high temperatures and in contact with various chemicals, their primary use is in sealing applications, where they function to prevent leakage after prolonged compression in hostile environments; rarely are they utilized under tension, and then only at modest elongations. Fluoroelastomeric parts are already employed extensively by the aircraft, aerospace, automotive, chemical, petroleum, and energy industries. The demanding high-performance requirements in these industries will continue to create great opportunities for fluoroelastomers.

Technologically important fluoroelastomers can be divided into three groups: the so-called fluorocarbon elastomers, fluorosilicones, and the recently developed fluoroalkoxylated poly(phosphazenes). Fluorocarbon elastomers are carbon-chain polymers, i.e., they are characterized by having all-carbon backbones; as a group they share some of the same monomers, crosslinking (vulcanization; curing) chemistry, and physical properties. Fluorosilicones and fluoropolyphosphazines have inorganic backbones; hence the chemistries are different from that of the fluorocarbon elastomers and are not described here. For some previous reviews of fluoroelastomers, see Refs. 1–6.

ANESTIS L. LOGOTHETIS • DuPont Elastomers, E.I. du Pont de Nemours & Co., Experimental Station, Wilmington, Delaware 19880-0328.

Organofluorine Chemistry: Principles and Commercial Applications, edited by R. E. Banks *et al.* Plenum Press, New York, 1994.

	CF_3 | $CF_2 = CF$	OR_F | $CF_2 = CF$	CH_3 | $CH_2 = CH$	Cl | $CF_2 = CF$
$CF_2 = CH_2$	X	X $(+CF_2 = CF_2)$		X
$CF_2 = CF_2$	X $(+CF_2 = CH_2)$	X $(+CH_2 = CH_2)$	X $(+CF_2 = CH_2)$	

Figure 1. Combination of two or more fluorocarbon monomers to give elastomeric structures. The sign X indicates commercially available fluoroelastomers.

16.1.1. Chemical Compositions

Homopolymers derived from tetrafluoroethylene (TFE) or vinylidene fluoride (VF_2) are long-chain macromolecules with high crystallinity which makes them unsuitable as elastomers (see Chapter 15). Copolymers of TFE or VF_2 with other suitable fluoromonomers are elastomeric, amorphous materials with low intermolecular forces (Figure 1 shows the monomers which constitute today's commercially important elastomers[7,8]). To obtain elastomeric properties, the basic monomers VF_2 and TFE are copolymerized with other monomers of topological similarity, i.e., having a bulky group attached to a vinyl functionality. In some cases an additional monomer is incorporated in small amounts to provide functional groups suitable for crosslinking.

16.1.2. Structural Considerations

Knowledge of rubber elasticity and of the solid-state behavior of organic polymers allows one to set out general structural characteristics necessarily associated with useful elastomers. Long linear molecules of high intrinsic flexibility are more desirable than those which are highly branched. Branching often leads to slow recovery, particularly at low temperatures, and large hysteresis loops in the stress–strain curves. It is important to note that groups which exhibit strong intermolecular forces and promote crystallinity must be avoided to minimize intermolecular attraction and to keep energy barriers low between conformational states of the chain. Since the polymers are amorphous, elastomers initially consist of randomly coiled macromolecules, which deform on stretching and recover upon removal of an applied force. To ensure complete recovery from large strains, crosslinking between chains to give a permanent three-dimensional network is required.[9]

At service temperatures above 0°C, uncrosslinked "raw" elastomers are highly viscous, incompressible liquids. Chemical crosslinking (vulcanization) is performed to produce thermoset elastomeric networks. However, physical crosslinking by crystallization of hard segments in segmented or grafted copolymers can also provide a

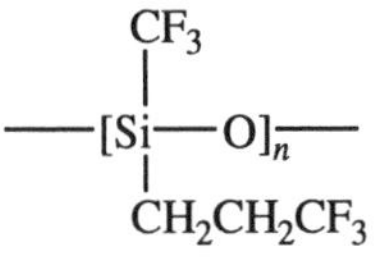

$$\begin{array}{c} CF_3 \\ | \\ -[Si-O]_n- \\ | \\ CH_2CH_2CF_3 \end{array} \qquad \begin{array}{c} OCH_2CF_3 \\ | \\ -[P=N]_n- \\ | \\ OCH_2CF_2CF_3 \end{array}$$

fluorosilicone 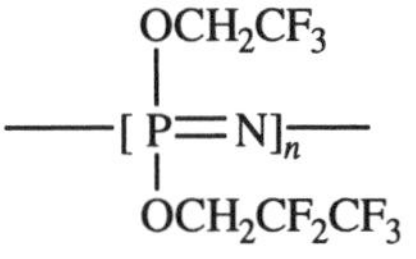   fluoropolyphosphazene

network structure in so-called thermoplastic elastomers. The few known thermoplastic elastomers have much lower upper-temperature service ceilings than the thermoset elastomers. Nevertheless, they possess advantages in processability, and they generally yield less extractable material because they contain no curatives. The restoring force that acts on a uniaxially or biaxially deformed elastomeric network is largely due to the lower entropy of the extended, and hence oriented, chains.

High fluorine content is desirable to achieve high solvent resistance and thermal stability in fluorocarbon elastomers. Low-temperature flexibility is governed by the glass transition temperature (T_g) which depends mainly on the conformational flexibility of individual chain elements. Therefore, if the various conformational states can be interconverted more freely, T_g is lower and the low-temperature performance is better. Thus, lack of crystallinity and a low T_g value is required for a serviceable elastomer. Poly(vinylidene fluoride) (polyVF$_2$), for example, has sufficiently low T_g (−40°C), but very strong intermolecular forces cause crystallization (latent heat of fusion ΔH_f, 6 kJ mol^{-1}; mp 165°C)[10] and thus restrict conformational freedom. Copolymerization with hexafluoropropylene (HFP), however, interrupts the crystallinity of polyVF$_2$ and takes advantage of its low T_g to give a fluoroelastomer with T_g in the minus 20°C range. Other similar combinations of monomers which give useful fluoroelastomers are shown in Figure 1.

Table 1.   Service Temperature Ranges and Fluorine and Hydrogen Contents of Fluoroelastomers

Type[a]	Service temp. (°C)[b]	F (wt%)	H (wt%)
Fluorocarbon elastomers			
VDF/HFP	−18 to 210	66	1.9
VDF/HFP/TFE	−12 to 230	66–69.5	1.1–1.9
VDF/PMVE/TFE	−27 to 230	64–66.5	1.1–1.7
TFE/Propylene	5 to 200	54	4.3
TFE/PMVE	0 to 260	73	0
TFE/PMVE/Ethylene	−15 to 230	66	1.1
Fluorosilicone	−65 to 175	37	4.5
Fluoropolyphosphazene	−65 to 175	55	1.4

[a]VDF, vinylidene fluoride; HFP, hexafluoropropylene; TFE, tetrafluoroethylene; PMVE, perfluoro(methyl vinyl ether).
[b]Temperature range for continuous service.

The low-temperature properties of fluorosilicones and fluoroalkoxylated poly(phosphazenes) are superior to those of any fluorocarbon elastomers. Their outstanding chain flexibility is a result of the oxygen or nitrogen backbone atoms. The low degree of substitution results in low activation energies for rotation about these atoms and adjacent chain atoms and, therefore, in low glass-transition temperatures. These polymers, however, contain less fluorine than the fluorocarbon elastomers, resulting in inferior solvent resistance and high-temperature performance (Table 1).

16.2. FLUOROELASTOMERS BASED ON VINYLIDENE FLUORIDE COPOLYMERS

16.2.1. General Description

Most commercially available fluoroelastomers consist of copolymers of vinylidene fluoride (VF_2) with hexafluoropropylene (HFP) and, optionally, tetrafluoroethylene (TFE). Worldwide, approximately 7,500 tonnes of such fluoroelastomers are sold annually at present. In very special applications, other monomers like perfluoro(alkyl vinyl ethers) ($R_FOCF{=}CF_2$), chlorotrifluoroethylene (CTFE), and 1H-pentafluoropropylene (PFP, $CHF{=}CFCF_3$) have been copolymerized with vinylidene fluoride to give specialty rubbers. These highly fluorinated polymers containing 62–70 wt% of fluorine have remarkable resistance to flame, chemicals, solvents, and oxidative attack. The stability has been attributed to the strength of the carbon–fluorine bond compared to that of the carbon–hydrogen bond, to steric hindrance due to the presence of fluorine, and to strong van der Waals forces between hydrogen and fluorine atoms.[11–14]

Copolymers of VF_2 and CTFE, prepared by M.W. Kellogg and Co. under contract to the United States Army Quartermaster Corps, were first described in a series of papers in 1955.[15–17] These fluoroelastomers (Kel-F[®]) exhibited solvent and heat resistance superior to any of the then existing elastomers. An even better fluoroelastomer, a dipolymer of VF_2/HFP (see top line in Figure 5 for the general formula), was described by Du Pont and M.W. Kellogg workers shortly afterward.[18–20] The Du Pont dipolymer was commercialized under the trade name Viton[®]A, while the 3M Company, which in the meantime had acquired M.W. Kellogg Co., introduced a similar fluoroelastomer under the trade name Fluorel[®]. A terpolymer, VF_2/HFP/TFE (Viton[®] B), was introduced[21,22] in the late 1960s with even better solvent and thermal resistance than the dipolymers. Also in the 1960s, elastomeric VF_2/PFP copolymers and VF_2/PFP/TFE terpolymers were introduced by Mont-Edison Co.[23–25] However, these polymers, sold under the Tecnoflon[®] name, lacked the stability of similar HFP-containing compositions and were replaced soon after with HFP analogs as the early patents were expiring. Daikin Kogyo Co. began production of VF_2-based fluoroelastomers (Daiel[®]) in 1970. A dipolymer, known as SKF-32 and produced in the Soviet Union, has not yet been traded on the world market. Asahi Chemical briefly entered the world market with VF_2-based fluoroelastomers, while in the last few years Nippon Mektron Co. has started to make such fluoroelastomers for internal use.

Table 2. Commercial Fluoroelastomers Based on
Vinylidene Fluoride

Copolymer[a]	Trade name	Manufacturer
VF₂/HFP	Viton	Du Pont
	Fluorel[b]	3M
	Tecnoflon	Montefluos
	Daiell[b]	Daikin
VF₂/HFP/TFE	Viton[b]	Du Pont
	Fluorel[b]	3M
	Tecnoflon	Montefluos
	Daiel[b]	Daikin
VF₂/PMVE/TFE	Viton[c]	Du Pont

[a]VF₂, vinylidene fluoride; HFP, hexafluoropropylene; TFE, tetrafluoroethylene; PMVE, perfluoro(methyl vinyl ether).
[b]Also available with proprietary cure site for peroxide curing.
[c]Contains proprietary cure site; can be cured only with peroxides.

In the mid-1970s, Du Pont Co. introduced VF₂-based compositions that contain small amounts of a bromine-substituted olefin as a cure-site monomer.[11,12,26] These polymers are curable either with peroxides or dinucleophiles and have the advantage that polymers with higher fluorine content (70 wt%) and better low-temperature performance can be made. Trademarks and manufacturers of commercial fluoroelastomers are shown in Table 2.

16.2.2. Production

16.2.2.1. Polymerizations

The useful compositional range for elastomeric structures from the three key monomers VF₂, HFP, and TFE is shown in Figure 2.[7,8] The region of interest is the one encompassing a composition of 20–70 wt% VF₂, 20–60 wt% HFP, and 0–40 wt% TFE. On a molecular scale, it is important to have about one HFP unit for every 2–4 of the other monomers present to prevent crystallization. These polymers are usually prepared in aqueous emulsion with free-radical initiators. Free-radical solution polymerizations are also possible, but chain transfer from the solvent keeps the molecular weights low. Typical aqueous emulsion polymerizations are conducted at 5–7 MPa and 100–120°C. Commonly used initiators are inorganic peroxides, such as ammonium persulfate; however, organic peroxides and redox systems are also used, especially at lower temperatures where the thermal decomposition of persulfate is slow. Emulsifying agents, when required, must be inert to the highly reactive fluorocarbon radicals of growing polymer chains, to minimize chain transfer: salts of long-chain fluorocarbon acids have proved satisfactory. Buffers control the pH and stabilize the emulsification system.

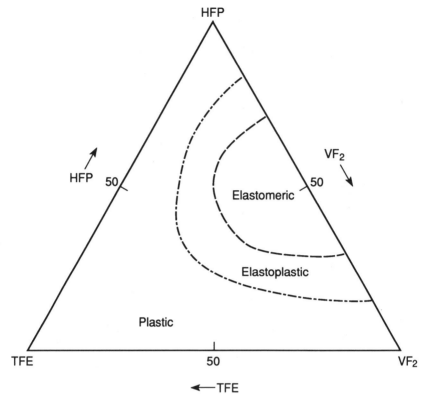

Figure 2. Fluoroelastomers prepared from vinylidene fluoride (VF_2), hexafluoropropylene (HFP), and tetrafluoroethylene (TFE). The elastomeric region lies in the monomer range of 20–70 wt% VF_2, 20–60 wt% HFP, and 0–40 wt% TFE.

In batch polymerizations, the ingredients are heated in the reactor and the reaction is allowed to proceed until the pressure drop corresponds to the desired extent of reaction. In semibatch polymerizations, monomers can be added continually under pressure, providing better control of copolymer composition and molecular weights. In a continuous polymerization, water, monomers, initiators and other components are fed to the reactor while polymer latex is removed at a corresponding rate. Thus monomers of widely ranging reactivities can be used effectively to give uniform polymeric structures. Unreacted monomers are removed from the latex and recycled.

16.2.2.2. Molecular Weight Control

In peroxide-initiated polymerizations, the molecular weight of the polymer usually increases with decreasing initiator concentration. The extent of the molecular weight increase may not be obvious if only solution viscosity measurements are utilized, because they show deviations from linearity in the higher ranges caused by gel formation. Use of chain-transfer agents is an effective way to eliminate the gel. Because of the high reactivity of fluorocarbon radicals, almost any compound, except a highly fluorinated one, can be used as a chain transfer agent, e.g., diethyl malonate, carbon tetrachloride, isopentane, or isopropanol. Recently, iodo-substituted fluorocarbons have been introduced as effective chain-transfer agents because they have the added advantage of giving reactive iodo-terminated polymeric chains.[27-31] The reactivity of the polymer's terminal iodine is about the same as that of the chain-transfer agent. Thus, successive chain transfers of the terminal iodine make the system behave like a "living" free-radical polymerization. The low-molecular-weight polymers formed initially are extended to higher ones by this process. To prepare high-molecular-weight species with iodine as the predominant terminal group, termination reactions not involving iodine are to be avoided as much as possible. Consequently, the least amount of iodo compound in relation to the initiator should be used throughout the polymerization to keep the kinetic chain as long as possible.

The regular unmodified co- and ter-polymers of VF_2 obtained at low initiator concentrations are of relatively broad molecular-weight distribution (MWD) and contain a significant amount of gel. With isopentane as modifier, gel formation is eliminated but the M_w/M_n ratio of 12.7 indicates that the MWD is still very broad. Modification by perfluoroalkyl iodides, on the other hand, makes the MWD considerably narrower, $M_w/M_n = 1.4$ (Figure 3). Di-iodo molecular-weight modifiers are

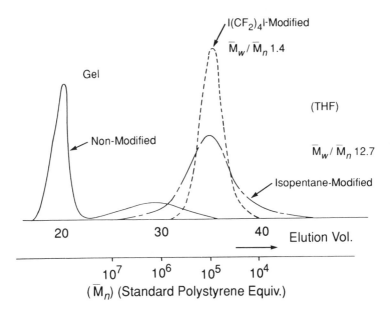

Figure 3. Gel permeation chromatography (GPC) of iodo- and isopentane-modified and nonmodified fluoroelastomers in THF.

preferred to their mono-iodo analogs, because they give narrow MWD polymers substituted with iodine at *both* ends of the macromolecular chain. These di-iodo-terminated fluoroelastomers cure more efficiently in the presence of peroxides, and are also capable of undergoing many reactions characteristic of the carbon–iodine bond. For example, olefins can be inserted at terminal carbon–iodine bonds, the iodine can be replaced with hydrogen, and it can participate in coupling reactions or polymerization sequences (in the presence of peroxides) with other monomers to give block copolymers. For example, di-iodo-terminated polymers have been treated with TFE, TFE/E mixtures, VF_2 or TFE/VF_2 mixtures to give ABA-type thermoplastic block fluoroelastomers, where A is the crystalline hard segment (polyTFE, polyTFE/E, polyVF_2) and B is the soft elastomeric segment.

16.2.3.   Curing (Vulcanization) of Fluoroelastomers Containing Vinylidene Fluoride

Fluoroelastomers, like all other thermosetting elastomers, require crosslinking in order to get useful properties. Fluoroelastomers can be cured (vulcanized) by nucleophiles, such as diamines or bisphenols, and with peroxides. Irrespective of the curing system, the best vulcanizate properties are obtained by a two-step process. The first step entails the application of both heat and pressure, by molding a specimen in a press (press-cure). The second step (post-cure) is carried out in air or under nitrogen at higher temperatures than the press cure and at atmospheric pressure. For thick sections, the temperature of the post-cure oven is usually raised stepwise to prevent fissuring of the part. A typical curing condition would be 5–15 minutes in a press at 150–180°C followed by 12–24 hours at 200–250°C in an air oven. Post-curing substantially improves the compression-set resistance of the vulcanizate, as well as tensile strength, and chemical and oxidative resistance.[12,32]

16.2.3.1.   Curing with Diamines

Diamine curing,[33–37] originally introduced in the late 1950s, was the predominant way of crosslinking raw fluoroelastomers until the late 1960s, when bisphenol curing was introduced. The most commonly used diamine is hexamethylenediamine, utilized conveniently as its carbamate salt ("Diak" 1) or *N,N'*-dicinnamylidene-1,6-hexanediamine ("Diak" 3). The chemistry involved is the initial removal by the amine of the elements of hydrogen fluoride from the polymer chain. The resulting polarized double bonds are susceptible to addition by the diamines to form a highly crosslinked network. It is advantageous to provide a metal oxide (MgO, PbO, CaO) as acid acceptor to react with the hydrogen fluoride generated during the curing step. These metals give insoluble, nonvolatile metal fluorides, which are thermally and chemically stable and are not detrimental to the physical properties of the cured fluoroelastomer. Typical compounding recipes contain 4–20 parts metal oxide, 10–30 parts carbon black, and 1–3 parts diamine per 100 parts of raw elastomer. Because of the processing difficulties and the relatively poor retention of physical properties at high temperatures (>200°C) after long exposures, the importance of the diamine cures has diminished.

16.2.3.2.   Curing with Bisphenols

Bisphenol cures were developed in the late 1960s and started replacing the diamine cure in the early 1970s.[38] Because of processing and property advantage, the most commonly used crosslinking agent is "Bisphenol AF" [2,2-bis(4-hydroxyphenyl)hexafluoropropane] (see Chapter 19). Others, like hydroquinone, substituted hydroquinones, and "Bisphenol A" also work well and are used commercially to a lesser degree. A typical curing recipe and curing curve are depicted in Figure 4. The curing system is very scorch safe, that is, no curing reaction is detected for 30 minutes at 121°C. However, at 177°C after a 2.5-minute induction period, the reaction is practically over after 5 minutes. The t_s2 is shortened and the cure rate increased with higher temperatures; however, the final state of cure remains essentially unchanged. The shape of the oscillating disk rheometer (ODR) curve changes little with copolymer composition. The curing behavior of this system indicates[10,37,38] that the bisphenol does not react with the polymer without an accelerator, which can be a phosphonium or tetra-alkylammonium salt, in combination with a metal oxide. The bisphenol reacts with the metal oxide to give the bisphenolate ion which, in turn, reacts with the salt to form a strong base that abstracts hydrogen fluoride from the polymer backbone to give a diene. The bisphenol then adds to the diene to give the crosslinked network (Figure 5).[39,40]

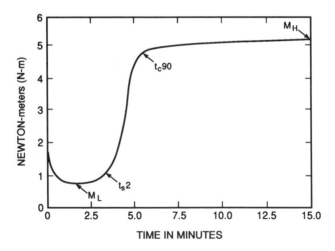

Figure 4.   Cure response by oscillating disk rheometry (ODR) at 177°C of a compound optimized for use in O-rings. The maximum cure rate is the initial slope of the curve: t_s2, the time to initiation (increase of the torque by two points from the minimum); t_c90, the time for 90% completion of the cure; M_L, the minimum torque; M_H, the maximum torque; and M_H–M_L, the degree of state of cure. The recipe consists of 100 parts polymer, 30 parts MT black, 6 parts calcium hydroxide, 3 parts magnesium oxide, 0.55 parts of benzyltriphenylphosphonium chloride (BTPPC), and 2 parts of "Bisphenol AF."

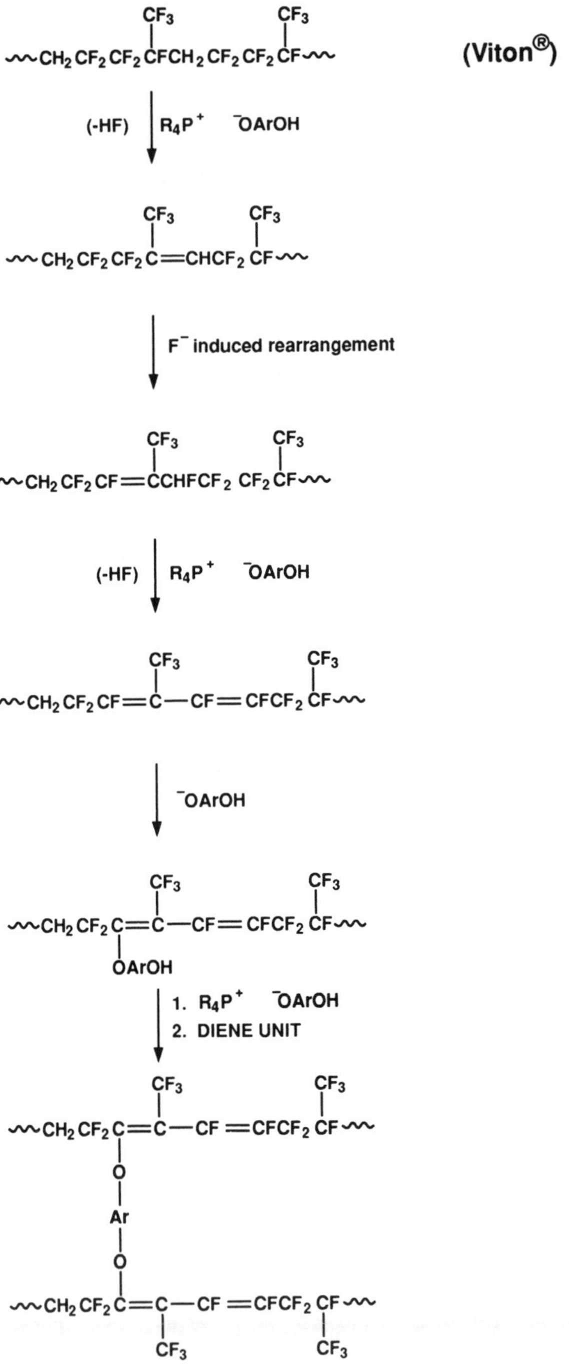

Figure 5. Proposed scheme for the crosslinking of VF$_2$/HFP polymers with bisphenols.

16.2.3.3. Curing with Peroxides

Incorporation of a monomer (known as a cure-site monomer) carrying a function susceptible to free-radical attack is required to make the vinylidene fluoride elastomeric copolymers peroxide-curable. Du Pont workers found that bromine-containing fluoro-olefins, such as bromotrifluoroethylene, 1-bromo-2,2-difluoroethylene, and 4-bromo-3,3,4,4-tetrafluorobut-1-ene, are excellent cure-site monomers, being incorporated readily. The resulting elastomeric ter-polymers then undergo attack by peroxide-generated radicals to give free-radical sites on the polymeric backbone; crosslinking to produce a network then occurs in the presence of a "co-agent".[11,12,41] The co-agents are usually di- or tri-functional electron-rich olefins, which act as radical traps. Some of the best are triallyl isocyanurate ("Diak" 7), triallyl cyanurate, and trimethallyl isocyanurate ("Diak" 8). Daikin workers have utilized modification of the fluorocarbon elastomers with perfluoroalkyl iodides (described earlier in Section 16.2.2.1) to introduce iodo end-groups.[27-31] These iodo-terminated polymers also give free-radical intermediates upon attack by peroxides. Some characteristics of this cure are: (1) the cure rate and final cure state are directly proportional to the cure-site concentration in the polymer; (2) the peroxide concentration has little effect on the final state of cure above a certain minimum level, but has a pronounced effect on the rate of cure [the radical trap (co-agent) concentration is important in maintaining a good cure rate and final cure state].

16.2.4. Properties

Fluoroelastomers are being used in spite of their high costs because of the combination of good physical properties (Table 3), as well as resistance to heat, chemical, and fluids.

16.2.4.1. Heat Resistance

Elastomers based on vinylidene fluoride can be used continuously at temperatures as high as 230°C, depending on the type of polymer and the formulation. Dipolymers retain over 50% of their tensile strength for more than 1 year at 200°C or 2 months at 260°C. They also retain over 95% of their elongation after 1 year at 200°C. By comparison, the fluid resistant nitrile and acrylate elastomers lose 50% of their elongation after aging at 150°C for 10 and 25 days, respectively.

16.2.4.2. Fluid and Chemical Resistance

Resistance to petroleum-based fuels and oils is excellent. However, certain lubricants containing additives that disperse sludge and inhibit metal corrosion, but are highly basic or decompose to basic products, will attack the fluoroelastomers.[42-44] Fluoroelastomers to be used in sealing applications involving hydrocarbon–alcohol blends (gasohol) must be chosen with care to minimize swelling. VF$_2$-based elastomers with 69.5% fluorine content are recommended for methanol and aromatic solvents, as

Table 3.   Physical Properties of Elastomers Containing
Vinylidene Fluoride

Property	Range
Specific gravity of raw polymer (g cm^{-3})	1.77–1.91
Tensile properties	
100% Modulus (MPa)	2–15
Tensile strength (MPa)	5–20
Elongation at break (%)	50–500
Hardness, durometer A	60–95
Compression set (%)	
70 h at 25°C	10–15
70 h at 200°C	10–40
1000 h at 200°C	50–70
Low-temperature properties	
Clash-Berg torsional-stiffness temperature (69 MPa) (°C)	–31 to –5
Retraction temperature (TR-10) (°C)	–31 to –6
Brittle point (°C)	–51 to –34

well as certain hydraulic fluids containing phosphate esters. Alcohols other than
methanol do not greatly swell these polymers, but blends of alcohols with other fluids
may act differently. Polar solvents tend to have a great affinity for these fluoroelastomers, and may actually dissolve raw polymer or destructively swell the vulcanizates.
For example, low-boiling ketones and esters (acetone, ethyl methyl ketone, ethyl
acetate), amides (*N,N*-dimethylacetamide, *N,N*-dimethylformamide), and ethers
(tetrahydrofuran, 2-methoxyethyl ether) cannot be used alone or as mixtures with other
fluids. Primary and secondary aliphatic amines severely attack VF$_2$-based fluoroelastomers, causing the formation of unsaturated bonds on the backbone (dehydrofluorination). Attack by nucleophiles in general leads to excessive crosslinking and
embrittlement. Resistance to aqueous alkali is good for short periods of exposure, but
eventually surface cracks appear and failure of the fluoroelastomer occurs. Phase-
transfer catalysts like quaternary ammonium or phosphonium salts accelerate degradation by aqueous alkali.

Resistance to aqueous acids is generally good, especially if acid acceptors such
as calcium hydroxide, magnesium oxide, or calcium oxide are present. Lead-based
acid acceptors, such as sublimed litharge and dibasic lead phosphite, are used for
optimal acid resistance.

16.2.4.3.   Other Types of Resistance

Ozone resistance and weatherability are excellent[45] and permeability to gases is
low. Compounding can be formulated to give nonflammable or self-extinguishing
vulcanizates. Abrasion varies according to the fillers in the formulation. Resistance to
β-radiation is moderate and independent of the polymer composition. Radiation doses

are cumulative and total exposure to 10^4–10^5 J kg^{-1} produces moderate to severe effects on physical properties. Failure is due to embrittlement rather than to chain scission.[46]

16.2.5. PROCESSING

16.2.5.1. Formulations

In O-ring applications, the compression-set resistance is of critical importance because sealing can be maintained only if the O-ring can retain some tendency to regain its uncompressed dimensions. Higher-molecular-weight polymers as well as tighter cures (lower elongation at break) give the best results. The bisphenol system is generally preferable to the peroxide curing systems in sealing applications because of the better compression sets. However, peroxide-cured stocks have better steam and acid resistance and are often specified for this reason.

In formulations for molded articles, elongation is more important. A lower cure state and greater elongation are achieved by using less crosslinking agent or lower-molecular-weight polymers. The tear strength at the demolding temperature, which increases with increasing elongation, becomes important in minimizing damage during removal of the finished parts from the mold. Specialized polymers containing perfluoromethyl vinyl ether can be compounded for exceptional resistance to fluids and low-temperature flexibility.[47,48]

16.2.5.2. Mixing

Standard rubber-compounding equipment and techniques are used. Rubber mills are suitable for most viscosity grades, although some reduction in molecular weight may occur when high-viscosity polymers are subjected to high shear (cold milled, and the clearance between the rolls small). Banbury and other internal mixers are also suited for mixing.

16.2.5.3. Molding

Compression molding is usually the most economical and flexible form of vulcanization. When extruded or cut preforms are used, the amount of flashing is kept to a minimum and higher viscosity polymers can be used because mold-flow requirements are low. Transfer molding requires lower compound viscosities because the compound must flow readily. Bisphenol-curable compounds are suited to transfer molding because of their processing safety. Injection molding requires higher capital investment and is usually less labor intensive. Injection molding is used increasingly and is justified in large-scale production, where labor savings and uniformity offset equipment costs and increased waste from mold filling channels. Compound formulation is critical because high flow-rates and compound safety are required. Thermoset elastomers are more demanding than thermoplastic polymers, because the former require irreversible chemical reactions that increase viscosity and create crosslinks

Table 4. Post-Cure Effect on Properties

Properties	Bisphenol		Peroxide	
	Press-cure[a]	Post-cure[b]	Press-cure[a]	Post-cure[b]
Tensile properties				
Modulus (M_{100}), MPa	5.0	7.9	5.0	7.9
Tensile (T_b), MPa	10.0	13.8	9.7	15.9
Elongation (E_b), %	225	175	165	150
Compression set (%) (70 h/200°C)				
O-rings	63	25	50	27
Pellets	85	20	52	20

[a]Press-cure, 10 minutes at 177°C.
[b]Post-cure, 24 hours at 232°C.

during the process. In the latter case, the thermoplastic polymer flows easily in the melt and is simply cooled or crystallized in the mold. Molding cycles are usually 1–5 minutes at 160–200°C and depend on the size of the article to be molded because of heat-transfer limitations.

16.2.5.4. Extrusion

Fluoroelastomers can be shaped and cured by extrusion, the fully compounded stock below the curing temperature being continuously forced through a shaping die. The extrudate, which may be a hose or tubing, is placed in a steam autoclave, microwave oven, or liquid curing medium for curing. Extrusion is also used to make preforms, rod stock for spliced O-rings, or profile stock for gaskets, which are cut and subsequently compression or transfer molded.

16.2.5.5. Post-Curing

Post-curing is utilized to develop optimum physical properties.[49] A VF$_2$/HFP/TFE elastomer containing a bromine functionality was cured with a bisphenol and with a peroxide. Physical tests on both products showed a 50% increase in modulus ($100M$) and tensile-at-break (T_b), a 50% decrease in elongation-at-break, and a two- to three-fold improvement in compression-set resistance after post-curing (Table 4).[11] The press-cured articles are usually placed in a circulating air oven at atmospheric pressure at 200–260°C for 12 to 24 hours, the conditions for a post-cure cycle being established rather arbitrarily. Each elastomeric system has a temperature range in air at which improvement of properties takes place. Upon heating longer or at higher temperatures, the physical properties deteriorate. Post-curing under nitrogen can be carried out at higher temperatures; consequently, the properties are better.[11]

It is not very clear what actually happens during post-curing. One of the essential functions is to remove water and other volatile products generated during the press-cure step.[50,51] In addition, the post-cure treatment contributes to the formation of additional crosslinks and to a thermally induced bond-breaking and bond-making process that results in a more stable network.[11,32,52–54] Thus, during post-curing an annealing process is occurring which relieves the stresses in the crosslinked network introduced during the press-cure.

16.3. FLUOROELASTOMERS BASED ON TETRAFLUOROETHYLENE/PROPYLENE COPOLYMERS

16.3.1. Introduction

In the early 1960s, it was found that TFE readily copolymerizes with propylene in a nearly alternating manner to give fluoroelastomers.[55–58] These polymers exhibit high T_g (–2°C), very good thermal stability and good chemical resistance to certain polar solvents, exceptional resistance to dehydrofluorination by amines, but inadequate resistance to aromatic hydrocarbons (most likely because of the low fluorine content, 54 wt%). They are cured by peroxide-radical trap systems,[59] unless a specific cure-site is present.[22,23,60,61]

Asahi Glass Co. introduced TFE/P fluoroelastomers in the mid-1970s under the "Aflas" trade name. It is estimated that about 200 tonnes per annum are sold worldwide.

16.3.2. Polymerizations

TFE/propylene copolymers have been prepared by emulsion polymerization, with perfluorocarbon surfactants and a modified persulfate redox system as initiator.[58–60,62] This redox system, consisting of ammonium persulfate, ferrous sulfate, ethylene diamine tetraacetic acid, and sodium hydroxymethanesulfinate, makes it possible to conduct the polymerizations close to room temperature and obtain high-molecular-weight copolymers (M_n 100,000–180,000). Disodium hydrogen phosphate (buffer) and sodium hydroxide are used to keep the pH between 5.5 and 10.0. At higher temperatures other redox systems can be utilized, but chain-transfer reactions begin to dominate and lower-molecular-weight polymers are produced (M_n about 70,000), which are not suitable as elastomers. The kinetic study indicates that the reduction of the ferric ion is the rate-determining step. NMR data indicate that nearly 70% of the TFE units are alternating with propylene units in terms of tetrads. The methyl groups appear to be in random configuration, which helps to give the polymers an amorphous morphology, in spite of the highly regular alternating structure. The amount of TFE in the monomer feed does not change the polymerization composition to any great extent, but affects the polymerization rate and the molecular weight of the polymer.[58]

Table 5. Properties of Standard Vulcanizates

Physical properties	"Bisphenol" AF or peroxide cured VF$_2$ elastomers	Peroxide cured C$_2$F$_4$/C$_3$H$_6$ elastomer
Specific gravity (g cm^{-3})	1.9	1.6
Tensile strength (MPa)	14–20	20–25
Elongation at break (%)	150–400	200–400
Tensile modulus at 100% (MPa)	4–6	2.5–3.5
Hardness, Shore A	60–90	70
Compression-set resistance (%)	15–40	35
Brittle point (°C)	–45	–40
Retraction temp. TR-10 (°C)	–20	3
Volume increase after immersion (%)		
95% H$_2$SO$_4$, 100°C, 3 days	5–25	4.4
60% HNO$_3$, 70°C, 3 days	12–47	10
H$_2$O, 150°C, 3 days	5–10	8.7
Steam, 160°C, 7 days	1–8	4.6
Fuel oil B, 25°C, 7 days	1	58.6
Benzene, 25°C, 7 days	5–12	40
Methanol, 25°C, 7 days	4–30	0.2
Ethyl acetate, 25°C, 7 days	>100	88

16.3.3. Curing Chemistry and Properties

TFE/propylene copolymers are cured by a peroxide/co-agent (radical trap) system similar to that described earlier for the bromine- or iodine-containing VF$_2$-based fluoroelastomers. Among the various peroxides tried, α,α'-bis(t-butylperoxy)diisopropylbenzene (Vul-Cup) is preferred; and of the co-agents, triallyl isocyanurate and triallyl cyanurate work the best. The properties of TFE–propylene fluoroelastomers are compared with those of vinylidene fluoride copolymers in Table 5. The results indicate that the two copolymers have similar tensile properties, but that TFE/propylene has poorer compression-set resistance, inferior resistance toward aromatic hydrocarbon solvents, and inadequate low-temperature performance; however, it displays superior resistance toward acids, bases, amines, and polar solvents.

TFE/propylene copolymers containing small amounts of cure-site monomers, such as an aryloxyalkyl vinyl ether,[61] vinylidene fluoride,[63,64] and glycidyl vinyl ether,[60] have also been prepared. These polymers can be cured either by peroxides or by nucleophiles. Depending on VF$_2$ content, the second type has improved low-temperature performance and chemical resistance toward aromatic hydrocarbons, but compromised resistance toward amines and polar solvents.

16.3.4. Processing

TFE/propylene polymers process similarly to, but less smoothly than, the VF$_2$-based fluoroelastomers on conventional rubber processing equipment. Processing aids such as carnauba wax or sodium stearate are widely used, and longer press-cure cycles are required followed by post-curing at 200°C for 16 hours to develop optimum properties.[65,66]

16.4. PERFLUOROELASTOMERS BASED ON TETRAFLUOROETHYLENE/PERFLUORO (ALKYL VINYL ETHER) COPOLYMERS

16.4.1. Polymer Description

Since the fluoroelastomers described earlier contain hydrogen, they have limited resistance to very hostile environments. In response to this deficiency, a perfluoroelastomer was developed that matches the oxidative, chemical, and thermal resistance of "Teflon." Its structure is based on copolymers of TFE with perfluoro(alkyl vinyl ethers); with 20–40 mol% of the vinyl ether incorporated, the copolymers are true elastomers with T_g below room temperature. The most readily copolymerizable homolog is perfluoro(methyl vinyl ether), which is the cornerstone of Du Pont's commercial Kalrez® elastomeric parts.[67-70]

A third monomer is usually added in small amounts (1 mol%) to make these polymers curable. It is very important that this cure-site monomer copolymerizes with TFE and perfluoro(alkyl vinyl ether) in a random fashion, that it does not cause chain transfer, and that it is capable of forming thermally and oxidatively stable networks inert to aggressive fluids. Three classes that meet the above requirements are shown in Figure 6. The perfluoro(alkyl vinyl ethers) carrying functional groups are the most useful and versatile. The substituted fluoroolefins are too unreactive in the copolymerization, and only the simplest types like vinylidene fluoride and trifluoroethylene can be used.[76] Bromine-containing olefins copolymerize well, but chain-transfer

(a) $CF_2 = CFOR'_FX$, $X = CO_2R$
                                $O\ C_6F_5$
                                $O\ (CF_2)_n\ CN$
                                CN

(b) $R^1CH = CR^2R^3$, $R^1, R^2, R^3 = H$
                                Br
                                F
                                R_F
                                $R'_F Br$

(c) $R_F I$ and $IR'_F I$, R_F = perfluoroalkyl
                                R'_F = perfluoroalkylene

Figure 6.   Classes of cure-site monomers. (a) Perfluoro(alkyl vinyl ether) carrying functional group; (b) simple substituted fluoroolefins; (c) perfluoroiodo alkyl and alkylenes.

reactions take place due to the presence of bromine. Perfluoroiodo compounds act predominantly as efficient chain-transfer agents.[27,28,71,72] The perfluoroelastomers containing bromine or iodine can be cured with peroxides and a co-agent.

16.4.2. Polymerizations

A persulfate/sodium sulfite redox system operating at 50–70°C is used when it is difficult to attain high-molecular-weight polymers. This initiating system gives sulfonic acid end groups which in turn form "strong" ionic domains, making the polymer very difficult to process. Thermally initiated free-radical polymerizations use persulfate initiation at 70–120°C and are suitable when chain-transfer reactions are less important. The end groups are mostly carboxylic acids which form weaker ionomers and are easier to process. All aqueous polymerizations require a surfactant, usually a salt of a perfluorinated carboxylic acid (C_8–C_{12}), and a buffer to maintain the pH close to neutral.

Nonaqueous polymerizations in chlorofluorocarbon ("Freon") solvents and with an organic peroxide as initiator have been described as well.[2,71] The advantage of such systems lies in the low concentrations of ionic end groups formed, and the subsequent good processing characteristics of the polymers. On the other hand, they suffer the disadvantage of the low solubility of the perfluorinated polymers and the formation of low-molecular-weight products. Use of a hybrid water–fluorocarbon solvent system at room temperature with a perfluoroacyl peroxide initiator [$Cl(CF_2)_n(CO)_2O_2$] has also been reported to give excellent results.[72] The polymerizations can be run in batch or continuous fashion.

16.4.3. Curing Chemistry

A number of curing systems have been developed for these materials, in spite of the small volume of total polymer used. The process is similar to that used with other fluoroelastomers and consists of a press-cure in a mold at 150–200°C for 10–60 minutes followed by post-curing in air or under nitrogen at 200–300°C for 24–120 hours. The post-cure improves the physical properties of the vulcanizates (tensiles properties, compression set, and chemical resistance) which are deficient after the press-cure. Fillers like carbon black, titanium dioxide, or silica have been used in amounts which vary in the range 10–30 parts per hundred of raw elastomer, depending on the kind of the filler and the desired properties. Typical properties of vulcanizates are given in Table 6.

16.4.3.1. Curing with Dinucleophiles

Aromatic or aliphatic diamines and, preferably, the more thermally stable bisphenols (utilized as their dipotassium salts) have been used to cure perfluoroelastomers containing small amounts of either perfluorophenoxy[68] or hydrogen[73] moieties on the polymeric backbone. In order to function well this system requires the presence of a

Table 6. Typical Properties of Perfluoroelastomers

Specific gravity (g cm^{-3})	2.05
100% Modulus (MPa)	9.0
Tensile strength (MPa)	16.0
Elongation at break (%)	160
Hardness, Shore A	75
Compression set (%)	
70 h at 200°C	20
70 h at 232°C	25
70 h at 288°C	35
Clash–Berg stiffness temp. (°C)	−2
(66.6 MPa torsional modulus)	
Brittle temp. (°C)	−38
Retraction temp. (°C)	
TR-10	−1
TR-50	+8

polyether accelerator, such as poly(ethylene glycol)dimethyl ethers (glymes) or "crown" ethers.

16.4.3.2. Curing of Nitrile-Containing Perfluoroelastomers

A typical nitrile-containing cure-site monomer is perfluoro-(8-cyano-5-methyl-3,6-dioxa-1-octene).[74] Crosslinking is brought about by the catalytic interaction of tetraphenyltin or silver oxide with the pendant nitrile groups. The structure of the crosslink is assumed to be mainly a triazinic,[1] as supported by infrared spectral data and the work of Henne's group[75] which showed that monomeric perfluoroalkanenitriles form 1,3,5-triazines under the influence of catalysts like those mentioned above.

16.4.3.3. Curing with Peroxides

Bromine- or iodine-containing perfluoroelastomers, like their analogs in the VF$_2$-based fluoroelastomer series, can be peroxide-cured satisfactorily in the presence of co-agents which act as radical traps. Both aromatic and aliphatic peroxides have been used and, among the co-agents, triallyl isocyanurate is preferred. The rate of cure is significantly faster than that of either the perfluorophenoxy- or nitrile-containing types, but post-cure is still required.

16.4.3.4. Curing by Irradiation

Exposure of perfluoroelastomers to high-energy irradiation (electron beam or γ-rays) produce clear cured parts.[84] The chemical resistance and compression set properties of these materials are excellent; the tensile properties, however, are a little inferior because they contain no reinforcing fillers.

16.4.4. Properties

16.4.4.1. Heat Resistance

Cured perfluoroelastomers show exceptional thermal stability and are differentiated by the robustness of their crosslinks. The perfluoroelastomers with pendant nitrile groups as cure sites retain their properties best after exposure to high temperatures. The triazine crosslinks, believed to be formed by trimerization of the nitrile functionality, are stable enough to show little or no change from the original post-cure properties after 18 days at 288°C in air. Even after short exposure to 316°C, useful properties are maintained.

Perfluoroelastomers cured with bisphenols are less stable to heat and, after 20 days at 288°C in air, the 100% modulus is reduced by about 25% and tensile strength by about 15%, while the elongation is increased by 120%. The peroxide-curable polymers are less thermally stable, starting to lose strength after but a few days exposure at about 230°C in air. In this respect, peroxide-cured perfluoroelastomers have about the same thermal resistance as fluoroelastomers based on vinylidene fluoride. The irradiation curable parts are thermally limited to about 200°C.

16.4.4.2. Fluid and Chemical Resistance

In general, cured perfluoroelastomers are very inert to chemicals or organic solvents, even at high temperatures. There are, however, some distinct differences in the amount of swell under aggressive conditions for perfluoroelastomer vulcanizates obtained with different cure systems,[1] as shown in Table 7. In glacial acetic acid and water, the bisphenol-cured stocks swell most while the peroxide-cured stocks swell

Table 7. Chemical and Fluid Resistance of Perfluoroelastomers[a]

Fluids	Temp. (°C)	Time (h)	Triazine (% swell)	Bisphenol AF (% swell)	Peroxide (% swell)
			Curing system		
Nitric acid	85	70	3	8	9
Acetic acid	100	70	3	32	16
Aniline	100	166	5	2	6
Butyraldehyde	70	70	13	10	14
Toluene	100	70	7	7	7
Skydrol 500B	125	70	9	5	6
Methyl ethyl ketone	70	70	6	4	4
Water	225	70	27	28	7

[a]Black-loaded stocks.

least. All perfluoroelastomers show fair to poor resistance toward chlorofluorocarbon solvents, so care must be exercised when using perfluoroelastomers in sealing applications with such solvents.

16.4.5. Processing

Perfluoroelastomers are difficult to process because of the stiffness of the fully fluorinated chains. Since the polymer is also difficult to crosslink, and expensive to produce, Du Pont commercialized these materials in the form of finished articles under the trade name Kalrez®.

The high cost of these parts is justified in applications where no other material can fulfill the requirements.[76] They are used in the chemical processing and transportation, aerospace, and oil exploration and production industries. Recently, Daikin commercialized a peroxide-curable perfluoroelastomer, "Perfluor," and Du Pont introduced a developmental product having a TFE/olefin/perfluoro(alkyl vinyl ether) composition that is peroxide curable.[77]

16.5. USES OF FLUORINATED ELASTOMERS

Commercial applications of fluoroelastomers reflect their rather special combination of properties: excellent resistance to heat, fluids, and oxidizing media combined with good physical characteristics. Elastomeric seals for industrial, aerospace, and automotive equipment constitute the largest applications market. These are mostly O-rings of all sizes, V-rings, flat or lathe-cut gaskets, and lip-type rotating or reciprocating shaft seals. Coated fabrics for diaphragms, sheet goods, expansion joints, chimney, and duct coatings account for considerable sales. Hose-linings, tubing, and industrial gloves for chemical service, extruded goods for such purposes as autoclave and oven seals, electrical connections, and wire coatings constitute examples of the varied end-uses. Their value in use has been confirmed in numerous applications that require a combination of chemical, thermal, fluid, and oxidation resistance.

Fluoroelastomers are relatively difficult to process and sometimes high temperatures and special techniques are required. Vigorous exothermic decompositions have occurred on rare occasions at high processing temperatures (>260°C) and when finely divided metals (aluminum or lead powder) were present in substantial amounts, or when a large excess of curing ingredients was added accidentally.[78]

Fluoroelastomers based on di- or ter-polymers of vinylidene fluoride have been compounded for optimum performance.[79] For better flex life, decreasing the carbon black content to lower the modulus (thus keeping the strain to low values) and increasing the molecular weight of the base polymer help to extend the useful life.[77] These fluoroelastomers can be compression-, transfer-, or injection-molded or extruded.[80–82] The peroxide-curable fluoroelastomers are extremely good in sealing applications where the parts are exposed to steam and acids at high temperatures and pressures. Raw polymers have recently found a very important use as processing aids in the extrusion of linear low-density poly(ethylene) (LLDPEs). At levels below 1 wt%, the fluoroelastomer acts as a die lubricant and reduces processing problems, such as draw resonance and melt fracture, and improves extrusion rates.[83]

16.6. REFERENCES

16.6.1. General References

1. A. L. Logothetis, *Prog. Polym. Sci. 14*, 251 (1989).
2. R. G. Arnold, A. L. Barney, and D. C. Thompson, *Rubber Chem. Technol. 46*, 619 (1973); S. Smith, in: *Preparation, Properties and Industrial Applications of Organofluorine Compounds* (R. E. Banks, ed.), Horwood–Wiley, Chichester–New York (1982).
3. D. P. Carlson and W. Schmiegel, in: *Ullman's Encyclopedia of Industrial Chemistry,* Vol. A11, p. 393, VCH Publishers, Weinheim (1988).
4. H. E. Schroeder, in: *Advances in Fluoroelastomers in High Performance Polymers: Their Origin and Development* (R. B. Seymour and G. S. Kirshenbaum, eds.), p. 389, Elsevier, Amsterdam (1986).
5. A. L. Logothetis, in: *Encyclopedia of Materials Science and Engineering*, p. 1809, Pergamon Press, Oxford (1986).
6. M. M. Lynn and A. T. Worm, in: *Encyclopedia of Polymer Science and Engineering 7*, p. 257, Wiley, New York (1987).

16.6.2. Specific References

7. D. C. England, R. E. Uschold, H. Starkweather, and R. Pariser, *Proceeding of the Robert A. Welch Conferences on Chemical Research, XXVI, Synthetic Polymers*, Houston, Texas (Nov. 15–17, 1982).
8. R. E. Uschold, *Polym. J. Jpn. 17*, 253 (1985).
9. H. F. Mark, *Chemtech 14*, 220, (1984).
10. G. J. Welch and R. L. Miller, *J. Polym. Sci., Polym. Phys. Ed. 14*, 1683 (1974).
11. D. Apotheker, J. B. Finlay, P. J. Krusic, and A. L. Logothetis, *Rubber Chem. Technol. 55*, 1004 (1982).
12. W. W. Schmiegel and A. L. Logothetis, *ACS Symposium Series. No. 260, Polymers for Fibers and Elastomers 10*, 159 (1984).
13. J. C. Montermoso, *Rubber Chem. Technol. 30*, 1521 (1961).
14. J. R. Cooper, *High Polymers, Vol. XXIII, Polymer Chemistry of Synthetic Elastomers*, p. 275, Wiley-Interscience, New York (1968).
15. M. E. Conroy, F. J. Honn, L. E. Robb, and D. E. Wolf, *Rubber Age 77*, 865 (1955).
16. C. B. Griffis and J. C. Montermoso, *Rubber Age 77*, 559 (1955).
17. W. W. Jackson and D. Hale, *Rubber Age 77*, 865 (1955).
18. A. L. Dittman, A. J. Passino, and J. M. Wrightson, U.S. Pat. 2,689,241 (1954) to M. W. Kellogg [*CA 49*, 11681a].
19. S. Dion, D. R. Rexford, and J. S. Rugg, *Ind. Eng. Chem. 49*, 1687 (1957).
20. J. S. Rugg and D. S. Rexford, *Rubber Age 82*, 102 (1957).
21. D. R. Rexford, U.S. Pat. 3,051,677 (1962) to DuPont.
22. J. R. Pailthorp and H. E. Schroeder, U.S. Pat. 2,968,649 (1961) to DuPont [*CA 55*, 13894g].
23. D. Sianesi, C. Bernardi, and A. Regio, U.S. Pat. 3,331,823 (1967) to Montecatini.
24. D. Sianesi, C. Bernardi, and G. Diotalleri, U.S. Pat. 3,335,106 (1967) to Montecatini [*CA 65*, 903a].
25. A. Miglierina and G. Ceccato, *4th Int. Sun. Rubber Symposium, No. 2*, 65 (1969).
26. D. Apotheker and P. J. Krusic, U.S. Pat. 4,035,565 (1977) to DuPont [*CA 85*, 193894f]; D. Apotheker and P. J. Krusic, U.S. Pat. 4,214,060 (1980) to DuPont [*CA 93*, 187547u].
27. M. Oka and M. Tatemoto, *Contemporary Topics in Polymer Science*, Vol. 4, p. 763, Plenum Press, New York (1984).
28. M. Tatemoto, *IX International Symposium on Fluorine Chemistry*, Avignon, France (1979).

29. K. Ishiwari, A. Sakakura, S. Yuhara, T. Yagi, and M. Tatemoto, *International Rubber Conference, Kyoto, Japan*, p. 407 (1985).

30. M. Tatemoto, T. Suzuki, M. Tomoda, Y. Furukawa, and Y. Ueta, U.S. Pat. 4,243,770 (1980) to Daikin [*CA 90*, 24564w].

31. M. Tatemoto and T. Nagakawa, U.S. Pat. 4,158,678 (1979) to Daikin [*CA 88*, 137374m]; M. Tatemoto and S. Morita, U.S. Pat. 4,362,678 (1982) to Daikin [*CA 93*, 27435c (1980)].

32. D. S. Ogunniyi, *Prog. Rubber Plast. Technol. 5*, 16 (1987).

33. J. F. Smith, *Rubber World 142*, 103 (1960); J. F. Smith and G. L. Perkins, *J. Appl. Polym. Sci. 5*, 560 (1961).

34. K. L. Paciorek, L. C. Mitchell, and C. T. Lenk, *J. Polym. Sci. 5*, 405 (1960).

35. K. L. Paciorek, W. G. Laginess. and C. T. Lenk, *J. Polym. Sci. 60*, 14 (1962).

36. D. K. Thomas, *J. Appl. Sci. 8*, 1415 (1960).

37. J. F. Smith, *J. Polym. Sci., Part A-2, 10*, 133 (1972).

38. A. L. Moran and D. B. Pattison, *Rubber Age 103* (1971).

39. W. W. Schmiegel, *Kautsch. Gummi, Kunstst. 31*, 137 (1971).

40. W. W. Schmiegel, *Angew. Makromol. Chem. 76/77*, 39 (1979).

41. J. B. Finlay, A. Hallenbeck, and J. D. MacLachlan, *J. Elastomers Plast. 10*, 3 (1978).

42. B. Frapin, *Rev. Gen. Caoutch. Plast. 672*, 125 (1987).

43. I. A. Abu-Isa and H. E. Trexler, *Rubber Chem. Technol. 58*, 326 (1985).

44. A. Nersasian, *Elastomerics 112*, No. 10, 26 (1980).

45. E. W. Bergstrom, *Elastomerics 109*, No. 10, 26 (1977).

46. Bulletin VT-515.1, *Radiation Resistance of "Viton"*, DuPont, Wilmington, DE.

47. Bulletin *VT-250.GLT, "Viton" GLT*, DuPont, Wilmington, DE.

48. Bulletin *VT-250.GF, "Viton" GF*, DuPont, Wilmington, DE.

49. Bulletin VT-440.1, *Effect of Oven Postcure Cycles on Vulcanizate Properties*, DuPont, Wilmington, DE.

50. A. W. Fogiel, H. K. Frensdorff, and J. D. MacLachlan, *Rubber Chem. Technol. 49*, 34 (1976).

51. J. D. MacLachlan and A. W. Fogiel, *Rubber Chem. Technol. 49*, 43 (1976).

52. A. W. Fogiel, *J. Polym. Sci. 53*, 333, (1975).

53. S. H. Kalfayan, R. H. Silver, and A. A. Mazzeo, *Rubber Chem. Technol. 49*, 1001 (1976); S. H. Kalfayan, R. H. Silver, and A. A. Mazzeo, *Rubber Chem. Technol. 48*, 944 (1975).

54. D. S. Ogunnyi, *Eur. Polym. J. 23*, No. 7, 577 (1987).

55. W. R. Brasen and C. S. Cleaver, U.S. Pat. 3,467,635 (1969) to DuPont [*CA 68*, 13885w].

56. Y. Tabata, K. Ishigure, and H. Sobue, *J. Poly. Sci., Part A-2* 2235 (1964).

57. G. Kojima and Y. Tabata, *J. Macromol. Sci., Chem. A5*, 1087 (1971).

58. G. Kojima, H. Kojima, and Y. Tabata, *Rubber Chem. Technol. 50*, 403 (1977).

59. G. Kojima and H. Wachi, *Rubber Chem. Technol. 51*, 940 (1978).

60. G. Kojima, H. Kojima, M. Morozuni, H. Wachi, and M. Hisasue, *Rubber Chem. Technol. 54*, 779 (1981).

61. R. Ro, U.S. Pat. 3,579,474 (1971) to DuPont [*CA 73*, 67217b].

62. G. Kojima and M. Hisasue, *Makromol. Chem. 182*, 1429 (1978).

63. J. R. Harrell and W. W. Schmiegel, U.S. Pat. 3,859,259 (1975) to DuPont [*CA 82*, 112956r].

64. G. Kojima and H. Wachi, *International Rubber Conference*, p. 242, Kyoto, Japan (1985).

65. D. E. Hull, *Elastomerics 115*, No. 10, 40 (1983).

66. D. E. Hull, *Elastomerics 114*, No. 7, 27 (1982).

67. A. L. Barney, W. S. Keller, and N. M. Van Gulick, *J. Polym. Sci., Part A-1, 8*, 1091 (1970).

68. G. H. Kalb, A. A. Kahn, R. W. Quareles, and A. L. Barney, *ACS Advances in Chemistry Series, No. 129*, pp. 13–26 (1973).

69. A. L. Barney, G. H. Kalb, and A. A. Khan, *Rubber Chem. Technol. 44*, 660 (1971).

70. G. H. Kalb, R. W. Quarles, and R. S. Graff, *Applied Polymer Symposium, No. 22*, p. 127 (1973).

71. M. Tatemoto and T. Amano, U.S. Pat. 4,487,903 (1984) to Daikin [*CA 99*, 39639j].

72. S. Nakagawa, T. Nakagawa, S. Yamaguchi, K. Ihara, P. Amano, M. Omori, and K. Asano, U.S. Pat. 4,499,249 (1985) to Daikin [*CA 100*, 52217d].

73. J. B. Finlay, U.S. Pat. 4,529,784 (1985) to DuPont [*CA 102*, 205254g].
74. A. F. Breazeale, U.S. Pat. 4,281,092 (1981) to DuPont [*CA 93,* 187543g].
75. A. Henne and R. L. Pelley, *J. Am. Chem. Soc. 74*, 1426 (1952).
76. S. Ogintz, *Elastomerics 119*, No.11, 21–23 (1987).
77. A. L. Moore, *Elastomerics 118*, No. 9, 14–17 (1986).
78. M. B. H. Simpson, *Kautsch. Gummi, Kunstst. 33*, 83 (1980).
79. J. B. Finlay, A. L. Moran, and A. L. Logothetis, *Proc. Int. Rubber Conf.*, pp. 93–100, Milan, Italy (1979); A. E. Hirsch, *Rubbercon 77, International Rubber Conference,* p. 12, London (1977).
80. J. E. Alexander and H. Omura, *Elastomerics 110*, 19 (1978); L. F. Pelosi and E. T. Hackett, *Elastomerics 109*, 31 (1977).
81. B. H. Spoo, *J. Elastomers Plast. 9*, 312 (1979).
82. R. Ferro, G. Fiorillo, and G. Restelli, *Elastomerics 121*, No. 4, 22 (1989); *121*, No. 3, 28 (1989).
83. A. Rudin, A. T. Worm, and J. E. Blacklock, *Plast Eng. 42*, No. 3, 63 (1986); *J. Plast. Film Sheeting 1*, No. 3, 189 (1985).
84. A. L. Logothetis, *Preprints, Second Pacific Polymer Conference*, p. 374, Otsu, Japan (1991); A. L. Logothetis, *Preprints, Third Pacific Polymer Conference,* p. 785, Gold Coast, Australia (1993).

17

Fluoropolymer Coatings

MASAAKI YAMABE

17.1. INTRODUCTION

Fluoropolymers are distinguished particularly by their extremely high thermal and chemical resistance, high electrical resistivity, low surface energy, and low refractive index. Accordingly, various kinds of fluoropolymers have been widely used in industrial applications where these characteristics are required.[1]

Fluoropolymers are expected to offer several advantages as coatings because they can impart their special properties to the surface of a variety of substrates. Major difficulties in using fluoropolymers for paints and coatings arise, however, owing to their poor solubilities in organic solvents and the need for high baking temperatures (above 200 °C). Thus, relevant technologies covering compounding, processing, and substrate pretreatment, as well as polymer design, have been investigated extensively and developed to optimize performance and economy.

Coatings of poly(tetrafluoroethylene) (PTFE), tetrafluoroethylene–hexafluoropropylene copolymer (FEP), or ethylene–tetrafluoroethylene copolymer (ETFE) are used mainly in antistick and anticorrosive applications. Among the conventional fluoropolymers, only poly(vinylidene fluoride) (PVdF) has been used for highly weather-resistant paints. In recent years, novel curable fluoropolymers have been developed and successfully commercialized, mainly as weather-resistant paints that can cure even at ambient temperature.[2]

17.2. THERMOPLASTIC FLUOROPOLYMER COATINGS

17.2.1. Tetrafluoroethylene Polymers

PTFE (e.g., Teflon®, Fluon®, etc.) is conveniently used for coatings in the form of an aqueous dispersion, which is produced by the emulsion polymerization of

MASAAKI YAMABE • Research Center, Asahi Glass Co. Ltd., Kanagawa-ku, Yokohama 221, Japan.

Organofluorine Chemistry: Principles and Commercial Applications, edited by R. E. Banks *et al.* Plenum Press, New York, 1994.

tetrafluoroethylene (see Chapter 15) followed by thermal concentration of as-polymerized latex up to 60 wt% of polymer. PTFE coatings provide the optimum effect where the special properties of fluoropolymers are concerned, especially nonstickiness and good lubricity—characteristics which are used to advantage in numerous applications, e.g., in the home (nonstick cookware) and the food-processing (bread, confectioneries and beer making, gardening implements, etc.), textiles (starching rolls, etc.), rubber and plastics (molds and lamination rolls, etc.) industries. A major drawback to PTFE as a coating is the polymer's very high melt viscosity at the baking (sintering) temperature (380 °C; see Chapter 15), which often causes the formation of pinholes in the final covering. Hence PTFE is not recommended as an anticorrosive coating, despite its excellent chemical resistance.

Pretreatment of a substrate to improve its adhesion to PTFE coating is very important. For example, when an iron or aluminum substrate is first treated with ceramic powder primers to form a coarse surface, and then subjected to PTFE coating, the finish becomes tough and more scratch resistant.[3] Another way to improve adhesion is to blend the PTFE dispersion with an engineering plastic like polyimide, poly(ether sulfone), or poly(phenylene sulfide). These blends are then used as a primer or one-coat enamel for the PTFE coating.[4]

Teflon[®] FEP and Teflon[®] PFA (C_2F_4/CF_2=$CFOC_3F_7$ copolymer) (see Chapter 15) also are produced in emulsion forms and used as dispersion coatings in the same way as PTFE. Pinhole-free coatings are obtained with these materials because of their lower melt viscosities at the baking temperature.

ETFE copolymer (e.g., Aflon COP[®] and Tefzel[®]; see Chapter 15), on the other hand, is most suitable for electrostatic powder coating for anticorrosive, antistick, and electrical applications in both the home (electrical hot plates, rice cooker, etc.) and industries (pump impellers and casings, pipe linings, valves, computer wire and cable coatings, etc.). One characteristic feature of ETFE electrostatic coating is the possibility of obtaining very thick finishes (reaching 1000 µm) by compounding about 20 wt% of carbon fiber to reduce the electrostatic repulsion of the coated powder.[5]

17.2.2. Poly(Vinylidene Fluoride)

PVdF is a crystalline polymer (e.g., Kynar[®]; see Chapter 15) but, being soluble to some extent in highly polar solvents, can be dispersed in a latent solvent such as dimethylformamide or dimethylacetamide. The resultant dispersion is then applied to a substrate and baked at a temperature of *ca* 300 °C. Kynar[®], PVdF paint, is usually formulated by blending in 20–30 wt% of an acrylic resin like poly(methyl methacrylate) to improve both melt-flow behavior at the baking temperature and adhesion characteristics. Long-term exposure tests have shown that the weather resistance of PVdF coatings is maintained for over 20 years.[6] Recently, a novel terpolymer composed of CH_2=CF_2, C_2F_4, and C_3F_6 (Kynar ADS[®]), has been developed as an air-drying paint, mainly for repair coating of an original PVdF finish.[7]

PVdF is now the preferred material for coil coating of galvanized steel and aluminum sheets used as maintenance-free protection for skyscraper side walls and industrial roofing.

17.3. CURABLE FLUOROPOLYMER COATINGS

Extensive studies have been conducted to develop crosslinkable fluoropolymers, which are particularly desirable for coatings and paints. Most of the research in this area has been concentrated on the choice of suitable cure sites and how to incorporate them into fluoropolymers.

A block ter-polymer [Fluoropolymer B] containing 65% VdF, 25% TFE, and 10% vinyl ester (e.g., vinyl butyrate) was developed several years ago by Du Pont.[8] This polymer can be cured by ultraviolet irradiation in the presence of a photosensitizer, and is especially useful as a weather-resistant, clear coating for wood substrates.

An excellent hard coating on plastic objects like acrylic sheets (Abcite® and Lucite®) was commercialized by Du Pont based on crosslinkable liquid mixtures composed of a hydroxy-loaded fluoropolymer (e.g., tetrafluoroethylene–hydroxyalkyl vinyl ether copolymer), hexa(methoxymethyl)melamine, and silica.[9]

Unique soluble fluoroolefin–vinyl ether copolymers (Lumiflon®) have been developed by Asahi Glass.[2] These polymers comprise alternating sequences of fluoroolefin and several specific vinyl monomer units (Figure 1), and are amorphous. The alternating sequence is responsible for the high weather resistance of the resultant paint finish. The combination of several kinds of vinyl monomers provides the polymer with various properties necessary for a paint, such as solubility, compatibility with pigments, crosslinking reactivity and good adhesion, hardness, and flexibility of the final finish.

The appealing characteristics of this polymer as a paint material are its ease of handling and processing just like conventional paint resins such as acrylic urethanes, and its excellent weatherability.[10] Since the hydroxyl group in the polymer functions as the cure site with polyisocyanates at room temperature, and with melamine resin or blocked isocyanates at higher temperature, Lumiflon® is easy to formulate not only for on-site coatings, but also for thermoset coatings in the factory. An aqueous dispersion of Lumiflon® has been developed recently by copolymerizing a macro-monomer containing a hydrophilic side chain with tetrafluoroethylene.[11]

Lumiflon® is convenient to use with various kinds of substrates—plastics, cements, metals, glass, etc.—and its field of application includes skyscraper side walls and other architectural structures, bridges (top-layer protective paint), and automobiles (no-wax brilliant top coat).

Several fluoroethylene (CF_2=CFX; X = F or Cl)/vinyl monomer [CH_2=CHOR, CH_2=CHOCOR, CH_2=CHCH$_2$OR, CH_2=CHSi(OR)$_3$, CH_2=CHOCH$_2$CF$_2$-CHF$_2$] co- and ter-polymers have been reported recently which are similar to the Lumiflon® structure (Figure 1).[12–15] Also, an entirely different family of fluorocoatings based on epoxy resin methodology has been developed.[16–18] The diglycidyl ether

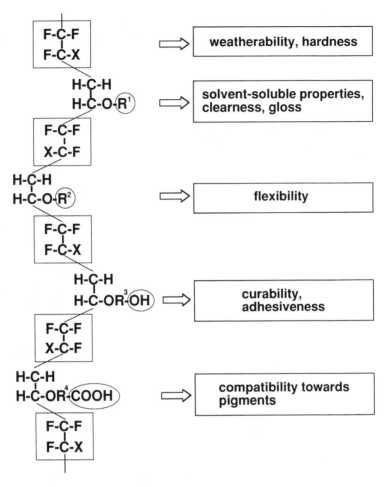

Figure 1. Structural features of Lumiflon® Polymer (X = F or Cl; R^1, R^2 = alkyl, cycloalkyl; R^3, R^4 = alkylene, cycloalkylene).

monomers (**1**) used are prepared from epichlorohydrin and diols synthesized from hexafluoroacetone (see Chapter 19). Once suitably crosslinked, they give coatings with superior weatherability, oil and water repellency, and resistance to the accumulation of dirt, ice, and marine fouling.[18]

1 (R = H or perfluoroalkyl)

17.4. REFERENCES

1. D. P. Carlson and W. Schmiegel, in: *Ullmann's Encyclopedia of Industrial Chemistry*, Vol. A11, pp. 393–427, VCH Publishers, Weinheim (1988).
2. M. Yamabe, H. Higaki, and G. Kojima, in: *Organic Coatings Science and Technology* (G. D. Parfitt and A. V. Pastis, eds.), Vol. 7, pp. 25–39, Marcel Dekker, New York (1984).
3. T. Concannon and E. Vassiliou, U.S. Patent 4,011,361 (to Du Pont) [*CA 86*, 74633 (1977)].
4. J. C. S. Fang, Brit. Patent 1,064,840 (to Du Pont) [*CA 66*, 47437 (1967)].
5. G. Kojima and M. Yamabe, in: *Proceedings of 11th International Conference in Organic Coatings Science and Technology* (A. V. Patsis, ed.), pp. 120–128, Technomic Publishing, Lancaster–Basel (1987).
6. J. E. McCarm, *Surf. Coating Aust. 27*, 8 (1990).
7. K. V. Summer, *Conf. Proc. Alum. Finish 19*, 36 (1986).
8. F. B. Stilmar, U.S. Patent 3,318,850 (to Du Pont) [*CA 64*, 9957 (1966)].
9. M. F. Bechtold, U.S. Patent 3,651,003 (to Du Pont) [*CA 73*, 78669 (1970)].
10. S. Munekata, *Prog. Org. Coat. 16*, 113 (1988).
11. N. Miyazaki and M. Kamba, Paper presented at First Pacific Polymer Conference, Hawaii, Abstract p. 577 (1989).
12. M. Ohka and Y. Murakami, Japanese Patent 84-102962 (to Dainippon Ink and Chemicals) [*CA 101*, 173148 (1984)].
13. T. Koishi and T. Yasamura, Japanese Patent 86-57609 (to Central Glass Co.) [*CA 105*, 8088 (1986)].
14. S. Honma, S. Murakami, and Y. Shimizu, Japanese Patent 86-141713 (to Mitsui Petrochemical Ind.) [*CA 105*, 192965 (1986)].
15. A. Ohmori, N. Tomihashi, and Y. Shimizu, Japanese Patent 84-189108 (to Daikin Industries) [*CA 102*, 47490 (1985)].
16. S. J. Shaw, D. A. Tod, and J. R. Griffith, *Chemtech* 290 (1982).
17. J. R. Griffith and S. Y. Lee, *Ind. Eng. Chem., Prod. Res. Dev. 25*, 572 (1986) .
18. R. F. Brady, *Chem Br. 26*, 427 (1990).

18

Fluorinated Membranes

MASAAKI YAMABE and HARUHISA MIYAKE

18.1. INTRODUCTION

The term fluorinated membranes generally refers to ion-exchange membranes composed of perfluorinated polymeric backbones. An overview of both the fundamental properties and the technological aspects of perfluorinated membranes is available.[1] The first perfluorinated membranes, Nafion® membranes, were developed and commercialized by Du Pont in the early 1970s. They were made of the perfluorinated sulfonic acid ionomer called XR resin.[2] Nafion® was first employed as a separator in fuel cells that were used in space exploration, and then as ion-exchange membranes that opened the way to the innovative electrolytic process for chlor-alkali production.[3]

Since the electrolytic performance of the early Nafion® membranes was unsatisfactory from the industrial point of view, extensive efforts have concentrated on improving current efficiencies of plain sulfonic acid membranes by chemical modification of their surfaces[4–6] or by film lamination.[7]

Flemion®, a new type of perfluorinated carboxylic acid membrane, was developed in 1975, and has been commercially produced since 1978 by Asahi Glass.[8–10] The different ion-exchange groups greatly affect the physical and electrochemical properties of the membranes.

The introduction of these high-performance membranes has significantly advanced electrolytic-cell technologies, and the membrane chlor-alkali process is now recognized worldwide as energy-saving and pollution-free, compared with the mercury or the diaphragm processes.

MASAAKI YAMABE and HARUHISA MIYAKE • Research Center, Asahi Glass Co. Ltd., Kanagawa-ku, Yokohama 221, Japan.

Organofluorine Chemistry: Principles and Commercial Applications, edited by R. E. Banks *et al.* Plenum Press, New York, 1994.

18.2. STRUCTURE AND PROPERTIES

Generalized structures (**1,2**) of the polymers used to manufacture perfluorinated membranes are shown below. Both **1** ($X = SO_2F$) and **2** ($Y = CO_2CH_3$) are melt-processable and can be fabricated into films by extrusion-molding. The resultant films can easily be converted to the corresponding ion-exchange membranes **1** ($X = SO_3^-Na^+$) and **2** ($Y = CO_2^-Na^+$) by alkaline hydrolysis.

Tensile properties of these membranes change drastically after hydrolysis,[9,10] and the ionic interaction or the formation of ionic clusters have been extensively investigated by small-angle X-ray, NMR, IR, and Mossbauer spectroscopies and by neutron-activation studies.[1] These studies reveal various different structural features of the membranes, but their ion-clustered morphologies remain unsolved because of their structural complexities.

Electrochemical characteristics, such as electric conductivity and current efficiency of the membrane, mainly depend upon the following inherent properties of the polymers: ion-exchange capacities, water content, and fixed-ion concentrations. The ion-exchange capacity represents the concentration of ion-exchange groups in a polymer, and the high conductivity of a membrane can be achieved by incorporating as many ion-exchange groups as possible without impairing polymer strength. Relationships between ion-exchange capacities and electric conductivities have been developed.[5,10,11]

These membranes have lower ion-exchange capacities than other ionomers in that ions present in the perfluorinated systems tend to aggregate and form domains. The superior performance of these membranes has been attributed to the presence of these domains. The diffusion of redox ions in Nafion® films has been discussed in connection with the two-phase structure of Nafion®.[12]

The membrane's water content or water uptake is determined by the balance of power based on the contraction of the polymer backbone and the swelling by water which associates with the counterion and the ion-exchange group in the polymer. Consequently, the concentration of ion-exchange groups in the water phase inside the

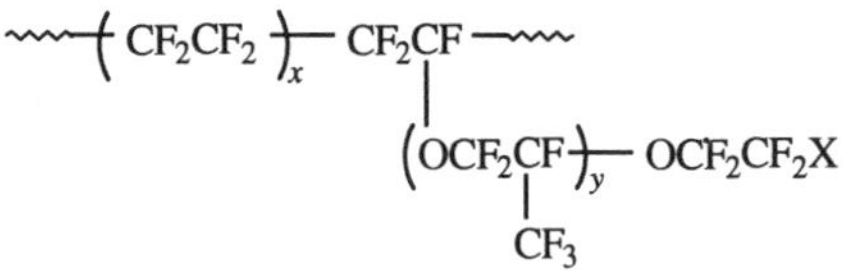

$$\sim\!\!\!\left(CF_2CF_2\right)_x\!\!-\!CF_2CF\!\!-\!\!\sim$$
$$\left(OCF_2CF\right)_y\!\!-\!OCF_2CF_2X$$
$$\qquad\quad CF_3$$

1

$$\sim\!\!\!\left(CF_2CF_2\right)_x\!\!-\!CF_2CF\!\!-\!\!\sim$$
$$\left(OCF_2CF\right)_m\!\!-\!O\!\!-\!\!\left(CF_2\right)_n\!\!Y$$
$$\qquad\quad CF_3$$

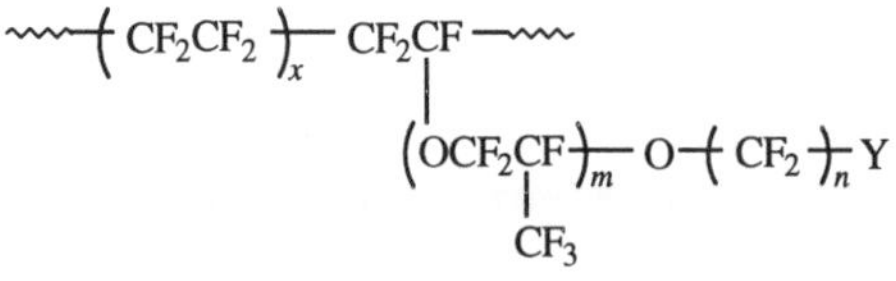

2 ($m = 0$ or 1, $n = 1\text{--}5$)

membrane is determined and expressed as the fixed ion concentration, which is an important factor governing the current efficiency in electrolysis. When the fixed-ion concentration is sufficiently high, the introduction of an ion with the same charge as that of the ion-exchange group is efficiently suppressed according to the theory of Donnan's equilibrium.

The CF_2CO_2H group is less acidic than the CF_2SO_3H group, and also has a lower energy of hydration. Therefore, the carboxylic acid membrane retains lower levels of water and higher fixed-ion concentrations than the sulfonic acid membrane through a wide range of ion-exchange capacities, which results in higher current efficiencies in the production of concentrated caustic soda.[13]

18.3. PREPARATION

The general procedure for manufacturing perfluorinated ionomer membranes includes synthesis of a functionalized perfluorovinyl ether, its copolymerization with tetrafluoroethylene, and fabrication of a membrane from the resulting copolymer. Usually, membranes are reinforced by fabrics or other materials to ensure the reliability of their mechanical properties over long-term use.

The commercial preparation of various perfluorovinyl ethers has become possible through Du Pont's pioneering work on the chemistry of hexafluoropropylene oxide (HFPO). The synthetic schemes for the key monomers used to make both Nafion® and Flemion® polymers are given in Schemes 1–3.[2,13,14] Alternative synthetic routes to perfluorinated vinyl ethers have also been disclosed.[8,16] New sulfonic acid membranes derived from perfluorovinyl ether monomers with shorter difluoromethylene chains (3) have been prepared.[5,15] These monomers are particularly useful for preparing membranes with higher ion-exchange capacities.

The functionalized vinyl ethers described above can be polymerized with tetrafluoroethylene in the presence of radical initiators in an aqueous medium or in an inert organic solvent.[17–19] A copolymer composition curve shows that random copolymerization of tetrafluoroethylene occurs with carboxylated perfluorovinyl ethers.[10]

The crystallinity of the copolymer depends upon the content of the functional comonomer wherein crystallinity decreases with increasing comonomer content.[10]

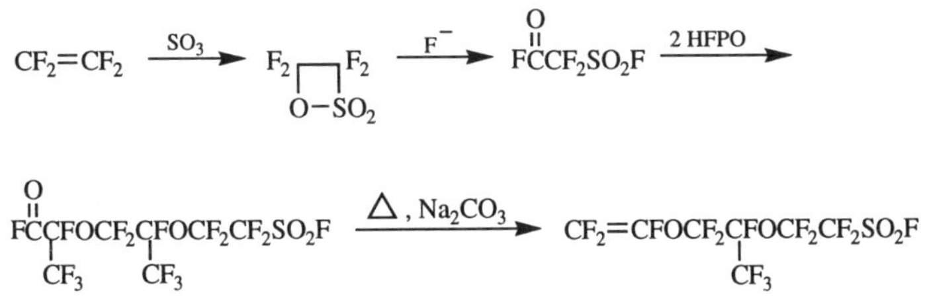

Scheme 1. Conversion of tetrafluoroethylene to a fluorosulfonylated perfluorovinyl ether monomer used in Nafion® production.

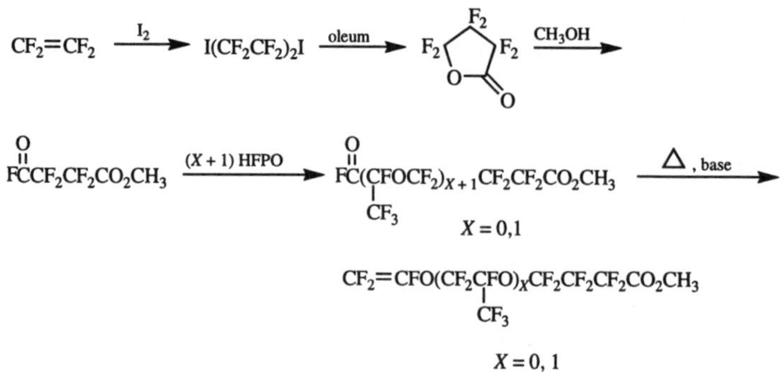

$$CF_2{=}CF_2 \xrightarrow{I_2} I(CF_2CF_2)_2I \xrightarrow{oleum} F_2\overset{F_2}{\underset{O}{\bigtriangleup}}F_2 \xrightarrow{CH_3OH}$$

$$\underset{O}{\overset{O}{\underset{\|}{F}}}CCF_2CF_2CO_2CH_3 \xrightarrow{(X+1)\ HFPO} \underset{\substack{|\\CF_3}}{\overset{O}{\underset{\|}{F}}}C(CFOCF_2)_{X+1}CF_2CF_2CO_2CH_3 \xrightarrow{\Delta,\ base}$$

$$X = 0,1$$

$$CF_2{=}CFO(CF_2\underset{\substack{|\\CF_3}}{C}FO)_XCF_2CF_2CF_2CO_2CH_3$$

$$X = 0,\ 1$$

Scheme 2. Conversion of tetrafluoroethylene to a methoxycarbonylated perfluorovinyl ether monomer used in Flemion® production.

$$CH_3O\overset{O}{\underset{\|}{C}}OCH_3 + CF_2{=}CF_2 \xrightarrow{NaOCH_3} CH_3O(CF_2)_2CO_2CH_3 \xrightarrow{oleum}$$

$$\overset{O}{\underset{\|}{F}}CCF_2CO_2CH_3 \xrightarrow{(m+1)\ HFPO} \underset{\substack{|\\CF_3}}{\overset{O}{\underset{\|}{F}}}C(CFOCF_2)_{m+1}CF_2CO_2CH_3 \xrightarrow{\Delta,\ Na_2CO_3}$$

$$CF_2{=}CFO(CF_2\underset{\substack{|\\CF_3}}{C}FO)_m(CF_2)_2CO_2CH_3$$

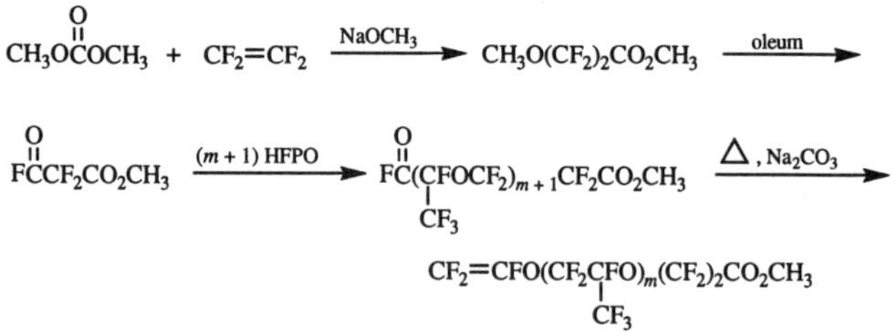

Scheme 3. The dimethyl carbonate route to a methoxycarbonylated perfluorovinyl ether monomer used in Flemion® production.

Amorphous or partly crystalline copolymers are fabricated into films with designed thickness (usually 100–250 µm) by conventional extrusion techniques. Two kinds of films with different ion-exchange capacities[20,21] or even with different ion-exchange groups[7] can be laminated, if necessary.

The polymer films are generally reinforced by poly(tetrafluoroethylene) (Teflon®) cloth but, in some cases, they can be reinforced by using a premixed copolymer with small amounts of fibrous materials.[22] The resulting films are converted to sulfonic or carboxylic acid salt ion-exchange membranes by alkaline hydrolysis.

A sulfonyl group in the membrane also can be converted to a carboxylic acid group (Scheme 4) by various chemical reactions. The sulfonyl halide group in the membrane is first converted to sulfinic acid by reduction, and thence to the carboxylic acid having one less CF_2 than the original chain of the sulfonyl halide.[5] Alternatively, the conver-

$$CF_2{=}CFO\left(CF_2\right)_n SO_2F$$

3 $(n = 2,3)$

$$\text{\textasciitilde\textasciitilde\textasciitilde} O(CF_2)_n SO_2 X \xrightarrow[80\,°C]{\text{aq. HI}} [-O(CF_2)_n SO_2 H] \xrightarrow{\Delta} \text{\textasciitilde\textasciitilde\textasciitilde} O(CF_2)_{n-1} CO_2 H$$

(X = Cl or F)

Scheme 4.   Conversion of a halogenosulfonylated membrane to a carbonylated analogue.

sion can be performed by treating the membrane sulfonyl chloride group with hot air in the presence of alcohol vapor.[6]

18.4.   APPLICATIONS

Very important features of fluorinated membranes are their excellent thermal and chemical stabilities, and retention of mechanical properties in corrosive and oxidative environments. Among fluorinated membranes, perfluorinated sulfonic and carboxylic acid ones have been widely used as separators in electrochemical processes such as brine electrolysis, water electrolysis, and fuel cells. In particular, the perfluorosulfonic acid membrane, which is highly acidic and has high hydrogen ion mobility when hydrated, is widely used for solid polymer electrolyte (SPE) cells.[23,24] The performance of the various types of perfluorinated ionic membranes in the chlor-alkali process will be emphasized in the following sections.

18.4.1.   Chlor-Alkali Process

The manufacture of chlorine and caustic is the most important application for fluorinated membranes.

The principle of brine electrolysis using an ion-exchange membrane is shown schematically in Figure 1. The membrane divides the cell into anode and cathode chambers. Saturated brine is fed to the anode chamber and chlorine is generated at the anode. Water is fed to the cathode chamber and produces hydrogen and hydroxide ions, the latter of which combine with the sodium ions to form sodium hydroxide. The electrochemical reactions for the membrane cell are:

anode: $2\,NaCl \rightarrow Cl_2 + 2\,Na^+ + 2e$
cathode: $2\,H_2O + 2e \rightarrow H_2 + 2OH^-$
overall: $2\,NaCl + 2\,H_2O \rightarrow Cl_2 + H_2 + 2\,NaOH$

The membrane permits only the passage of sodium ions from the anode chamber to the cathode chamber and prevents the migration of hydroxide ions to the anode chamber, which would reduce the current efficiency if it occurred.

The essential requirements for the membrane used in the chlor-alkali process are high conductivity, selectivity, and mechanical strength. Therefore, various technologies have been devised depending on the composition of the membrane. To attain high

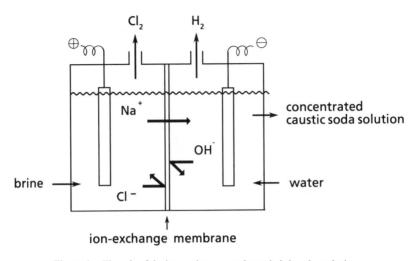

Figure 1.   The role of the ion-exchange membrane in brine electrolysis.

electrolytic performance, the membranes are usually composed of layers of both sulfonate and carboxylate perfluoropolymers. To endow the membranes with necessary strength, the membranes are generally reinforced by fluoropolymer cloth.

18.4.1.1.   Nafion® Membranes

Du Pont's Nafion® perfluorosulfonic acid resin membrane has a long history of improvements to its electrolytic performance. The initial Nafion® 400 series was composed of plain sulfonic acid membranes that were not suitable for caustic soda production because of low current efficiencies. The Nafion® 300 series was then introduced for the purpose of producing 10–20% caustic soda. These are laminated membranes with two sulfonic-acid-type polymers having different ion-exchange capacities,[4] one with ca 0.7 meq g^{-1} and the other with 0.91 meq g^{-1}. For the production of 20–28% caustic soda, the Nafion® 200 series was developed, and is characterized by having surfaces chemically modified with diamines.[4]

The latest Nafion® membranes have been designed to be carboxylate–sulfonate two-layer membranes with unique reinforcement. Nafion® 90209, 954, and 961, for example, show 96% current efficiency at 32% caustic soda production.[7]

18.4.1.2.   Flemion® Membranes

Various types of Flemion® membranes have been commercialized in response to user's requirements. Flemion® has an organic and inorganic composite texture. With these types of membranes, gas bubbles can be removed easily from the membrane surfaces. Flemion® can thus be used with zero gap between the membrane and electrode.

A new electrolytic process with a zero-gap cell, called the AZEC system, has attained drastic reduction in energy consumption when Flemion® and a new electrode system are employed. The energy consumption of this membrane process is now about 1000 kWh per tonne of NaOH lower than that of the mercury or the asbestos diaphragm process. Flemion® 795 is the membrane which has a low ohmic drop and is mainly used in AZEC electrolyzers. Flemion® 865 membranes are mechanically stronger and are less sensitive to operational upsets, so they are mainly used for general filter-press-type electrolyzers. Flemion® 854 has high resistance to damage from folding and is mainly used in retrofitting diaphragm cells. These membranes are used for the production of 32–35% caustic soda with a current efficiency of ca 96%. Flemion® 737 is designed for 35% caustic potash with a current efficiency of 97–98% and excellent product purity. Flemion® 795, 865, 854, and 737 are carboxylate-polymer based membranes. Recently, the Flemion® 890 series, which are sulfonate/carboxylate laminated polymer membranes, have been developed. They are used in both AZEC electrolyzers and general filter-press-type electrolyzers for the production of 32–35% caustic soda with a current efficiency of ca 96%.

Asahi Glass is now developing an entirely new membrane that is capable of directly producing 50% caustic soda without need of further evaporation.[25,26] Present membranes cannot afford to be applied to 50% caustic production because of a decrease in current efficiency. To overcome this technological difficulty, a specific layer is attached to the surface of the conventional membrane at the cathode side.

18.4.1.3. Aciplex® Membrane

Asahi Chemical has developed a two-layer membrane which is characterized by having a carboxylic acid layer on the cathode side of the sulfonate membrane.[27,28] A woven Teflon® web is embedded in the membrane. For the production of 21–23% caustic soda, Aciplex® F2200 is used with a current efficiency of ca 96%. Aciplex® F4100 and F4200 are used for the production of 30–32% caustic soda. Aciplex® F4200 and the latest F5200 version are surface-treated membranes for the prevention of gas-bubble retention on the membrane surfaces. Aciplex® F5200 is used for the production of 30–35% caustic soda and shows low cell voltage. At a current density of 4.0 kA m^{-2} and caustic soda concentration of 30–35%, 2150–2230 kWh per tonne NaOH has been achieved in commercial operation.

18.4.2. Outlook

Besides the membranes described above, several attempts to develop a fluorinated membrane with high electrolytic performance have been made.[29] Salt brine electrolysis using ion-exchange membranes has progressed remarkably in this decade and is now recognized worldwide as the most energy efficient process for the production of chlorine and caustic alkali.

Particularly in Japan, the mercury process was completely closed down in 1986 and has been converted almost entirely to the membrane process. The diaphragm process is also being supplanted by membrane technology. As a result, about 80% of total chlor-alkali is currently produced by the membrane process in Japan. The membrane process drastically reduces not only electrical power consumption but also pollution problems.

A commercial membrane process came into operation in 1975 and, since then, it has been increasingly adopted in many countries. At present, world caustic production by the membrane process amounts to 7.4 million tonnes per year, of which nearly 60% is shared in Asia, and 30% between North America and Western Europe. In 1980, the membrane process accounted for only about 3% of the world's chlor-alkali production, but since then it has gradually increased to the point where it now represents about 16% of installed worldwide caustic capacity. In the year 2000, total production capacity is estimated to reach 55 million tonnes, and the membrane process is expected to account for half of this.

18.5. REFERENCES

1. A. Eisenberg and H. L. Yeager (eds.), *Perfluorinated Ionomer Membranes*, ACS Symposium Series 180, American Chemical Society, Washington, DC (1982).
2. D. J. Vaughan, *Du Pont Innovation 4*, 10 (1973).
3. *Chem. Week 35* (Nov. 17, 1982).
4. W. Grot, U.S. Patent 3,969,285 (to Du Pont) [*CA 83*, 132591 (1975)].
5. M. Seko, H. Miyauchi, J. Ohmura, and K. Kimoto, in: *Modern Chlor-Alkali Technology* (C. Jackson, ed .), Vol 3, p. 76, Ellis Horwood, Chichester (1983).
6. T. Sata and Y. Onoue, in: *Perfluorinated Ionomer Membranes*, ACS Symposium Series 180, pp. 411–425, American Chemical Society, Washington, DC (1982).
7. S. M. Ibrahim, E. H. Price, and R. A. Smith, in: *Modern Chlor-Alkali Technology* (C. Jackson, ed.), Vol. 3, p. 53, Ellis Horwood, Chichester (1983).
8. H. Ukihashi, T. Asawa, and H. Miyake, in: *Ion Exchange Membranes* (D. S. Flett, ed.), p. 165, Ellis Horwood, Chichester (1983).
9. H. Ukihashi, M. Yamabe, and H. Miyake, *Prog. Polym. Sci. 12*, 229 (1986).
10. H. Ukihashi, *Chemtech* 118 (1980).
11. W. G. Grot, G. E. Munn, and P. N. Walmsley, *J. Electrochem. Soc. 119*, 108C (1972); Paper presented at the 141st National Meeting, The Electrochemical Society, Abstract No. 154, Houston, TX (1972).
12. D. A. Bultry and F. C. Anson, *J. Am. Chem. Soc. 105*, 685 (1983).
13. H. Ukihashi and M. Yamabe, in: *Perfluorinated Ionomer Membanes*, ACS Symposium Series 180, pp. 427–457, American Chemical Society, Washington, DC (1982).
14. D. C. England, U.S. Patent 4,131,740 (to Du Pont) [*CA 90*, 72627 (1979)].
15. B. R. Ezzel, W. P. Carl, and W. A. Mod, U.S. Patent 4,358,412 (to Dow Chemical Co.) [*CA 96*, 143514 (1982)].
16. D. C. England, U.S. Patent 4,138,426 (to Du Pont) [*CA 90*, 138599 (1979)].
17. D. P. Carlson, U.S. Patent 3,528,954 (to Du Pont) [*CA 71*, 39639 (1969)].
18. H. Ukihashi, T. Asawa, M. Yamabe, and H. Miyake, U.S. Patent 4,138,373 (to Asahi Glass Co.) [*CA 89*, 25341 (1978)].
19. H. Ukihashi, T. Asawa, M. Yamabe, and H. Miyake, U.S. Patent 4,116,888 (to Asahi Glass Co.) [*CA 89*, 25341 (1978)].
20. E. H. Price, E. J. Peter, and D. R. Pulver, *J. Electrochem. Soc. 124*, 319C (1977); Paper presented at the 152nd National Meeting, The Electrochemical Society, Abstract No. 443, Atlanta, GA (1977).
21. M. Nagamura, H. Ukihashi, and O. Shiragami, in: *Modern Chlor-Alkali Technology* (M. O. Coulter, ed.), Vol. 1, p. 221, Ellis Horwood, Chichester (1980).

22. H. Ukihashi, T. Asawa, and T. Gunjima, U.S. Patent 4,255,523 (to Asahi Glass Co.) [*CA 90*, 169829 (1979)].

23. R. S. Yeo, in: *Perfluorinated Ionomer Membranes*, ACS Symposium Series 180 (A. Eisenberg and H. L. Yeager, eds.), pp. 454–473, American Chemical Society, Washington, DC (1982).

24. B. Kippling, in: *Perfluorinated Ionomer Membranes*, ACS Symposium Series 180 (A. Eisenberg and H. L. Yeager, eds.), pp. 475–487, American Chemical Society, Washington, DC (1982).

25. M. Miyake, I. Kaneko, and A. Watakabe, European Patent 229,321 (to Asahi Glass Co.) [*CA 112*, 27264 (1990)].

26. K. Suzuki, Y. Sugaya, A. Watakabe, and T. Shimohira, PCT Int. Appl. 80-09, 799 (to Asahi Glass Co.) [*CA 110*, 20584 (1988)].

27. M. Seko, A Yomiyama, and S. Ogawa, in: *Ion Exchange Membranes* (D. S. Flett, ed.), p. 121, Ellis Horwood, Chichester (1983).

28. T. Hiyoshi, H. Wakamatu, and H. Shiraki, *Soda To Enso* (Japan Soda Industry Association) *41*, 126 (1990).

29. S. C. Stinson, *Chem. Eng. News 22* (March 15, 1982).

19

Monomers and Polymers from Hexafluoroacetone

WOLFGANG K. APPEL, BERND A. BLECH, and MICHAEL STÖBBE

19.1. INTRODUCTION

It has been known for some years that the incorporation of trifluoromethyl groups, particularly in the form of the 1,1,1,3,3,3-hexafluoroisopropylidene function, $(CF_3)_2C$ (HFIP), into suitable monomers confers a rather unique property profile on the corresponding polymers.[1,2] Among the main advantages are enhanced solubility and processability, higher thermo-oxidative stability, and superior electrical properties. This has led to the synthesis of a large variety of such monomers and polymers, as disclosed in recent reviews.[3,4] The aim of this chapter is to focus on the synthesis of basic intermediates and important monomers, and to highlight important polymer classes containing the 1,1,1,3,3,3-hexafluoroisopropylidene moiety.

19.2. SYNTHESIS OF HEXAFLUOROACETONE (HFA)

Hexafluoroacetone (HFA) can be synthesized in several ways,[5] the most important being halogen-exchange fluorination[6] (HF/chromium oxide catalyst) of hexachloroacetone and oxidative transformations of hexafluoropropene (HFP) (Scheme 1).

It seems that the latter approach is economically and ecologically superior and becoming increasingly more dominant. The starting material for this sequence, HFP, can be oxidized either directly to HFA[7,8] or first to hexafluoropropene oxide (HFPO),[9,10] which is then rearranged to HFA.[11-13]

WOLFGANG K. APPEL and MICHAEL STÖBBE • Hauptlaboratorium, Hoechst AG, D-6230 Frankfurt-am-Main 80, Germany.   BERND A. BLECH • RLMTC, Hoechst Celanese Corporation, Summit, New Jersey 07901.

Organofluorine Chemistry: Principles and Commercial Applications, edited by R. E. Banks *et al.* Plenum Press, New York, 1994.

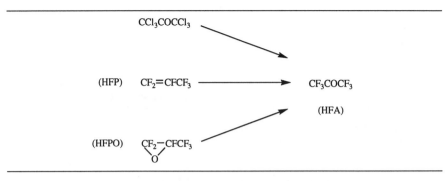

Scheme 1. Routes to hexafluoroacetone (HFA).

Handling HFPO has advantages over handling HFA, because HFPO is considerably less toxic[14,15] and its conversion to HFA can easily be performed in the vapor phase over a metal oxide catalyst as needed,[17] or *in situ* in the reaction mixture if HF is used as solvent. HFA is not only highly toxic but also a suspect teratogen.[14] Thus, the transportation and handling of HFA are strictly regulated and require the utmost care. HFA boils at –27.4°C and freezes at –122°C. It combines exothermally with water to form hydrates, e.g., hexafluoroacetone mono- and tri-hydrate. The monohydrate, $(CF_3)_2C(OH)_2$ is a solid, mp 49°C; higher hydrates are liquids, the trihydrate distilling at 105°C.

19.3. SYNTHESIS OF INTERMEDIATES CONTAINING THE HEXAFLUOROISOPROPYLIDENE GROUP

19.3.1. The Overall Reaction Sequence

The conversion of HFA to a hexafluoroisopropylidene (HFIP) bridged intermediate is a two-step process. First, an arylbis(trifluoromethyl) carbinol* is formed by treating the ketone either with a metallated or Grignard-type arene, or with the parent arene in the presence of a Friedel-Crafts catalyst. Subsequently, the carbinol is condensed with a second aromatic unit to yield a 2,2-diarylhexafluoropropane (Scheme 2).

19.3.2. Formation of Arylbis(trifluoromethyl)carbinols (Step 1)

19.3.2.1. Carbinol Formation via Electrophilic Aromatic Substitution

In general, the electrophilic reaction of HFA with aromatic compounds follows the pattern typical of Friedel-Crafts reactions. Arenes bearing *ortho,para* directing

---

*In the carbinol system of nomenclature, alcohols are named as derivatives of the first member of the series, methanol (CH_3OH). For this purpose, methanol is called carbinol, and other alcohols are named as derivatives produced by replacement of hydrogen(s) of the methyl group.—Eds.

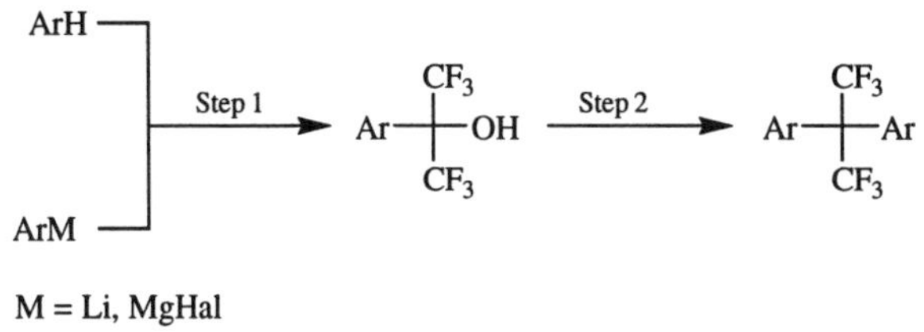

Scheme 2. Reaction sequence leading to diarylhexafluoropropanes.

groups, including alkyl, aryl, aryloxy and halogen, are easily attacked by HFA in the presence of $AlCl_3$.[16] Phenols and anilines react with HFA even without a catalyst.[18,19] To our knowledge, there is no example in the literature where HFA reacts in an electrophilic manner with aromatic compounds bearing only strongly deactivating substituents, such as nitrobenzene.

With alkyl, aryl, aryloxy or halogen substituted benzenes, the entering incipient hexafluoro-2-hydroxy-2-propyl group is directed predominantly to the *para* position. Only if this position is occupied will the *ortho* position be attacked. In the case of phenols the picture is more complicated and greatly influenced by the catalyst itself. When heated with HFA under pressure without catalyst and solvent, phenol gives predominantly the *ortho* substitution product (**1**; see Scheme 3).[18] The same reaction run under atmospheric pressure with xylene as solvent and *para*-toluenesulfonic acid (PTSA) as catalyst also gives the *ortho*-substituted phenol **1**, yet faster and in better yields; this holds for other phenols, except when *ortho* positions are blocked or when formation of an *ortho* isomer is sterically disfavored, as with 2-*t*-butylphenol.[18] Boron trifluoride and also anhydrous hydrogen fluoride, on the other hand, convert phenol

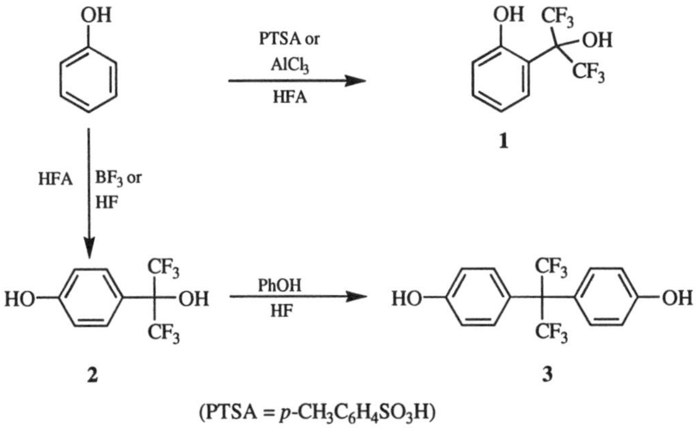

Scheme 3. Reactions of hexafluoroacetaone (HFA) with phenol.

4a: X = H
4b: X = OH
4c: X = F

5a: X = H
5b: X = OH

preferentially to the *para*-carbinol 2; with an excess of phenol, both catalysts promote the formation of the condensation product Bisphenol AF (3). Like *p*-toluenesulfonic acid, aluminum chloride promotes *ortho* substitution at the expense of *para*.

Anilines combine immediately with HFA to give unstable "salt-like" addition compounds which slowly rearrange to carbinols when heated[19]; positions *para* to NH_2 are predominantly attacked, if available, otherwise *ortho* isomers are formed. Although a catalyst is not generally necessary, reaction rates and yields are improved if a small amount of *p*-toluenesulfonic acid is added, and use of a solvent such as xylene is advantageous too.

Under more severe conditions it is possible to replace up to three hydrogens on the same aromatic nucleus. The second and third $(CF_3)_2COH$ groups enter the ring primarily *meta* to the first one. Thus, treatment of benzene with a two-molar proportion of HFA in the presence of $AlCl_3$ gives an 86:14 mixture of *meta* (4a) and *para* isomers.[18,20,21] Under very high pressure (20,000 atmospheres), 1,3,5-tris(2-hydroxyhexafluoro-2-propyl)benzene (5a) has been prepared in 90% yield from benzene and HFA in the presence of boron trifluoride.[22] When 1,3-bis(2-hydroxyhexafluoro-2-propyl)benzene (4a) was used as starting material, the same product (5a) was obtained in 91% yield under milder conditions (20 atmospheres, 225°C).[23]

Reactions between an excess (at least two-molar proportions) of HFA and phenol(ate)s[24] or halobenzenes[25] lead to formation of 2,4-disubstituted derivatives (4b,4c). Starting from sodium phenoxide a threefold substitution can be achieved (→5b).[23] Some of these oligocarbinols have been incorporated into epoxy resins, resulting in materials with hydrophobic and oleophobic properties.[26]

When carrying out these electrophilic aromatic substitutions, HFA can be introduced into reaction mixtures at atmospheric pressure, either as a neat gas or in the form of one of its hydrates[27]; alternatively, it can be preformed in an autoclave from HFPO, and this is the preferred method for most large-scale technical reactions. Anhydrous hydrogen fluoride is then, for several reasons, the reaction medium of choice: it serves as solvent, it catalyzes the isomerization of HFPO to HFA, it catalyzes carbinol formation, and (usually at elevated temperatures) it can bring about the condensation reaction leading to 2,2-diarylhexafluoropropanes (e.g., see Scheme 3) if required. As a matter of fact, in most known cases, the condensation reaction only proceeds in HF, as will be discussed in more detail later.

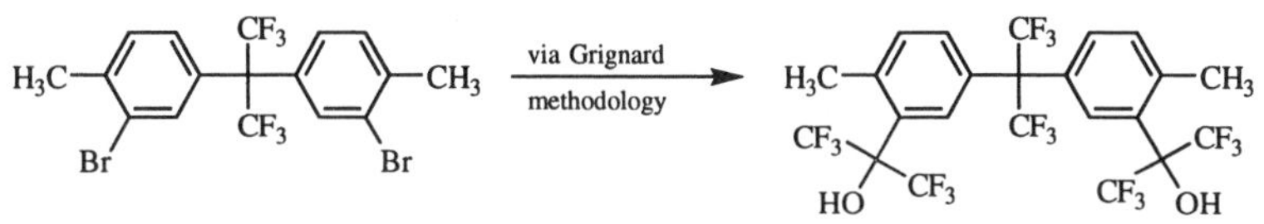

Scheme 4. An example of the Grignard route to hexafluoro-2-hydroxy-2-propyl derivatives.

19.3.2.2. Via Grignard or Metallation Procedures

The introduction of hexafluoroisopropyl substituents into arenes via organometallic methodology, while not exactly the production chemist's delight, is certainly the method of choice for systems which are not accessible via Friedel-Crafts reactions. The method is especially useful for the selective multiple introduction of HFIP groups. A recent example is the synthesis of an "18-F" compound [i.e., containing three $(CF_3)_2C$ groups] (**6**, Scheme 4) starting from 2,2-bis(4-tolyl)hexafluoropropane via double bromination and subsequent treatment of the corresponding Grignard derivative with HFA at low temperatures.[28]

An interesting example of a multiple metallation involves double deprotonation of 1,2,4,5-tetrafluorobenzene with two equivalents of butyllithium, followed by addition of HFA at −78°C to yield the dicarbinol $1,4-[(CF_3)_2C(OH)]C_6F_4$ after an aqueous work-up.[29]

19.3.3. Properties of Arylbis(trifluoromethyl)carbinols

Carbinols obtained by the reaction of HFA and aromatic compounds are stable liquids or solids which can be purified by vacuum distillation or recrystallization. The hydroxyl group is rather acidic, hence some of the carbinols are soluble in aqueous sodium hydroxide.[16] Under drastic alkaline conditions (KOH in diethylene glycol at *ca* 175°C) haloform cleavage takes place and the corresponding arene carboxylic acids are isolated as final products after acidification.[30] This degradation can be used to determine the orientation of the 2-hydroxy-HFIP group in products from carbinol syntheses,[30] although NMR analysis is less tedious. ^{1}H NMR data for a number of mono- and bis-carbinols can be found in Refs. 16–18. Ring protons *ortho* to a $(CF_3)_2COH$ group and *ortho*-methyl protons suffer characteristic deshielding. Known chemical transformations of the carbinol hydroxyl group include exchange for fluorine[31] and chlorine[32] by treatment with sulfur tetrafluoride or thionyl chloride, respectively, ether synthesis, formation of mesylates and tosylates, esterification, and lactone formation.[28] Cassidy and his co-workers have also described a number of attempted or only marginally successful transformations, including preparation of a polyether and a polyester using a di-carbinol as a comonomer.[28]

19.3.4. Formation of Diphenylhexafluoropropanes (Step 2)

The only viable route to HFIP-bridged diaryls is the condensation of arylbis(tri-fluoromethyl)carbinols with substituted or unsubstituted arenes, e.g., toluene, chlorobenzene, benzene, etc. Clearly, both symmetrical and unsymmetrical diphenylhexafluoropropanes can be prepared by this method.[27,33] Unfortunately though, the scope of this condensation is limited to a rather small selection of substituents that are able to survive the severe reaction conditions and/or do not deactivate either reactant too much to allow the reaction to proceed: satisfactory results are obtained only with alkyl, aryl, aryloxy, hydroxyl, and halogen substituents; in all cases *para*-alkylation is the major process. No direct access to interesting monomers such as diamines[34] or dianhydrides has been reported so far.

Therefore, almost all syntheses of HFIP-bridged monomers comprise two quite distinct stages: preparation of basic HFIP-bridged systems bearing substituents stable to the conditions involved, followed by transformation of those substituents to groups which confer monomer status on the final products (e.g., see Scheme 5). The only noteworthy exception from this plight of devising routes to appropriately functionalized monomers is the synthesis of Bisphenol AF (**3**), where the reaction between phenol and HFA leads directly to an important monomer.[35] Bisphenol AF itself can act as a precursor of other types of monomers, as exemplified in Scheme 5.

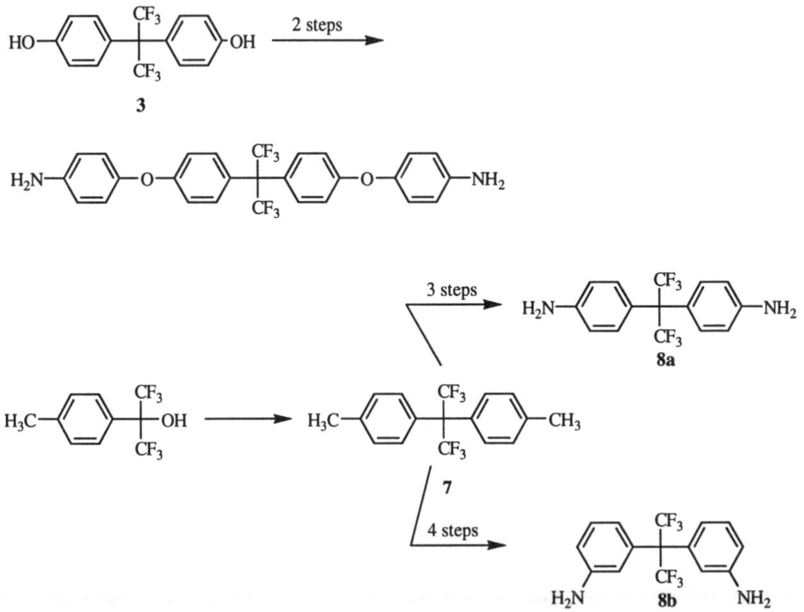

Scheme 5.   Functionalization of directly accessible HFIP-bridged compounds.

$$\text{H}_3\text{C}-\text{C}_6\text{H}_4-\overset{\text{CF}_3}{\underset{\text{CF}_3}{\text{C}}}-\text{F} \qquad \text{H}_3\text{C}-\text{C}_6\text{H}_4-\overset{\text{CF}_3}{\underset{\text{CF}_3}{\text{C}}}-\text{H}$$

9   10

Most references to the actual condensation step are found in industrial patents, reflecting the commercial interest in HFIP-based monomers for high-performance polymers, and also the rather "technical" nature of the reaction itself. Typically, the carbinols are treated with a suitable aromatic compound in an 8–40 molar excess of anhydrous HF, using a stainless steel autoclave. The HF serves both as solvent and catalyst. In the case of symmetrical HFIP-bridged diaryls, the rearrangement of HFPO to HFA, formation of the carbinol, and condensation to the final product can all be accomplished in one pot.[11]

Reaction temperatures are rather specific for each system. They range from 80 to 180°C and depend mainly on the nature of the substituents in the carbinol. It is therefore assumed that formation of an intermediate carbenium ion is the rate-determining step; some ionization rates have been studied.[36] Carbinols derived from phenols begin to react at 80°C, while those bearing aryl or alkyl groups need temperatures of at least 110°C. Reactions involving the 4-halogen substituted and the unsubstituted phenylcarbinol require temperatures of 150 to 170°C; the yields are typically lower than for carbinols containing methyl, aryl, or hydroxyl groups. Improvements in yields and conversions have recently been reported employing an even larger excess (40 moles) of HF.[37] For condensations leading to HFIP-bridged bisphenols, sulfonic acids have also been used successfully as both reaction medium and catalyst.[27]

Apart from the desired condensation, three major side-reactions are observed. Their extents depend on reaction conditions and the particular reactants used. The first one is reversible formation of heptafluoroisopropyl compounds, such as 9; these do still give the desired condensation products, but more slowly. The second process is irreversible reduction of carbinols to 2H-hexafluoroisopropyl derivatives, such as 10; this is mainly observed when benzylic hydrogens, e.g., as in methyl groups, are present in the arene. Finally, extensive intra- and inter-molecular methyl-scrambling has been detected in certain cases.

19.4. DIRECTLY ACCESSIBLE HFIP-BRIDGED MONOMERS AND INTERMEDIATES

It is beyond the scope of this review to list all the known hexafluoroiso-propylidene-bridged aromatic compounds. Their number is already quite large and increasing steadily, and a dozen or so are available commercially.[3] HFIP-bridged monomers and intermediates that are obtained directly via the condensation route discussed in Section 19.3.4 are displayed in Table 1.

Among the relatively large number of HFIP-bridged aromatic monomers known, there are but two fairly large-volume products at this time, and just a handful with the

Table 1.   Directly Accessible HFIP-Bridged Monomers and Intermediates

3: R = H	R′ = H
R = H	R′ = CH$_3$
R = CH$_3$	R′ = CH$_3$

7: R = H	
11: R = CH$_3$	

R = H	
R = halogen	

R = OH	R′ = H
R = CH$_3$	R′ = H
R = R′ = CH$_3$	

R = OH	R′ = H
R = CH$_3$	R′ = H
R = R′ = CH$_3$	

potential of getting into that league within the next few years. The most important product by far is Bisphenol AF (**3**; Scheme 3), with a world production of around 150 tonnes per annum. The runners-up are 6F-dianhydride (**12**; 6FDA) and diesters of the corresponding tetracarboxylic acid. 6FDA (**12**) is synthesized from the 6F-tetramethyl precursor **11** by air oxidation and subsequent dehydration (Scheme 6). It offers a very versatile entry into fluorinated high-performance polymers.

Another noteworthy pair of compounds are the HFIP-bridged aromatic diamines **8a** and **8b** (Scheme 5). They are interesting not only for their contribution to polymer properties,[1,2] but also because they are less toxic than the currently used aromatic diamines, such as *meta*- and *para*-phenylenediamine (preliminary results indicate that both **8a** and **8b** give a negative response in mutagenicity tests).

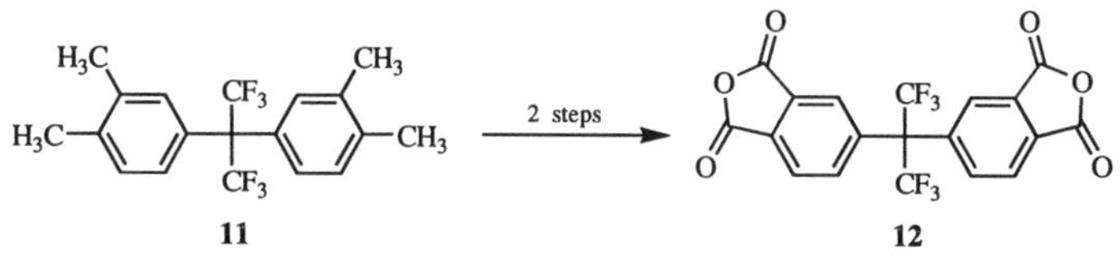

Scheme 6. Final steps in the production of 6FDA.

19.5. POLYMERS CONTAINING HEXAFLUOROISOPROPYLIDENE GROUPS

Unfortunately, the 6F-diamines (Scheme 5) as well as all the other monomers containing the hexafluoroisopropylidene function are rather expensive materials due to the technically demanding nature of their multistep synthesis. Therefore, it is not surprising that the practical use of such compounds can seldom be justified in areas other than the production of high-performance polymers. Since two comprehensive reviews on polymers derived from HFA have been published recently,[3,4] the following short discourse about selected condensation-type polymers is designed to highlight general trends where structure–property relationships of such materials are concerned.

19.5.1. Aromatic 6F-Polyesters

Like their nonfluorinated counterparts, which are commercially available engineering resins, linear HFIP-containing polyesters are usually synthesized via the classical polycondensation reaction of dicarboxylic acids or derivatives thereof with diols.

Bisphenol AF (**3**) has been condensed with isophthalic and terephthalic[38–41] acid derivatives to yield the corresponding polyesters and copolyesters (**13**) with enhanced thermal and thermo-oxidative stabilities.

These HFIP-bridged polyarylates were amorphous, and transparent tough films could be cast from solutions; they were much more soluble in halogenated hydrocarbon solvents than their isopropylidene analogues.

In the early 70s, the synthesis of a Bisphenol A/Bisphenol AF/iso- and tere-$C_6H_4(COCl)_2$ copolyarylate, useful as a coating material in aerospace applications, was patented.[42] Obviously, the inventor balanced the mechanical properties of the coating by adjusting the ratios of isophthalic acid dichloride and terephthalic acid dichloride, as well as the solubility of the copolymer by optimizing the ratio of

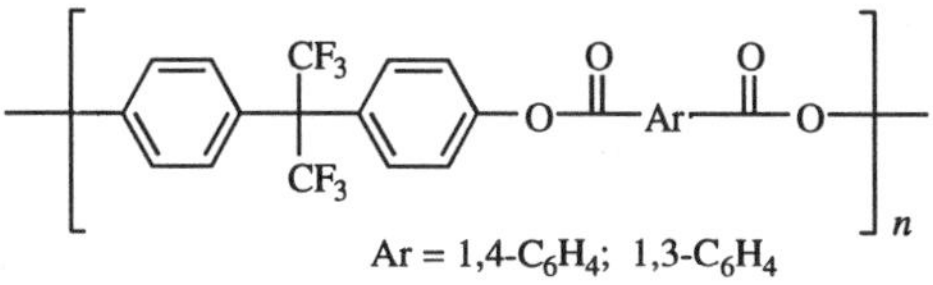

$Ar = 1,4\text{-}C_6H_4;\ 1,3\text{-}C_6H_4$

13

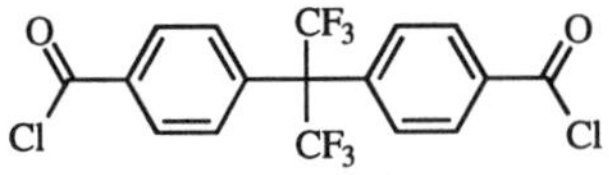

14

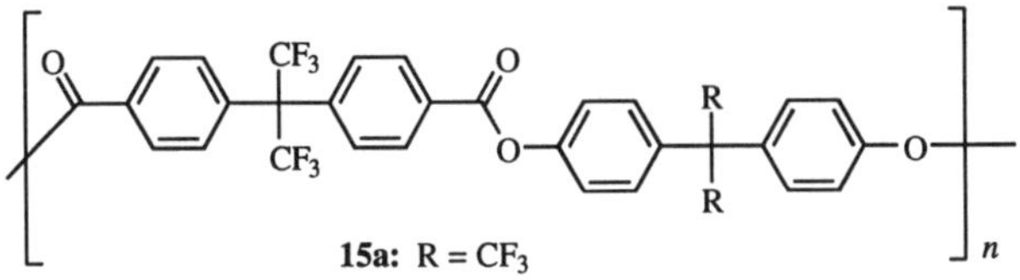

15a: R = CF$_3$
15b: R = CH$_3$

Bisphenol AF to its acetone analogue Bisphenol A. Product molecular weights were controlled in interfacial polymerization reactions by the addition of monofunctional termination agents (such as C$_6$H$_5$COCl), and polymers with excellent fuel/weather resistance were obtained.

Similar trends regarding the solubility, color, and transparency of HFIP-containing polyarylates versus 6H-polyesters were encountered when 2,2-bis(4-chloroformylphenyl)hexafluoropropane (4,4'-6F-diacid chloride) (**14**) was used in polyester synthesis together with comonomers such as Bisphenol AF (→**15a**)[43] and Bisphenol A (→**15b**).[43,44] The presence of the hexafluoroisopropylidene group in the main chain of polyarylates seems to lead to lower degrees of crystallinity than found in polyesters containing the isopropylidene group.[45,46]

19.5.2. Aromatic 6F-Polyamides

Aromatic polyamides serve most often in applications where high abrasion resistance, strength, and solvent resistance at elevated temperatures are required. The presence of amide bonds in the polymer backbone results in an increase of chemical stability compared to polyarylates. Interchain hydrogen bonding in these polymers leads generally to high transition temperatures, therefore high processing temperatures are often required.

In the late 60s, the first examples of hexafluoroisopropylidene-linked polyamides of type **16** were reported by Du Pont[47] and Russian scientists.[48,49] A variety of aliphatic and aromatic diamines, including *m*-phenylenediamine, *p*-phenylenediamine, and 2,2-bis(4-aminophenyl)propane, were polycondensed with the 4,4'-6F-diacid chloride **14**; and related polyamides (**17**) were synthesized from 2,2-bis(4-aminophenyl)hexafluoropropane (**8a**, Scheme 5) and diacid chlorides such as **14**, isophthaloyl chloride, terephthaloyl chloride, and mixtures thereof to evaluate the impact on polymer performance of siting the HFIP groups in the diamine residues.

Film-forming polyamides were obtained which exhibited comparably higher solubilities and lower glass-transition temperatures. This property combination greatly facilitated fiber-spinning processes and orientation drawing at lower temperatures.

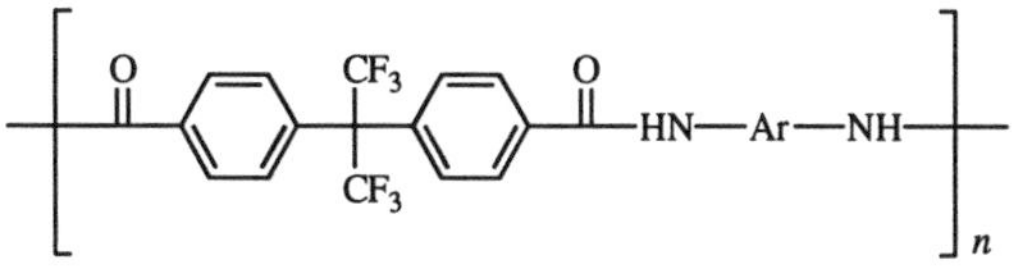

16

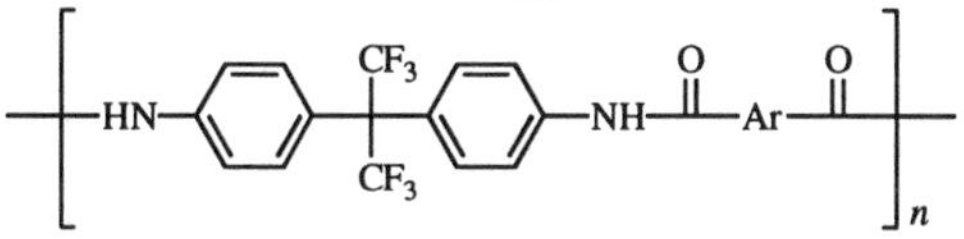

17

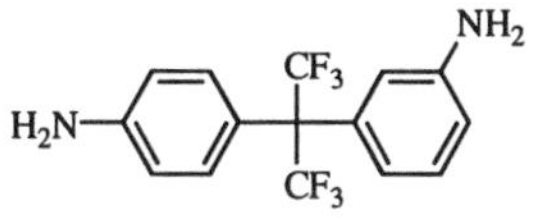

18

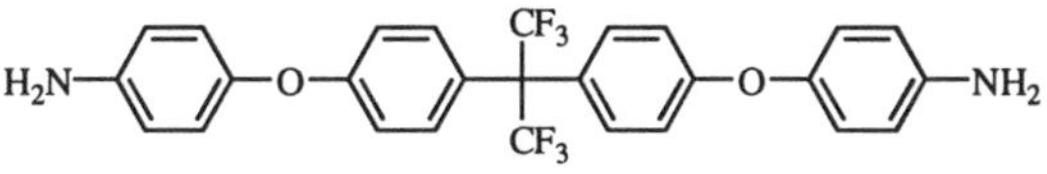

19

Additionally, it was found that these HFIP-derived polymers showed enhanced thermal stability and a substantially improved resistance to soiling when compared to nonfluorinated aromatic polyamides.

Lower abrasion losses and improved melt flow behavior can be achieved compared with isopropylidene-linked polymers, by condensing 2-(3-aminophenyl)-2-(4-aminophenyl)hexafluoropropane (**18**)[50] or 2,2-bis[4-(4'-aminophenoxy)phenyl]-hexafluoropropane (**19**)[51] with selected aromatic diacid chlorides.

19.5.3. Aromatic 6F-Polyimides

Polymers with cyclic secondary amide, i.e., imide, groups in the main chain are generally referred to as polyimides. It is especially in this polymer class that "6F-chemistry" has made its mark, and polyimide products containing HFIP links (spacers) have already been commercialized.

High-molecular-weight aromatic polyimides were first synthesized around 1960.[52] Since then, polymers of this type have been recognized as members of the most important class of high-performance polymers, exhibiting unparalleled versatility by combining excellent mechanical and electrical performance with processability and exceptionally high thermo-oxidative stability. Based on about 20 dianhydride monomers and a much larger number of diamines, an enormous number of materials, especially of the fully aromatic polyimide type, has been synthesized, evaluated, and

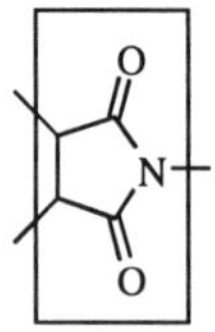

Imide unit

reviewed.[53,54] Today, the estimated worldwide consumption of polyimides is 4.5 million lb. With an average selling price of about 110 $ per lb, the superb performance and not the market volume of polyimides clearly determines their strategic importance.

The already attractive performance profile of aromatic polyimides was even further enhanced by the introduction of the hexafluoroisopropylidene function into the polymer main chain. In 1989, 6000 lb of 6FDA (**12**; Scheme 6) were used to produce advanced composites, hot-melt adhesives, high-performance coatings, and other high-value-added products. Further applications of 6F-polyimides as high-performance functional and structural materials are sure to be established in the future.

Historically, Du Pont was the first company to recognize the commercial value of polyimides and 6F-polyimides. In 1967, a patent was filed by Rogers,[55] describing the polycondensation reaction products of 2,2-bis(4-aminophenyl)hexafluoropropane **8a** (Scheme 5) with one or more dianhydrides, including 6FDA **12** (→**20**), in an inert solvent. The resulting soluble polymers could be used as coatings, films, fibers, filaments, foams, and melt-fusible powders. The glass-transition temperatures ranged from 310–340°C, and some of the 6F-polyimides appeared to be nearly colorless. In the case of 6F-polyimide **20**, degradation in air was reported to start at about 475°C.

The first linear aromatic HFIP-derived polyimide products were introduced as structural high-temperature adhesives in 1972 by Du Pont.[56–58] This product line, called NR-150, utilized the improved thermoplastic behavior and extraordinary high thermostability of compositions based upon 6FDA (**12**), phenylenediamines (NR-150 B: T_g = 360°C), and 4,4′-diaminodiphenyl ether (NR-150 A: T_g = 290°C). The processing of the monomer blend polyimide precursor products proved to be difficult, because volatile condensation byproducts and residual high-boiling solvents had to be removed very carefully or would have caused voids and cracks in the reinforced laminates. Improved high-temperature structural adhesives are available nowadays and account for the bulk of 6FDA monomer consumed in the USA. Still applying the difficult polycondensation processing of monomer blend impregnated prepregs, 6F-tetraacid **21** or its dimethyl esters **22** and phenylenediamines are used to manufacture tools and high-performance carbon-fiber-reinforced composites for space and aircraft

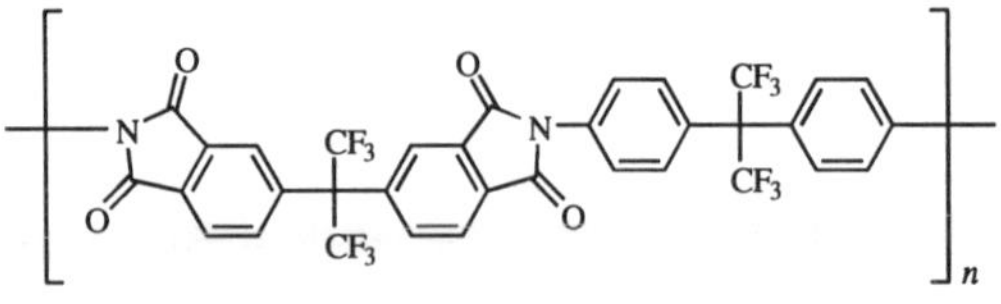

20

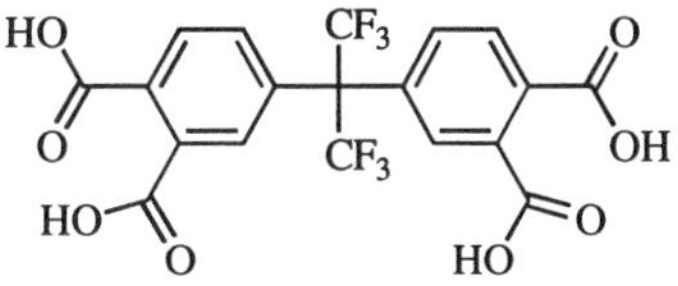

21 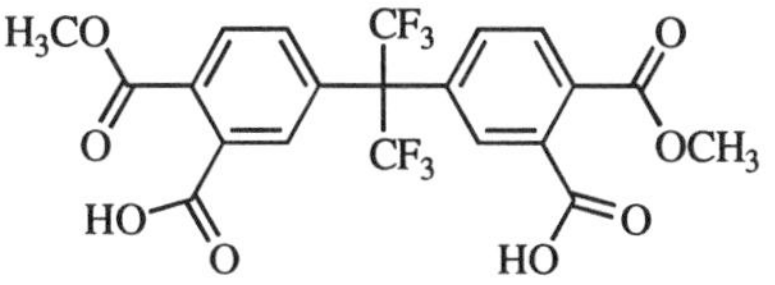   **22** + Isomers

applications under the trade name Avimid®.[59,60] These materials still exhibit excellent properties even after 100 h exposure in air at 371°C, and are utilized also as molded jet engine parts.

In 1972, NASA had developed[61,62] a thermosetting polyimide product, PMR-15, based upon the PMR-technology (*in situ* polymerization of *m*onomeric *r*eactants). New versions,[63] called PMR-II (**24**), incorporate the bis esters **23**, *p*-phenyl-enediamine, and 5-norbornene-2,3-dicarboxylic acid methyl ester (NADIC-ester, NE) (Scheme 7). This modification increased the thermo-oxidative stability impressively: graphite-fiber-reinforced PMR-II laminates showed a weight loss of only 6% after 500 h at 343°C in air.

The beneficial influence of the hexafluoroisopropylidene function in 6F-poly-imides on thermoprocessability and thermal stability can also be detected in high-performance polymer blends with polybenzimidazole (PBI).[64] The thermo-

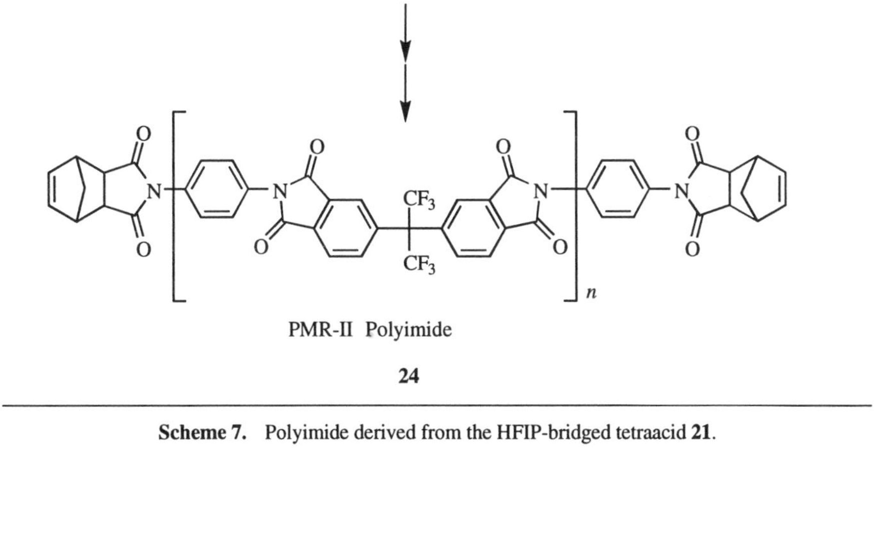

23 + Isomers

PMR-II Polyimide

24

Scheme 7. Polyimide derived from the HFIP-bridged tetraacid **21**.

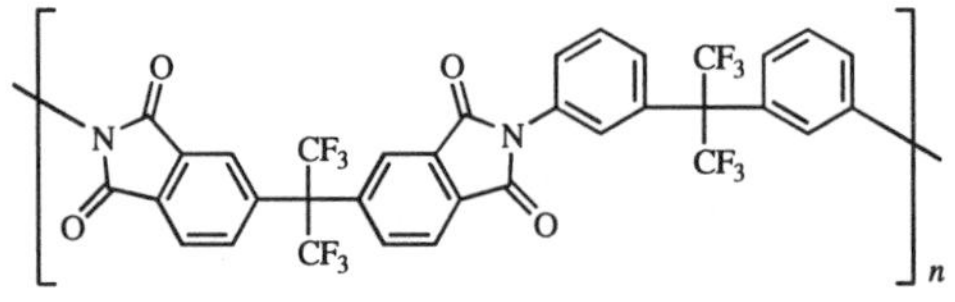

25

oxidative performance of such blends was much better than that of PBI and clearly a function of the 6F-polyimide content.

There has been a great deal of interest in the research and development of polyimides for electronic and microelectronic applications. The miniaturization of multilevel metallization IC devices requires the use of ultrapure, high-temperature-stable (*ca* 450°C/15 min), mechanically tough, processable and film-forming dielectrics as planarizing and insulating layers between conductor lines.[65] Since conventional polyimides had been used extensively in this application, fluorinated competitors promised to enable even higher switching speeds in such devices based upon their lower permittivity. The incorporation of the HFIP function into the polyimide main chain seems to be one of the most important prerequisites for the synthesis of polymers with low dielectric constants.[66] When the HFIP-group-based fluorine content of the polymer repeat unit is maximized, as represented by structure 25 (from 12 and 8b) and its isomer 20, the permittivity can be decreased to values as low as ε = 2.7 at RT and 10 MHz.[67] The electrical performance is quite unaffected by relative humidity changes of the environment. The fully imidized polymers 20 (T_g = 320°C) and 25 (T_g = 250°C) are obtainable in high molecular weights (> 200,000) through polycondensation reactions in polar aprotic solvents, and form soluble, transparent, and colorless films with superb long-term thermo-oxidative stability at temperatures as high as 340°C in air.

Polyimide 25 was also patented as a protective coating for the manufacture of solar cells,[68,69] and further studies conducted by NASA[70] clearly indicated once more the most beneficial effect of the 6F-group in 6F-polyimides upon color, thermal stability, and solubility when being used as a bulky, electronic interaction hindering moiety within the polymer backbone. Polymers 20 and 25 were evaluated recently, among others, as planar lightguide materials. As the number of HFIP functions per polymer repeat unit increased, the optical losses of the corresponding lightguides decreased to values below 0.1 dB cm^{-1}.[74,75]

The number of HFIP-derived polyimide products continues to increase steadily, and such polymers serve as high-temperature-stable photoresists, LCD alignment layers, protective coatings, fibers, films, and moldings. Separation processes via membrane technology represent one of the most recently developed applications of HFIP polyimides.[71,72] Membranes made from specially designed polymers appear to offer an excellent combination of high gas permeability and selectivity for the purification and production of industrial gases.[73]

Table 2. Introduction of the 6F-Spacer into the
Polymer Main Chain: Generalized Trends and Effects

Polymer property	Increase ■/Decrease ⊖
Thermo-oxidative stability:	■
Flame resistance:	■
Processibility:	■
Thermal flow	■
Solubility	■
Abrasion resistance:	■
Electrical performance:	■
Dielectric constant	⊖
Tracking resistance	■
Humidity sensitivity	⊖
Optical properties:	■
Refractive index	⊖
UV-cutoff wavelength	⊖
Color	⊖
Radiation resistance	■
Crystallinity:	⊖
Surface tension:	⊖
Adhesion	⊖

19.6. CONCLUSION

Based upon hexafluoroacetone as the starting material, a large number of intermediates and monomers containing the $(CF_3)_2C$ (HFIP or 6F) group has been developed in the past few years. The use of these monomers in the synthesis of polymers, especially polyimides, has created a new group of high-performance materials with extremely useful properties. A summary of general effects which might be encountered when 6F-functions (spacers) are built into a polymer main chain is given in Table 2.

The use of trade names or manufacturers does not constitute an official endorsement of such products or manufacturers, either expressed or implied, by Hoechst AG and Hoechst Celanese Corporation.

19.7. REFERENCES

1. D. A. Scola and J. H. Vontell, *Chemtech* 112 (1989).
2. K. S.Y . Lau, A. L. Landis, W. J. Kelleghan, and C. D. Beard, *J. Polym. Sci. 20*, 2381 (1982).
3. P. E. Cassidy, T. M. Aminabhavi, and J. M. Farley, *J. Macromol. Sci., Rev. Macromol. Chem. Phys. C29*, 365 (1989).
4. V. V. Korshak, I. L. Knunyants, A. L. Rusanov, and B. R. Livshits, *Russ. Chem. Rev. 56*, 288 (1987).
5. C. G. Krespan and W. J. Middleton, *Fluorine Chem. Rev. 1*, 145 (1967).
6. F. W. Swamer, French Patent 1,372,549 (to Du Pont) [*CA 62*, 6397 (1965)].

428   Wolfgang K. Appel *et al.*

7. T. Tozuka and O. Yonosuke, European Patent 17,171 (to Daikin Kogyo Co.) [*CA 94*, 120870 (1980)].
8. A. Kurosaki and S. Okazaki, *Nippon Kagaku Kaishi 11*, 1800 (1988); A. Kurosaki and S. Okazaki, *Chem. Lett.* 17 (1988).
9. H. Millauer, German Patent 2,460,468 (to Hoechst) [*CA 85*, 177239 (1976)]; H. Millauer, German Patent 2,658,328 (to Hoechst) [*CA 90*, 22787 (1979)]; H. Millauer and W. Lindner, German Patent 2,658,382 (to Hoechst) [*CA 89*, 75641 (1978)]; H. Millauer, *Chem. Ing. Tech. 52*, 53 (1980).
10. H. Millauer, W. Schwertfeger, and G. Siegemund; *Angew. Chem. 97*, 164 (1985); *Angew. Chem., Int. Ed. Engl. 24*, 161 (1985).
11. P.-P. Rammelt and G. Siegemund, European Patent 54,227 (to Hoechst) [*CA 97*, 181732 (1982)].
12. Du Pont, French Patent 1,416,013 [*CA 64*, 6502 (1965)].
13. D. E. Morin, U.S. Patent 3,213,134 (to 3M Co.) [*CA 63*, 18029 (1965)].
14. M. R. Brittelli, R. Culik, O. L. Dashiell, and W. I. Fayerweather, *Tox. Appl. Pharmacol. 47*, 35 (1979).
15. G. N. Tkacuk. N. A. Minkina, and E. G. Berliner, *Nek. Vopr. Eksp. Prom. Toksikol.* p. 87 (1977); [*CA 93*, 38837].
16. E. P. Moore and A. S. Milan Jr., British Patent 1019788 (to Du Pont) [*CA 64*, 15745 (1966)].
17. B. S. Farah, E. E. Gilbert, and J. P. Sibilia, *J. Org. Chem. 30*, 998 (1965).
18. B. S. Farah, E. E. Gilbert, M. Litt, J. A. Otto, and J. P. Sibilia, *J. Org. Chem. 30*, 1003 (1965).
19. E. E. Gilbert, E. S. Jones, and J. P. Sibilia, *J. Org. Chem. 30*, 1001 (1965).
20. D. C. England, French Patent 1,325,204 (to Du Pont) [*CA 59*, 11339 (1963)].
21. K. Schneider and G. Siegemund, European Patent 254,219 (to Hoechst) [*CA 109*, 8396 (1988)].
22. D. von der Brück, R. Bühler, C. Heuck, H. Plieninger, K. E. Weale, J. Westphal, and D. Wild, *Chemiker-Ztg.-Chem. Apparatur 94*, 183 (1970).
23. E. S. Jones, U.S. Patent 3,304,334 (to Allied Chemical Corp.) [*CA 67*, 21630 (1967)].
24. V. I. Dyachenko, A. F. Kolomiets, and A. V. Fokin, *Izv. Akad. Nauk SSSR. Ser. Khim.* 2849 (1987); V. I. Dyachenko, M. V. Galakhov, A. F. Kolomiets, and A. V. Fokin, *Izv. Akad. Nauk SSSR Ser. Khim.* 923 (1989),
25. M. Masamichi and N. Takayuki, German Patent 3,920,518; Japanese Patent 88/155,717 (to Central Glass Co.) [*CA 113*, 6984 (1990)].
26. R. L. Soulen and J. R. Griffith, *J. Fluorine Chem. 47*, 195 (1989).
27. V. Mark and C. V. Hedges, World Patent (to General Electric Co.) 82/02380 [*CA 98*, 71666 (1983)].
28. D. W. Reynolds, P. E. Cassidy, C. G. Johnson, and M. L. Cameron, *J. Org. Chem. 55*, 4448 (1990).
29. C. Tamborski, W. H. Burton, and L. W. Breed, *J. Org. Chem. 31*, 4229 (1966).
30. B. S. Farah, E. E. Gilbert, E. S. Jones, and J. A. Otto, *J. Org. Chem. 30*, 1006 (1965).
31. W. A. Sheppard, *J. Am. Chem. Soc. 87*, 2410 (1965).
32. W. A. Sheppard, *J. Org. Chem. 33*, 3297 (1968).
33. J. Lau, G. Siegemund, and F. Röhrscheid, German Patent 3,833,338 (to Hoechst) A1 [*CA 113*, 116054 (1990)].
34. K. L. Paciorek, T. I. Ito, J. H. Nakahara, and R. H. Kratzer, NASA CR159403, SN-8320-F (August 1978).
35. I. L. Knunyants, T-Y. Chen, N. P. Gambaryan, and E. M. Rokhlin, *Zh. Vses. Khim. Obshchestva im. D.I. Mendeleeva 5*, 114 (1960) [*CA 54*, 20962 (1960)].
36. A. D. Allen, V. M. Kanagasabapathy, and T. T. Tidwell, *J. Am. Chem. Soc. 105*, 5961 (1983); A. D. Allen, V. M. Kanagasabapathy, and T. T. Tidwell, *J. Am. Chem. Soc. 108*, 3470 (1986).
37. T. Maruta and T. Wada, Japanese Patent 01 50,833 (to Central Glass Co.) [*CA 111*, 77622 (1989)].
38. V. V. Korshak, S. V. Vinogradova, and V. A. Pankratov, *Dokl. Akad. Nauk SSSR 156*, 880 (1984); [*CA 61*, 8419a (1964)].
39. V. V. Korshak, S. V. Vinogradova, and V. A. Pankratov, *Izv. Akad. Nauk SSSR, Ser. Khim.* 1649 (1965); [*CA 64*, 8321h (1966)].
40. M. A. Kakimoto and Y. Imai, *J. Polym. Sci., Part A, Polym. Chem. Ed. 24*, 3555 (1986).
41. V. V. Korshak, S. V. Vinogradova, and V. A. Pankratov, British Patent 1,122,201 (to Institute of Elemental-Organic Compounds, Academy of Sciences, USSR) [*CA 69*, 77985g (1968)].
42. W. W. Howerton, U.S. Patent 3,824,2111 [*CA 82*, 18780g (1975)].
43. C. G. Johnson and P. E. Cassidy, *J. Macromol. Sci., Rev. Macromol. Chem. Phys. C29*, 365 (1989).

44. V. V. Korshak, *Heat Resistant Polymers*, Chapter 3, Israel Program for Scientific Translations Ltd., Jerusalem (1971).
45. V. V. Korshak, S. V. Vinogradova, M. M. Dzhanashvili, V. A. Vasnev, and R. G. Keshelava, *Deposited Doctorate 1980*, VINITI 4140-80 [*CA 96*, 69684a (1982)].
46. V. V. Korshak, V. A. Vasnev, S. V. Vinogradova, T. M. Babchinister, M. M. Dzhanashvili, Y. V. Genin, and R. G. Keshelava, *Vysokomol. Soedin, Ser. A 23*, 2573 (1981) [*CA 96*, 105005r (1982)].
47. S. L. Kwolek, U.S. Patent 3,328,352 (to Du Pont) [*CA 68*, 60466v (1968)].
48. R. M. Gitina, E. L. Zaitseva, and A. Ya. Yakubovich, *Russ. Chem. Rev. 40*, 679 (1971).
49. I. L. Knunyants, S. V. Vinogradova, and B. R. Livshits, USSR Patent 226,845 [*CA 70*, 48054t (1969)].
50. R. H. Vora, U.S. Patent 4,914,180 (to Hoechst Celanese Corp.) [*CA 114*, 7455 (1991)].
51. Japanese Patent 58,149,944 (to Hitachi Chemical Co.) [*CA 100*, 104486n (1984)].
52. C. E. Strong, A. L. Endrey, S. V. Abramo, C. E. Berr, W. M. Edwards, and K. L. Olivier, *J. Polym. Sci. A1 3*, 1373 (1965).
53. M. I. Bessonov, M. M. Koton, V. V. Kudryavtsev, and L. A. Laius, in: *Polyimides; Thermally Stable Polymers* (W. W. Wright, ed.), Consultants Bureau, Plenum Publ. Corp., New York (1987).
54. F. W. Harris, T. Takekoshi, T. L. St. Clair, H. D. Stenzenberger, P. R. Young, R. Escott, P. Hergenrother, D. Wilson, H. Satou, H. Suzuki, D. Makino, and S. C. Sroog, in: *Polyimides* (D. Wilson, H. D. Stenzenberger, and P. M. Hergenrother, eds.), Chapman and Hall, Inc., New York (1990).
55. F. E. Rogers, U.S. Patent 3,356,648 (to Du Pont) [*CA 68*, 30419q (1968)].
56. H. H. Gibbs, *17th Int. SAMPE Symposium* III-B6, 1 (1972).
57. H. H. Gibbs, German Patent 2,223,807 (to Du Pont) [*CA 78*, 125449v (1973)].
58. C. E. Rogers, U.S. Patent 3,959,350 (to Du Pont) [*CA 85*, 78937k (1976)].
59. Product Brochures, E.I. Du Pont de Nemours & Co., Composites Center, Wilmington, Delaware (1989).
60. H. H. Gibbs and D. E. Myrik, *33rd Int. SAMPE Symposium* 1473 (1988).
61. T. Serafini, P. Delvigs, and G. Lightsey, *J. Appl. Polym. Sci. 16*, 905 (1972).
62. P. J. Cavano and W. E. Winters, *NASA TM CR 135377* (1978).
63. R. D. Vanucci and D. Cifani, *NASA TM 100923* (1988).
64. M. Jaffe, "Tailoring Polyimide Performance Through Blending," *Interdisciplinary Symposium on Recent Advances in Polyimides and Other High Performance Polymers*, Am. Chem. Soc., Pol. Chem., San Diego (January 22–25, 1990).
65. N. Kinjo, "Polyimides For Electronic Applications," *Interdisciplinary Symposium on Recent Advances in Polyimides and Other High Performance Polymers*, Am. Chem. Soc., Pol. Chem., San Diego (January 22–25, 1990).
66. A. St. Clair, Technical Support Package for Tech Brief *LAR-13769*, "Low-Dielectric Polyimides," NASA Langley RC, Virginia (1987).
67. W. H. Mueller and R. H. Vora, Product Data Sheet American Hoechst Corp., Coventry, Rhode Island (1987).
68. P. S. Dupont and N. Bilow, PCT Int. Appl. World Patent 8402,529 (1984) (to Hughes Aircraft Co.) [*CA 102*, 80466u (1985)].
69. A. L. Landis and A. B. Naselow, U.S. Patent 4,645,824 (to Hughes Aircraft Co.) [*CA 107*, 7845 (1987)].
70. A. K. St. Clair, T. L. St. Clair, W. S. Slemp, and K. S. Ezzell, *NASA TM 87650* (1985).
71. C. E. Sroog, in Ref. 54, p. 278.
72. H. H. Hoehn and J. W. Richter, German Patent 2,336,870 (to Du Pont) [*CA 81*, 78937 (1974)].
73. R. A. Hayes, U.S. Patent 4,717,394 (to Du Pont) [*CA 109*, 130426 (1988)].
74. H. Franke, G. Knabke, and R. Reuter, "Molecular and Polymeric Optoelectronic Materials: Fundamentals and Applications," *Proceedings of SPIE, The International Society for Optical Engineering*, San Diego (August 21–22, 1986).
75. R. Reuter, H. Franke, and C. Feger, *Appl. Opt. 27*, 4565 (1988).

20

Perfluoropolyethers (PFPEs) from Perfluoroolefin Photooxidation

Fomblin® and Galden® Fluids

DARIO SIANESI, GUISEPPE MARCHIONNI, and RALPH J. DE PASQUALE

20.1. INTRODUCTION

The development of perfluoropolyethers (PFPEs) represents a major contribution of organofluorine chemistry to advanced technology. These water-white mellifluous fluids, benign toxicologically and environmentally, exhibit an impressive range of physicochemical properties that are already legendary in certain application areas. The aim of this chapter is to substantiate the above statements by dissection and analysis of PFPEs from the viewpoints of their intriguing synthetic chemistry, properties, derivatives, and applications.

PFPEs have for some time now been commercially available worldwide in the form of neutral end-capped fluids based on work pioneered by Du Pont (Krytox®) and Montedison (Fomblin®) in the 1960s. More recently, commercial-scale plants have been inaugurated by Nippon Mektron (Aflunox®), Daikin (Demnum®), Hoechst (Hostinert®), and Montedison (Galden®); the last two have extended the available range to low-boiling fractions. Here the discussion focuses on Fomblin–Galden fluids, and relates information gleaned from first-hand experience. Such a format gains merit from a synthetic/chemical development basis as well. PFPE synthesis can be based on

DARIO SIANESI, GUISEPPE MARCHIONNI, and RALPH J. DE PASQUALE • Ausimont/Montefluos, Montedison/Montefluos Group, 20021 Bollate, Milano, Italy. *Present address for R. J. D. P.*: 5500 Atlantic View, St. Augustine, Florida 32084.

Organofluorine Chemistry: Principles and Commercial Applications, edited by R. E. Banks *et al.* Plenum Press, New York, 1994.

the isolation of either a perfluorinated or highly fluorinated cyclic ether, with subsequent ionic ring-opening polymerization to give a fluoropolyalkylene ether intermediate containing mono-structural repeating units, as in the production of Krytox® fluids from hexafluoropropene oxide or Demnum® from 2,2,3,3-tetrafluorooxetane (see Chapter 21). By contrast, Montedison chemistry involves the light-assisted oxidation of perfluoroolefins to yield a polymer directly without isolation of an intermediate. Radical rather than ionic species intervene throughout the reaction schemes. It will become apparent that this chemistry is quite versatile by affording a variety of repeating units in the polymer backbone that can be regulated not only by changing the composition of the perfluoroolefin feedstock, but also the reaction conditions. The interplay between these two factors serves as the basis for the ongoing development of neutral and organofunctional PFPE products.

20.2. THE PHOTOOXIDATION ROUTE TO NEUTRAL PFPEs

The empirical evolution of the reaction between perfluoroolefins and oxygen can be traced to offsetting factors: the ease of reaction of fluorinated olefins relative to their hydrogen counterparts, and the often hazardous nature of the reaction products.[1] Early work in the authors' laboratories with "dark" reactions based on tetrafluoroethylene (TFE) and oxygen in the liquid phase produced extremely shock-sensitive liquid products with high peroxide contents. [The enhanced stability of the O—O bond in perfluoroalkyl peroxides (BDE *ca.* 50 kcal mol^{-1}) renders highly energetic perfluorinated polyperoxides isolable.[2]] Motivated by survival instincts, attention was transferred first to gas-phase then to liquid-phase oxidation with photolytic assistance. The formation of PFPEs via liquid-phase photooxidation of perfluoroolefins proved to have both industrial applicability and a fascinating chemistry.

20.2.1. Oxidative Photopolymerization of Fluoroolefins

This type of reaction was probably first encountered serendipitously during an attempted photooligomerization of hexafluoropropene (HFP, bp –29°C) when monomer contaminated with oxygen gave rate enhancement accompanied by a series of unexpected products.[3] In fact, the introduction of oxygen into UV-irradiated (≤300 nm) liquid HFP at –40°C yields an oligomeric mixture of random peroxidic polyethers

$$CF_3CF{=}CF_2 \ + \ O_2$$

(1) $\Big\downarrow$ UV light $<$–40 °C

$$AO(CF_2\overset{\underset{\displaystyle CF_3}{|}}{C}FO)_h(CF_2\overset{\underset{\displaystyle CF_3}{|}}{C}FOO)_i(\overset{\underset{\displaystyle CF_3}{|}}{C}FCF_2O)_j(\overset{\underset{\displaystyle CF_3}{|}}{C}FCF_2OO)_k(CF_2O)_l CF_2OO)_m(CFO)_n B$$

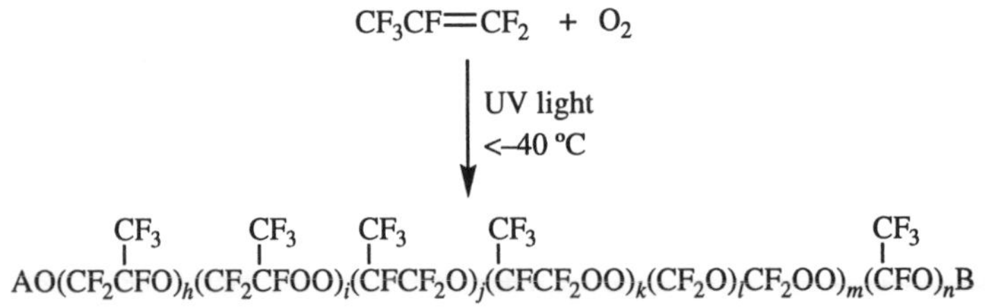

1   A, B = CF$_3$, COF (major), CF$_2$COF (minor)

("polyperoxides"; **1**, Eq. 1).[4] Following a short induction period, rapid oxygen uptake occurs. Initially, the degree of polymerization (DP) and oxygen content of the polymer are both high, then level off as monomer conversion increases and the viscosity of the medium rises. Under favorable conditions the yield of **1** exceeds 95% (based on HFP), the predominant byproducts being COF_2 and CF_3COF. Hexafluoropropene oxide (HFPO) becomes an increasingly important reaction product if the temperature rises from $-40°C$. Provided oxygen is in slight excess, reaction rate and product composition are independent of oxygen concentration, while temperature, HFP concentration, and light intensity are sensitive variables. Quantum yields based on HFP or oxygen vary according to conditions, but rarely fall below 20, indicative of free-radical chain processes.

Spectroscopic analysis (notably ^{19}F NMR), supported by thermal degradation data, indicates that structurally the polyperoxide intermediate (**1**) consists of randomly distributed branched C_3, branched C_2 (minor) and C_1 fluoroether repeating units (h, n, and l, respectively) interspersed with C_3 (i, k) and C_1 (m) peroxy units [only traces of the C_2 unit $CF(CF_3)OO$ are present] joined in tail-to-head (e.g., $h-h, h-i$) with a minor amount of head-to-head ($j-h, k-i$) sequences. Tail-to-tail sequences ($h-j, h-n$) are barely detectable. Two types of end group (A, B) are present: neutral (or inert), namely CF_3, and functional (or reactive), namely COF (major) and CF_2COF (minor), the neutral/functional ratio being nearly one. The ratio of neutral–functional to neutral–neutral to functional–functional end-group pairings is approximately 2:1:1, as determined by tedious isolation and characterization of low-to-intermediate molecular-weight-range hydrolyzed products having low peroxide contents.

A comprehensive study on the effects of systematically changing the reaction variables has established the following empirical relationships. The average molecular weight of **1** falls in the range 700–10,000 and depends on reaction temperature, higher temperatures favoring the formation of low-molecular-weight oligomers; the molecular-weight distribution is monomodal and shows a polydispersity of *ca.* 1.5. The peroxide content varies directly with HFP concentration and inversely with light intensity, and normally is $1 \pm 0.25\%$ by weight (3% max for safety); typically, neat HFP (as reactant and solvent) and a light intensity of 3×10^{-10} einstein cm^{-3} sec^{-1} are employed. A linear (negative slope) relationship is found between the yield of hexafluoropropene oxide versus difluoromethylene oxide units and temperature at constant peroxide content; at low temperature (below $-60°C$), h/l approaches 1000, while at $20°C$ (liquid phase) h/l is in the vicinity of 5. The microstructure of **1** is affected by temperature, HFP concentration, and light intensity; regioselectivity (tail-to-head) varies inversely with temperature. Reactor residence time is adjusted to reflect a manageable balance between monomer conversion and the viscosity of the system, which changes as a function of the desired product.

Replacement of HFP by tetrafluoroethylene (TFE; bp $-76°C$) in the oxidative photopolymerization process yields a viscous peroxidic perfluoropolyether having structure **2** (Eq. 2)[5] in which distribution of the difluoromethylene oxide and tetrafluoroethylene oxide units is random. The reaction is run at a low temperature ($\leq -60°C$) in an inert solvent (CFC-12, a low-boiling perfluoroalkane, or a PFPE)

$$CF_2\!\!=\!\!CF_2 \;+\; O_2$$

(2) $\Big\downarrow$ UV light $<-60\,°C$
inert solvent

$$AO(CF_2CF_2O)_p(CF_2CF_2OO)_q(CF_2OO)_r(CF_2O)_sB$$

2 A, B as in Eq. 1

and the gaseous byproduct comprises COF_2 containing small amounts of tetrafluoroethylene oxide, perfluorocyclopropane, and CF_3COF. Yields of **2** are extremely dependent on temperature as well as TFE concentration and, under the preferred conditions, exceed 70%. The ratio p/s falls in the range 0.1–20, and careful inspection of the ^{19}F NMR spectrum has also revealed the presence of $O(CF_2)_nO$ units ($n = 3,4$), which increase in importance with radiation intensity and TFE concentration. These units are thought to arise from a combination of degenerative cage reactions associated with peroxide fission (see later) and addition of radical sites of type $—O(CF_2)_n^•$ to TFE.

As discussed subsequently, peroxy units of type CF_2OO are more stable than CF_2CF_2OO and can reach substantial concentrations given time (in excess of the 3% by weight recommended limit); irradiation intensity is a controlling factor and should be closely monitored. The most easily discernible and distinctive feature between HFP/O_2 and TFE/O_2 photopolymers is the higher product average molecular weight reached with TFE (values in excess of 10^4 are common).

Photooxidative polymerization of HFP/TFE mixtures provides oils with structures (**3**) comprising randomly distributed C_1, linear C_2, and branched C_3 ether and peroxide units, according to Eq. (3).[6] It is convenient to run the reaction by bubbling TFE and oxygen separately but simultaneously into liquid HFP under UV irradiation at temperatures in the range -80 to $-40°C$. In general, the principles established for the HFP and TFE photooxidations hold when considering the effects of reaction variables on structure **3**. By adjusting mainly the olefin concentrations and temperature, one controls the relative proportions of $t,v,$ and z units in the polymer. Typically, a t/v range of 0.5–30 applies, with $(t+v)/z$ lying between 20 and 100; however, reaching both (CF_2CF_2O) and (CF_2O) contents of 65 and 30 mol%, respectively, are within the practical limits of the reaction. Notably z is the more easily adjusted to zero, even with v approaching a value corresponding to 50 mol%, owing to HFP concentration effects.

$$CF_3CF\!\!=\!\!CF_2 \;+\; CF_2\!\!=\!\!CF_2 \;+\; O_2$$

(3) UV light $\Big\downarrow$ $<-40\,°C$

$$AO(CF_2\overset{\underset{\displaystyle CF_3}{|}}{C}FO)_t(CF_2\overset{\underset{\displaystyle CF_3}{|}}{C}FOO)_u(CF_2CF_2O)_v(CF_2CF_2OO)_x(CF_2OO)_y(CF_2O)_zB$$

3 A, B as in Eq. 1

Oxidative photopolymerization methodology has been extended to a variety of other fluorinated olefins including perfluorocyclobutene,[7] higher homologues of HFP (α-olefins),[8] perfluorobuta-1,3-diene,[9] and various substituted olefins (usually in conjunction with TFE and HFP), like perfluoro(alkyl vinyl ethers),[10] and chloro- and bromo-trifluoroethylene.[11]

Currently, however, only oxidative photopolymerization of HFP, TFE, and HFP–TFE mixtures are of industrial significance; hence only these systems are discussed in detail here.

20.2.2. Removal of Peroxide Linkages

Both the fragile peroxide links and the reactive fluoroformate/acid fluoride end-groups in polymers **1–3** must be removed to produce stable PFPEs. Chemical peroxide removal is feasible but has drawbacks, notably excessive lowering of average molecular weights and the inconvenience of using solvents and reagents; such procedures are, however, practiced when alternatives are limited, as in the preparation of difunctional PFPEs. Thermal and photochemical peroxide-elimination reactions reduce average molecular weights to a lesser extent than chemical methods, and therefore are preferred; the mechanisms involved are exemplified in Scheme 1, using the major HFP-based peroxide unit (**4**) occurring in photopolymer **1** (*i–h* units). Heat- or light-driven homolysis of TFE-based peroxide links (see **5** and **6**) found in **2** and **3**, followed by analogous β-scissions of resultant alkoxy radicals and subsequent radical cage-combination reactions, produces the PFPE structure units shown in Scheme 2. Weight loss during thermal or photochemical removal is clearly a function of peroxide content, and is less than 15% under favorable conditions. Byproducts consist predominantly of COF_2 (bp $-83°C$; from **2**) and COF_2 plus CF_3COF (bp $-59°C$; from **1** and **3**). Perfluorinated cyclic ethers account for a small percent of the volatile products, indicating the occurrence of radical cyclization processes. Multiple carbonyl fluoride loss per peroxide group (in some cases COF_2:O–O ratios exceed 2) occurs during decomposition of **5** and **6** (Scheme 2), as determined by mass balance measurements and further supported by an increase in the frequency of $O(CF_2)_nO$ ($n = 3,4$) units in the final peroxide-free product.

Peroxide removal from type **2** polymers gives products with molecular weights only slightly lower (5%) than expected from the experimentally determined evolution of COF_2. By contrast, polyperoxides containing HFP-based type **4** units (i.e., **1** and **3**) decompose to PFPEs with molecular weights much reduced but predictable from the number of randomly distributed type **4** units per polymer molecule. These observations indicate that CF_3 transfer leading to permanent chain scission of type **4** units (pathway 2 in Scheme 1) are competitive with those nearly preserving molecular weight, i.e., β-scission of an alkoxy radical with loss of COF_2 (pathway 3) or CF_3COF (pathway 1) followed by cage recombination (Scheme 1). Formation of the tail-to-tail ether units $OCF_2CF(CF_3)OCF(CF_3)CF_2O$ during peroxide elimination from polymers **1** and **3** is extremely small, as found in the initial production of these peroxidic PFPE prepolymers from HFP or HFP–TFE mixtures and oxygen (Section 20.2.1). During thermal

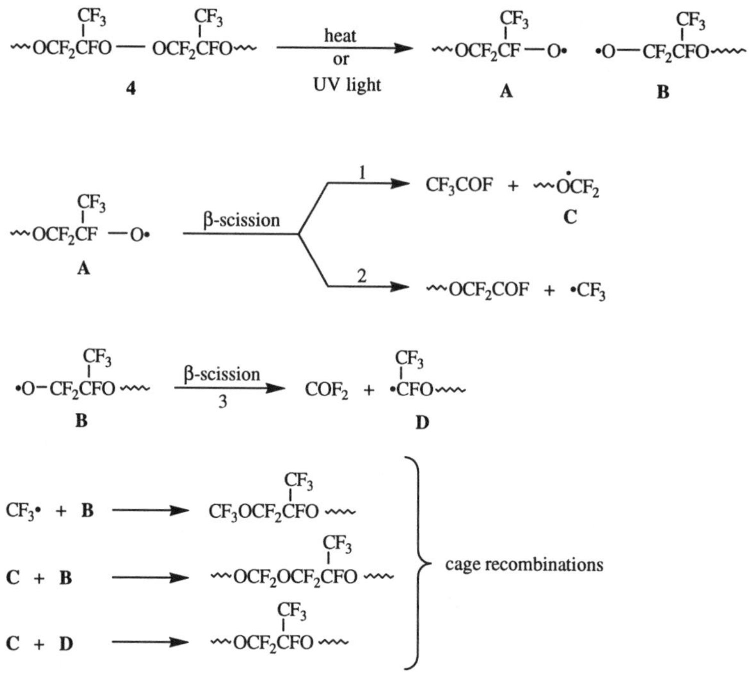

$$\overset{CF_3}{\underset{|}{\sim\!\!\sim OCF_2CFO}} \!\!-\!\! \overset{CF_3}{\underset{|}{OCF_2CFO}}\!\!\sim\!\!\sim \quad \xrightarrow[\text{UV light}]{\text{heat}\atop\text{or}} \quad \overset{CF_3}{\underset{|}{\sim\!\!\sim OCF_2CF}}\!\!-\!\!O\bullet \quad \bullet O\!\!-\!\!\overset{CF_3}{\underset{|}{CF_2CFO}}\!\!\sim\!\!\sim$$

$$\mathbf{4} \qquad\qquad\qquad\qquad\qquad\qquad\qquad \mathbf{A} \qquad\qquad\qquad \mathbf{B}$$

$$\overset{CF_3}{\underset{|}{\sim\!\!\sim OCF_2CF}}\!\!-\!\!O\bullet \quad \xrightarrow{\beta\text{-scission}} \quad \begin{cases} \xrightarrow{\ 1\ } CF_3COF \;+\; \sim\!\!\sim O\overset{\bullet}{C}F_2 \\[2pt] \qquad\qquad\qquad\qquad\quad \mathbf{C} \\[6pt] \xrightarrow{\ 2\ } \sim\!\!\sim OCF_2COF \;+\; \bullet CF_3 \end{cases}$$

$$\mathbf{A}$$

$$\overset{CF_3}{\underset{|}{\bullet O\!\!-\!\!CF_2CFO}}\!\!\sim\!\!\sim \quad \xrightarrow[3]{\beta\text{-scission}} \quad COF_2 \;+\; \overset{CF_3}{\underset{|}{\bullet CFO}}\!\!\sim\!\!\sim$$

$$\mathbf{B} \qquad\qquad\qquad\qquad\qquad\qquad\qquad\qquad \mathbf{D}$$

$$CF_3\bullet \;+\; \mathbf{B} \quad\longrightarrow\quad \overset{CF_3}{\underset{|}{CF_3OCF_2CFO}}\!\!\sim\!\!\sim$$

$$\mathbf{C} \;+\; \mathbf{B} \quad\longrightarrow\quad \overset{CF_3}{\underset{|}{\sim\!\!\sim OCF_2OCF_2CFO}}\!\!\sim\!\!\sim \qquad \left.\vphantom{\begin{array}{c}1\\1\\1\end{array}}\right\} \text{cage recombinations}$$

$$\mathbf{C} \;+\; \mathbf{D} \quad\longrightarrow\quad \overset{CF_3}{\underset{|}{\sim\!\!\sim OCF_2CFO}}\!\!\sim\!\!\sim$$

Scheme 1. Pathways for removal of peroxide linkages from HFP/O_2 peroxidic polyethers of type 1.

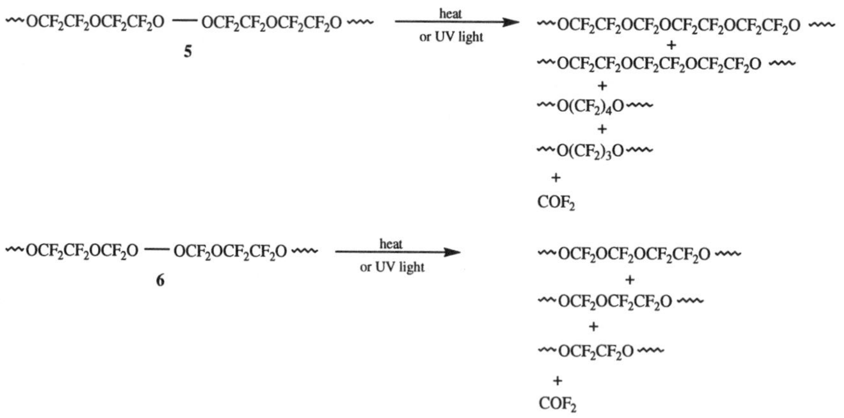

$$\sim\!\!\sim OCF_2CF_2OCF_2CF_2O\!\!-\!\!OCF_2CF_2OCF_2CF_2O\!\!\sim\!\!\sim \quad \xrightarrow[\text{or UV light}]{\text{heat}} \quad \begin{array}{l} \sim\!\!\sim OCF_2CF_2OCF_2OCF_2CF_2OCF_2CF_2O\!\!\sim\!\!\sim \\ + \\ \sim\!\!\sim OCF_2CF_2OCF_2CF_2OCF_2CF_2O\!\!\sim\!\!\sim \\ + \\ \sim\!\!\sim O(CF_2)_4O\!\!\sim \\ + \\ \sim\!\!\sim O(CF_2)_3O\!\!\sim\!\!\sim \\ + \\ COF_2 \end{array}$$

$$\mathbf{5}$$

$$\sim\!\!\sim OCF_2CF_2OCF_2O\!\!-\!\!OCF_2OCF_2CF_2O\!\!\sim\!\!\sim \quad \xrightarrow[\text{or UV light}]{\text{heat}} \quad \begin{array}{l} \sim\!\!\sim OCF_2OCF_2OCF_2CF_2O\!\!\sim\!\!\sim \\ + \\ \sim\!\!\sim OCF_2OCF_2CF_2O\!\!\sim\!\!\sim \\ + \\ \sim\!\!\sim OCF_2CF_2O\!\!\sim\!\!\sim \\ + \\ COF_2 \end{array}$$

$$\mathbf{6}$$

Scheme 2. Pathways for removal of peroxide linkages from TFE/O_2 peroxidic polyethers of type 2. The type of peroxide bond in **6** is more stable by *ca.* 7 kcal mol^{-1} than that in **5**; hence it has proved possible to effect controlled degradation of polymers of type **2** to new peroxides (*ca.* 1.5% by wt.) containing >90% of units of the type OCF_2OOCF_2O (peroxide bond energy *ca.* 47 kcal mol^{-1}).

deperoxidation, randomization of the chain length occurs to some extent, so molecular-weight distributions are broadened by *ca.* 10%. In typical conversions, a TFE-based polyperoxide (**2**) containing 1.5 wt% peroxide (in units of type **5** and **6**, Scheme 2) has a peroxide half-life of 30 minutes at 200°C; similarly, a type **1** polyperoxide with 0.8 wt% peroxide content has a half-life of approximately 30 minutes at 140°C.

The trends encountered in photolytic peroxide decomposition follow those found with the thermolytic procedure. Conversions can be performed at much lower temperatures (–40 to 50°C), where the reaction parameters depend almost exclusively on irradiation wavelength and intensity. Light-induced decomposition enjoys the benefit of better reaction control, so that higher yields are attained owing to a diminution of multiple COF_2 expulsions per peroxide unit, especially in structures **2** and **3**. The low reaction temperature range, complimented by solvent dilution, allows for the collection and interpretation of data correlating medium viscosity to yield and product structure.[12] Notably, the introduction of co-reagents during peroxide decomposition (e.g., aromatics and quinones,[13] olefins, halogens[14]) that scavenge perfluoroalkoxy radicals or carbon-centered radicals derived from their β-scission, produces results rich in chemistry and surprises which continue to inspire experimentation at Montefluos.

20.2.3. End-Group Modification to Give Neutral PFPEs

Oxidative polymerization of HFP and TFE produces oligomers (**1–3**) in which acid fluoride and fluoroformyl moieties comprise nearly 50% of the end groups. Thermal deperoxidation effects a partial redistribution: in reactions generally commencing at 150°C, fluoroformate groups are converted in part to acid fluorides and ketones at rates dependent on structure, as exemplified in Eqs. (4) and (5). Photochemi-

$$(4) \quad \text{\textasciitilde\textasciitilde OCF}_2\text{CF}_2\text{OCF}_2\text{OCOF} \xrightarrow{\text{heat}} \text{\textasciitilde\textasciitilde OCF}_2\text{COF} + 2COF_2$$

$$(5) \quad \overset{\overset{\textstyle CF_3}{\textstyle |}}{\text{\textasciitilde\textasciitilde OCF}_2\text{CFOCOF}} \xrightarrow{\text{heat}} \text{\textasciitilde\textasciitilde OCF}_2\text{COCF}_3 + COF_2$$

cal peroxide decomposition has little effect on functional terminals under normal processing conditions; however, prolonged irradiation leads to chain extension of acid fluorides through radical coupling with loss of CO and COF_2.[15] Early manufacturing procedures for "pacifying" the functional acid fluoride, ketone, and remaining fluoroformate end-groups relied on treatment with base at elevated temperature; this produces hydrogen-containing end-groups (Eqs. 6 and 7) which have little influence on the overall properties of relatively high-molecular-weight PFPEs, but whose presence

$$(6) \quad \text{\textasciitilde\textasciitilde OCF}_2\text{COF} + \text{HO}^- \longrightarrow \text{\textasciitilde\textasciitilde OCHF}_2 + CO_2 + \text{F}^-$$

$$(7) \quad \text{\textasciitilde\textasciitilde OCF}_2\text{COCF}_3 + \text{HO}^- \longrightarrow \text{\textasciitilde\textasciitilde OCHF}_2 + CO_2 + [\text{CF}_3^-]$$

(8) $\overset{CF_3}{\underset{|}{\sim\!\!\sim\!OCFCOF}}$ + F$_2$ $\longrightarrow$ $\left[\overset{CF_3}{\underset{|}{\sim\!\!\sim\!OCFCF_2OF}} \right]$ $\longrightarrow$ $\overset{CF_3}{\underset{|}{\sim\!\!\sim\!OCF_2}}$ + COF$_2$

(9) $\sim\!\!\sim\!OCF_2CF_2OCOF$ + F$_2$ $\longrightarrow$ $\left[\sim\!\!\sim\!OCF_2CF_2OCF_2OF \right]$ $\longrightarrow$ $\sim\!\!\sim\!OCF_3$ + 2COF$_2$

(10) $\sim\!\!\sim\!OCF_2COCF_3$ + F$_2$ $\longrightarrow$ $\left[\sim\!\!\sim\!OCF_2CF(OF)CF_3 \right]$ $\longrightarrow$ $\sim\!\!\sim\!OCF_3$ + CF$_3$COF

is undesirable. This led to the development of methods for replacing the hydrogen by fluorine, and in the 1970s the manufacturing process was switched to a fluorination procedure which converts the carbonyl functionals directly to completely fluorinated end-caps (Eqs. 8–10).[16]

Such "fluorine-finishing" reactions are performed under a slight positive pressure of the halogen either thermally at temperatures often exceeding 200°C or photochemically below 100°C. It is quite a challenge to find organic fluids other than the PFPE final products which are stable under these conditions. The reactions proceed through hypofluorite intermediates (see Eqs. 8–10) that are detectable and even isolable in some instances with photochemical initiation. These PFPE hypofluorites have exciting chemistries in their own right.[17]

To summarize, three sequential steps—photooxidaton, peroxide removal and fluorination—are needed to convert the perfluoroolefins HFP and TFE to neutral PFPEs. Reaction variables allow the PFPE structure to be modified to contain CF$_2$O, CF$_2$CF$_2$O, and CF(CF$_3$)CF$_2$O ether repeating units distributed randomly in the polymer chains. An ambitious objective still evolving is to tailor PFPE structures to coincide with optimum performance properties required of fluids for specific applications. On the industrial scale,[18] photooxidation is performed continuously in reactors where residence time and recycle streams (including byproduct removal) are adjusted to conform to a desired composition of the polyperoxide. Design factors (like mass transport, light intensity, and mean distance of the reaction zone from the irradiation source) are critical to the avoidance of broad concentration gradients of peroxidic material. Once formed, the polyperoxides are metered to a continuous thermal or photochemical peroxide degradation unit, again equipped to separate and scrub volatile byproducts. The resulting functional polyethers are fluorinated in a batch reactor under slight pressure, yielding end-capped PFPEs; these are separated into various commercial fractions (see later) by standard or molecular distillation.

20.3. MECHANISM OF HFP AND TFE OXIDATIVE PHOTOPOLYMERIZATION

The oxidative photopolymerizatons of hexafluoropropene and tetrafluoroethylene are complicated processes. However, information collected over the years concerning, for example, product structure,[18] and reaction parameter structure

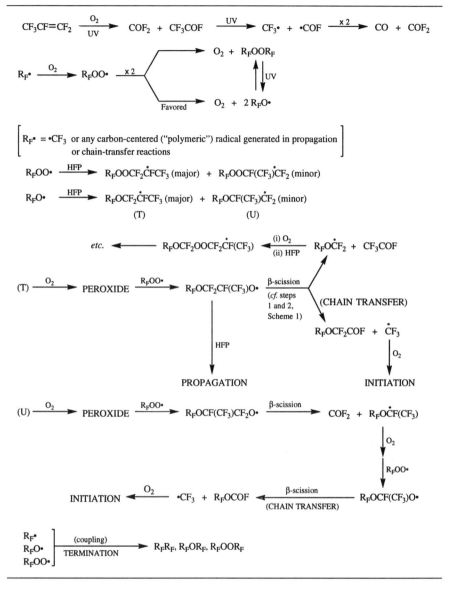

Scheme 3. Hexafluoropropene–oxygen photocopolymerization mechanism: preliminary initiation sequence and generalized important propagation, chain-transfer, and termination steps.

relationships, has provided the basis for an internally consistent interpretation of the major steps involved. In short, radical copolymerization of the olefins and oxygen occurs, with simultaneous decomposition of growing radicals, as exemplified in Scheme 3. At the outset, trifluoromethyl radical is the initiating species,[19] being formed by photolysis of CF_3COF generated by the oxidation of HFP or TFE[20]; support for this proposal is provided by the abundance of CF_3 end-groups in products 1–3, and the disappearance of induction periods upon adding CF_3COF to the monomer feed. During propagation, CF_3 radicals, then subsequently R_F radicals, are oxidized to peroxy radicals at diffusion-controlled rates that are orders of magnitude faster than $R_F^{\bullet}$ addition to monomers (HFP, TFE).[21] At this critical stage of the sequence, peroxy radicals react with monomer affording peroxides whose subsequent fission constitutes chain initiation. Alternatively, peroxy radicals competitively disproportionate to either dialkyl peroxides and oxygen (an important chain-termination step) or two alkoxy radicals and oxygen in bimolecular, diffusion-controlled reactions,[22] presumably via tetroxides; the latter mode ($R_FO^{\bullet}$ formation) is favored by a factor of 40 (maximum kinetic chain length) based on retrofit of kinetic simulation data.[23,24] The fluorinated alkoxy radicals ($R_FO^{\bullet}$) are important chain-carrying intermediates that react with monomer to afford new carbon-centered radicals which enter fresh oxidation cycles.[25]

In the photooxidation of HFP (see Scheme 3) the β-scissions[26,27] are chain-transfer reactions only when CF_3 radical expulsion occurs, leaving an acid fluoride end-group. There is scant support for the occurrence of β-scission of C—F bonds (i.e., F atom liberation) during photooxidation [not unexpectedly, since the C—F bond is approximately 20 kcal mol^{-1} stronger than the C—C bond (see Chapter 3)[28]]. This explains why product molecular weights are significantly higher in TFE versus HFP photooxidations, molecular-weight-limiting chain-transfer reactions with the former monomer being either structurally impossible (no β-CF_3) or improbable (failure to expel F$^{\bullet}$). The additional termination steps $R_FO^{\bullet} + {}^{\bullet}COF \rightarrow R_FOCOF$ and $R_FOCF_2^{\bullet} + {}^{\bullet}COF \rightarrow R_FOCF_2COF$ are invoked in the case of TFE to account for the acid fluoride end-groups.[29] Measurements taken during oxidative copolymerization of the TFE with HFP indicate that overall at –60°C fluoroalkoxy radicals favor TFE over HFP addition by a factor of roughly 300.[24]

20.4. CHEMICAL AND PHYSICAL PROPERTIES OF FOMBLIN® AND GALDEN® FLUIDS

Structures 7–9 represent neutral PFPEs distributed commercially.[30] Fomblin® K, an experimental series, derived from a type 3 polyperoxide, is related structurally to Fomblin Y® (8), the difference being the presence of randomly dispersed CF_2CF_2O units. The ranges of n, $m+n$, and $p+q$ are representative but may vary slightly at the extremes, depending on the application. Relative to hydrocarbon materials, these and other PFPEs enjoy enhanced thermodynamic stabilities and kinetic invulnerability stemming from very high C—F, C—C, and C—O bond strengths and the steric and electronic shielding provided by the fluorine atoms (see Chapter 3).[25]

20.4.1. Chemical Properties and Solubilities

PFPEs are chemically inert to the majority of reagents that eventually, if not instantaneously, degrade hydrocarbon-based fluids and polymers. These include mineral acids, alkalis, organic acids and bases, oxidizing and reducing agents and halogens,

$$CF_3$$
$$\overset{|}{R_FO(CF_2CFO)_nR_F} \qquad R_F = CF_3, C_2F_5, C_3F_7$$

7 $n = 1\text{--}10$

$$CF_3$$
$$\overset{|}{CF_3O(CF_2CFO)_m(CF_2O)_nR_F} \qquad R_F = CF_3, C_2F_5, C_3F_7$$

8

$$m + n = 8\text{--}45$$

$$m/n = 20\text{--}1000$$

$$CF_3O(CF_2CF_2O)_p(CF_2O)_qCF_3 \qquad p + q = 40\text{--}180$$

9

$$p/q = 0.5\text{--}2$$

as illustrated in Table 1. Even though the oxygen atom in fluorinated ethers is essentially nonbasic and nonnucleophilic,[31] latent characteristics can be stimulated by several aggressive reagents. Thus PFPEs containing an α-fluorine substituent undergo halogen exchange and ether cleavage on contact with strong Lewis acids at elevated temperature.[32] Reaction rates depend on substituent effects (steric and electronic) and reaction conditions (homogeneity and stirring rates), but generally proceed at measurable rates according to Lewis acid strength in the series SbF_5 (at *ca.* 100°C), $AlCl_3$ (*ca.* 120°C), $SbCl_3$ (*ca.* 200°C), $FeCl_3$ (*ca.* 202°C).[33] Mechanisms involving carbocation intermediates stabilized by resonance of the type $\ddot{O}\!-\!C^+ \leftrightarrow {}^+O\!\!=\!\!C$ and $F\!-\!C^+ \leftrightarrow {}^+F\!\!=\!\!C$ have been suggested, such as

$$-CF_2CF(CF_3)OCF_2CF(CF_3)- + SbF_5 \rightarrow -CF_2CF(CF_3)O\overset{+}{C}FCF(CF_3)- \quad SbF_6^-$$

$$\rightarrow \quad -CF_2CF_2CF_3 + SbF_5 + FOCCF(CF_3)-$$

PFPEs are degraded by strong electron-transfer reducing agents as well, e.g., Na, K, Mg, Al at temperatures above 100°C, especially in equipment where shear produces freshly exposed metal surfaces.[34] Reactions proceed exothermically and can be capricious.

Due to their mutual incompatibility with hydrocarbons, PFPEs are neither solutes nor solvents where typical organics are concerned. They are soluble in highly halogenated solvents, e.g., other PFPEs, perfluoroalkanes, chlorofluorocarbons, HCFCs, and HFCs. Accordingly, the physical properties of conventional ("hydrocarbon") thermoplastics, thermosets, and elastomers undergo little if any modification on immersion in PFPE fluids.

Table 1. Compatibility of PFPE Fluids (Fomblin® Y 25 and Z 25) with Chemical Agents

Agents	Temp. (°C)	Conc. (wt%)	Time (h)	Result[a]
Solvents				
Hydrocarbons, aromatics, acids, alcohols, amides, esters, ketones, dipolar aprotics	25	50	—	immiscible
Mineral acids				
Conc. H_2SO_4, HNO_3, H_3PO_4, HCl	70	50	24	unchanged
Alkalis				
KOH, NaOH, Na_2CO_3	200	20	24	unchanged
Hydrogen halides				
HF, HCl, HBr	200	flow[b]	24	unchanged
Halogens				
F_2, Cl_2, Br_2,	200	flow[b]	24	unchanged
Cl_2 (liquid)	15	30	24	unchanged
Oxidizing agents				
$KMnO_4$, $K_2Cr_2O_7$	200	50	24	unchanged
$KMnO_4$ aqueous (6% wt)	70	5	24	unchanged
Nonmetallic halides				
BCl_3, UF_6, BF_3	80	5	24	unchanged
PBr_3, PCl_5, $SiCl_4$	150	5	24	unchanged
Sulfur anhydrides				
SO_2, SO_3	200	flow[b]	24	unchanged
Organic acids				
(various)	200	50	24	unchanged
Organic bases				
Tributylamine, quinoline	200	50	24	unchanged
Ammonia				
300 mm Hg at 20°C	200	gas	168	unchanged

[a]No significant change in acidity of the recovered Fomblin® fluid; viscosity change ±2%.
[b]Flow rate of the gaseous agent: 1 liter per hour.

20.4.2. Thermal Properties—Influence of Metals

The thermooxidative stability of PFPEs is outstanding and perhaps their hallmark. Thermogravimetric analysis (TGA) of Fomblin Y and Z fluids shows onset of decomposition at temperatures exceeding 350°C.[35,36] The independent isoteniscope method gives similar values; both methods indicate little dependence on molecular weight. TGA in oxygen reveals a modest shift, approximately 10°C to lower temperatures, consonant with the notion that radical abstractions induced by O_2 in hydrocarbons have little relevance in PFPE chemistry. The presence of metals and oxygen negatively influence decomposition temperatures as a function of metal and fluid structure.[37] In the most reactive combinations, Fomblin® Y-titanium/aluminum alloy and Fomblin® Z-aluminum, the TGA-determined threshold temperatures in oxygen reduce to 325 and 270°C, respectively. The greater susceptibility of Z fluids was verified by the microoxidation test,[38] whereby aggressive metal/oxygen combinations reduce operating temperatures to 240 (Z), 260 (M), and 315°C (Y). As discussed later, additives show a prophylactic influence on this behavior. During thermal or metal/oxygen induced decomposition, volatile gases containing perfluoroolefins plus toxic COF_2 and CF_3COF are evolved, hence in open systems undecomposed PFPE fluid retains its physical properties.

20.4.3. Oxidative Stability, Radiation Resistance, and Bioinertness

Extreme resistance to oxidants, for example, O_2 and F_2, is another notable characteristic of PFPEs. At elevated pressure, an indicative ignition temperature for Y fluid of mol. wt. 3000 is 395°C at 3370 psi O_2 The value varies slightly with molecular weight and oxygen concentration.[39] PFPEs are not flammable and show no self-ignition temperature in air at atmospheric pressure.

PFPEs show good radiation resistance. At absorbed γ-ray levels of 50 Mrad from a ^{60}Co source, weight loss and viscosity changes are minimal at 0.1 and 0.5%, respectively. Decomposition increases at higher dose levels and is accompanied by the formation of lower-molecular-weight fragments containing acid fluoride end-groups.[40]

Extensive biological evaluations of neutral PFPEs indicate high mammalian tolerance, e.g., at a level of 25 g kg^{-1} no pathological effects were detected in acute oral rat resting.[41] Data have been collected from oral acute and subacute, interperitoneal acute, skin and eye irritation, mutagenic and teratogenic testing. Based on such findings, neutral Galden® and Fomblin® fluids appear to be innocuous.

20.4.4. Physical Properties

Like perfluorocarbons[42] (see Chapter 3), PFPEs show extremely low intermolecular forces which, coupled with low polarizability, give rise to low surface energies, low dielectric constants, and high coefficients of expansion (see Table 2). In sharp contrast to perfluorocarbons, the oxygen atoms within the PFPE chain allow for a greater degree of intramolecular rotational and bending freedom, resulting in improved chain flexibility; this confers better low-temperature and migratory properties, as well as slightly lower specific gravities.

Galden® PFPE (7) fractions are available over a near-continuous range of boiling points and viscosities (see Table 3). The availability of these dual-character (perfluoro-carbon/ether) multicomponent fractions, in which the high polyether content (C—O—C) enhances low-temperature properties by comparison with perfluorocarbons, have aided instrument design in electronic testing applications where a low level of fluid consumption is an important economic criterion.[43] The chemical structure (7) of Galden® fractions ensures against the release of harmful quantities of highly toxic perfluoroisobutene during thermal decomposition (cf. Chapter 4).[44]

Table 2.   General Properties of Neutral PFPFs (Fomblin® and Galden® Fluids)

Property	Values	Comments
Molecular weight	450–15,000	Broad range according to grade
Kinematic viscosity (cSt) (20°C)	<1–2000	Broad range according to grade
Viscosity index	70–370	Good viscoelasticity and lubricating properties
Density (g cm^{-3})	1.7–1.9	Good heat-transfer properties
Pour point (°C)	< –100 to –20	Low
Surface tension (mN m^{-1})	16–23	Low
Refractive index	*ca.* 1.3	Low
Specific heat (cal g^{-1}°C^{-1})	0.23	Good heat-transfer properties
Heat of vaporization (cal g^{-1})	<17	
Dielectric constant (20°C, 1 kHz)	2.1	Good dielectric properties
Vol. resistivity at 25°C (ohm cm)	10^{15}	Good dielectric properties
Coefficient of expansion (cm^3°C^{-1})	0.001	Good heat-transfer properties
Thermal/oxidative stability		High thermooxidative stability
(°C glass)	>300	
(°C metal)	>240	
Self ignition temperature or flash point	none	Nonflammable
Solubility of air cm^3 (gas) per 100 cm^3 (25°C)	25	High gas solubility
Solubility of water (ppm)	<15	Hydrophobic

Fomblin® is customized according to market segment into working fluids for vacuum-generating devices or as lubricants. Vacuum fluids have evolved into three classifications[45]: general vacuum fluids, designated GV, which are available in different viscosity grades [80, 140, 200 and 250 cSt at 20°C (GV 80, etc.)] to serve needs in the chemical, food, and related industries; and Y-L VAC (06/6, 14/6, 16/6, and 25/6) and Y-H VAC (18/8, 25/9, 40/11, and 140/13) fluids designed for the ever-increasing clean-vacuum demands of the electronics industry. To interpret the notations in the Y-VAC series, multiply by 10 the first number before the slash to get the viscosity at 20°C; the number following the slash refers to the negative exponent to base 10 of the guaranteed vapor pressure (mm Hg) at 20°C; e.g., 06/6 is a 60 cSt fluid with equal to or less than 10^{-6} mm Hg vapor pressure, both at 20°C. Vapor pressure and evaporation-weight-loss decrease in the series order GV > L-VAC > H-VAC, reflecting narrower molecular-weight distributions achieved through fractionation. Ultimate total pressures are determined using rotary pumps (pump limit approximately 10^{-3} mm Hg) for GV and L-VAC grades, then with diffusion pumps for H-VAC fluids.

Fomblin® lubrication fluids (**8, 9**) cover a broad spectrum of physical properties and molecular composition. The oxygen:carbon ratio in these PFPEs increases in the order Y < M < Z; accordingly, one- and two-carbon ether units in the backbone increase chain flexibility, allowing shifts to higher-molecular-weight fractions at constant viscosity. As illustrated in Table 4, these structural modifications accommodate higher viscosity indices and lower evaporation weight losses—both crucial properties for lubricants.

Table 3. Physical Properties of Galden® Fractions[a]

Properties (units)	HT 70	HT 90	DET	HT 110	HT 135	D02 TS	D 02	HT 200	HT 230	HT 250	HT 270
Boiling point (°C)	70	90	91	110	135	165	175	200	230	250	270
Pour point (°C)	<-110	-110	-110	<-110	<-110	-97	-97	-85	-77	-72	-66
Density at 25 °C (g cm^{-3})	1.68	1.69	1.70	1.72	1.79	1.77	1.77	1.79	1.82	1.84	1.85
Kinematic viscosity at 25 °C (cSt)	0.5	0.75	0.6[b]	0.83	1.0	1.7[c]	1.8	2.4	4.4	9	14
Vapor pressure at 25 °C (mm Hg)	165	100	84	60	50	<1	<1	<1	<1	<1	<1
Heat of vaporization at boiling point (cal g^{-1})	17	17	17	17	16	17	17	15	15	15	15
Surface tension at 25 °C (mN m^{-1})	14	16	16	16	17	17	17	19	20	20	20
Refractive index at 25 °C	1.280	1.281	1.280	1.280	1.280	1.280	1.280	1.281	1.283	1.283	1.283
Dielectric constant at 25 °C (1 kHz)	2.1	2.1	2.1	2.1	2.1	2.1	2.1	2.1	2.1	2.1	2.1
Water solubility (ppm) (weight)	14	14	14	14	14	14	14	14	14	14	14
Air solubility cm^3 (gas) per 100 cm^3 (liquid)	26	26	26	26	26	26	26	26	26	26	26
Average molecular weight	410	460	400	580	610	750	760	870	1020	1320	1550

[a] Members of this family of fluids show essentially the same values for each of the following properties: specific heat at 25 °C (0.23 cal g^{-1} °C^{-1}), thermal conductivity at 25 °C (7 × 10^{-4} W cm^{-1} °C^{-1}), coefficient of expansion per °C (1.1 × 10^{-3}), dielectric strength at 25 °C (40 kV, 2.54 mm gap), dissipation factor at 25 °C [2 × 10^{-4} (1 kHz)] and volume resistivity at 25 °C (1 × 10^{-15} ohm cm).
[b] 6.0 at -54 °C.
[c] 45 at -54 °C.

Table 4.   Fomblin® Lubricants—Typical Properties

Fomblin® grade	Kinematic viscosity (cst)			Viscosity index	Pour point (°C)	Evaporation wt loss % (22 h)		Specific gravity 20°C (g cm⁻³)	Surface tension 20°C	Mol. wt.
	20°C	40°C	100°C			149°C	204°C			
Y 06	60	22	3.9	70	−50	20	—	1.88	21	1800
Y 25	250	81	10.4	108	−35	2	15	1.90	22	3200
Y 45	470	147	16.5	117	−30	0.7	1.7	1.91	22	4100
Y R	1200	345	33	135	−25	0.5	1.2	1.91	24	6250
Y PL 1500	1500	420	40	135	−25	0.3	0.9	1.91	24	660
Y R 1800	1850	510	46.5	135	−20	*b*	0.5	1.92	24	7250
M 03*ᵃ*	30	17	5	253	−85	15	—	1.81	23	4000
M 15*ᵃ*	150	85	22	286	−75	0.8	3	1.83	24	8000
M 30*ᵃ*	280	159	310	338	−65	*b*	0.7	1.85	25	9800
M 60*ᵃ*	550	310	86	343	−60	*b*	0.4	1.86	25	12500
Z 03	30	18	5.6	317	−90	6	—	1.82	23	4000
Z 15	160	92	28	334	−80	0.2	1.2	1.84	24	8000
Z 25	260	159	49	358	−75	*b*	0.4	1.85	25	9500
Z 60	600	355	98	360	−63	*b*	0.2	1.85	25	13000

*ᵃ*Preliminary, new grade.
*ᵇ*Below the limit of detection.

In brief, PFPE fluids not only possess lubrication/load-carrying capabilities comparable to conventional mineral oils or synthetic fluids, but also display exceptional thermooxidative stability and chemical resistance, with service temperatures rising to 250°C and beyond in hostile environments; no wonder they are said to provide "lubrication for life". The new M series (**9**) captures many of the features of Z fluids, shows improved thermal stability in the presence of metals, and addresses the cost/performance issue.

20.4.5.   Greases

PFPE greases[46] are prepared by blending Fomblin® fractions with up to 30% by weight of fluoroplastic thickening agents, predominantly PTFE. Some formulations incorporate antirust additives, e.g., sodium nitrite, benzotriazoles, and succinimides. The substitution of PTFE by a mineral thickener (talc) increases the oxygen impact stability threefold, from *ca.* 70 to 220 bar at 200°C. These greases display a formidable range of performance. At one extreme an M-based grease shows a low-temperature torque (ASTM D 1478) value of 360 g cm⁻¹ at 59°C, while a Y grease shows hundreds of lifetime hours in a high-speed bearing test (DIN 51821) at 230°C. The latter performance is unachievable with a mineral oil or synthetics-based grease.

20.4.6.   Tribology

Fomblin® PFPEs have been subjected to a multiplicity of lubricity tests.[35–37,47–50] Depending on the lubrication region and conditions addressed in a specific test, the relative performances of Z and Y vary, but generally are similar, excellent overall, and far superior to conventional lubricants at high temperature (>200°C). Good-to-high viscosity indices support favorable hydrodynamic lubrication,[49,50] while their propensity to form insulating boundary films reduces wear/coefficient of friction throughout elastohydrodynamic lubrication.[50-52]

Extreme pressure evaluation indicates that PFPE are among the best lubricants for preventing metal seizure and welding.[49,52] Under the high temperatures generated by metal-to-metal contact (>400°C), decomposition gradually occurs producing a fluorided metal wear surface that retards the welding process. Investigation of wear track surfaces by us and others shows extensive metal fluoriding, as deep as 50 Å.[53] The metal fluorides, while deterring the welding process, are not nearly as effective as conventional hydrocarbon EP agents under boundary conditions,[54] as pointed out in several studies.[55] Information gleaned from past experience with mineral oils and synthetics strongly suggests that additives can enhance the lubrication characteristics as well as the corrosion protection in this class of fluid. With regard to the latter, even though PFPEs are hydrophobic and satisfactory film-formers on metals, improved barrier properties (toward oxygen and water vapor) are warranted via stronger association between the medium and metal surfaces—physiabsorption or chemiadsorption. For some time now work has focused on soluble additives that retain the thermooxi-

Table 5. Wear, Lubricity, Corrosion Testing of Fomblin®

Test[e]	Method	Y 25	M 15	M 30	Z 25	Min. oil + antiwear/EP
4 Ball wear (scar, mm)	ASTM D 4172B	0.75, 0.45[a]	0.9, 0.6[a]	1, 0.8[a]	0.9, 0.6[a]	0.35
4 Ball EP welding load (kg)[a]	IP 239	400	580	631	562	282
FZG (stage)[c]	DIN 51354	12[d]	12	12	12	7–10
Low temp. torque	ASTM D 1478					
start (g cm^{-1})		2440	358	585	488	not tested
1 h (g cm^{-1})		1500	130	293	130	not tested
Corrosion	D 665	5,5; 0,0[a]	5,5; 0,1[a]	5,5; 0,0[a]	5,5; 0,0[a]	0,0
	D 665 B	fail; pass[a]	fail; pass[a]	fail; pass[a]	fail; pass[a]	pass
	d 665 A	5,5; 0,1[a]	5,5; 0,0[a]	5,5; 0,0[a]	5,5; 0,0[a]	0,0
Microoxidation[f] corrosion (temp.°C)	Internal	315	not tested	260, 315[g]	240, 290[g]	<200

[a]With *ca.* 4% soluble, experimental additive.
[b]Initial seizure and mean hertz for PFPE range from 250–500 and 104–126 kg, respectively.
[c]Specific wear loss was less than 0.2 mg, kWh^{-1} for all Fomblin tested.
[d]Test was carried out using Y 45 not Y 25.
[e]Greases were tested that were prepared from based fluid shown with *ca.* 27 %wt PTFE to LGI 1. Test performed at –59°C. With Y 25 as base fluid, the grease contained *ca.* 22 % wt PTFE, LGIO, –40°C.
[f]Metal specimen consisted of 30455 containing 6% Ti–Al, 4% Valloy using 1 liter h^{-1} air flow. At the temperature tabulates, one obtains <2% fluid loss after 72 h; see also Ref. 9 for a test description.
[g]With 1 %wt soluble experimental additive.

dative properties of PFPEs[56]; recent results with a series of antirust, antiwear, and thermal stabilizers suggest a ray of light at the end of a labyrinthed tunnel. Tribological data comparing fractions containing additives are listed in Table 5.

20.4.7. Rheology

For information concerning the fundamental fluid/rheological properties of PFPEs, see Refs. 57–63. From plots of glass-transition temperature (T_g) versus molecular weight, T_g at infinite molecular weight is approximately 70°C lower for Fomblin® Z than for Y fluids (**8** and **9**), indicating that a higher carbon/oxygen ratio and branching pendant trifluoromethyl groups cause stiffening of the polymer chain [$T_g = -65$ and -131°C, respectively for Fomblin Y R (5 mol% OCF_2 units) and Z 25 (60 mol% OCF_2), which have $M_n = 2 \times 10^4$]. By contrast, poly(propylene oxide) and poly(ethylene oxide) possess similar T_g values at infinite molecular weight (−62 versus

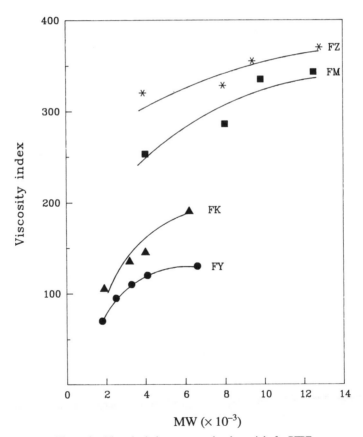

Figure 1. Viscosity index versus molecular weight for PFPEs.

−55°C), but neither of these hydrocarbon analogs of the Fomblins® contains oxymethylene units [so important in Z fluids (as —OCF_2— groups), see **9**] which would have a pronounced T_g lowering effect.

The concept of free volume has been used successfully to interpret flow characteristics of PFPE fluids.[58,59] The values for excess end-group free volumes of Fomblin Y and Z have been computed to be among the smallest known, and are an indication that free volume is little affected by chain ends. This is a tribute to the flowability of these fluids over the ranges investigated.

Density versus temperature plots are nearly linear from −33 to 67°C for Y and Z fluids; expansion coefficients increase with temperature and are roughly 10^{-3} per°C, actual values being *ca.* 15% higher for Z than for type Y fluids (Fomblin® Y R, 0.856 × 10^{-3}°C^{-1}; Fomblin® Z 25, 1.006 × 10^{-3}°C^{-1}).

Viscosity/viscosity index has been studied as a function of molecular weight and composition as illustrated in Figure 1. The greater dependence on temperature for Y types is a notable distinction. Critical molecular weights, above which chain entanglement occurs, are 5000–6000 for Y and 8000 for Z PFPEs. Activation energy for viscous flow increases with molecular weight and levels at 49.2 and 21.4 kJ mol^{-1} for Y R and Z 25, respectively, the lower of the two being indicative of more favorable viscoelasticity and superior low-temperature properties. Collectively, the rheological parameters available are indicators of the lower intermolecular attractive forces and intramolecular rotational barriers that the Z exhibit over the Y PFPEs. The influence of

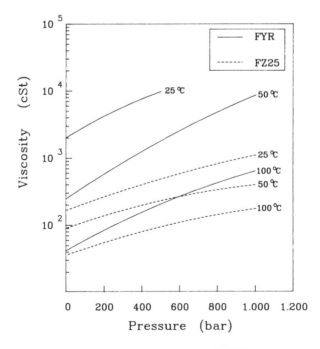

Figure 2. Viscosity versus pressure for PFPEs.

pressure on viscosity (Figure 2) exemplifies this behavior, Z fluids being noticeably more resistant to perturbation (increase in viscosity as pressure rises at constant temperature) than the Y types. Y and Z fractions are essentially Newtonian, independent of shear rate gradients.

20.5. FUNCTIONAL PFPEs FROM PHOTOOXIDATION

Mono- and difunctional PFPE acid fluorides and derived products based on fluoride ion-initiated ring-opening polymerization of perfluoroepoxides have been known for some time.[20b,30b] Analogs, including polyfunctionals, can be prepared via fluoroolefin photooxidation, as described here. Photooxidation of HFP followed by peroxide removal provides PFPE material in which acid fluoride and fluoroformyl functions comprise roughly 50% of the end-groups (see Section 20.2.2). Hydrolysis and further processing enables the material to be separated into two fractions, the major of which is rich in PFPE monoacids, while the minor comprises diacid and completely neutral (perfluoroalkyl end-groups) PFPE molecules.

Alternatively, monofunctional PFPEs can be prepared by addition of bromotrifluoroethylene (BTFE), for example, to the HFP feed during photooxidation.[64] This causes the production of oligomeric PFPE polyperoxides characterized by the presence of a high proportion of monofunctional molecules, i.e., carrying one COF end-group. Appropriate work-up generates PFPE monoacids. Mechanistically, the reaction probably proceeds via regiospecific attack of perfluoroalkoxy radicals ($R_FO^\bullet$, see Scheme 3) on BTFE, followed by conversion of the resultant α-bromo radical $R_FOCF_2CFBr^\bullet$

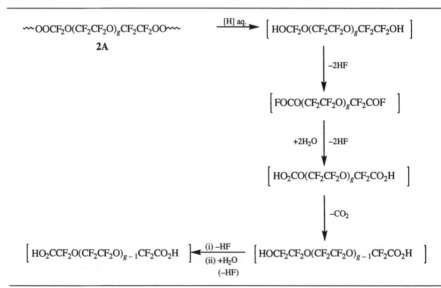

Scheme 4. Conversion of Z-type polyperoxides to dicarboxylic acids via reductive cleavage of peroxide linkages.

Table 6. Fomblin® Derivatives: Typical Properties of $XCF_2O(CF_2CF_2O)_m(CF_2O)_nCF_2X$

Product	X = CO_2H	X = CO_2Me	X = CH_2OH	X = $CH_2(OCH_2CH_2)_lOH$ $l = 1.5$	X = $CH_2OCH_2CH(OH)CH_2OH$	X = 3-NCO,4-$CH_3C_6H_3NHC(O)$	X = Aromatic
Code	Z DIAC	Z DEAL	Z DOL	Z DOL TX	Z TETRAOL	Z DISOC	AM 2001
Color	colorless	colorless	colorless	colorless	colorless	yellow	yellow
Molecular weight	2000	2000	2000	2200	2000	3000	2300
Difunctional content (wt%)	>94	>94	>94	>94	>95[a]	>94	>94
Kinematic viscosity at 20 °C (cSt)	60	20	80	140	2000	160[b]	80
Density (g cm^{-3})	1.82	1.77	1.81	1.73	1.74	1.72	1.72
Refractive index	1.300	1.298	1.300	1.316	1.324	—	1.345
Vapor pressure							
at 20 °C (mm Hg)	5×10^{-5}	3×10^{-4}	2×10^{-5}	6×10^{-6}	1×10^{-7}	—	2×10^{-6}
at 100 °C (mm Hg)	2×10^{-3}	2×10^{-3}	2×10^{-3}	1×10^{-3}	5×10^{-3}	—	2×10^{-4}

[a]Tetrafunctional.
[b]75 wt% solution in CFC-113 ($CF_2ClCFCl_2$).

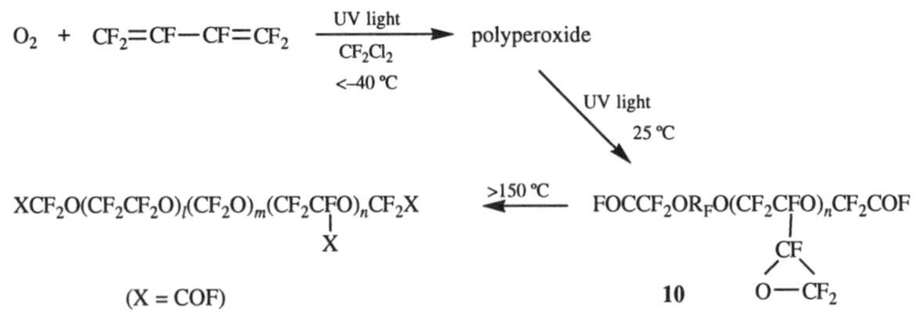

$$O_2 \; + \; CF_2{=}CF{-}CF{=}CF_2 \quad \xrightarrow[\substack{CF_2Cl_2 \\ <-40\,^\circ C}]{UV\ light} \quad \text{polyperoxide}$$

$$\xrightarrow[\substack{25\,^\circ C}]{UV\ light}$$

$$XCF_2O(CF_2CF_2O)_l(CF_2O)_m\underset{\underset{X}{|}}{(CF_2CFO)_n}CF_2X \quad \xleftarrow{\;\geq 150\,^\circ C\;} \quad FOCCF_2OR_FO(CF_2CFO)_n CF_2COF$$

$$(X = COF) \qquad\qquad\qquad\qquad\qquad\qquad \mathbf{10} \qquad \underset{O{-}CF_2}{CF}$$

Scheme 5. Production and thermal decomposition of PFPE epoxides derived from perfluorobuta-1,3-diene. Thermal elimination of difluorocarbene (in last step) from trifluoroepoxy groups in fluorocarbon systems is well known (see Chapter 3.).

to the new alkoxy species $R_FOCF_2CFBrO^\bullet$ (cf Scheme 3) which then loses bromine via β-scission ($R_FOCF_2CFBrO^\bullet \rightarrow R_FOCF_2COF + Br^\bullet$). The known photolytic oxidation of R_FCHX_2 (X = Cl, Br) to R_FCOX, e.g., the conversion of HCFC-123 (CF_3CHCl_2) to trifluoroacetyl chloride,[65,66] provides support for this proposal. Narrow-molecular-weight-range PFPE acids obtained via these methods have been converted into salts, amides, nitriles, alcohols, ethoxylated alcohols, and alcohol methacrylates; these are now available in limited quantities for experimental purposes.

Polyperoxide intermediates procured via olefin photooxidation not only serve as precursors to neutral PFPEs (see Sections 20.2.2 and 20.2.3), but additionally show a diversified chemistry in their own right. Thus, the Z polyperoxide series (**2**) can be converted directly to diacid products by chemically induced reductive cleavage of the peroxide bonds,[67] as exemplified in Scheme 4, where structure **2A** is a specific version of the generalized formula **2**. When a CF_2O unit is adjacent to the peroxide linkage, reduction is rapidly followed by decomposition of unstable OCF_2OH groups (see Chapter 3) and hydrolysis of the resulting fluoroformate intermediate, with accompanying loss of a CO_2 and HF. If a run of two or more CF_2O units lies next to the peroxide link, the above process repeats until a CF_2CF_2O unit is reached, ultimately forming a CF_2CO_2H end-group. Clearly, the average molecular weight and difunctional content of the PFPE diacid product depend on the peroxide content and average molecular weight of the precursor (**2**), respectively. These are controlled by adjusting the conditions during the photooxidative polymerization reaction.

Modification of the carboxyl end-groups provides a family of Z derivatives. Typical properties of representative "difunctionals" now available on a semicommercial scale are displayed in Table 6. Photooxidation of perfluorobuta-1,3-diene gives a polyperoxide containing trifluoroepoxide groups.[19a] Extension of this early observation to TFE/perfluorobutadiene oxidative copolymerization has led to the development of a route to polyfunctional PFPEs,[9b] as illustrated in Scheme 5. By varying the polymerization conditions, both the number of pendant epoxide groups in the unique polyfunctional PFPE **10** and its average molecular weight can be regulated; 3–5 and 2000, respectively, are representative values. Investigation of the chemistry and properties of PFPE epoxides (**10**) and their derivatives continues in our laboratories.[68]

20.6. APPLICATIONS OF PFPEs

PFPEs are exemplary specialty chemicals that enjoy myriad applications due to their performance properties. The volume and use requirements vary over a broad spectrum from providing lubrication for the delicate and precise mechanisms of Rolex® watches through to the ultracentrifuge bearings in UF_6 enrichment plants. In this section certain applications are highlighted with references to further reading; no attempt has been made to produce a comprehensive account.

Historically, PFPEs first found major acceptance as working fluids in mechanical vacuum pumps used in various semiconductor manufacturing processes.[45,69] In this application traditional hydrocarbon fluids often succumb to the degrading action of aggressive chemicals by "polymerization," which translates into high system mainte-nance. The inherent stability of PFPEs* reduces safety risks associated with extremely reactive process gases and chemicals, e.g., oxygen, chlorine, fluorine, silane, boron trifluoride, and boron trichloride.

20.6.1. Testing of Electronic Devices and Equipment

During the 1980s PFPEs were introduced as vapor-phase soldering media as alternatives to perfluorinated amines and cycloaliphatics (see Chapter 4).[71] Besides showing good thermooxidative stability, low surface tension, good heat-transfer characteristics, low volatility, and low levels of toxicity, an ample selection of PFPE grades are available according to boiling points (i.e., reflow vapor temperatures), to more nearly satisfy the diverse needs of surface mount assembly.[72] Similarly,[73] carefully designed PFPE fractions have gained acceptance in electronic testing, a market which perfluorocarbon fluids once dominated unchallenged. Thus PFPE counterparts find use as thermal shock,[47] hermetic seal (gross leak),[74] and liquid burn-in testing fluids,[75] as heat transfer/cooling media,[76] and in blood analysis,[77] often with performance advantages when compared with perfluorocarbon equivalents. Cases in point include NID (Negative Ion Detection,[78] an automated gross leak test), where PFPEs show higher response sensitivity and lower consumption than perfluorocarbon analogs,[79] and single-fluid thermal-shock testing, where use of PFPE fluids with a broader liquid range allow test-procedure simplification (one versus two baths) that minimizes fluid consumption.

---

*Du Boisson and Sellers have compared the rates of $AlCl_3$-catalyzed decomposition between structure **8** and an alternative commercial PFPE containing slightly different terminal groups and devoid of OCF_2 units in the polymer chain.[70a] Under the test conditions (160°C for 1–7 h, 1–3 wt% $AlCl_3$) in open or closed systems, decomposition of the former dramatically exceeded that of the latter. These test results contrast strongly with those obtained in our own laboratory[70b] under test conditions which more nearly simulated semicon plasma etch (Al by $BCl_3/Cl_2/CCl_4$) conditions. Periodically the vacuum pumps were examined for oil consumption, oil viscosity change, and ultimate total pressure. Over the test period (744 h), "oil consump-tion" for *both fluids* was *ca.* 14% of total pump charge, and the visosity of each dropped by only *ca.* 8%, but fluid of structure **8** slightly outperformed the alternate fluid in all three categories mentioned above. The results of our test have been corroborated by a major PFPE user.

20.6.2. Polymer Curing

An embryonic technique for polymer curing comprises vapor (similar to vapor-phase soldering) or even liquid immersion techniques using PFPEs as the medium. Resins, elastomers, and adhesives have been crosslinked by this method,[80] which offers the advantages of efficient heat transfer (precise temperature control), air exclusion, and reduced curing times over standard procedures.

20.6.3. Lubrication

In the lubrication field, the use of PFPEs extends well beyond working fluids for pumps[81] into general industrial engineering and transportation practices. For example, they find use in lubricating valves and bearings associated with aggressive gas handling (including UF_6)[82] and (a more recent application) with refrigerant compressors.[83] Also, there is a gradual but distinct trend to partially substitute hydrocarbons and silicones with PFPE lubricants for demanding, long-life applications in conveyor belts, gears, roller bearings, and vibrating blast-furnace screens associated with food processing, metal foundry work, and textile and paper manufacturing.

In the light mechanical arena, PFPEs show superior performance in plastic (to metal or plastic) lubrication in typewriters or photocopying machines,[84] and impregnated (self-lubricating) sintered metal bearings for rotating shaft supports.[85]

The trend toward space-saving miniaturization, electronic gadgetry, and drag minimization, among others, in the automotive industry is associated with higher overall operating temperatures. Compounded by long-term guarantees/maintenance-free aspirations, PFPEs offering life-lubrication possibilities have stimulated end-user attention in fan clutch, transmission, anti-brake lock, shock absorber and differential lubrication systems.

Low volatility and good lubricity characteristics support PFPE use in aerospace projects.[86] As noted previously (Section 20.4.6) attempts to further improve properties through incorporation of additives are being explored. Despite lower bulk modulus compared to hydrocarbons, synthetic and CTFE (chlorotrifluoroethylene-based) fluids, PFPEs are suitable nonflammable hydraulic fluids,[87] which, however, may require system redesign rather than retrofit.

The Z neutral and functional PFPEs[88] have captured the major market share of lubricants for floppy and rigid disks, and also show potential on magnetic tapes.[89] PFPEs effectively lubricate push-button and sliding-switch contacts, offering performance and cost savings by minimizing precious metal coatings on the contacts.[90] Related use as dielectric fluids, for example in cable insulation, has been reported.[91]

20.6.4. Surfactants, Emulsions, and Cosmetics

PFPEs are distinguished from perfluorocarbons by their ability to maintain fluidity even at high molecular weights, a feature which broadens their scope of application. Structural synergy between fluid and surfactant,[92] when both are PFPEs, allows for reduced aqueous interfacial tension (to several mN m^{-1}). Critical micelle concentrations of photochemically generated PFPE surfactants[93] approach 10^{-5} mol L^{-1}; their aqueous solutions show surface tensions < 20 mN m^{-1}. Applications of PFPE emulsions, microemulsions, and surfactants are at a preliminary stage of development, thus far encompassing unusually stable three-phase (mineral oil/neutral PFPE/H$_2$O) emulsions for cosmetics,[94] surfactants for aqueous dispersion polymerization of fluoroolefins,[95] conductive microemulsions,[96] defoamers,[97] and water-borne systems for stone protection (see below).

20.6.5. Polymer Modification

The difunctional Fomblin® Z derivatives are versatile condensation-type step-growth prepolymer intermediates, convertable through appropriate reactions into thermoplastics, thermosets, or elastomers. Polyurethanes, epoxies, polyesters, polyacrylates, polyamides, polyimides, liquid crystal polymers,[98] and the like can be prepared with different fluorine contents. Reaction conditions may be varied systematically to modify the properties of the final polymer, for example, T_g, hydrophobicity, oleophobicity, chemical resistance, refractive index, surface energy, and mechanical strength; although the condensation techniques often are experimentally difficult, such modification of properties to approach those of conventional fluoropolymers has obvious performance advantages. Hence, use of this technique, in conjunction with polymer blending, continues to be an area of interest in polymer property improvement for coatings, adhesives, and sealants.

Miscellaneous uses of PFPEs as a pendant group or in a polymer backbone include elastomers,[99] gas-permeable membranes[100]—specifically in opthalmics,[101] release coatings,[102] and textile finishes.[103]

Within the realm of polymer technology, the lubrication properties of PFPE in combination with their low surface energies (migration occurs to air–substrate interfaces) make them good candidates as plastomer extrusion aids[104] and elastomer-mold release agents.[105] Coupling agents with PFPE tails improve compatibility between fluoropolymers and fillers.[106]

20.6.6. Conductive Films

Hard, self-lubricating, electrically conductive films are formed on stainless steel or silicon substrates by first applying a thin layer of PFPE grease followed by rf sputtering.[107]

20.6.7. Mass Spectrometry

In analytical chemistry, PFPEs offer expansive ranges of volatility and molecular weight, hence their attractiveness as mass markers in mass spectrometry.[108]

20.6.8. Environmental Applications

During the present century, corrosive air-borne chemicals have defaced stone architectural structures and monuments at an accelerated rate. The facts that the historic landmarks of Italy are both abundant and part of the national heritage prompted Montefluos to develop a penetrating surface-treatment agent for conserving such treasures which provides a high degree of weatherability and protection, yet binds only reversibly to the substrates (therefore is easy to remove).

In collaboration with the Research Centre of Artwork Conservation (University of Florence, Italy), a neutral PFPE fraction, Fomblin® MET, was customized for this application and successfully evaluated in long-term artificial then natural aging-in-field tests.[110] Water and solvent-based formulations of functional PFPEs are also being developed to protect masonry and concrete[111]; not surprisingly, these confer oil as well as water-repellant properties. Similarly, solvent-borne or water-dispersed fluoroelastomers are effective aggregating and protective agents for limestone or aged sandstone, i.e., porous substrates.[112]

The CFC–HCFC/ozone issue has created a potential opportunity for nonchlorine-containing substitutes. Several PFPE fractions have been customized for niche solvent applications where recycle allows the fluoroether alternative to approach cost parity.

20.7. FUTURE PROSPECTS

From 1985 through 1989, PFPE citations in the literature increased annually, the 1989 number being approximately twice that of 1985; during the same period the world market for PFPEs doubled to roughly 700 tons per annum. Both are positive indicators. The high-performance properties of PFPEs seem well-aligned with the projected requirements in several high-growth commercial sectors, which is not an unusual scenario in fluorine chemistry. The market place previously has risen to the technology, as with fluoropolymers and fluorochemicals for electronics. Over the past 20 years attention has been focused on developing applications for neutral PFPE; in this light, if the encouraging trend of the relatively new functional PFPEs continues, the prognosis is favorable indeed.

Several global organizations are well entrenched in PFPE manufacture...enough to sustain a healthy spirit of competition, and the commitment to the marketplace is fierce. Newcomers are welcome but capital investment costs are high: well-integrated technology to basic feedstocks is complex, challenging to refine, and shows lengthy learning curves.

The price of PFPEs is declining, as expected, in keeping with the realization of new volume applications. In general, PFPE technology is well-suited to economy of scale, and an analogy can be drawn between the current situation and the position of silicones 35 years ago. Potential PFPE applications based, for example, on turbine-engine lubricants for aircraft, electrorheological fluids, magnetic fluids, gas/oil additives, and aqueous emulsion/microemulsion lubrication will—if successfully commercialized—accelerate these trends.

ACKNOWLEDGMENTS

Thanks are due to our colleagues, whose endeavors established the foundation and development of PFPEs at Montedison. Special recognition is given to the scientists, engineers, and field representatives actively engaged in the continued evolution of PFPEs: S. Briggs, E. Calloni, A. Chittofrati, C. Corti, U. De Pato, A. Faucitano, L. Flabbi, M. Gastaldi, G. Gavazzi, P. Gavezotti, Q. Hakim, D. Higgins, K. Johns, B. Lim, A. Locaspi, M. Malavasi, M. Meucke, G. Moggi, L. Montagna, W. Navarrini, C. Pace, M. Paganin, S. Panerati, M. Pianca, D. Sama, P. Savelli, A. Schultheiss, S. Schwartz, S. Shaffer, R. Silvani, H. Soma, P. Srinivasan, A. Staccione, E. Strepparola, A. Tasca, G. Tommasi, C. Tonelli, and M. Visca.

20.8. REFERENCES

1. D. Sianesi, A. Passetti, and C. Corti, *Macromol. Chem. 86*, 308 (1965); V. Caglioti, A. Dell Site, M. Lenzi, and A. Mele, *J. Chem. Soc.* 5430 (1964); F. Gozzo and G. Camaggi, *Tetrahedron 22*, 2181 (1966); J. Heickler, *Advances in Photochemistry*, Vol. 7, Interscience, New York (1969).
2. L. Batt and R. Walsh, *Int. J. Chem. Kinet. 14*, 933 (1982).
3. R. N. Haszeldine, *J. Chem. Soc.* 3559 (1953).
4. D. Sianesi, A. Pasetti, and C. Corti, British Patent 1,104,482 (1968); U.S. Patent 3,704,214 (to Montedison) [*CA 70*, 29557t (1969)]; D. Sianesi and R. Fontanelli, British Patent 1, 226,566 (to Montedison) [*CA 70*, 20513t (1969)].
5. D. Sianesi, A. Pasetti, and G. Belardinelli, British Patent 1,217,871 (to Montedison) [*CA 78*, P125179g (1973)]; F. Gozzo and V. Oprandi, U.S. Patent 3,622,635 [*CA 73*, 110319a (1970)]; K. Kobiashi, M. Fkazawa, and S. Ishikawa, U.S. Patent 4,859,299 [*CA 109*, 129862f (1988)].
6. D. Sianesi, A. Pasetti, and C. Corti, U.S. Patent 3,722,792 (1973); G. Marchionni and U. De Patto, European Patent 344,547 (to Ausimont) [*CA 112*, 180167 (1990)].
7. G. Marchionni and U. De Patto, unpublished results.
8. D. Sianesi, A. Pasetti, and C. Corti, U.S. Patent 3442,942 (to Montedison) [*CA 64*, 14360g (1966)].
9. (a) D. Sianesi, G. C. Bernardi, and A. Pasetti, U.S. Patent 3,451,907 (to Montedison) [*CA 69*, P51649a (1968)]; (b) G. Marchionni, U. De Patto, E. Strepparola, and G. T. Viola, European Patent 244,839 (to Ausimont) [*CA 108*, 168200 (1988)].
10. D. Sianesi, A. Marracini, and G. Marchionni, Italian Patent 20207A/89; S. P. Brindue, A. B. Clinch, D. K. McIntyre, A. L. Noreen, and M. J. Pellertite, European Patent 259,980 (to 3M Co.) [*CA 109*, 74137 (1988)].
11. G. Marchionni and A. Staccione, European Patent 340,739 (to Ausimont) [*CA 112*, 140100 (1990)].
12. G. Marchionni, F. Spataro, and R. J. De Pasquale, *J. Fluorine Chem. 49*, 217 (1990).
13. D. Sianesi, A. Grazioli, and R. Fontanelli, U.S. Patent 3,720,646 (to Montedison) [*CA 76*, 26193j (1972)]; G. Caporiccio, G. T. Viola, and C. Corti, U.S. Patent 4,500,793 (to Montedison) [*CA 99*, P213089s (1988)].
14. G. Marchionni and G. T. Viola, U.S. Patent 4,668,357 (to Ausimont) [*CA 105*, 206914j (1986)].
15. See, for example, J. F. Harris, *J. Org. Chem. 30*, 2182 (1965).
16. D. Sianesi and R. Fontanelli, U.S. Patent 3,665,041 (1972); G. Marchionni and G. T. Viola, U.S. Patent 4,664,766 (to Ausimont) [*CA 106*, 33664e (1987)]; G. Marchionni, A. Staccione, and G. Gregorio, *J. Fluorine Chem. 47*, 515 (1990).
17. G. Marchionni and A. Staccione, European Patent 308,905 (to Ausimont) [*CA 111*, 78902 (1989)].
18. D. Sianesi, *Chim. Ind. (Milan) 50*, 206 (1968).
19. (a) D. Sianesi, A. Pasetti, R. Fontanelli, G. C. Bernardi, and G. Capriccio, *Chim. Ind. (Milan) 55*, 208 (1973); (b) N. M. Baranova and V. A. Poluektov, *Khim. Vys. Energ. 21*, 446 1987; (c) N. M. Baranova, V. A. Poluektov, and V. G. Vereskunov, *Khim. Vys. Energ. 29*, 1299 (1987).
20. (a) M. Lenzi and A. Mele, *J. Chem. Phys. 43*, 1974 (1965); (b) H. S. Eleuterio, *J. Macromol. Sci., Chem. A 616*, 1027 (1972).

21.   K. J. Ryan and I. C. Plumb, *J. Phys. Chem.* 78, 4678 (1982).
22.   A. Faucitano, A. Buttafava, F. Martinotti, G. Marchionni, and R. J. De Pasquale, *Tetrahedron Lett.* 29, 5557 (1988).
23.   A. Faucitano, A. Buttafava, F. Martinotti, V. Comincioli, and G. Marchionni, IUPAC Int. Symp. on Free Radical Polymerizations: Kinetics and Mechanism, Santa Margherita, preprints p. 280 (1987).
24.   G. Marchionni, U. De Pato, and R. J. De Pasquale, 12th Int. Symp. Fluorine Chem., Santa Cruz, Ca., Abstr. No. 333 (1988); A. Faucitano, A. Buttafava, V. Comincioli, G. Marchionni, and R. J. De Pasquale, *J. Phys. Org. Chem. 4*, 293 (1991).
25.   For a review of the Mayo–Sianesi scheme for the radical copolymerization of oxygen and terminal perfluoroalkenes, see L. F. Sokolov, P. I. Valov, and S. V. Sokolov, *Russ. Chem. Rev. 53*, 711 (1984).
26.   A. Faucitano, A. Buttafava, F. Martinotti, A. Staccione, and G. Marchionii, 13th Int. Conf. on Photochem., Budapest, Hungary, 1987, Vol. II, p. 662; A. Faucitano, A. Buttafava, F. Martinotti, and G. Marchionii, *Tetrahedron Lett. 29*, 4611 (1988).
27.   A. Faucitano, A. Buttafava, G. Caporiccio, and G. T. Viola, *J. Am. Chem. Soc. 106* (1984).
28.   B. E. Smart, in: *Molecular Structure and Kinetics* (J. F. Liebman and A. Greenberg, eds.), Vol. 3, p. 141, VCH Publishers, Weinheim (1986).
29.   A. Faucitano, A. Buttafava, E. Martinotti, G. Marchionni, A. Staccione, and R. J. De Pasquale, ACS 9th Winter Fluorine Conference, St. Petersburg, Fl, Abstr. No. 52 (1989).
30.   For general references on PFPE, see: (a) R. E. Danielson, *Kirk-Othmer Encyclopedia* (3rd edn.) *10*, 1874 (1980); (b) P. R. Resnick, *ibid.,* p. 956.
31.   A. L. Henne and S. B. Richter, *J. Am. Chem. Soc. 74*, 5420 (1952).
32.   For examples, see G. V. D. Tiers, *J. Am. Chem. Soc. 77*, 4837 (1955); *ibid.*, 6703 (1955); *ibid.*, 6704 (1955); D. Sianesi and R. Fontanelli, *Macromol. Chem. 102*, 115 (1967).
33.   G. Marchionni, unpublished results.
34.   ASTM 3115/73 (modified).
35.   D. Sianesi, V. Zamboni, R. Fontanelli, and M. Binaghi, *Wear 18*, 85 (1971).
36.   C. Corti, 10th International Symposium on Fluorine Chemistry, Vancouver, 1982, Abstr., p. 7.
37.   G. Caporiccio, L. Flabbi, G. Marchionni, and G. T. Viola, *J. Synth. Lubr. 6*, 133 (1989).
38.   See Refs. 6 and 7, and C. E. Snyder and C. Tamborski, U.S. Patent 4,643,976 (to U.S. Department of the Air Force) [*CA 89*, P182232v (1978)]; C. E. Snyder Jr. and R. E. Dolle, Jr., ASLE Trans. *19*, 171 (1976).
39.   BAM tests, Bundesanstalt für Materialprufung, Berlin 45 Abteilung 4 (Chemische Sicherhertstechnik).
40.   A. Faucitano, A. Buttafava, F. Martinotti, G. Caporiccio, and G. T. Viola, *J. Fluorine Chem. 29*, 211 (1985); A. Faucitano, A. Buttafava, F. Martinotti, G. Marchionni, and R. J. De Pasquale, *J. Rad. Phys. Chem. 37*, 493 (1991); for E-beam, see J. Pacansky, R. J. Waltman, and C. Wang, *J. Fluorine Chem. 32*, 283 (1986) and J. Pacansky, R. J. Waltman, and M. Maier, *J. Phys. Chem. 91*, 1225 (1987); J. Pacansky and R. J. Waltman, *Polym. Mater. Sci. Eng. 60*, 6148 (1989).
41.   E. C. Vigliani and G. Cavagna, "Toxicity tests of PFPE oligomers", Work Clinic, Milan Montedison report (1969); U. Dal Re and F. Volagerna, "Determination of acute oral toxicity of PFPEs", Montedison report (1981): G. Chiappino, P. Castano, and O. Picchi, "Dermal Acute Toxicity of PFPEs", Toxicological Laboratories, BIOLAB/ITALY (1981); J. A. Cuthbert and E. Eackesham, "Human Repeat Insult Patch Test", Inveresk Research International, Report 2715 (1983); E. Santoro, E. Martinelli, and G. Malinverno, "Analysis of Fomblin in the Biological Fluids of Orally Treated Rats", Institute Donegani, Report 003 (1989).
42.   For a general reference, see D. S. L. Slinn and S. W. Green, in: *Preparations, Properties and Industrial Applications of Organofluorine Compounds* (R. E. Banks, ed.), p. 57, Ellis Horwood, Chichester (1982).
43.   J. S. Shaffer and A. Schultheiss, *Test and Measurement World* 26 (Nov. 1988).
44.   L. Flabbi and R. S. Briggs, "PFPEs—Safe Primary Fluids for VPH", Nepcon East, Boston, 302 (1987); L. Flabbi and R. S. Briggs, *Brazing and Soldering 7*, 6 (1984).
45.   G. Caporiccio, C. Corti, S. Soldini, and G. Carniseli, *Ind. Eng. Chem., Prod. Res. Dev. 21*, 515 (1982).

46. J. T. Skenan, NLGI Preprint, 37th meeting, MLGI, Kansas City (1969); L. Messina, *Lubr. Eng. 25,* 459 (1969).
47. E. Cosmacini and V. Veronesi, *Wear 108,* 269 (1986).
48. L. Flabbi and C. Corti, *Tribologia e Lubrificazione 20,* 109 (1985).
49. C. Corti and P. Savelli, *Proc. Cont. Synth. Lubr.,* Sapron, Hungary, p. 128 (1989); *Magy. Kem. Lapja 45,* 486 (1990).
50. J. A. Spikes and G. Caporiccio, *J. Synth. Lubr. 1,* 23 (1984).
51. W. R. Jones, R. J. Johnson, W. D. Winer, and D. M. Sanborn, *ASLE Trans. 18,* 249 (1975); A. Plagge, *Schmierungstechnik 15,* 338 (1984).
52. C. E. Snyder, L. J. Gschwender, and C. Tamborski, *Lubr. Eng. 37,* 344 (1981); W. R. Jones and C. E. Snyder, Jr., *ASLE Trans. 23,* 253 (1980).
53. D. J. Carrè, *ASLE Trans. 29,* 121 (1986).
54. See, for example, 5th International Colloquium, "Additives for Lubricants and Operational Fluids", Vols. 1 and 2, Technische Akademie Esslingen, Germany (Jan 1986).
55. D. J. Carrè, *ASLE Trans. 31,* 447 (1988); W. Morales and D. H. Buckley, *Wear 123,* 345 (1988); S. Mori and W. Morales, *Wear 132,* 111 (1989).
56. See Ref. 6 and C. Tamborski, U.S. Patent 3,499,041 (to U.S. Department of the Air Force) [*CA 72,* 100879, (1970)]; K. J. L. Paciorek, D. H. Harris, M. E. Smythe, and R. H. Kratzer, *NASA CR-168224* (1983); W. R. Jones, J. K. L. Paciorek, J. H. Nakahara, M. E. Smythe, and R. H. Kratzer, *Ind. Eng. Chem. Res. 26,* 1930 (1987); J. B. Christian and C. Tamborski, U.S.Patent 4,324,673 (to U.S. Department of the Air Force) [*CA 94,* 86992a (1981)] and 4,132,660 (1979); E. Strepparola, M. Alfieri, and P. Gavezotti, European Patent 322,916 (to Ausimont) 1987; E. Strepparola, P. Gavezotti, and C. Corti, European Patent 337,425 (to Ausimont) 1989; E. Strepparola, C. Corti, P. Gavezottii, and A. Chittofrati, European Patent 382,224 (to Ausimont) 1990; C. Corti, P. Savelli, L. Montagna, and R. J. De Pasquale, U.S. Patent 5,124,058 (to Ausimont) [*CA 115,* 139493z (1991)]; M. Sato, A. Mori, and M. Aizawa, Japanese Patent 0133165 (to Nippon Soda Co.) [*CA 111,* 156031 (1989)].
57. A. C. Quano, B. Appelt, and T. D.Watson, *Org. Coat. Appl. Polym. Sci. Proc. 46,* 230 (1981).
58. B. A. Wolf, M. Klimiuk, and M. J. R. Cantow, *J. Phys. Chem. 93,* 2672 (1989).
59. G. Marchionni, G. Ajroldi, and G. Pezzin, *Eur. Polym. J. 24,* 1211 (1988); G. Marchionni, G. Ajroldi, P. Cinquina, E. Tamepllini, and G. Pezzin, *Polym. Eng. Sci. 30,* 829 (1990).
60. M. J. R. Cantow, T. Y. Ting, E. M. Barrall, R. S. Porter, and E. R. George, *Rheol. Acta 25,* 69 (1986); M. J. R. Cantow, E. M. Barrall, B. A. Wolf, and H. Geerissen, *J. Polym. Sci. Phys. 25,* 603 (1987); M. J. R. Cantow, R. B. Larrabee, E. M. Barrall II, R. S. Butner, P. Cotts, F. Fvy, and T. Y. Ting, *Macromol. Chem. 187,* 2475 (1986).
61. Y. Ohasaka, *Petrotech 8,* 840 (1985).
62. Y. Tanaka, N. Nojiri, K. Onta, H. Kubota, and T. Makita, *Int. J. Thermophys. 10,* 857 (1989).
63. S. Yasutomi, S. Blair, and W. O. Winter, *J. Tribology 106,* 291 (1984).
64. G. Marchionni and A. Staccione, European Patent 340,740 (to Ausimont) [*CA 112,* 159247 (1990)].
65. R. N. Haszeldine and F. Nyman, *J. Chem. Soc.* 420 (1950).
66. A. L. Dittman, German Patent 2,418,676 (to Halocarbon Products Corp.) [*CA 82,* 97680 (1975)].
67. D. Sianesi, A. Pasetti, and G. Belardinelli, U.S. Patent 4,451,646 (to Montedison, Italy) [*CA 101,* 91718 (1984)].
68. G. Marchionni, E. Strepparola, and F. Spataro, European Patent 337, 346 (to Ausimont) [*CA 112,* 8373 (1990)].
69. L. Holland, *Vacuum 22,* 234 (1972); L. Holland and L. Laurenson, *Vacuum 23,* 139 (1973); G. Caporiccio and R. A. Steenrod, *J. Vac. Sci. Technol. 15,* 775 (1978); E. Falcone, *Chimica Oggi. 8,* 27 (1983); E. Calloni, A. Tasca, and L. Stoppa, European Patent 223,251 (to Ausimont) [*CA 107,* 80813 (1987)].
70. (a) R. A. Du Boisson and S. F. Sellers, *Macroelectronic Manufacturing and Testing,* p. 84 (May 1986); (b) Montefluos Bulletin, "The Resistance of PFPEs to Lewis Acids in Vacuum Pumping Systems for Semiconductor Production".
71. S. Briggs, *Brazing and Soldering 7,* 6 (1984).
72. C. Lea and K. Johns, *Brazing and Soldering 12,* 34 (1987).

73. L. Flabbi and S. Briggs, *European Semiconductor Design and Production*, p. 24 (May 1986); G. Bargigia, G. Caporiccio, C. Tonelli, L. Flabbi, and G. Marchionni, European Patent 203,348 (to Montedison, Italy) [*CA 106*, 139420 (1987)]; see also Ref. 42.

74. P. Srinivasan, "Fluid Performance in Gross Leak Bubble Testing", Proc. Internepcon., Singapore (Sept. 1988).

75. L. Flabbi and G. Bargigia, *Test and Measurement World 32*, (Nov. 1986).

76. P. Srinivasan and S. Landonio, Int. Microelectronics and Systems Conference, Malysia (July 1989).

77. L. B. Anderson and J. D. Madsen, *Sand. J. Chem. Lab. Invest. 47*, 503 (1987).

78. B. Evans, Web Technology Inc., Dallas, Texas, Personal communication.

79. P. Srinivasan, V. Frattini, and M. Scapin, Internepcon Semiconductor Asia Pacific Conference, Abstr. No. 251 (1990).

80. S. Briggs, P. Newton, and P. Srinivasan, *Hybrid Circuits 18*, 42 (1989); K. Johns, A. Re, and G. Bargigia, European Patent 287,398 (to Ausimont) [*CA 110*, 96941, (1989)]; G. Bargigia, A. Re, L. Corbelli, K. Johns, L. Flabbi, and P. Srinivasan, European Patent 282,973 (to Ausimont) [*CA 110*, 155462 (1989)].

81. M. Soei, S. Shimazaki, and T. Totsuka, Japanese Patent 61,113,695 (to Daikin Industries) [*CA 106*, 7816 (1987)]; R. Lowrie, *ASTM Spec. Tech. Publ. 812*, 84 (1983).

82. A. Ciancia, A. Ascensioni, C. Corti, and G. Caporiccio, *Nucl. Sci. Eng. 86*, 232 (1984); J. Hennig and H. Lotz, *Vacuum 27*, 171 (977).

83. K. Honma, S. Komasuzaki, F. Nakano, and T. Iizuka, Japanese Patent 62,288,692 (to Hitachi Co.) [*CA 108*, 153457 (1988)].

84. J. C. Anderson, L. Flabbi, and G. Caporiccio, *J. Synth. Lubr. 5*, 199 (1988).

85. V. D'Agostino, V. Niola, and G. Caporiccio, *Tribol. Int. 21*, 105 (1988).

86. K. E. Demorest and E. L. McMurtrey, *Lubr. Eng. 34*, 137 (1978); G. F. Brouwer, "Lubrication and Wear of a Stainless Steel Actuating Cable", *Proc. Second Space Tribology Workshop ESTL*, Risley UK (Dec. 1980); see also Ref. 55.

87. C. E. Snyder and R. E. Dolle, *ASLE Trans. 19*, 171 (1976); C. E. Snyder and L. Gschwender, *Lubr. Eng. 36*, 458 (1980); C. E. Snyder Jr., L. Gschwender, and W. B. Campbell, *Lubr. Eng. 38*, 41 (1982); M. J. Fifolt and J. Forcucci, U.S. Patent 4,528,109 (to Occidental Chemical Corporation) [*CA 103*, 144615 (1985)]; T. Sato, K. Yamamoto, and T. Nakajina, Japanese Patent 63,295,810 (to NTN Toyo Bearing Co.) [*CA 111*, 137353 (1989)].

88. A. M. Scarati and G. Caporiccio, *IEEE Trans. Magn. 23*, 106 (1987) and European Patent 239,123 (to Ausimont), [*CA 108*, 132991 (1988)]; P. Butafava, V. Bretti, L. G. Ciardiello, G. Caporiccio, and A. M. Scarati, *IEEE Trans. Magn. 21*, 1533 (1985); C. H. Tsai, Y. Memandost, H. Samani, and A. Eltoukhy, *J. Vac. Sci. Technol. 7*, 2491 (1989); G. Caporiccio, E. Strepparola, and A. M. Scarati, European Patent 165,650 (to Montedison, Italy) [*CA 104*, 186986 (1986)]; M. Hoshino, A. Terada, and S. Sugarawa, German Patent 3,440,361 (to Nippon Telegraph and Telephone Corp., Japan) [*CA 103*, 97747 (1985)].

89. U. Bagatta, A. R. Corradi, L. Flabbi, and L. Salvioli, *IEEE Trans. Magn. 20*, 16 (1984). For a recent review on magnetic media, see A. Berkowizz (ed.), *MRS Bulletin 15*, 23–72 (1985).

90. M. Antler, *IEEE Trans. Compon. Manuf. Technol. 10*, 24 (1987); Y. Tsuchiya, Y. Miazawa, and K. Mogi, Japanese Patent 60, 251, 285 (to Hirose Electric Co.) [*CA 105*, 10506 (1986)]; G. L. Lilvestrand, *Electr. Contacts 28*, 11 (1982); G. Kovacs, *Proc. Int. Conf. Electr. Contact Phenom.* 10th, Budapest, p. 475 (1980).

91. B. Uccellio and C. Basisio, British Patent 2,064,579 (to Industre Pirelli, Italy) [*CA 96*, 70085 (1982)]; A. Luches and L. Provenanzo, *J. Phys. D 10*, 339 (1977).

92. Ref. 19a and G. Caporiccio, F. Burzio, C. Carniselli, and V. Biancardi, *J. Colloid Interface Sci. 98*, 202 (1984); M. Visca and D. Lenti, European Patent 315,078 (to Ausimont) [*CA 111*, 135482 (1989)].

93. A. Chittofrati, D. Lenti, A. Sanguinati, M. Visca, C. Gambi, D. Senatra, and Zhou Zhen, *Colloids Surf. 41*, 45 (1989) and *Prog. Colloid Polym. Sci. 79*, 218 (1989). For a review of fluorocarbon surfactants, see H. C. Fielding, *Organofluorine Chemicals and Their Industrial Application* (R. E. Banks, ed.), p. 214, Ellis Horwood, Chichester (1979).

94. S. Bader, F. Brunetta, G. Pantini, and M. Visca, *Cosmetic Toiletries 101*, 45 (1986); G. Pantini and A. Antonini, *Drug Cosmet. 143*, 34 (1988); M. Pianca and G. Pantini, *Technol. Chim. 9*, 116 (1989); F. Brunetta, S. Bader, and G. Pantini, European Patent 196,904 (to Ausimont) [*CA 106*, 38237 (1987)].

95. E. Gianetti and M. Visca, European Patent 250,767 (to Ausimont) [*CA 108*, 222277 (1988)]; E. Gianetti, A. Rotasperti, and E. Marchesi, European Patent 247,379 (to Ausimont) [*CA 108*, 187440 (1988)].

96. A. Chittofrati and D. Lenti, European Patent 315, 841 (to Ausimont) [*CA 111*, 121624 (1989)].

97. Japanese Patent 60, 022,909 (to Nippon Mectron Co.) [*CA 103*, 8022 (1985)].

98. F. Pilati, M. Toselli, A. Re, F. A. Bottino, A. Pollicino, and A. Recca, *Macromolecules 23*, 348 (1989); S. Birkle, D. H. Feucht, and M. E. Rissel, European Patent 322,624 (to Siemens, Germany) [*CA 111*, 116973 (1989)]; F. Pilati, V. Bonora, P. Manaresi, A. Munari, M. Toselli, A. Re, and M. De Giorgi, *J. Polym. Sci. A, Polym. Chem. 27*, 951 (1989); A. Re and G. Donati, Eur. Pat. Appl. EP 249,048 (to Ausimont) [*CA 108*, 151587w (1988)]; A. Re, E. Strepparola, and P. Gavezotti, European Patent 192,190 (to Ausimont) [*CA 105*, 192658 (1986)]; G. Caporiccio, E. Strepparola, G. Bargigia, G. Novaira, and E. Peveri, *Macromol. Chem. 184*, 935 (1983); regarding biomedical elastomers, see R. J. Zdrahala and M. A. Strand, U.S. Patent 4,841,007 (to Becton Dickinson and Co.) [*CA 112*, 8596 (1990)]; V. V. Zuev, T. I. Zhukova, and S. S. Skorokhodov, *Vysokomol. Soedin., Ser B 31*, 406 (1989).

99. G. Moggi and G. Marchionni, European Patent 320,005 (to Ausimont) [*CA 111*, 154676 (1989)]; R. E. Uschold, *Polym. J. (Tokyo) 17*, 253 (1985); W. R. Griffin, U.S. Patent 4,201,876 (to U.S. Department of the Air Force) [*CA 93*, 133669 (1980)].

100. M. Ikeda and A. Aoshima, Japanese Patent 61,166,834 (to Asaki Chemical Industry Co.) [*CA 106*, 19780 (1987)].

101. D. E. Rice and J. V. Ihlenfeld, Canadian Patent 1,201,247 (to Texaco Development Corp.) [*CA 105*, 46081 (1986)].

102. J. M. Larson, U.S. Patent 4,830,910 (to 3M Co.) [*CA 111*, 136167 (1989)]; J. M. Larson and A. Noreen, European Patent 98,698 (to 3M Co.) [*CA 100*, 157905 (1984)]; P. E. Olson, Eur. Pat. Appl. EP 98, 699 [*CA 100*, 105217n (1984)].

103. K. F. Mueller and R. A. Falk, U.S. Patent 4,046, 944 (to Ciby-Geigy Corp.) [*CA 87*, 153384 (1977)].

104. C. Tonelli, G. Marchionni, P. Gavezotti, and R. J. De Pasquale, Fluoropolymer Division, SPI Conference, Asheville, NC, September (1992).

105. V. Arcella, G. Brinati, and F. Barbieri, European Patent 310, 966 (to Ausimont) [*CA 111*, 155652 (1989)]; Y. Goto, K. Shimamura, and T. Tsutsumi, Japanese Patent 6,354, 463 (to Nitsui Toatsu Chemicals Inc.) [*CA 109*, 232160 (1988)]; G. Tommasi, R. Ferro, and G. Cirillo, European Patent 222, 201 (to Ausimont) [*CA 107*, 116783 (1987)].

106. E. Strepparola, T. Terenghi, E. Monza, and F. Felippone, European Patent 227,103 (to Ausimont) [*CA 107*, 177271 (1987)].

107. I. Sugimoto and S. Miyake, *Thin Solid Films 158*, 51 (1988).

108. G. A. Warburton, R. A. McDowell, K. T. Taylor, and J. R. Chapman, *Adv. Mass Spectrom. 8B*, 1953 (1980).

109. G. Moggi, *Proc. XV Int. Cont. Org. Coat.* Athens, p. 283 (July 1989).

110. F. Piacenti, L. Manganelli Del Fà, U. Matteoli, P. Tiano, and A. Scala, *Proc. VI Int. Cong. Deterioration and Conservation of Stone*, Torun, Poland, p. 881 (1988).

111. M. Visca and D. Lenti, Eur. Pat. Appl. EP 337,311; 337,312; 337,313 (to Ausimont) [*CA 112*, 11253s (1990)]; Ref. 110, p. 515.

112. Ref. 110, p. 492.

21

Perfluoropolyether Fluids (Demnum®) Based on Oxetanes

YOHNOSUKE OHSAKA

21.1. INTRODUCTION

Until the 1980s, perfluoropolyether (PFPE) oils had only limited, specialized uses, but since then their applications have expanded very rapidly. Major applications now include working fluids for vacuum pumps designed for semiconductor manufacturing and base oils for high-temperature lubricating greases.

This chapter describes the features of poly(hexafluorooxetane), a recently developed PFPE which differs from other PFPEs with respect to its structure and manufacturing process. Other types are described in Chapter 20.

21.2. PREPARATION

21.2.1. Monomers

The synthesis of poly(hexafluorooxetane) begins with 2,2,3,3-tetrafluorooxetane (1), which is prepared from tetrafluoroethylene and paraformaldehyde in anhydrous HF (Eq. 1).[1,2] The oxetane 1 and HF form an azeotrope from which the HF can be removed by washing with water. More practical purification methods for large-scale manufacture involve extracting the 1 · HF azeotrope with a halocarbon solvent, such as $CF_2ClCFCl_2$, followed by distillation,[3] and passing the distillate in the gas phase over alumina or silica.[4]

YOHNOSUKE OHSAKA • Daikin Industries, Ltd., Chemical Division, 1-1 Nishi Hitotsuya, Settsu-shi, Osaka 566, Japan. *Present address*: 1-16-5 Shirakawa, Ibaraki, Osaka, Japan.

Organofluorine Chemistry: Principles and Commercial Applications, edited by R. E. Banks *et al.* Plenum Press, New York, 1994.

(1) $\qquad$ $CF_2{=}CF_2 + (CH_2O)_n \xrightarrow{\text{HF}}$

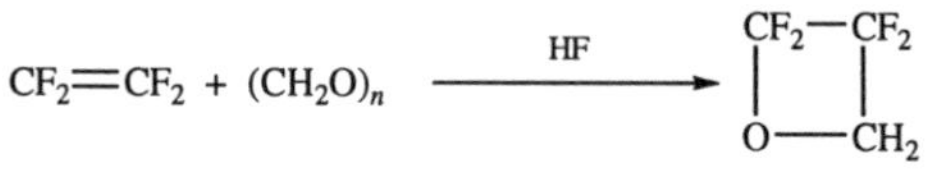

$$\begin{array}{cc} CF_2{-}CF_2 \\ | \quad\quad | \\ O{-\!-}CH_2 \end{array}$$
$$\mathbf{1}$$

The gas-phase oxidation of 1,1,2,2-tetrafluoropropane (Eq. 2) has been reported recently to be an alternative, high-yield preparation of $\mathbf{1}$.[5]

(2) $\qquad$ $HCF_2CF_2CH_3 + O_2 \xrightarrow{500\,°C} \mathbf{1} + H_2O$

21.2.2. Oligomers

The oxetane monomer $\mathbf{1}$ is treated with catalytic nucleophilic reagents in non-protic polar solvents such as poly(ethylene glycol) dimethyl ethers (glyme or diglyme, for example) to produce oligomers $\mathbf{2}$ (Eq. 3).[1,2] Alkali metal halides (KF, KBr, KI, CsF) are particularly useful catalysts. Lewis acids such as SbF_5 also can be employed to initiate ring-opening oligomerization in the absence of solvent.

(3) $\qquad$ $\mathbf{1} + X^- \rightarrow X{-}[CH_2CF_2CF_2O]_n{-\!-}CH_2CF_2COF$
$$\mathbf{2}\ (n = 2{-}200)$$

Oligomer $\mathbf{2}$ is then treated with fluorine gas to substitute fluorine for hydrogen atoms (Eq. 4).[1,2] The process can be accelerated by UV irradiation. Little decomposition occurs and the yield is over 90%, even though this direct fluorination is performed at temperatures above 100°C. The reaction proceeds in a stepwise manner, $({-}CH_2CF_2CF_2O{-\!-}) \rightarrow ({-}CHFCF_2CF_2O{-\!-}) \rightarrow ({-}CF_2CF_2CF_2O{-\!-})$, whereby the regularly increasing level of fluorine content likely stabilizes the oligomer toward the fragmentation processes commonly encountered in direct fluorinations.

(4) $\qquad$ ${-}[CH_2CF_2CF_2O]_n{-\!-} + F_2 \xrightarrow{100{-}120\,°C} {-}[CF_2CF_2CF_2O]_n{-\!-}$

The fluorination reaction is usually prolonged to insure that the unstable end groups in the oligomer are removed. The resulting PFPE oligomers ($\mathbf{3}$) are fractionally distilled to provide commercial products with the required physical properties for specific applications. They are sold by Daikin Industries, Ltd. under the "Demnum" tradename.

$$F{-}[CF_2CF_2CF_2O]_n{-\!-}CF_2CF_3$$
$$\mathbf{3}$$

21.3. PROPERTIES

21.3.1. General

The general properties of some commercial poly(hexafluorooxetane) products whose average molecular weights have been adjusted by fractional distillation are

Table 1. Typical Properties of Poly(hexafluorooxetane) (Demnum®)

Property		Grade			
		S-20	S-65	S-100	S-200
Average molecular weight		2700	4500	5600	8400
Kinematic viscosity (cSt) at	20°C	53	150	250	500
	40°C	25	65	100	200
	60°C	14	33	50	95
Viscosity index		150	180	200	210
Pour point (°C)		−75	−65	−60	−53
Density at 20°C (g cm^{-3})		1.860	1.873	1.878	1.894
Refractive index n_D^{20}		1.290	1.295	1.296	1.298
Surface tension (dyn cm^{-1})		17.7	18.0	18.5	19.1
T_g (°C)		−115	−111	−107	−104

given in Table 1. These products are all oils at room temperature and, among those listed, Demnum grades S-20 and S-65 are used mainly as vacuum pump oils while S-100 and S-200 are used in general lubricating applications.

The poly(hexafluorooxetane) materials are noncombustible and are thermally stable up to about 400°C. They withstand NaOH and H_2SO_4 up to 300°C, but are decomposed by strong Lewis acids at temperatures below 100°C. This lability, which is common to most PFPEs, is a disadvantage of the ether functionality (see p. 441).[6,7]

The Demnum PFPEs are insoluble in water and in almost all organic solvents. Their solubility is similar to that of CFC-113, $CF_2ClCFCl_2$.

21.3.2. Toxicity

The LD_{50} for Demnum PFPEs in mice is over 40 g kg^{-1} in acute toxicity tests.[2] This result, coupled with the finding that Demnum PFPEs do not accumulate in animal tissues or organs, indicates that these PFPEs are virtually nontoxic. Their volatile acid fluoride decomposition products, however, are highly toxic and adequate ventilation is required when these PFPEs are used under conditions where decomposition is likely.

21.3.3. Comparative PFPE Properties

In addition to poly(hexafluorooxetane), three other types of PFPEs are commercially available: —{ [CF(CF$_3$)CF$_2$O]$_m$—(CF$_2$O)$_n$}— (Fomblin®Y, Montedison), —[(CF$_2$CF$_2$O)$_p$—(CF$_2$O)$_q$]— (Fomblin®Z, Montedison), and — [CF(CF$_3$)CF$_2$O]$_n$— (Krytox®, Du Pont). Their viscosity characteristics differ somewhat, but otherwise they generally have similar physical properties.

Some properties of Demnum®, Krytox®, and Fomblin®Y oils of similar molecular weight are compared in Table 2. Demnum is distinguished by its (a) lower viscosity, (b) higher viscosity index (smaller change in viscosity with temperature), (c) lower

Table 2. Comparison of PFPE Properties

Chemical structure	$-[CF_2CF_2CF_2O]_n-$	CF_3 $\vert$ $-[CFCF_2O]_n-$	CF_3 $\vert$ $-[(CFCF_2O)_m(CF_2O)_n]-$
Average molecular weight	2700	2450	2500
Kinematic viscosity at 20°C (cSt)	53	85	140
Viscosity index	150	85	70
Pour point (°C)	−75	−50	−45
T_g (°C)	−115	−75	−72

pour point, and (d) lower glass-transition temperature. These features can be attributed to the linear Demnum structure versus the branched structures of the other two PFPEs.

The intrinsic thermal stabilities of the various PFPEs do not differ significantly,[8] but those containing —OCF$_2$O— acetal units are less chemically resistant, especially toward Lewis acids, metals, and metal oxides.[6,7,9,10]

21.4. APPLICATIONS

21.4.1. Base Oil for Greases

Demnum, with its good viscosity characteristics and outstanding thermal stability, can work very effectively as a base oil for greases over a wide range of temperatures from low to high. To best take advantage of Demnum's thermal and chemical resistance, fluoroplastics are the preferred thickening agents for the PFPE greases. Most commercial products currently available use poly(tetrafluoroethylene) for this purpose.[2,11] The characteristics of greases made from Demnum base oils with different average molecular weights (see Table 1) are compared in Table 3.

The performance of a grease depends on more than just the characteristics of its base oil, but it has been shown that Demnum offers several advantages in comparison with other PFPE base oils: (a) wider serviceable temperature range, (b) lower apparent viscosity, which provides the ability to work at lower torque and higher torque stability, and (c) smaller evaporation loss at high temperature.

21.4.2. Semiconductor Manufacturing

This application represents the largest market for PFPE oils. Demnum has been extensively evaluated as a working fluid for vacuum pumps used in the semiconductor manufacturing process.[2] Its good properties as a vacuum pump oil include (a) high average molecular weight and serviceability with low viscosity at room temperature and (b) high chemical resistance. These properties offer the following practical advantages: (a) low vapor pressure to permit high vacuum, (b) low operating temperature of pump, and (c) long service life of oil.

Table 3. Properties of Greases Made from Demnum

Property	Condition	Grade[a]		
		L65	L100	L200
Penetration	0 strokes	280 ± 15	280 ± 15	280 ± 15
	60 strokes	280 ± 15	280 ± 15	280 ± 15
Worked stability	10^4 strokes	260–290	260–290	260–290
	10^5 strokes	260–290	260–290	260–290
Apparenty viscosity (mPa s)	25°C, 300 s^{-1}	2500	3300	5300
Evaporation	200°C, 22 h	< 1 wt%	< 1 wt%	< 0.1 wt%
Oil separation	100°C, 30 h	< 3 wt%	< 3 wt%	< 2 wt%
	200°C, 30 h	< 12 wt%	< 11 wt%	< 10 wt%
	300°C, 30 h	—	< 17 wt%	< 16 wt%

[a]Appearance: All white, opaque.

21.4.3. Miscellaneous

Poly(hexafluorooxetane) can work effectively as a lubricant for pumps, valves, and compressors for oxygen and halogens. It also can be employed in various measuring devices used with corrosive liquids. The PFPE oils have been thoroughly evaluated in other diverse applications such as magnetic recording medium (mainly hard disk lubricants[12]), impregnating oils, magnetic fluid medium, and grease for high-temperature heating furnaces.

Perfluoropolyether oils and greases also have been used in high-vacuum systems and cleanrooms for various industries, including semiconductor manufacturing.

21.5. REFERENCES

1. Y. Ohsaka, T. Tozuka, and S. Takaki, U.S. Patent 4,845,268 (to Daikin Industries) [*CA 104*, 69315 (1986)].
2. Y. Ohsaka, *Petrotech (Tokyo) 8*, 840 (1985).
3. Y. Ohsaka, S. Takaki, and H. Sakai, European Patent 252,454 (to Daikin Industries) [*CA 108*, 188896 (1988)].
4. Y. Ohsaka and S. Takaki, European Patent 326,054 (to Daikin Industries) [*CA 112*, 98358 (1990)].
5. Y. Ohsaka and S. Khono, Japanese Patent 88-22,572 (to Daikin Industries) [*CA 109*, 93808 (1988)].
6. W. R. Jones, Jr., in: *New Directions in Lubrication, Materials, Wear, and Surface Interactions* (W. R. Loomis, ed.), pp. 402–437, Noyes Publication, Park Ridge, NJ (1985).
7. W. R. Jones, Jr., K. J. L. Paciorek, T. I. Ito, and R. H. Kratzer, *Ind. Eng. Chem., Prod. Res. Dev. 22*, 166 (1983).
8. L. S. Helmick and W. R. Jones, Jr., *NASA Tech. Memo*, NASA-TM-102493 (1990).
9. S. Mori and W. Morales, *Wear 132*, 111 (1989).
10. M. J. Zehe and O. D. Faut, *NASA Tech. Memo*, NASA-TM-101962 (1989).
11. S. Fukui, S. Shimasaki, and T. Tozuka, European Patent 180,996 (to Daikin Industries) [*CA 105*, 82018 (1986)].
12. J. F. Moulder, J. S. Hammond, and K. L. Smith, *Appl. Surf. Sci. 25*, 446 (1986).

22

Surface Fluorination of Polymers

MADHU ANAND, J. P. HOBBS, and IAN J. BRASS

22.1. INTRODUCTION

This chapter describes the existing commercial applications of direct fluorination of polymer surfaces and briefly discusses some emerging applications that have not yet been commercialized. To those familiar with the highly oxidizing and toxic nature of gaseous elemental fluorine, its acceptance by industries that have traditionally avoided the use of potentially hazardous substances may be surprising. This acceptance in industries such as blow-molding is a tribute to the engineers, materials scientists, and chemists who have worked to develop safe and reliable methods for the production, storage, transport, and application of this powerful fluorinating agent.

22.2. REACTIONS OF ELEMENTAL FLUORINE WITH ORGANIC POLYMERS

The violent nature of reactions between elemental fluorine and hydrocarbons is notorious (see Chapter 1). It arises as a consequence of the weakness of the F—F bond, the strength of the C—F bond, the stability of the HF leaving group, and the extremely low activation energy for hydrogen abstraction and fluorine substitution. The reaction of fluorine with organic polymers is so exothermic that it generally leads to fragmentation and charring of the substrate if the reaction is not controlled. Moderation of the reaction may be effected by (1) diffusion control in the gas phase arising from the presence of diluent gases, (2) thermal control of the gas or substrate, and (3) diffusion control in the solid phase. The last phenomenon can be achieved by creation of a

---

MADHU ANAND and J. P. HOBBS • Air Products and Chemicals, Inc., Allentown, Pennsylvania 18195-1501.   IAN J. BRASS • Air Products PLC, Walton-on-Thames, Surrey KT12 4RZ, England.
Organofluorine Chemistry: Principles and Commercial Applications, edited by R. E. Banks *et al.* Plenum Press, New York, 1994.

relatively passive surface layer produced by exposing the surface to a low partial pressure of fluorine.

Early studies examined the fluorination of simple polymers, such as polyethylene, to prepare perfluorinated materials.[1-6] Polymer surface fluorination proceeds via a free radical mechanism,[7,8] where fluorine abstracts hydrogen atoms from the hydrocarbon and fluorine atoms are substituted. A substituent fluorine deactivates a geminal hydrogen atom toward removal, so that the halogenation becomes progressively more difficult as it proceeds.[9,10]

The process can be illustrated as follows: $F_2 \rightarrow 2F\cdot$ (Initiation); $F\cdot + >CH_2 \rightarrow HF + >CH\cdot$, then $>CH\cdot + F_2 \rightarrow >CHF + F\cdot$ (hydrogen abstraction, fluorine addition repeating sequence or Propagation); $2F\cdot \rightarrow F_2$, $2>CH\cdot \rightarrow >CH—CH<$, or $>CH\cdot + F\cdot \rightarrow >CHF$ (Termination steps). In this simplified reaction scheme, initiation is attributed to the equilibrium dissociation of molecular fluorine; at room temperature and pressure the partial pressure of atomic fluorine is approximately 10^{-6} Pa. The initiation may be accelerated by the heat of reaction, by preheating the polymer, or by heat or radiation applied to the gas.[11] Clark *et al.* studied the surface fluorination of polyethylene and discussed possible reaction pathways and the probabilities for these reactions occurring.[12] A wide range of reaction products is predicted. These include partially fluorinated, perfluorinated, oxygenated (see below), and unfluorinated species that may be linear, branched, or crosslinked.

Following these early studies, fluorination of a wide range of polymers has been investigated.[1,2,4,13-15] In a large number of polymers, complete fluorination was often achieved by gradually increasing the partial pressure of fluorine to compensate for the progressively decreasing ease of hydrogen abstraction.

Several efforts have been made to model the kinetics of polymer surface fluorination reactions, although none has been completely successful in representing the chemical and mass transport aspects of the process. Shimada and Hoshino concluded from experiments on polycarbonate poly(methyl methacrylate), and polystyrene that at ambient temperatures the depth of fluorination is controlled by diffusion of gas into the polymer.[16] Al-Hussaini concluded that in surface fluorinated low-density polyethylene (LDPE) the reaction is rate-limited at low temperatures and mass transport or diffusion of the reacting gas limited at high temperatures.[17] This was based on the cross-sectional profile of the fluorine concentration against depth. Similar work has been done on LDPE by Corbin *et al.*[18] Volkmann and Widdecke also concluded that, across a wide temperature range, the fluorination rate is limited by the diffusion rate of molecular fluorine to the reaction front.[19] Anand and Hobbs have examined cross-sections of high-density polyethylene (HDPE) fluorinated at room temperature and 180 °C,[20] and found fluorine incorporation profiles consistent with those reported by Volkmann and Widdecke.[19]

It is difficult to draw broad conclusions on the overall kinetics from the results reported above. Clearly the amount of fluorine incorporated into a cross-section of the polymer surface is entirely dependent on the flux of fluorine into the polymer. The flux of fluorine into the polymer surface, and the consequent profile of the incorporated fluorine, will depend on the temperature, the thermal history of the substrate, and the diffusivities and solubilities of the reactants in the range of partially fluorinated compositions present. The fluorine flux will also depend on the rate of fluorine removal from the concentration gradient by reaction. All of these factors are changing dramatically throughout the course of the fluorination process. As a result of these many factors, the concentration of incorporated fluorine through a cross-section of the polymer surface may decrease steadily with depth or it may approximate to a step function.

Even though the kinetics of surface fluorination reactions have not been elucidated fully, general methods to control the reactions are well understood: specifically, increased extent and depth of fluorination may be achieved by increasing the fluorine partial pressure, the substrate temperature, or the reaction time. The exact conditions used will depend on the nature of the polymer in question and the surface properties required.

Fluorine reactions with polymer substrates may be further complicated by the presence of additives or dissolved gases in the polymer or by reactive contaminants in the fluorine. Several researchers have investigated the role of oxygen in the surface fluorination of polymers.[21-23] The key reactions are thought to be: $>CH\cdot + O_2 \rightarrow >CHOO\cdot$; $>CHOO\cdot + >CH_2 \rightarrow >CHOOH + >CH\cdot$; $2CHOO\cdot \rightarrow >CHOOCH< + O_2$; $>CHOO\cdot + >CH\cdot \rightarrow >CHOOCH<$; $F\cdot + O_2 \rightarrow FOO\cdot$; $FOO\cdot + -CH_2-CH_2- \rightarrow -C(O)F + -CH_2 + H_2O$. Hence the presence of oxygen in the reaction environment results in the incorporation of oxygen-containing groups in the polymer backbone, which tend to impart an acidic character to the surface. The presence of oxygen also results in a reduction in the overall depth and degree of fluorine incorporation. Oxygen incorporation can beneficially alter the polymer surface by generating functional groups required for specific applications such as adhesion (see below).[24] Coreactant gases such as sulfur dioxide, sulfur trioxide, and chlorine have been used to incorporate other moieties into polymer surfaces to obtain novel properties. This area has not been explored fully and remains fertile for further developments.

In summary, the surface fluorination of polymers can result in surfaces having a wide range of fluorine content, and differing cross-sectional profiles for any given incorporated fluorine content. The required surface properties may be obtained by controlling parameters such as the fluorine partial pressure and dilution, reaction time, temperature, and the presence of coreactants. This control has been successfully exercised in the production of surfaces for many specific applications, which will be discussed in the following sections.

22.3. PROPERTIES AND APPLICATIONS OF SURFACE-FLUORINATED POLYMERS

22.3.1. Barrier Properties

The most commercially significant use of polymer surface fluorination is the creation of barriers to hydrocarbon permeation. The effectiveness of such barriers has never been in doubt, reductions in permeation rates by two orders of magnitude having been observed routinely on thinner HDPE substrates. The origin of the effect, however, has been the subject of much discussion. The barrier is generally believed to arise from at least three cooperative effects.

First, the application of elemental fluorine to polyolefin surfaces generally raises their surface energy. This differs from plasma-fluorinated polyolefins and polymers made from fluorine-containing monomers such as PVF (—CHF—CH$_2$— monomer unit), PVdF (—CF$_2$—CH$_2$—), and PTFE (—CF$_2$—CF$_2$—), that exhibit a general trend toward lower surface energies with increasing fluorine incorporation.[25] This increase in the surface energy can be demonstrated by an increasing water-wettability of the fluorinated surfaces. It is evidence of an increase in the effective solubility parameter of the modified polymer, which in turn reduces the solubility of limited-polarity organic liquids in that material.[26] While this behavior can often be attributed to the incorporation of contaminant oxygen, which causes marked increases in both the dispersive and polar components of the surface energy, the authors have also observed increasing surface energies on very lightly fluorinated surfaces in the absence of oxygen incorporation.

The second effect of fluorination is to reduce the free volume of the polymer. Interestingly, for polyolefins, while the free volume is reduced, there is also a 34% to 44% decrease in carbon atom density when the fluorination is carried to completion. The reduction in free volume is accompanied by a further reduction in solubility.

The crosslinking effect of fluorine exposure provides the third contribution to the barrier. Crosslinked polymers resist much of the swelling that characterizes exposure to hydrocarbon liquids. Swelling is comparable with plasticization, which is well known to increase diffusivities of larger permeant molecules. In addition, crosslinking directly affects diffusivity by restricting the freedom of motion available to polymer segments. Depending on the fluorination conditions, significant crosslinking of the polymer beneath the heavily fluorinated surface layer may also occur. This lightly fluorinated, crosslinked material also contributes to the permeation barrier.

The effects of incorporation of contaminant or adjuvant gases, particularly oxygen, can be complex. The positive or negative effects on the barrier depend upon a number of factors; however, the nature of the permeant is often dominant. It has been found that the barrier to a nonpolar hydrocarbon permeant can be improved through the incorporation of oxygen, although some loss in long-term barrier stability may occur.[27,28] Oxygen incorporation can have a devastating impact on barrier performance when polar components, such as alcohols or surfactants, are included in the hydrocarbon permeant.[29]

An economically attractive improvement in hydrocarbon barrier has been demonstrated on HDPE. Fluorinated blow-molded HDPE containers have found widespread application in the packaging of industrial and consumer chemicals such as mineral spirits and creosote. One should anticipate growth for similar packaging applications given the increasing governmental focus on hydrocarbon emissions and air pollution. Fluorinated containers are also widely used for the packaging of agricultural chemicals, where the product may benefit from improved retention of active ingredients, in addition to a reduction in solvent permeation.

Similarly, in food contact applications the fluorinated barrier layer can prevent the loss of flavor and fragrance components to the surroundings or to the container wall through absorption.[30,31] The barrier may also prevent the migration of unwanted flavors or odors from the plastic into the food.

Some applications of fluorinated containers depend upon improved retention of the container shape. In many of the above applications, and also in the packaging of hand cleaners, the most visible evidence of a permeation problem is distortion of the container. Hobbs *et al.* have shown that this "paneling" can be reduced or eliminated by the incorporation of a fluorinated barrier layer.[32]

Finally, an area where fluorinated barriers is increasingly important is in manufacturing automotive plastic fuel tanks (PFTs) from HDPE. Initially, the goal was elimination of objectionable odors entering the passenger compartment of the vehicle from the PFT. Now, especially in the U.S. and Germany, the primary reason for fluorinating PFTs is to comply with increasingly stringent statutory efforts to reduce air pollution: (1) hydrocarbon emission limits for vehicles sold in California are scheduled to be reduced by more than an order of magnitude starting in 1996; and (2) the American Environmental Protection Agency has mandated the use of oxygenated (alcohol and MTBE containing) fuels during the winter months in many localities to reduce carbon monoxide pollution. Such regulatory actions are stimulating further research into the production of PFTs with surfaces modified by fluorination to have enhanced barriers. Attainment of the 0.1 gram/day emission standard targeted by most automotive manufacturers will be impossible for a PFT without such a barrier.

22.3.2. Adhesion and Surface Energy

Although high surface energy and adhesion are by no means synonymous, or even necessary and sufficient conditions each for the other, they are related enough to justify their being dealt with together. When a polymer is exposed to elemental fluorine, particularly in combination with one or more adjuvant gases, the surface energy of the polymer almost invariably increases. Surface energies as high as 72 mN m^{-1} have been reported immediately following exposure to fluorine-containing gases,[26] although stable levels in the range 45 to 55 mN m^{-1} are perhaps more representative. Several ways to exploit this effect have been suggested. For example, Dixon and Hayes proposed that synthetic resins treated with mixtures of fluorine, oxygen, and sulfur dioxide find application for their enhanced water-wicking and moisture transport properties.[33] However, the effect on adhesive and coating properties has provoked the greatest interest in surface energy enhancement by fluorine.

Theories explaining how surface fluorination promotes adhesion are diverse and numerous. Crosslinking of the surface is believed to play a significant role, particularly on polyolefins, by eliminating oligomeric material and additives that contribute to a weak boundary layer. Crosslinking will also impede further "blooming" of these materials to the surface. Surface microetching may also play a role by increasing the adhesive bond strength through mechanical interlocking or by an increase in the surface area available for bonding.

Incorporating surface functionalities certainly plays a role in most if not all instances of improved adhesion. As discussed earlier (Section 22.2), surface functionalities arise from: (1) the partial substitution of fluorine for hydrogen; (2) free-radical-initiated chain scission or rearrangement; and (3) the fluorine-assisted incorporation of oxygen or other contaminant or adjuvant gas(es). The interactions of the functionalized polymer surface and the adhesive or matrix material may be expressed in terms of: (1) improved wetting; (2) polar–polar interactions; (3) dispersive interactions promoted by the incorporation of polarizable groups such as carbonyl; and (4) the formation of covalent bonds. The role of surface functionalities may also be explained by the general theory of surface acid–base interactions proposed by Fowkes.[34]

The improvements in adhesion achieved through fluorination can be significant and are being explored on a commercial scale. Proposed applications include: (1) enhancement of the paint receptivity of molded plastic articles for automotive and other applications; (2) enhanced adhesion between polyester yarn, cord, or fabric and rubber as might be employed in the construction of automobile tires[35]; (3) increased resistance to delamination in coated flexible films[36]; (4) improved strength in fiber-reinforced composite parts through better matrix-to-fiber energy transfer; and (5) increased coating integrity on molded thermoplastic containers.[37]

One application that has achieved commercial acceptance is the functionalization of ultrahigh-molecular-weight polyethylene (UHMWPE) particles. PRIMAX® powders, which are of this type, were recognized by Industrial Research Magazine in the U.S. as one of the hundred most significant new technical products introduced in 1989. Incorporating these UHMWPE powders into paints can reduce friction and improve abrasion resistance.[38] Composites with thermoset polymers, such as polyurethanes, have shown wear resistance superior to the already exceptional performance of UHMWPE alone.[39]

22.3.3. Other Properties of Fluorinated Surfaces

Fluorination can enhance the gas-barrier properties of polymers,[40,41] but the improvements have not been found adequate for food packaging. It is somewhat paradoxical that in the field of gas separation, surface fluorination can induce commercially significant selectivity effects.[42] Selectivity, the preferential permeation of one component of a mixture over another, is observed for multicomponent permeation through all nonporous polymeric membranes. The application of surface-fluorinated polymers to effect liquid-phase separations by selective permeation has not been investigated.

Tribological changes can be achieved by means of surface fluorination. Although the typical effect of fluorination on polymers is to increase the coefficient of friction, a reduction in stick–slip behavior results when the technique is applied to materials such as butyl, halobutyl, and isoprene rubbers. This has been used to ease the handling of a variety of artifacts including medical disposables.

Fluorination has also been found to: (1) reduce the refractive indices of a number of polymers of interest in optical applications, (2) increase the resistance of polyolefins to chemical attack in acidic media, and (3) reduce particle shedding from HDPE containers. The latter two points are useful in the packaging of high-purity liquids used in semiconductor fabrication.

To date, the commercial exploitation of most of these properties of surface-fluorinated polymers has been limited.

22.4 TECHNIQUES FOR THE DIRECT FLUORINATION OF POLYMER SURFACES

Elemental fluorine may be applied to the surfaces of polymeric materials by two main methods: (1) post-forming exposure ("post-treatment"); (2) simultaneous forming and exposure (*in situ* treatment). These methods differ not only in terms of principle of operation but also in terms of economics, flexibility, and applicability and may differ in terms of the ultimate properties of the product.

22.4.1. Post-Treatment

The advantage of applying fluorine-containing gases to polymer substrates after completion of the forming process is that it permits separation of the forming and treatment processes, with resulting improvement in operational flexibility and better economics for smaller production runs. The disadvantage is usually a significant inventory of partially processed material, and higher operating costs. Depending on the physical form of the substrate, the fluorination reactor may fit a number of general descriptions.

Reactors for the treatment of continuous webs of film or fiber consist of closed vessels fitted with substrate entry and exit ports, which are in turn fitted with means to exclude entrained air and retain the reactive atmosphere (see Figure 1).[36,43] Fluorine-containing gas is continuously injected into the reactor and gaseous byproducts are removed. Clearly, given the toxic nature of the gases in question, steps must be

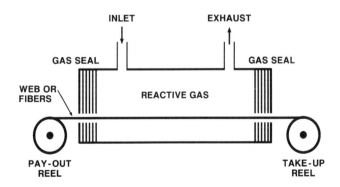

Figure 1. Post-treatment of films or fibers.

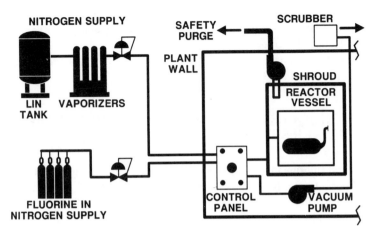

Figure 2. Schematic diagram of a post-treatment system.

taken to avoid worker exposure during start-up when the substrate must first be threaded through the reactor. Such reactors may be employed to improve the adhesion of inks, adhesives, and other coatings to polymeric films and sheets and to increase the bond strengths obtainable between reinforcing fibers and matrix materials in polymer composites.[35]

Reactors for treatment of articles that can be conveniently handled individually or on pallets operate in batch mode and consist of a suitably sized vacuum chamber provided with means of evacuation and gas injection (see Figure 2). They may also be fitted with systems for reactive atmosphere analysis,[44] heating and recycle,[45] pressure pulsing and multiple chambers.[46] Generally, reactors of this type operate at subambient pressures, with the desired fluorine concentration(s) being obtained by pressure blending a fluorine-containing gas with a diluent within the reactor. Infiltration of air and the degree of gas mixing are often critical factors to the treatment quality. Cycle times of up to one hour have been reported.

Reactors of this type are employed to impart enhanced hydrocarbon barriers to HDPE containers such as automotive fuel tanks and containers for hydrocarbon solvent-based products, as described in Section 22.3.1. With correct choice of gas composition they may also be used to provide enhanced adhesion and paint receptivity on parts such as molded automotive parts, as described in Section 22.3.2.

Reactors treating small articles or powders are generally constructed in the form of a rotating vacuum vessel. Operating in batch mode, these reactors also require the use of vacuum pumping equipment at the beginning and end of the process cycle. Examples of the commercial use of such equipment include modification of the frictional characteristics of medical rubber articles and enhancement of the compatibility of ground polymeric material in thermosetting matrices,[39] as also described in Section 22.3.2.

An area with potential is the application of a liquid or a high-density vapor rather than nitrogen or air as the diluent for the fluorine.[47,48] The higher thermal capacitance of such a carrier can assist in controlling the reaction.

22.4.2. *In Situ* Treatment

The principle underlying *in situ* treatment is that one uses existing polymer-processing equipment to apply the desired fluorine-containing gas to the polymer in question. Of course, the polymer-processing equipment and the operating cycle may require modification and additional equipment will be necessary, but the intention is to take advantage of some or all of the following: (1) reduced capital cost, particularly at higher production levels; (2) reduced product handling; (3) reduced equipment "footprint"; (4) access to polymer in the melt with consequent reduction in required exposure time and fluorine concentration.

Several reductions to practice of *in situ* treatment have been proposed. Of these, only one, the Airopak® process[27] developed by Air Products and Chemicals, Inc., has achieved commercial significance. Operation of this process involves replacing the air, or nitrogen, normally employed to form the product in the blow-molding process by a dilute mixture of fluorine in nitrogen. The purpose of the substitution is to create on the inside surface of the formed container a layer resistant to permeation of hydrocarbon liquids. This has been found useful in automotive and packaging applications, as described in Section 22.3.1. above.

The products of the Airopak® process differ to a degree from those of processes depending on post-treatment for the creation of a hydrocarbon barrier. Airopak®-fluorinated surface material is both amorphous and more crosslinked because the process exposes the molten or only partially solidified polymer to fluorine. This differs significantly from the alternative process where the already-formed surface crystalline regions are generally more resistant to the action of fluorine than the surrounding amorphous matrix.

A simplified representation of the equipment required for the operation of the Airopak® process is shown in Figure 3. The major equipment difference between the *in situ* and post-treatment process systems is that: (1) the *in situ* systems invariably use a large supply reservoir and a blender system, because of the quantities of parts being made and the need to dilute the fluorine from 10% or 20% to the lower concentrations required for treatment; and (2) the post-treatment systems, which can operate directly from the supply concentration, require separate vacuum vessels and pumps.

In order to comply with the more stringent emission standards being adopted for automotive fuel tanks, it has been necessary to go beyond optimizing the current technologies and develop a new generation of fluorination processes. One of these, the Airoguard™ Process,[49] developed by Air Products and Chemicals, Inc., has been shown to produce barriers comparable to those achievable with multilayer PFTs and approaching those attainable from metal fuel tanks.

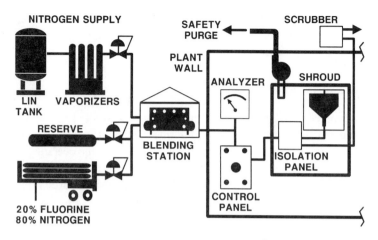

Figure 3. Schematic diagram of an Airopak® system.

Further ideas based on the principle of *in situ* treatment that have not yet become commercial realities include: (1) passage of a fluorine-containing gas through the bubble in the film-blowing process to improve the resulting film's printability, and adhesion[50]; (2) prefilling the mold cavity of an injection molding machine with a fluorine-containing gas to increase the surface energy of the resulting molding or to influence the point of failure when the molding is placed under stress[51,52]; (3) passage of fluorine-containing gas into the bore of an extruding polymer tube to create a hose with a barrier layer on the inside surface.[53,54]

22.5. SAFETY AND ENVIRONMENTAL CONCERNS

Those wishing to embark on the use of elemental fluorine in experimental investigations or commercial ventures are advised to consult with appropriate fluorine or equipment suppliers.

Elemental fluorine, even when diluted with an inert gas, is toxic and corrosive: exposure to it can be fatal! Its sharp penetrating odor provides a useful warning of an acutely toxic atmosphere, but instrumental detection methods are recommended. Manifestations of overexposure include irritation or burns to the eyes, skin, or respiratory tract. Prompt medical attention is indicated in all cases of overexposure. Systemic absorption of fluorides associated with exposures in excess of permissible levels over the course of many years may lead to fluorosis, a disease of the bones and connective tissues.

The correct choice of materials of construction and appropriate cleaning, preparation, and maintenance of those materials are vital for prevention of fluorine fires. Fires involving elemental fluorine as oxidant generally cannot be extinguished by the application of water or chemicals. These may act as additional fuels. The correct response to a fluorine fire is to isolate the source of the fluorine and to tackle any subsequent conventional fire by conventional means. Byproducts of fluorine reactions, such as fires, may include substances that are reactive, toxic, or corrosive in their own right.

Also, waste gas streams from fluorine processes must be scrubbed prior to release into the environment. Commercial fluorination systems generally rely on the use of either caustic solution scrubbers (potassium hydroxide is preferred to sodium hydroxide) or solid scrubbers containing calcium carbonate or alumina chips. The disposal of the fluoride wastes from these scrubbers must be carried out in an environmentally sound manner.

Careful consideration of the safety and environmental implications of system design and operation enable fluorine to be tamed. Neumark and Siegmund have provided a useful first reference,[55] but again those wishing to embark on the use of elemental fluorine in experimental investigations or commercial ventures are advised to consult with appropriate fluorine or equipment suppliers.

22.6. CONCLUSIONS

Clearly, the effects of exposing polymer surfaces to fluorine are numerous. Not all of these effects have made the transition from laboratory to industrial practice. The Airopak® process and the post-treatment of HDPE fuel tanks and containers are commercial. The post-treatment of UHMWPE powders, polymer films, and rubber parts for adhesion, printability, and frictional modification, respectively, are the current commercial examples known to the authors. Fluorine will remain a highly effective tool for polymer surface modification, but its commercial use will be limited to the small number of applications where the value of the modified product can justify the cost.

22.7. REFERENCES

1. A. J. Rudge, British Patent 710,523 to ICI [*CA 49*, 2778c (1955)].
2. H. Schonhorn, P. K. Gallagher, J. P. Luongo, and F. J. Padden, *Macromolecules 3*, 800 (1970).
3. S. P. Joffre, U.S. Patent 2,811,468 to Shulton Inc. [*CA 31*, 2453g (1958)].
4. R. J. Lagow and J. L. Margrave, *J. Polym. Sci., Polym. Lett. Ed. 12*, 177 (1974).
5. D. W. Brown, R. E. Florin, and L. A. Wall, *Appl. Polym. Symp. 22*, 169 (1973).
6. K. Tanner, *Chimia 22*, 197 (1968).
7. W. T. Miller and A. L. Dittman, *J. Am. Chem. Soc. 78*, 2793 (1956).
8. L. A. Bigelow, *Chem. Rev. 40*, 51 (1947).
9. P. S. Fredricks and J. M. Tedder, *J. Chem. Soc.* 144 (1960).
10. J. M. Tedder, *Adv. Fluorine Chem. 2*, 104 (1961).
11. G. A. Corbin, R. E. Cohen, and R. F. Baddour, *J. Appl. Polym. Sci. 30*, 1407 (1985).
12. D. T. Clark, W. J. Feast, W. K. R. Musgrave, and I. Ritchie, *J. Polym. Sci., Polym. Chem. Ed. 13*, 857 (1975).
13. G. Jorgensen and P. Schissel, *Sol. Energy Mater. 12*, 491 (1985).
14. H. Shinohara, M. Iwasaki, S. Tsujimura, K. Watanabe, and S. Okazaki, *J. Polym. Sci., Part A-1 10*, 2129 (1972).

15. G. E. Gerhardt, E. T. Dumitru, and R. J. Lagow, *J. Polym. Sci., Polym. Chem. Ed. 18*, 157 (1979).

16. J. Shimada and M. Hoshino, *J. Appl. Polym. Sci. 19*, 1439 (1975).

17. H. S. A. Al-Hussaini, *Penetrant Permeation through Fluorinated and Untreated Polyethylene Films*, Thesis submitted to North Carolina State University, Dept. of Chemical Engineering (1983).

18. G. A. Corbin, R. E. Cohen, and R. F. Baddour, *Polymer 23*, 1546 (1982).

19. T. Volkmann and H. Widdecke, *Makromol. Chem., Macromol. Symp. 25*, 243 (1989).

20. J. P. Hobbs and M. Anand, Air Products and Chemicals, Inc., unpublished results (1989).

21. R. E. Florin and L. A. Wall, *J. Chem. Phys. 57*, 1791 (1972).

22. J. L. Adcock, S. Inoue, and R. J. Lagow, *J. Am. Chem. Soc. 100*, 1948 (1978).

23. T. Volkmann and H. Widdecke, *Kunststoffe 79*, 743 (1989).

24. H. Schonhorn and R. H. Hansen, *J. Appl. Polym. Sci. 12*, 1231 (1968).

25. D. G. Rance, in: *Industrial Adhesion Problems* (D. M. Brewis and D. Briggs, eds.), pp. 48–86, Orbital Press, Oxford (1985).

26. W. J. Koros, V. T. Stannett, and H. B. Hopfenberg, *Polym. Eng. Sci. 22*, 738 (1982).

27. D. D. Dixon, D. G. Manly, and G. W. Recktenwald, U.S. Patent 3,862,284 to Air Products and Chemicals [*CA 82*, 157405 (1975)].

28. M. Eschwey, R. VonBonn, and H. Neumann, U.S. Patent 5,073,231 to M. G. Ind. [*CA 113*, 116739 (1990)].

29. J. P. Hobbs and M. Anand, *Eng. Plastics 5*, 247 (1992).

30. C. J. Farrell, *Ind. Eng. Chem. Res. 27*, 1946 (1988).

31. M. Anand, in: *Proceedings of the 4th Annual High Performance Blow Molding Technical Conference*, October 16–17 (1989).

32. J. P. Hobbs, M. Anand, and B. A. Campion, in: *Barrier Polymers and Structures* (W. J. Koros, ed.), *ACS Symp. Ser. 423*, Chapter 15 (1989).

33. D. D. Dixon and L. J. Hayes, U.S. Patent 3,940,520 to Air Products and Chemicals Inc. [Derwent WPI No. 75-37578W/23 (1976)].

34. F. M. Fowkes, "Acid–Base Interactions", in: *Encyclopedia of Polymer Science and Engineering* (J. Kroschwitz, ed.), Suppl. Vol., pp. 1–11, Wiley, New York (1990).

35. D. D. Dixon and W. M. Smith Jnr., U.S. Patent 4,009,304 to Air Products and Chemicals Inc. [*CA 85*, 19391 (1976)].

36. R. Milker and A. Koch, *Adhesion 6*, 33 (1989).

37. B. D. Bauman, R. K. Mehta, and M. A. Williams, U.S. Patent 4,764,405 to Air Products and Chemicals Inc. [*CA 109*, 232831 (1988)].

38. B. D. Bauman, U.S. Patent 4,972,030 to Air Products and Chemicals Inc. [*CA 113*, 117135 (1990)].

39. B. D. Bauman, U.S. Patent 4,880,879 to Air Products and Chemicals Inc. [*CA 112*, 159676 (1990)].

40. V. C. McGinniss and F. A. Sliemers, U.S. Patent 4,491,653 to Battelle Development Corporation [*CA 102*, 114526 (1985)].

41. C. L. Kiplinger, D. F. Persico, R. J. Lagow, and D. R. Paul, *J. Appl. Polym. Sci. 31*, 2617 (1986).

42. M. Langsam, U.S. Patent 4,657,564 to Air Products and Chemicals Inc. [*CA 107*, 116637 (1987)].

43. T. R. Bierschenk, U.S. Patent 4,743,419 to Dow Chemical Co. [*CA 109*, 94370 (1988)].

44. M. Eschwey and R. van Bonn, U.S. Patent 4,701,290 to Messer Griesheim GmbH [*CA 106*, 19466 (1987)].

45. G. Tarancon, U.S. Patent 4,484,954 to Union Carbide Corp. [*CA 101*, 8179 (1984)].

46. G. Tarancon, E. Acevedo, and A. Saud, U.S. Patent 4,576,837 to Tarancon Corporation [*CA 104*, 226007 (1986)].

47. C. Bleifert, H. M. Boldhaus, and M. Hoffmann, U.S Patent 4,536,266 to Hewing GmbH [*CA 101*, 111606 (1984)].

48. G. Tarancon, U.S. Patent 4,994,308 to Tarancon Corporation [*CA 115*, 9684 (1991)].

49. J. P. Hobbs, Patent Pending to Air Products and Chemicals, Inc.

50. C. Matsushita, S. Tsukamoto, and Y. Odama, Japanese Patent Application 281965 (1985), 62/140281-A, 91/060304-B to Toyo Soda Mfg. KK. [Derwent No. 87-21607/31 (1987)].

51. M. A. Williams, B. D. Bauman, D. R. Ruprecht, and P. D. Marsh, U.S. Patent 4,752,428, to Air Products and Chemicals, Inc., [Derwent WPI No. 88-190273/27 (1988)].

52. B. D. Bauman, D. R. Ruprecht, P. D. Marsh, and M. A. Williams, U.S. Patent 4,800,053, to Air Products and Chemicals, Inc. [Derwent WPI No. 89-046761/06 (1989)].
53. H.-D. Finke, B. Hoeffker, and M. E. Hoffmann, European Patent 0267441 A2 to Hewing GmbH [*CA 110*, 25015 (1989)].
54. K. Olbrich, German Patent Application DE 3820254 to Stramax AG [Derwent WPI No. 89-333544 (1989)].
55. H. R. Neumark and J. M. Siegmund, in: *Kirk-Othmer Encyclopedia of Chemical Technology* (2nd edition), Vol. 9, pp. 506–526, Wiley, New York (1963).

23

Fluorinated Carbon

GEORGE A. SHIA and GANPAT MANI

23.1. INTRODUCTION

Fluorinated carbon or graphite fluoride (CAS# 51311-16-2) is a nonstoichiometric solid made by the reaction of carbon with fluorine.[*] Its empirical formula is CF_x, where x can vary between almost 0 and about 1.3 and depends on the crystallinity of the starting carbon, reaction temperature, and time. Materials having x-values less than about 0.3 are composed of fluorinated and virgin carbon.[1]

The material was first prepared by Ruff *et al.* in 1934,[2] and in some later work by the Rüdorffs compounds ranging from $CF_{0.676}$ to $CF_{0.988}$ were obtained.[3] The latter materials were prepared by exposing graphite to fluorine at temperatures between 410 and 550°C. Independently, $CF_{1.04}$ was prepared by Palin and Wadsworth using similar conditions.[4]

Commercial interest in these materials was awakened in the late 1960s when applications utilizing their unique properties began to be developed. Today, fluorinated carbon is a specialty product used in lithium batteries and lubricants. Other applications are under development.

There are three producers of fluorinated carbons. Allied-Signal Inc. in the United States has six types of material sold under the Accufluor® CF_x trademark: two fluorinated petroleum cokes, one fluorinated graphite, and three grades of fluorinated carbon black. The Central Glass Co., Ltd. in Japan produces three types of material under the Cefbon® trademark; and Daikin Industries, Ltd., also in Japan, produces three types of fluorinated carbon. Worldwide production capacity for all products is estimated to be 50 to 75 tonnes per annum.

---

[*]Such solids are usually referred to as *graphite fluoride*, $(CF_x)_n$, in the Japanese literature, regardless of which starting carbon is used to make the fluorinated material. We use the term *fluorinated carbon* to encompass materials made from all forms of carbon.

GEORGE A. SHIA • Allied-Signal Inc., Fluorine Products Division, Buffalo Research Laboratory, Buffalo, New York 14210. GANPAT MANI • Allied-Signal Inc., Fluorine Products Division, Morristown, New Jersey 07962-1139.

Organofluorine Chemistry: Principles and Commercial Applications, edited by R. E. Banks *et al.* Plenum Press, New York, 1994.

23.2. PHYSICAL PROPERTIES

By fluorinating different carbons, and by fluorinating the same carbon to different degrees, products with differing physical and chemical properties can be made. Table 1 contains the physical properties of five commercially available fluorinated carbons. Some physical properties, such as electrical conductivity and color, depend on the fluorine content of the material; other properties, such as decomposition temperature, are much more a function of the nature of the starting carbon. By choosing the right starting material, a CF_x can sometimes be customized to suit a particular application.

While the various fluorinated carbons are approximately pseudomorphic with their respective starting carbons, the insertion of fluorine into the structure causes substantial buckling of the carbon rings, separation of the carbon layers, and a decrease in crystallinity. These changes are exemplified in Figures 1 and 2, which provide a comparison of virgin and fluorinated carbons, and are also apparent in other properties. For example, the interlayer spacing is roughly doubled and the surface area may be increased more than 10 times in going from virgin carbon to CF_1 (the interlayer spacing, d_{002}, and surface area of a typical coke are 3.3 Å and 10 m^2 g^{-1}, respectively, while for CF_1 made from that coke the same parameters are 7.1 Å and 130 m^2 g^{-1}). The structure of this nonstoichiometric material has not been definitely identified because large single crystals of CF_x cannot be made. It is generally accepted that the fluorine is covalently bonded to the carbons, and that the carbons are arranged in cyclohexane "chairs" with alternating fluorines above and below the carbon plane.[5,6] The aromaticity of the virgin carbon is lost, explaining the very poor electrical conductivity of CF_1. Another structure ("boat" conformation) has been proposed,[2,7] but is inconsistent with all the experimental data. For $CF_{x<1}$ the structure is even less

Table 1. Physical Properties of Some Commercially Available Fluorinated Carbons[a,b]

	Carbon source				
	graphite	petroleum coke	carbon black	carbon black	carbon black
Fluorine content (wt%)	62	62	65	28	10
X-Value	1.03	1.03	1.17	0.25	0.07
True density (g cm^{-3})	2.6	2.7	2.5	2.1	1
Bulk density (g cm^{-3})	0.6	0.6	0.1	0.1	0.09
Median particle size (µm)	5	8	<1	<1	<1
Surface area (m^2/g^{-1})	250	130	340	130	170
Decomposition temperature (°C)	600	630	500	450	380
Thermal conductivity (cal/cm^{-1} sec^{-1} °C–1)	N.A.	10^{-3}	10^{-3}	10^{-3}	N.A.
Resistivity (ohm-cm)	$>10^{10}$	10^{11}	10^{11}	10^8	<10
Color	white	gray	white	black	black

[a]Data Sources: CEFBON® Graphite Fluoride Technical Information, Central Glass Co., Ltd. (graphite based material); ACCUFLUOR® CF_x Technical Brochure, Allied-Signal Inc. (1985) (all other materials).
[b]N.A.—not available.

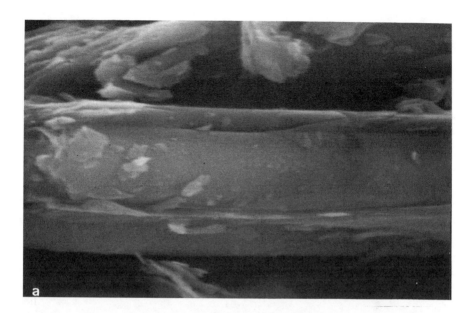

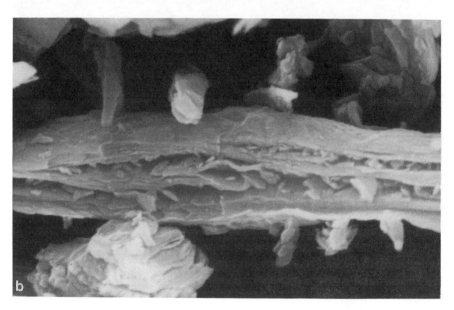

Figure 1. (a) Scanning electron micrograph of virgin petroleum coke, on-edge view. (b) Scanning electron micrograph of fluorinated petroleum coke (62 wt% F), on-edge view. Note the dramatic change in morphology as compared to (a). The introduction of fluorine into this graphitic material causes a marked increase in the interlayer spacing between the carbon layers and much greater surface area and porosity. × 10,000. Scale bar = 1 μm.

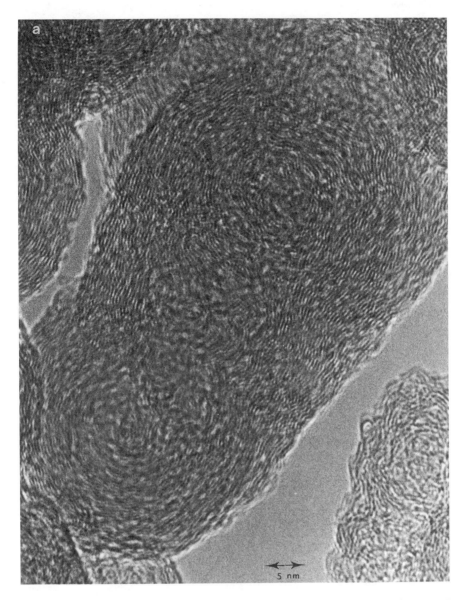

Figure 2. (a) Transmission electron micrograph of virgin carbon black. Note the roughly concentric arrangement of ordered, graphitic regions. (b) Transmission electron micrograph of fluorinated carbon black, fluorine content 65 wt%. Much of the order seen in (a) is no longer observed.

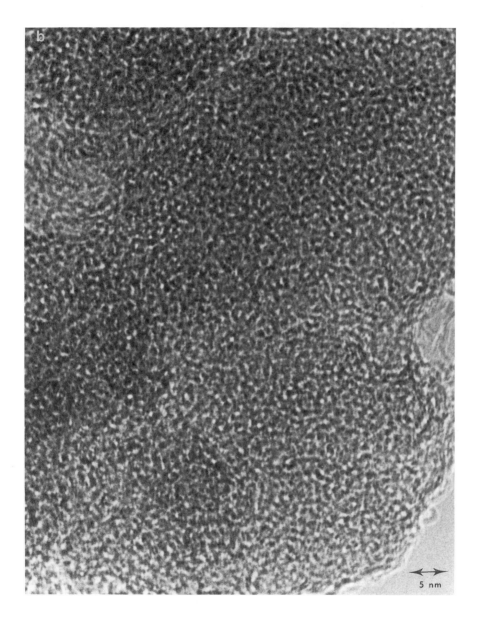

Figure 2.   (*Continued*)

clear. A distinct phase, C_2F, has been proposed by several workers, but this material has been identified only in samples made from highly crystalline graphite.[6,8] It seems very likely that all fluorinated carbons are nonhomogeneous and are composed of CF_x mixtures of varying stoichiometries.

The surfaces of fluorinated carbon particles have some unique properties. Even materials having carbon-to-fluorine ratios substantially less than 1 generally have highly fluorinated surfaces.[1,8,9] Furthermore, the carbons along the edges of the graphite planes are bound to only two other carbons, leading to the formation of CF_2 groups on the surface.[9] Therefore, the reported very low values for the surface energy of this material are not surprising.[8] The highly fluorinated surface also gives rise to some of the other useful properties of fluorinated carbon: very marked hydrophobicity, excellent lubricity, and interesting triboelectricity.

23.3. CHEMICAL PROPERTIES

CF_x is generally stable in the presence of oxidizing agents, and inorganic and organic acids, and reacts slowly to moderately fast with reducing agents, bases, and amines. In the latter case, these reactions may be catalyzed or accelerated by UV light.[8,10] The material is compatible with many engineering resins except for polyacetal, which readily decomposes in its presence at typical processing temperatures. A guide to the stability of fluorinated carbon in different solvents, reagents, and resins can be found in Allied-Signal's ACCUFLUOR® CF_x Technical Brochure.

Fluorinated carbons made at low temperatures (below about 350°C), or those that are not fluorinated to nearly $CF_{x\geq1}$, or those that are made from amorphous starting carbons (or made under combinations of these conditions), will contain high levels of adsorbed fluorine. Activated carbons (which are completely amorphous) represent an extreme case and can adsorb large amounts of fluorine, even at ambient temperatures.[11] The presence of adsorbed fluorine may be detected by slurrying the sample in a water/isopropanol solution containing potassium iodide: liberation of iodine indicates the presence of an oxidizing agent, and the amount can be quantified. The two carbon black-based samples listed in Table 1 having the lowest levels of combined fluorine (28% and 10%) have levels of adsorbed fluorine of 2 and 3%, respectively. CF_x made at higher temperatures using more crystalline carbon has very little oxidizing power with respect to iodide. The fluorinated petroleum coke sample noted in Table 1 contains less than 0.03% adsorbed fluorine.

Fluorinated carbons having high levels of adsorbed fluorine may degrade some materials and several patents describing chemical treatments to reduce this problem have been filed.[12–14] Useful reactions with these materials have also been demonstrated, such as the oxidation of alcohols and organic sulfides. However, commercial applications may be limited because reaction rates are low.[8,11]

Mass spectrometric analysis of gases evolved during the thermal decomposition of CF_x *in vacuo* showed that the predominant products were a variety of perfluorocarbon fragments; CO, CO_2 and SiF_4 were also detected, and probably resulted from attack of some of the hot fluorocarbon fragments with the glass walls of the apparatus.[15] Thermal decomposition of CF_x in moist air produces small amounts of HF.[16] Unlike the situation with poly(tetrafluoroethylene), formation of the highly poisonous olefin perfluoroisobutylene has not been reported to occur.

When the fluorine is removed from CF_x, via chemical reaction or thermal decomposition, the resulting carbon bears little resemblance to the parent carbon. With samples made from crystalline carbons, e.g., calcined cokes and graphites, a reduction in particle size is usually observed. Furthermore, the microstructure of the parent carbon is irrevocably destroyed and the resulting product is an amorphous soot.[17,18]

23.4. MANUFACTURE

The preparation of CF_x seems rather straightforward, but controlling any reaction with elemental fluorine and dealing with the concomitant heat produced are never simple matters. The heterogeneous reaction between solid carbon and gaseous fluorine, $2nC$ (s) $+ xnF_2$ (g) $\rightarrow 2$ $(CF_x)_n$ (s), is strongly exothermic, so careful temperature control is required to achieve reproducible results. Fluorine is almost never used undiluted, concentrations of up to 50 vol% with the balance made up of an inert gas (e.g., N_2 or Ar) being typical. The heats of formation $[\Delta H_f^{\theta}]$ for $CF_{1.25}$ and $CF_{0.597}$ determined by fluorine bomb calorimetry are -195.73 and -101.74 kJ mol^{-1}, respectively.[19,20]

Even more highly exothermic reactions may proceed under some conditions and further aggravate heat-management problems on a commercial scale, namely: "burning" of the carbon in the fluorine to give CF_4, C_2F_6, and other perfluorocarbons; thermal decomposition of CF_x, to finely divided amorphous carbon ("soot"), and gaseous perfluorocarbons (CF_4, C_2F_6, etc.), followed by burning of the soot in the fluorine.

Production of small amounts of CF_4 and other volatile perfluorocarbons usually occurs during the whole course of the fluorination process. Carbons having low crystallinity, e.g., carbon blacks and activated carbons, are more prone to this problem. Excessively high temperatures and fluorine gas concentrations will also exacerbate the production of perfluorocarbons. The thermal decomposition of CF_x is somewhat suppressed by the presence of fluorine,[20] but if the temperature in a localized area gets high enough (conditions depend on the nature of the carbon), the CF_x will decompose. The amorphous carbon produced when CF_x decomposes is highly reactive, and can react further to produce yet more perfluorocarbon gases.[17,18] Taken together, these reactions lead to hot spots in the carbon bed during the fluorination process, if the heat of reaction is not adequately removed. In the worst case, a localized hot spot can spread as more and more heat builds up, and eventually cause an explosion, due to the rapid production of gaseous perfluorocarbons.[18]

The fluorination kinetics for several different types of carbon have been reviewed recently, and a detailed discussion will not be repeated here.[8] However, some generalized comments regarding the effects of the reaction kinetics on the manufacturing process are warranted. The speed at which the fluorination reaction proceeds is greatly dependent on the crystallinity of the carbon, its particle size, the fluorination temperature, and, to a somewhat lesser degree, the concentration of fluorine gas. Initially the reaction between fluorine and carbon proceeds rapidly, but as the particle's CF_x layer grows, the diffusion of fluorine into the particle is impeded and the reaction rate begins to decrease. To make the most highly fluorinated CF_x materials, long fluorination times are required (more than 8 hours). Highly crystalline carbons or graphite tend to fluorinate more slowly, but tend to give better yields of the desired product (i.e., production of perfluorinated side products is minimal).

The critical design requirements for any commercial-scale reactor used in the manufacture of CF_x are: (1) efficient removal of heat generated by the process; (2) provision of sufficient contact time to make the desired product; and (3) *precise* control of all reaction variables (e.g., temperature and fluorine concentration). Examination of the patent literature reveals that these requirements have spurred the development of several types of reactor, including static bed,[21] vertical multiple tray ("tray-tower"),[22] rotating and vibrating inclined tubular[23-25] flat moving bed,[26] and fluidized bed.[27]

These various reactors represent engineering or mechanical approaches to mitigating the hazards and long reaction times associated with the manufacture of fluorinated carbon. Some chemical approaches, based on the intercalation of certain halides into carbon, have shown promise in reducing reaction temperatures and times. Nakajima *et al.* have fluorinated graphites at 400°C (about 200°C lower than the temperature normally employed) by intercalating with AlF_3 or MgF_2.[28,29] Nalewajek has achieved similar results using antimony or arsenic pentahalides.[30] Another and very promising approach, because it is not limited to graphitic carbons, is fluorination with Cl_2 and F_2.[31] More recently, the room-temperature fluorination of graphite has been accomplished using volatile fluorides of I, Br, Cl, W, Mo, B, or Re and a catalyst (HF). However, these materials are not entirely the same as those made at higher temperatures because they retain small amounts of the intercalate and contain ionic fluorine species.[32] A more complete discussion regarding fluorine intercalates can be found in Section 23.6.

23.5.   APPLICATIONS

Fluorinated carbon's unique chemical and physical properties have led to quite diverse applications for this material. In this section the major commercial applications, in batteries and for lubrication, will be covered. Imaging—an emerging applications area—will also be discussed.

23.5.1.   Batteries

The use of CF_x as the cathode in lithium anode batteries {discharge reaction nLi + $[CF]_n \rightarrow C_n(LiF)_n$} remains the largest commercial application for the material. These batteries have many desirable traits: high operating voltage (3 V nominal), high specific energy (360 Wh kg^{-1}), excellent shelf life (capacity losses of 0.5% per annum are typical), very good safety and reliability, and the capability to operate over a wide temperature range (–20 to +60°C).

The mass production and commercialization of lithium/CF_x batteries was pioneered by the Matsushita Electric Co., Ltd, of Osaka, Japan. Under the Panasonic® trade name, the company currently manufactures coin, pin, and cylindrical batteries in a variety of sizes for use in electronic watches and calculators, IC memory protection circuits, illuminated fishing floats, cameras, and many other devices. Another manufacturer of these batteries for consumer products is the RayOVac Corporation of Wisconsin (U.S.A.). Lithium/CF_x batteries for industrial and military applications are manufactured in the U.S.A. by Eagle-Picher Industries of Missouri, and Wilson Greatbatch, Ltd., of New York. IBM and Kodak are large users of lithium/CF_x batteries in some of their products.[33,34]

The best type of CF_x for use in lithium batteries is the variety made from high-grade, calcined petroleum coke.[35,36] Metals contamination (particularly by iron, nickel, and copper) and the level of adsorbed fluorine should be maintained below 500 ppm. These impurities can greatly affect shelf life. Materials having x-values about equal to or slightly greater than 1 are used to maximize the energy content of the battery. The petroleum coke-based sample listed in Table 1 is specifically manufactured for this use.

Since others have already reviewed the great wealth of literature and patents available regarding the lithium/CF_x system, attention here is confined to some of the most recent work, particularly that which demonstrates the capabilities of this battery system beyond the range of the commonly available commercial products. For those requiring more background information regarding this system, Refs. 8, 37, and 38 are recommended.

A comparison of the performance of three very popular lithium systems, CF_x, MnO_2, and SO_2, can be found in Marple's 1987 paper (see Figure 3).[39] This work is noteworthy for two reasons: one, a very fair comparison is made using cells of the same size; and two, the performance of a promising mixed cathode system, CF_x and MnO_2, is discussed.[39] Batteries made with this combination of cathode materials have higher operating voltage and longer life than those made with either material separately. The performance of the lithium/CF_x system exceeds that of the other popular solid cathode system, lithium/MnO_2, by a very wide margin. But it is even more remarkable that the CF_x-based batteries out-perform those based on lithium/SO_2, a system used by the military and noted both for its performance and its safety hazards.

Other improvements in lithium/CF_x battery performance have also been demonstrated. Eagle-Picher manufactures very large batteries (0.9 kg, 225 Ah) with specific energies over 600 Wh kg^{-1}, nearly twice that of the mass-produced products.[40] High-rate pulse discharge performance, an important criteria for some applications like charging a camera's flash, can be greatly improved through changes in cell construction, organic electrolyte, and cathode formulation.[41,42] Low-temperature performance can be improved by a change in organic electrolyte.[43] Specially designed lithium/CF_x batteries, that can perform safely at temperatures that range from –20 to 180°C, are available commercially.[44] These batteries are well suited for remote sensing equipment that will be used in severe environments. Even greater improvements in

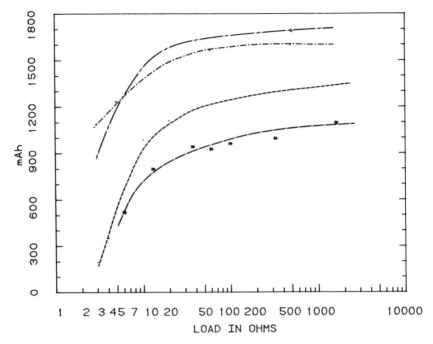

Figure 3 Comparison of output capacities versus discharge loads for 2/3A size cells. Li/MnO₂, -----; LiCF$_x$, ·····; LiMnO₂–CF$_x$, —·—·—; LiSO₂, – – –. J. W. Marple, *J. Power Sources 19*, 325 (1987).

operating voltage and discharge rate performance may be expected by using some of the intercalates described in Section 23.6.

23.5.2. Lubrication

Many of the attributes of fluorinated carbon make it an excellent solid lubricant. Its highly fluorinated surface makes this material unlikely to adhere or bond under pressure to other materials. Its lamellar structure allows the material to deform and coat irregular or rough surfaces. Unlike poly(tetrafluoroethylene) and some other polymers, fluorinated carbon is not subject to "cold flow". CF$_x$'s thermal stability, hydrophobicity, low toxicity, and light color are also advantages.

The earliest cited use in lubrication dates back to a patent issued in 1961,[45] but it was not until the early 70s, when NASA developed an interest in new lubricants for space applications, that the full potential of CF$_x$-based lubricants began to be recognized.[46,47] Comparisons of fluorinated carbon with the common solid lubricants graphite and molybdenum disulfide, showed important advantages over each. Graphite is a good lubricant in normal environments, but fails miserably in a vacuum due to the absence of moisture[46–48]; MoS₂, on the other hand, performs very poorly in the presence of moisture, because abrasive oxides form through its reactions with water.[46,47] CF$_x$ is not affected by moisture nor requires the presence of moisture to

Table 2. CF_x Performance in a Lubricating Grease[a]

	Four ball wear scar (mm)	
	Run #1	Run #2
Grease with 0.05 wt% CF_x	0.60	0.62
Grease without CF_x	0.73	0.76

[a]Aluminum complex grease containing anti-wear and extreme pressure additives. Data supplied by Certified Laboratories, Division of NCH Corp.

provide lubrication. Other advantages include its light color (almost all other solid lubricants are black), its low toxicity (see Section 23.7), its effectiveness at low concentrations (see Table 2), and its good thermal conductivity. The last property is particularly advantageous because heat removal is a problem in many bearing applications. CF_x's thermal conductivity is about two orders of magnitude greater than poly(tetrafluoroethylene)'s and comparable to that of amorphous carbon.

Under most conditions, fluorinated carbon will meet or exceed the performance of graphite or MoS_2. Contradictory results regarding the load-carrying capability and wear life of polymer-bonded films containing MoS_2 or CF_x can be found, but this confusion may be related to performance differences between materials.[49,50] Fusaro has evaluated several types of fluorinated carbon and found major differences in performance.[51] These results were obtained with the CF_x sample providing the only source of lubrication (no binders were used). In applications where CF_x is used in conjunction with other lubricating material, e.g., oils or polymeric binders, certain types of CF_x may be preferred.

Bonded polymer films incorporating CF_x have been developed with several resin systems.[50,52,53] Polyimide binders are very attractive because of their very high thermal stability (>300°C).[53] This system has found application in foil gas bearings.[54] More recently, a fluorinated polyimide resin containing up to 50% CF_x by weight was used to coat the skirt area of aluminum pistons used in gasoline engines; enhanced lubrication between the piston and the cylinder wall was cited.[55]

Metal/CF_x films can be made by electrolytic or electrodeless plating from baths containing the dispersed solid. Surfactants must be used to disperse the hydrophobic solid into these aqueous systems. Several multicomponent systems have been patented and rely on combinations of nonionic and cationic surfactants.[56–59] It is also important to choose a material with a particle size below 10 μm but not smaller that about 1 μm. With the larger materials, settling is a problem, while with smaller materials, gas bubbles, mainly encountered in electrodeless plating, tend to carry the particles to the surface of the bath. Graphite and coke-based fluorinated carbons can be used in this application, although coke-based materials have an advantage in high-load situations because of their greater hardness.

Fluorinated carbons can be added to lubricating oils and greases to reduce friction, decrease wear under normal loads, and greatly increase the load required to cause bearing seizure.[60–62] Evaluations in gasoline and diesel engines have demonstrated improvements in fuel economy.[61] For engine oils, very fine materials (<1 μm particle size) are required to allow the solids to pass through typical oil filters. Fluorinated carbon blacks or milled fluorinated graphites can be used. CF_x's use in engine oil is currently under development, and could result in a substantial market for the material.

Several other smaller and/or developmental lubrication applications involving fluorinated carbons are known. Thus, use of CF_x has been demonstrated in composites for bearings,[63,64] in coatings for magnetic recording media (tapes and disks),[65] and in lubricants for metal forming and wire drawing.[66–68] Using fluorinated carbons in some cases can be very advantageous. For example, in coatings used for magnetic recording, carbon blacks are added to the formulation containing the magnetic particles and binders to lubricate and control static charging. By replacing the carbon black with a partially fluorinated carbon black (<50 wt% fluorine), a 30% reduction in friction can be achieved.[69]

23.5.3. Imaging

Fluorinated carbons are being evaluated in a variety of imaging processes such as electrostatic and electroerosion technologies. Interest in using CF_x for these diverse uses is due to the unique combination of properties exhibited by these materials and the ease with which many of these properties can be varied—like color (black to white) and electrical resistivity (conductive to insulating).

Most of the literature in the public domain deals with use of CF_x in dry two-component electrostatic imaging, commonly known as xerography. Carbon black-based CF_x materials ranging greatly in degree of fluorination and used in concentrations of 1 to 40% by weight of the toner, with or without additional ingredients such as PTFE, enable improved machine operation and copy quality. The benefits of using CF_x include reduced blemishing of the transferred image, and toner buildup on the photoconductor plate and carrier particles.[70,71]

Another patent claims the use of CF_x as either an internal additive to the toner or as an external developer-additive for color electrophotography.[72] Using a fully fluorinated graphite fluoride at a concentration of 0.001 to 10 wt% of the toner provides improved charge control and control of the density of color toner (since the CF_x is white or light in color).

Use of CF_x as an additive to single-component developers for electrographic imaging also provides benefits important for this process, such as improved flow of the developer due to reduced sensitivity to changes in humidity, and images of high resolution, because the developer containing CF_x is more negatively charged.[73]

Fluorinated carbon coatings on the carrier core particles of two-component developer systems have been investigated by several workers. Such coatings either include binder resins that are fluoropolymer based,[74,75] or specifically exclude resins containing fluorine.[76] Addition of the CF_x provides stable triboelectric charge control to the developer, enhances fluidity of developers, and prevents toner from fusing to the carrier cores or photosensitive surfaces. Fluorinated carbons made from various feedstocks (carbon black, crystalline graphite, and petroleum coke) and varying fluorination levels (x-values from 0.25 to 1.2) have been used to achieve the desired triboelectric and resistivity properties of the carrier core.

Fluorinated carbons can also be used in electroerosion recording. In this method an arc is used to erode a thin film of conductive material (typically vapor-deposited aluminum on a paper or plastic substrate) from the surface of the recording material, in accordance with digitally coded image information. Less than 5% by weight of graphite fluoride incorporated into an intermediate polymer layer between the aluminum film and the substrate has been found to improve the scratch resistance of the recording medium, as well as its handling and writing characteristics.[77,78]

23.6. FLUORINE INTERCALATION COMPOUNDS OF GRAPHITE

At low temperatures (below about 300°C), the reaction between fluorine and graphite and other crystalline forms of carbon is almost nonexistent. However, intercalation compounds can be made with fluorine and graphite even at room temperature. These materials are quite different from those made at elevated temperatures.

Early reports of intercalation of graphite by fluorine aided by HF were made by W. Rudörff and G. Rudörff.[79] Similar compounds have been made from highly oriented pyrolytic graphite.[80] In both cases the materials contained less than 50 wt% of fluorine. Characterizations of the latter materials indicated enhanced electrical conductivity (1.5 to 4.5 times that of the virgin pyrolytic graphite) and higher discharge voltage when incorporated into a lithium anode battery.[80,81]

Other fluorine intercalates have been made using volatile fluorides of I, Br, Cl, W, Mo, B, or Re and a catalyst (HF). Owing to the higher fluorine content of these samples (typically >50 wt%) their electrical conductivity is not greater than that of the virgin graphite, but is several orders of magnitude greater than the conductivity of fluorinated carbons made via the high-temperature process.[32] The use of these materials in lithium anode batteries appears quite promising owing to their high discharge voltage and good electrochemical capacity.[82,83]

Structurally, all of these intercalates are quite different from materials made at higher temperatures. X-ray diffraction and NMR data indicate that the fluorine is not covalently bonded to the carbon, but is ionically or "semi-ionically" bound. Furthermore, the planar nature of the graphene layers appears to be preserved.[32,80]

23.7. HEALTH/SAFETY

23.7.1. Toxicity

An early study of the acute oral toxicity of poly(carbon monofluoride) (inferred to be $CF_{1.0}$) using male mice reported the LD_{50} at not less than 4 g kg^{-1} and concluded that its oral toxicity is extremely mild.[84] More recently, fluorinated carbons of varying fluorine contents and from different carbon feedstocks have been similarly evaluated.[85] Fully fluorinated carbons, i.e., those having fluorine contents in the range 59–66 wt%, were found to have very low oral toxicity, no deaths or signs of systemic poisoning being observed in rats at dosages of 5 g kg^{-1}. The partially fluorinated carbons were found to have slightly greater health effects owing to the higher content of surface-adsorbed fluorine (see Section 23.3). Thus, fluorinated carbons made with carbon black and containing 26–30 wt% fluorine caused no deaths at the 5 g kg^{-1} (rat) level, but some reversible symptoms such as increased activity, ataxia, and diarrhea were observed. Fluorinated carbon black containing between 7 and 14 wt% fluorine, while somewhat more active, was still of low oral toxicity. The LD$_{50}$ values for these materials fell between 1.5 and 3.0 g kg^{-1} for male rats, and exceeded 3 g kg^{-1} for females.

23.7.2.   Dermal and Ocular Effects

Fully fluorinated petroleum cokes and partially fluorinated carbon blacks have been evaluated for their potential to cause irritation to the skin, using rabbits and guinea pigs. No detectable skin irritation or sensitization was found for the fully fluorinated carbon, but the partially fluorinated variety had the potential to cause mild irritation.[86]

Acute eye-irritation studies conducted on rabbits established that neither the fully fluorinated carbons nor the partially fluorinated ones are likely to be an eye irritant in humans.[87]

23.7.3.   Cytotoxicity

Lung macrophage studies of fully and partially fluorinated carbons using rabbit alveolar cells have revealed that neither grade is likely to cause fibrotic lung disease. The estimated LC$_{50}$ for both materials was greater than 100 µg cm^{-3}.[88]

23.7.4.   Handling of Fluorinated Carbons

The key point in the safe handling of these materials is to note their dusty character arising from the small particle size. To avoid discomfort from inhalation, operatives are advised to wear dust respirators; normal work clothing and safety glasses are usually sufficient to protect other body parts. In case of contact, washing with water is appropriate.

In case of fire, any standard fire extinguishing agent can be used. While CF$_x$ does not burn, it does undergo thermal decomposition which, in the presence of moist air, may produce hydrofluoric acid vapor (see Section 23.3).

23.8.   REFERENCES

1.   M. H. Luly, *J. Mater. Res. 3*, 890 (1988).
2.   O. Ruff, C. Bretschneider, and F. Ebert, *Z. Anorg. Allg. Chem. 217*, 1 (1934).
3.   W. Rüdorff and G. Rüdorff, *Z. Anorg. Allg. Chem. 253*, 281 (1947).

4. D. E. Palin and K. D. Wadsworth, *Nature 162*, 925 (1948).
5. Y. Kita, N. Watanabe, and Y. Fujii, *J. Am. Chem. Soc. 101*, 3832 (1979).
6. H. Touhara, K. Kadono, Y. Fujii, and N. Watanabe, *Z. Anorg. Allg. Chem. 544*, 7 (1987).
7. L. B. Ebert, J. I. Brauman, and R. A. Huggins, *J. Am. Chem. Soc. 96*, 7841 (1974).
8. N. Watanabe, T. Nakajima, and H. Touhara, *Studies in Inorganic Chemistry 8, Graphite Fluorides*, Elsevier Science Publishers, New York (1988).
9. D. T. Clark and J. Peeling, *J. Polym. Sci., Polym. Chem. Ed. 14*, 2941 (1976).
10. Japanese Patent 59 086,155 (to Matsushita Electric Industrial Co.) [*CA 101*, 76036 (1984)].
11. N. Watanabe and K. Ueno, *Bull. Chem. Soc. Jpn. 54*, 127 (1981).
12. Japanese Patent 56 038,774 (to Matsushita Electric Industrial Co.) [*CA 96*, 69917 (1982)].
13. Japanese Patent 59 087,762 (to Matsushita Electric Industrial Co.) [*CA 101*, 76034 (1984)].
14. M. C. C. Brendle French Patent 2,327,308 (to Agence Nationale de Valorisation de la Recherche) [*CA 88*, 173332 (1978)].
15. D. Devilliers, M. Vogler, F. Lantelme, and M. Chemla, *Anal. Chim. Acta 153*, 69 (1983).
16. H. Tung, *Thermal Decomposition of Fluorinated Carbon*, Allied-Signal unpublished report (1986).
17. N. Watanabe, Y. Chong, and S. Koyama, *J. Jpn. Chem. Soc. 2*, 228 (1981).
18. D. T. Meshri. *Polym. News 6*, 200 (1980).
19. P. Karmarchik, Jr. and J. L. Margrave, *Therm. Anal. 11*, 259 (1977).
20. P. Kamarchik, Jr. and J. L. Margrave, *Acc. Chem. Res. 11*, 296 (1978).
21. J. L. Margrave, R. B. Badachhape, J. L. Wood, and R. J. Lagow, U.S. Patent 3,674,432 (to RI Patents, Inc.) [*CA 77*, 141956 (1972)].
22. T. Komo, S. Mizobata, R. Ukaji, and T. Kamihigoshi, U.S. Patent 3,929, 920 (to Daikin Kogyo Co.) [*CA 84*, 124106 (1976)].
23. T. Aikyama, T. Kamihigoshi, S. Takagi, and T. Maeda, U.S. Patent 4,348,363 (to Daikin Kogyo Co.) [*CA 94*, 33097 (1981)].
24. T. Komo, S. Mizobata, R. Ukaji, and T. Kamihigoshi, U.S. Patent 4,447,663 (to Daikin Kogyo Co.) [*CA 94*, 33097 (1981)].
25. J. Metzger, U.S. Patent 4,855,121 (to Atochem) [*CA 109*, 172997 (1988)].
26. Y. Kita, S. Moroi, M. Aramaki, and H. Nakano, British Patent 2,104,884 (to Central Glass Co.) [*CA 98*, 200743 (1983)].
27. H. Staab, European Patent 236,882 (to Hoescht) [*CA 107*, 219949 (1987)].
28. T. Nakajima, M. Kawaguchi, and N. Watanabe, *Chem. Lett.* 1045 (1981).
29. T. Nakajima, M. Kawaguchi, and N. Watanabe, *Carbon 20*, 287 (1982).
30. D. Nalewajek, U.S. Patent 4,795,624 (to Allied-Signal, Inc.) [*CA 111*, 99872 (1989)].
31. D. Nalewajek, U.S. Patent 4,886,921 (to Allied-Signal, Inc.) [*CA 112*, 39259 (1990)].
32. A. Hamwi, M. Daoud, and J. C. Cousseins, *Synth. Met. 26*, 89 (1988).
33. P. S. Clark, *Design News 43* (May 23, 1983).
34. L. J. Hart and T. Ciobanu, *Batteries Today 8* (Summer 1984).
35. A. Morita, T. Iijima, T. Fujii, and H. Ogawa, *J. Power Sources 5*, 111 (1980).
36. Y. Toyoguchi, T. Iijma, and M. Fukuda, U.S. Patent 4,271,242 (to Matsushita Electric Industrial Co.) [*CA 90*, 171536 (1979)].
37. M. Fukuda and T. Iijima, in: *Lithium Batteries* (J. P. Gabano, ed.), Academic Press, New York (1983).
38. D. Linden, in: *Handbook of Batteries and Fuel Cells* (D. Linden, ed.), McGraw-Hill Book Co., New York (1984).
39. J. W. Marple, *J. Power Sources 19*, 325 (1987).
40. Eagle-Picher Industries, Joplin, MO, Lithium Batteries Technical Brochure.
41. T. W. Martin, P. J. Muehlbauer, and K. Pan, European Patent 165,794 (to Eastman Kodak Co.) [*CA 104*, 98046 (1986)].
42. M. A. Faust and H. W. Osterhoudt, U.S. Patent 4,565,751 (to Eastman Kodak Co.) [*CA 104*, 92188 (1986)].
43. T. Iijima and K. Nakamura, *Denki Kagaku 54*, 274 (1986).

44. M. F. Pyszczek, S. J. Ebel, and M. A. Zelinsky, to be published in the *Proceedings of the 34th Power Sources Symposium* (June 1990).
45. H. Schachner, U.S. Patent 2,933,567 (to Inst. Dr. Ing. Reinhard Straumann Akt.-Ges.) [*CA 55*, 19224 (1961)].
46. R. L. Fusaro and H. E. Sliney, *NASA Tech. Note D-5097* (1969).
47. R. L. Fusaro and H. E. Sliney, *ALSE Trans. 13*, 56 (1970).
48. R. L. Fusaro, *ALSE Trans. 20*, 15 (1977).
49. B. D. McConnell, C. E. Snyder, and J. R. Strang, *ASLE Lub. Eng. 33*, 184 (1977).
50. R. L. Fusaro and H. E. Sliney, *NASA Tech. Note D-6174* (1972).
51. R. L. Fusaro, *Tribology Trans. 32*, 121 (1989).
52. R. L. Fusaro, *NASA Tech. Note D-8033* (1975).
53. R. L. Fusaro, *NASA Tech. Memorandum 82968* (1982).
54. R. C. Wagner and H. E. Sliny, *NASA Tech. Memorandum 83596* (1984).
55. J. R. Presswood, U.S. Patent 4,831,977 (to Ethyl Corp.) [*CA 111*, 26150 (1989)].
56. H. Takeuchi, Y. Okamoto, S. Kurosaki, and K. Nakamoto, U.S. Patent 3,756,295 (to Uemura Kogyo Co., Monolon Co., and Nippon Carbon [*CA 80*, 29109 (1974)].
57. J. T. Kim, U.S. Patent 4,716,059 (to Allied Corp.) [*CA 109*, 214784 (1988)].
58. J. F. Paulet, J. C. Puippe, and H. Steup, U.S. Patent 4,728,398 (to Fluhmann Werner) [*CA 106*, 164581 (1987)].
59. J. R. Henry and E. M. Summers, U.S. Patent 4,830,889 (to Wear-Cote International Inc.) [*CA 111*, 82779 (1989)].
60. H. Gisser, M. Petronio, and A. Shapiro, *Lubr. Eng. 28*, 161 (1972).
61. F. Defretin, J. P. Eudeline, E. Schoch, and A. Voisin, U.S. Patent 4,314,907 (to Ugine Kuhlmann) [*CA 93*, 242422 (1980)].
62. T. Ishikawa and T. Hori, U.S. Patent 3,607,747 (to Japan Carbon Co.) [*CA 74*, 89497 (1971)].
63. T. Hiratsuka and T. Shimada, U.S. Patent 3,717,576 (to Japan Carbon Co.) [*CA 78*, 125469 (1973)].
64. L. Abrahams and T. P. J. Izod, U.S. Patent 4,333,977 (to Waters Associates, Inc.) [*CA 97*, 56899 (1982)].
65. A. Kawata, Y. Morino, H. Okajima, and S. Miya, U.S. Patent 4,621,016 (to Kanto Denka Kogyo Co.).
66. Belgian Patent 829,843 (to Ugine Kuhlmann) [*CA 86*, 20450 (1977)].
67. Japanese Patent 57 65,795 (to Nippon Steel Corp.) [*CA 97*, 165883 (1982)].
68. Japanese Patent 58 185,694 (to Central Glass Co.) [*CA 100*, 194884 (1984)].
69. S. Miya, T. Niinuma, A. Kawata, and S. Suzuki, paper presented at Intermag, St. Paul, MN (April 29–May 2, 1985).
70. T. Nozaki, Japanese Patent 51 101,536 (to Fuji-Xerox Co.) [*CA 86*, 148812 (1977)].
71. T. Ushimaya, Japanese Patent 53 147,542 (to Ricoh Co.) [*CA 90*, 178170 (1979)].
72. T. Hasegawa, H. Toma, and K. Satomi, U.S. Patent 4,141,849 (to Canon K.K.) [*CA 88*, 129022 (1978)].
73. R. H. Helland and C. A. Burton, U.S. Patent 4,681,830 (to 3M Co.) [*CA 96*, 113485 (1982)].
74. Japanese Patent 51 140,635 (to Daikin Kogyo Co.).
75. Japanese Patent 57 147,648 (to Fujitsu Co.) [*CA 99*, 222363 (1983)].
76. M. H. Luly, G. D. Lockyer, R. E. Eibeck, and J. Gaynoer, U.S. Patent 4,524,119 (to Allied Corp.) [*CA 102*, 195146 (1985)].
77. A. Afzali-Ardakani, M. S. Cohen, K. S. Pennington, and K. G. Sachdev, U.S. Patent 4,554,562 (to IBM Corp.) [*CA 104*, 13120 (1986)].
78. A. Afzali-Ardakani, M. S. Cohen, K. S. Pennington, and K. G. Sachdev, U.S. Patent 4,567,490 (to IBM Corp.) [*CA 104*, 59462 (1986)].
79. W. Rüdorff and G. Rüdorff, *Chem. Ber. 80*, 417 (1947).
80. T. Mallouk and N. Barlett, *J. Chem. Soc., Chem. Commun.* 103, (1983).
81. R. Hagiwara, M. Lerner, and N. Barlett, *J. Electrochem. Soc. 135*, 2393 (1988).
82. A. Hamwi, M. Daoud, and J. C. Cousseins, *Synth. Met. 30*, 23 (1989).
83. A. Hamwi, M. Daoud, and J. C. Cousseins, *J. Power Sources 27*, 81 (1989).

84. Y. Yoshida, A. Harada, T. Okamura, K. Kono, M. Watanabe, S. Toyota, and K. Iwasaki, *Bull. Osaka Med. Sch. 23*, 14 (1977).

85. *Acute Oral Toxicity Study of Fluorinated Carbon*, Allied-Signal unpublished reports (1983, 1985).

86. *Primary Dermal Irritation of Fluorinated Carbon*, Allied-Signal unpublished reports (1983, 1985).

87. *Acute Eye Irritation Study of Fluorinated Carbon*, Allied-Signal unpublished reports (1984, 1985).

88. *Determination of Cytotoxicity of Fluorinated Carbons,* Allied-Signal unpublished reports (1985).

24

Uses of Fluorine in Chemotherapy

PHILIP NEIL EDWARDS

24.1. GENERAL INTRODUCTION

This account of uses of fluorine in chemotherapy is the sometimes idiosyncratic view of a practicing medicinal chemist who has no special expertise or commitment to organofluorine chemistry, but who, like many or perhaps most medicinal chemists, has become increasingly keen to investigate fluorine-containing compounds in the search for improved drugs. Previous approaches to this subject have been based on the selection of compounds from particular biochemical or therapeutic areas mainly because they happen to contain fluorine.[1–5] In this author's opinion, such an approach inevitably tends to produce a very unbalanced view of the selected areas from a medicinal chemist's standpoint; and clearly the one-time considerable value of such a compilation to fluorine chemists is increasingly limited as computer-based retrieval of complete structures and accompanying abstracts—based on part structures (e.g., C—F!) and keyword searches—becomes a routine matter. Additionally, a recent publication which is fully computer searchable provides a compilation of over 5500 compounds that have been used or studied as medicinal agents in man.[6–7] The output includes full structures, references, and, if available, therapeutic classification and some physical data. Another recent book provides information on the physical and chemical properties, structures, and pharmacological actions of all the most significant drugs in use today.[8]

What seemed to be lacking in the literature to-date was an in-depth attempt to look at the principles behind the uses of fluorine in drugs and to assess the possible value of fluorine, relative to other atoms or groups, which have been, or could become, targets of a typical analog synthesis or program. This review aims to partially correct

PHILIP NEIL EDWARDS • Zeneca Pharmaceuticals, Alderley Park, Macclesfield, Cheshire SK10 4TG, England.

Organofluorine Chemistry: Principles and Commercial Applications, edited by R. E. Banks *et al.* Plenum Press, New York, 1994.

that omission. The approach is generally from a mechanistic and physicochemical standpoint, but some synthetic routes to compounds, particularly those of (potential) commercial importance, are included. For a wider coverage of synthetic methodology in drug synthesis, the reader should consult the recent reviews that deal more fully with such matters.[4-5]

The chapter is largely organized along physicochemical lines, with compounds drawn from various chemotherapeutic areas in order to exemplify particular points such as, for example, substituent size, partition coefficient effects, hydrogen bonding potential, and metabolism. But since neat divisions of this type almost never exist in real life, the same or similar compounds will appear in a number of sections. The major exception to this organizational scheme is Section 24.2, on the fluoroquinolone antibacterials. This section to some extent sets the scene for the chapter as a whole in showing up our often poor understanding of why fluorine can be so special and how, through the subsequent generation of multiple hypotheses, one might be able to gain further insights into its roles in this increasingly important niche of chemotherapy. Indeed in the area of fluorine-containing drugs, this is by far the most active field of current and projected research, and since essentially all recent agents contain fluorine, the entire area falls comfortably within the intended scope of this book.

24.2.   FLUOROQUINOLONE ANTIBACTERIAL AGENTS

24.2.1.   Introduction

The fluoroquinolone carboxylic acids have, over the past decade, become a major class of clinically important antibacterial agents, displaying broad-spectrum oral and parenteral activity with usually mild side effects, which are reversible upon discontinuation of therapy if that should be needed. They thus seem set to rival the β-lactams in terms of research effort deployed worldwide and, despite over 5000 compounds having already been reported, there is reason to expect the flow to increase in the future.[9] The vast majority of recently reported compounds contain a fluorine at position-6 of 1-substituted-4-quinolone-3-carboxylic acids, or at the equivalent position in closely related ring systems; many contain one or more additional fluorine atoms as part of the 1-substituent and some have fluorines at position-6 and -8. A recent, extensive and well balanced review, covering all aspects except synthetic methodology and some matters of the type herein, is strongly recommended.[10] This section will gloss over many aspects of the topic in order to sharpen the focus on those facts which are relevant to later sections.

The progenitor of this class of synthetic antimicrobials, nalidixic acid (**1**, Figure 1), was first reported in 1962; it shows only moderate activity versus some Gram-negative organisms and its use is largely confined to infections of the urinary tract (UTIs)—in which fluid it becomes strongly concentrated and thereby reaches the levels required to achieve its bactericidal effects. Doses of about four grams per day are required!

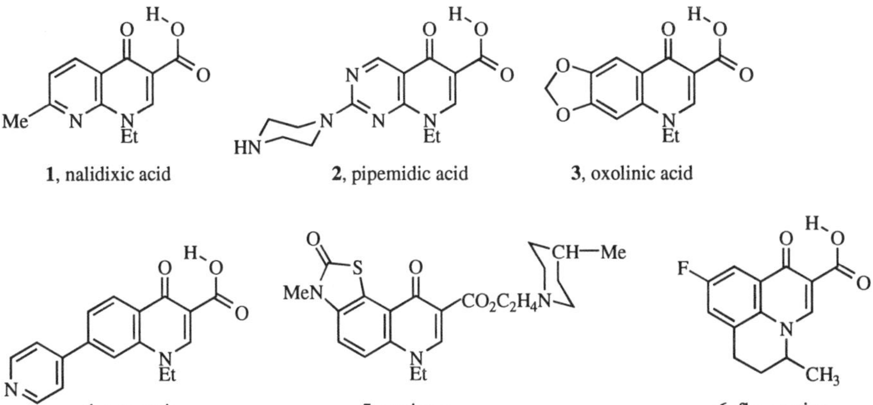

Figure 1. Some early quinolones.

There were a great many useful advances in the following two decades in terms of development of structure–activity relationships (SAR), but only moderate gains in terms of antibacterial potency and spectrum: the hydrophilic pipemidic acid (2, Figure 1) is notable for the presence of a piperazine ring and a somewhat better spectrum, while the much more lipophilic oxolinic acid (3) showed some of the largest advances in terms of both potency and spectrum. Rosoxacin (4) and metioxate (5) exemplify other early nonfluorinated varients. Flumequine (6) is of particular relevance here as an early example of a compound containing a 6-fluoro substituent. These compounds and several others from the same period are limited (mainly) to the treatment of selected UTIs due to a lack of potency against relevant bacteria.

24.2.2. Early 6-Fluoro-7-piperazino-quinolones

It was the systematic investigation in Japan by Koga and co-workers of multisubstituted derivatives, and the application of multiparameter quantitative structure-activity relationships (QSAR), that led to the synthesis (Scheme 1) and subsequent development of the more potent and much broader spectrum agents, norfloxacin (7a) and (later) pefloxacin (7b), which can be used against many systemic infections as well as those of the urinary tract.[11] The QSAR analysis of 1-substituted versions of 7 ($R^6 = R^7 = R^8 = H$), and benz-monosubstituted compounds with $R^1 = Et$, had indicated that antibacterial properties were correlated parabolically with a variety of steric parameters for positions 1, 6, and 8. The optimum substituents were $R^1 = Et$, $R^6 = F$, and $R^8 = Cl$ or CH_3. At position-7, all of a moderate variety of substituents [NO_2, Cl, Me, C(O)Me, OMe, NMe_2, piperazino] enhanced the potency, relative to hydrogen, by the remarkable amount of 10- to 30-fold. The physicochemical parameters responsible for such effects were unclear, so a further set of 7-substituted compounds was made in the 1-ethyl-6-fluoro series. Also, a range of 1-ethyl-7-piperazino-6-substituted compounds was made to help understand any possible interaction effects—a likely

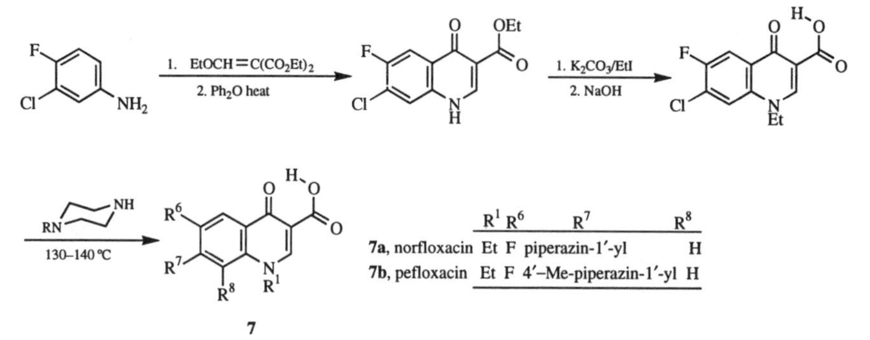

Scheme 1. The synthesis of norfloxacin and pefloxacin.

possibility for *ortho*-disubstituted derivatives due to conformational effects if for no other reason. Some of the results from the latter set are displayed in Table 1.

Clearly, the steric correlation found for R^6 in the parent series is similarly manifested in the much more potent 7-piperazino series, with fluorine being the best substituent by a considerable margin.

The choice of norfloxacin (**7a**) from the large number of compounds that had been made involved far more than a simple consideration of such *in vitro* potencies (compounds up to four times as potent were available), but its very broad spectrum

Table 1. Antibacterial Activity of Some
6-Substituted-7-piperazin-1′-yl-quinolones

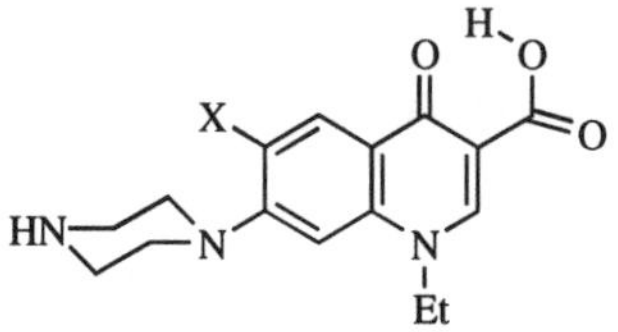

	Minimum inhibitory conc. (MIC/μg ml^{-1})		
X	*S. aureus* 209 P	*E. coli NIHJ* JC-2	*P. aeruginosa* V-1
H	12.5	0.78	3.13
F	0.39	0.05	0.39
Cl	1.56	0.20	3.13
Br	3.13	0.39	12.5
Me	3.13	0.39	6.25
MeS	25	0.78	12.5
CH$_3$C(O)	100	100	>100
nalidixic acid	>100	3.13	100
pipemidic acid	25	1.56	12.5

and 16- to 500-fold improvement over nalidixic acid, depending upon bacterial species and strain, were complemented by ease of synthesis, adequate absorption (oral bioavailability *ca* 40% in man), and a half-life in patients of 3–7 hours.

Thus this compound represents the first of the new generation of fluoroquinolones and it rapidly attracted a great deal of attention, particularly in Japan. It has been followed by a succession of competitors which recently have been entering clinical studies at the rate of several new agents per year.

24.2.3. Mode of Action

This high level of interest has been fueled by the identification of the target of the quinolones as DNA gyrase (bacterial topoisomerase II) which was first isolated from *Escherichia coli* in 1976. This enzyme catalyzes bacterial DNA supercoiling and appears to be involved in several other reactions; the supercoiling is necessary if the very long, *ca* 1300 µm, chromosomes are to be fitted into a cell with dimensions of only *ca* 2×1 µm. The enzyme, an A_2B_2-tetramer, achieves the process by breaking the DNA, in both strands, through transfer of phosphate ester links to tyrosine residues on the A subunits. This allows supercoiling to occur and rejoining of the strands completes the process. It is of some concern that the concentrations necessary for inhibition of isolated enzyme activity can be up to 100-fold greater than those necessary for antibacterial effects, and that their correlation is far from good. This latter aspect has been ascribed to the compounds differing abilities to penetrate the organisms. Thus, from such data, enoxacin (**8a**, Table 2) was deduced to have superior cell permeability relative to norfloxacin (**7a**) and the authors of the study concluded that both gyrase inhibition (2- to 17-fold) and cell penetration (1- to 70-fold) are enhanced by the presence of a fluorine at position-6. However, such analyses should be treated

Table 2.   Biological Properties of Naphthyridones

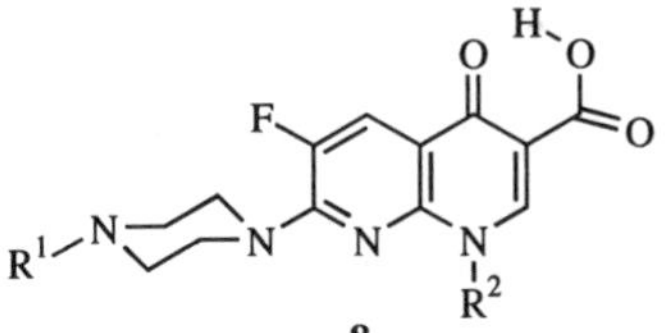

8

			ED$_{50}$ (p.o.) (mg kg^{-1})			
Compd.	R1	R2	*S. aureus* 50774	*E. coli* P-5101	*P. aeruginosa* 12	LD$_{50}$ (p.o.) (mg kg^{-1})
8a	H	Et	10	1.8	9.0	>2000
8b	Me	Et	4.8	1.2	10.6	210
8c	H	CH$_2$=CH	33.4	1.3	2.4	>2000
8d	Me	CH$_2$=CH	10.5	1.1	3.7	354
8e	H	FCH$_2$CH$_2$	11.5	3.0	27.2	>2000
8f	Me	FCH$_2$CH$_2$	1.4	0.52	4.2	1866

with considerable caution since gyrase enzymes differ between species and even strains, and enzyme in the cellular environment may differ substantially from that as used in these tests. Also, the intracellular free-drug concentration (chemical potential) is the end result of many rate processes,[12] of which penetration is only one (multicomponent) part. There is no reason to doubt that the presence of a fluorine atom at position-6 will influence the majority of these rate processes in some degree and will have a variable effect on the compound affinity for enzymes from different sources. It will also influence nonspecific adsorption to various bacterial constituents, especially phospholipids, and change the proportions of the drug existing as anion, cation, zwitterion, and neutral forms.

24.2.4. New Structural Types

The discussion thus far has concentrated on *in vitro* antibacterial effects, but the objective is to find agents for human and veterinary applications and the correlation of *in vitro* and *in vivo* potency is normally poor, thus *in vivo* testing is essential. The usual first step on the *in vivo* road for antibacterial agents is to test compounds in mice and hope that the ranking of compounds will correspond to that in man (almost always the major target species). This hope is usually given better grounding by testing a very few compounds in other laboratory species, but examples of both better and much worse than expected results in man can be found in the fluoroquinolones, as in most other areas of chemotherapy. With this background and given the data in Table 2, and with many millions of dollars at stake, which compound should be chosen for development? The compounds with R^1 = CH_3 are generally somewhat less potent *in vitro* but more potent *in vivo* when dosed orally, which is the commercially most attractive route. *In vivo* efficacy is very important but **8b** and **8d** are much more acutely toxic than their nonmethylated counterparts (the basicity-reducing and lipophilicity-enhancing effects of the N-methyl groups will probably cause more rapid absorption and distribution and therefore higher peak tissue levels). But what is the relevance of acute toxicity at even this lower level to a drug's normal use at a very small fraction of these quantities? Certainly compound **8f** would seem an excellent choice on these (limited) data and its *in vivo* performance outstrips that of **8a**. Despite its apparent inferiority, compound **8a**, enoxacin (Dainippon), was chosen. In man it is well absorbed with a bioavailability of roughly 80%, twice that of norfloxacin, and a half-life of 3.4–6.4 hours, which is typical for nonmethylated piperazines. Why was **8f** not chosen? It is unlikely that we will ever know the answer, but the risk of fluoroacetate poisoning from synthetic intermediates, or indeed from well precedented (but not in this series) oxidative nitrogen de-alkylation of the drug itself (by the human liver), must have been considerations.

The same concerns may have been weighed in a subsequent decision to develop fleroxacin (Roche) which is the 8-fluoroquinolone analog of **8f**, but if that was part of the gamble then it seems to have paid off, since no fluoroacetate-type toxicity has been reported, and indeed this was probably always a very small risk at normal therapeutic doses; the high (maximum tolerated) doses used in toxicological studies may present

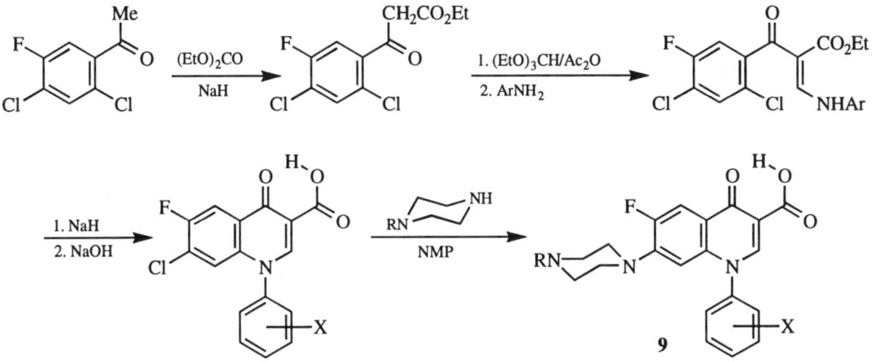

Scheme 2. Synthesis of 1-arylquinolones.

a different picture. Workers at Roche have investigated the toxicology of fluoroacetate in rats and dogs. One of the distinguishing features of this compound is the very low frequency of development of resistance *in vitro*—only 10^{-10} to 10^{-11} mutants per colony forming unit—it is therefore very much superior to the early nonfluorinated quinolones, for which development of resistance has been a clinical problem.[13] Most recent quinolones share this type of advantage, but in varying degrees.

Metabolic transformation to fluoroacetate, or indeed oxidation in any manner, is not a problem for the compound difloxacin (**9a**, Scheme 2), a 1-(4′-fluorophenyl)qui-nolone which has a much longer half-life in man than any of the 1-alkyl types. Such 1-aryl compounds were latecomers to this scene, probably because all prior SAR at the 1-position had indicated that groups bigger than ethyl were deleterious, but also because the normal route of synthesis of quinolones (Scheme 1) was not applicable to this new type. The synthesis of difloxacin is shown in Scheme 2.[14] Two further examples of this type are temafloxacin (**10**) and tosufloxacin (**11**). These fluorinated 1-aryl types are broadly similar to their aliphatic counterparts against Gram-negative

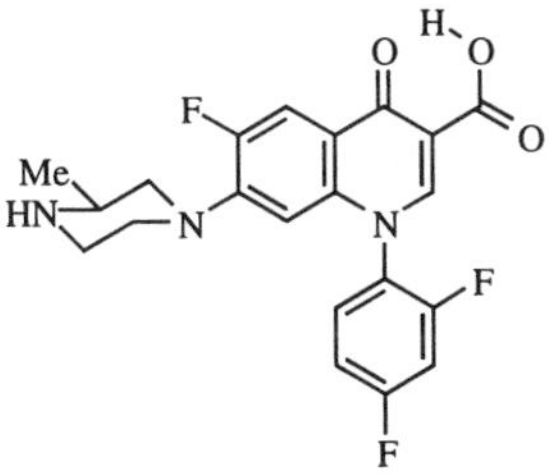

10, temafloxacin 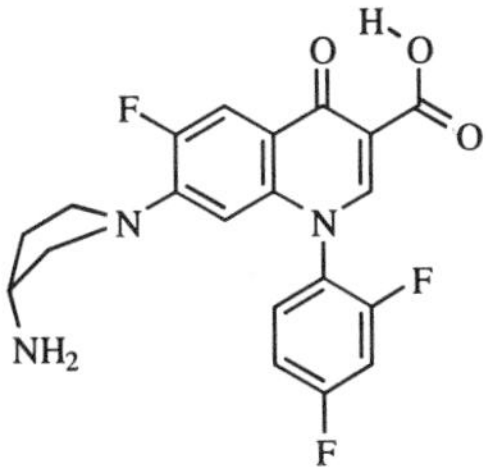   11, tosufloxacin 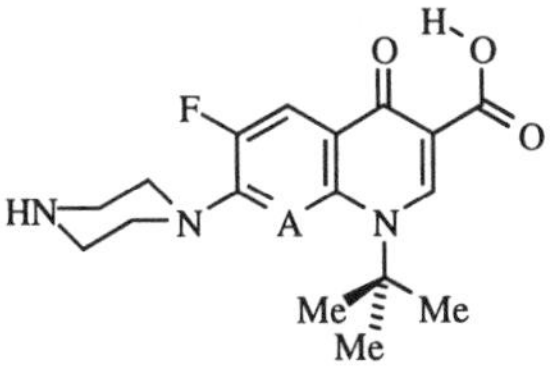   12

13   14, lomefloxacin   15, sparfloxacin

organisms, but some have advantages against *S. aureus*. *In vivo* in mice, the subcutaneous-to-oral dose ratios indicate better bioavailability and/or longer duration of action as a result of the presence of the fluorinated aryl ring. Methylation of the piperazine nitrogen produces a related *in vivo* enhancement for both series, so that difloxacin, with both effects, has very high bioavailability after oral dosage to mice. Norfloxacin (**7a**), on the other hand, seems very poor by the oral route, probably as a consequence of poor rates of intrinsic absorption (these rates vary along the length of the gut) combined with low solubility in the intestinal fluids.

The problems that arise from low aqueous solubility are a commonplace headache for medicinal chemists and the way in which structural changes, including fluorine substitution might be expected to modify this important property are discussed later. The phrase "might be expected" is carefully chosen since prediction in this regard is notoriously poor, depending as it does on both partition coefficients and the efficiency and energy associated with crystal packing. The melting point of a compound provides a (usually) good guide to crystal energies and on this basis the quinolones have problems: their average melting point (sometimes decomposition point due to thermal decarboxylation) is about 250°C, but can be much higher: oxolinic acid (**3**) has a melting point of 314–316°C (dec.)! Aqueous solubility ranges from extremely poor [tosufloxacin (**11**) is reported at 0.008 mg ml^{-1}] to fairly good for compound **12** at 0.82 mg ml^{-1}. Compound **13** is one of a series of analogs of tosufloxacin which had improved solubility as a major goal: it retains the excellent antibacterial properties of tosufloxacin and achieves its other main target with a twentyfold improvement in aqueous solubility. Extremely poor solubility was presumably behind the choice of metioxate (**5**), with its water solubilizing basic side-chain (which will be hydrolyzed off *in vivo*) rather than the probably very sparingly soluble parent acid. This pro-drug approach to improved absorption has been tried many times in the quinolone area, sometimes with useful results.

Compounds like **12** with a 1-*tert*-butyl group were, like the 1-aryl type, latecomers to the scene and unexpectedly potent in the light of the marked drop in potency in changing from 1-ethyl to 1-isopropyl. The best antibacterial effects are seen with A = nitrogen, closely followed by A = CH and A = CF a poor third. This is similar to the pattern seen in the 1-aryl types, but elsewhere, as in fleroxacin (the 8-fluoroquinolone analog of **8f**), lomefloxacin (**14**), and sparfloxacin (**15**), the 8-fluoro compounds can be slightly better *in vitro* but significantly better in mouse protection tests. The latter

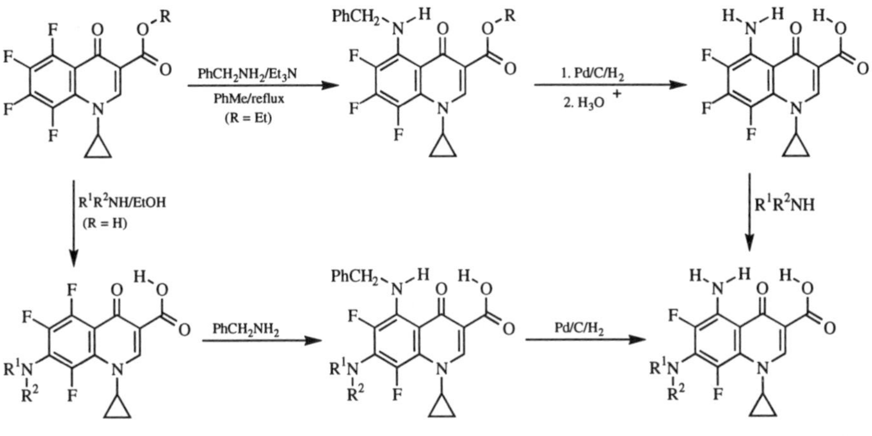

Scheme 3. Synthetic routes to sparfloxacin and analogs.

compound revives an old theme in containing a 5-amino group, which in this case produces a quinoline with seven out of the eight positions carrying substituents! Two synthetic routes to such compounds are shown in Scheme 3.[15] The compound **15** is of great interest from many points of view: it has an excellent antibacterial spectrum and is very potent *in vitro* and *in vivo*; it was designed to avoid some of the side effects of previous fluoroquinolones and appears in animal studies to be successful in this regard. Since it incorporates features which modulate its physical properties and conformational characteristics, it will reappear in later sections.

24.3. ELECTRONEGATIVITY AND ITS CONSEQUENCES

Essentially all previous attempts to appreciate or explain the utility of fluorine in biochemistry and chemotherapy have placed great emphasis on its extreme electronegativity. However, it is now clear that this property is only indirectly responsible for the vast majority of the observations concerned with utility, since field effects rather than inductive effects are by far the most important means of transmission of substituent effects in compounds of medicinal interest.

Electronegativity effects tend to be very localized; they are relevant to the sizes of attached atoms, bond lengths, bond angles, hybridization preferences, hyperconjugation/σ-resonance, double bond tautomerism/isomerism and conformational analysis, but space limitations exclude their further discussion here.

24.3.1. What is Electronegativity and Can "It" Be Quantified?

In any discussion of electronegativity one should state Pauling's definition of the concept as "the power of an atom in a molecule to attract electrons to itself".[16] The existence of his (thermochemically derived) scale must surely be known to every chemist. It is atom based and no doubt provides widely usable values for monovalent atoms. Many people have modified and extended the concept and the interesting work of Nagle,[17] based on atomic polarizability, is the latest of many alternative approaches which generate similar but more extensive sets of numbers, χ-values, for atoms. In most of chemistry, however, one is interested in polyatomic systems and thus there might arise the need for χ-values for groups of atoms, including groups which incorporate changes in hybridization, if the atomic values were shown to be unsatisfactory in some way. But what way?

Quantifiable properties that correlate well with χ-values are extremely rare and mainly involve NMR chemical shifts and coupling constants—which are difficult or impossible to relate to chemical energetics. A recent exception to this is the demonstration that differences in the heats of formation between hydrogen and methyl derivatives correlate linearly with V_X, where $V_X = n_X/r_X$; n_X is the number of valence electrons in atom X and r_X is its covalent radius. The "fit" for hydrogen is a complete fudge, but despite this the data allow of another new χ-scale. Significantly, two lines are required to fit the data, one for the halogens, the other for composite groups (OH, NH_2, SH, CH_3, SiH_3).[18]

The most significant recent approach to χ-values for substituent groups is that of Taft, Topsom, and co-workers who have used an *ab initio* molecular orbital approach to calculate atomic charge densities on hydrogen in compounds HX, and thereby produce an extensive and easily extendable set of σ_χ values.[19] Precise details of the procedures are not available, but it seems inevitable that where X is a polyatomic array, there must be changes in charge density on H due to field effects from the bonds and lone pairs of the more distant atoms. Thus the separation of field effects and electronegativity effects may be substantial but incomplete.

The values obtained depart very significantly from a curvilinear correlation with the atomic values of Nagle only when the atom bonded to H is sp-hybridized, or carries a high level of charge, as in the nitro group and the oxyanion. Smaller deviations clearly indicate the operation of σ-resonance effects in both decreasing (X = CH_2CN) and increasing [X = RC(O), CH_2OH, CH_2F, CF_3] the charge density on H in HX. The implication of this is that another series Y—X, in which the σ-resonance demand of Y is substantially greater (or less) than for H, will show systematic changes from the presently defining series. Thus, as with the related σ_F-values, one should understand the possible limitations for correlation analysis, while making use of the values available.[20] Values for the interesting series $CH_{(3-n)}(CF_3)_n$ ($n = 1$–3) and $C(O)CF_3$ were not calculated.

In a later re-analysis of these data, and with particular regard to the greater than normal deficiencies of Mulliken population analysis for hydrogen bonded to carbon, Topsom used a variety of arguments to support changing the relative position of hydrogen, toward that of carbon, finally choosing to make them equal.[21] This is clearly at odds with all other χ-scales so, if this new and differently scaled set is used in correlation analysis, the residuals for the parent compound should be of particular concern.

24.4. DIPOLE MOMENTS AND ELECTRON DENSITY DISTRIBUTIONS IN HALOGENO COMPOUNDS

The electric field complementarity of a drug with its binding site can be roughly broken down into mutually induced fields, which are always attractive, and the interactions of permanent fields, which one aims to improve through structural change. The former are normally significant because proteins, and indeed all biological macromolecules, contain extremely polar repeating units, some of which almost inevitably are exposed within binding sites. The greater polarizability of "aromatic" rings compared to saturated alicyclic rings is one reason why so many drugs contain the former type of ring. Fluorine substitution in any situation reduces the ability of molecules to respond to complex fields. This is apparent in many ways. For example, (1) boiling points of related nonpolar molecules are normally well correlated with surface area, but the larger surface area of hexafluorobenzene compared to benzene fails to produce any such increase; (2) the wavelengths of the charge-transfer bands of tetracyanoethylene complexes with toluene and 2,3,4,5,6-pentafluorotoluene are 406 and 325 nm, respectively[22]; (3) fluorobenzene is a weaker hydrogen-bond acceptor than benzene (see Section 24.6). As the electronegativity of a substituent decreases, these effects are reduced and in some cases reversed. For the halogens at least, the permanent fields associated with C—Hal bonds can be much more important than the above secondary effects, but if maximum insight into any electrostatic interaction is to be achieved it is necessary to appreciate that the simplifying concept of bond dipoles is likely to be inadequate. Indeed, the simplification that only electrostatic forces are involved breaks down for the heavier halogens, especially iodine.

Starting with some facts, the dipole moments of the hydrogen halides and some alkyl halides are recorded in Table 3.[23,24] The initial observation must be that the concept of a constant sp^3 C—Hal bond dipole moment is untenable at other than a superficial level. For the alkyl fluorides, the slight increase in dipole moment with increasing size of the R group can be understood in terms of small contributions from increased ionic character, σ-resonance, and polarizability in the larger members; the magnitudes of the changes are intuitively reasonable. Jumping to the iodides, it is clear that additional factors are required to explain these profound changes. The C—I bond should be polarized in the sense $C^{\delta-}$—$I^{\delta+}$ if assessed purely from respective electronegativities, but contributions from ionic resonance structures and σ-resonance effects

Table 3.   Dipole Moments (debye^a) of Hydrogen and Alkyl Halides (R—Hal)

R	F	Cl	Br	I
H (gas phase)	1.74	1.03	0.83	0.45
Me (in CCl₄)	1.7	1.8₅	1.8₅	1.5
t-Bu (in CCl₄)	2.0₅	2.1₅	2.2	2.15
adamant-1-yl (in CCl₄)	2.1	2.4	2.5	—

^{a}1 debye (D) = 3.336×10^{-30} coulomb meter.

should, as for fluorine but possibly in a somewhat greater degree, produce field effects in the "expected" direction. In total, only a small dipole moment in the observed direction might be expected (compare the value for H—I). Thus a major part of the dipole moment of tertiary-butyl iodide remains to be explained.

Potentially relevant to the explanation is the analysis of intermolecular oxygen / iodine contacts in crystals.[25] Some of these "contacts" are very considerably closer than conventional van der Waals' distances would predict and the shorter the distance, the closer the C—I---O angle is to 180°. These findings go precisely contrary to simple expectations based on dipole effects and are best understood as a combination of two effects: $n_O \rightarrow \sigma^*_{IC}$—partial covalent bonding, "halogen bonding," equivalent to hydrogen bonding (see later), and an electrostatic term in which the partially negatively charged oxygen is repelled by a toroidal-like region of negative charge and attracted toward the positive central axis. The pear shape of CPK iodine atoms provides an approximate representation of this model. Such a distribution of electron density is expected from consideration of how the very large, loosely held cloud of electrons around iodine will be repelled by the more rigidly held electrons of the atoms and bonds of the attached group. All atoms in a 1,2- and 1,3-relationship and some in a 1,4-relationship to iodine are (well) within the normal van der Waals' contact distance and so will contribute most to this distortion. The asymmetry of the distortion is proposed to be the main source of the large dipole moment and contributes to the generation of a positive axial field. Nature uses this special property of iodides to increase the association constant of thyroxine for its prealbumin binding site.[26] Medicinal chemists will no doubt try similar tricks.

The effect is not restricted to iodine: bromine and sulfur show very similar but less dramatic asymmetry in their electron density distributions and potential for interactions with "nucleophiles."[27] Chloro compounds need geminal electron-attracting atoms or groups to be present, as in CCl_4 for instance, before weak intermolecular interactions with nucleophiles become observable *in solution*.[28] Nitrogen, sulfur, selenium, and a variety of other electron donors including aromatic systems have been shown to be effective partners in these types of interactions. The subject has been reviewed.[28] In fluoro compounds, the dipole is expected to arise mainly from the electronegativity-induced asymmetry of the C—F bonding electrons, with smaller contributions from lone pairs. Distortion of the small, tightly held electron cloud should be minimal, so the axial field beyond fluorine remains negative. However, the experimental and theoretical geometry of the hydrogen fluoride homo-dimer indicates that electron density is most available in the lone-pair regions, so yet again a simplistic bond dipole model would be misleading.[29]

24.5. FLUORINE SUBSTITUTION IN
ACID-SENSITIVE COMPOUNDS

Acid sensitivity has been a problem for orally administered drugs almost since the advent of modern chemotherapy: penicillin G is easily destroyed at the low pH (*ca.* pH 1.0) of parietal cell secretions. Fortunately this acidity is usually buffered and diluted to some degree by other secretions and food, with the result that this drug usually gives effective blood levels from large oral doses. Simply increasing the dose is not always possible, however, and in these circumstances the introduction of electron-attracting atoms or groups, such as fluorine, has often been successful in reducing acid sensitivity, while retaining adequate potency and a selectivity profile not too different from that of the parent. The success of penicillin V, with its phenoxyacetyl side-chain in place of the phenylacetyl side-chain of penicillin G, is largely due to its improved resistance to stomach acid.

The chemotherapy of human HIV (human immunodeficiency virus) infections provides a good recent example of this approach. The problem is that $2',3'$-dideoxy-inosine (ddI), a compound currently under clinical investigation as an anti-HIV agent, has a half-life at pH 1 and 37°C of only 30 seconds. Under these circumstances oral administration cannot be expected to provide useful effects, yet oral dosing of a simple formulation is highly desirable for the long-term therapy of a large patient population. Special formulations are available that overcome such problems, e.g., by coating tablets in a film that only dissolves in the less acidic environment beyond the stomach, but this increases costs and can be unreliable. The alternative is to modify the drug chemically in such a way as to improve its acid stability. This was achieved in this case by the introduction of a $2'$-β-fluorine which drastically destabilizes the transition state leading to the intermediate oxacarbenium ion shown in Scheme 4 (X = F). This fluoro analog is stable under the aforementioned acidic conditions and retains activity against the virus *in vitro*, but it has some disadvantages which may preclude its development. The corresponding adenine analog shows essentially identical improvements in acid stability over its nonfluorinated counterpart—again a known anti-HIV agent—but in this case there are potentially advantageous changes in the complex enzymological profile that characterizes such agents.[30] Perhaps both agents should be tested in man in order to help determine the relevance, if any, of the several *in vitro* effects.

Coated tablets are a possible answer to the problems encountered by ddI, but the acid sensitivity of the natural prostaglandin PGI_2 (16) is so great, half-life 5–10 minutes at 37°C and pH 7.4, that the development of useful drugs based on its vasodilatory and platelet aggregation inhibiting properties is only feasible if the "acid" sensitivity can be vastly improved. The problem arises through intramolecular general acid catalysis of the hydrolysis of the vinyl ether.[31] No doubt Nature has chosen to use this molecule and the still more unstable thromboxane A_2 (TXA_2) as a means of restricting their distribution and duration of action. Man has now synthesized acid-stable analogs of

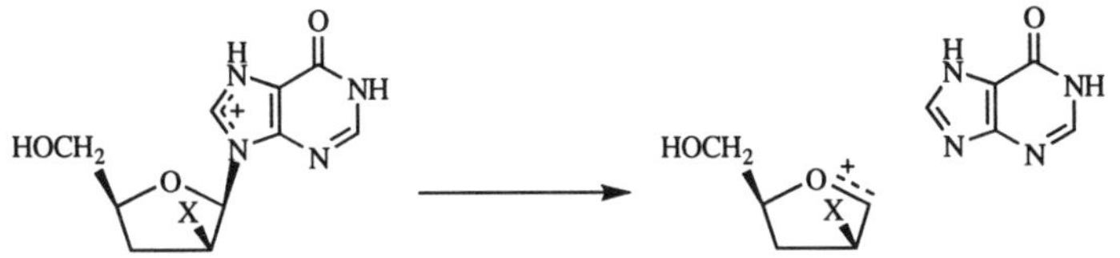

Scheme 4. Improving the stability of ddI towards acids.

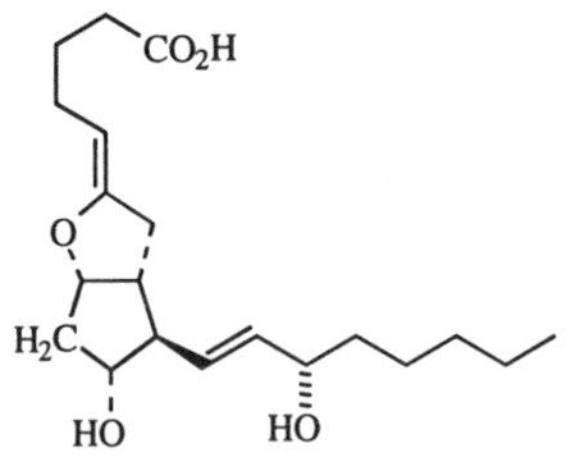

16 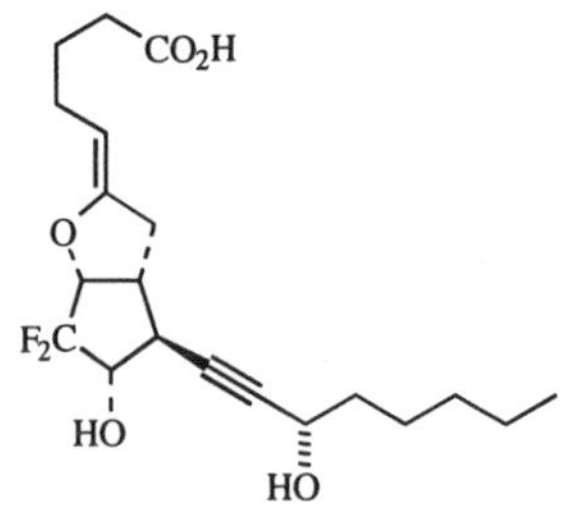   17

PGI$_2$ by changes involving the double bond, such as hydrogenation, which is very deleterious to potency, its incorporation into an aromatic ring, replacing hydrogen at C-5 by a chloro or cyano group, replacing the ether oxygen by sulfur or a methylene group, and, of course, by introducing fluorine.[31] As in the case of fluoro-ddI, a degree of success can be demonstrated in most of these approaches and, as expected, fluorination near to the vinyl ether (viz at positions 5, 7, and 10) improves acid stability dramatically. 10,10-Difluoro-13,14-dehydro-PGI$_2$ (**17**) at pH 7.4 and 37°C is only half decomposed in 24 hours, but unfortunately its duration of effect in dogs is essentially unchanged, with a half-life of only 0.6 minute. And this despite the incorporation of the acetylenic function to slow oxidation of the 15-OH group—a rapid mode of removal for olefinic analogs.

This demonstrates that in chemotherapy there are normally many obstacles on the road to a potential drug. In this case it is likely that β-oxidation of the acid, the normal catabolic route of fatty acids, is limiting activity. Difluorination at C-2 has been shown to block this type of degradation, but this device was not used here. Even if it had been, it is possible that the duration of effect would still be insufficient because of oxidation of C-19 and C-20—another very common oxidative pathway for linear saturated aliphatic chains. A terminal trifluoromethyl group should block that pathway (see section on metabolism), but even seven fluorine atoms will not guarantee success and the production costs hardly bear thinking about! Despite the lack of biological success in this last example, the use of fluorine to improve acid stability in *carefully selected* cases remains as one of the valuable tools in the medicinal chemical toolchest.

24.6. HYDROGEN BONDING IN FLUORO COMPOUNDS

The importance of hydrogen bonding to interactions in the biological sphere can scarcely be overstated. Many extensive reviews are available but a very recent paper is recommended to medicinal chemists.[32] The following discussion is broken down into two subsections, one dealing with fluorine as hydrogen bond acceptor—it is much weaker than many people suppose—the other with the influence of fluorine and chlorine substitution on the hydrogen-bonding properties of other acceptor and donor groups.

24.6.1. Fluorine as Hydrogen-Bond Acceptor

Some of the confusion over the hydrogen-bonding abilities of organofluorine compounds likely comes from the facts that hydrogen fluoride is a very powerful hydrogen-bond donor and fluoride anion is an outstanding hydrogen-bond acceptor. Given iso-basic comparisons in the gas phase, it is the strongest of all acceptors![33] The clue as to why covalently bonded fluorine is such a weak acceptor comes from a comparison of the basicities of NH_3, H_2O, and HF, or, more conveniently, the pK_a values of their conjugate acids. That of NH_4^+ is +9.3 and for H_3O^+ it is −1.7, values which are widely recorded. Nowhere can I find a value recorded for H_2F^+ in solution; but, as with water, one can easily derive the required number from the pH (in this case H_0 value) of the pure liquid, plus its molarity and self-ionization constant.[34] The answer is roughly −23; so, despite atom-type dependence in hydrogen bonding being considerable (due to changes in field *gradients* and orbital effects), it is clear that the weakness of hydrogen bonds to covalent fluorides is a reflection of how tightly held are the lone pairs in such structures: the $n_F \rightarrow \sigma^*_{Hx}$ charge transfer at a given distance is weak, and the electrostatic component from the moderately polar C—F bond produces too little attraction to cause much change in the internuclear distance and therefore to the orbital overlap. Thus the normal synergy between these contributions to hydrogen-bonding energetics is particularly weak for fluorine.

Determination of the quantitative acceptor ability of aliphatic fluoro compounds has been almost totally neglected—only one good study (involving cyclohexyl fluoride) has been found.[35] This contrasts with the vast amount of work that has been devoted to the field as a whole. One part of this work has culminated in the immediate past with the publication of large amounts of data and the development of a level of understanding that allows the calculation of hydrogen-bonding equilibrium constants in tetrachloromethane (or 1,1,1-trichloroethane) for the majority of donor–acceptor pairs that are likely to be of interest to medicinal chemists.

For solutions in tetrachloromethane, the acceptors have been scaled from a zero point defined by multiple regression, to unity for hexamethyl phosphoric triamide (HMPA). These numbers define the β_2^H scale and individual values are related to hydrogen-bonding K_B^H constants by Eq. (1).

(1) $$\beta_2^H = (\log K_B^H + 1.1)/4.636$$

Note that for hydrogen bonding to phenol, a change in β_2^H of 0.068 corresponds to a 2-fold change in equilibrium constant, so even changes of 0.04 are of interest and perhaps of relevance to medicinal chemists.

A selection of β_2^H values for halogeno compounds is recorded in Table 4, along with a selection of other data to allow them to be put into perspective.[32] The value for tetrachloromethane is necessarily zero since this is the defining solvent and, for dilute solutes, only positive changes from this baseline can be observed. It does *not* imply that tetrachloromethane is incapable of being a hydrogen-bond acceptor. In the gas phase at 0 K, where the baseline truly is zero, *ab initio* calculations indicate that even the much weaker acceptor, difluorine, forms a "hydrogen bond".[96]

Table 4. Hydrogen-Bond Acceptor Values (β_2^H) for Compounds in CCl_4

Solute	β_2^H	Solute	β_2^H	Solute	β_2^H
CCl_4	0.00	RBr	0.17	PhCl/Br/I	0.09
$CHCl_3$	0.02	RI	0.18	PhF	0.10
CH_2Cl_2	0.05	c-$C_6H_{11}F$	0.24	PhH	0.14
n-BuCl	0.11	Et_2S	0.28	PhOMe	0.26
t-BuCl	0.19	Et_2O	0.45	Ph_3PO	0.92

In the real world at 37°C (310 K) there are overwhelming entropic limitations (three degrees each of rotational and translational freedom) to the effectiveness of weak intermolecular interactions.

From the data in Table 4 we observe that fluorobenzene, $\beta_2^H = 0.10$, is a weaker acceptor than benzene itself. This fact, plus statistical considerations (six carbon atoms to one fluorine), plus the change in β_2^H in going from diethyl ether to anisole ($\delta\beta_2^H = 0.19$), leads to the conclusion that *in this system* aromatic fluorine has no detectable acceptor ability. However, there is evidence from infrared studies on *ortho*-fluorophenol in tetrachloromethane of some sort of interaction of the hydrogen-bond type. Much more relevantly, the cyclohexane/water partition coefficient of *ortho*-fluorophenol is sevenfold greater than that of its *para*-isomer, while the octanol/water values show no difference.[36] The conventional wisdom derived from the latter fact would have it that these isomers are isolipophilic, yet lipids, and even more so the central regions of natural phospholipid bilayers are much closer in character to cyclohexane than they are to octanol. If easy passage through cell membranes is relevant to the activity of a phenolic drug, then its *ortho*-fluorinated derivative might have considerable advantage; or disadvantage if that membrane restricts access to catabolic enzymes.

The difference between the above observations in nonpolar solvents arises through the intramolecularity of the second example: entropic considerations are obviously very important, as is the poor geometry of the intramolecular "hydrogen bond"—see structure **18**. Another important consideration is that the alternative low-energy conformer of the phenol **19** has lone-pair/dipole repulsion problems which might contribute as much, perhaps even more, to the energetics of this system than does the "hydrogen bond". (Trifluoromethyl)phenols might be expected to show more extreme differences between the positional isomers because of the potentially better geometry in a 6-membered intramolecular hydrogen bond in the *ortho*-isomer. Molecular modeling actually shows up a problem—two fluorines are forced to straddle the oxygen and, in the resonance-preferred *syn*-planar conformation, the hydroxylic hydrogen would point toward the positively charged carbon, rather than at a fluorine atom—see structure **20**. The alternative conformer **21** evidently suffers from even worse repulsion problems than **19** because, in tetrachloromethane, the *syn*-conformer is slightly preferred, despite the implication from the frequency of the O—H stretching vibration (3625 cm^{-1}) that the interaction is actually repulsive relative to the anti-conformer ($v_{OH} = 3605$ cm^{-1})![37] At higher concentrations in the same solvent,

18   19   20   21

^{19}F NMR analysis shows that hydrogen bonding self-association is stronger in the *ortho-* than in the *meta-* or *para-*isomers, despite the intramolecular competition. This shows that the lone-pair/dipole repulsions in conformer **21** very significantly increases the hydrogen-bond acceptor ability of the oxygen atom.[38] Partitioning (c-C_6H_{12}/H_2O) shows a twofold greater "lipophilicity" for the *ortho-* over the *para-*isomer.[36]

The extremely favorable entropic factors in the above compounds clearly apply to drugs containing similar part-structures; rather less obviously, they also apply to the pseudointramolecularity of tightly bound (regions of) drugs. In the above examples the hydrogen bonds are partly forced to occur, and similar situations can be envisaged within a drug/binding-site complex. Depending on the degree of forcing, the difference in binding energy between a fluoro compound and its hydrogen equivalent can be expected to vary between being beneficial to fluorine to being slightly beneficial to hydrogen (ignoring all other factors).

The hydrogen-bond donors considered up to this point have been (implied to be) hydroxyl groups which have the conformational freedom to present, or not present, the acidic hydrogen toward an acceptor atom. When forming a hydrogen bond, they have to give up significant amounts of freedom (entropy) in terms of the number of energy minima occupied, and the low-frequency torsional and "stretching" modes are constrained. So even the (pseudo-) intramolecular bond has its limitations. These limitations are reduced if the donor system is rigidified. The serine proteinases to be considered later provide an excellent example of how enzymes use this energetic principle to stabilize alkoxide anions: two rigidly held main-chain amidic-NH groups point from the base of a cleft in the enzyme toward an oxy-anion. Let us apply this principle to the design of a hypothetical binding site for the fluoroquinolones. No extreme measures are called for, so a single main-chain amidic-NH group is positioned to be in contact with the fluorine; much of the SAR can be rationalized because this NH$^{\delta+}$···F$^{\delta-}$ "contact" must be better than the NH$^{\delta+}$···H$^{\delta\delta+}$ contact of the parent. Other halogens are weaker acceptors and possibly also too long or too big. Size may also exclude other substituents to variable extents. On this hypothesis what would one do next? I would replace fluorine by a carbonyl group and make the necessary ring-atom changes to maintain "aromaticity."

There are many possible explanations for the potency-enhancing effect of the 6-fluoro group in the quinolone antibacterial agents, but similar enhancement of potency or binding is occasionally seen in other series for which only hydrogen-bonding or dipolar explanations seem to be viable. Two good examples of this type come from research concerned with dopamine β-hydroxylase (DBH). The first exam-

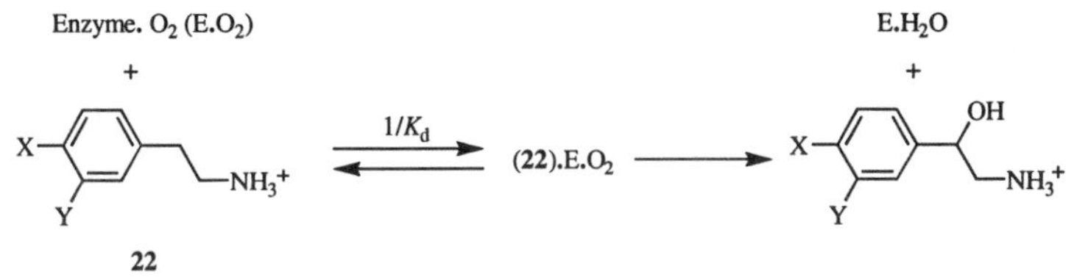

Enzyme. O_2 (E.O_2)   E.H_2O

Scheme 5.   Enzymic hydroxylation of substituted β-phenylethylamines.

ple concerns a detailed study of the enzyme mechanism shown in abbreviated form in Scheme 5, which involved the dissection of kinetic parameters for a variety of phenylethylamine derivatives.[39] Some of the equilibrium and kinetic parameters are shown in Table 5 and clearly demonstrate the superior binding of *para*-fluoro and *para*-hydroxy derivatives over the remainder. This commonality does not extend to the V_{max} values.

The second example involves the inhibition of this copper-containing enzyme by a series of *N*-arylmethylimidazolidin-2-thiones which have therapeutic potential in the treatment of cardiovascular disorders.[40] The thioamide group is believed to bind to the metal center(s) that normally activate dioxygen, while the aryl group binds, perhaps rather approximately, in the manner of the natural substrates aryl ring. The best compound (**23.i**), shown in Table 6, binds 10^6-fold more tightly than tyramine, but is not active in reducing blood pressure in rats when dosed orally. Rapid conjugation of the phenol is the likely cause, however, the still very tight binding 3,5-difluoro compound shows good oral activity. In this series it is possible that hydrogen bonding to both fluorine and the "acidic" *para*-hydrogen atom is involved in enzyme binding.

Aromatic fluoro and aromatic or aliphatic trifluoromethyl groups may, on the above analysis, possibly be involved in hydrogen bonds, but this is likely to constitute a small minority. Aliphatic monofluoro groups, on the other hand, are much better contenders for this role, with those where the fluorine is on tertiary carbon and distal

Table 5.   Equilibrium and Kinetic Constants
for Substrates of DBH

Substrate **22**			
X	Y	K_D (mM)	V_{max} (s^{-1})
OH	OH	0.96 ± 0.64	13
OH	H	0.5 ± 0.1	15
H	H	14 ± 5.3	18
CH$_3$	H	19 ± 1.4	25
OMe	H	61 ± 0.8	27
F	H	0.2 ± 0.1	50
Cl	H	9.1 ± 1.0	41
CF$_3$	H	300 ± 57	2

Table 6. Physical Properties and DBH Inhibitory Potency of
Arylmethylimidazolidin-2-thiones

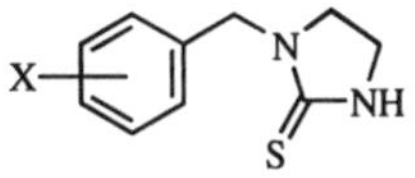

23

Compd. 23	Aryl substit.	IC_{50} (µM)	mp (°C)	log P (octanol)	Aqueous solubility[a] (mg ml^{-1} at 37°C)
a	4-CF$_3$	>1000	148	2.4[a]	1
b	4-Br	>1000	192	2.4[a]	0.3
c	4-Cl	96	189	2.25[a]	0.5
d	4-OH	2.6	188	1.02	10
e	4-H	32	145	1.55[a]	8
f	3-OH	148	167		
g	3-Cl	12	131		
h	3-F	5.6	114	1.70	14
i	3-F,4-OH	1.5	172	1.12	13
j	3,5-Cl$_2$	2.4	209	3.1[a,b]	0.04
k	3,5-F$_2$	1.2	143	1.98	3
l	3,5-F$_2$,4-OH	0.074	215	1.08	5

[a]Estimated values (this work). See Section 24.9.2.
[b]The recorded value of 1.78 is clearly erroneous.

to other functionality having the best chances of achieving strong bonds (compare n-BuCl with t-BuCl in Table 4, and dipole moments in Table 3). The use of fluorine as a replacement for oxygen probably provides more clearly defined target structures than any other category of fluorine use, and it sometimes allows one to assess the importance of hydrogen-bond donor versus acceptor functions in natural products (or drugs) containing hydroxyl groups. Fluorine can of course only be an acceptor. Several fluoro prostaglandins have already been mentioned in other connections, where the fluorine atom(s) was not potentially replacing an oxygen atom, but such F for O replacement is commonplace in this area as in most other chemotherapeutic areas and biochemical types. Some examples show the unpredictability of such work.

The prostaglandin flunoprost (24), has fluorine at position-9, but with β-orientation, unlike the α-hydroxy group of F-types or the planar 9-carbonyl of E-types. It is being clinically tested as a nasal decongestant.[41] Volunteer studies showed it to be highly potent and free from the side effects of previously used agents. The stability of formulations is important for successful development, and this consideration may

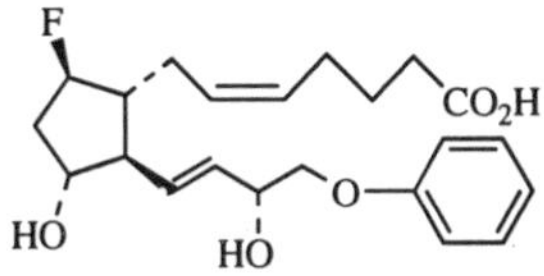

24

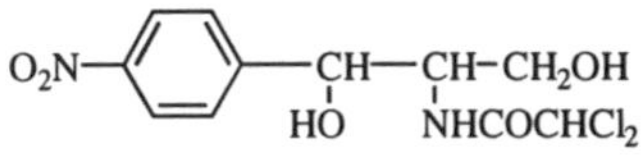

25

prevent the use of E-type agents because they readily dehydrate to give conjugated ketones.

Chloroamphenicol (25) has two aliphatic hydroxyl groups. Replacement of that on the benzylic carbon by fluorine (not the synthetic route!) to give both *erythro* and *threo* isomers produced compounds with no antibacterial or antifungal activity. The three nearby electron-attracting groups p-$O_2NC_6H_4$, dichloroacetamido, and hydroxymethyl guarantee that this fluoro compound will have, in β_2^H terms, very low hydrogen-bond acceptor ability. Conversely, the hydroxyl group in the parent drug will be a particularly powerful hydrogen-bond donor (see next subsection) so this result is not surprising. The alternative monofluoro analog, in which the primary alcohol is replaced, is reported to be active.[5]

A study of the binding of deoxy and deoxyfluoro analogs of glucose to the enzyme glycogen phosphorylase provided equilibrium constants which can almost certainly be interpreted in terms of hydrogen bonding of the fluorine substituents.[42] Together with the X-ray determined structure of the enzyme/glucose complex, this work shows that, where a hydroxyl group acts only as an acceptor, i.e., at C-1 and C-2, its replacement by fluorine (contrast hydrogen) leaves the binding constant essentially unchanged. Thus in moving from the very polar aqueous phase to the very polar binding site, the C—O and C—F bonds/lone pairs in these compounds can behave in quantitatively very similar ways; but those hydroxyl groups that are both donors and acceptors cannot be replaced with impunity.

At the qualitative level of X-ray determined structures, hydrogen bonding to fluorine almost never competes successfully with other co-present acceptor atoms.[43] The best it achieves is a bifurcated interaction, except for a single example in which an aromatic fluorine seems to be the only recipient of a hydrogen bond from water coordinated to a calcium ion.[44] Whether bifurcated interactions are generally involved in the binding of the above fluoro sugars is not known, but, if such bonding is present, the generalization of those results could be quantitatively misleading.

24.6.2. Halogeno-Substituent Effects on Hydrogen-Bond Donor and Acceptor Ability

Attention has already been drawn to the results of successive geminal chlorination of alkyl chlorides (Table 4), which is to progressively reduce β_2^H values. This is precisely what would be expected from the combination of decreasing lone-pair orbital energies and diminishing electrostatic effects. The reduction seen for the first additional chlorine is larger than for subsequent additions, and this effect is even greater if statistical corrections are applied (there are several assumptions built into this analysis, one of which is to focus on the number of like atoms rather than lone pairs; thus a factor of two for dichloromethane corresponds to a "correction" of –0.06 in β_2^H—which apparently takes the value *off* the scale, but to maintain correct comparisons one should apply corresponding "corrections" of –0.10 for trichloromethane and –0.13 for tetrachloromethane). Lone-pair repulsions may contribute to the diminishing effects. Additional examples of the weakening effects of halogen substitution on donor

Table 7.   Values of α_2^H and β_2^H for Halogeno Compounds in CCl_4

Solute	α_2^H	β_2^H	Solute	α_2^H	β_2^H
CH_3CCl_3	0.01		FCH_2CH_2OH	0.40	0.36
CH_2Cl_2	0.13	0.05	F_3CCH_2OH	0.57	0.18
$CHCl_3$	0.20	0.02	$(F_3C)_2CHOH$	0.77	0.03
$CHBrClCF_3$	0.22		PhOH	0.60	0.22
RSH	0		2-MeOPhOH	0.26	
RNH_2	0		2-ClPhOH	0.65	
$PhNH_2$	0.26	0.38	$2,6-Cl_2PhOH$	0.32	
MeCONHMe	0.38		F_5PhOH	0.76	0.02
EtOH	0.33	0.44	Cl_5PhOH	0.55	
H_2O	0.35	0.38	$4-NO_2,3-CF_3PhOH$	0.95	
RNH_2		0.70	pyridine		0.62
$CF_3CH_2NH_2$		0.36	2-F-pyridine		0.43
$MeCONEt_2$		0.71	2-Cl-pyridine		0.45
CF_3CONEt_2		0.47	3-Cl-pyridine		0.49

ability are shown in Table 7. Again, progressively weakening effects of multiple substitution are evident.

Also shown in Table 7 are α_2^H values for proton donor ability.[32] The values are related to equilibrium constants in a precisely identical manner to that which applies to acceptors: one simply substitutes α_2^H for β_2^H in Eq. (1). This exact correspondence of zero points is vital to the ability to predict a range of equilibrium constants through the application of Eq. (2).

$$(2) \qquad \log K = 7.354\ \alpha_2^H \cdot \beta_2^H - 1.094$$

Returning to the data in Table 7, the α_2^H and β_2^H values for hexafluoroisopropanol are particularly interesting since the extreme separation of α_2^H and β_2^H values can potentially be of great interest in medicinal chemistry. Also of particular note are the zero α_2^H values for aliphatic amines and thiols and the clearly significant values of many C—H donors when activated by nearby halogens. This will come as a rude shock to many chemists, medicinal and otherwise.

These are qualitative aspects—we have quantitative data here and would hope to apply them in that way to drug research. What are the prospects? Well, according to two highly respected chemists, very poor! Their view, basically, is that it is the release of solvating water molecules, from the unbound drug and binding site, which produces an *entropic* advantage for the bound complex.[45,46] This then results in each hydrogen bond contributing roughly 1.0 ± 0.5 kcal mol^{-1} to the binding energy, irrespective of its nature.[46] However, α_2^H and β_2^H values relate to free energy, not entropy, so there are problems still to be solved.

Let us return to Eq. (2) and examine it in more detail; the fact that $\log K$ is related to the *product* of α_2^H with β_2^H has profound consequences for medicinal chemistry

26   27

and biology. In that order, because as medicinal chemists we can manipulate properties—through the use of fluorine substitution, for example—more easily and over a wider range than Nature can normally manage. A very neat demonstration of these consequences is to take separate solutions of tetrahydrofuran (THF, bp 65°C) and hexafluoroisopropanol (HFIP, bp 59°C) in water—they are both fully miscible—and mix them. Two phases are formed; the lower one can be separated and distilled at 99–100°C, but it is not water—it is a hydrogen-bonded 1:1 complex between THF and HFIP.[47] This example is far removed from dilute solutions in tetrachloromethane and, in any discussion of the phenomenon, many factors should be considered in addition to "intrinsic" hydrogen-bonding abilities, but examination of the data in Table 7 shows that such a three-solute system would be predicted to associate in the above manner, since Eq. (2) shows that strong donors bond preferentially to strong acceptors. This principle can be extended to the biological sphere.

Prostatic cancer, the second most common cause of death in American males in 1974, can be treated with flutamide if tumor growth is androgen-dependent.[48] The anti-tumor effects of the drug are the result of rapid and selective oxidative metabolism at its tertiary aliphatic C—H bond to give hydroxyflutamide (26), which is a pure androgen antagonist working through competitive inhibition of binding of testosterone and dihydrotestosterone to androgen receptors on tumor cells.[49] Unfortunately, concurrent antagonism of hypothalamic receptors results in a feedback stimulation of testicular androgen synthesis and an increased agonist challenge. A tissue-selective antagonist might avoid this problem. The discovery of such selective antagonists at ICI Pharmaceuticals involved solution studies of the conformational and hydrogen-bonding properties of hydroxyflutamide and various analogs.[50] These revealed that the affinity for the receptor was, *inter alia*, strongly dependent on the hydrogen-bond donor ability of the hydroxyl group. Flutamide itself had no measurable affinity, while the good affinity of hydroxyflutamide ($\alpha_2^H = ca$ 0.65) was increased a further threefold (racemate) by changing one of the methyl groups to trifluoromethyl ($\alpha_2^H = ca$ 0.77). A similar structural change in a more complex series resulted in the compounds displaying partial agonist properties—a highly deleterious change.[51] It is possibly relevant that this structural change results in a greater tendency to assume a conformation in which the hydroxyl group behaves as a donor to the amidic oxygen atom; this conformation is dominant (60% in 1,1,1-trichloroethane) in compound 27 in which the hydroxyl group will be less basic and more acidic than a des-fluoro analog. All compounds with strong electron-withdrawing groups in the aryl ring exist essentially exclusively (in 1,1,1-trichloroethane) in conformations corresponding to 26. In

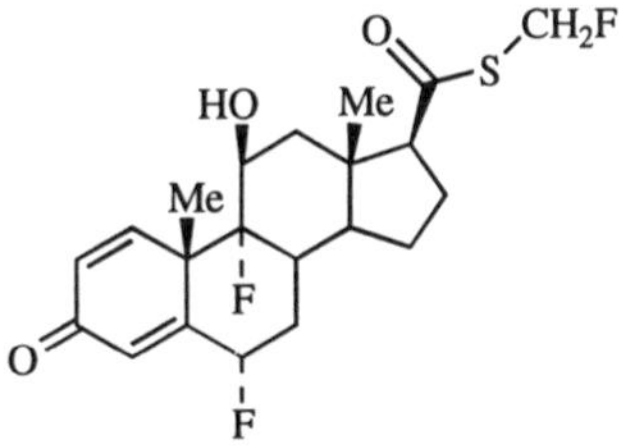

28 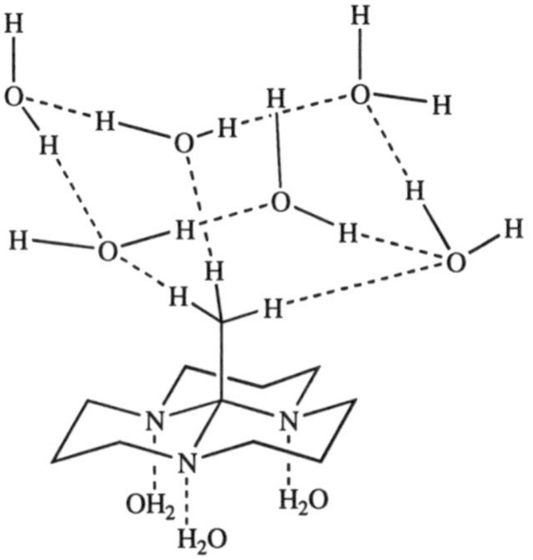   29

this conformation the weak NH⋯O hydrogen bond is augmented by strong dipolar effects which render the hydroxyl group an excellent donor and a very weak acceptor: the measured donor properties of hydroxyflutamide are midway between those of trifluoroethanol and HFIP. X-ray crystallography of hydroxyflutamide shows the same NH⋯O interaction in the solid state, and furthermore reveals that the nitro and amide groups are almost coplanar with the aromatic ring. This conformation displays a convincing resemblance to the natural agonists when the structures are overlaid (with the CF_3 group filling the space near C-1, C-2 in the agonists).[52]

Fluorinated steroids and particularly 9α-fluorocortisol and related compounds have been the subject of much study and speculation regarding the changes in potency and selectivity which result from the presence of the halogen atom. Among these suggestions, that concerning increased hydrogen-bond donor ability of the axial 11β-hydroxyl group is relevant here.[1] As can be seen from the data in Table 7 for fluoroethanol and ethanol, the changes in α_2^H and β_2^H could, in combination with a receptor containing a good acceptor and no donor for the 11-hydroxyl group, produce *in vitro* potency changes of about twofold for each such effect. However, fluoroethanol in a nonpolar environment exists in a *gauche–gauche* conformation with a weak intramolecular hydrogen bond, so both $\Delta\alpha_2^H$ and $-\Delta\beta_2^H$ values should be larger for the app-conformation in the steroid and differences in potency of an order of magnitude could conceivable arise from hydrogen-bonding effects alone. However, possibilities are not probabilities and the suggestion that improved resistance to oxidative metabolism of the 11β-hydroxyl function increases potency has excellent chemical precedent and some pharmacokinetic backing.[3] Also, the observation of considerable changes in the conformation of the A-ring due to 9α-fluorination might be expected to influence both potency and selectivity.[53,54]

Systemic selectivity between anti-inflammatory and mineralocorticoid activities is also achieved by substitution in ring D, but a recent trend has been toward topical versus systemic, i.e., distributional selectivity. Compounds containing 17-carboxylic acid esters can achieve such selectivity through their rapid hydrolytic inactivation by enzymes in the blood and/or liver.[55] A recent compound of this type containing a fluoromethylthioester—fluticasone (**28**)—is particularly intriguing because of the extremely unusual functional group and the possibility that the terminal fluorine may be involved as a hydrogen-bonding mimic of the natural C-21 hydroxyl group.[56] This fluorine will also make the ester much more susceptible to hydrolysis.

The acidic hydrogens of many volatile anesthetics are poor-to-moderate hydrogen-bond donors, but this property can be a major determinant of both potency and specificity, as shown in the first and so far only convincing published example of the application of hydrogen-bonding data to QSAR studies.[57] These hydrogen bonds are assumed to occur within a very low polarity environment, in this case cell membranes. The latest volatile anesthetic to be tried in humans, desflurane $CHF_2OCHFCF_3$,[58] contains two acidic hydrogens and these can be estimated from Ref. 57 to enhance its potency by almost twentyfold. In contrast, the methyl groups in CH_3CFCl_2 and related compounds appear not to be hydrogen-bond donors *in vivo*: this certainly relates to the enthalpy versus entropy balance referred to in Section 24.6.1. The pseudointramolecularity of most drug/binding site interactions, which overcomes many of the entropic limitations on hydrogen bonding, has parallels in the environments of low-unit molecular mass crystals, wherein "weak" hydrogen-bonding potential is more frequently satisfied than in solution. The orthoamide **29** as its crystalline trihydrate provides a remarkable example of this phenomenon, in which hydrogen bonding occurs to all three hydrogens of a methyl group[59]: this description does not derive from the $H_2O\cdots H$ distance, which is that expected for a normal van der Waals contact, 2.67 Å, but from the fact that the methyl group eclipses the CN_3 unit (N—C—C—H torsion angles are *ca* 8°). The enthalpy of each $CH\cdots O$ hydrogen bond (the C—H$\cdots$O angle is 170°) is estimated to be $\geq$ 1.5–2 kcal mol^{-1}. The enhanced $C^{\delta-}$—$H^{\delta+}$ bond dipole due to σ-resonance effects ($\sigma_{CH} \rightarrow \sigma^*_{CN}$ and $n_N \rightarrow \sigma^*_{CC}$), repulsion effects[60] ($n_N \leftrightarrow \sigma_{CC}$) and increased water basicity due to hydrogen bonding to nitrogen lone pairs, will contribute to this unique deviation from the expected staggered geometry, but the main effect is probably due to the rigid orientation of all three CN: $\cdots$H—OH fields producing near to their maximum possible polarization of the CH_3 group.

In the crystalline state it is generally the case that aliphatic and olefinic/aromatic C—H bonds adjacent to nitrogen atoms (and other electron-attracting groups) have much greater than average probability of being involved in hydrogen bonds if the H$\cdots$X (X = O,N,S,Cl atoms) distance is the main defining criterion of such bonds.[61] If this criterion is changed to one of preferred orientation, then aromatic C—H bonds seem to hydrogen-bond to the π-clouds of adjacent molecules, and this finding extends to Ar$\cdots$Ar contacts in proteins, where near-perpendicular contacts are a very common finding and overlaid coplanar orientations are very rare.[62] This contrasts with the 1:1 complexes formed between various polyfluorinated and nonfluorinated aromatics, which crystallize with their rings essentially parallel and often face-to-face.[22] This is not due to intermolecular $\pi \rightarrow \pi^*$ charge transfer in the ground state (a concept which is overused by medicinal chemists in particular) but rather to quadrupole/multipole attractions and $\pi:\sigma$ interactions[63]: benzene tends to behave as a ring of negative charge surrounded by a ring of positive hydrogens, while hexafluorobenzene does the reverse and face-to-face contact matches these fields. The good anesthetic characteristics of hexafluorobenzene are consistent with its behavior as a hydrogen-bond donor, which is obviously impossible for a compound lacking hydrogen, but theoretical calculations show that hydrogen-bond acceptor atoms are attracted to the positive field at the ring center, with interaction energies similar to those found for trichloromethane.[64]

These characteristics of polyfluorinated versus hydrocarbon rings represent important differences for the design and understanding of drug interactions, and such differences will extend to electron-attracting (both π and σ) groups other than fluorine. Returning to speculate yet again on the example of sparfloxacin (**15**), its aromatic ring might be a "hydrogen-bond donor," or the whole bicyclic system may willingly lie parallel to an aromatic protein residue; a tyrosine anion could fulfil both functions.

24.7. FLUOROKETONE ENZYME INHIBITORS

The chronic incapacity and premature death due to severe rheumatoid arthritis and pulmonary emphysema are, in part, due to an imbalance between proteinase and endogenous antiproteinase activities and the resulting excessive destruction of connective tissue by neutrophile-derived proteolytic enzymes. Inhibitors of these enzymes with appropriate *in vivo* characteristics are expected to be useful drugs.[65] Various aldehydes and chloromethyl ketones have long been known to be inhibitors of such serine- and cysteine-based enzymes, but none of these is promising *in vivo*. Both types of inhibitors can act by reversibly forming tetrahedral adducts, hemiacetals, or hemiketals, with the reactive enzymic serine-OH or cysteine-SH groups, i.e., as transition-state analog inhibitors. 3-Chloro-1-phenoxypropan-2-one powerfully inhibits acetylcholine esterase *only* by this mechanism despite its high chemical reactivity, but the same and related compounds are good irreversible inhibitors of chymotrypsin, another serine-based enzyme.[66] The problem with chloromethyl ketones is that they generally lack specificity and the high (*ca* 4 mM) concentration of glutathione in cells tends to rapidly destroy such compounds *in vivo*. Fluoromethyl ketones should be less reactive, and indeed they react 500 times more slowly with glutathione; in spite of this, some cysteinyl proteinases, but not serine proteinases, are readily alkylated by appropriate peptidyl fluoromethyl ketones, because increased binding (K_i) can more than offset their only somewhat reduced intracomplex alkylating ability (k_i).[67] Acyloxy and sulfonium groups are effective replacements for fluorine despite their size and charge differences; the very large 2,6-bis-(trifluoromethyl)benzoyloxy group is particularly effective *in vitro* and should be sterically shielded from inactivation by esterases *in vivo*. It was then potentially surprising that peptidylmethyl and especially the trifluoromethyl ketones were found to be essentially inactive against cysteine-based enzymes.

These latter structural changes produce very different results in serine proteinases, where the methyl ketones are again ineffective, but some trifluoromethyl ketones bind much more strongly than previously described aldehydes and exhibit much greater

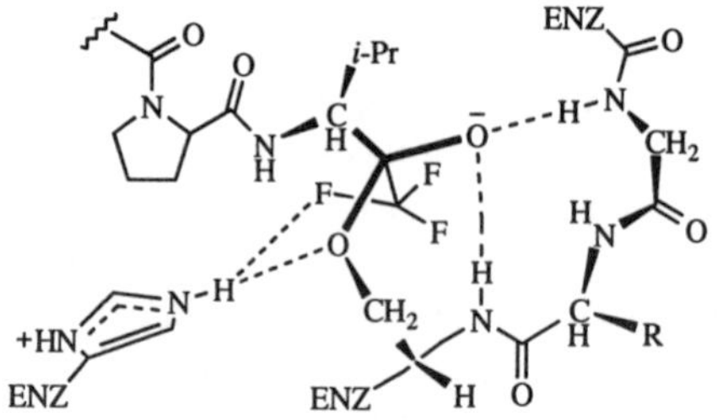

30

duration of action *in vivo*.[68] X-Ray data and NMR studies reveal that such compounds bind as transition state inhibitors in the truest sense, since it is the anion of the hemiketal that binds tightly in the "oxyanion hole" (see structure **30**).[69] The pK_a of this group is substantially below 4.9 and therefore more than 4.2 units lower than the pK_a of model trifluoromethyl hemiketals in water.[70] This reflects the strong stabilization of the anion by the amide groups forming the oxyanion hole, the poor compatibility of the site with an hydroxyl group, and the potentially large effects of CF_3 and other strongly dipolar groups inside proteins, where field reductions through dipole reorientation of surrounding groups is very limited. One of the fluorine atoms of the CF_3 group is weakly involved, along with the more strongly interacting serine oxygen, in a bifurcated hydrogen bond with one of the NH groups of the cationic imidazolium ring of His-57. In a related α,α-difluoro-β-ketoamide inhibitor complex, one of the fluorine atoms interacts very strongly, as judged by the NH···F heavy-atom distance of 2.58 Å, but even here it is bifurcated with the serine oxygen atom.[71]

The advantages produced by electron-withdrawing groups close to the carbonyl function can be analyzed approximately according to Scheme 6. Given that the hemiketal anion binds much the most strongly, and ignoring the thiol adducts and the potentially very slow kinetics of inhibition for the present, it should be advantageous

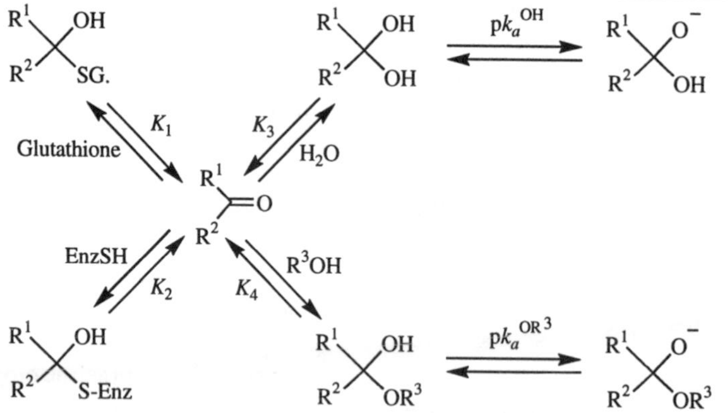

Scheme 6. Generalized mechanisms for hemiketal formation and for inhibition of enzymes by ketones.

to increase the electron withdrawal until [hydrate anion]/[hydrate]+[ketone] = 1 (approximately); further increases will only serve to weaken the hydrogen bonding in the oxyanion hole and further reduce the kinetics of inhibition. Compounds that achieve this target will have $K_3 >> 1$ in all likely candidates, so the objective becomes that of $pK_a^{OH} = ca$ 7.5. Other points worth noting are that: (1) the result of replacing aldehydic hydrogen by a trifluoromethyl group is to increase K_3 much more than K_4 (R_3 = e.g. CH_3),[72] due to both steric and other effects: the nonsteric effects are probably associated with the strong hydrogen bonds to the two acidic hydrogens of such gem-diols, in comparison with only one for the hemiketals, and the stronger solvation of alcohol lone pairs than of water lone pairs combined with the much reduced solvation of these in their carbonyl adducts; (2) thiols add 10^4-fold more easily to some carbonyl groups than does water,[73] so glutathione inside cells (ca 4×10^{-3} molar) will compete, but not severely so, with 55 molar water for reaction with such functionalities; and (3) the fact that the cysteinyl proteinases examined so far are not inhibited by trifluoromethyl ketones implies an extraordinarily severe steric/elecrostatic problem, which may not extend to all such enzymes nor to other important thiols in other proteins. Selectivity, as always, has to be the name of the game.

The tetrahedral carbonyl adducts with cysteine and serine proteinases are particularly effective inhibitors because they resemble the transition states of normal substrate hydrolysis. For the acid proteinases in which no covalently bound enzyme intermediates are formed, one might still expect ketone hydrates to be good inhibitors. Such is the case and several fluorinated ketones show good inhib: on of enzymes in this category. Various α,α,-difluoro-β-ketoamides are good inhibitors of renin and one compound of this type inhibits pepsin at 6×10^{-11} M![74] Even though binding to the enzyme is fast, this implies that the off rate could be in the region of one day, so inhibition is effectively irreversible.[75]

24.8. INHIBITION OF PYRIDOXAL-DEPENDENT ENZYMES

This fascinating and extremely versatile group of enzymes has been the proving ground for a wide array of mechanism-based inhibitor types, many of which rely on fluoride "ion" being eliminated for the generation of electrophilic olefins. Inhibitor design relies on the fact that the mechanisms of action of many of the enzymes are moderately well established, with X-ray crystallography providing protein organizational details in the case of aspartate transaminases. They all proceed through a transaldimination of an enzyme lysine-linked pyridoxal 5'-phosphate (see Scheme 7) and the loss of a proton, carbon dioxide or the group YCX_2 (e.g., serine $\rightarrow$ formaldehyde) to give quinonoid-like species with the potential to react further in many and diverse ways.

The entire field has been the subject of an excellent recent review,[76] so only two target enzymes will be discussed—hopefully enough to give some insight into the chemically interesting complexities that abound in the area.

The alanine racemases are a group of pyridoxal 5'-phosphate-containing bacterial enzymes that catalyze the racemization of alanine, providing D-alanine for cell-wall

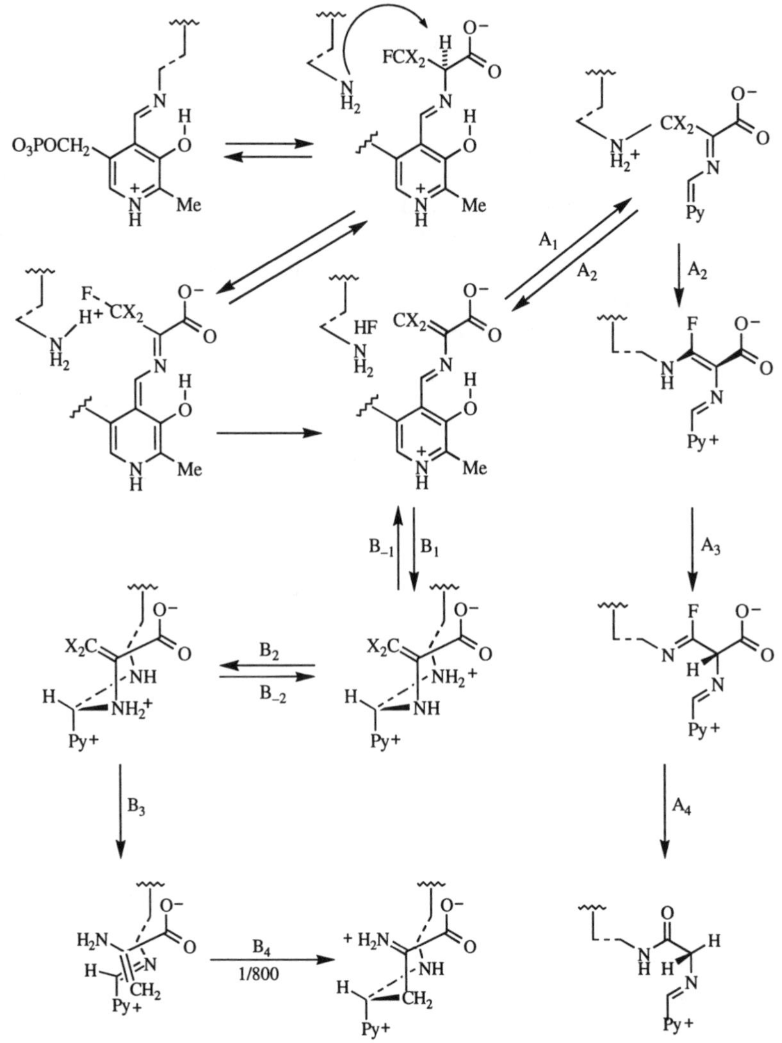

Scheme 7. Mechanisms of inhibition of pyridoxal-dependent enzymes.

synthesis and growth of the organism. As such they have long been a target of antibacterial research: 2-deutero-3-fluoro-D-alanine, in combination with a pro-drug of D-cycloserine, has excellent broad-spectrum antibacterial effects in mice.[77] Enzyme inactivation by this agent, and other Cα-fluoromethyl amino acids and amines, was originally considered to be due to capture of intermediate α,β-unsaturated imines, in Michael fashion, by the active site lysine, path A, but it now seems likely that (all?) these agents make use of the more complex pathway B1 → B2 → B3 → B4 in which

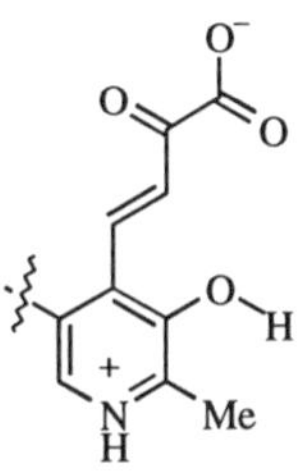

31 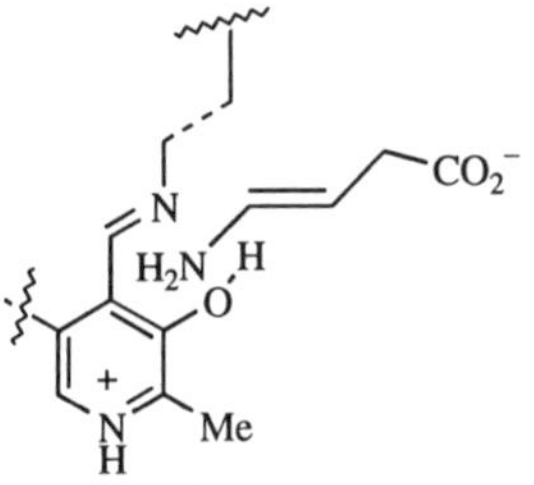   32 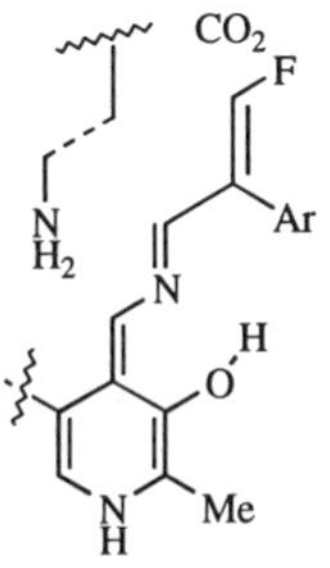   33

a very stable C—C bond is eventually formed. Certainly such a bond is present in a pyridoxal derived product (31) after enzyme denaturation. It may well be that a Michael adduct is formed, but this then reverses too easily and essentially irreversible capture depends on the alterative chemistry. For fluoroalanine the process can be very inefficient: the enzyme from *S. typhimurium (dad B)* turns over on average 800 times (to give pyruvate anion) prior to the inactivation event (partition ratio = 800). Other enzyme / inhibitor combinations are more efficient with partition ratios ranging even to zero. One example of a partition ratio of zero (reaction always results in inactivation) is for trifluoroalanine in combination with alanine racemase from *B. stearothermophilus*. Here reaction seems to occur exclusively via attack at the doubly activated (carboxylate-arginine ion pair//pyridinium ring) difluoromethylene group, but the very unfavorable equilibrium B_2/B_{-2} may be largely responsible for this apparent selectivity. The initial, extensively conjugated product is formed with the loss of only two fluorides, but subsequent partial denaturation of the enzyme results in the rapid formation of a less conjugated system, followed by the slower loss of CO_2 and the final fluoride. The active site lysine was shown to be the covalently modified group[78]—probably in the manner shown on the right of Scheme 7.

The loss of fluoride within the active sites of most of these enzymes is likely to occur by general acid catalysis leading to the direct formation of HF, rather than the extremely solvation-dependent fluoride anion.

The second enzyme to be considered, γ-aminobutyrate transaminase (GABA-T), is the key enzyme controlling the catabolism of GABA (a neurotransmitter) in the brain. Of various halogenated GABA analogs tested, the (S)-4-fluoromethyl and 4-trifluoromethyl derivatives irreversibly inactivate the enzyme, whereas (R,S)-3-chloro- and (R,S)-3-fluoro-GABA do not. The first-named compound has been shown to use the enamine/C—C bond-forming inactivation mechanism, but in this case the partition ratio is zero, probably because the enamine is of a very reactive type and is anchored to the enzyme by both its carboxylate group and its lipophilic chain, thus reducing its tendency to diffuse away.[79] Dehydroalanine, on the other hand, is anchored mainly by its carboxylate group in an ion pair with an arginine residue—a combination that should lead to much reduced reactivity of the enamine and its relatively easy loss to the medium. The failure of the 3-halogeno-GABA compounds to inactivate this type of enzyme probably relates to the lack of proximity of the appropriate groups, as implied in **32**, thus allowing alternative chemistry or product loss to dissipate the threat. A recent approach to the selective inhibition of monoamine oxidase A (MAO-A) in the brain[80] probably relies for its success on a similar lack of proximity of a lysine amino group to a potentially reactive fluoromethylene group (see structure **33**). The latter group is transiently formed when the enzyme, aromatic amino-acid decarboxylase, produces the mechanism-based MAO-A inhibiting fluoromethylene amine from its amino-acid prodrug. The approach depends on the decarboxylase enzyme not being irreversibly inhibited.

24.9. DRUG ABSORPTION AND DISTRIBUTION

Two misconceptions are apparent in most reviews concerning fluorine in a biological context. The more general one is that through enhancing lipophilicity, fluorine substituents assist absorption and tissue penetration; the second is that a CF_3 substituent is extremely lipophilic. The author is reluctant to accept such views even as the generalizations they clearly represent. Indeed the use of perfluoro compounds in blood substitutes is only possible because of their lipophobic character. Were they not immiscible with lipids, they would kill cells by disrupting their membranes. Such compounds are weakly lipophilic but extremely hydrophobic, hence their partition coefficients are typically very large. Because of this fact their initial rates of diffusional transfer through tissues, i.e., across a series of aqueous compartments separated by a series of membranes, will be very slow: each transition from a lipoidal membrane to the next aqueous phase is highly unfavorable. Very hydrophilic drugs have problems in the reverse sense, so the early maximization of diffusional transport of any drug—fluorinated or otherwise—requires that "the" partition coefficient should be close to unity,[81] i.e., $\log P \approx 0$ (larger values—at constant size/diffusivity—are not deleterious if and when an idealized steady-state is achieved). The question that must follow such an analysis is: which of the common solvents might adequately represent a membrane?

Cell membranes are highly organized entities and one needs to distinguish between equilibrium partitioning data *into* undefined regions of membranes or liposomes, as against *through* natural membranes. A more complete analysis[82] of permeability data generated by Gutknecht and co-workers[83] allows a decision for lipid bilayer membranes formed from egg phosphatidylcholine, which is that a saturated hydrocarbon, e.g., hexadecane or, even better, a very soft hydrocarbon polymer, is the best model of such a kinetic barrier.

Natural membranes are much more complex, but they are not expected to differ substantially from this picture except perhaps for those of intestinal epithelial cells under some circumstances and for interactions with compounds possessing extreme hydrogen-bond donor or acceptor characteristics. Therefore the influence of fluorine substitution on hydrocarbon/water partitioning of compounds would be very relevant to their kinetics of absorption and distribution, while octanol data would be more appropriate to their nonspecific equilibrium binding to tissues generally. Only the latter type of data are currently available for other than a trivial number of fluoro compounds;

Table 8.   Inhibition of Rat Embryo Limb Cell Differentiation (*in Vitro*) by
Some Antifungal Agents

	X	IC_{50} (µM)	log $P_{octanol}$
	2,4-Cl_2	18	1.53
	4-Cl	115	0.93
	2,4-F_2	131	0.44
	2-Cl,4-CF_3	174	1.85
	4-CF_3	414	1.15
	2-F,4-CF_3	617	1.25

but, before discussing the octanol data, it is relevant to record that the low cohesive energy density/polarizability, which causes fluorocarbons to be immiscible with hydrocarbons,[84] is likely to be a relevant parameter for correlating the teratogenicity of the series of antifungal agents shown in Table 8. Replacing chlorine by the much less polarizable fluorine, together with the effects of fluorine and particularly CF_3 groups on the polarizability of the benzene π-system, results in reduced teratogenicity *in vitro* and *in vivo* (data not shown).[85] Note that there is no correlation of teratogenicity with the octanol log P values, and hydrocarbon log P values, with these substituents, will be very strongly correlated with those in octanol. An impression of the potential magnitude of such polarizability effects can be gained from inverted systems: the log P values for 4-ethylphenol in n-C_6F_{14}, n-C_6H_{14}, CH_2ClCH_2Cl, and octanol are –0.26, 1.12, 2.33, and 3.33, respectively.[86] Much smaller but still significant effects are likely to apply to the interactions of the compounds discussed above with a hypothetical enzyme or receptor involved in teratogenesis.

24.9.1.   Fluorine Substitution in Relation to log $P_{octanol}$ Values

In the absence of nearby hydrogen-bonding groups, which includes double and triple CC bonds, the replacement of up to at least three aliphatic hydrogens by fluorine results in reduced partition coefficients (see Table 9).[36]

However, if hydrogen-bonding atoms or groups have their pK_a values substantially reduced, then positive changes are seen; and the larger the dipole moment and hydrogen-bonding solvation number of the parent, the larger can be the increment. The change of 2.2 units between the π-values of CH_3SO_2 and CF_3SO_2 attached to simple aromatics is a well known example of the effect. Pentafluorophenyl groups have much smaller effects, but the log P values for pentafluorophenol (3.23), hexafluorobenzene (2.22), and benzene (2.13) reveal an anomaly. These values do not indicate that a phenolic hydroxyl group can somehow become extremely lipophilic, rather they demonstrate that some minor species in water-saturated octanol [possibly $O(H\cdots octanol)_2$] is a substantially stronger hydrogen-bond acceptor than octanol-saturated water, and *very* strong donors (which includes some heterocyclic NH groups) give the impression of being lipophilic. In a biological environment such

Table 9. Octanol log P Values for Some Fluoro Compounds

R–CH$_{3-n}$F$_n$	Value of n			
	0	1	2	3
H—	1.09	0.51	0.20	0.64 (CF$_4$ = 1.18)
PhCH$_2$CH$_2$—	3.63	2.95		3.31
HOCH$_2$—	–0.31	–0.8^a		0.36^a
HOCH(CF$_3$)—	0.70 (Pr^{i}OH = 0.05)			1.66
Me$_2$NCH$_2$—	0.70			1.06

^{a}Average of two or more values.

(e.g., antiseptic) phenols will (if not ionized) try to find acceptors much more powerful than those present in water or fats: phospholipid membranes with their phosphate anions and embedded proteins should provide an ideal environment for both their "non"-polar and strong donor characteristics. In crossing the membrane, however, the hydrocarbon-like center brings into view the other side of the coin—the absence of significant acceptor properties in hydrocarbons results in pentafluorophenol having slightly negative log $P_{\text{hydrocarbon}}$ values (–0.30 in n-hexane; –0.52 in cyclohexane), while the values for benzene (and probably hexafluorobenzene) remain similar to or slightly greater than the octanol value given above.

The enhanced hydrogen-bond acceptor properties of 2-(trifluoromethyl)phenol have been mentioned previously (Section 24.6.1). Similar effects are reflected in the partitioning of 8-substituted quinolines; thus the trifluoromethyl compound, log P = 2.50, is slightly less lipophilic than the methyl compound, log P = 2.60, despite greater steric screening of the nitrogen lone pair. The 8-chloro compound is also more hydrophilic than predicted (log P = 2.33, quinoline = 2.04) by 0.4 of a unit—so the prediction must be that 8-fluoroquinoline will be more hydrophilic than its parent compound.

24.9.2. Solubility Effects

In order to be absorbed after oral dosage, a drug must dissolve either in the aqueous phase or the bile micelle phase of the gut contents, and it must do so rather rapidly if the flow is not to carry it from the body or even from the small intestine where the great majority of absorption normally takes place. The aqueous route is by far the most common and poor solubility in water is a frequent limitation on drug bioavailability. The approximate relationship of Eq. (3) for nonionized drugs[87] shows that where solubility is known to be a problem, reducing melting point and/or log P should be beneficial.

(3) $$\log S_w = 1.6 - 1.1 \log P_{\text{octanol}} - 0.012 \,(MP - 37)$$

(S_w is the molar aqueous solubility at 37°C; MP is the melting point in °C)

The relative values of the latter parameter can be estimated with confidence for most projected target compounds and a substantial advantage for fluoro over heavier halogeno and CF_3 substituents will be the norm. Melting points are far less predictable, but those shown in Table 6 (Section 24.6.1) for a set of aromatic compounds are typical in that the presence of symmetry elements, e.g., *para* substitution, usually increases melting point relative to the *meta* equivalents; the relatively free C—CF_3 rotor will normally be restricted in the solid so this helps reduce melting points; the greater dispersion forces and polarizability of the heavier halogens tends to progressively increase melting points.

Overall, fluorine is much superior to its common rivals (but not to hydrogen) in terms of its effects on aqueous solubility and, since relative oil (bile micelle) solubilities are expected to be dominated by melting point differences (for compounds which vary only in their nonhydrogen bonding regions), it will usually have substantial advantages in extremely lipophilic, oil-soluble series as well.

24.10. FLUORINATION EFFECTS ON METABOLISM AND ELIMINATION

By far the most common reason for incorporating fluorine into a potential drug is to reduce the rate of oxidative metabolism of a constituent aromatic ring. The ploy is frequently successful, hence its popularity, but the heavier halogens and some other electron-withdrawing groups can provide much greater protection of exposed aryl residues than that obtainable from a *single* fluorine substituent. The use of these alternatives, however, is sometimes limited by potency, selectivity, or solubility constraints as discussed in the previous Section. In such cases, two fluorine atoms in the one ring will usually provide excellent protection against oxidation.

In saturated systems the higher chemical reactivity of most substituents leads to greater reliance on fluorine, and particularly on the almost inert polyfluorinated groups CF_3, C_2F_5, n-C_3F_7, and CHF_2CF_2. The origin of this increased oxidative stability has essentially nothing to do with the greater strength of the carbon–fluorine bond relative to the carbon–hydrogen bond, despite frequent assertions to the contrary. Biological oxidations do not involve the isolated homolysis of C—H or C—F bonds, so their strengths cannot be directly related to oxidation rates; much more relevant are the high bond energies and heats of formation of H—O and C—O bonds relative to those of F—O bonds. The latter are so grossly unfavorable that alternative mechanisms which avoid "attack" at fluorine and instead lead to the loss of fluoride or HF will probably always apply in the biological sphere. The protection provided by an aliphatic CF_3 or related group extends to adjacent CH or CH_2 groups, and even somewhat to β-related hydrogens. This results from field-effect destabilization of the transition state (**34**; R_F = per- or poly-fluoroalkyl) and is somewhat analogous to the selectivity observed in the radical chlorination of fluoroalkanes.[88] Most of the sp^3C—H oxidations carried out by the ubiquitous cytochrome P_{450} family of enzymes proceed through radical intermediates, with radical cations frequently involved if the carbon is attached to an easily oxidized atom or group. The rate retardation results from the electron deficiency

$$CH_3CH_2CH_2CH_3 \qquad CF_3CH_2CH_2CH_3$$

34

$$\downarrow Cl_2 \qquad\qquad \downarrow Cl_2$$

$CH_3CH_2CH_2CH_2Cl$	1	
$CH_3CH_2CHClCH_3$	4	

$CF_3CH_2CH_2CH_2Cl$	5	
$CF_3CH_2CHClCH_3$	4	
$CF_3CHClCHCH_3$	0	

at the radical or radical-cation center and the field due to the developing very acidic H—OFe group being opposed by the field due to the per- or poly-fluoroalkyl (R_F) group.

Benzylic hydrogens are particularly sensitive to radical attack, so it is interesting that among the *para*-substituents investigated, only fluorine destabilizes radical character (relative to hydrogen), $\sigma_{\alpha\cdot} = -0.011$, and CF_3 is only marginally stabilizing.[89] However, the oxidation of benzylic hydrogens will generate very significant electron deficiency at carbon and rate constants are expected to correlate with a combination of $\sigma_{\alpha\cdot}$, σ_F, and σ_R parameters in an idealized series where steric effects and enzyme fit are invariant. In practice, steric access to the active site often dominates other factors and some of the protection provided by the larger halogens and CF_3 groups comes from this source.

The objective in slowing oxidative metabolism can be to avoid or reduce the extent of formation of reactive species: many epoxides, quinone-methides, etc., are likely or proven toxophores, and even extreme measures to reduce their formation may be appropriate.[90] Such is not the case if the objective is simply to extend *in vivo* half-life. In most cases, the optimum for successful development and freedom of use is a drug with a half-life in man and laboratory animals of 12–24 hours, and which makes use of both renal and biliary elimination. This ideal is almost never achieved, but the quinolone antibacterial difloxacin (**9a**) comes close and provides useful lessons. Its single fluorine in the *N*-1 aryl group provides entirely adequate protection here, since oxidation occurs mainly in the *N*-methyl piperidine ring, and conjugation—in this case glucuronide ester formation—does not require the oxidative introduction of, e.g., hydroxyl group. Its 26-hour half-life in man (half-life typically varies ±2-fold from the mean) is much longer than the class norm, 3–7 hours, due to its reduced excretion in urine—which in turn reflects its increased ability to passively diffuse out of the incipient urine (an increasingly concentrated plasma ultrafiltrate), through the surrounding cell membranes (smaller % zwitterion and more favorable log $P_{(membrane\ interior)}$), and back into the bloodstream. Its nearly quantitative absorption from an oral dose also depends on its good membrane permeability in conjunction with adequate water solubility.[91]

Too long a half-life, most particularly at the high doses used in toxicology testing where catabolic enzyme saturation is commonplace, can result in massive drug accumulation with adverse effects which will contribute to the high failure rate at this stage of development. Such was the fate of the antifungal agents **35a** and **35b** shown in Table 10. One may deduce that their difluorophenyl rings are very resistant to

Table 10. Manipulation of Metabolism of Antifungal Agents

35 a–i

35a–i	R^1	R^2	C. albicans in vitro IC_{50} ($\mu g\ ml^{-1}$)	C. albicans in vivo ED_{50} ($mg\ kg^{-1}$) (p.o.)	Half-life in rat (days)
a	2,4-F_2	CF_3	0.09	1–0.5	v. long
b	2,4-F_2	OCF_3	0.05	0.25	6.5
c	2,4-Cl_2	H	0.06	>50	
d	2,4-F_2	F	0.05	10	
e	2,4-F_2	OCH_3	0.001	>25	
f	2,4-F_2	OCH_2CH_2F	0.003	25	
g	2,4-F_2	OCF_2CF_2H	0.004	0.25	9
h	2,4-F_2	OCH_2CF_3	0.03	0.5	1.5
i	2,4-F_2	$OCH_2C_2F_4H$	0.003	0.25	1

oxidation, and conjugation of the very hindered hydroxyl groups is also very slow or reversible. The poor oral activity of the monofluorostyryl compound **35d**, and the absence of oral activity in **35c** and **35e**, is almost certainly due to their rapid or very rapid oxidative elimination, and this pointed the way to tuning the half-life to the desired range—at least for the rat! In the fluorinated alkoxy compounds, **35f–i**, one fluorine is inadequate, four is even more extreme than OCF_3, but trifluoroethoxy is almost optimum and tetrafluoropropoxy provided the desired one-day half-life. This became the development compound, ICI 195739.[92]

The site of metabolism—aromatic ring or methylene group—in the above development compound was not stated, but, as all other sites were concluded to be robust, the choice is limited to these two. In a series of oestrogen antagonists, four of which are shown as **36a–d**, there seems to be almost unlimited opportunity for oxidation; and conjugation can act directly to speed elimination. This work uncovered the first pure antagonists ever to be described, but, surprisingly, both the profile and the potency of many compounds in this series were clearly related to metabolism occurring in the terminal region of the long amidic side chains. Dosed to rats by the subcutaneous route, which avoids first-pass metabolism in the liver, all four compounds shown were pure antagonists of the uterotrophic activity of oestradiol, with **36b** having the greatest potency. By the oral route, however, **36a** was a partial agonist and **36d** was 3–4 times as potent as its nonfluorinated equivalent, which probably indicates substantial first-pass oxidation of the nonfluorinated chain terminii.[93] Pure antagonists may be superior to the currently used partial agonists in the treatment of breast cancer and other oestrogen-dependent tumors and conditions.

a: $R^1 = H$; $R^2 = n\text{-}C_4H_9$
b: $R^1 = CH_3$; $R^2 = n\text{-}C_4H_9$
c: $R^1 = H$; $R^2 = CH_2C_3F_7$
d: $R^1 = CH_3$; $R^2 = CH_2C_3F_7$

36 a–d

Related effects on profile and half-life due to fluorination have been extensively investigated in analogs of vitamin D3—another molecule with multiple potential sites of oxidation; very substantial changes can result from well located fluorine atoms, with up to twentyfold improvements in potency *in vivo* despite reduced receptor affinity.[5]

Conversely, fluorine-related inhibition of metabolism can slow or prevent the formation of active metabolites from less active or inactive parent drugs: fluorinated analogs of cyclophosphamide were much inferior anticancer agents than the parent compound,[94] and inhibiting the aromatic ring hydroxylation of antioestrogenic triarylethylenes, such as tamoxifen, prevents their conversion to vastly more potent phenols, which are the likely mediators of most of their endocrinological and anticancer effects in women.[95]

24.11. THE SIZE AND SHAPE OF FLUORO SUBSTITUENTS

The widespread ability of the CF group to mimic both CH and COR groups in bioactive entities is widely known and several examples have been quoted in previous sections. Thus from these observations alone, it is likely that, in terms of size and polarity/hydrogen-bonding ability, the CF group may lie somewhere in between the other pair. With regard to size, the common use of Pauling's atomic van der Waals radii (r_w), $H = 1.20$, and $F = 1.35$ Å (*all distances here are in angstrom units*),[23] as a single method of comparison, leads to a differential, $\Delta(F\text{-}H)$, of only 0.15 and the inappropriate conclusion that the CF group is very similar to the CH group. Widthwise this would be true, but the inclusion of bond lengths and the use of radii derived by Bondi during his study of van der Waals volumes[96] leads to estimates of length (L) which show that CF and CO groups are very similar to each other and significantly different from the CH group, especially if aromatic CX comparisons are used: the values for (bond length[23] + radius) in atomic number order are $(1.08 + 1.0) = 2.08$, $(1.36 + 1.5) = 2.86$, $(1.33 + 1.47) = 2.80$, which leads to $\Delta(F\text{-}H) = 0.72$. The value for chlorine, $(1.70 + 1.77) = 3.47$, shows that $\Delta(Cl\text{—}F) = 0.67$ is a slightly smaller differential; but, as discussed in Section 24.4, the anisotropy around CBr and CI groups renders further comparisons inappropriate.

Bondi did not detect any anisometry around the heavier halogens and, in order to carry out his investigation, he made the simplifying assumption that all bonded atoms are spherical, despite clear evidence to the contrary for doubly bonded oxygen and nitrogen. He also found it necessary to use different radii for the same atom, depending on its local environment. As an extreme example, aliphatic hydrogen was found to require $r_w = 1.17$, while aromatic hydrogen required $r_w = 1.00$. For fluorine, the value was 1.47 for all environments except CH_2F, which gave the best data fit with $r_w = 1.40$. Clearly, there are problems in the assignment of r_w values. These problems arise in many ways: all bonded atoms are anisotropic to variable degrees and in various ways, e.g., in their electric fields, their electron iso-density surface shapes, and the density gradient at particular points on those surfaces. Furthermore, the fitting of such calculated surfaces to atom-centered spheres demonstrated that the average radius of a bonded atom depends on its hybridization state (the size order is opposite to that given by Bondi), the presence of angle strain, and the electronegativity of attached atoms: the radii of carbon in CH_2F_2 and CH_2Me_2 are calculated to be 1.760 and 1.944, respectively.[97] Add to these complications the variable nature of charge transfer,[98] exchange repulsion, and correlation effects with particular contacting molecules, and it is remarkable that averaged radii are useful to about ± 0.1 Å.

Even more remarkably, molecular mechanics programs, such as MM2, which have sphericity and uniform atom sizes as central characteristics, can, while using a limited number of fudge factors, reproduce the shapes of X-ray crystallographically determined structures with reasonable precision. In this program, two parameters are used to describe atoms—a radius (not equivalent to r_w) and a "hardness" factor. By these criteria, fluorine is slightly smaller than oxygen, 1.65 versus 1.74 Å, but is considerably "harder", 0.078 versus 0.05 (kcal mol^{-1}).[99]

Two distances, r_w and L, may be enough to characterize a very few simple substituents, but steric effects due to even fairly simple groups such as CH_3 and CF_3 pose major problems for correlation analysis. Numerous approaches to the problem have been tried, but none is generally useful.[100] Those that are based on changes in chemical reactivity tend to sense only local steric effects, but they can incorporate entropic effects due to increased restriction of rotation around, e.g., $C—CX_3$ bonds. Geometrical characterization can theoretically allow even distant atoms/groups to be incorporated into the analysis, but this approach produces a confusing array of numbers which have no clearcut significance (due to the need to define arbitrary axes) and entropic effects pose severe problems. Extensive tables of steric parameters are available,[100] but some of the numbers seem very strange and all values should ideally be checked and considered before use.

Bondi-derived volumes are not normally used in qualitative or quantitative analysis, but those in Figure 2 provide food for thought, especially when combined with other qualitative descriptors—such as cyano being long and thin, while CF_3 is short and fat. This "fat" quality produces in 2-(trifluoromethyl)-2-adamantyl tosylate such severe steric problems that its solvolysis rate is almost equal to that of 2-adamantyl tosylate itself.[101] The investigators had expected a rate reduction of many orders of magnitude. In view of this indication of large steric demand from a CF_3 group, it is somewhat surprising that a CF_2 group can successfully replace a linking oxygen in the triphosphate group of adenosine triphosphate (ATP).[102] Fluorine may here be mimicking the oxygen lone pairs to produce an isopolar, if barely isosteric, analog of ATP.

Finally, we return for the last time to the 6-fluoro-7-piperazino-quinolones which demonstrate such a marked advantage for the fluoro analog (Table 1). Is it possible

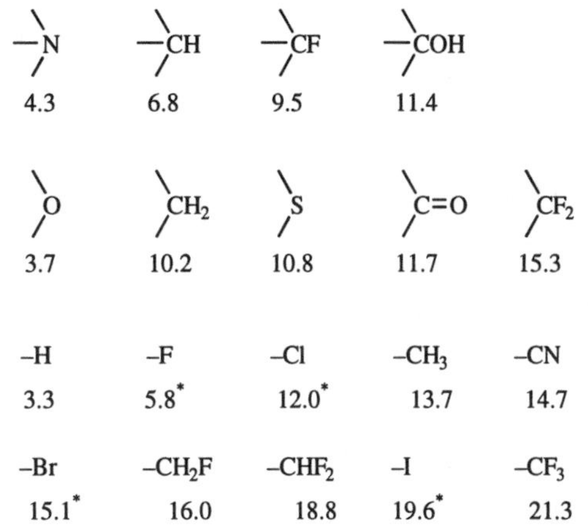

$-N$	$-CH$	$-CF$	$-COH$
4.3	6.8	9.5	11.4

O	CH_2	S	$C=O$	CF_2
3.7	10.2	10.8	11.7	15.3

$-H$	$-F$	$-Cl$	$-CH_3$	$-CN$
3.3	5.8*	12.0*	13.7	14.7

$-Br$	$-CH_2F$	$-CHF_2$	$-I$	$-CF_3$
15.1*	16.0	18.8	19.6*	21.3

Figure 2. Bondi volumes ($cm^3 mol^{-1}$) of atoms and simple groups bonded to carbon (* = values for Ar—Hal).

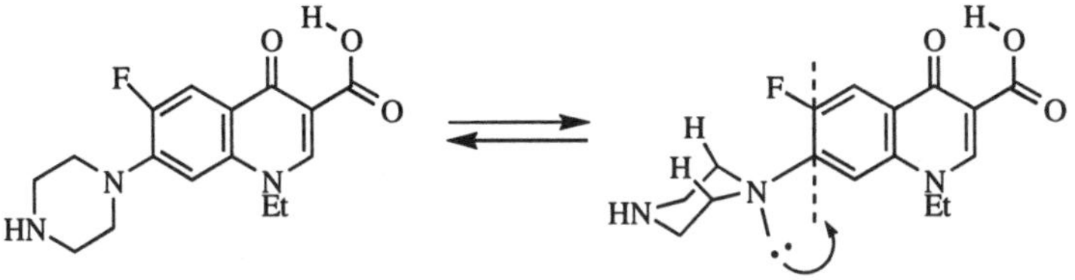

37a   37b

that this advantage is (in part) a consequence of conformational energetics? In the notional equilibrium **37a** $\rightleftharpoons$ **37b**, all groups at position-6 will destabilize the planar conformer **37a** since even hydrogen generates a strong steric clash with the equatorial hydrogen of the piperazine ring. This will be relieved by twisting the rings around the aryl–nitrogen bond; some concurrent shallow pyramidalization of the nitrogen atom is also likely. A fluorine atom at position-6 is more demanding of space than hydrogen and renders carbon-6 electron-deficient. The combination of these effects will help to stabilize conformer **37b** in which the fluorine atom can just fit between the two axial hydrogens of the now fully chair-like piperazine ring: the fields of the CF unit and the highly pyramidized nitrogen are nicely complementary and σ-resonance ($n_N \rightarrow \sigma^*_{CC}$) is maximized, so this conformer is likely to correspond to a local minimum in conformational space (the electronegative nitrogen atom attached to carbon-7 will decrease its electron density and thereby increase σ-resonance as shown in **37b**: the donated electron density is attracted by the two electron-deficient carbon atoms[103]). Thus another hypothesis regarding the 6-fluoro effect is that this orthogonal conformer has good complementarity to the gyrase target site(s) and/or the bacterial permease systems.

24.12. REFERENCES

1. R. Filler, in: *Organofluorine Chemicals and their Industrial Applications* (R. E. Banks, ed.), p. 123, Ellis Horwood, Chichester (1979).
2. R. Filler and Y. Kobayashi, *Biomedicinal Aspects of Fluorine Chemistry*, Kodansha, Tokyo and Elsevier Biomedical, Amsterdam (1982).
3. C. Walsh, *Adv. Enzymol. Relat. Areas Mol. Biol. 55*, 197 (1983).
4. J. Mann, *Chem. Soc. Rev. 16*, 381 (1987).
5. J. T. Welch, *Tetrahedron 43*, 3123 (1987).
6. C. Hansch, P. G. Sammes, and J. B. Taylor, *Comprehensive Medicinal Chemistry: The Rational Design, Mechanistic Study & Therapeutic Applications of Chemical Compounds*, Pergamon Press, Oxford (1990).
7. P. N. Craig, in Ref. 6, Vol. 6, p. 237.
8. J. Elks and C. R. Ganellin, *Dictionary of Drugs*, Chapman and Hall, London (1990).
9. D. T. W. Chu and P. B. Fernandes, *Antimicrob. Agents Chemother. 33*, 131 (1989) and references cited therein.
10. T. Rosen, *Prog. Med. Chem. 27*, 235 (1990) and references cited therein.
11. H. Koga, A. Itoh, S. Murayama, S. Suzue, and T. Irikura, *J. Med. Chem. 23*, 1358 (1980).
12. S. P. Cohen, D. C. Hooper, J. S. Wolfson, K. S. Souza, L. M. McMurry, and S. B. Levy, *Antimicrob. Agents Chemother. 32*, 1187 (1988).
13. J. S. Chapman, A. Bertasso, and N. H. Georgopapadakou, *Antimicrob. Agents Chemother. 33*, 239 (1989).
14. D. T. W. Chu, P. B. Fernandes, A. K. Claiborne, E. Pihuleac, C. W. Nordeen, R. E. Maleczka, and A. G. Pernet, *J. Med. Chem. 28*, 1558 (1985).
15. T. Miyamoto, J-i. Matsumoto, K. Chiba, H. Egawa, K-i. Shibamon, A. Minamida, Y. Nishimura, H. Okada, M. Kataoka, M. Fujita, T. Hirose, and J. Nakano, *J. Med. Chem. 33*, 1645 (1990).
16. L. Pauling, *The Nature of the Chemical Bond*, 3rd edn., Cornell University Press, New York (1960).
17. J. K. Nagle, *J. Am. Chem. Soc. 112*, 4741 (1990) and references cited therein.
18. Yu-Ran Luo and S. W. Benson, *J. Am. Chem. Soc. 111*, 2480 (1989).
19. S. Marriott, W. F. Reynolds, R. W. Taft, and R. D. Topsom, *J. Org. Chem. 49*, 959 (1984).
20. S. Marriott and R. D. Topsom, *J. Am. Chem. Soc. 106*, 7 (1984).
21. R. D. Topsom, *Prog. Phys. Org. Chem. 16*, 125 (1987).
22. R. Filler, in: *Fluorine-containing Molecules: Structure, Reactivity, Synthesis and Applications* (J. F. Liebman, A. Greenberg, and W. R. Dolbier, eds.), p. 19, VCH, Weinheim (1988).
23. A. J. Gordon and R. A. Ford, *The Chemists Companion*, Wiley, New York (1972).
24. O. Exner, *Dipole Moments in Organic Chemistry*, Georg Thieme, Stuttgart (1975).
25. P. Murray-Rust and W. D. S. Motherwell, *J. Am. Chem. Soc. 101*, 4374 (1979).
26. J. M. Blaney, E. C. Jorgensen, M. L. Connolly, T. E. Ferrin, R. Langridge, S. J. Oatley, J. M. Burridge, and C. C. F. Blake, *J. Med. Chem. 25*, 785 (1982).
27. R. E. Rosenfield, R. Parthasarathy, and J. D. Dunitz, *J. Am. Chem. Soc. 99*, 4860 (1977).
28. J-M. Dumas, M. Gomel, and M. Guerin, in: *The Chemistry of the Functional Groups, Supplement D: The Chemistry of Halides, Pseudo-halides and Azides* (S. Patai and Z. Rappoport, eds.), Part 2, p. 985, Wiley, Chichester (1983).
29. J. D. Dill, L. C. Allen, W. C. Topp, and J. A. Pople, *J. Am. Chem. Soc. 97*, 7720 (1975).
30. R. Masood, G. S. Ahluwalia, D. A. Cooney, A. Fridland, V. E. Marquez, J. S. Driscoll, Z. Hao, H. Mitsuya, C-F. Perno, S. Broder, and D. G. Johns, *Mol. Pharmacol. 37*, 590 (1990).
31. N. H. Andersen, C. J. Hartzell, and B. De, *Adv. Prostaglandin Thromboxane Leukotriene Res. 14*, 1 (1985).
32. M. H. Abraham, P. P. Duce, D. V. Prior, D. G. Barratt, J. J. Morris, and P. J. Taylor, *J. Chem. Soc., Perkin Trans. 2*, 1355 (1989) and references cited therein.
33. J. W. Larson and T. B. McMahon, *J. Am. Chem. Soc. 104*, 5848 (1982).
34. A. W. Jache, in Ref. 22, p. 165.

35. R. West, D. L. Powell, L. S. Whatley, M. K. T. Lee, and P. von R. Schleyer, *J. Am. Chem. Soc. 84*, 3222 (1962).
36. C. Hansch and A. L. Leo, *Log P Database*, Pomona College Medicinal Chemistry Project, Claremont, California 91711.
37. A. W. Baker and A. T. Shulgin, *Nature 206*, 712 (1965).
38. D. Doddrell, E. Wenkert, and P. V. Demarco, *J. Mol. Spectrosc. 32*, 162 (1969).
39. S. M. Miller and J. P. Klinman, *Biochemistry 24*, 2114 (1985).
40. L. I. Kruse, C. Kaiser, W. E. DeWolf, J. S. Frazee, S. T. Ross, J. Wawro, M. Wise, K. E. Flaim, J. L. Sawyer, R. W. Erickson, M. Ezekiel, E. H. Ohlstein, and B. A. Berkowitz, *J. Med. Chem. 30*, 486 (1987).
41. *Drugs of the Future 14*, 942 (1989).
42. S. G. Withers, I. P. Street, and M. D. Percival, in: *Fluorinated Carbohydrates, Chemical and Biochemical Aspects* (N. F. Taylor, ed.), ACS Symposium Series, No. 374 (1988).
43. P. Murray-Rust, W. C. Stallings, C. T. Monti, R. K. Preston, and J. P. Glusker, *J. Am. Chem. Soc. 105*, 3206 (1983).
44. A. Karipides and C. Miller, *J. Am. Chem. Soc. 106*, 1494 (1984).
45. W. P. Jencks, *Catalysis in Chemistry and Enzymology*, McGraw-Hill, New York (1969).
46. A. R. Fersht, J-P. Shi, J. Knill-Jones, D. M. Lowe, A. J. Wilkinson, D. M. Blow, P. Brick, P. Carter, M. M. Y. Waye, and G. Winter, *Nature 314*, 235 (1985).
47. W. J. Middleton and R. V. Lindsey, *J. Am. Chem. Soc. 86*, 4948 (1964).
48. R. O. Neri, K. Florance, P. Koziol, and S. van Cleave, *Endocrinology 91*, 427 (1972).
49. B. Katchen and S. Buxbaum, *J. Clin. Endocrinol. 41*, 373 (1975).
50. J. J. Morris, L. R. Hughes, A. T. Glen, and P. J. Taylor, *J. Med. Chem. 34*, 447 (1991).
51. H. Tucker, J. W. Crook, and G. J. Chesterson, *J. Med. Chem. 31*, 954 (1988).
52. W. L. Duax and J. F. Griffen, *Adv. Drug. Res. 18*, 115 (1989).
53. C. M. Weeks, W. L. Duax, and M. D. Wolff, *J. Am. Chem. Soc. 95*, 2865 (1973). Molecular mechanics calculations fail to reproduce the (solid state) changes in the conformation due to fluorination. It may be that deficiencies in the MM2 parameterization, especially the lack of any terms corresponding to $n_F \rightarrow \pi^*$, are limiting the reproduction of the observed changes: allowance for such an effect somewhat increases ring-A distortion.[82]
54. T. C. Wong, V. Rutar, and J-S. Wong, *J. Am. Chem. Soc. 106*, 7046 (1984).
55. H. J. Lee and M. R. I. Soliman, *Science 215*, 989 (1982).
56. *Scrip.* No 1551, p. 26 (Sept. 21st, 1990).
57. R. H. Davies, R. D. Bagnall, W. Bell, and W. G. M. Jones, *Int. J. Quantum Chem. Quantum Biol. Symp. No. 3*, 171 (1976).
58. R. M. Jones, J. N. Cashman, E. I. Eger, M. C. Damask, and B. H. Johnson, *Anesthesia and Analgesia 70*, 3 (1990).
59. P. Seiler, G. R. Weisman, E. D. Glendening, F. Weinhold, V. B. Johnson, and J. D. Dunitz, *Angew. Chem., Int. Ed. Engl. 26*, 1175 (1987).
60. K. B. Wiberg, *J. Am. Chem. Soc. 112*, 3379 (1990).
61. R. Taylor and O. Kennard, *J. Am. Chem. Soc. 104*, 5063 (1982).
62. S. K. Burley and G. A. Petsko, *Science 229*, 23 (1985).
63. C. A. Hunter and J. K. M. Sanders, *J. Am. Chem. Soc. 112*, 5525 (1990).
64. Reference 22 in Ref. 57.
65. W. C. Groutas, *Med. Res. Rev. 7*, 227 (1987).
66. A. Dafforn, J. P. Neenan, C. E. Ash, L. Betts, J. M. Finke, J. A. Garman, M. Rao, K. Walsh, and R. R. Williams, *Biochem. Biophys. Res. Commun. 104*, 597 (1982).
67. E. Shaw, *Adv. Enzymol. 63*, 271 (1990).
68. P. D. Edwards, B. Hesp, D. A. Trainor, and A. K. Willard, in: *Enzyme Chemistry: Impact and Applications* (C. J. Suckling, ed.), 2nd. ed., p. 171, Chapman and Hall, London (1990).
69. L. H. Takahashi, R. Radhakrishnan, R. E. Rosenfield, E. F. Meyer, D. A. Trainor, and M. Stein, *J. Mol. Biol. 201*, 423 (1988).
70. T-C. Liang and R. H. Abeles, *Biochemistry 26*, 7603 (1987).

71. L. H. Takahashi, R. Radhakrishnan, R. E. Rosenfield, E. F. Meyer, and D. A. Trainor, *J. Am. Chem. Soc. 111*, 3368 (1989).
72. J. P. Guthrie, *Can. J. Chem. 53*, 898 (1975).
73. E. G. Sander and W. P. Jencks, *J. Am. Chem. Soc. 90*, 6154 (1968).
74. W. J. Greenlee, *Med. Res. Rev. 10*, 173 (1990).
75. J. F. Morrison and C. T. Walsh, *Adv. Enzymol. 61*, 201 (1988).
76. D. Gani in Ref. 6, Vol. 2, p. 213.
77. J. Kollonitsch, *Isr. J. Chem. 17*, 53 (1978). G. K. Darland, R. Hajdu, H. Kropp, F. M. Kahan, R. W. Walker, and W. J. A. Vandenheuvel, *Drug Metab. Dispos. 14*, 668 (1986).
78. W. S. Faraci and C. T. Walsh, *Biochemistry 28*, 431 (1989).
79. R. B. Silverman and B. J. Invergo, *Biochemistry 25*, 6817 (1986).
80. M. G. Palfreyman, M. Zreika, and I. A. McDonald, in: *Enzymes as Targets for Drug Design* (M. G. Palfreyman, P. P. McCann, W. Lovenberg, J. G. Temple Jr., and A. Sjoerdsma, eds.), p. 139, Academic Press Inc., San Diego (1989).
81. J. C. Dearden in Ref. 6, Vol. 4, p. 375.
82. P. N. Edwards, unpublished work.
83. A. Walter and J. Gutknect, *J. Membr. Biol. 90*, 207 (1986). See also: W. D. Stein, *Transport and Diffusion Across Cell Membranes,* Academic Press, London (1986).
84. A. F. M. Barton, *Chem. Rev. 75*, 731 (1975).
85. O. P. Flint and F. T. Boyle, *Handb. Exp. Pharm. 96*, 231 (1990).
86. N. H. Anderson, S. S. Davies, M. James, and I. Kojima, *J. Pharm. Sci. 72*, 443 (1983).
87. S. H. Yalkowski, S. C. Valvani, and T. J. Roseman, *J. Pharm. Sci. 72*, 866 (1983).
88. J. M. Tedder, *Quart. Rev. 14*, 336 (1960); A. L. Henne and J. B. Hinkamp, *J. Am. Chem. Soc. 67*, 1197 (1945).
89. J. M. Dust and D. R. Arnold, *J. Am. Chem. Soc. 105*, 1221 (1983).
90. D. J. Pope, A. P. Gilbert, D. J. Easter, R. P. Chan, J. C. Turner, S. Gottfried, and D. V. Parke, *J. Pharm. Pharmacol. 33*, 302 (1981).
91. G. R. Granneman, K. M. Snyder, and V. S. Shu, *Antimicrob. Agents Chemother. 30*, 689 (1986).
92. F. T. Boyle, D. J. Gilman, M. B. Gravestock, and J. M. Wardleworth, *Ann. N.Y. Acad. Sci. 554*, 86 (1988).
93. J. Bowler, T. J. Lilly, J. D. Pitlam, and A. E. Wakeling, *Steroids 54*, 71 (1989).
94. A. Foster, M. Jarman, R. Kinas, J. Van Maaren, G. Taylor, J. Gaston, A. Parkin, and A. Richardson, *J. Med. Chem. 24*, 1399 (1981).
95. R. McCague, G. Leclerq, N. Legros, J. Goodman, G.M. Blackburn, M. Jarman, and A. B. Foster, *J. Med. Chem. 32*, 2527 (1989), and references cited therein.
96. A. Bondi, *J. Phys. Chem. 68*, 441 (1964).
97. M. M. Francl, R. F. Hout, and W. J. Hehre, *J. Am. Chem. Soc. 106*, 563 (1984).
98. A. E. Reed, L. A. Curtiss, and F. Weinhold, *Chem. Rev. 88*, 899 (1988).
99. N. L. Allinger and V. Burkert, *Molecular Mechanics*, American Chemical Society, Washington, D.C. (1982).
100. C. Silipo and A. Vitloria in Ref. 6, Vol. 4, p. 153.
101. A. D. Allen, R. Krishnamurti, G. K. S. Prakash, and T. T. Tidwell, *J. Am. Chem. Soc. 112*, 1291 (1990).
102. G. M. Blackburn, F. Eckstein, D. E. Kent, and T. D. Perree, *Nucleosides Nucleotides 4*, 165 (1985).
103. A. E. Read and P. v. R. Schleyer, *J. Am. Chem. Soc. 109*, 7362 (1987).

25

Fluorinated Inhalation Anesthetics

DONALD F. HALPERN

25.1. INTRODUCTION

No class of chemical compounds has contributed more toward the elimination of hospital trauma than anesthetics. These drugs have changed the operating room from a chamber of horrors to a place where medical care is provided in a tranquil atmosphere to some 50 million patients every year. When inhaled, anesthetics enter the brain and induce profound sleep (hypnosis), sedation (a passive state), muscle relaxation (flaccidity) and analgesia (the absence of pain) to the level required to perform surgery. Inhalation anesthetics leave the brain and the body chemically unchanged because they are exhaled and need not be metabolized to be eliminated. In addition, the newer inhalants are not metabolized to any great extent.

Although early medical literature contains several references to drugs and methods to relieve pain, provide sedation, and induce sleep, few agents were actually used to prevent the pain and suffering associated with surgical procedures. The so-called anesthesia of our ancestors was more of a philosophical treatment than a practical one. Patients who underwent operations suffered indescribable agonies.

There are Chinese references as early as 220 A.D. describing hashish as an analgesic. The writings of Dioscorides, a celebrated first-century A.D. physician who coined the word *anesthesia*, include a recipe for an oral dosage form. The near 2000-year delay before anesthesia was accepted by the rest of the world may be ascribed to tradition and religious opposition which blocked the popularization of these medical advances.

Although Joseph Priestly discovered nitrous oxide in 1771, an obscure doctor in New York City, Samuel Latham Mitchell, in *Remarks on the Gaseous Oxyd of Azote*

DONALD F. HALPERN • Ohmeda, P.P.D. Murray Hill, New Jersey 07974. *Present address*: 55 Murray Hill Square, Murray Hill, New Jersey 07974.

Organofluorine Chemistry: Principles and Commercial Applications, edited by R. E. Banks *et al.* Plenum Press, New York, 1994.

or Nitrogene, etc. (New York, T & J Swords, 1795), presented a description of the effects of inhaling nitrous oxide and what might well be the first documentation of a person being anesthetized.[1] In 1799, Humphry Davy recognized the anesthetic and analgesic qualities of this gas, and recorded that nitrous oxide was capable of "destroying pain," and that it "may be used with advantage during surgical operations."[2]

Diethyl ether was described by Valerius Cordus in the 16th century, and its anesthetic properties were recognized in the same century by Paracelsus. At this point, science had the necessary drugs for the relief of suffering, but humanity resisted this advance. Several attempts to introduce anesthetics into the practice of surgery failed.[2]

In 1842, in Danielsville, Georgia, Crawford Long used ether as an anesthetic during surgery and did so for seven years before publishing an account of his work.[3] By that time, three other physicians were arguing over who had discovered it as an anesthetic.

As inhalation anesthesia became established in the late 19th and early 20th centuries, carbon dioxide, chloroform, cyclopropane, ether, nitrous oxide, and trichloroethylene were important anesthetic agents at one time or another. Any inhalation agent available to the medical community offered major advantages over previous agents; each agent had at least one serious disadvantage: odor, narrow margin of safety, degree of metabolism, flammability, or side effects. Although ether was the dominant inhalant for over 100 years, with its last recorded use as an anesthetic for humans in the United States in 1982,[4] its odor, flammability, and side effects left a lot of room for improvement. In spite of these drawbacks, the practice of anesthesia was built around the anesthetic profile of ether. Even today, the often used "Guedel's classical signs of anesthesia" are essentially the effects of diethyl ether on the central nervous system.[5]

25.2. THE FLUORINE REVOLUTION

The history of organofluorine compounds as pharmaceuticals is not measured in centuries like the history of the opioids or other natural products; they are just a few

Table 1. Major Inhalation Anesthetics

Generic name	Proprietary name	Formula	Commercial introduction (date, company)
Diethyl ether		$(C_2H_5)_2O$	1842 Long
Chloroform		$CHCl_3$	1847 Simpson
Trilene		$CCl_2{=}CHCl$	1935 Jackson
Halothane	Fluothane	$CF_3CHBrCl$	1956 ICI
Fluroxene	Fluoromar	$CF_3CH_2OCH{=}CH_2$	1960 Anaquest
Methoxyflurane	Penthrane	$CH_3OCF_2CHCl_2$	1962 Abbott
Enflurane	Ethrane	$CHFClCF_2OCHF_2$	1972 Anaquest
Isoflurane	Forane	$CF_3CHClOCHF_2$	1981 Anaquest
Sevoflurane		$(CF_3)_2CHOCH_2F$	1990 Maruishi
Desflurane	Suprane	$CF_3CHFOCHF_2$	1993 Anaquest

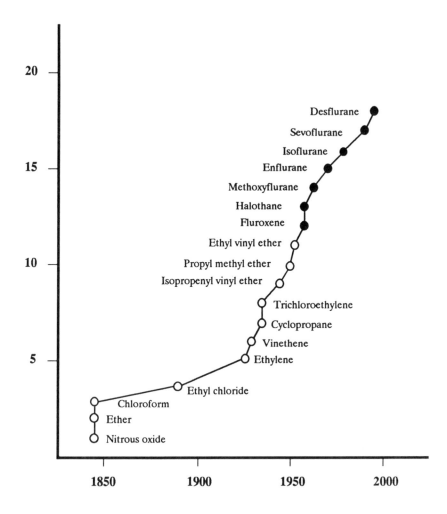

Figure 1. Timeline for the introduction of inhalation anesthetics. ○, Nonfluorinated anesthetics; ●, fluorinated anesthetics. [Adapted from R. N. Rosenberg, J. R. Guest, and B. E. Etsten, *Anesthesiol. Rev. 1*, 11 (1974).]

decades old. After World War I, many chlorofluorocarbons were synthesized and used as refrigerants. During World War II, new methods of synthesizing organofluorine compounds were developed and many fluorinated compounds that once were laboratory curiosities became commercially available.[6] Robbins published the first paper describing the pharmacology of fluorinated hydrocarbons. It included the tri-fluoromethyl compounds CF_3CHCl_2 and CF_3CHBr_2.[7] During this period Shukys at Air Reduction Co. was synthesizing various compounds that might have potential as inhalation anesthetics. He intended to explore the effect of fluorination on anesthetic properties and to eliminate the fire hazard posed by some of the anesthetics in use at the time. In 1951 two compounds were synthesized in different laboratories. The first was fluroxene ($CF_3CH_2OCH{=}CH_2$), synthesized by Shukys and modeled after Krantz's ethyl vinyl ether work.[8] The other was halothane ($CF_3CHBrCl$), synthesized by Suckling and McGinty at ICI. Halothane was bromochlorotrifluoroethane, the missing $CF_3CHBrCl$ from Robbins's series.

Fluroxene (bp 43.2°C) showed excellent anesthetic properties when tested on several species of small laboratory animals, but it did not eliminate fire hazards. Before any new chemical entity can be evaluated on people, it must be tested in a variety of animals to demonstrate its safety as well as its efficacy. Such testing was particularly important in the case of fluroxene (Fluoromar®) because it was the first fluorinated substance to be considered as an anesthetic for people. When the data from the animal studies indicated that a human trial was justified, John Krantz, MD did the honors in 1953: "I anesthetized Dr. Max Sadove of the University of Illinois for a period of 20 minutes. The anesthesia was smooth and the recovery was uneventful. It appeared that our extrapolation of animal data to man was valid. That same afternoon, Dr. Sadove, a distinguished anesthesiologist, anesthetized three patients requiring surgery with Fluoromar."[9]

This was the beginning of the fluorine revolution. Fluroxene was later submitted to and approved by the U.S. Food and Drug Administration. Other fluorinated anesthetics followed: halothane, methoxyflurane, enflurane, isoflurane, and sevoflurane (Table 1).[10] All of the anesthetics have been extensively reviewed.[5,6,10–12] Figure 1 shows the total number of inhalants introduced into clinical practice and their year of introduction.

25.2.1. Halothane ($CF_3CHClBr$)

Halothane, 2-bromo-2-chloro-1,1,1-trifluoroethane (bp 50.2°C), was the first nonflammable inhalant synthesized and commercialized because of its predicted chemical and physical properties. It can be manufactured by two routes. The liquid-phase ICI synthesis by Suckling and Raventos starts with trichloroethylene (Eq. 1)[13] while the Hoechst process uses trifluorochloroethene (Eq. 2).[14]

(1) $\quad CCl_2{=}CHCl \xrightarrow{HF/SbCl_5} CF_3CH_2Cl \xrightarrow{Br_2/h\nu} CF_3CHBrCl$

(2) $\quad CF_2{=}CFCl \xrightarrow{HBr/h\nu \text{ or peroxide}} CF_2BrCHFCl \xrightarrow{AlBr_3} CF_3CHBrCl$

25.2.2. Methoxyflurane ($CH_3OCF_2CHCl_2$)

Methoxyflurane, 2,2-dichloro-1,1-difluoroethyl methyl ether (bp 105°C), was the first fluorine containing ether used as an inhalation anesthetic. It was initially synthesized by Gowland (Eq. 3)[15] and later in the pure form by Larsen, who removed an unsaturated impurity in Gowland's product by ozonolysis (Eq. 4).[16]

(3)

$$CF_2{=}CCl_2 \xrightarrow[\text{MeOH}]{\text{MeONa}} CH_3OCF_2CHCl_2 + CH_3OCF{=}CCl_2$$

(4)

$$CH_3OCF{=}CCl_2 + O_3 \rightarrow CH_3OCOF + COCl_2$$

Initially methoxyflurane enjoyed a period of commercial success. Although it is five times more potent than halothane, there is a dose-related fluoride nephrotoxicity due to its metabolites that has eliminated its utility in the hospital operating room. Additionally, induction with and recovery from methoxyflurane is slow. At present, methoxyflurane is used only as a veterinary anesthetic.

25.2.3. Enflurane ($CHFClCF_2OCHF_2$)

Enflurane, 2-chloro-1,1,2-trifluoroethyl difluoromethyl ether (bp 56.5°C), is prepared commercially from chlorotrifluoroethylene (Eq. 5).[17] In addition to the Anaquest manufacturing facility in Puerto Rico and Abbott's plant in the United Kingdom, a 1986 report suggested that enflurane manufacture in mainland China was then at the pilot plant stage with at least 17 successful runs.[18]

(5)

$$CF_2{=}CFCl \xrightarrow[\text{CH}_3\text{OH}]{\text{CH}_3\text{ONa}} CH_3OCF_2CHFCl \xrightarrow{Cl_2/h\nu} CHCl_2OCF_2CHFCl$$

$$CHCl_2OCF_2CHFCl \xrightarrow{HF/SbCl_5} CHF_2OCF_2CHFCl$$

Reductive dechlorination of CFC-113 with aqueous alkaline methanolic triethanolamine containing $CuCl_2$ provides an alternate route to the enflurane precursor (Eq. 6):

(6)

$$CF_2ClCFCl_2 \rightarrow CH_3OCF_2CHFCl$$

Similar treatment of CFC-112a gives methoxyflurane[19]:

(7)

$$CF_2ClCCl_3 \rightarrow CH_3OCF_2CHCl_2$$

During the large-scale synthesis of enflurane, a number of byproducts are formed which lend themselves to various recycling methods.[6] One such compound is $CFCl_2CF_2OCHF_2$, which can be converted to enflurane by reductive dechlorination with aqueous methanolic $NaOH/CuCl_2$/triethanolamine.

The chronology of the registration of enflurane with the FDA was not unusual. The only major regulatory delay centered on the need for a thorough evaluation of atypical EEG patterns to ensure that this was not a deleterious side effect.

25.2.4. Isoflurane ($CF_3CHClOCHF_2$)

In the present process, isoflurane, 1-chloro-2,2,2-trifluoroethyl difluoromethyl ether (bp 48.8°C), is manufactured by the *in situ* trapping of difluorocarbene, generated

(8)

$$CHF_2Cl \xrightarrow{\text{NaOH}} :CF_2 \xrightarrow{\text{CF}_3\text{CH}_2\text{OH}} CHF_2OCH_2CF_3$$

$$CHF_2OCH_2CF_3 \xrightarrow{\text{Cl}_2/h\nu} CHF_2OCHClCF_3 + CHF_2OCCl_2CF_3$$

from HCFC-22 and NaOH with trifluoroethanol, followed by chlorination to isoflurane (Eq. 8).[21] One of the byproducts of the chlorination, $CF_3CCl_2OCHF_2$, can be selectively converted to isoflurane by reductive dechlorination with aqueous methanolic $NaOH/CuCl_2$/triethanolamine.[20]

The registration of isoflurane proved to be a more complex procedure than for enflurane. During the regulatory review of isoflurane, the results of a small pilot study without adequate controls were submitted to the FDA. It suggested that isoflurane might be a hepatocarcinogen. Extensive animal studies were conducted using a wider range of isoflurane concentrations in a more controlled manner. These studies failed to show any evidence of carcinogenicity. Subsequent examination of the conditions under which the pilot study was conducted revealed that the mice had been fed grain contaminated with polybrominated biphenyls which are potent teratogens and mutagenic carcinogens.[22] FDA approval of isoflurane was delayed for more than

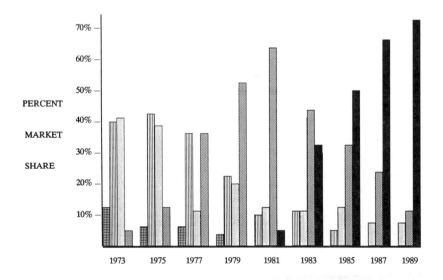

Figure 2. Percent U.S. market share of inhalation anesthetics (equivalent units). ▦, ether; ▥, methoxyflurane; ◼, halothane; ▨, enflurane; ◼, isoflurane.

five years because of this invalid study. In order to correct this inequity, the Congress of the United States granted Anaquest patent restoration for a five-year period to compensate for the lost years of patent exclusivity.

A new agent like isoflurane does not immediately displace the more established inhalants. It gradually becomes popular as professors in teaching hospitals introduce it to their students, the anesthesiologists of the future. Isoflurane's popularity is the result of its pharmacological profile, which includes low metabolism and no suggestion of hepatitis after repeated exposures. The emergence of enflurane and isoflurane over the past 15 years has reduced halothane's market position in the United States (Figure 2).[23] The change in the proportion of units of halothane sold with respect to units of enflurane and isoflurane is less dramatic in selected European countries (Table 2).[23-25]

Just as inhalation agents have evolved over the decades, so have the needs of the anesthesiologist. When enflurane was launched in 1972 neither outpatient surgery nor the use of inhalation anesthetics for outpatient surgery was an important issue for the patient, the anesthesiologist, the surgeon, the hospital administrator, or the insurance group that paid the bills. Today there is a tendency to perform minor surgical procedures in outpatient facilities rather than by admitting the patient to a hospital. Outpatient surgery increases the importance of rapid recovery. Its goal is to allow the

Table 2. Percent of Each Inhalant Sold 1985–88 (Equivalent Units)[a]

Agent	1985	1986	1987	1988
Germany				
Halothane	40%	34%	29%	28%
Enflurane	49%	52%	52%	51%
Isoflurane	9%	13%	18%	20%
Italy				
Halothane	20%	16%	14%	9%
Enflurane	67%	50%	44%	42%
Isoflurane	12%	33%	41%	47%
United Kingdom				
Halothane	54%	42%	29%	27%
Enflurane	30%	41%	49%	44%
Isoflurane	14%	15%	21%	28%
France				
Halothane				11%
Enflurane				54%
Isoflurane				34%
United States				
Halothane	11%	10%	7%	6%
Enflurane	32%	30%	27%	23%
Isoflurane	51%	60%	73%	71%

[a]See Ref. 23.

patient to become "street fit" as rapidly as possible. A street fit patient can go home in a matter of hours after surgery and enjoy a return to health in the friendly and cost-effective surroundings of his or her home rather than in an expensive hospital bed. There is a correlation between rapid recovery and the blood–gas partition coefficient of the inhalant. With all other factors being equal, a lower blood–gas solubility results in a shorter recovery time for the patient.[26]

25.2.5. Sevoflurane [$(CF_3)_2CHOCH_2F$]

Compounds like aliflurane (2-chlorotetrafluorocyclopropyl methyl ether) from W. R. Grace, synthane (difluoromethyl 1,2,2,3,3-pentafluoro-n-propyl ether) and sevoflurane [fluoromethyl 2,2,2-trifluoro-1-(trifluoromethyl)ethyl ether], both from Baxter Travenol, were synthesized, evaluated in man, and "put on the shelf".[6] Sevoflurane was taken off the shelf and reevaluated by BOC and then by Mariushi.

For a time sevoflurane, a nonpungent fluorinated ether (bp 58.5°C) with a blood–gas partition coefficient of 0.69 (isoflurane = 1.43), had the lowest recorded blood–gas partition coefficient.[10] It is stable to light, oxygen, and metals. An early route to the synthesis of sevoflurane uses 2H-hexafluoroisopropanol[29]:

(9) $(CF_3)_2CHOH + (CH_2O)_x + HF + H_2SO_4 \rightarrow (CF_3)_2CHOCH_2F$

Another approach first methylates 2H-hexafluoroisopropanol, then photochlorination of the derived methyl hexafluoroisopropyl ether, followed by halogen exchange with KF in hot sulfolane, gives sevoflurane (Eq. 10).[30]

$$(CF_3)_2CHOH \xrightarrow{(CH_3)_2SO_4/NaOH} (CF_3)_2CHOCH_3 \xrightarrow{Cl_2/h\nu} (CF_3)_2CHOCH_2Cl$$

(10)
$$(CF_3)_2CHOCH_2Cl \xrightarrow{NaF \text{ or } KF} (CF_3)_2CHOCH_2F$$

A 75% yield of sevoflurane is obtained by treating $(CF_3)_2CHOCH_2Cl$ with NaF or KF in a pressure vessel at 185°C for 19 hours.[31] Since $(CF_3)_2CHOCH_2Cl$ is both the solvent and reactant, it is believed that the reaction proceeds under supercritical conditions. If $(CF_3)_2CHOCH_2Cl$ is allowed to react with NaF at 190–220°C for 17 hours, an 87% yield of $(CF_3)_2CHOCH_2F$ is obtained. Alternatively, $(CF_3)_2CHOCH_2Cl$ can be fluorinated with BrF_3 to provide $(CF_3)_2CHOCH_2F$.[30] The use of expensive 2H-hexafluoroisopropanol can be avoided by the substitution of 2H-hexachloroisopropanol [$(CCl_3)_2CHOH$] as the starting material followed by the replacement of chlorine by fluorine (Eq. 11).[32] Another route to sevoflurane involves the synthesis of the alkoxyacetic acid, $(CF_3)_2CHOCH_2COOH$, using $ClCH_2COOH$ followed by fluorodecarboxylation with XeF_2[33] or with BrF_3.[34]

$$(CCl_3)_2CHOH + (CH_3)_2SO_4 \xrightarrow{NaOH} (CCl_3)_2CHOCH_3 \xrightarrow{BrF_3} (CF_3)_2CHOCH_2F$$

(11) or
$$(CCl_3)_2CHOCH_3 \xrightarrow{Cl_2/h\nu} (CCl_3)_2CHOCH_2Cl \xrightarrow{BrF_3} (CF_3)_2CHOCH_2F$$

Several research groups have reported that sevoflurane degrades when it comes in contact with soda-lime, the carbon dioxide scavenger used in anesthetic rebreathing circuits (used to absorb CO_2 from exhaled anesthetic gases before recirculation to the patient).[26,27,35,36] However, in the clinic only small amounts of the degradation products appear to be involved.[28] Products vary depending on the conditions used to degrade sevoflurane. Two important ones are 2-fluoromethoxypentafluoropropene [$CF_2{=}C(CF_3)OCH_2F$] and 1-methoxy-2-fluoromethoxy-1,1,3,3,3-pentafluoropropane [$CH_3OCF_2CH(OCH_2F)CF_3$]. Using an anesthetic circuit with soda-lime connected to a model lung in a closed system, Hanaki's group found five fluorinated degradation products with $CF_2{=}C(CF_3)OCH_2F$ and $CH_3OCF_2CH(OCH_2F)CF_3$ present in appreciable amounts. In a semiclosed system $CF_2{=}C(CF_3)OCH_2F$ was the major degradation product.[35] This propene can be obtained by dehydrofluorination of sevoflurane under anhydrous conditions. It needs to be removed rapidly from the reaction mixture under reduced pressure in addition to maintaining the temperature below −20°C throughout the reaction (Eq. 12).[37]

(12) $\quad (CF_3)_2CHOCH_2F + [(CH_3)_3Si]_2NLi \rightarrow CF_2{=}C(CF_3)OCH_2F$

An unusual addition–elimination results from an aqueous workup of the previous reaction with K_2CO_3. It leads to the formation of the cyano-ether $CF_3CH(CN)OCH_2F$.[37] If $NaOCH_3$ is used as the base, $CH_3OCF_2CH(OCH_2F)CF_3$ is formed.[37]

During the 1980s only sevoflurane and desflurane were evaluated in large numbers of patients as new inhalation anesthetics. Sevoflurane was approved in January 1990 for drug use in Japan. The product launch took place in May 1990. It is now in phase III clinical trials in the United States.[28]

25.2.6. Desflurane ($CF_3CHFOCHF_2$)

The search for an anesthetic closer to the present ideal led to the reevaluation of desflurane, difluoromethyl 1,2,2,2-tetrafluoroethyl ether (bp 23.5°C). It was first synthesized in 1971 by Russell. When compared to the current anesthetics, its rapid recovery, stability to oxygen, light, metals, soda-lime, low flammability limits (Table 3), and extremely low rate of biodegradation (<0.03%) made it an attractive drug candidate.[10,26,39]

Table 3. Lower Flammability Limits of Inhalation Agents

Halothane	5% (70% N_2O, 30% O_2)[a]
Enflurane	6% (70% N_2O, 30% O_2)[a]
Isoflurane	7% (70% N_2O, 30% O_2)[a]
Sevoflurane	12% (100% O_2)[b]
Desflurane	20% (70% N_2O, 30% O_2)[c]

[a]See Ref. 38; [b]Ref. 10; [c]Ref. 22.

Table 4. Rate of Recovery

Inhalant	Blood–gas partition coefficient	Recovery of muscle coordination (min.)[a]
Halothane	2.3^b	47.2 (± 4.7)
Isoflurane	1.4^b	23.2 (± 7.6)
Sevoflurane	0.69^c	14.2 (± 8.1)
Desflurane	0.42^b	4.7 (± 3.0)

[a]See Ref. 40; [b]Ref. 26; [c]Ref. 41.

It was felt that desflurane's rapid recovery was due to its very low blood–gas partition coefficient (0.42).[26] The rate of awakening of rats exposed to desflurane, sevoflurane, isoflurane, or halothane was measured to determine the significance of the differences in blood–gas partition coefficients. Upon exposure to 1.2 times MAC (*Minimum Alveolar Concentration*) for two hours, the recovery of muscle coordination was measured (rotarod test) and tabulated (Table 4). The results show a significant reduction in the rat's recovery time when desflurane, which has the lowest blood–gas partition coefficient, was used.

Russell *et al.* synthesized desflurane by direct fluorination of the isoflurane precursor $CF_3CH_2OCHF_2$ (cf Eq. 8).[42] Other syntheses based on the exchange of fluorine for chlorine in isoflurane utilized NaF,[31] KF,[43,44] or BrF_3.[45] An approach independent of isoflurane starts with the trichlorination of methoxyacetyl chloride followed by fluorination with SF_4 to produce desflurane (Eq. 13).[46]

(13) $CHCl_2OCHClCOCl + SF_4 \rightarrow CF_3CHFOCHF_2$

There are several ways of converting trifluoroacetaldehyde via its methyl-hemiacetal $[CF_3CH(OH)OCH_3]$ to methyl-1,2,2,2-tetrafluoroethyl ether,

$$CF_3CH(OH)OCH_3 \xrightarrow{EtNCF_2CHClF} CF_3CHFOCH_3{}^{47}$$

(14) $CF_3CH(OH)OCH_3 \xrightarrow{PCl_5} CF_3CHClOCH_3 \xrightarrow{BuNH_2MeF} CF_3CHFOCH_3{}^{48}$

$$CF_3CH(OH)OCH_3 \xrightarrow{TsCl} CF_3CH(OTs)OCH_3 \xrightarrow{MF\ (M = Cs, K)} CF_3CHFOCH_3{}^{49}$$

$CF_3CHFOCH_3$, a desflurane precursor (Eq. 14). The two steps needed to convert $CF_3CHFOCH_3$ to desflurane parallel the enflurane process (cf. Eq. 5).

Desflurane, whose proprietary name is Suprane®, received Federal Drug Administration approval for use in the United States as an inhalation anesthetic in September 1992.

25.3. CONCLUSIONS AND ACKNOWLEDGMENTS

Standing on the shoulders of giants in the field of inhalation anesthesia, I wish I could say that my view of the future of anesthesia is clearer than it may have been to previous reviewers. In truth, it is just as obscure. The observed progression over the last 50 years has been to replace the hydrogen atoms of inhalants first with chlorine atoms and now with fluorine atoms. Although potency is reduced, blood solubility is also reduced thereby shortening the recovery time. In addition, the degree of metabolism of desflurane is lower than found with its analogues enflurane and isoflurane. At one time the potency of an agent was very important. The dose of the more potent inhalant would be smaller, thereby leaving less metabolites for *in vivo* clearance. With an inhalant that is not metabolized, potency is less of an issue.

I am indebted to Ross C. Terrell for suggesting that I write this chapter, to Mark Robin for being an enthusiastic and dedicated laboratory partner, and to Gerald G. Vernice and Chialang Huang for their help. Portions of this chapter were originally published as a historical review: *Magic in a Vial—Development of New Inhalation Anesthetics* in the BOC Group Technology Magazine, June 1985 (The BOC Group, Chertsey Road, Windlesham, Surrey, GU 206 HJ, UK). This was later revised and published as *Inhalation Anesthetics, The New Generation* in *Chemtech* (Ref. 11).

25.4. REFERENCES

1. N. A. Bergman, *J. Am. Med. Assoc. 253*, 675 (1935).
2. F. Boland, *The First Anesthetic*, p. 3ff, University of Georgia Press, Athens (1950).
3. C. W. Long, *Southern Medical and Surgical Journal* (New Series), 5 (1849).
4. C. Carlsson and S. Cooper, *Anesth. Analg. 70*, 339 (1990).
5. A. Dobkin, *Development of New Volatile Inhalation Anesthetics, Vol. 6*, p. 6, Excerpta Medica, New York, (1979).
6. W. G. M. Jones, in: *Preparation, Properties and Industrial Applications of Organofluorine Compounds* (R. E. Banks, ed.), pp. 157–192, Ellis Horwood, Chichester (1982); E. R. Larsen, *Fluorine Chem. Rev. 3*, 3 (1969).
7. B. H. Robbins, *J. Pharmacol. Exp. Ther. 86*, 197 (1946).
8. J. G. Shukys, U.S. Patent 2,830,007 (to Air Reduction) [*CA 52*, 14561 (1958)].
9. J. C. Krantz, *Profiles of Medical Science and Inspired Moments*, 87, Lucas, Baltimore (1967).
10. R. C. Terrell, *Br. J. Anaesth. 56* (Supplement), 3 (1984).
11. D. Halpern, *Chemtech 19*, 304 (1989).
12. A. K. Barbour, in: *Organofluorine Chemicals and Their Industrial Applications* (R. E. Banks ed.), p. 49, Ellis Horwood, Chichester (1979).
13. C. W. Suckling and J. Raventos, British Patent 767,779, U.S. Patent 2,921,098 (to ICI) [*CA 51*, 15547 (1957)].
14. O. Scherer and H. Kuhn, German Patent 1,161,249 (to Hoechst) [*CA 60*, 9147 (1964)].
15. T. B. Gowland, British Patent 523,499 (to ICI) [*CA 35*, 6265 (1941)].
16. E. R. Larsen, U.S. Patent 3,264,356 (to Dow) [*CA 65*, 16865 (1966)].
17. R. C. Terrell, U.S. Patent 3,469,011 (to Airco) [*CA 72*, 3025 (1970)].
18. W. Wang, S. Chen, W. Chang, and Z. Sun, *Yaowu Fenxi Zazhi 5*, 322 (1985) [*CA 104*, 95600 (1986)].
19. R. C. Terrell and K. Hansen, U.S. Patent 4,365,097 (to BOC) [*CA 99*, 5203 (1983)].
20. R. C. Terrell and K. Hansen, U.S. Patent 4,346,246 (to BOC) [*CA 94*, 83604 (1981)].
21. L. S. Croix, U.S. Patent 3,637,477 (to Airco). Note: This patent was not abstracted by *Chemical Abstracts* and is incorrectly listed in the patent concordance.
22. E. I. Eger II, A. E. White, C. L. Brown, C. G. Biava, T. H. Corbett, and W. C. Stevens, *Anesth. Analg. 57*, 678 (1978).
23. Anaquest, Internal Data.
24. J. Tarpley and P. Lawler, *Anesthesia 44*, 596 (1989).

25. D. Noble and L. Martin, *Anesthesia 45*, 339 (1990).
26. E. I. Eger II, *Anesth. Analg. 66*, 971 (1987).
27. R. Wallin, B. Regan, M. Napoli, and J. Stern, *Anesth. Analg. 54*, 758 (1975).
28. Personal communication from Maruishi.
29. C. Coon and R. Simon, U.S. Patent 4,250,334 (to Baxter) [*CA 94*, 208538 (1981)].
30. B. Regan and J. Longstreet, U.S. Patent 3,683,092 (to Baxter) [*CA 77*, 156346 (1972)].
31. D. Halpern and M. L. Robin, U.S. Patent 4,874,901 (to BOC) [*CA 112*, 157680 (1990)].
32. C. Huang and G. G. Vernice, U.S. Patent 4,874,902 (to BOC) [*CA 112*, 157679 (1990)].
33. T. Patrick, K. Johri, D. White, W. Bertrand, R. Mokhtar, M. Kilbourn, and M. Welch, *Can. J. Chem. 64*, 138 (1986).
34. D. Halpern and M. L. Robin, U.S. Patent 4,996,371 (to BOC) [*CA 115*, 28690 (1991)].
35. C. Hanaki, K. Fujii, M. Morio, and T. Tashima, *Hiroshima J. Med. Sci. 36*, 61 (1987) [*CA 107*, 242577 (1987)].
36. D. Strum, B. Johnson, and E. Eger II, *Anesthesiology 67*, 779 (1987).
37. C. Huang, V. S. Venturella, A. L. Cholli, F. M. Venutolo, A. T. Silbermann, and G. G. Vernice, *J. Fluorine Chem. 45*, 239 (1989).
38. P. F. Leonard, *Anesth. Analg. 54*, 238 (1975).
39. E. I. Eger II, *Anesth. Analg. 66*, 983 (1987).
40. E. I. Eger II and B. Johnson, *Anesth. Analg. 66*, 977 (1987).
41. D. Strum and E. I. Eger II, *Anesth. Analg. 66*, 654 (1987).
42. J. Russell, A. Szur, and R. C. Terrell, U.S. Patent 3,897,502 (to Airco) [*CA 83*, 178306 (1975)].
43. R. C. Terrell, U.S. Patent 4,762,856 (to BOC) [*CA 109*, 189832 (1988)].
44. K. Toshikazu, U.K. Patent Appl. GB 2,219,292 (to Central Glass, Japan) [*CA 112*, 234804 (1990)].
45. M. L. Robin and D. Halpern, Eur. Patent Appl. EP 341,004 (to BOC) [*CA 112*, 178046 (1990)].
46. D. Halpern and M. L. Robin, U.S. Patent 4,855,511 (to BOC), [*CA 112*, 54963 (1990)].
47. G. Siegemund, *Chem. Ber. 106*, 2960 (1973).
48. H. Muffler and R. Franz, Ger. Offen. 28 23 969 (to Hoechst) [*CA 92*, 197322 (1980)].
49. M. L. Robin and D. Halpern, Eur. Patent Appl. EP 352,034 (to BOC) [*CA 113*, 39937 (1990)].

26

Properties and Biomedical Applications of Perfluorochemicals and Their Emulsions

KENNETH C. LOWE

26.1. INTRODUCTION

During the past 20 years, much attention has focused on the potential applications of emulsified perfluorochemicals (PFCs) in medicine and biology. Of particular interest has been their use as vehicles for respiratory gas transport, primarily as O_2-carrying resuscitation fluids designed to supplement conventional blood transfusion. For this reason, PFC emulsions have been frequently, but somewhat inaccurately, referred to as "blood substitutes". However, important clinical applications for PFCs and their emulsions also exist in areas as diverse as hemodilution and microcirculatory management, cancer therapy, diagnostic tissue imaging, and ophthalmologic surgery. Moreover, PFCs appear to be valuable in both basic and applied research for perfusing isolated organs and for regulating the growth of cell cultures.

This chapter is an overview of the development and assessment of PFCs and their emulsions for biomedical uses. Following an outline of the basic physicochemical properties of PFCs, and of the selection criteria for biological applications, consideration is given to problems relating to PFC emulsification and the choice of acceptable surfactants. This is followed by a discussion of the *in vivo* biocompatibility testing of PFC emulsions and their components in man and other species, together with details of complementary cell culture studies. Finally, some of the possible applications for PFCs and their emulsions in both clinical medicine and basic biomedical research are described.

KENNETH C. LOWE • Mammalian Physiology Unit, Life Science Department, University of Nottingham, University Park, Nottingham NG7 2RD, England.

Organofluorine Chemistry: Principles and Commercial Applications, edited by R. E. Banks *et al.* Plenum Press, New York, 1994.

Table 1. Some Physicochemical Properties of PFCs of Medical Interest[a]

Compound	Structure	Acronym	Molecular weight	Boiling point (°C)	Vapor pressure at 37°C	Stability of emulsions[b]
Perfluorodecalin	$c\text{-}C_{10}F_{18}$[c]	FDC	462	142	12.5	Poor
Perfluoro-octyl bromide	$C_8F_{17}Br$	PFOB	499	141	14.0	Excellent
Perfluorotripropylamine	$(C_3F_7)_3N$	FTPA	521	129	18.5	Good
Perfluorotributylamine	$(C_4F_9)_3N$	FTBA	671	177	1.1	Excellent

[a]Relates to emulsions of individual PFCs prepared with *Pluronic*® F-68 and/or lecithins as emulsifying agent(s).
[b]Data taken from Refs. 10, 12, and 33.
[c]

(*cis* and *trans*)

26.2. PROPERTIES OF PFCs

PFCs are cyclic or straight-chain hydrocarbons or their derivatives in which all hydrogen atoms attached to carbon have normally been replaced with fluorine. Those commonly used for biomedical purposes are saturated perfluorinated hydrocarbons, usually referred to simply as perfluorocarbons or fluorocarbons,* or analogs containing other atoms, such as oxygen, nitrogen, and bromine, in addition to carbon and fluorine.

The PFCs of interest as O_2 carriers are dense, colorless liquids with specific gravities about twice that of water; some of the basic physicochemical properties of important compounds are listed in Table 1. In general,** they are very stable thermally and essentially chemically inert, due to the great strengths of carbon–fluorine bonds and the excellent steric and electronic protection the fluorine substituents provide for the carbon skeletons (see Chapters 3 and 4). PFCs should not, of course, be confused with CFCs, the commercial chlorofluorocarbons, which have low boiling points and contain potentially reactive chlorine atoms (see Chapter 28).[1]

PFC liquids have very high capacities for dissolving gases (Table 2). It was this property, coupled with their biological inertness, that led to their assessment as vehicles for respiratory gas transport[2-7] and for regulating gas supply to cell cultures.[8,9] The solubility of O_2 in PFCs pertinent to intravascular (i.v.) use is typically 40–50 vol%, while CO_2 solubility can be 3–4 times greater (Table 2). This solubility depends upon the molecular volume of the dissolving gas, and decreases in the order $CO_2 > O_2 > N_2$.[10] By contrast with the characteristic sigmoid binding curve of O_2 to hemoglobin, O_2 solubility in PFCs and their emulsions increases linearly with partial pressure approximating to Henry's Law (Figure 1). The amount of gas dissolved depends upon the PFC concentration and its solubility coefficient. Moreover, gas solubility in PFCs is inversely related to temperature.[10] It appears that the molecular structures and shapes

---

*For a discussion of fluorocarbon nomenclature, see Chapter 1.
**Perfluoroalkyl bromides like $C_8F_{17}Br$ (PFOB) are more reactive than perfluorocarbons (e.g., perfluorodecalin) owing to the relative strengths of C—F and C—Br bonds (see Chapter 8).

Table 2.   Solubilities (vol% at 1 atm and 37°C) of
Oxygen and Carbon Dioxide in Water and Selected
Liquid PFCs[10,12]

Compound	O_2	CO_2
Water	2.5	65
Perfluorotributylamine	40.0	140
Perfluorodecalin	40.3	142
Bis(perfluorohexyl)ethene[a]	43.0	180
Perfluorotripropylamine	45.0	166
Perfluoro-octyl bromide	50.0	210
Bis(perfluorobutyl)ethene[b]	50.0	230

[a,b]*Trans*-1,2-bis(perfluoro-*n*-alkyl)ethenes, $R_fCH=CHR_f$, where $R_f =$
n-C_4F_9[a] or n-C_6F_{13}.[b] These are known as F-44E and F-66E, respectively.

of PFCs are important determinants of their gas-dissolving properties, since gas molecules are believed to occupy "cavities" within a PFC liquid.[10]

Fluorine-containing compounds are used extensively in modern medicine as inhalation anesthetics (see Chapter 25), analgesics, anti-inflammatory agents, antibacterials, anticancer agents, and so forth.[11] Even fluoropolymers, such as poly(tetrafluoroethylene), are also widely used as orthopedic implants and replace-

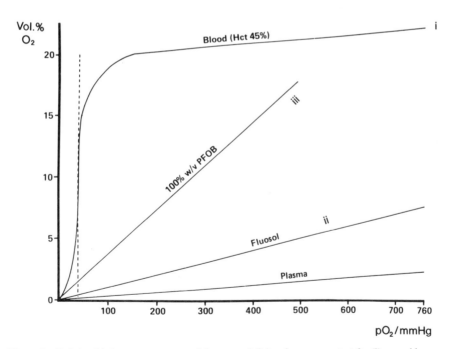

Figure 1.   Relationship between oxygen partial pressure (pO_2) and oxygen content for (i) normal human blood (hematocrit *ca* 45%), (ii) *Fluosol*® emulsion (Green Cross, Japan), and (iii) a 100% (w/v) perfluorooctylbromide (PFOB) emulsion. Note that the PFOB emulsion carries over 3 times more oxygen than *Fluosol*® at a tissue pO_2 of *ca* 45 mm Hg as indicated by the broken line (from Ref. 6).

ment vascular structures.[11] Thus, the development of PFC emulsions for i.v. use is but a further example of the increasing medical exploitation of organofluorine compounds.

26.3. SYNTHESIS OF PFCs

Methodologies for the synthesis of organofluorine compounds have been discussed in Chapter 2, and information bearing directly on the preparation of PFCs having importance as O_2 carriers can be found in Chapters 4 and 5. Briefly, there are two main approaches to PFC synthesis: (1) substitution, and (2) oligomerization. In the substitution method, hydrogen atoms in a hydrocarbon analog of the desired compound are replaced progressively by fluorine. This can be achieved in one of several ways, including electrochemical fluorination (see Chapter 5), use of high-valency metal fluorides (e.g., CoF_3), or directly with gaseous fluorine.[12] One widely used PFC, perfluorodecalin (FDC), is produced by the CoF_3 fluorination method (Chapter 4). Unfortunately, this method is somewhat unselective, and can produce mixtures of unpredictable composition due to incomplete fluorination, bond-breaking, and isomerization.[12]

The oligomerization approach to PFC synthesis involves combining small, already-fluorinated compounds in a building-block manner. The great advantage of this technique, however, is that pure PFCs can be made in a reproducible manner.[12] For example, perfluorooctyl bromide (perflubron; PFOB) can be synthesized by the oligomerization of tetrafluoroethylene:

$$C_2F_5I + 3CF_2{=}CF_2 \rightarrow C_2F_5(CF_2CF_2)_3I \rightarrow \text{(with } Br_2) \, n\text{-}C_8F_{17}Br$$

26.4. EMULSIFICATION OF PFCs

PFC liquids are immiscible in aqueous systems and are poor solvents for most physiological solutes. Thus, PFCs destined for i.v. use must be emulsified in an electrolyte solution containing an appropriate oncotic component, together with one or more surface-active agents (surfactants). PFCs are normally emulsified either by ultrasonic vibration (sonication) or by high-pressure homogenization.[10,12,13] Sonication is essentially for emulsification on a small scale, and is much more vigorous than homogenization methods, which can produce larger volumes of emulsion.[13] Sonication can, however, cause PFCs to degrade with release of potentially toxic fluoride ion (F^-).[10,12,13] Consequently, laboratory preparation of PFC emulsions requires a supplementary "clean-up" procedure in which emulsions are subjected to ion-exchange and resin dialysis to remove unwanted F^-.[13,14]

26.4.1. "First-Generation" Emulsions

The first commercial PFC emulsion developed for i.v. use was *Fluosol®* (formerly known as *Fluosol®-DA 20%*), produced in the late 1970s by the Green Cross Corporation, Japan.[15] *Fluosol®* is available in the USA and UK through the Alpha

Table 3. Composition[a] of *Fluosol*® and *Oxypherol*®

	Fluosol®	*Oxypherol*®
Perfluorodecalin (FDC)	14.0	—
Perfluorotripropylamine (FTPA)	6.0	—
Perfluorotributylamine (FTBA)	—	20.0
Pluronic® F-68	2.72	2.56
Yolk phospholipids	0.40	—
Potassium oleate	0.032	—
Glycerol	0.8	—
NaCl	0.6	0.6
NaHCO$_3$	0.21	0.21
Glucose	0.18	0.18
MgCl$_2$	0.043	0.043
CaCl$_2$	0.036	0.036
KCl	0.034	0.034

[a]All values are w/v (%).

Therapeutic Corporation. It was the first PFC formulation to be tested in man,[16–18] and was also the first emulsion to be subsequently accepted for clinical use (see below). *Fluosol*® is a 10% (v/v) emulsion of FDC and perfluorotripropylamine (FTPA) in a 70:30 weight ratio (Table 3). FTPA is added because FDC emulsions alone are relatively unstable (Table 1).[10,12,15] Hence, the stem emulsion of *Fluosol*® must be stored deep-frozen prior to use,[15] an obvious practical disadvantage. Other deficiencies of *Fluosol*®, discussed in detail elsewhere,[6,7,12] include a relatively low O$_2$-carrying capacity relative to its low PFC content; inadequate purity and prolonged tissue retention of its FTPA component; and acute, transient, adverse reactions in some patients.

Green Cross also produce a second commercial preparation, *Oxypherol*® (FC-43; formerly *Fluosol*®-43), consisting of 10% (v/v) emulsified perfluorotributylamine (FTBA; Table 3). Unlike *Fluosol*®, FC-43 is not advocated for clinical use owing to the prolonged retention time of FTBA in the body.[10,15] It has, however, been widely used in animals,[2–7] and has been employed for the perfusion of a variety of isolated organs (see below).

While *Fluosol*® is still the most widely used PFC emulsion, it is, nevertheless, now regarded as one of several "first-generation" formulations. Two additional first-generation PFC emulsions are "Emulsion No. II" and *Ftorosan*® produced in China and Russia, respectively (Table 4).[19,20] Emulsion No. II is a mixed emulsion of FDC and FTPA prepared with *Pluronic*® F-68 and is thus broadly similar in composition to *Fluosol*®. Ftorosan contains 7.6% (v/v) FDC and 3.8% (v/v) perfluoro[*N*-(4-methylcyclohexyl)piperidine] with the same surfactant. Both *Ftorosan*® and Emulsion No. II have been extensively used clinically, but only in their respective countries of origin.[19–21] During the past 5 years or so, these first-generation formulations have been superseded by improved "second-generation" emulsions (Table 4).

Table 4. Composition and Origin of PFC-Based Emulsions of Medical Interest

Name	Country of origin	PFC(s)	Surfactant(s)
"First-generation" formulations			
Fluosol[®]	Japan	Perfluorodecalin	Pluronic[®] F-68
		Perfluorotripropylamine	Egg yolk phospholipids
			Potassium oleate
Oxypherol[®]	Japan	Perfluorotributylamine	Pluronic[®] F-68
Ftorosan[®]	Russia	Perfluorodecalin	Pluronic[®] F-68
		Perfluoro-[(4-methylcyclohexyl)piperidine]	
Perflucol[®]	Russia	Perfluorodecalin	Pluronic[®] F-68
		Perfluorotripropylamine	
"Emulsion No II"	China	Perfluorodecalin	Pluronic[®] F-68
		Perfluorotripropylamine	
"Second-generation" formulations			
Addox[®]	USA	Perfluoro(trimethyl-adamantanes)	Lecithins
Hemagen[®]	USA	Perfluorodecalin	Lecithins
Oxygent[®]	USA	Perfluoro-octyl bromide	Lecithins
Imagent[®]	USA	Perfluoro-octyl bromide	Lecithins
—	France	1,2-bis(perfluorohexyl)ethene	Pluronic[®] F-68
			Lecithins
		1,2-bis(perfluorobutyl)ethene	Fluorinated surfactants[a]
—	Japan	Perfluoro(methylocta-hydroquinolizine)	Pluronic[®] F-68
			Egg yolk phospholipids
—	Japan	Perfluoro-(perhydro-*N*-methylisoquinoline)	Egg yolk phospholipids
			Potassium oleate
—	U.K.	Perfluorodecalin	Pluronic[®] F-68[b]
			Soya oil

[a]Perfluoroalkylated polyhydroxylated compounds.[7,12,47]
[b]Purified material.[39]

26.4.2. Improved Emulsions

Efforts to develop improved PFC emulsions have included: (1) identification of more readily emulsified PFCs of high purity and possessing acceptable *in vivo* half-life characteristics; (2) preparation of emulsions with greater PFC content, and correspondingly increased potential gas-carrying properties; and (3) the use of well-characterized and highly purified surfactants.

The most promising second-generation emulsions are those based on either FDC or PFOB (Table 4). The molecular weights of both compounds fall well within the range 460–520 (Table 1), which is recognized as that giving acceptable tissue retention times.[12] Other formulations, such as emulsified perfluoro(trimethyladamantanes),[22,23] perfluoromethyloctahydroquinolizine (FMOQ), and perfluoro(perhydro-*N*-

methylisoquinoline) (FMIQ) (Table 4),[24,25] are not regarded as "front runners," primarily because they are expensive and difficult to purify.

FDC is still retained as the sole or major PFC in newer emulsions currently being developed in the USA and other countries (Table 4) because its excellent biocompatibility characteristics outweigh its generally poor emulsifying properties. One solution to this problem has been to protect FDC emulsions from degradation attributable to molecular diffusion (Ostwald Ripening) by adding small quantities of high-boiling perfluorinated naphthenes of the fused-ring polycyclic type (e.g., perfluoroperhydrofluoranthene).[26–28] Emulsions containing such "higher-boiling-point-oil" (HBPO) additives are markedly more stable than emulsions such as *Fluosol*®, and appear to possess comparable biocompatibility characteristics.[27,29–31]

PFOB is another very good candidate for i.v. use since it can readily be emulsified with lecithins, and shows exceptionally fast excretion characteristics, owing to the high lipophilicity conferred by the bromine substituent.[12,32,33] Furthermore, it has one of the highest O_2-dissolving capacities of commonly used PFCs (Table 1). Neat PFOB was initially used in humans as a contrast agent in radiographic imaging of the gastrointestinal tract and lungs.[32,33] More recently, PFOB emulsions have been used for tissue oxygenation,[32,33] and for sensitizing tumors to radiation and cytotoxic drugs. Clearly, the combined imaging and O_2-carrying characteristics of PFOB make it a potentially very valuable compound for both diagnostic and therapeutic purposes.

A major advance has been the development of emulsions with increased PFC concentrations, thereby substantially increasing their O_2-transport characteristics (Figure 1). One such emulsion, a 50% (v/v) (*ca* 100% w/v) PFOB-based system, produced commercially by the Alliance Pharmaceutical Corporation in the USA, has been tested in dogs and has been evaluated in preliminary clinical trials.[34] Similarly, emulsions containing up to 30% (v/v) FDC, plus *Pluronic*® F-68 and soya oil, have been tested in rats,[35] while a 27% (v/v) emulsion based on 1,2-bis(perfluorohexyl)ethane (F-66E) has been used experimentally in mice as an adjunct to anticancer therapy.[36]

26.4.3. Surfactants

Surfactants are an essential constituent of any emulsion system and stabilize the emulsion by preventing flocculation or coalescence.[37] The surfactant commonly used for PFC emulsification is the nonionic, polyoxyethylene–polyoxypropylene block co-polymer, *Pluronic*® F-68 (*Poloxamer 188*). Commercial-grade material is a polydisperse preparation with an average molecular weight of 8400. It has numerous industrial and pharmaceutical applications,[38] and is present, for example, in both *Fluosol*® and FC-43 (Table 3).

Despite the acceptable emulsifying properties of *Pluronic*®, much concern has been expressed about the purity and biocompatibility of commercial samples. These can contain variable amounts of low-molecular-weight undesirable impurities, including aldehydes, formic and acetic acids, and the antioxidant, O,O'-di-*t*-butylcresol.[12] Evidence exists that *Pluronic*® or its components may be responsible for many of the reported adverse effects of first-generation PFC emulsions, especially *Fluosol*® (see below). This concern has prompted the preparation of highly purified fractions of *Pluronic®* for PFC emulsification, using either silica-Amberlite resin filtration[39] or supercritical fluid fractionation.[40] In both cases, the purified fractions showed markedly reduced toxicity, both *in vivo*[39] and *in vitro,*[40] and are now used routinely in many of the second-generation PFC emulsions under development (Table 4).[26–28]

One further problem with *Pluronic®*-containing emulsions is sterilization: such preparations cannot be autoclaved to the conventional temperature of 121°C because this exceeds the surfactant's "cloud point" (*ca* 115°C), above which its effectiveness drops markedly.[38] However, Johnson *et al.*[41] claim that the addition of 2% (w/v) soya oil to FDC (10% v/v) emulsified with 4% (w/v) *Pluronic®* F-68 raises the mean cloud point to 128°C, thus enabling nondisruptive sterilization to be carried out.

Concern about toxicity problems with *Pluronic®* F-68 has led to the use of alternative emulsifiers, including lecithins (phospholipids) and perfluoroalkylated compounds, including a novel class of "fluorophilic" compounds based on modified sugars and amino acids.

Egg-yolk lecithins are an obvious choice for emulsifying PFCs since they are already widely used in injectable lipid emulsions for parenteral nutrition.[37] Stable PFC emulsions containing lecithins as surfactant have been prepared,[22,34,42] and preliminary biocompatibility studies with a PFOB-lecithin emulsion in dogs have been encouraging.[34] One problem with lecithins, however, is that they are poorly defined, sensitive to degradation, and can contain potentially toxic contaminants, especially lysophospholipids. Indeed, studies using rat peritoneal mast cells showed that a petrol soluble extract of *Fluosol®*—in which trace quantities of lysophospholipids were detectable— stimulated histamine release in a dose-dependent manner.[43]

An alternative strategy has been to develop a range of fluorophilic surfactants and/or co-surfactants specifically for use with PFCs. Only limited biocompatibility data are available, yet Clark *et al.* have used a perfluoroalkylated amine oxide, XMO-10, and related compounds to prepare room-stable PFC emulsions[44]; a similar approach has been adopted by others.[45,46] One promising line of work has been the development of a series of well-defined, monodisperse, fluorinated co-surfactants derived from naturally occurring compounds, such as sugars or amino acids. Stable emulsions have been prepared with such compounds and the results of preliminary biocompatibility studies have been satisfactory.[12,47]

Overall, substantial progress has been made in the selection and assessment of surfactants for PFC emulsification. More biocompatibility data are needed on the fluorophilic surfactants in particular, but both purified *Pluronic®* F-68 and lecithins have been used successfully in the preparation of PFC emulsions for many biomedical applications.

26.5. BIOCOMPATIBILITY ASSESSMENT

A detailed consideration of the biological effects of PFC emulsions and their components, in both *in vivo* and *in vitro* systems, is well beyond the scope of this chapter. The interested reader is therefore referred to recent specialist reviews in this area.[2–7,48–50]

26.5.1. Tissue Uptake and Excretion

PFCs can accumulate in organs such as the liver and spleen, and the retention time depends on the molecular weight of the compound involved.[12] Emulsion droplets are readily visible in tissue macrophages, giving the cells a characteristic foam-like appearance.[51] This accumulation of PFCs produces marked increases in the weight of the tissues concerned, generally in proportion to the dose administered.[48]

PFCs are excreted primarily by expiration from the lungs,[52] and at rates which depend mainly on the molecular weights, and hence volatilities, of the compounds used.[10,12] Clearance rates appear to be little influenced by other factors, such as the presence of cyclic systems or heteroatoms in PFC molecular structures.[12] Small losses of PFCs from the body can also occur by transpiration through the skin.[52]

26.5.2. Effects on Lymphoid Tissues and the Reticuloendothelial System

The average droplet diameter of PFC emulsions prepared for i.v. use is generally less than 0.25 μm and this allows uptake into lymphoid tissue macrophages and other reticuloendothelial system (RES) cells; PFCs are visible in such cells as "foamy" membrane-bound vesicles.[51] Much attention has been focused on RES uptake of PFCs and, in particular, the consequences of this for immunological responsiveness. Earlier work showed that *Fluosol*® could induce transient impairment of RES phagocytic function as assessed, for example, by colloidal carbon and heterologous red cell clearance methods.[53,54] The temporary inhibition of RES clearance was intermediate between that produced by the "mild" RES blocker, *Intralipid*®, and the more potent agent, ethyl palmitate.[53] Later experiments revealed a biphasic pattern of RES clearance induced by *Fluosol*® in rats and it was suggested that this was a consequence of cyclic regeneration of RES cells.[55] An alternative explanation is that the secondary blockade may be due to re-uptake of PFCs from the circulation from which they had been released as a consequence of macrophage senescence.[56]

26.5.3. Effects on Immunological Competence

Since many of the proposed uses of PFCs as resuscitation fluids could involve their administration to the immunocompromised patient, it is particularly important that the effects of emulsified PFCs and their components on immune system function in the recipient should be studied in detail.[48] Indeed, experiments have shown that PFC effects on the humoral immune responses to immunological "challenges" are highly variable and dependent on: (1) emulsion composition; (2) dose and route of administration; (3) tissue and species involved; and (4) timing of emulsion administration relative to immune challenge.[48,57–60] Such variability has inevitably introduced difficulties in assessing the effects of different PFC emulsions on immunological compe-

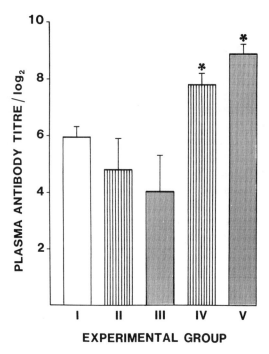

Figure 2. The mean hemagglutination response (expressed as $\log_2$ titer) at 7 days after i.p. injection of SRBC in female Wistar rats also receiving injections (10 cm^3 kg^{-1} body weight) of the following: Group I—saline (controls); Group II—*Fluosol*[®] (i.v.); Group III—*Fluosol*[®] (i.p.); Group IV—novel perfluorodecalin emulsion containing perfluoroperhydrofluoranthene (i.v.); Group V—novel perfluorodecalin emulsion containing perfluoroperhydrofluoranthene (i.p.). Vertical bars represent S.E.M.; *$p < 0.05$ (from Ref. 29).

tence. Moreover, the fact that responses are highly species-specific makes it difficult to extrapolate from laboratory animals into man.

Injection of PFC emulsions can certainly produce alterations in immunological responsiveness. For example, injection of rats with either *Fluosol*[®57] or a novel FDC emulsion containing the HBPO additive, perfluoroperhydrofluoranthene, to enhance stability increased the hemagglutination response to sheep red blood cells (SRBC) (Figure 2)[29,30]; similar results were obtained using mice.[59] An additional important finding was that changes in antibody titers produced by *Fluosol*[®] depended upon the timing of emulsion injection relative to immunization.[59,60]

It has been suggested that PFC emulsions act as adjuvants when injected into the peritoneal cavity of rodents immunized with SRBC.[48,57,58,60] This possibility is supported by results from related studies showing that *Fluosol*[®] can form complexes with antigens in hormone radioimmunoassay systems,[61] a property which has been effectively employed in an immunological agglutination assay.[62] While the precise nature of any adjuvant properties of PFCs or other emulsion components remain to be determined, one possibility is that the intracellular mediator, interleukin-1 (IL-1), is involved. However, the effects of *Fluosol*[®] or other PFC emulsions on IL-1 release have not been reported, although other particulate material, such as silica, can enhance its release from rabbit macrophages in culture.[63]

Further evidence of altered immune system function in response to injection of emulsified PFCs is that near-total transfusion of rats with *Fluosol*® inhibited the afferent (induction) phase of a specific immune response, but did not affect the efferent phase[64]; this was attributed to PFC-induced RES blockade as described above. Subsequent work showed that transfusion with *Fluosol*® also inhibited the humoral immune response to pneumococcal antigen.[65] These results contrasted markedly with those in which animals had been immunized with SRBC[64] showing that alterations in immunological responsiveness produced by emulsified PFCs varied according to whether a T-lymphocyte dependent (e.g., SRBC) or T-lymphocyte independent (e.g., pneumococcal polysaccharide) was involved.

26.5.4. Effects on Tissue Biochemistry

An important area currently under active investigation concerns the effects of PFCs on liver biochemistry and the consequences of their use on the metabolism of other drugs, especially anesthetics and analgesics.[50] Some PFCs can transiently alter the activity of enzymes, such as the microsomal cytochromes (P-450) complex[66,67] (Figure 3) and aryl esterase.[68] The responses, which included induction of liver P-450, depended on the nature of the compound(s) administered, together with both sex and endocrine status of the recipient.[68,69] The observation that pentobarbital-induced "sleeping time" in rats was decreased markedly by prior injection of a novel FDC emulsion suggests that PFC-mediated liver P-450 induction enhances drug metabolism, thereby reducing its anesthetic properties.[69] This is supported by earlier work showing that partial transfusion of rats with *Fluosol*® produced a marked increase in the clearance of antipyrine, a drug commonly used to assess hepatic metabolism.[70,71] While further detailed studies are required, these effects occur independently of any metabolism of the PFC molecules themselves which are known to be highly unreactive in the body.[10,12,50] A further recent development in this area concerns possible applications for PFC emulsions as controlled inducers of metabolic enzymes.[72] The finding that some PFCs can induce hepatic enzymes without apparently generating pharmacological side-effects[50,66–69] could be of considerable therapeutic benefit in the treatment of drug overdose and poisoning.

26.5.5. Other *in vivo* Responses

Several studies have investigated the effects of PFC emulsions and their components on the hematopoietic and endocrine systems, although in the latter case these have been far from extensive. Early experiments showed that *Fluosol*® was able to produce complement activation via the alternate pathway.[17,18,73] *Fluosol*® can also inhibit both plasma and liver phospholipase A_2 activities leading to alterations in prostaglandin (PG) metabolism.[74] However, the changes in PG biosynthesis by PFCs

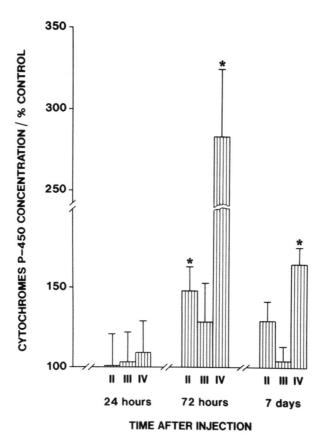

Figure 3. Liver cytochromes P-450 concentrations in male Wistar rats at 24 hr, 72 hr, and 7 days after injection of either *Fluosol®* (Group II), FC-43 (Group III), or a novel perfluorodecalin emulsion containing perfluoroperhydrofluoranthene and *Pluronic®* F-68 (Group IV). Values are percents of mean control value (Group I—data not shown); vertical bars represent S.E.M.; *$p < 0.05$. (From Ref. 67.)

were not of sufficient magnitude to offset the increase in immunoreactive thromboxane B_2 which followed endotoxin injection in FC-43-transfused rats.[75]

Experiments have also shown that near total blood replacement with FC-43 in anesthetized rats is followed by alterations in circulating catecholamine concentrations, as reflected by increases in arterial free dopamine, adrenaline, and nor-adrenaline.[76] Further studies on the endocrine responses to emulsified PFCs and their components are required.

26.5.6. Cellular Effects *in vitro*

The finding that *in vivo* use of emulsified PFCs could alter immune system function prompted studies on the direct effects of different emulsions and their components on immune cells *in vitro*. Thus, incubation of human blood with either *Fluosol®* or FC-43 inhibited the phagocytic activity of both neutrophils and monocytes, as assessed by their uptake of fluorescent polystyrene beads.[77,78] In addition, exposure to emulsified PFCs produced alterations in cell structure and function characterized by increased cytoplasmic vacuolation, decreased chemotaxis and reduced aggregation, adherence and induced superoxide (O_2^-) release.[77,78] Similar findings of reduced cell adherence have been seen in both mouse peritoneal macrophages and nondepleted splenocytes cultured with FC-43 or *Fluosol®*, respectively.[79,80]

The *in vitro* cellular responses to emulsified PFCs are dependent on the cell type under investigation and the species of origin. For example, FC-43 has adverse effects on mouse macrophages, but not lymphocytes, and it has been proposed that this selective toxicity was caused by either disruption of phospholipid membrane or alterations in the intracellular oxygen detoxification system, or a combination of both.[79] Emulsion composition is also an important variable since *Fluosol®* is more effective than FC-43 in promoting procoagulant generation and inhibiting O_2^- generation in human mononuclear leucocytes.[81] Further details of the effects of PFC emulsion components on cells *in vitro* are given in specialist reviews.[4,6,8,9,48,49]

26.5.7. Effects of Surfactant(s)

Attention has also focused on the biological effects of the surfactant component(s) of PFC emulsions. For example, *Pluronic®* F-68 can mimic the complement-activating effects of *Fluosol®*,[17,18,73] and has thus been proposed as the active principle responsible for the acute, transient, hematological disturbances seen in some patients receiving the emulsion. *Pluronic®* can inhibit neutrophil functions both *in vivo* and *in vitro*,[81,82] although there is speculation that some of these effects may be caused by contaminants in commercial grade *Pluronic®* fractions rather than by the surfactant itself.[39,40] In addition, peroxide derivatives of *Pluronic®*, formed during steam sterilization or long-term storage of *Fluosol®*, have been implicated in adverse physiological effects of this emulsion.[83] While some of the unwanted side-effects of *Pluronic®* can be prevented by using highly purified fractions,[39,40] it has been noted already that there is a view that other, less toxic, surfactants should be used preferentially for emulsifying PFCs for *in vivo* use. For example, experiments using a PFOB-lecithin emulsion in anesthetized dogs showed less pronounced hemodynamic disruption compared with *Fluosol®*.[34]

While the arguments over choice of surfactant(s) continue, it is somewhat paradoxical that the neutrophil-inhibiting effects attributable to *Pluronic®* may be advantageous when PFC emulsions are used, for example, in the ischemic myocardium to protect against neutrophil-induced reperfusion injury.[5] In support of this, experiments have shown that the effects of *Fluosol®* on myocardial salvage in dogs appeared to involve suppression of neutrophil function[84] (see below).

Table 5. Potential Biomedical Applications for PFCs
and Their Emulsions

- Resuscitation fluids
- Pre-operative hemodilution
- Perfusion of ischemic tissues
    myocardium
    brain
    intestine
    placental insufficiency
- Adjuncts to cancer treatment
    radiotherapy
    chemotherapy
- Respiratory applications
    liquid ventilation
    decompression sickness
    whole body oxygenation
- Contrast agents for *in situ* tissue imaging and diagnosis
    X-ray
    magnetic resonance
    ultrasound
- Management of blood diseases
- Ophthalmologic surgery
- Organ perfusion and preservation for transplants
- Cell cultures
    regulation of gas supply
    surface for cell growth

26.6. BIOMEDICAL APPLICATIONS

PFC emulsions were initially proposed as O_2-carrying resuscitation fluids or, more dramatically, as "blood substitutes". It is now clear, however, that PFCs have great potential in more specific areas of medicine and biotechnology (Table 5). Some of these applications are outlined below.

26.6.1. Microcirculatory Support in Ischemic Tissues

PFC emulsions have properties which make them very attractive for providing microcirculatory support in ischemic tissues. In addition to O_2 transport, PFCs have very low viscosities which are almost independent of shear rate (Figure 4).[3,6] Thus, PFC emulsions can be used as hemodiluents in ischemic disease to reduce blood viscosity and improve perfusion through available collaterals. Additionally, the particle size in emulsions for i.v. use is generally < 0.25 μm, and this allows penetration into hypoxic tissue beds otherwise inaccessible to red blood cells. Emulsions such as *Fluosol*® and FC-43 have been used in several species for ischemic tissue rescue with

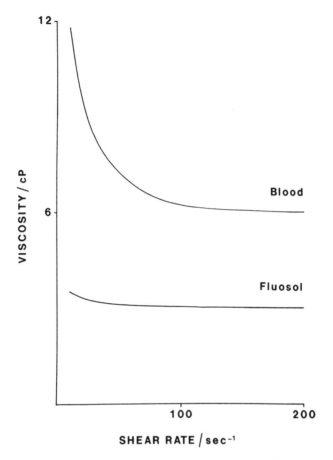

Figure 4. Relationship between viscosity (centipoise) and shear rate (sec^{-1}) of normal human blood (hematocrit *ca* 45%) and *Fluosol*® emulsion (from Ref. 6).

varying degrees of success.[3,5,85–88] There is evidence that PFC emulsions may be preferred over other O_2-carrying perfusates since they probably affect the outcome of ischemia by direct inhibitory actions on leukocytes, thereby protecting against reperfusion injury.[5,84,88] However, the extent to which the individual PFCs or surfactant components(s) are involved requires further study.

26.6.2. Coronary Angioplasty

One area currently attracting much interest is the clinical use of PFC emulsions as distal O_2-carrying perfusates to overcome ischemic myocardial damage during percutaneous transluminal coronary angioplasty. The remarkable success of preliminary trials[89–94] contributed significantly to the recent decision by the United States Federal Drug Administration Committee to approve *Fluosol*® for this purpose.

Fluosol® has also been used during balloon angioplasty in the UK and the first operation employing it (January 1991) received wide news media coverage.

26.6.3. Cancer Therapy

Solid tumors generally contain hypoxic cells and this decreases their sensitivity to the cytotoxic effects of ionizing radiation and anticancer drugs. Such conditions arise as a result of abnormal vasculature and consequent inadequate blood flow. PFC emulsions have been used with varying degrees of success to increase tumor oxygenation and thus improve sensitivity to either radiation or chemotherapy.[95–104] PFC emulsions developed for their cancer therapy-enhancing properties (i.e., increased i.v. persistence providing more prolonged therapeutic effects) are substantially different from those developed as, for example, resuscitation fluids, which ideally have low vascular retention times.[2] This illustrates the point that future generations of PFC emulsions will need to be "tailor-made" for specific requirements.

A related area of study concerns the effects of emulsified PFCs as adjuncts to photodynamic therapy (PDT), which depends on the use of tumor-localizing porphyrins that are activated by visible light in the presence of O_2 to produce tumor destruction.[105] Experiments have shown that incubation of porphyrin derivatives with either FC-43 or FDC, emulsified with *Pluronic*® F-68, increased the efficiency of photosensitized oxidation of histidine.[106] However, while PFCs appear to be of no immediate benefit in animals receiving PDT while breathing air,[107–109] there is evidence that emulsified FDC may nevertheless protect the skin against PDT-induced damage.[108,109] Further work is in progress to confirm this.

26.6.4. Contrast Media and Diagnostic Imaging

Another valuable clinical application for PFCs and their emulsions is as contrast enhancement agents for *in situ* tissue imaging and diagnosis. PFCs are radio-opaque, especially if they contain bromine atoms, and compounds such as PFOB have been used extensively for X-ray imaging of the lungs, gastrointestinal tract, and RES tissues.[32,33,110,111] PFOB and other PFCs have also been used effectively as contrast agents for both proton and ^{19}F NMR imaging studies of various tissues.[32,33,111,112] PFOB has additionally been used as a contrast agent in computed tomography and sonography.[32] The uptake of PFC emulsion particles by macrophages of malignant tissues has provided a convenient method for localizing tumors.[113,114] As noted previously,[6,33] the dual combination of tumor imaging and cancer therapy-enhancing properties of PFOB and related compounds makes them potentially very powerful clinical tools.

26.6.5. Respiratory Distress Syndrome

Respiratory distress syndrome (RDS) is one of the most common diseases of the human premature newborn and is associated with a deficiency in alveolar surfactant.

Studies using fetal lambs and miniature pigs have shown that ventilation of the respiratory airways with PFC liquids can improve lung compliance and increase arterial O_2 tensions.[115,116] In addition, the small quantity of PFC remaining in the alveoli after evaporation acted as a wetting agent enabling subsequent ventilation to occur at lower pressures.[116] There is great optimism that such use of PFCs will assist the immature lungs of premature babies in making the transition to air breathing, and preliminary trials to this effect have been encouraging.[117]

The impetus for this work was provided by the classical experiments of Leland Clark and Frank Gollan in the mid-1960s which demonstrated that rats and mice could survive for extended periods while completely submerged in liquid perfluoro(2-butyltetrahydrofuran) saturated with O_2.[118] Their experiments were prompted by the earlier paper by Kylastra et al. entitled "Of mice as fish," in which it was reported that mice had been kept alive while "breathing" from oxygenated saline solutions.[119]

26.6.6. Decompression Sickness

N_2 has a high solubility in PFCs.[10,12] This led to the suggestion that administration of PFCs could be used to alleviate the potentially fatal effects of decompression sickness, by "mopping-up" gas bubbles from the bloodstream. Initial studies in hamsters showed that ventilation with fluorocarbon liquid had a protective effect against subsequent exposure to hyperbaric compression.[120] Related experiments in cats, however, revealed the thermoregulatory problems associated with fluorocarbon breathing[121]: under hypothermic conditions comparable to those which exist at great depth, there was a marked decline in cardiovascular function that could not be maintained by normal regulatory mechanisms. While this approach is clearly limited, and remains to be tested systematically in humans, it has nevertheless received popular acclaim in the movie "The Abyss" (Twentieth Century Fox, 1989).

More recent experiments in rats showed that pre-injection with emulsified PFCs increased survival of animals with experimental decompression sickness.[122] In addition, a combination of i.v. FC-43 emulsion and 100% O_2 breathing was effective in providing both hemodynamic and neurologic protection against decompression sickness.[123]

26.6.7. Lung Damage and Respiratory Failure

One possible application for PFC emulsions is for oxygenating and removing CO_2 from the body in cases of respiratory failure. Intraperitoneal perfusion of experimental animals with oxygenated *Fluosol®* has been shown to increase arterial pO_2 and decrease pCO_2.[124-126] The success of this procedure is limited by the relatively small surface area of the peritoneum compared with that available in the lungs; nevertheless, this preliminary work provides good evidence that it may be useful for supplementing whole body oxygenation in patients with respiratory disease.

26.6.8. Blood Diseases

Experiments have shown that PFC emulsions may have therapeutic value in patients suffering from sickle-cell crisis. Reindorf et al.[127] found that oxygenated FC-43 was effective in reducing the incidence of sickling when cultured with deoxygenated red cells from patients with sickle-cell anemia. Subsequent studies showed that the loss of erythrocyte deformability induced by FC-43 was due to adsorption of the emulsion onto red cell membranes.[128] In related work, the Pluronic® F-68 surfactant component of Fluosol® was found to improve the rheology of sickled red cells independently of changes in oxygenation provided by the PFCs. The experiments also showed that the surfactant reduced the adherence of cells to endothelial monolayers, possibly by lubricating effects on cell surfaces. This work raises the question of whether Pluronic®-containing PFC emulsions may be useful in the management of other blood diseases, such as cerebral malaria, the onset of which involves adhesion of infected erythrocytes to endothelium.[130]

26.6.9. Applications in Ophthalmology

Low-viscosity PFC liquids such FDC, FTBA, and perfluorooctane have been used as intraoperative tools during vitreoretinal surgery for trauma-induced retinal detachment.[131,132] The greater specific gravities of PFCs make them much more effective than, say, sodium hyaluronate or fluorosilicone oils, for hydrokinetic retinal manipulation during eye surgery.

26.6.10. Organ Perfusion and Preservation

PFC emulsions have been employed for the perfusion of isolated organs including the heart, liver, kidney, lung, pancreas and testis.[133-143] The successful reimplantation of ex-vivo PFC-perfused kidneys has also been reported.[137] Additionally, Fluosol® has been used to maintain various traumatically amputated human extremities for up to 72 hours prior to their successful reimplantation.[144] The specific inhibitory effects of Pluronic®-containing PFC emulsions on leukocytes may be especially valuable in transplantation studies, since such preparations could be used to "flush out" sensitized cells and thereby reduce subsequent immunocompatibility problems.

26.6.11. Cell Culture Studies

PFCs have value in cell culture systems for regulating gas supply, leading to improved growth and productivity.[8,9] Such use of PFCs for oxygenation of fragile cell cultures can reduce or eliminate mechanical damage caused by conventional aeration methods. An additional approach has been to grow animal cells at the interface between PFC oil and aqueous culture media.[8,9] This provides support for anchorage-independent cell growth, additional to any alterations in O_2 supply. More recently, the use of an FC-43-supplemented medium for the culture of early chick embryonic heart tissue has been reported.[145]

PFCs have also been used to supply other gases, including CO_2, to microbial cells.[146] This suggests that the potential use of such compounds for regulating gas supply in cell cultures can be extended beyond aerobic systems. One possible application for PFCs in *in vitro* systems would be as scavengers of toxic gaseous cellular byproducts.

26.7. CONCLUDING REMARKS

Not surprisingly, initial attention in both academia and industry focused inevitably on the possible value of PFC emulsions as O_2-carrying "blood substitutes," but it is now abundantly clear that this was both misleading and highly restrictive: biomedical applications—established and potential —for PFCs are extensive, involving areas as diverse as tissue imaging and perfusion, organ preservation, cancer therapy, respiratory support, eye surgery, and cell culture. PFCs are thus rapidly becoming recognized as one of the most versatile classes of compounds yet exploited by man.

26.8. REFERENCES

1. J. P. Cohn, *Bioscience 37*, 647 (1987).
2. G. P. Biro and P. Blais, *CRC Crit. Rev. Oncol. Haem. 6*, 311 (1987).
3. N. S. Faithfull, in: *Blood Substitutes: Preparation, Physiology and Medical Applications* (K. C. Lowe, ed.), pp. 130–148, Ellis Horwood, Chichester (1988).
4. K. C. Lowe, in: *Blood Substitutes: Preparation, Physiology and Medical Applications* (K. C. Lowe, ed.), pp. 149–172, Ellis Horwood, Chichester (1988).
5. K. C. Lowe, in: *Ischemic Diseases and the Microcirculation—New Results* (K. Messmer, ed.), pp. 100–107, Zuckschwerdt, Munich (1989).
6. K. C. Lowe, *Vox Sang. 60*, 129 (1991).
7. J. G. Riess, *Vox Sang. 61*, 225 (1991).
8. A. T. King, B. J. Mulligan, and K. C. Lowe, *Bio/Technology 7*, 1037 (1989).
9. A. T. King, B. J. Mulligan, and K. C. Lowe, in: *Oxygen Transport to Tissue* (J. Piiper, T. K. Goldstick, and M. Meyer, eds.), Vol. XII, pp. 283–290, Plenum Press, New York (1990).
10. J. G. Riess and M. Le Blanc, *Pure Appl. Chem. 54*, 2383 (1982).
11. G. D. Chase, R. A. Deno, A. R. Gennaro, M. R. Gibson, C. K. Stewart, R. E. King, A. N. Martin, E. A. Swinyard, C. T. VanMeter, A. Osol, B. Witlin, and J. E. Hoover, *Remington's Pharmaceutical Sciences* (14th edn.), Mack, Easton (1970).
12. J. G. Riess and M. Le Blanc, in: *Blood Substitutes: Preparation, Physiology and Medical Applications* (K. C. Lowe, ed.), pp. 94–129, Ellis Horwood, Chichester (1988).
13. S. K. Sharma, K. C. Lowe, and S. S. Davis, *Drug Dev. Ind. Pharm. 14*, 2371 (1988).
14. C. Chubb, *Biol. Reprod. 33*, 854 (1985).
15. R. Naito and K. Yokoyama, *Perfluorochemical Blood Substitutes*, Tech. Inform. Ser. No. 5, Green Cross, Osaka (1978).
16. K. K. Tremper and S. T. Anderson, *Ann. Rev. Med. 36*, 309 (1985).
17. D. E. Hammerschmidt and G. M. Vercellotti, in: *Blood Substitutes* (T. M. S. Chang and R. P. Geyer, eds.), pp. 431–438, Marcel Dekker, New York (1989).
18. G. M. Vercellotti and D. E. Hammerschmidt, in: *Blood Substitutes: Preparation, Physiology and Medical Applications* (K. C. Lowe, ed.), pp. 173–182, Ellis Horwood, Chichester (1988).
19. R. Xiong, R. A. Zhang, H. F. Chen, W. Y. Huang, C. P. Luo, and W. T. Cao, *Chin. J. Surg. 19*, 213 (1981).

20. F. F. Beloyartsev, E. I. Mayevsky, and B. I. Islamov, *Ftorosan Oxygen Carrying Perifluorochemical Plasma Substitute*, Acad. Sci. USSR, Pushchino (1983).
21. H. S. Chen and Z. H. Yang, in: *Blood Substitutes* (T. M. S. Chang and R. P. Geyer, eds.), pp. 403–409, Marcel Dekker, New York (1989).
22. R. E. Moore, in: *Blood Substitutes* (T. M. S. Chang and R. P. Geyer, eds.), pp. 443–445, Marcel Dekker, New York (1989).
23. P. Menasch, C. Grousset, C. Mouas, R. E. Moore, and A. Piwnica, in: *Blood Substitutes* (T. M. S. Chang and R. P. Geyer, eds.), pp. 607–616, Marcel Dekker, New York (1989).
24. K. Yokoyama, T. Suyama, H. Okamoto, M. Watanabe, H. Ohyanagi, and Y. Saitoh, *Artif. Org.* 8, 34 (1984).
25. T. Mitsuno, H. Ohyanagi, K. Yokoyama, and T. Suyama, in: *Blood Substitutes* (T. M. S. Chang and R. P. Geyer, eds.), pp. 365–373, Marcel Dekker, New York (1989).
26. S. S. Davis, K. C. Lowe, and S. K. Sharma, *Br. J. Pharmacol.* 89, 665P (1986).
27. S. K. Sharma, A. D. Bollands, S. S. Davis, and K. C. Lowe, in: *Oxygen Transport to Tissue* (I. A. Silver and A. Silver, eds.), Vol. IX, pp. 97–108, Plenum Press, London (1987).
28. S. K. Sharma, K. C. Lowe, and S. S. Davis, in: *Blood Substitutes* (T. M. S. Chang and R. P. Geyer, eds.), pp. 447–450, Marcel Dekker, New York (1989).
29. A. D. Bollands, K. C. Lowe, S. K. Sharma, and S. S. Davis, *J. Pharm. Pharmacol.* 39, 1021 (1987).
30. A. D. Bollands, K. C. Lowe, S. K. Sharma, and S. S. Davis, in: *Blood Substitutes* (T. M. S. Chang and R. P. Geyer, eds.), pp. 451–453, Marcel Dekker, New York (1989).
31. K. C. Lowe, A. D. Bollands, and P. D. Raven, *Comp. Biochem. Physiol.* 93C, 377 (1989).
32. R. F. Mattrey, *Am. J. Radiol.* 152, 247 (1989).
33. D. M. Long, D. C. Long, R. F. Mattrey, R. A. Long, A. R. Burgan, W. C. Herrick, and D. F. Shellhamer, in: *Blood Substitutes* (T. M. S. Chang and R. P. Geyer, eds.), pp. 411–420, Marcel Dekker, New York (1989).
34. R. F. Mattrey, P. L. Hilpert, C. D. Long, D. M. Long, R. M. Mitten, and T. Peterson, *Crit. Care Med.* 17, 652 (1989).
35. P. K. Bentley, O. L. Johnson, C. Washington, and K. C. Lowe, *J. Pharm. Pharmacol.* 45, 182 (1993).
36. C. Thomas, F. L. De Vathaire, E. Lartigau, E. P. Malaise, and M. Guichard, *Radiat. Res.* 118, 476 (1989).
37. S. S. Davis, J. Hadgraft, and K. J. Palin, in: *Encyclopedia of Emulsion Technology* (P. Becher, ed.), Vol. 2, pp. 159–238, Marcel Dekker, New York (1985).
38. I. R. Schmolka, *J. Am. Oil Chem. Soc.* 54, 110 (1977).
39. P. K. Bentley, S. S. Davis, O. L. Johnson, K. C. Lowe, and C. Washington, *J. Pharm. Pharmacol.* 41, 661 (1989).
40. T. A. Lane and V. Krukonis, *Transfusion* 28, 375 (1988).
41. O. L. Johnson, C. Washington, and S. S. Davis, *Int. J. Pharmaceutics* 59, 131 (1990).
42. S. Magdassi and A. Siman-Tov, *Int. J. Pharmaceutics* 59, 69 (1990).
43. K. C. Lowe, D. C. McNaughton, and J. P. Moore, *Br. J. Pharmacol.* 82, 276P (1984).
44. L. C. Clark, E. W. Clark, R. E. Moore, D. G. Kinnett, and E. I. Inscho, in: *Advances in Blood Substitute Research* (R. B. Bolin, R. P. Geyer, and C. J. Nemo, eds.), pp. 169–180, Liss, New York (1983).
45. C. M. Sharts, A. A. Malik, J. C. Easdon, and L. Khawli, *J. Fluorine Chem.* 34, 365 (1987).
46. B. M. Fung, E. A. O'Rear, J. Afzal, C. B. Frech, D. L. Mamrosh, and M. Gangoda, in: *Blood Substitutes* (T. M. S. Chang and R. P. Geyer, eds.), pp. 439–440, Marcel Dekker, New York (1989).
47. J. G. Riess, C. Arlen, J. Greiner, M. Le Blanc, A. Manfredi, S. Pace, C. Varescon, and L. Zarif, in: *Blood Substitutes* (T. M. S. Chang and R. P. Geyer, eds.), pp. 421–430, Marcel Dekker, New York (1989).
48. K. C. Lowe, in: *Oxygen Transport to Tissue* (M. Mochizuki, C. R. Honig, T. Koyama, T. K. Goldstick, and D. F. Bruley, eds.), Vol. X, pp. 655–663, Plenum Press, New York (1988).
49. K. C. Lowe, *Clin. Hemorheol.* 12, 141 (1992).
50. W. R. Ravis, J. F. Hoke, and D. L. Parsons, *Drug Metab. Rev.* 23, 375 (1991).
51. L. Nanney, L. M. Fink, and R. Virmani, *Arch. Pathol. Lab. Med.* 108, 631 (1984).

52. Y. Tsuda, K. Yamanouchi, K. Yokoyama, T. Suyama, M. Watanabe, H. Ohyanagi, and Y. Saitoh, in: *Blood Substitutes* (T. M. S. Chang and R. P. Geyer, eds.), pp. 473–483, Marcel Dekker, New York (1989).

53. J. Lutz and P. Metzenauer, *Pflügers Arch. 387*, 175 (1980).

54. O. Castro, A. E. Nesbitt, and D. Lyles, *Am. J. Hematol. 6*, 15 (1984).

55. J. Lutz, in: *Perfluorochemical Oxygen Transport* (K. Tremper, ed.), *Int. Anesth. Clin.*, Vol. 23, pp. 63–93, Little Brown, Boston (1985).

56. F. Pfannkuch and N. Schnoy, in: *Advances in Blood Substitute Research* (R. B. Bolin, R. P. Geyer, and C. J. Nemo, eds.), pp. 209–219, Liss, New York (1983).

57. A. D. Bollands and K. C. Lowe, *Comp. Biochem. Physiol. 85C*, 309 (1986).

58. A. D. Bollands and K. C. Lowe, *Comp. Biochem. Physiol. 86C*, 431 (1987).

59. A. D. Bollands and K. C. Lowe, *Comp. Biochem. Physiol. 89C*, 127 (1988).

60. K. C. Lowe and A. D. Bollands, in: *Blood Substitutes* (T. M. S. Chang and R. P. Geyer, eds.), pp. 495–504, Marcel Dekker, New York (1989).

61. M. Hammarstrom, R. Mullins, and D. Sgoutas, *Clin. Chem. 29*, 1418 (1983).

62. T. L. Prather, J. Grane, C. R. Keese, and I. Giaever, *J. Immun. Methods 87*, 211 (1986).

63. R. F. Kampschmidt, M. L. Worthington, and M. I. Mesecher, *J. Leukocyte Biol. 39*, 123 (1986).

64. G. R. Hodges, S. E. Worley, J. M. Kemner, N. I. Abdou, and G. M. Clark, *J. Leukocyte Biol. 39*, 141 (1986).

65. N. C. Molina, G. R. Hodges, S. E. Worley, and N. I. Abdou, *Clin. Immun. Immunopath. 42*, 211 (1987).

66. N. V. Adrianov, A.I. Archakov, and M. Zigler, *Bull. Exp. Biol. Med. (USSR) 108*, 164 (1989).

67. F. H. Armstrong and K. C. Lowe, *Comp. Biochem. Physiol. 94C*, 345 (1989).

68. F. H. Armstrong and K. C. Lowe, *Comp. Biochem. Physiol. 99C*, 561 (1991).

69. K. C. Lowe and F. H. Armstrong, *Biomat. Art. Cells Immob. Biotechnol. 20*, 993 (1992).

70. R. P. Shrewsbury, L. G. White, G. M. Pollack, and W. A. Wargin, *J. Pharmacol. 38*, 883 (1987).

71. R. P. Shrewsbury and L. G. White, *Res. Commun. Chem. Pathol. Pharmacol. 62*, 137 (1988).

72. K. C . Lowe and C . Washington, *Nature 358*, 717 (1992).

73. G. M. Vercellotti, D. E. Hammerschmidt, P. R. Craddock, and H. S. Jacob, *Blood 59*, 299 (1982).

74. K. M. M. Shakir and T. J. Williams, *Prostaglandins 23*, 919 (1982).

75. J. A. Cook, W. C. Wise, G. E. Tempel, and P. V. Halushka, *Circ. Shock 15*, 193 (1985).

76. J. X. Wilson, N. H. West, M. Amies, and E. Michalska, *Neurosci. Lett. 62*, 329 (1985).

77. R. Virmani, D. Warren, R. Rees, L. M. Fink, and D. English, *Transfusion 23*, 512 (1983).

78. R. Virmani, L. M. Fink, K. Gunter, and D. English, *Transfusion 24*, 343 (1984).

79. R. Bucala, M. Kawakami, and A. Cerami, *Science 220*, 965 (1983).

80. A. D. Bollands and K. C. Lowe, *Biotechnol. Lett. 11*, 265 (1989).

81. R. L. Janco, R. Virmani, P. J. Morris, and K. Gunter, *Transfusion 25*, 578 (1985).

82. T. A. Lane and G. E. Lamkin, *Blood 64*, 400 (1984).

83. L. E. McCoy, C. A. Becker, T. H. Goodin, and M. I. Barnhart, *Scanning Electron Microsc. 16*, 311 (1984).

84. A. K. Bajaj, M. A. Cobb, R. Virmani, J. C. Gray, R. T. Light, and M. B. Forman, *Circulation 79*, 645 (1989).

85. R. D. Bell, J. L. Osterholm, S. W. Duckett, K. Anderchek, P. L. Gorsuch, and L. A. Schneider, in: *Cerebrovascular Diseases* (M. D. Ginsberg and W. D. Dietrich, eds.), pp. 265–271, Raven Press, New York (1989).

86. A. J. Triolo, J. L. Osterholm, G. M. Alexander, R. D. Bell, and G. D. Frazer, *Neurosurgery 26*, 480 (1990).

87. R. Virmani, M. B. Forman, and F. D. Kolodgie, *Circulation 81* (Suppl. IV), 57 (1990).

88. M. B. Forman, D. A. Ingram, and J. D. Murray, *Clin. Hemorheol. 12*, 121 (1992).

89. H. V. Anderson, D. L. Nelson, P. P. Leimgruber, G. S. Roubin, and A. R. Gruentzig, in: *Transfusion Medicine: Recent Technological Advances* (K. Murawski and F. Peetoom, eds.), pp. 3–20, Liss, New York (1986).

90. M. Cleman, C. C. Jaffe, and D. Wohlgelernter, *Circulation 74*, 555 (1986).

91. C. C. Jaffe, D. Wohlgelernter, H. A. Highman, and M. Cleman, in: *Transfusion Medicine: Recent Technological Advances* (K. Murawski and F. Peetoom, eds.), pp. 21–27, Liss, New York (1986).

92. M. J. Cowley, F. R. Snow, G. DiSciascio, K. Kelly, C. Guard, and J. V. Nixon, *Circulation 81* (Suppl. IV), 27 (1990).

93. K. M. Kent, M. W. Cleman, M. J. Cowley, M. B. Forman, C. C. Jaffe, M. Kaplan, S. B. King, M. W. Krucoff, and T. Lassar, *Am. J. Cardiol. 66*, 279 (1990).

94. M. B. Forman, J. M. Perry, B. H. Wilson, M. S. Verani, P. R. Kaplan, F. A. Shawl, and G. C. Friesinger, *J. Am. Coll. Cardiol. 18*, 911 (1991).

95. B. A. Teicher and S. A. Holden, *Cancer Treat. Rep. 71*, 173 (1987).

96. S. Rockwell, in: *Blood Substitutes* (T. M. S. Chang and R. P. Geyer, eds.), pp. 519–531, Marcel Dekker, New York (1989).

97. R. A. Lustig, C. M. Rose, and N. L. McIntosh-Lowe, in: *Blood Substitutes* (T. M. S. Chang and R. P. Geyer, eds.), pp. 511–518, Marcel Dekker, New York (1989).

98. G. E. Kim and C. W. Song, *Cancer Chemother. Pharmacol. 25*, 99 (1989).

99. B. A. Teicher, N. L. McIntosh-Lowe, and C. M. Rose, in: *Blood Substitutes* (T. M. S. Chang and R. P. Geyer, eds.), pp. 533–546, Marcel Dekker, New York (1989).

100. S. A. Holden, T. S. Herman, and B. A. Teicher, *Radiother. Oncol. 18*, 59 (1990).

101. M. Guichard, *Radiother. Oncol. Suppl. 20*, 59 (1991).

102. D. J. Chaplin, M. R. Horsman, and D. S. Aoki, *Br. J. Cancer 63*, 109 (1991).

103. C. Thomas, J. G. Riess, and M. Guichard, *Int. J. Radiat. Oncol. Biol. Phys. 59*, 433 (1991).

104. B. A. Teicher, *Biomat. Art. Cells Immob. Biotechnol. 20*, 875 (1992).

105. B. W. Henderson and T. J. Dougherty, *Photochem. Photobiol. 55*, 145 (1992).

106. R. K. Chowdhary, R. B. Cundall, and C. G. Morgan, *Photochem. Photobiol. 51*, 395 (1990).

107. V. H. Fingar, T. S. Mang, and B. W. Henderson, *Cancer Res. 48*, 3350 (1988).

108. M. C. Berenbaum, S. L. Akande, F. H. Armstrong, P. K. Bentley. R. Bonnett, R. D. White, and K. C. Lowe, in: *Oxygen Transport to Tissue* (J. Piiper, T. K. Goldstick, and M. Meyer, eds.), Vol. XII, pp. 277–282, Plenum Press, New York (1990).

109. K. C. Lowe, S. L. Akande, R. Bonnett, R. D. White, and M. C. Berenbaum, *Biomat. Art. Cells Immob. Biotechnol. 20*, 925 (1992).

110. P. M. Joseph, Y. Yuasa, H. L. Kundel, B. Mukherji, and H. A. Sloviter, *Invest. Radiol. 20*, 504 (1985).

111. R. F. Mattrey, D. J. Schumacher, H. T. Tran, Q. Guo, and R. B. Buxton, *Biomat. Art. Cells Immob. Biotechnol. 20*, 917 (1992).

112. J. E. Fishman, P. M. Joseph, M. J. Carvlin, M. Saadi-Elmandjra, B. Mukherji, and H. A. Sloviter, *Invest. Radiol. 24*, 65 (1989).

113. R. F. Mattrey, F. W. Scheible, B. R. Gosink, D. M. Long, and C. B. Higgins, *Radiology 145*, 759 (1982).

114. N. J. Patronas, J. Hekmatpanah, and K. Doi, *J. Neurosurg. 58*, 650 (1983).

115. V. K. Bhutani and T. H. Shaffer, *Biol. Neonate 44*, 257 (1983).

116. B. Widjaja, J. Wuthe, A. Schmidt, B. Seitz, and R. Rufer, *Res. Exp. Med. 188*, 425 (1988).

117. J. S. Greenspan, M. R. Wolfson, S. D. Rubenstein, and T. S. Shaffer, *J. Pediatr. 117*, 106 (1990).

118. L. C. Clark and F. Gollan, *Science 152*, 1755 (1966).

119. J. A. Kylastra, M. O. Tissing, and A. van der Maen, *Trans. Am. Soc. Artif. Int. Org. 8*, 378 (1962).

120. P. R. Lynch, J. S. Wilson, T. H. Shaffer, and N. Cohen, *Undersea Biomed. Res. 10*, 1 (1983).

121. T. H. Shaffer, D. L. Forman, and M. R. Wolfson, *Undersea Biomed. Res. 11*, 287 (1984).

122. J. Lutz and G. Herrmann, *Pflugers Arch. 401*, 174 (1984).

123. B. D. Spiess, R. J. McCarthy, K. J. Tuman, A. W. Woronowicz, K. A. Tool, and A. D. Ivankovich, *Undersea Biomed. Res. 15*, 31 (1988).

124. N. S. Faithfull, J. Klein, H. T. Van Der Zee, and P. J. Salt, *Br. J. Anaesth. 56*, 867 (1984).

125. N. S. Faithfull, P. J. Salt, J. Klein, H. Van Der Zee, H. Soini, and W. Erdmann, *Adv. Exp. Med. Biol. 191*, 463 (1985).

126. J. Klein, N. S. Faithfull, P. J. Salt, and A. Trouwborst, *Anesth. Analg. 65*, 734 (1986).

127. C. A. Reindorf, J. Kurantsin-Mills, J. B. Allotey, and O. Castro, *Am. J. Hemat. 19*, 229 (1985).

128. V. V. Tuliani, E. A. O'Rear, B. M. Fung, and B. D. Sierra, *J. Biomat. Res. 22*, 45 (1988).

129. C. M. Smith, R. P. Hebbel, D. P. Tukey, C. C. Clawson, J. G. White, and G. M. Vercellotti, *Blood* 69, 1631 (1987).
130. K. P. W. J. McAdam (ed.), *New Strategies in Parasitology*, Churchill-Livingstone, Edinburgh (1989).
131. S. Chang, H. Lincoff, N. J. Zimmerman, and W. Fuchs, *Arch. Ophthalmol. 107*, 761 (1989).
132. S. Chang, V. Reppucci, N. J. Zimmerman, M. H. Heinemann, and D. J. Coleman, *Ophthalmology 96*, 785 (1989).
133. L. D. Segel, J. L. Ensunsa, and W. A. Boyle, *Am. J. Physiol. 252*, H349 (1987).
134. K. Ueda, T. Genda, I. Hirata, M. Shimada, T. Shibata, T. Ueda and R. Omoto, *J. Heart Lung Transplant. 11*, 646 (1992).
135. J. L. Skibba, J. Sonsalla, R. J. Petroff, and P. Denor, *Eur. Surg. Res. 17*, 301 (1985).
136. T. E. Felker, D. Gantz, A. M. Tercyak, C. Oliva, S. B. Clark, and D. M. Small, *Hepatology 14*, 340 (1991).
137. S. Fuchinoue, K. Takahashi, S. Teraoka, H. Toma, T. Ashishi, and K. Ota, *Transplant. Proc. 18*, 566 (1986).
138. T. S. Kurki, A. L. Harjula, L. J. Heikkila, A. L. Lehtola, P. Hammainen, E. Taskinen, and S. P. Mattila, *J. Heart Transplant. 9*, 424 (1990).
139. V. P. Malley, D. M. Keyes, and R. G. Postier, *J. Surg. Res. 40*, 210 (1986).
140. H. Ohyanagi, O. Ohashi, S. Nakayama, M. Yamamoto, S. Okumura, and Y. Saitoh, in: *Blood Substitutes* (T. M. S. Chang and R. P. Geyer. eds.), pp. 585–594, Marcel Dekker, New York (1989).
141. T. Urushihara, K. Suminoto, M. Ikeda, K. Yamanaka, H. Q. Hong, H. Ito, Y. Fukuda, and K. Dohi, *Biomat. Art. Cells Immob. Biotech. 75*, 227 (1992).
142. C. Chubb and P. Draper, *Am. J. Physiol. 248*, E432 (1985).
143. C. Chubb and P. Draper, *Proc. Soc. Exp. Biol. Med. 184*, 489 (1987).
144. A. R. Smith, W. van Alphen, N. S. Faithfull, and M. Fennema, *J. Plastic Reconst. Surg. 75*, 227 (1985).
145. A. R. Scialli and G. C. Goeringer, *In Vitro Cell Dev. Biol. 26*, 507 (1990).
146. C. Ceschlin, M. C. Malet-Martino, G. Michel, and A. Lattres, *C.R. Acad. Sci. Paris 300 Ser. III*, 669 (1985).

27

The Fluorochemical Industry

A. The Fluorochemical Industry in the United States

RICHARD A. DU BOISSON

27A.1. THE MANUFACTURE OF HYDROGEN FLUORIDE IN THE UNITED STATES

27A.1.1. Industrial Preparation of Hydrogen Fluoride

Hydrogen fluoride (HF) is the largest volume synthetic fluorine compound manufactured in America, and virtually all of the myriad fluorine products that surround us have their origins in HF. In 1991, U.S. domestic sales of fluorochemicals approached $3 billion. Annual production of HF throughout the world is close to one million tons, and it is manufactured almost exclusively by interaction of refined, or acid grade, fluorspar (CaF_2) with sulfuric acid. Attempts to manufacture HF from "phosphate rock" [fluorapatite, $3Ca_3(PO_4)_2 \cdot CaF_2$] have largely been shelved as uneconomic. Fortunately, in view of the increasing demand for fluorine-containing products, known reserves of fluorspar are increasing.

State-of-the-art production of HF involves feeding a slurry of refined fluorspar and sulfuric acid into a revolving kiln at about 200°C ($CaF_2 + H_2SO_4 \rightarrow CaSO_4 + 2HF$). The gaseous products are cooled, scrubbed, and condensed to afford a 98–99% grade acid which is further purified by distillation to 99.9% assay. For every 1 lb of HF manufactured, 4 lb of calcium sulfate are formed, and environmental and economic concerns have prompted development of new outlets for this co-product. For example, Allied-Signal uses the $CaSO_4$ as a feedstock for synthetic gypsum for flooring screeds,[1] and Du Pont has developed a use as a soil stabilizer. In the past few years Du Pont has

---

RICHARD A. DU BOISSON • PCR Inc., Gainesville, Florida 32602.
Organofluorine Chemistry: Principles and Commercial Applications, edited by R. E. Banks *et al.* Plenum Press, New York, 1994.

sold more calcium sulfate than it produced, although estimates indicate it will take another 20 years to eliminate the stockpile![2]

27A.1.2. U.S. Manufacturers of Hydrogen Fluoride

There are four major manufacturers of HF in the U.S.,[3] namely Allied-Signal, Atochem North America (formerly Pennwalt), Du Pont, and Aluminum Company of America (Alcoa), although in the case of the last company the HF produced is not isolated. Dow Chemical ceased manufacture of HF at their Essex Chemical facility in 1987, and Henley Manufacturing, California, produces small volumes of ultrapure HF for the electronics industry. Often included in the reported statistics for production volumes are Allied-Signal's Amhurstburg plant in Ontario, and four HF production facilities in Mexico. In the case of the Mexican facilities, namely Fluorex SA, Industrias Quimicas de Mexico SA (IQM), Quimbasicos SA de CV, and Quimica Fluor SA de CV, the bulk of the production is destined for markets in the U.S. Indeed, Du Pont owns a 30% share in the Quimica Fluor SA de CV plant in Matamoros, Mexico. Table 1 details the North American manufacturers of HF, plant locations, and capacities.

27A.1.3. Hydrogen Fluoride Production Statistics

Although anhydrous HF was first made in the 1850s by the French chemist Frémy, it was only when the U.S. company Sterling Products commenced production in 1931 that significant commercial quantities became available. Up until that time HF (aqueous) had been mainly used for etching glass (Section 27A.2.5.), and it was demand for chlorofluorocarbon refrigerants that prompted large-scale industrial production. The development of new uses for HF during the war years, particularly as an alkylation catalyst for aviation fuel production and as an intermediate in the manufacture of uranium fluorides, dramatically increased demand, as can be seen in Figure 1. The

Table 1.   North American HF Producers and Capacities

Company	Location	Annual capacity (10^3 tons)
Allied-Signal Inc.[4]	Amherstburg, Ontario, Canada	52
Allied-Signal Inc.[3]	Geismar, LA	105
Aluminium Company of America[3]	Port Comfort, TX	50
Atochem[3] (formerly Pennwalt)	Calvert City, KY	26
Dow Chemical[3]	Paulsboro, NJ	11
E I Du Pont de Nemours & Co Inc.[3]	La Porte, TX	75
Fluorex SA[5]	Mexico	22
Henley Manufacturing Inc.[3]	Bay Point, CA	NA
Industrias Quimicas de Mexico SA[5]	San Luis Potosi, Mexico	14
Quimica Fluor[5]	Matamoros, Mexico	68
Quimbasicos[5]	Mexico	5

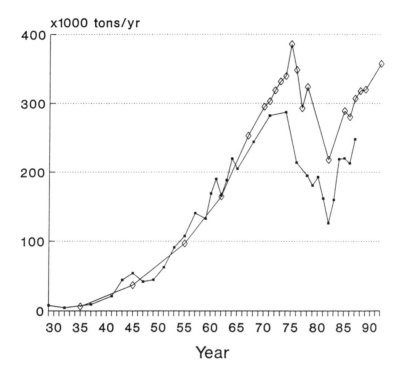

Figure 1. Hydrogen fluoride in the United States.—■—, U.S. production; —◇—, U.S. consumption.

subsequent rapid growth of these industries and new requirements for fluorine-containing polymers, fine chemicals, aluminum production, and steel pickling stimulated HF production capacity to peak levels in the mid 1970s. Demand for HF decreased in the late 1970s when the commissioning of nuclear power stations virtually ceased, the stockpiling of atomic weapons was slowed, aluminum production was optimized, HF recovery in stainless-steel pickling operations was implemented, and fear of destruction of the ozone layer slowed growth in CFC manufacture.

Figure 1 shows the changes in U.S. production and consumption of HF from 1930 to 1990, the data being gathered from a number of sources. The gap between U.S. production and consumption which opened in the early 70s is attributable to importation of HF, principally from Mexico. Figure 2 shows the current sources of HF imported by the U.S.[4,5]; European and Japanese material accounts for less than 1% of the HF consumed (U.S. exports are similarly insignificant).

In the early eighties, demand for HF started to increase with the upswing in the world economy, and in 1989 Allied increased the capacity of its Geismar plant by 10% to 142,000 tons per year.[6] In 1989, HF demand was at, or near, capacity[7]; however, the outlook for the 1990s is mixed. The most likely replacements for ozone-depleting chlorofluorocarbons (CFCs) require up to four times as much HF per mole to produce,

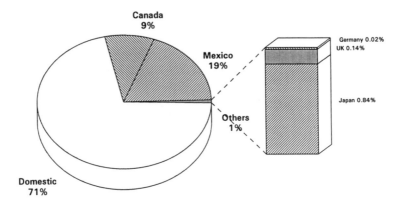

Figure 2. Sources of hydrogen fluoride in the United States.

but this is countered by the emergence of alternative technologies for certain applications that markedly reduces, or in some cases completely obviates, the need for any fluorine-containing materials. For example, AT&T announced in 1990 that its communications and computer products facility at Little Rock, Ark., has completely eliminated the use of CFCs, reducing its emissions from 330,000 lb in 1988 to zero.[8] More efficient use and recovery of CFCs for recycling caused a slackening of HF demand in 1990,[9] and until CFC replacements have been defined, it is difficult to forecast trends for demand through the 1990s.

27A.2. INDUSTRIAL USES OF HYDROGEN FLUORIDE

27A.2.1. Introduction

Hydrogen fluoride is the primary raw material for virtually all other fluorine-containing products and is used directly in the manufacture of many fluorocarbons (FCs), hydrofluorocarbons (HFCs), chlorofluorocarbons (CFCs), and hydrochlorofluorocarbons (HCFCs). Such materials are not only prime building blocks in organofluorine chemistry, but are themselves valuable end-products with applications ranging from refrigerants to solvents, and foam-blowing agents to dielectric fluids. Another major use for HF is in the manufacture of aluminum; indeed Alcoa manufactures, but does not isolate, some 45,000 tons of HF every year for this sole purpose. Chemical intermediates, including herbicides, fluoroborates, surfactants, anesthetics, and chemicals for the electronics industry, account for a large proportion of HF consumption. Other uses include steel pickling, glass etching, and specialty metal processing, all of which make use of aqueous HF, that is, hydrofluoric acid. Anhydrous HF is also used in petroleum alkylation and uranium refining processes.

The U.S. market for HF is reported[10,11] at 320,000 tons annually of which only 2% is apportioned to aluminum production.[11] This figure does not include the HF

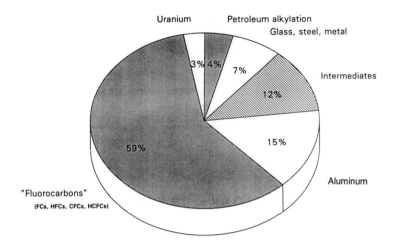

Figure 3. Market distribution of hydrogen fluoride in the United States.

manufactured by Alcoa specifically for the production of synthetic cryolite (Na_3AlF_6), but its inclusion gives a more meaningful market breakdown (see Figure 3).

27A.2.2. Chlorofluorocarbons and Related Halogenoalkanes (See Chapters 6 and 7)

The largest volume of organofluorine compounds are simple (C_1 and C_2) chlorofluoroalkanes, which are generally referred to by a numerical code (see Chapter 1) devised by the American Society of Refrigerating Engineers[12] and unofficially extended to include other fluoroaliphatics.[13] The numerical code may be preceded by an alphabetical designator, e.g., R- (Refrigerant), FC- (Fluorocarbon, a nonspecific term for all alkanes containing C—F bonds), CFC- (Chlorofluorocarbon), HCFC- (Hydrochlorofluorocarbon), HFC- (Hydrofluorocarbon), or by a trade name. For example, Du Pont uses *Freon*® and *Suva*®*; Allied-Signal uses *Genetron*®; Atochem (formerly Pennwalt) uses *Isotron*® and *Racon*® but may replace these with *Forane*®, which is currently used in Europe; ICI uses *Arcton*®; and Rhône-Poulenc (formerly ISC) uses *Isceon*®.

From 1986 to 1988 the consumption of these chlorofluorocarbons and related halogenoalkanes in the U.S. increased by 40 million lb to 1111 million lb. However, by 1993–94 consumption is expected to decline by about 27% in the U.S. This reduction is in response to the Montreal Protocol, an international treaty that mandates complete phaseout of CFC 11, 12, 113, 114, and 115 by the year 2000. In 1988 these

---

*The *Suva* products will not have a numerical designation; instead Du Pont will use initials to describe each product use area. *Suva* products will include *Suva Trans A/C*; *Suva Centri-LP* (FC-123); *Suva Cold-MP* (FC-134a); *Suva Chill-LP* (FC-124); *Suva Freez-HP* (FC-125).

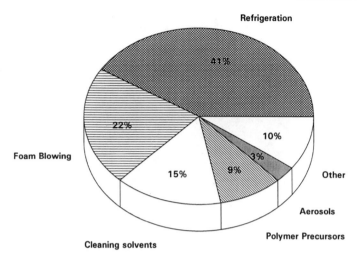

Figure 4. Use of CFCs by demand.

five restricted chlorofluorocarbons accounted for 72% of the total U.S. demand for all fluorocarbons and Figure 4 shows the chlorofluorocarbon demand by end-use.[14]

27A.2.2.1. Refrigeration and Air Conditioning

Cooling and foam-blowing dominate U.S. CFC uses, with refrigeration and air conditioning consuming the largest proportion (41%). CFCs 11, 12, and 114 are extensively used as refrigerants, and even CFC-113 and CFC-115 have some utility, although their use is somewhat limited. The inert nature of CFCs allows materials released in the lower atmosphere to diffuse into the stratosphere, where intense UV radiation liberates chlorine atoms that destroy ozone. Finding environmentally acceptable alternatives within the given time-frame is an enormous undertaking (see Chapter 28). With $135 billion worth of equipment in the U.S. that use CFCs, finding replacement refrigerants that can be substituted in existing equipment is essential. Replacement refrigerants must be nontoxic and nonozone depleting, contribute insignificantly to global warming, and possess suitable thermodynamic properties. There are two approaches to meet these requirements, namely to use materials that contain no chlorine and/or design products that are degraded in the troposphere before they can reach the stratosphere. Inclusion of hydrogen in chlorofluorocarbons gives products that will break down in the lower atmosphere, and complete replacement of chlorine with hydrogen provides a number of potentially useful products that have zero calculated ozone-depletion potential. Table 2 lists the major contenders, and, with the exception of HFC-125, viable commercial processes are in existence, or in the final stages of development, for all of them.

Hydrogen fluoride is used in the manufacture of all these materials, but their synthesis is far from straightforward since materials with the greatest potential require

Table 2. Chlorofluorocarbon Alternatives

FC-Number	Formula	BP (°C)	Toxicology	Ozone depleting potential[a]	Greenhouse potential[b]	Potential application
FC-22	CHF_2Cl	−40.8	Low	0.05	0.07	Refrigerant, blowing agent
FC-32	CH_2F_2	−51.7	Incomplete	0.0		Refrigerant
FC-123	CF_3CHCl_2	28.7	Low	<0.05	<0.1	Refrigerant, foam, solvent
FC-124	CF_3CHFCl	−12	Low	<0.05	<0.1	Refrigerant, blowing agent
FC-125	CF_3CHF_2	−48.5	Incomplete	0.0	<0.2	Refrigerant
FC-134a	CF_3CH_2F	−26.5	Low	0.0	<0.1	Refrigerant
FC-141b	CH_3CFCl_2	32	Low	<0.05	<0.1	Blowing agent, solvent
FC-142b	CH_3CF_2Cl	−9.2	Low	<0.05	<0.1	Propellant
FC-143a	CH_3CF_3	−47.6	Incomplete	0.0	<0.3	Refrigerant
FC-152a	CH_3CHF_2	−24.7	Low	0.0	<0.1	Refrigerant, propellant

[a]Estimated value with FC-11 set at 1.0.
[b]Estimated value with FC-12 set at 1.0.

three or four process steps. The most promising replacements contain hydrogen, and as such have the desirable property of being susceptible to oxidative degradation in the troposphere. However, this inherent instability leads to manufacturing problems when their breakdown under the necessarily harsh processing conditions causes deactivation of catalyst beds and generation of unwanted byproducts.

HFC-134a is probably the most likely candidate for *new* domestic and automobile air conditioners, since its thermodynamic properties are very close to those of CFC-12. However, it is not a "drop in" replacement and its use will require modified compressor design, different lubricants and gasket materials. Manufacture of HFC-134a has proven difficult, although it is now being produced by Du Pont (Corpus Christi, Texas) and ICI (Runcorn, U.K.) by unspecified routes; ICI entered the U.S. market in 1993 with 134a production at St. Gabriel, La. There are a number of viable chlorinated starting materials available for the synthesis of 134a, but it is far from obvious which is the easiest to use or most cost effective. However, no matter which route is chosen, every mole of 134a will have consumed at least 4 moles of HF, twice as much as required for CFC-12, and will cost three to five times as much when it is in full commercial production.

As with HFC-134a, none of the alternative hydrofluorocarbons is a "drop in" replacement because their thermodynamic properties do not match those of any of the refrigerants currently in use. Du Pont has recently patented a mixture of HCFC-124, HFC-22, and HFC-152a as a "drop in" replacement suitable for refrigerators, food freezers, automobile airconditioners, water coolers, and food display cases. This mixture will be sold as *Suva Blend-MP*, and is already available. HFC-22 and

HFC-152a are commercially available, and HCFC-124 has only recently (1992) become commercially available from new plants constructed by Du Pont and Allied-Signal.

27A.2.2.2. Foam-Blowing Agents

CFC-11 is extensively used to expand polyurethane and polyisocyanate foams, which are widely used as insulating material for buildings; the trapped blowing agent contributes to their insulating capacity. CFCs 12, 113, and 114 are also used as blowing agents for both rigid and nonrigid foams. There are several alternative materials that may be used as blowing agents; see Table 2. The manufacturers of rigid insulating foam already face stiff competition from other types of insulation, such as glass fiber, and the increased cost of replacement blowing agents will cause great difficulty in this already fragile market. The most promising replacements for CFC-11, of which about 200 million lb are used annually as a blowing agent, are HCFCs 123 and 141b. Preliminary toxicity tests of these two replacements are promising, but the results of long-term toxicity tests are not yet completed. However, the U.S. Environmental Protection Agency (EPA) has "approved" these two replacements, which are available commercially.

Manufacture of expanded polystyrene food containers for the *fast food* industry has largely switched from CFCs to HCFCs and/or hydrocarbon products as blowing agents. Notwithstanding, public opinion is firmly against polystyrene packaging, which is perceived as a nonbiodegradable, nonrecyclable material and is fast being replaced by paper-based products.

27A.2.2.3. Solvents and Cleaning Fluids

1,1,2-Trichlorotrifluoroethane (CFC-113) is widely used as a solvent for degreasing and removing soldering flux residues from electronic circuit boards. Du Pont and Allied-Signal manufacture CFC-113 under the trade names Freon® 113 and Genetron® 113, respectively, but its excellent solvent properties, inertness, and nontoxicity are marred by its high ozone-depletion potential, hence it will be phased out under the terms of the Montreal Protocol. Both Allied-Signal and Du Pont are developing HCFC-123 as one alternative to CFC-113, and commercial quantities are available from their Geismar and Maitland (Canada) plants, respectively. Du Pont has also introduced a nonfluorinated replacement for both CFC-113 and 1,1,1-trichloroethane, called Axarel® 52. It is a hydrocarbon compound suitable for semiaqueous cleaning systems and is effective for removing heavy greases, oils, waxes, and other impurities soluble in organic solvents from metals, and most plastics. Allied-Signal has developed a solvent system based on HCFC-141b which is available commercially. The plant to manufacture HCFC-141b is located at their Geismar facility. There is also interest in fluorinated propanes as potential replacements in solvent applications, but toxicity testing and development of viable commercial processes are incomplete.

Other companies outside the fluorochemicals area are offering promising replacements. For example, Union Camp is developing an environmentally sound, nontoxic substitute for CFC solvents based on terpenes. Jones Chemical Inc. of LeRoy, NY, has introduced an electronic circuit board defluxer called *Sunny Sol ESC-10* which is made from byproducts of pulp and paper manufacturing. It is estimated that only about 40% of chlorofluorocarbon demand in the year 2000 will be met by HCFC or HFC replacements; the remainder will be met by conservation and nonfluorocarbon alternatives. The CFC-based solvent industry will be adversely effected more than other segments of the industry, since already there are moves to plan a phase-out of HCFCs, which are regarded by some as stopgap alternatives merely awaiting the commercialization of suitable HFCs.

27A.2.2.4. Halons

Bromine-containing fluoroalkanes are generally known as Halons, usually being designated by a four-digit number where the first digit corresponds to the number of carbon atoms, the second to the number of fluorine atoms, the third to the number of chlorine atoms, and the fourth to the number of bromine atoms. For example, 1,2-dibromotetrafluoroethane is often referred to as Halon 2402. The bromine-containing fluoroalkanes are chemically quite reactive and are useful intermediates in organic synthesis (see Chapter 8).

Halons 1211 and 1301 have been widely used as fire extinguishing agents in the U.S., since their dense vapors quickly smother fires without causing additional damage such as when water is used. Halon 2402 has seen use as a fire extinguisher outside the U.S. However, weight-for-weight these bromine-containing materials have the potential to destroy 3–10 times more atmospheric ozone than CFC-11 (see Chapter 8); hence, under the terms of the Montreal Protocol, they will be phased out before the year 2000. Proposed alternative fire-extinguishing media include bromodifluoromethane (Great Lakes, Firemaster® 100), 2,2-dichloro-1,1,1-trifluoroethane to replace Halon 1211 (HCFC-123, ICI and Du Pont), and pentafluoroethane to replace Halon 1301 (HFC-125, Du Pont). The current world market for Halons is about 60 million lb annually, but this is likely to change dramatically as their use as fire-extinguishing agents is phased out. Production of Halon 2402 is already restricted to the synthesis of intermediates which are consumed at the site of manufacture.

27A.2.2.5. Intermediates

Many of the CFCs that find use as refrigerants, blowing agents, and solvents are important intermediates in the production of fluorinated monomers, anesthetics, fire-extinguishing media, and chemical intermediates. The market for fluorinated plastics and elastomers is expanding, and will help sustain the demand for HF during the 1990s.

Tetrafluoroethylene (TFE) is the most important of the fluoroalkenes, and is manufactured in the U.S. by Du Pont (for in-house consumption) and by ICI. TFE is

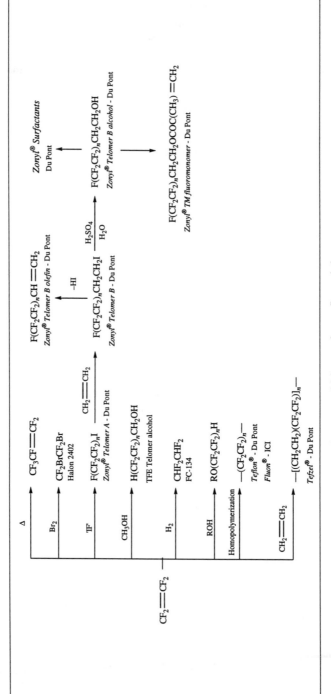

Scheme 1. Major uses for tetrafluoroethylene.

synthesized by pyrolysis of HCFC-22: $2\ CHF_2Cl \rightarrow CF_2{=}CF_2 + 2HCl$. This seemingly simple equation belies the hazards associated with the manufacture, handling, and use of this alkene. For TFE is readily polymerized with the evolution of much heat, and if the reaction is not properly controlled, or takes place inadvertently, the monomer disproportionates into carbon and CF_4 with the explosive force, weight-for-weight, of black powder! However, the large number of unique products synthesized from TFE justify the effort required for its safe manufacture. Scheme 1 shows the major uses of TFE.

Homopolymerization of TFE affords poly(tetrafluoroethylene), PTFE (see Chapter 15), which is easily the largest-volume commercial fluoropolymer. It is manufactured in the U.S. by Du Pont at their Parkersburg, WV, plant which has an annual capacity of 20 million lb, and is sold under the famous Teflon® trade name. It is also manufactured by ICI at their Bayonne, NJ plant, which has a 6 million lb per annum capacity; that material is sold under the trade name Fluon®. Available to fabricators as a powder or aqueous dispersion, PTFE finds use in many different industries: in the electrical/electronic industry for wire and cable insulation; in the chemical/petroleum industry for chemical processing equipment; in the clothing industry for textile laminates; in the printing and rubber/plastics industry for lubricants; in the consumer goods industry for cookware coatings; and in numerous other applications. Growth of PTFE consumption is projected at 4% per annum from 1991–1993; by contrast, demand for melt-processible fluoropolymers is expected to grow at 7% per annum. Copolymerization of TFE with HFP gives a fluorocarbon resin with superior melt processing characteristics over PTFE. Du Pont began commercial production of this material in 1960 under the name Teflon® FEP (Fluorinated Ethylene Propylene). A 1:1 copolymer of TFE with ethylene modified with a proprietary monomer is sold as Tefzel® by Du Pont.

Hexafluoropropene (HFP), manufactured by Du Pont by low-pressure pyrolysis of TFE, is another important monomer and intermediate. Scheme 2 shows some of the important uses of HFP. It is difficult to form a high-molecular-weight homopolymer from HFP. However, copolymerization of HFP and 1,1-difluoroethylene affords a useful elastomer sold under the Viton A® and Viton E® name by Du Pont, and as Fluorel FC® by 3M. Inclusion of TFE as a third monomer gives another elastomer sold as Viton B® by Du Pont and as Fluorel FT® by 3M (see Chapter 16).

Hexafluoropropene oxide (HFPO) is manufactured by Du Pont by oxidation of HFP; telomerization of HFPO followed by fluorodecarbonylation affords perfluoropolyether (PFPE) fluids sold as synthetic vacuum pump oils under the trade names Krytox® (Du Pont) and Aflunox™ (PCR). Co-oxidation of HFP and TFE affords a different type of PFPE fluid, also useful as a vacuum pump oil, sold as Fomblin® by Montedison (see Chapter 20). Chlorotrifluoroethylene (CTFE) manufactured by Allied-Signal by dechlorination of CFC-113 is another versatile intermediate. Low-molecular-weight telomers are useful oils for nonflammable hydraulic fluids, and higher-molecular-weight polymers, with and without the inclusion of 1,1-difluoroethylene units, are clear, colorless materials sold as Kel-F® by 3M. Poly(1,1-

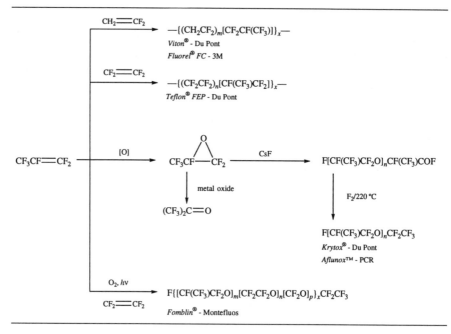

Scheme 2. Some important uses of hexafluoropropene.

difluoroethylene) is sold by Atochem as Kynar®, and poly(vinyl fluoride) as Tedlar® by Du Pont.

27A.2.3. Aluminum Production

Hydrogen fluoride is important to the aluminum industry for the synthesis of cryolite (Na_3AlF_6) and aluminum fluoride from bauxite (hydrated Al_2O_3), which are used in the electro-winning of aluminum by the classic Hall–Héroult process. The annual production of aluminum in the United States in 1991 stood at nearly 4×10^6 ton.

27A.2.4. Petroleum Alkylation

HF is used as an alkylation catalyst in the petroleum industry to make *alkylate*, a blending component for high-octane gasoline. The reaction is carried out at 20–35°C using either sulfuric acid or anhydrous hydrogen fluoride, and the product recovered from the acid in a settler. The strong catalytic activity of anhydrous hydrogen fluoride is attributed to its ability to protonate alkenes. The resultant carbenium ions produced in olefin-rich gas streams (from cracking processes) then alkylate other hydrocarbons present in the mixture. The resultant product can have octane numbers as high as 95. About 16,000 tons of HF are used annually in the U.S. petroleum industry.

27A.2.5. Aqueous Hydrofluoric Acid

Hydrogen fluoride dissolves readily in water with evolution of much heat to give hydrofluoric acid, which is available in polyethylene containers in 48, 60, and 70% concentrations. The use of hydrofluoric acid for etching and polishing glass goes back to the eighteenth century and the effect can be varied, depending on what the acid is mixed with. For example, hydrofluoric acid with sulfuric acid produces on lead glass a high gloss, but, neutralized with ammonia, will leave a frosted effect. The use of hydrofluoric acid in the glass industry has not diminished, rather, it has been overshadowed by the ever increasing requirements of other industries.

Increasing demand for stainless steel has kept pace with changes in HF production. Approximately 3–4% of annual HF output is used in pickling acids; however, steel manufacturers are now recycling spent acids, thus reducing their demands for HF per unit of stainless steel produced. This can save considerable quantities of HF; for example, Washington Steel, which uses Allied Corp's *Aquatech* Process, in which bipolar electrodialysis membranes are used to provide recovery of spent hydrofluoric acid and nitric acid, saves $0.4 million every year.

Other uses of aqueous hydrofluoric acid include cleaning metal castings (removal of adhering sand particles from the mold); as a reagent in inorganic syntheses, including the preparation of metal fluorides, fluoroborates, and fluorosilicon products; as an etchant in semiconductor fabrication; to prepare laundry sours and stain removers; and in mineral analysis.

27A.2.6. Fluorine

Fluorine is manufactured by electrolysis of the molten salt $KF \cdot 2HF$ in cells equipped with carbon anodes. The major U.S. producers are Air Products, who both market it and use it to prepare a range of fluorine-containing fine chemicals and for surface fluorination of polymers (see Chapter 22). Air Products *Multifluor* products are manufactured by cobalt fluoride technology (see Chapters 2 and 4), the CoF_2 being oxidized back to CoF_3 continuously with elemental fluorine. Both Kerr-McGee and Allied manufacture and use fluorine for uranium processing. Elemental fluorine is not widely used in commercial organic synthesis as a fluorinating agent, since it is difficult to control the reactions (see Chapter 1). Perhaps the largest-volume fluoroorganic produced by direct fluorination is the anticancer agent 5-fluorouracil, manufactured by PCR Inc. at their Puerto Rico facility.

27A.2.7. Uranium

Enrichment of uranium-235 by gaseous diffusion requires a volatile compound of uranium, the most suitable being uranium hexafluoride. It is produced from uranium oxide in two stages: first UO_2 is fluorinated with HF at 400°C to afford uranium tetrafluoride, which is then oxidized to UF_6 by treatment with fluorine at 300°C. The enriched uranium hexafluoride is then hydrolyzed in aqueous ammonia to precipitate ammonium diuranate, $(NH_4)_2U_2O_7$, which is then reduced with hydrogen and calcined to UO_2. Uranium processing uses some 3–4% of the HF consumed in North America.

27A.2.8. Chemical Intermediates

Passing an electric current through a solution of an organic compound in anhydrous HF in cells equipped with nickel anodes results in perfluorination of the substrate (see Chapter 5). Such electrochemical fluorination (Simons ECF) is an energetic process and yields are generally moderate at best, with side reactions and fragmentation occurring, especially with larger molecules. 3M is the only U.S. company to practice this technology, offering a range of perfluorinated products under the *Fluorinert*® trade name. Such materials are used as dielectric fluids and heat-transfer agents, for vapor-phase soldering, and as inert fluids. Electrochemical fluorination of alkanesulfonic acids affords perfluoroalkanesulfonyl fluorides, which are valuable raw materials for the manufacture of *Scotchguard*® and fluorinated surfactants sold under the *Fluorad*® trade name by 3M.

Fluoroaromatics are important intermediates to the pharmaceutical, agrochemical, and engineering resins industries. Mallinckrodt, a St. Louis based subsidiary of International Minerals and Chemical, has been involved in the production of fluoroaromatic intermediates since 1980. This involvement culminated in the world's first continuous production facility for fluoroaromatic intermediates and the first commercial U.S. source for such products. Mallinckrodt produces the most basic fluoroaromatic intermediate, fluorobenzene, at the rate of 1200 tonnes per year, the diazotization step of this fluoro-de-diazoniation chemistry being carried out in HF (see Chapter 9). Du Pont started operation of a fluorobenzene facility at their Chambers works late in 1990, with a capacity of 1360 tonnes per year. The world market for fluoroaromatics is estimated at 3000–4000 tonnes per year, but Mallinckrodt projects that demand will rise to 10,000 tonnes per annum by 1994.

The world's largest producer of benzotrifluoride is Occidental in the U.S., with production heavily centered on 4-chlorobenzotrifluoride for Eli Lilly. Other important benzotrifluoride derivatives are 3,4-dichlorobenzotrifluoride and 3-aminobenzotrifluoride, both of which are used in the manufacture of agrochemicals (see Chapters 10 and 11).

27A.3. REFERENCES

1. *Can. Chem. Process.* 14 (1980).
2. "Calcium Sulphate runs out again for Du Pont", *Chem. Mark. Rep. 235(8)*, 13 (1989).
3. *Directory of Chemical Producers*, SRI International, 697 (1990); 691 (1989).
4. "Trade and Production Statistics", *Chem-Intel*, Dialog on-line database, file 318 (1990).
5. "Mexico's Industrial Minerals: Hydrofluoric Acid", *Ind. Miner. (London) 250*, 24 (1988).
6. "Allied-Signal Expands in Hydrofluoric Acid", *Chem. Eng. (Int. Ed.) 95(15)*, (1988).
7. "Commercial Fluorine Compounds", Market Research Report, Business Communications Co. Inc. (Publisher), Norwalk, CT, USA (1989).
8. "AT&T Eliminates CFCs", *Chem. Mark. Rep. 238*, 9 (1990).

9.   "The US Market for Performance Fluorine Chemicals", Report A2154, Frost & Sullivan (Publishers), NY (1990); "Montreal Protocol Big Factor in Fluorine Chemicals Future", *Chem. Mark. Rep. 238(11)*, 24 (1990).

10. "Hydrofluoric Acid Stays on Track; CFC Shift Drives Growth", *Chem. Mark. Rep. 256(5)*, 22 (1989).

11. "Hydrogen Fluoride Production and Markets", *Ind. Miner. (London) 254*, 77 (1988).

12. *Refrigerating Engineering 65*, 49 (1957); ANSI/ASHRAE 34-1989, "Number Designation and Safety Classification of Refrigerants" (1989).

13. "Fluorocarbon Numbering", *PCR Research Chemicals Catalog*, 162 (1992).

14. R.F. Bradley, A. Leder, and Y. Sakuma, *Chemical Economics Handbook*, 543.7000A (1990).

27

The Fluorochemical Industry

B. Organofluorine Products and Companies in Western Europe

PETER FIELD

27B.1. INTRODUCTION

The objective of this chapter is to provide a broad picture of the organofluorine chemicals industry in Western Europe as seen at the beginning of the 1990s. Commercialization worldwide of organofluorine chemistry has been driven by the demand for "speciality" and "effect" chemicals, and continues to be so. Although a surprisingly high proportion of the global effort devoted to the development of organofluorine chemistry has been expended in Western Europe, the U.S. and Japan feature more strongly in the commercial exploitation of the area. For convenience here, organofluorine products have been divided into three categories, namely, fluoroaliphatics, fluoroaromatics, and fluoropolymers.

27B.2. FLUOROALIPHATICS

27B.2.1. Fluorocarbons

27B.2.1.1. Fluoroalkanes

This class includes the partially fluorinated alkanes, i.e., hydrofluoroalkanes (HFAs) and perfluoroalkanes (PFAs). HFAs, which CFC producers in the rest of the world refer to as HFCs (hydrofluorocarbons) and HCFCs (hydrochlorofluorocarbons), are currently the favored products to replace CFCs now regulated under the terms of

---

PETER FIELD   •   PGF Associates Ltd., Buxton, Derbyshire SK17 6EW, England.

Organofluorine Chemistry: Principles and Commercial Applications, edited by R. E. Banks *et al.* Plenum Press, New York, 1994.

the Montreal Protocol (see Chapters 7 and 28). The most important products undergoing extensive toxicological testing in 1990 were HFAs 125 (CF_3CHF_2), 134a (CF_3CH_2F), 143a (CF_3CH_3) and 152a (CHF_2CH_3). HFA-134a is the replacement for CFC-12 (CF_2Cl_2) in refrigeration applications, and the first commercial plant was commissioned by ICI at Runcorn (U.K.) in October 1990. While CFCs can be produced in a single-stage reactor system, HFAs in general have to be produced in multistage reactor systems with attendant increases in costs. Plants are dedicated rather than multipurpose or swing, and purification of the products to the requisite high level is difficult.

Perfluoroalkanes are exceptionally stable, both thermally and chemically, are nonflammable, possess excellent dielectric properties, and have very low toxicities. Products such as perfluorooctane, perfluorodecalin, perfluorohexane, etc., find extensive application as testing fluids in the electronics industry (see Chapter 4).

Perfluoroalkanes are produced either by fluorination of hydrocarbons with CoF_3 [as carried out by Rhône-Poulenc UK (ISC Division) at Avonmouth until recently (see Chapter 4 and Section 27B.5.)] or by electrochemical fluorination [as used by 3M (Chapter 5) at Antwerp in Belgium and Miteni SRL at Trissino in Italy].

27B.2.1.2. Fluoroalkenes

The most significant fluoroalkenes produced are tetrafluoroethylene (TFE-monomer), hexafluoropropylene (HFP), and vinylidene difluoride (VDF or VF_2). Used mainly as monomers for the production of fluoropolymers and functional telomers, their manufacture (see Chapter 15) involves materials regulated by the Montreal Protocol, hence concern exists about the future of current routes. TFE monomer, for example, is produced by pyrolysis of HCFC-22 (CHF_2Cl). The main producers therefore are companies which produce HCFC-22, such as ICI, Du Pont, Hoechst, and Ausimont.

HFP is used for co-polymer production (see Section 27B.4), and as the raw material for the manufacture of perfluoropolyethers and hexafluoropropylene oxide (HFPO) (Chapters 20 and 19). Formed in around 88% yield by the low-pressure pyrolysis of TFE, its production lies in the hands of the same companies that manufacture TFE.

Vinylidene difluoride (VDF) is produced by the dehydrohalogenation of HCFC-142b (CH_3CF_2Cl) by the PVDF producers, Solvay et Cie and Atochem SA. Chlorotrifluoroethylene (CTFE), produced by dechlorination of CFC-113 ($CF_2ClCFCl_2$), and vinyl fluoride, from $CH{\equiv}CH + HF$, are also used in polymer production (see Section 27B.4.).

27B.2.2. Halogenofluorocarbons (See Chapters 6–8)

Production and consumption of the five major CFCs [11 ($CFCl_3$), 12 (CF_2Cl_2), 113 ($CF_2ClCFCl_2$), 114 (CF_2ClCF_2Cl), and 115 (CF_3CF_2Cl)] are regulated under the

Table 1. West European CFC Production (1989)[a]

CFC	11	12	113	114	115	Total
Volume (10^3 tonnes)	178	172	70	5	8	433

[a]Source: PFG Associates Ltd.

Table 2. West European Production of CFCs and HCFCs (1989)[a]

Company	Country	Volume (10^3 tonnes)
Akzo NV	Netherlands	20.0
Atochem SA	France	130.0
Du Pont Europe	Netherlands	36.0
Hoechst AG	Germany	76.0
ICI PLC[b]	United Kingdom	102.0
Ausimont	Italy	80.0
Rhône-Poulenc (ISC)	United Kingdom	25.0
SICNG	Greece	21.0
Solvay/Kali Chemie[b]	Germany	50.0
TOTAL		540.0

[a]Source: PFG Associates Ltd.
[b]See Section 27B.5.

terms of the Montreal Protocol, which became effective in its initial form on July 1, 1989 (see Chapter 28).

All CFC production in Western Europe occurs within the EEC, which in 1989 contributed 35% to total world production (see Table 1). Production in Western Europe of the most important hydrochlorofluorocarbons CHF_2Cl (HCFC-22) and CF_2ClCH_3 (HCFC-142b) via partial halogen-exchange (with HF) in chloroform and 1,1,1 trichloroethane, respectively, totalled 107,000 tonnes in 1989 (HCFC-22, 95000; HCFC-142b, 12000 tonnes). The combined production of CFCs and HCFCs in Western Europe in 1989 amounted to 540,000 tonnes, estimated tonnages by company being listed in Table 2. Some companies, such as Atochem, have several plants located in different countries.

Halons also are now regulated by the Montreal Protocol. Only Halons 1211 and 1301 are commercially significant in Europe, being produced by thermal bromination of HCFC-22 and HFA-23, respectively ($CHF_2Cl + Br_2 \rightarrow CF_2ClBr + HBr$; $CHF_3 + Br_2 \rightarrow CF_3Br + HBr$). Atochem SA (France), Solvay (Germany), and ICI (United Kingdom) manufacture Halon 1211, but only the first two companies produce Halon 1301. Halons possess exceptional fire-extinguishant ability and have found widespread application in both fixed and mobile systems.

27B.2.3. Trifluoroacetic Acid and Its Derivatives

Trifluoroacetic acid (TFA) and several of its derivatives, such as $CF_3CO_2C_2H_5$, CF_3CONH_2, and CF_3COCl, are used increasingly in small-volume applications.

Partially fluorinated (e.g., CHF_2CO_2H) and mixed fluorohalo acids (e.g., CF_2ClCO_2H) are currently being investigated as speciality items.

The largest producer worldwide of TFA is Rhône-Poulenc SA (France), other producers being Halocarbon Products (USA) and Kali-Chemie (Germany). The most important process for the production of TFA is based on perchloroethylene, and has been patented by both Rhône-Poulenc and Hoechst: $CCl_2{=}CCl_2$ (with O_2/UV) $\rightarrow$ CCl_3COCl $\rightarrow$ (with HF/chromium oxyfluoride) CF_3COF $\rightarrow$ (with CCl_4/catalyst) CF_3COCl $\rightarrow$ (with H_2O) TFA].[1–3] The use of carbon tetrachloride to convert CF_3COF to the corresponding acid chloride solves the difficult problem of separating the acid fluoride from unreacted HF and the co-product HCl in the first halogen-exchange step.[3] Industrially important synthetic applications of TFA and its derivatives continue to increase in number, notably in the area of trifluoromethylated heterocycles; two examples in the herbicide area are the production of 2-trifluoromethyl-5-methylamino-1,2,4-thiadiazole for treatment with methyl isocyanate or phosgene/methylamine to produce thiazafluron (Ciba-Geigy),[4] and condensation of ethyl trifluoroacetate with benzyl acetate/sodium hydride en route to flurazole (benzyl 2-chloro-4-tri-fluoromethylthiazol-5-yl carboxylate), a Monsanto herbicide (MONN-4606).[5] The antimalarial drug mefloquine can be synthesized from ethyl trifluoroacetate and 2-aminobenzotrifluoride.[6]

Trifluoroacetic anhydride cannot conveniently be produced from TFA on a large scale using phosphorus pentoxide, hence the following indirect route is used industri-ally: $CHCl_2CO_2H$ $\rightarrow$ $(CHCl_2CO)_2O$ $\rightarrow$ (with TFA) $(CF_3CO)_2O$ + $CHCl_2CO_2H$ (recycled).[7] 2,2,2-Trifluoroethanol (TFE), which currently is produced by catalytic reduction of TFA using a precious metal catalyst,[8] is the key raw material for the production of the inhalation anesthetic isoflurane (see Chapter 25). Interest exists in the development of processes for the production of both TFA and TFE from trichlo-roethylene via CF_3CH_2Cl (HCFC-133a), which is the key intermediate in the manu-facture of halothane, another important anesthetic.

The estimated total European market for TFA and its derivatives (including TFE) is about 500 tonnes per annum, with the price of TFA being around DM25 per kg.

27B.2.4. Trifluoromethanesulfonic Acid and Its Derivatives

Trifluoromethanesulfonic acid (triflic acid) is an extremely acidic hygroscopic liquid (bp 162°C). The most important producer is 3M, which uses electrofluorination (ECF) of methanesulfonyl chloride. However, other methods of production are being researched, with emphasis on a route involving CF_3Br (Halon 1301). A process patented[9] by Rhône-Poulenc involves the reactions: CF_3Br $\rightarrow$ CF_3SO_2Na $\rightarrow$ (with H_2O_2) CF_3SO_3Na $\rightarrow$ (with H_2SO_4) CF_3SO_3H (see Section 8.3.2.3.).

A potentially vast range of catalytic, synthetic and speciality applications exists for triflic acid, but in many cases the present cost (>FF 500/kg) precludes industrial use. However, owing to its relatively low boiling point, it can, in theory at least, easily be recovered; undoubtedly it will be widely used in the future.

27B.2.5. Higher (>C$_2$) Fluoroalkane Derivatives

27B.2.5.1. Telomer Alcohols

2,2,3,3-Tetrafluoropropan-l-ol and 2,2,3,4,4,4-hexafluorobutan-1-ol are produced in high yield (>85%) by benzoyl peroxide-initiated free-radical addition of
methanol to tetrafluoroethylene and hexafluoropropylene, respectively: $CF_2{=}CF_2$ +
$CH_3OH \rightarrow CHF_2CF_2CH_2OH$; $CF_3CF{=}CF_2$ + $CH_3OH \rightarrow CF_3CHFCF_2CH_2OH$.
Higher homologs of the propanol [$H(CF_2CF_2)_nCH_2OH$, n = 2–12] are also produced
from methanol and tetrafluoroethylene.

Acrylate and methacrylate esters of fluoroalcohols find application as speciality
monomers for fluoropolymers, e.g., in optical waveguides and oxygen-permeable contact
lenses. Currently small in volume terms, the production in Europe is dominated by Hoechst
AG, although a number of companies are actively investigating applications of the esters.

27B.2.5.2. Hexafluoroacetone

Hoechst produces hexafluoropropylene oxide via epoxidation of HFP and converts it to a range of derivatives [$(CF_3)_2CO$ and hence $(CF_3)_2CHOH$,
$R_FCF_2OCF{=}CF_2$ (R_F = F, CF_3, etc.), and perfluoropolyethers (Hostinert)].
Hexafluoroacetone is also used to produce fluorinated Bisphenol A derivatives,
including hexafluoroisopropylidene-2,2-bis(phthalic acid anhydride) (6FDA) used as
a monomer in the production of high-performance polymers (see Chapter 19).
Hexafluoroacetone can also be produced from hexachloroacetone by halogen-
exchange methodology, using HF and chromium-based catalysts (Section 19.2).

27B.2.5.3. Telomer Iodides

Functionalized fluoroalkanes containing six or more carbon atoms have generally
been produced by three main routes, which lead to products with different structural
characteristics: telomerization, electrochemical fluorination (ECF), and oligomerization. Telomerization technology is based on the reaction of a telogen such as pentafluoroethyl iodide, with a polymerizable ethylenic compound such as
tetrafluoroethylene: 5 $CF_2{=}CF_2$ + 2 I_2 + $IF_5 \rightarrow$ 5 C_2F_5I; n $CF_2{=}CF_2$ + $C_2F_5I \rightarrow$
$C_2F_5(CF_2CF_2)_nI$ (R_FI). The resultant telomer iodides, R_FI, are then combined with
ethylene via free-radical addition, providing a mixture of iodides: R_FI + $CH_2{=}CH_2 \rightarrow$
$R_FCH_2CH_2I$. The perfluoroalkyl group is predominantly straight chain with an even
number of carbon atoms, and the chain length varies widely (see Section 8.4.2).

Functionalization of the perfluoroalkyl iodides by nucleophilic substitution is not
possible, but, following free-radical ethylenation, conventional synthetic manipulation
of the ethyl iodide moiety in $R_FCH_2CH_2I$ can be achieved. In this way a wide range
of products with valuable surface-modification properties, e.g., carboxylic acids and
alcohols, acrylate and methacrylate monomers, amides, amines, quaternary ammonium salts, etc., is produced (see Chapter 14). The manufacturing technology devel-

oped independently by Hoechst AG in West Germany, Atochem in France, and Du Pont in the U.S., each with variations on the main theme, is difficult and expensive to operate. The estimated total Western Europe telomerization capacity lies somewhere between 1100 and 1300 tonnes per annum (see Chapter 14).

27B.2.5.4. ECF-Based Products

Electrochemical fluorination (ECF), pioneered by Simons (see Chapter 5), has been commercialized in Europe by 3M (Belgium) and Miteni (Italy). Reduction of the perfluoroalkanecarboxylic acid fluorides thus produced is carried out by catalytic hydrogenation (e.g., $C_7H_{15}COCl \rightarrow C_7F_{15}COF \rightarrow C_7F_{15}CO_2H \rightarrow C_7F_{15}CH_2OH$). Acrylate and methacrylate monomers based on electrochemical fluorination products available from 3M include FX-13 [$C_8F_{17}SO_2N(C_2H_5)CH_2CH_2OCOCH=CH_2$], FX-14 [$C_8F_{17}SO_2N(C_2H_5)CH_2CH_2OCO(CH_3)=CH_2$], L-9186 [$C_7F_{15}CH_2O-COCH=CH_2$] L-9187 [$C_7F_{15}CH_2OCOC(CH_3)=CH_2$] and L-11913 [cyclo-$C_6F_{11}CH_2OCOC(CH_3)=CH_2$].

27B.3. FLUOROAROMATICS

27B.3.1. General Comments

Fluoroaromatics are mainly used as intermediates in the production of pharmaceuticals, agrochemicals, advanced polymers, reactive dyes, and liquid crystals (see Chapters 9–13, 24). Their utilization in pharmaceutical and agrochemical circles has been driven by the ever-increasing demand for products containing more active, lower-dosage ingredients with fewer side effects.

Analysis of a recent compilation of fluorine-containing agrochemicals revealed the statistics shown in Table 3. The diversity of products of commercial significance in the pharmaceutical industry is much greater, but the volumes are much smaller; drugs marketed fall mainly into three therapeutic categories: antibiotics and antibacterials; antidepressants; and nonsteroidal anti-inflammatories (NSAIDS).

Table 3.   Structural Characteristics of
Fluorine-Containing Agrochemicals (1990)[a]

Main structural characterstics	No. of products
CF_3 group on benzene ring	63
One or more F atoms on benzene ring	34
CF_3 group and F atom(s) on benzene ring	5
Aliphatic fluorine atoms	42
CF_3 group attached to a pyridine or other heteroaromatic nucleus	9
Inorganic and others	11
Total	164

[a]Source: PFG Associates Ltd.

27B.3.2. Nuclear-Fluorinated Aromatics

Table 4 shows a list of the important pharmaceuticals and agrochemicals available in 1991 which contain a fluorinated aromatic ring, together with the primary fluoroaromatic raw materials involved in their manufacture.

27B.3.2.1. Manufacturing Processes for Nuclear-Fluorinated Aromatics

Basically, only two technologies are used industrially to introduce fluorine into the aromatic nucleus: Halex and diazotization. A great deal of interest exists in the development of other methodologies that are wider in scope and involve more selective, less hazardous, and more environmentally acceptable reactions.

Table 4. A Selection of the More Important Nuclear-Fluorinated Aromatics[a]

Fluoroaromatic raw material	End product	Company[b]
Fluorobenzene	Flusilazol	Du Pont
	Flutriafol	ICI
	Haloperidol	Janssen
	Trifluperidol	Janssen
2-Fluoroaniline	Flurbiprofen	Boots
4-Fluoroaniline	Flumequine	Riker
	Flumazenil	Roche
	Clinoril	M.S.D.
3-Chloro-4-fluoroaniline	Flamprop isopropyl	Shell
	Ciprofloxacin	Bayer
	Norfloxacin	Kyorin
	Perfluoxacin	Rhône-Poulenc
	Amifloxacin	Sterling Drug
2,4-Difluoroaniline	Diflunisal	M.S.D.
	Diflufenican	Rhône-Poulenc
	Temafloxacin	Abbott Labs
2,3,4-Trifluoroaniline	Fleroxacin	Roche
	Lomefoxacin	Searle
	Ofloxacin	Daiichi
1,3-Difluorobenzene	Fluconazole	Pfizer
3-Fluorotoluene	Ciprofloxacin	Bayer
4,4'-Difluorobenzophenone	PEEK	ICI
4-Hydroxy-4'-fluorobenzophenone	PEEK, PEK	ICI
2-Fluorobenzoyl chloride	Flutriafol	ICI
	Flurazepam	Roche
	Flunitazepam	Roche
2,3,5,6-Tetrafluorophthalonitrile	FORCE	ICI
2-Chloro-6-fluorobenzaldehyde	Flucloxacillin	S.K. Beecham
2,6-Difluorobenzonitrile *or*	Diflubenzuron	Duphar
2,6-Difluorobenzoyl chloride	Teflubenzuron	Shell
	Flufenoxuron	Shell
2,4,6-Trifluoro-3,5-dichloropyridine	Fluroxypyr	Dow

[a]Source: PFG Associates Ltd.
[b]See Section 27B.5.

27B.3.2.2. The Halogen-Exchange ("Halex") Process

The Halex reaction (see Chapter 9) is a nucleophilic aromatic substitution in which a chlorine or bromine atom, activated by an electron-withdrawing group in an *ortho* or *para* position, is replaced by fluorine using a metal fluoride (usually spray-dried KF) in a polar aprotic solvent. A commercially attractive manufacturing process which can be operated in multipurpose equipment, it does pose certain hazards, particularly a tendency to "runaway". The generally severe reaction conditions required have prompted an intensive search for ways to facilitate exchange under milder conditions.

27B.3.2.3. Diazotization-Based Processes

Two diazotization-based processes are used to effect nuclear fluorination (see Chapter 9). The first, known as the Balz–Schiemann Process, involves the conversion of an aromatic amine to the corresponding diazonium fluorborate, which is thermally decomposed subsequently to yield the fluoroaromatic:

$$ArN_2^+ \, BF_4^- \rightarrow ArF + N_2 + BF_3$$

Clearly, only one-quarter of the available fluorine is utilized in the fluorination step but the boron trifluoride released can be absorbed in 30–40% aqueous hydrogen fluoride, or with sodium fluoride, and thus quantitatively recovered as aqueous fluoroboric acid or sodium fluoroborate, respectively, which can be recycled. This process was not known to have been commercialized until 1986, when a Laporte company [Wendstone Chemicals (U.K.)] published details of their process for the production of 4,4'-difluorodiphenylmethane from the corresponding diamine using a continuous decomposition stage.

The second diazotization-based process uses anhydrous hydrogen acid (AHF) as solvent and reagent. The water liberated in diazonium salt formation dilutes the anhydrous HF; but the fluoride values can be reclaimed, for example, by using the aqueous HF in fluoroboric acid production. The intermediate diazonium fluoride is not isolated but decomposed *in situ* under very carefully controlled conditions to yield the fluoroaromatic:

$$ArNH_2 + NaNO_2 + 2HF \rightarrow 2H_2O + NaF + ArN_2^+ \, F^- \rightarrow ArF + N_2$$

One of the major problems associated with this HF-diazotization process is the potentially hazardous thermal decomposition step. This has tended to limit the scale of batch processes to $< 5 \text{ m}^3$. To circumvent this problem, companies such as Riedel-de Haen (Germany) and Mallinckrodt (U.S.) have developed continuous HF-diazotization processes where the diazonium fluoride is decomposed as soon as it is formed. The Riedel-de Haen process uses a falling-film decomposer;[10] Mallinckrodt employ a cascade of well-agitated reactors.[11]

The reaction can be applied to a wide range of substituted anilines, with yields varying from 60 to 90%. Yield improvements have been reported for reactions in which the separately prepared nitrous acid solution is continuously added to the anilinium fluoride solution. The process is operated on a continuous basis for the higher-volume products like fluorobenzene, where HF consumption and unit costs are reduced and throughput can be increased if necessary. According to producers operating the HF-diazotization process, the key to efficient operation is the economic use of HF and the correct choice of constructional materials to combat the highly corrosive properties of aqueous HF.

The choice of diazotization method depends mainly on the thermal stability of the diazonium salt relative to the boiling point of the HF/salt mixtures. Solids-handling in the Balz–Schiemann process represents a cost increase over the HF solution process. The HF-diazotization process is considered to be a more capital intensive process than the Halex process, but to have lower operating costs. Halex technology requires that the chlorines be activated toward nucleophilic substitution; but in an increasing number of fluoroaromatic intermediates, this prerequisite for halogen exchange cannot be met, so HF-diazotization or Balz–Schiemann technology must be used.

27B.3.2.4.   Fluorodenitration

The nitro function is not only an activating group in the Halex reaction: it is also readily activated toward nucleophilic displacement by the presence of other electron-withdrawing groups on the aromatic nucleus (Chapter 9). In Halex reactions, the activating groups should be located on specific carbon atoms. An unsuitable configuration such as in 2,6-dichloronitrobenzene can lead to fluorodenitration (see Scheme 1).

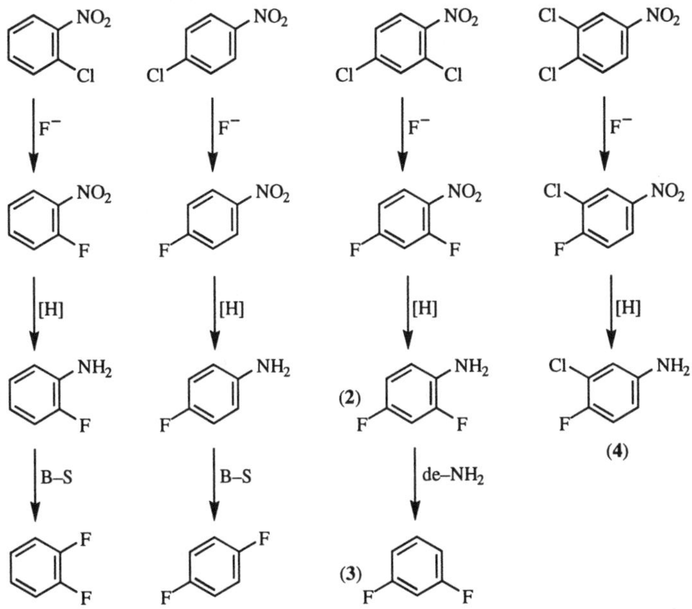

Scheme 1.   Halex products and their primary derivatives based on chloronitro-aromatics: B–S denotes the Balz–Schiemann process.

27B.3.3. Production of Nuclear-Fluorinated Aromatics in Europe

Nine companies in Western Europe produce ring-fluorinated aromatics on a significant scale. Formerly, companies tended to specialize in either halogen exchange or diazotization, but with the increasing demand for polyfluorinated products, several have diversified their production capability to use both processes. Halex capacity in Western Europe in early 1990 was about 3000 tonnes per annumn, distributed as shown in Table 5. The structural diversity of market demand is illustrated in Scheme 1.

The largest single outlet for 2-fluoroaniline (1) is the production of flurbiprofen by Boots Chemicals at Nottingham in the UK. The major uses of 2,4-difluoroaniline (2) are for the production of the Rhône-Poulenc herbicide diflufenican (JAVELIN) (Chapter 11, product 12), the Merck Sharpe & Dohme NSAID diflunisal (DOLOBID) and 1,3 difluorobenzene (3) via deamination. Shell used to produce 3-chloro-4-fluoroaniline[b] (4) for captive manufacture of flamprop-isopropyl (BARNON) but there are several small-volume applications related to the production of fluoroquinolone antibiotics. 2,3,5,6-Tetrafluoroterephthalonitrile produced by the Halex reaction (p. 199) of the tetrachloro analog is used by ICI in the manufacture of FORCE, a synthetic pyrethroid.

Fluorobenzene is currently produced in Europe by four companies: Riedel-de Haen, Rhône-Poulenc, Miteni, and ICI. ICI's 2000 tpa continuous diazotization plant at Grangemouth in Scotland also produces 2-fluorobenzoyl chloride and 4,4′-difluorodiphenylmethane.

The availability of cheap fluorobenzene from a low-cost HF-diazotization process presents the opportunity of making the economic switch of technology for the production of 4,4′-difluorobenzophenone (BDF) from the Balz–Schiemann process to a Friedel–Crafts process based on fluorobenzene and 4-fluorobenzoyl chloride. BDF is used as a monomer in the production of polyetheretherketone (PEEK) polymer, one of the ICI range of engineering polymers.

Table 5.   Halex Capacity in Western Europe (1990)[a]

Company	Plant type	Location	Capacity (tonnes)
Rhône-Poulenc (ISC)	Multipurpose	Avonmouth (U.K.)	1000
Shell Chemicals[b]	Dedicated	Stanlow (U.K.)	1000
Duphar BV	Dedicated	Weesp (Netherlands)	250
Angus Fine Chemical	Multipurpose	Ringaskiddy (Eire)	300
ICI FCMO	Multipurpose	Grangemouth (U.K.)	100
Boots Chemicals	Dedicated	Nottingham (U.K.)	200
Hoechst AG[c]	Multipurpose	Frankfurt (Germany)	50
		Total	2900

[a]Source: PFG Associates Ltd.
[b]See Section 27B.5.
[c]Hoechst produce three products on a campaign basis: 4-fluoroaniline, 2,4-difluroaniline, and 3-chloro-4-fluoroaniline. Capacity has been expanded to 500 tonnes per annum.

Miteni, Riedel-de Haen, and ICI produce 2-fluorobenzoyl chloride in Western Europe. Most of the ICI production is used for the production of 2,4-difluorobenzophenone, the flutriafol intermediate. Riedel-de Haen uses 2-fluorobenzoyl chloride as a precursor for other 2-substituted benzophenones: 2-amino-5-chloro-2'-fluorobenzophenone for the production of flunitrazepam, 2-chloro-4'-fluorobenzophenone, and 4-fluoro-4'-hydroxybenzophenone.

The two most important fluorinated benzaldehydes are 2-chloro-6-fluorobenzaldehyde and 4-fluoro-3-phenoxybenzaldehyde. The first is used by Smith-Kline-Beecham to produce flucloxacillin (antibiotic), and the second to produce synthetic pyrethroids. Rhône-Poulenc (France), Hoechst (Germany), and Miteni (Italy) manufacture the chlorofluoroaldehyde.

2,6-Difluorobenzamide, produced from 2,6-difluorobenzonitrile, is the key fluoroaromatic intermediate in the synthesis of diflubenzuron and several other fluorinated benzoyl ureas. Shell have developed new biochemical technology to produce this amide. 2-Chloro-4,5-difluorobenzoic acid is a favored raw material for the synthesis of a number of fluoroquinolone antibiotics (see Chapter 24).

27B.3.4. Side-Chain Fluorinated Aromatics

World demand for side-chain fluorinated aromatics is much larger than that for nuclear analogs. However, though there are many development products available, very few are in high-volume regular production.

Side-chain fluorination can be effected using very similar technology to that used for aliphatic halogen exchange (see Chapter 10). The four main end-uses for benzotri-

Table 6. Side-Chain Fluorinated Aromatic Products
and Their Fluoroaromatic Precursors[a]

Side-chain fluorinated precursor	Commercial product	Company
Benzotrifluoride	Flumeturon	Ciba-Geigy
	Diflufenican	Rhône-Poulenc
	Flufenamic acid	Park Davis
	Trifluperidol	Janssen
	Norflurazon	Sandoz
	Fluridone	Dow-Elanco
	Flurochloridone	ICI
4-Chlorobenzotrifluoride	Trifluralin	Dow-Elanco
	Benfluralin	Dow-Elanco
	Ethalfluralin	Dow-Elanco
	Fluchloralin	BASF
3,4-Dichlorobenzotrifluoride	Lactofen	PPG
	Acifluorfen	Rohm & Haas
	Oxyfluorfen	Rohm & Hass
	Fomesafen	ICI
2-Chloro-5-trifluoromethylpyridine	Fluazifop butyl	ICI
2,3-Dichloro-5-trifluoromethylpyridine	Haloxyfop	Dow

[a]Source: PFG Associates Ltd.

fluoride rely on its transformation to 3-aminobenzotrifluoride via mononitration of benzotrifluoride to produce a mixture of isomers containing a high proportion (>90%) of the *meta* isomer; after hydrogenation of the mixture, 3-aminobenzotrifluoride is separated by distillation. This process is used by Rhône-Poulenc, the largest producer of side-chain fluorinated aromatics in Europe. The manufacture of diflufenican, the Rhône-Poulenc herbicide, requires 3-trifluoromethyl phenol; this is produced from 3-aminobenzotrifluoride, as are several small-volume pharmaceutical products such as flufenamic acid and trifluperidol.

4-Chlorobenzotrifluoride and 3,4-dichlorobenzotrifluoride are made by Hoechst, Rhône-Poulenc and Miteni. Trifluoralin, the largest-volume product based on 4-chlorobenzotrifluoride, is manufactured in Europe by Dow-Elanco (France) and IPICI (Italy). The four products based on 3,4-dichlorobenzotrifluoride are all relatively recent, highly successful herbicides (Table 6).

Table 7. Major Fluoropolymers Available in Europe[a]

Polymer	Acronym	Manufacturer	Trade name
Poly(tetrafluoroethylene)	PTFE	Atochem	Soreflon
		Du Pont	Teflon
		Hoechst	Hostaflon
		ICI	Fluon
		Montefluos	Algoflon
Poly(vinylidene fluoride)	PVDF	Atochem	Foraflon
			Kynar
		Solvay	Solef
Tetrafluoroethylene–hexafluoropropylene copolymer	FEP	Du Pont	Teflon
Ethylene–tetrafluoroethylene copolymer	ETFE	Du Pont	Tefzel
Poly(chlorotrifluoroethylene)	PCTFE	Allied Signal	Aclar
		Atochem	Voltalef
		3M Corp	Kel-F 81
		Montefluos	Edifren
		Halocarbon Products	Halocarbon Oil
Chlorotrifluoroethylene copolymer		Allied-Signal	Aclon
			Aclar
		3M	Kel-F 82
			Kel-F 800
		Occidental	Fluorolube
Copolymers of perfluorovinyl ethers and tetrafluoroethylene	PFA	Du Pont	Teflon
		Du Pont	Kalrez
		Hoechst	Hostaflon
Poly(vinyl fluoride)	PVF	Du Pont	Tedlar
Chlorotrifluoroethylene–ethylene copolymer	ECTFE	Allied-Signal	Halar
Tetrafluoroethylene/hexafluoropropylene/		Hoechst	Hostaflon
vinylidene fluoride/ethylene copolymers		Du Pont	Viton
Perfluoropolyether fluids	PFPE	Montefluos	Fomblin
			Galden

[a]Source: PFG Associates Ltd.

Table 8.   Other Fluoropolymers Available in Europe[a]

Polymer	Manufacturer	Trade name
Amorphous Fluoropolymers		
Copolymer of TFE and 2,2-bistrifluoromethyl- 4,5-difluoro-1,3-dioxole	Du Pont	Teflon AF
Cyclopoly[perfluoro(allyl vinyl ether)]	Asahi Glass	Cytop
Fluoroelastomers		
Vinylidene fluoride–hexafluoropropylene copolymer	Ausimont Du Pont 3M	Technoflon Viton A Fluorel
Vinylidene fluoride-hexafluoropropylene– tetrafluoroethylene terpolymers	Du Pont	Viton B Viton C
Vinylidene fluoride–chlorotrifluoroethylene copolymer	3M	Kel-F
Perfluoroelastomers		
Tetrafluoroethylene–perfluoro(methyl vinyl ether) copolymer	Du Pont	Kalrez
Fluorosilicone Elastomers		
1,1,1-Trifluoropropylsiloxy-based products	3M	Sylon
Perfluoropolyether (PFPE) Fluids		
Products based on photooxidation of perfluoroolefins	Ausimont	Fomblin Galden
Products based on ROP polymerization of		
(i) perfluoropropylene oxide	Du Pont	Krytox
and	Hoechst AG	Hostinert
(ii) 2,2,3,3-tetrafluorooxetane	Daikin	Demnum
Surface-Fluorinated High Density Poly(ethylene)	Air Products	Airopak
Perfluoro-Ionomers		
Copolymers of tetrafluoroethylene with sulfonated or carboxylated perfluorovinyl ethers	Du Pont Asahi Glass Dow	Nafion Flemion

[a]Source: PFG Associates Ltd.

27B.4.   FLUOROPOLYMERS

In 1989, the total European production of fluoropolymers was estimated at around 20,000 tonnes of which PTFE accounted for about 70%. Some of the fluorine-containing polymers available in Europe are listed in Tables 7 and 8.

27B.5.   NOTE ADDED IN PROOF: DEVELOPMENTS CONCERNING COMPANIES

The Shell plant at Stanlow (U.K.) was seriously damaged by explosion and fire in mid-March 1990, and Shell has effectively withdrawn from production of primary fluoroaromatic feedstocks by fluorination.

Kali Chemie AG, Fluor Products Division, became known as Solvay Fluor und Derivate GmbH in January 1991.

The early 1993, BNFL Fluorochemicals acquired the FLUTEC perfluorocarbon business of Rhône-Poulenc and is rebuilding the plant at its Springfields site (Preston, U.K.).

In June 1993, the original ICI Plc demerged into two separate companies, the 'new' ICI Plc and Zeneca Plc, the latter comprising the bioscience and speciality activities. The CFC replacement business, i.e., KLEA, remains within ICI, whereas the fluoroaromatics activities are now part of the Fine Chemicals Division of Zenea.

27B.6. REFERENCES

1. Ramanadin, German Patent 2,221,849 (to Rhône-Progil) [*CA 78*, 42858 (1973)].
2. French Patent 1,343,392 (to Hoechst) [*CA 60*, 9147 (1964)].
3. P. P. Rammelt and G. Siegemund, German Patent 2,203,326 (to Hoechst) [*CA 79*, 14606 (1973)].
4. K. Hoegerle, P. Rathgeb, H. J. Cellarius, and J. Rumpf, German Patent 1,816,694 (to Agripat S. A.) [*CA 72*, 12734 (1970)].
5. R. K. Howe and L. F. Lee, German Patent 2,919,511 (to Monsanto) [*CA 92*, 110998 (1980)].
6. G. Grethe and T. Mitt, German Patent 2,806,909 (to Hoffmann-La Roche) [*CA 90*, 22838 (1979)].
7. L. Amiet and C. Disdier, European Patent 168,293 (to Rhône-Poulenc) [*CA 105*, 114606 (1986)].
8. G. Cordier, French Patent 2,544,712 (to Rhône-Poulenc) [*CA 103*, 37090 (1985)].
9. M. Tordeux, B. Langlois, and C. Wakselman, French Patent 2,593,808 (to Rhône-Poulenc) [*CA 108*, 166975 (1988)]; European Patent 278,822 (1988) [*CA 110*, 94514 (1989)].
10. E. Begemann and H. Schmand, German Patent 3,520,316 (to Riedel-de Haen AG) [*CA 106*, 175936 (1987)].
11. N. J. Stepaniuk and B. J. Lamb, U.S. Patent 4,822,927 (1989) (to Mallinckrodt Inc.) [*CA 111*, 59987 (1989)].

27

The Fluorochemical Industry

C. Manufacturers of Organic Fluoro Compounds in Japan

NOBUO ISHIKAWA[†]

Asahi Chemical Industry Co., Ltd.

[(1-1-2, Yurakucho, Chiyoda-ku, Tokyo 100. Tel: 03-3507-2730; Fax: 03-3507-2005)

1. Fluorinated Membranes

Aciplex-F Ion-exchange membrane.
Acilyzer-ML Chlorine–alkali electrolyzing membrane.

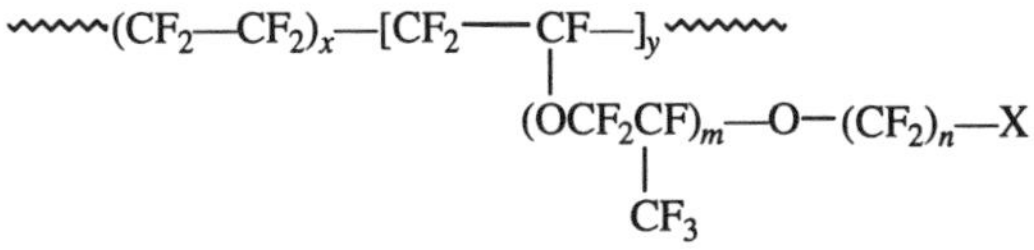

$$\sim\!\!\sim\!\!\sim(CF_2\!-\!CF_2)_x\!-\![CF_2\!-\!\!-\!CF\!-\!]_y\!\sim\!\!\sim\!\!\sim$$
$$(OCF_2CF)_m\!-\!O\!-\!(CF_2)_n\!-\!X$$
$$CF_3$$
$$[X = CO_2^-Na^+, SO_3^-Na^+]$$

2. Fluorinated Elastomer

Miraflon (vinylidene fluoride-based).

3. Anticancer Drug

Sunfural S (1-(2-tetrahydrofuryl)-5-fluorouracil, tegafur).

[†]Deceased.

NOBUO ISHIKAWA • F & F Research Center, Shizushin Building, Minato-ku, Tokyo 107, Japan.

Organofluorine Chemistry: Principles and Commercial Applications, edited by R. E. Banks *et al.* Plenum Press, New York, 1994.

Asahi Glass Co., Ltd.

(2-1-2 Marunouchi, Chiyoda-ku, Tokyo 100. Tel: 03-3218-5566; Fax: 03-3215-8840)

1. Fluorinated Gases and Liquids

Asahi Flons
AF-11, 12, 13, 14 (CF_4), 21 ($CHCl_2F$), 22, 23 (CHF_3), 13B1 ($CBrF_3$) 112, 113, 114, 115, 116, 122 ($CHCl_2$—$CClF_2$), 123 ($CHCl_2$—CF_3), 142b ($CClF_2$—CH_3), 152a (CHF_2—CH_3), 218 ($CF_3CF_2CF_3$), 1113 ($CClF{=}CF_2$), 1114 ($CF_2{=}CF_2$), 1216 ($CF_3CF{=}CF_2$), c-318 ($\overline{CF_2CF_2CF_2C}F_2$), 500 ($CCl_2F_2/CHF_2$-$CH_3$), 502 ($CHClF_2/CClF_2$—$CF_3$), 503 ($CHF_3/CClF_3$).

2. Fluorinated Fire Extinguisher

Asahi Halons
1301, 1211, 2402.

3. Fluorinated Plastics and Elastomers

Fluon   PTFE plastic, in the form of molding powder, fine powder, dispersion, and lubricant.

Aflon-COP   Ethylene/TFE copolymer (ETFE). Easy moldability, high chemical resistance, good mechanical properties, etc.

Aflon-PFA   Perfluorovinyl ether/TFE copolymer.   —[CF_2-CF_2-CF_2-CF]$_n$—
Easy processability. Excellent resistance to heat, |
chemicals and solvents.   O—R_f

AFLAS   Propylene/TFE copolymer-based fluoroelastomer. Excellent properties including nonflammability, high solvent and oil resistance.

Technoflon   Vinylidene fluoride/HFP, and vinylidene fluoride/HFP/TFE-based fluorinated elastomers.

4. Fluorinated Coating Material

LUMIFLON   Solvent-soluble fluoropolymer which is used for coating various types of outdoor structures. Characteristic properties are: high weather resistance, wide range of coating, excellent gloss and beautiful appearance, easy removal of stains, and economical long-term maintenance, etc.

5. Functional Membrane and Materials

Flemion   Fluorinated membranes used for electrolysis of sodium chloride solution for the alkali and chlorine production. Low-energy consumption, high-quality durability, and easy adaptability are the features.

HISEP   Vapor separation film with selective permeability. Medical use for oxygen enrichment and industrial use for gas separation.
CYTOP   Noncrystalline-type perfluorinated polymer with high transparency.

6. Fluorinated Surface-Active Materials

Asahi-Guard   Fluorocarbon-containing water and oil repellent for apparel and upholstery use. Emulsion type: AG-310, 410, 430, 460, 530, 550, 710, 740, 800 etc. Solvent type: AG-610, 640, 670, 804, 805, etc.
SURFLON   Fluorocarbon-based surfactant. Used for emulsion polymerization, fire extinguisher, and others. Emulsion type (S-111, 112, 121, 131, 141, etc.) and solvent type (S-381, SC-101, SR-100) are provided.
AG-LUB   Lubricant based on TFE telomers.

7. Fluorinated Intermediates

Trifluoroacetic acid and its derivatives   Trifluoroacetic anhydride, ethyl trifluoroacetate, ethyl trifluoroacetoacetate, trifluoroacetamide, etc.
Chlorobenzotrifluoride and its derivatives   4-Chloro-, 3,4-dichloro-, 2,4-dichloro-, 3-chloro-4-fluoro-, and 4-fluoro-benzotrifluorides.
Fluoroanilines   2- and 4-Fluoroanilines, 2,4- and 3,4-difluoroanilines, 3-chloro-4-fluoroaniline, 2,3,4-trifluoroaniline.
Fluorobenzonitriles   2- and 4-Fluoro-, 2,4- and 2,6-difluorobenzonitriles.
Fluorophenols   4-Fluoro-, 2-chloro-4-fluoro- and 3-chloro-4-fluoro-phenols.
Fluorobenzoic acids   2- and 4-Fluorobenzoic acids, 2,4- and 2,6-difluorobenzoic acids, 2,3,4,5-tetrafluorobenzoic acid, pentafluorobenzoic acid.

Central Glass Co., Ltd.

(3, Kandanishikicho, Chiyoda-ku, Tokyo 101. Tel: 03-3259-7324; Fax: 03-3293-2145)

1. Hexafluoroacetone and Its Derivatives

HFA   Hexafluoroacetone, $(CF_3)_2C{=}O$, colorless gas, bp 27.5°C, nonflammable.
HFA Trihydrate   colorless liquid, bp 105°C.
HFIP   Hexafluoroisopropanol, $(CF_3)_2CHOH$, bp 58.6°C. Strongly corrosive to skin and eyes. Good solvent for polyamides, polyesters and other organic compounds. *BIS-AF* 2,2-bis(4-hydroxyphenyl)hexafluoropropane.
Bis-AF derivatives   BIS-A-AF, BIS-AT-AF, BIS-B-AF, BIS-AF-G, etc.

$$HO{-}C_6H_4{-}\underset{\underset{CF_3}{|}}{\overset{\overset{CF_3}{|}}{C}}{-}C_6H_4{-}OH$$

2. Triflic Acid (Trifluoromethanesulfonic Acid)

    Triflic acid, CF_3SO_3H, colorless liquid, bp 162°C, freezing point –40°C.
    Derivatives: Li, Na, K, Ag, Sn salts and anhydride, $(CF_3SO_2)_2O$.
    Uses: Electric cell, synthetic catalyst for polymers, medicines and others.

3. Benzotrifluoride Derivatives

    1,3- and 1,4-Bis(trifluoromethyl)benzene.
    o- and *p*-Trifluoromethylbenzaldehyde.
    o-Trifluoromethylbenzoyl chloride.

Daikin Industries, Ltd.

(2-4-12, Nakazaki-nishi, Kita-ku, Osaka. Tel: 06-373-1201; Fax: 06-373-4390)

1. Fluorinated Gases and Liquids

    Daiflon R-11, -12, -13, -13B1, -22, -112, -113, -114, -115, -142b, -500, -502,
        -502P, $(CHClF_2/C_2ClF_5/C_3H_8)$, *c*-318($C_4F_8$).
    DAIFLON S1(CCl_3F), S2($C_2Cl_4F_2$), S3($C_2Cl_3F_3$).
    Dryflon (for dry cleaning) -11, -113, -A(113/surfactant).

2. Fluorinated Plastics and Elastomers

    Daikin-Polyflon TFE   PTFE plastic, in the form of molding powders, filled
molding powders, fine powders, dispersions, enamels, and tough-coat enamel papers.
    Neoflon PFA   Copolymer of TFE/PFA (perfluoroalkyl vinyl ether). Pellets for
wafer baskets, electric wire, film tubing, lining, etc.
    Neoflon FEP   Copolymer of TFE/HFP (hexafluoropropene). Pellets for films,
tubes, and electrical wiring. Corrosion-resistant linings, insulated covers, etc.
    Neoflon ETFE   Copolymer of TFE/ethylene. Molding pellets for electrical
wires, linings, tubings, etc.
    Neoflon CTFE   Poly(chlorotrifluoroethylene). Molding powder for packings,
gaskets, etc.

$$\left[\begin{array}{cc} \underset{|}{\overset{|}{F}} & \underset{|}{\overset{|}{F}} \\ -C & -C- \\ \underset{|}{F} & \underset{|}{Cl} \end{array} \right]_n$$

    DAI-EL   Fluoroelastomer, based on the copolymer of vinylidene fluoride and
HFP. Molding powder for O-rings, gaskets, fuel hoses, etc.

3. Surface-Active Materials

    Texgard   Fluorocarbon-based oil and water repellent, used for textile and paper
finishing.

Daifree   Fluorocarbon-based releasing agent for rubber and plastic molding.
Unidyne   Fluorocarbon-based surfactant used for wetting, leveling, fire extinguishing, etc.

4. Fluorinated Intermediates

Fluoroalcohols $H(C_2F_4)_nCH_2OH$.
Fluorocarboxylic acids $H(C_2F_4)_nCO_2H$.

5. Bioactive Compounds

Halothane $CHBrCl-CF_3$ Anesthetic.
Tetrapion $CHF_2CF_2CO_2Na$ Herbicide, used for forest and nonagricultural area.

Du Pont-Mitsui Fluorochemicals Co., Ltd.

(1-2-3, Otemachi, Chiyoda-ku, Tokyo 100. Tel: 03-3214-5241; Fax: 03-3215-0064)

This company is a joint venture established by Mitsui Petrochemicals and Du Pont (U.S.A.) in 1963. They are manufacturing fluorochemicals such as Freons and Teflons, being supported by the technology of Du Pont (U.S.A.).

1. *Freons*   CFC-11, -12, -113, -114, -115, HCFC-22.

2. *Teflons*   TFE, FEP, PFA, TEFZEL.

Japan Halon Co., Ltd.

(3-2-5, Kyobashi, Chuo-ku, Tokyo 107. Tel: 03-3274-1301; Fax: 03-3274-1211)

1. Fluorinated Intermediates

HEF-9   (Perfluoro-*n*-butyl)ethylene, $n\text{-}C_4F_9CH{=}CH_2$ (for fluorosilicone oils).
TFP   (3,3,3-Trifluoropropene), $CF_3CH{=}CH_2$ (modifier of fluorosilicone elastomer).
TFEC   (1-Chloro-2,2,2-trifluoroethane), CF_3CH_2Cl (for TFEA and HFC-134a).
TFEA   (2,2,2-Trifluoroethanol), CF_3CH_2OH (for anesthetics).

2. Halons

Halon-1211 (Bromochlorodifluoromethane), $CBrClF_2$ (fire extinguisher, bp −2.5°C).
Halon-1301 (Bromotrifluoromethane), $CBrF_3$ (fire extinguisher, bp −58°C).

Mitsubishi Materials Corporation

(4-6-23, Takanawa, Minato-ku, Tokyo 108. Tel: 03-3213-2111; Fax: 03-3280-8155)

1. Fluorinated Surfactants

   EFTOP anionic (EF-102, -103, -104, -105, -112, -123a, -123b, -306a, -501, -201,-204),
   cationic (EF-132),
   nonionic (EF-121, -122a, -122b, -122c, -122a3, -301, -303, -305, -351, -352, -801, -802).

2. Perfluorinated Inert Liquids

   EF-L -102 (bp 102°C), -155 (155°C), -174 (174°C), -215 (215°C).

3. Perfluoroalkylating Agents

   MF-100   3-(2-perfluorohexyl)ethoxy-1,2-dihydroxypropane   (liquid, bp 125°C/0.2 mm Hg).
   MF-110   N-n-propyl-N-2,3-dihydroxypropyl(perfluorooctyl)sulfonamide (solid, mp 57°C).
   MF-120   3-(2-perfluorohexyl)ethoxy-1,2-epoxypropane   (clear liquid, bp 71°C/0.15 mm Hg).
   MF-140   (perfluorohexyl)ethylene   (bp 107°C/760 mm Hg) etc.

4. Fluorinated Aliphatic Compounds

   TFE (trifluoroethanol), *TFEA* (2,2,2-trifluoroethyl acrylate).
   EF-101 (perfluorooctanesulfonic acid).
   EF-201 (perfluorocaprylic acid).
   etc.

5. Fluorinated Aromatic Compounds

   p-Fluorobenzonitrile, 2- and 4-fluorobenzoic acid, p-fluorophenol, o-, m-, and p-difluorobenzene.

Morita Kagaku Kogyo Co., Ltd.

(2-6-10, Korai-bashi, Chuo-ku, Osaka, 541. Tel: 06-203-3571; Fax: 06-203-3570)

1. Fluorinated Aromatic Compounds

    o- and p-Fluoroaniline, o-, m-, and p-fluorophenol, 3,4-difluorophenol, 1,2-difluorobenzene, 2,6-difluorobenzonitrile, 2,6- and 2,4-difluoroaniline, 2,3,4-trifluoroaniline, 5-fluorouracil.

Neos Co., Ltd.

(6-2-1 Kanomati, Chuo-ku, Kobe 650. Tel: 078-331-9381; Fax: 078-391-0448)

1. Fluorinated Surfactants

    Neos Ftergent   Surfactants containing highly branched fluorocarbon chains (hexafluoropropene dimers and trimers). They have high surface activity even at very low concentration.
    anionic *Ftergent 100* (sulfonate), *150* (carboxylate).
    nonionic *Ftergent 250* (EO adduct), *251* (EO adduct).
    cationic *Ftergent 300* (quat. ammonium salt).
    amphoteric *Ftergent 400s* (betaine).

Tokuyama Soda Co., Ltd.

(1-4-5, Nishishinbashi, Minato-ku, Tokyo 105. Tel: 03-3597-5049; Fax: 03-3597-5188)

1. Perfluorinated Inert Liquids

    Perfluord IL-310   Inert liquid for vapor-phase soldering process at high temperature, bp 216–218°C, d_{25} 1.93.
    Perfluord IL-270   Inert liquid for vapor-phase soldering process at low temperature, bp 176–178°C, d_{25} 1.89.
    Perfluord IL-260   Inert liquid used for thermal shock-test and leak-test, etc., in electronics industry, bp 160°C, d_{25} 1.87.

28

CFCs and the Environment: Further Observations

RICHARD L. POWELL and J. HUGO STEVEN

28.1. INTRODUCTION

The introduction of the chlorofluorocarbon (CFC) fluids in the early thirties marked the inception of the organofluorocarbon industry. Both in terms of tonnage produced and product value, they have dominated the industry and provided feedstocks for the development of other products, such as fluoropolymers (see Chapter 6).

The possible implication of chlorofluorocarbons—and to a lesser extent hydrochlorofluorocarbons (HCFCs)—in the depletion of stratospheric ozone, postulated in 1974[1] and reinforced in the late 1980s by the discovery of a possible link to the thinning of the ozone layer over Antarctica during its springtime,[2] continues to have profound effects on the fluorocarbon fluids industry. The situation is changing so rapidly that whatever is written now will be superseded, either scientifically or politically, within 6 months. This chapter reflects the situation at the end of 1991.*

---

*In 1992, at an intergovernmental conference in Copenhagen, the Montreal Protocol was substantially revised, bringing forward the date of the deadline for the phase-out of CFCs and, for the first time, introducing a specific timetable for the phase-out of HCFCs. The European Union has decided subsequently to phase out CFCs at the end of 1994, one year ahead of the date required by the Montreal Protocol, and has proposed an accelerated timetable for HCFC phase out. Table 1 summarizes the current situation at the end of 1993.

RICHARD L. POWELL and J. HUGO STEVEN • ICI Klea, Runcorn Technical Centre, Runcorn, Cheshire WA7 4QD, England.

Organofluorine Chemistry: Principles and Commercial Applications, edited by R. E. Banks *et al.* Plenum Press, New York, 1994.

Table 1. Regulatory Reduction in the Consumption of CFCs/HCFCs in %

	Copenhagen amendments to Montreal Protocol			European Community's regulation		
	Agreed and in force			Enacted and in force		Proposed
	CFCs[a]	'Other CFCs'[b]	HCFCs	CFCs[a]	'Other CFCs'[b]	HCFCs
Cap level[c]			3.1%			2.6%
Base year	1986	1989	1989	1986	1986	1989
Year						
1993	—	−20	—	−50	−50	—
1994	−75	−75	—	−85	−85	—
1995	−75	−75	—	−100	−100	Freeze and use controls
1996	−100	−100	Freeze			
2000			Freeze			Freeze and use controls
2004			−35			−35
2007			−35			−60
2008			−35			−60
2010			−65			−80
2012			−65			−80
2013			−65			−95
2015			−90			−100
2020			−99.5			
2030			−100			

[a]11, 12, 113, 113a, 114, 114a, 115.
[b]13, 112, 112a, 111 and the fully halogenated C_3 CFCs.
[c]The cap is defined as the percentage of the 'calculated' level of chlorofluorocarbons consumed in the base year plus the calculated level of hydrochlorofluorocarbons consumed in the same base year. It applies only to HCFCs. 'Calculated' in this context means that the amount of each substance is adjusted by its ozone depletion potential (ODP), a measure of its potential to deplete stratospheric ozone relative to that of CFC-11.

28.2. REFRIGERATION

The history of commercial refrigeration can be usefully divided into three periods. The first runs from the 1870s until the 1930s, when progress was based on flammable or toxic refrigerants. Despite the obvious attendant problems, the technology developed allowed the transportation of perishable foodstuffs over thousands of miles by land and sea via the newly developed railways and steamships to feed the large populations of industrial cities in Europe and the USA.[3] The introduction of public electricity supplies in the 1880s made the domestic refrigerator a realistic proposition; however, it was not until 1913 that the first reliable model, the "Dolmere," based on sulfur dioxide, appeared in the USA.

The U.S. market for domestic refrigerators expanded rapidly in the 1920s as households abandoned their old-fashioned ice boxes, which required daily deliveries of fresh ice by the local ice-making companies. However, sensational newspaper reports of fatal incidents involving the leakages of sulfur dioxide and methyl chloride resulted in local laws restricting refrigerants, and renewed competition from the ice distributors. Frigidaire, a leading domestic refrigerator producer, realized that its business would prosper only if a suitable safe refrigerant was identified, and it asked the research laboratory of its parent company, General Motors, to find such a product. To seek it among aliphatic fluorine compounds was an act of vision, considering the limited developments in that field at that time.[4]

The project team was led by an engineer, Thomas Midgley, and a chemist, Albert Henne, who is now respected[5] as one of the pioneers of organofluorine chemistry. After a systematic search of possible compounds they selected CFC-12, dichlorodifluoromethane (CF_2Cl_2), as the most promising candidate (see Chapter 1). The work was started in 1928, and the first announcement of CFC-12 (bp -30 °C) as a refrigerant was made in 1930 at the Atlanta meeting of the American Chemical Society. Production commenced in 1931 at Kinetic Chemicals, a joint GM/Du Pont company using halogen-exchange reactions, still the preferred manufacturing route (see Chapter 6).

Dichlorodifluoromethane was first used in small Frigidaire ice cream cabinets in 1931, and rapidly found service in other commercial units and room coolers. Carrier adopted trichlorofluoromethane ($CFCl_3$, CFC-11) as the refrigerant in centrifugal air conditioners in 1933. When Frigidaire introduced the first domestic fridge, the "Meter-Miser", it was based on CFC-114 (CF_2ClCF_2Cl), not the lower-boiling CFC-12.

A graphic illustration of the impact caused by the chlorofluorocarbon refrigerants is attributed to Midgley. It is said that in public lectures he would inhale a sample of CF_2Cl_2 and then gently exhale it over a candle flame, which was duly extinguished. He claimed that this showed two important facts about CFC-12: first, it was nontoxic, and second, it was nonflammable. Attempting the demonstration with other refrigerants available at that time (sulfur dioxide, ammonia, methyl chloride, propane) would obviously have had dire consequences!

The second period in the history of refrigeration, which commenced with the introduction of the chlorofluorocarbon refrigerants, runs from the 1930s to the 1990s, again a span of about 60 years. It is characterized by the dominance of chlorofluorocarbons in refrigeration. Airconditioning also grew rapidly in industrialized countries, especially those which have high summer temperatures and humidities, like the U.S. HCFC 22 (CHF_2Cl), a hydrochlorofluorocarbon, became the dominant refrigerant in this application area. Also HCFC-22 and its azeotrope with CFC-115 (CF_3CF_2Cl), code numbered R-502, have dominated supermarket refrigeration.

This period is notable, then, for the spread of refrigerators into almost every home in the industrialized world. These devices are no longer luxuries for the rich but necessities for all, ensuring that everyone can enjoy a varied, safe, and pleasant diet free from the taint of pathogenic bacteria. Not surprisingly, citizens of developing countries such as India and China see the domestic "fridge" as a key aspect of their aspirations to a better standard of living.

CFC-11 has also found uses as a working fluid in large airconditioning units, but mainly it has been employed in insulation technology. This latter application has arisen because it is an excellent blowing agent for polyurethane foam owing to its low vapor thermal conductivity, convenient boiling point (24 °C), nonflammability, and low

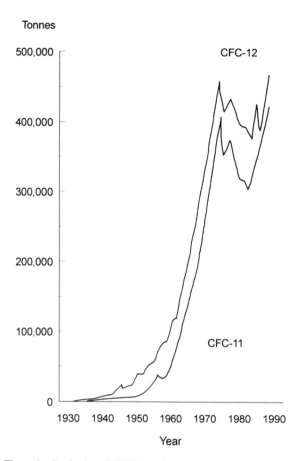

Figure 1. Production of CFCS 11 and 12 (excluding the Eastern Bloc).

toxicity. The chlorofluorocarbon, dissolved in the polyol component of the formulation, is evaporated by the heat of the polyol/isocyanate reaction and generates the cells
of the foam. Foams blown with CFC-11 have 30–40% better insulating properties than
equivalent foams blown with carbon dioxide and are used to insulate refrigerator
cabinets, replacing the rockwool fiber common up to the end of the 1950s. Foam blocks
blown with CFC-11 are used to insulate the walls and roofs of frame construction
buildings, thus reducing heat losses and helping to conserve fossil fuels. In this area,
polyurethane foam competes with polystyrene foam blown with CFC-12.

   The production of chlorofluorocarbon refrigerants expanded rapidly after 1945,
reaching about one million tonnes per annum in 1986 (Figure 1) despite a drop in
production in the late 70s when the U.S. and several other countries banned the use of
chlorofluorocarbons as aerosol propellants. However, environmental concerns have
led to the rapid decline in production and consumption of chlorofluorocarbons in the
early 1990s.

Until 1989 the single biggest use for chlorofluorocarbons was aerosol propellants, an application which grew rapidly from the 1950s to the mid-1970s. The low toxicity, nonflammability and good solvent properties of CFC-11 and CFC-12 made them especially attractive for this purpose. Of the other CFCs, 113 ($CF_2ClCFCl_2$, bp 48 °C) has found widespread application as a nonaggressive cleaning agent, and is particularly valued in the electronics industry for removing solder flux residues; it is also the feedstock for producing the important monomer $CF_2{=}CFCl$.

Brominated fluorocarbons derived from chlorofluorocarbons are highly effective fire extinguishants (see Chapter 8). This property, coupled with their volatility and the fact that they leave no residues, has made them attractive for use in a wide variety of situations ranging from aircraft to computer installations, industrial plants, and domestic dwellings.

28.3. THE OZONE LAYER PROBLEM

In 1974,[1] it was conjectured by Rowland and Molina that the high stability of the chlorofluorocarbons which made them so attractive as refrigerants, etc., also enabled them to pass unchanged through the troposphere to the stratosphere[*]; once there, it was pointed out, the intense UV radiation from the sun could cause the carbon–chlorine bonds to break homolytically, giving chlorine atoms (e.g., $CF_2Cl_2 + UV \rightarrow Cl\cdot + \cdot CF_2Cl$) which were known from laboratory experiments to catalyze the destruction of ozone: $Cl\cdot + O_3 \rightarrow ClO\cdot + O_2$; $ClO\cdot + O \rightarrow Cl\cdot + O_2$ (net: $O_3 + O \rightarrow 2\,O_2$). The trace gas ozone, which is present throughout Earth's atmosphere, reaches its peak concentration (10 ppmv) within the stratosphere. Hence talk of the *ozone layer*, which plays a role critical for life on Earth by absorbing biologically harmful UV radiation in the shorter-wavelength regions below 320 nanometers (nm), i.e., in the UV-C (<290 nm) and UV-B (290–320 nm) regions. Thus stratospheric chlorine catalyzes the reversal of the photochemical generation of ozone (*trioxygen*, O_3) from the more familiar respirable form of oxygen (O_2, *dioxygen*): $O_2 + UV$ (<240 nm) $\rightarrow 2\,O$; $O + O_2 \rightarrow O_3$; $O_3 + UV$ (210–320 nm) $\rightarrow O + O_2$. One million ozone molecules may be decomposed by just ten chlorine atoms before the latter are eliminated, mainly through encounters with natural methane ($Cl\cdot + CH_4 \rightarrow HCl + \cdot CH_3$). The hydrogen chloride produced ultimately precipitates as hydrochloric acid in rain ("rains out") in quantities negligible compared with natural background concentrations.

The refrigerant producers themselves first questioned the fate of chlorofluorocarbons released into the atmosphere; thus, in 1972, some two years before publication of the Rowland–Molina hypothesis implicating chlorofluorocarbons in stratospheric ozone depletion,[1] they initiated the industry research program into the environmental impact of chlorofluorocarbons through an organization called the "Fluorocarbon Program Panel" (FPP). This organization and its variously named successors sponsored much of the fundamental research into atmospheric chemistry through the 1970s.

---

[*]The troposphere extends roughly for the first 10–17 km of the atmosphere, and the stratosphere through the next 30–40 km.

28.4. REPLACEMENTS FOR THE CHLOROFLUOROCARBONS

Initial evidence for the effect of chlorofluorocarbons on the ozone layer was very limited and based mainly on computer models of the atmosphere. In the 1970s and until 1988 there was no observational support for ozone depletion, but, contrary to the impression sometimes gained from the media, the refrigeration manufacturers responded positively to the challenge. By 1976, within 18 months of the original Rowland and Molina hypothesis being published, most companies had initiated substantial research programs to identify environmentally acceptable replacements for chlorofluorocarbons. Patents with priority dates from 1976 onward, claiming manufacturing routes for candidate compounds, bear testimony to this activity.

Alternatives to the chlorofluorocarbons needed to retain the attractive properties of chlorofluorocarbons, but avoid any adverse environmental impact. Low toxicity, nonflammability, good thermodynamic properties, and accessibility via economically viable manufacturing routes were key factors in the selection. The most promising candidates are still to be found among fluids containing fluorine and carbon atoms (see Chapter 7). The inclusion of one or more hydrogen atoms in their molecules results in them being largely destroyed in the lower atmosphere by naturally occurring hydroxyl radicals (see Chapter 3), ensuring that relatively little of the material survives to enter the stratosphere.

The candidates which emerged from the research programs are listed in Table 2. Of these only 22, 23, 152a, and 142b were available commercially in the mid-1970s

Table 2.   Candidate Replacements for CFCs

Candidate compound	Formula	bp (°C)	Compound(s) replaced	Flammability[c]
134a	CF_3CH_2F	−27	12	N
123	CF_3CHCl_2	28	11	N
125	CF_3CHF_2	−48	502[a] and 22	N
32	CH_2F_2	−51	502[a] and 22	F
143a	CF_3CH_3	−48	502[a] and 22	F
134	CHF_2CHF_2	−20	12	N
22	CHF_2Cl	−41	502	N
152a	CHF_2CH_3	−24	12	F
124	CF_3CHFCl	−10	114 and 12	N
142b	CF_2ClCH_3	−9	114 and 12	F
133a	CF_3CH_2Cl	6	11	N
141b	$CFCl_2CH_3$	32	11	F
31	CH_2FCl	−9	11	F
21	$CHFCl_2$	9	11	N
23	CHF_3	−82	503[b]	N

[a]Azeotropic mixture of CF_3CF_2Cl and CHF_2Cl.
[b]Azeotropic mixture of CHF_3 and CF_3Cl.
[c]N = nonflammable; F = flammable.

when concern over the adverse effect of chlorofluorocarbons on the ozone layer was first postulated; 22 alone was an established major refrigerant.

In the first stage of the research, producers were primarily seeking alternatives for the major chlorofluorocarbon refrigerants and foam blowing agents CFC-12 and CFC-11, which were also the main aerosol propellants. This search was crucially influenced by the major customer: the refrigeration and airconditioning industry. Between 1931 and 1974 the industry owed its rapid development to the availability of the chlorofluorocarbon and hydrochlorofluorocarbon fluids, designing its equipment around their specific thermodynamic properties. By careful, evolutionary development, the industry had acquired an enviable reputation for reliability. Not surprisingly the refrigerant manufacturers were expected to develop substitutes whose physical properties were similar to the chlorofluorocarbon fluids which were being replaced. In particular, similar vapor pressures were needed, since pressure was a key factor in refrigerator and airconditioning design. Essentially this meant searching for fluids with similar boiling points to CFC-12 and CFC-11 which also met the other criteria outlined above.

By 1976, HFC-134a (bp –27 °C) had emerged from the research program as the fluid closest in physical properties to CFC-12 (bp –30 °C) and therefore the most acceptable to the refrigeration and airconditioning industry. Its isomer HFC-134 (CHF_2CHF_2; bp –20 °C) was not favored because of its significantly lower vapor pressure. HFC-152a (bp –24 °C) was rejected, primarily because of its flammability.

The search for a CFC-11 replacement proved more difficult; but by the end of the 1970s, HCFC-123 and HCFC-141b had been identified as potential substitutes. Other candidates, such as HCFC-133a, HCFC-31, and HCFC-21, had failed on toxicity grounds. To hasten the development program, manufacturers shared data about the toxicity of candidate compounds to avoid needless expense and time-consuming duplication.

In parallel with their own research programs, the manufacturers, through the FPP, also jointly funded research to study the atmospheric chemistry of chlorofluorocarbons in order to assess the extent of any risk they might pose. Independent research workers at universities and research institutes worldwide were contracted to measure the rates of reactions which were essential input data for the complex computer models needed to predict the rate of ozone depletion. This value could not be measured directly in the 1970s because the large daily and seasonal fluctuations in stratospheric ozone concentrations swamped the modest depletion expected from chlorofluorocarbons.

By 1980, the consensus of informed scientific opinion based on the best available evidence was that, although chlorofluorocarbons could cause depletion of stratospheric ozone, the overall effect at the current production levels would be less than 3%. This suggested that action to replace chlorofluorocarbons would be required, but this could be taken over a sufficiently long period to minimize the cost to the refrigeration industry and ultimately the customers.

As a follow-up to the 1978 ban in the U.S. on the use of CFCs as aerosol propellants, the U.S. Environmental Protection Agency undertook in 1980 to promote legislation restricting the manufacture and other uses of chlorofluorocarbons. However, the perceived low rate of ozone depletion coupled with changing U.S. political priorities removed the urgency to enact the appropriate legislation.

28.5. DISCOVERY OF ANTARCTIC OZONE DEPLETION

The discovery in 1984 of ozone depletion over Antarctica during the spring period provided the first observational support for the possible effect of chlorofluorocarbons on stratospheric ozone.[2] However, the observation of ozone loss did not indicate its cause. From 1984 to 1988 several theories competed—ranging from CFC chemistry, to atmospheric dynamics or even cosmic electron fluxes. It was not until 1988, with the results of the 1987 Airborne Antarctic Ozone Expedition, that a probable link with CFCs was established. This prompted the Natural Resources Defense Council, an American pressure group, to sue the EPA to fulfill its 1980 promise to seek legislation to control further the manufacture and use of chlorofluorocarbons in the U.S.

Industrial research programs to develop chlorofluorocarbon substitutes were renewed with vigor, and the Refrigerant Manufacturers set up in 1988 the jointly funded "Programme for Alternative Fluorocarbon Toxicity Testing" (PAFT) to initiate and manage the toxicity testing of the preferred candidates. The Refrigerant Manufacturing Industry also renewed its research into the atmospheric aspects of fluorocarbon fluids by forming in 1989 and funding an organization called AFEAS ("Alternative Fluorocarbons Environmental Acceptability Study"). Continuing the work started in the 1970s, this organization contracts independent research scientists to develop the understanding of the complex interactions of chemical species in the atmosphere.

An important outcome of this work has been the formulation of the ozone depletion potentials (ODPs) for chlorine-containing fluids.[6] Arbitrarily, CFC-11 is

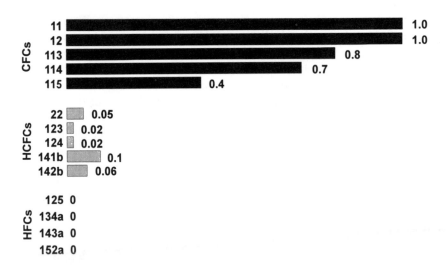

Figure 2.   Ozone depleting potentials.

assigned an ODP of unity; the ODPs of other fluids are normalized to that of CFC-11 on a mass-for-mass basis. Important factors in determining the ODP of a fluid include its atmospheric lifetime and the quantity of chlorine it contains. As Figure 2 clearly shows, hydrochlorofluorocarbons have much lower ODPs than chlorofluorocarbons—a consequence mainly of the former's lower atmospheric lifetimes. Hydrofluorocarbons, which by definition contain no chlorine, have zero ODPs.

28.6 THE MONTREAL PROTOCOL[*]

Worldwide concern over the depletion of the ozone layer was discussed at the Vienna Convention in 1985 under the auspices of UNEP. This culminated in the Montreal Protocol, the first international agreement made to protect the global environment. Signed in September 1987 and revised in June 1990, the Protocol now controls, throughout much of the world, use and production of chlorofluorocarbons, Halons (firefighting agents), carbon tetrachloride, and 1,1,1-trichloroethane (base solvent for degreasing formulations used in the engineering industry). In detail, the products currently (December 1991) controlled are the following:

- all fully-halogenated chlorofluorocarbons containing 1, 2, or 3 carbon atoms, including all isomers (i.e., molecules containing the same types and numbers of atoms but in different arrangements);
- Halons 1211 (CF_2ClBr), 1301 (CF_3Br), and 2402 (CF_2BrCF_2Br);
- carbon tetrachloride (CCl_4, CTC);
- 1,1,1-trichloroethane (methyl chloroform, CCl_3CH_3).

The 1990 Montreal Protocol mandates a total phase-out of production and consumption of chlorofluorocarbons, Halons, and CTC to be achieved by 2000, and a total phase-out of 1,1,1-trichloroethane by 2005. The intermediate stages are set out in Table 3. The first column shows the stages for the five chlorofluorocarbons in the original Protocol, and the second column shows the stages for the newly controlled chlorofluorocarbons. Provision—in the form of funding and agreements to ensure access to appropriate technologies—is made for the special needs of the developing countries.

Hydrofluorocarbons, which contain hydrogen as well as chlorine and fluorine (e.g., HCFC-22, CHF_2Cl), are not controlled under the revised Protocol. They have considerably lower Ozone Depletion Potentials (ODPs) than chlorofluorocarbons, and therefore have an important role to play in speeding the transition away from chlorofluorocarbons in applications where no suitable alternatives are currently available. The Parties to the Revised Montreal Protocol passed a Resolution which classed them as "Transitional Substances" and set guidelines for their use, namely, they are to be

---

[*]This section is based (with permission) on ICI's Information Note (August 1991) dealing with the Montreal Protocol and represents the position when the chapter was completed in December 1991.

Table 3. Phase-Out Timescales for Controlled Compounds[a]

	CFCs	New CFCs	Halons	CTC	111-Tri
Freeze	b	—	1992	—	1993
–20%	—	1993	—	—	—
–50%	1995	—	—	—	—
–85%	1997	—	—	—	—
–100%	2000	2000	2000	2000	2005

[a]All dates shown apply from 1 January of the year shown.
[b]The freeze applies only to the 5 chlorofluorocarbons (11,12,113, 114, 115) controlled by the original Protocol (1987), and is already in force.

used only as substitutes for chlorofluorocarbons, i.e., their range of applications cannot be extended; they are to be recovered and recycled during the transitional period in order to limit the quantity manufactured; and when their useful life is finished they are to be destroyed.

Scientific data on the hydrochlorofluorocarbons will be reviewed regularly with the intention that they should be replaced as and when suitable alternatives can be identified and made available. On the basis of current evidence, this suggests phase-out would be some decades into the next century; and the legislation recommends specific dates of 2040, or if possible 2020, for this to be achieved.

The Parties also passed a resolution regarding the production and use of certain fully halogenated Halons not controlled by the Protocol. Data will be monitored regularly, and a working group will study the definition of "essential uses" for these substances. A separate working group is to define essential uses which will continue beyond the year 2000 for the three Halons controlled.

A key aspect of the Protocol is its provision of a review process, so that new information on the various issues may be evaluated, and the Protocol amended if necessary. The first such revision occurred in June 1990, when the original control measures were strengthened and new substances became subject to controls.

28.7. REPLACEMENTS FOR THE CHLOROFLUOROCARBONS: A SECOND LOOK

Why did HFC-134a (CF_3CH_2F) continue to be the preferred replacement for CFC-12 as a refrigerant when the industrial chlorofluorocarbon replacement programs resumed in 1985? Could HFC-134 (CHF_2CHF_2), HFC-152a (CH_3CHF_2), or even propane have been reconsidered? Some reports give the impression that the refrigerant producers have forced HFC-134a onto the refrigerator and auto-airconditioner industry. In fact the truth is more prosaic. Consumer product-orientated industries rejected flammable refrigerants, such as HFC-152a and propane which were openly available; their use was deemed unacceptable to companies which had striven hard to improve the reliability and safety of their products. HFC-134, on the other hand, could not meet

Table 4. Additional Candidate Replacements

Fluid	Formula	bp (°C)	Fluid replaced	Flammability[a]
E134	CHF_2OCHF_2	5	114	N
E125	CF_3OCHF_2	−35	22	N
227ea	CF_3CHFCF_3	−18	12	N
225ca	$CF_3CF_2CHCl_2$	50	113	N
225cb	CF_2ClCF_2CHFCl	54	113	N
218	$CF_3CF_2CF_3$	−38	22 and 502[b]	N

[a] N = nonflammable; F = flammable.
[b] 502 = azeotropic mixture of CF_3CF_2Cl and CHF_2Cl.

the deadlines set by the Montreal Protocol for the reduction and ultimate phase-out of CFC-12. The compound had been subjected to minimal toxicity testing and would need a lengthy program to establish its safety to modern standards, a program which could not be accelerated. Only HFC-134a, benefitting from the headstart given by work in the late 1970s, could simultaneously meet the requirements of the refrigeration industry, complete its toxicity testing, and meet the timescale set by the Protocol.

Although the major replacements for CFC-12 and CFC-11 emerged during the 1970s, substitutes were still required for solvent CFC-113 and the refrigerants CFC-114 and CFC-115. Renewed efforts over the last 6 years have identified the additional fluids listed in Table 4. Some of these still contain chlorine, but the presence of hydrogen ensures that the compounds will decompose in the troposphere via attack by hydroxyl radicals. Of course, under the 1990 revision of the Montreal Protocol, these hydrochlorofluorocarbons (HCFCs) are viewed as transitional substances.

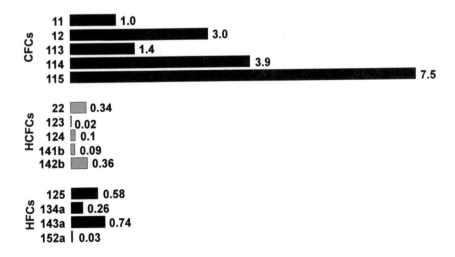

Figure 3. Global warming potentials.

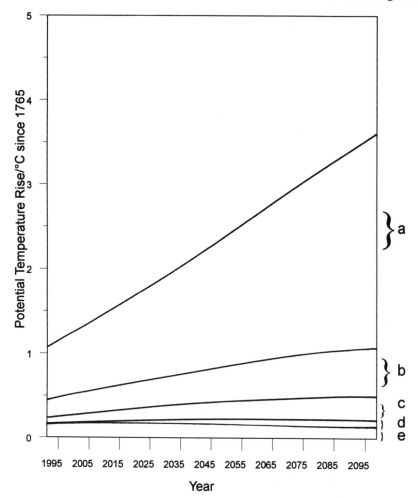

Figure 4. Relative contributions of greenhouse gases to global warming[6-8]: (a) From carbon dioxide, assuming the "business as usual" scenario for emissions from IPCC (1990).[7] (b) From methane, assuming emissions as above.[7] (c) From nitrous oxide, assuming emissions as above.[7] (d) From alternative fluorocarbons, assuming the "aggressive substitution" scenario for emissions from UNEP (1992).[6] (e) From chlorofluorocarbons (CFCs), assuming the scenario from UNEP (1992).[6]

The replacement of chlorofluorocarbons by hydrofluorocarbons will considerably reduce that part of the potential global warming contribution attributable to refrigerants (see Figure 3). An important contribution to this reduction will come from minimizing the leakage rate of refrigeration equipment, and from refrigerant recycling. Already refrigerant manufacturing companies have instituted recycling schemes for chlorofluorocarbons. For HFCs, low loss-rates and recycling will be normal. Only sufficient new refrigerant will be manufactured to "top up" the supply system and allow for the wider use of refrigeration and airconditioning. Most important of all, HFCs are at least as energy efficient as the refrigerants they replace—and it is energy efficiency (via CO_2) which totally dominates the global warming picture (see Figure 4). On this basis the contribution of HFC refrigerants to global warming is calculated to be much less than 1% of the total effect.

28.8. COMMERCIAL PRODUCTION OF CHLOROFLUOROCARBON REPLACEMENTS: THE THIRD PERIOD OF REFRIGERATION

The extensive research and development by the refrigerant manufacturers (see Chapter 7) is now being realized as full-scale production plants. HFC-134a went into production in October 1990 with the start-up by ICI C&P Ltd in the U.K. of the world's first commercial scale plant and of the first U.S. plant by Du Pont in February 1991; these will be followed by further plants in Europe and Japan between now and the year 2000 as other companies bring plants onto line and capacity is expanded to meet the growing demand. HCFC-123 and HCFC-141b may be introduced in 1992/1993; they are mainly intended for blowing polyurethane foam insulation, but HCFC-123 will also find application in centrifugal chillers.

In the longer term, the complete replacement of all chlorine-containing refrigerants by hydrofluorocarbons is anticipated. Hence research by the refrigerant manufacturers into the development and commercialization of new products of this type will continue for the foreseeable future.

ACKNOWLEDGMENTS

The authors would like to acknowledge the considerable advice and assistance from their colleagues in the Klea Business of ICI Chemicals and Polymers Ltd.

28.9. REFERENCES

1. M. J. Molina and F. S. Rowland, *Nature 810*, 249 (1974).
2. J. C. Farman, B. G. Gardiner, and J. D. Shankin, *Nature 315*, 207 (1985).
3. B.A. Nagengast, CFCs Time of Transition Pub. *ASHRAE*, 3 (1989) (ISBN 0-910-110-58-1).
4. R. E. Banks and J. C. Tatlow, in: Fluorine: The First Hundred Years (R. E. Banks, D. W. A. Sharp, and J. C. Tatlow, eds.), Chapter 4, Elsevier Lausanne (1986).
5. Ref. 4, p. 100.
6. Scientific Assessment of Stratospheric Ozone 1992; United Nations Environment Programme/World Meteorological Organization Global Ozone Research and Monitoring Project UNEP (1992).
7. F. Bertnthal (ed.), Climate Change; the Intergovernmental Panel for Climate Change Response Strategies, United Nations Environment Programme/World Meteorological Organization IPCC (1990).
8. T. M. L. Wigley, T. Holt, and S. C. B. Raper, STUGE (an Interactive Greenhouse Model) User's Manual, Climate Research Unit, Norwich, U.K. CRU (1991).

Index

Accufluor®, 483, 488
Acidity, 67
  of fluoroalcohols, 67
  of fluoroalkanesulfonic acids, 67
  of fluoroamides, 68
  of fluorocarboxylic acids, 67
  of fluorohydrocarbons (C–H acidity), 68, 76–77
  of fluorophenols, 68
  of FSO_3H, 68
  of HF, 68
  superacids, 68
Acids: *see* Fluorinated derivatives
Acifluorfen, 241
Acilyzer®, 609
Aciplex®, 409, 609
Addition reactions of fluoro-olefins: *see*
    Fluoroalkenes
Additives (HBPO) for PFC emulsions, 561
Addox®, 560
Adhesives, 424
Aerosol propellants, 151, 584
Aflas®, 387, 610
Aflon COP®, 398, 610
Aflunox®, 431, 589
Agrochemicals
  acifluorfen, 241
  acylureas, 246
  benzylamines, 256
  bifenox, 241
  bifenthrin, 248
  bioallethrin, 248
  bioresmethrin, 248
  bromethalin, 257
  chlorazifop, 243
  chlorfluazuron, 246
  cyfluthrin, 248, 250

Agrochemicals (*cont.*)
  cyhalothrin, 248
  cypermethrin, 247
  diclofop methyl, 243
  diflubenzuron, 199, 216, 246
  diflufenican, 216, 242, 604, 606
  dinitroanilines, 240
  diphenyl ethers, 240
  fenvalerate, 249
  flampropisopropyl, 217
  flocoumafen, 257
  fluazifop-butyl, 243
  flucythrinate, 248
  flufenoxuron, 246
  flumetralin, 256
  fluoroglycofen ethyl, 241
  fluotrimazole, 253
  fluoxypyr 1-methylheptyl, 244
  fluridone, 215, 242
  flurochloridone, 242
  flurprimidol, 253, 256
  flusilazole, 215, 253, 601
  flutolamil, 255
  flutriafol, 215, 253, 605
  fluvalinate, 249
  fomesafen, 241
  fungicides, 252*ff*
  Gibberellin biosynthesis, 255
  haloxyfop ethoxyethyl, 243
  herbicides, 240*ff*, 613
  hydramethylnon, 252
  imazalil, 252
  imidazoles, 254
  insecticides, 246*ff*
  lactofen, 241
  mefluidide, 245

Agrochemicals (*cont.*)
  MONN-4606, 598
  names—common and trade, 238
  nitrofen, 241
  norflurazon, 242
  nuarimol, 253
  oxyfluorfen, 241
  perfluidone, 246
  permethrin, 247
  plant growth regulators, 255
  primisulfuron methyl, 245
  pyrethroids, 247
  pyrimidines, 254
  rodenticides, 257*ff*
  sterol biosynthesis, 252
  structural types, 240
  sulfometuron methyl, 245
  sulfonylureas, 245
  teflubenzuron, 246
  tefluthrin, 217, 248, 250
  thiazafluron, 598
  triadimefon, 252
  triarimol, 252
  1,2,4-triazoles, 253
  triflumizole, 253
  trifluoromethanesufonanilides, 245
  trifluralin, 21, 221, 227, 240
  U.S. industry, 592
Aim®, 238
Air-conditioning, 584
Airoguard™ process, 477
Airopak®, 477, 607
Algoflon®, 349
Algofrene®, 146
Aliphatic fluorides, early history, 7
Alkali-metal fluorides: *see* Fluorinating agents
Alkyl fluorides, reactivity of, 70
Alternatives to CFCs, 159*ff*, 584, 617*ff*
Aluminum production, 590
Ambush®, 239
Amdro®, 238
Amines: *see* Fluorinated derivatives
3-Aminobenzotrifluoride, 221, 225
4-Aminotetrafluoropyridine, 319
Anesthetics: *see* Inhalation anesthetics
Angioplasty, 569
Anhydrous hydrogen fluoride (AHF)
  fluorination with: *see* Fluorinating agents;
    HF-diazotization/fluorodediazoniation
  hazards, 6
  history, 5, 6
  production, 5, 6, 579–582
  uses, 582–583, 590–592

Anionic processes in fluorocarbon chemistry, 78
Anodes
  for electrochemical fluorination,122, 134
  in $[CF]_n$/Li batteries, 490
Antarctic ozone hole, 621
Antimony halides: *see* Fluorinating agents;
    Swarts, F.
Aquaflow®, 238
Arcton® 4, 146, 583
Ardent®, 238
Aromatics: *see* Fluoroaromatics; Polyfluoroarenes
Arylbis(trifluoromethyl)carbinols, 413*ff*
Asahi Frons®, 610
Asahiflon, 146
Asahigard®, 332
Astonex®, 238
Atabron®, 238
Atomic bomb: *see* Manhattan Project
Atomic physical properties, 58
Avimid®, 425
Axarel®, 586
Azo dyes, 315*ff*

Balz–Schiemann reaction, 9,40, 203–207
  comparison with alternative methods, 213
  effects of ring substituents, 204–205
  industrial use and manufacturers, 217–218, 602
  plant procedures, 205
  reaction mechanisms, 204–205
  use of diazonium hexafluorophosphates, 205
Barnon®, 604
Basicity of fluorinated amines, 68–69
Batteries ($[CF]_x$/Li type), 490
Bayleton®, 239
Baylon®, 239
Baythroid®, 238
Beacon®, 239
Bellmark®, 238
Benzotrifluoride derivatives, 222*ff*, 240*ff*, 271*ff*,
    592, 600, 605, 611, 612
  alternative synthesis, 229–232
  hydrolysis of the trifluoromethyl group, 225, 228
  nucleophilic displacement of nuclear halogen in,
    227, 228
Benzotrifluoride, 10, 12, 27, 221–225, 228, 592,
    600, 605
  bromination, 227
  chlorination, 226
  electrophilic substitution, 225
  nitration, 225
Bioallethrin®, 238
1,2-Bis(perfluoroalkyl)ethenes, 27, 557, 560
Bisphenol AF, 362, 381, 415*ff*, 611

Blazer®, 238
Blood substitutes, 555ff
  biocompatibilities, 562
  biomedical applications, 568
  compositions, 559
  emulsification, 558
  gas solubilities in, 557
  history, 559
  perfluorochemicals (PFCs) for use in, 556
  retention times, 563
  surfactants for, 561
Blowing agents, 103, 133, 152, 172, 586
Bond dipoles, 58, 61
Bond energies, 70–73
Bond strengths, effect of adjacent F, 71–72
Brake®, 238
Brigade®, 238
Bromination reactions, 178, 186, 546
Bromochlorodifluoromethane (Halon 1211), 27
  as a fire extinguishant, 185–188
  reactions with carbon-centered nucleophiles, 181
  reactions with enamines, 181
  reactions with sulfur-centered nucleophiles, 182
  reactions with zinc, 184
  See also Halons; Perfluoroalkyl bromides
Bromochlorofluorocarbons (BCFCs), 177ff; see
  also Halons
4-Bromofluorobenzene, 215
Bromofluorocarbons (BFCs), 177ff; see also
  Halons; Perfluoroalkyl bromides
Bromofluoromethanes, preparation, 178
Bromotrifluoromethane (Halon 1301), 27
  as a fire extinguishant, 185–188
  electrochemical reactions, 185
  reaction with phosphorus-centered nucleophiles, 181
  reaction with sulfur-centered nucleophiles, 182,
    598
  reactions with activated arenes, 179, 231
  See also Halons; Perfluoroalkyl bromides
Building block approach to synthesis, 17, 26–29,
  240ff, 271ff, 405, 406, 413ff, 546ff

C—C bond stabilities in PFCs, 72
C—F bond strengths, 70
C=O bond energies, 73
Cancer chemotherapy: see Drugs
Capture®, 238
Carbanions (fluoro), 76
Carbenes (fluoro), 19, 71–72, 80, 180, 184, 245,
  249, 341, 548
Carbocations (fluoro), 74
  resonance destabilization of, by β-F, 75
  resonance stabilization of, by α-F, 74

Carbon monofluoride, 14, 483ff
Carbon tetrafluoride, 14, 89, 91, 93, 95, 150
Cascade®, 238
Catalysts (for fluorination): see Fluorinating agents
Cefbon®, 483
CFCs
  11 (CFCl₃), 146, 150ff, 583, 596
  12 (CF₂Cl₂), 146, 150ff, 583, 596
  113 (CF₂ClCFCl₂), 97ff, 146, 150ff, 164–167,
    547, 583, 596
  113a (CF₃CCl₃), 164, 249
  114 (CF₂ClCF₂Cl), 146, 150ff, 164 583, 596
  114a (CF₃CFCl₂), 164, 168, 172
  115 (CF₃CF₂Cl), 146, 150ff, 168
  See also Chlorofluorocarbons
Chemotherapeutic agents: see Drugs
Chloro-alkali cells, 403ff
Chlorination reactions, 222–225
4-Chlorobenzotrifluoride, 221–222, 226–227, 592
Chlorodifluoromethane: see HCFC-22
5-Chloro-2,4-difluoro-6-methylpyrimidine
  (MFCP), 289
Chlorodinitrobenzotrifluorides, 226
Chlorofluorocarbons (CFCs), 145ff
  alternatives, 159ff, 584, 595, 622
  applications, 147, 151ff, 162, 583ff, 596
  blowing agents, 152, 586
  chemical intermediates, 155–156, 230
  chemical properties, 155
  disproportionation and isomerization, 164, 167,
    230
  environmental aspects, 145, 159, 617ff
  GWPs, 160, 585, 618, 627
  history, 13, 145, 618
  hydrogenolysis of C—Cl bonds in, 164–165,
    167–168
  manufacture, 147
  manufacturers, 146
  nomenclature, 4, 160, 583
  ODPs, 160, 587, 624
  of high molecular weight, 147, 149,
  physical properties, 151–154
  production figures, 147, 583, 597, 620
  propellants (aerosol), 151, 621
  purity specifications, 150
  refrigerants, 152, 584, 618
  replacements: see alternatives
  safety, 145, 619
  solvents, 153, 621
  toxicities, 150, 586, 619
  tradenames, 146, 583
  uses as lubricants, 153
  See also CFCs

3-Chloro-4-fluoronitrobenzene, 198
Chloropentafluorobenzene, 19, 78, 212
Chlorotrifluoroethylene, 73, 77, 155, 165, 169,
    339, 341, 350, 546–547
5-Chloro-2,4,6-trifluoropyrimidine (FCP), 289,
    297
Chrysron®, 238
Cibacron®, 289*ff*
Coatings (fluoropolymer), 397*ff*, 421
Cobalt trifluoride: *see* Fluorinating agents
Cobra®, 238
Combat White Fly Insecticide®, 238
Combat®, 238
Commercial aspects
    fluorinated carbons, 483
    fluorinated ion-exchange membranes, 403, 409
    inhalation anaesthetics, 548, 549
Complexes of perfluoroaromatics, 511
Composites, 424
Condensation polymers, 362
Coopex®, 239
Cosmetics, 454
Cougar®, 238
Covalent radii, 80
Critical surface tensions, 321, 347*ff* (fluoroplas-
    tics)
Crosslinking reactions: *see* Fluoroelastomers
Cryolite, 590
Cure-site monomers: *see* Fluoroelastomers
Cutlass®, 238
Cyanuric fluoride: *see* Trifluoro-*s*-triazene
Cybolt®, 238
Cyclization reactions, 18, 73, 129
Cycloalkanes: *see* Perfluorocarbons
Cymbush®, 238
Cythrin®, 238
Cytop®, 349, 611

Daiflon®, 146, 612
Daifree®, 613
Dart®, 239
Defluorination reactions, 19, 165, 211
Dehalogenation (not F), 12, 155
Dehydrohalogenation, 12, 78, 165
Demnum®, 431, 463*ff*
Desflurane, 524, 551
Destun®, 239
Deuteriated compounds, 528
Diarect®, 239
Diarylhexafluoropropanes, 413*ff*
Diazotization, 280, 319: *see also* Fluorination
    methods
    fluorodediazoniation: *see under* HF

Dibromochlorotrifluoroethane,179, 183
Dibromodifluoromethane, 27; *see also*
    Perfluoroalkyl bromides
    reactions with carbon-centered nucleophiles, 181
    reactions with oxygen-centered nucleophiles, 183
    reactions with phosphorus-centered nucleophiles,
        181
    reactions with selenium-centered nucleophiles,
        182
    reactions with sulfur-centered nucleophiles, 182
    reactions with zinc, 184
1,2-Dibromotetrafluoroethane (Halon 2402)
    as a fire extinguishant, 185–187
    preparation, 178
    reactions with oxygen-centered nucleophiles,
        182–183
    reactions with sulfur-centered nucleophiles, 182
    See also Halons; Perfluoroalkyl bromides
Dicarbon monofluoride, 488
3,4-Dichlorobenzotrifluoride, 221
3,5-Dichloro-2,4,6-trifluoropyridine, 199, 216,
    244
2,6-Difluorobenzamide, 198, 605
Difluorobenzenes, 213, 276, 281, 601, 603
2,4′- and 4,4′-Difluorobenzophenone, 215, 602
2,4-Difluoronitrobenzene, 198
α,ω-Diiodoperfluoroalkanes, 379, 406
Dimilin®, 238
Dipole moments, 61, 511
Direct fluorination, 363, 438, 469*ff*, 483; *see also*
    Fluorinating agents, F_2
Directional effects in additions to fluoroalkenes,
    75–79
Dolobid®, 604
Drimalan®, 293
Drimaren®, 289
Drugs
    absorption and distribution, 530
    antibacterial agents, 502
    anticancer agents, 522, 536, 609
    antifungal agents, 535
    antimalarial agents, 598
    cardiovascular agents, 518
    ciprofloxacin, 601
    deoxyfluoro analogs of glucose, 520
    2-deutero-3-fluoro-D-alanine, 528
    difloxacin, 507
    diflunisal, 216, 604
    electronegativity effects, 509
    fleroxacin, 506, 508
    floxacillin, 216
    flucloxacillin, 601, 605
    flumequine, 503

Drugs (*cont.*)
  flunitrazepam, 605
  flunoprost, 519
  flurbiprofen, 216, 604
  fluorimide, 216
  fluoroacetate formation, 506
  fluoroalanine, 529
  9α-fluorocortisol, 523
  fluoroketone enzyme inhibitors, 525
  fluoroquinolones, 502*ff*, 517, 537, 604
  flurazepam, 216, 601
  flutamide, 522
  fluticasone, 524
  GABA analogs, 529
  history, 20
  hydrogen bonding, 514*ff*
  hydroxyflutamide, 522
  ICI 195739, 535
  lomefloxacin, 508
  mefloquine, 598
  metabolism and elimination, 533
  naphthyridones, 505
  norfloxacin, 216, 503
  NSAIDS, 604
  ofloxacin, 216, 601
  pefloxacin, 216, 503
  prostaglandins, 513
  pyridoxal-dependent enzymes, 527
  quinolones, 502
  radiopharmaceuticals: *see* [18]F
  sparfloxacin, 508, 509
  steric effects, 536*ff*
  steroids, 523, 536
  temafloxacin, 507
  trifluperidol, 606
  tosufloxacin, 507
  vitamins, 536
Dumas–Péligot experiment, 7, 35
Duo Top®, 239
Dyes
  reactive, 287*ff*
  others, 315*ff*

ECTFE, 354
Eksmin®, 239
E1*cb* reactions, 78
Elastomers: *see* Fluoroelastomers
Electrochemical reactions, 50–51, 121*ff*
  anodic fluorination of arenes, 44
  Phillips (CAVE) process, 51, 133*ff*
  Simons (ECF) electrochemical fluorination, 50–51, 92, 121*ff*, 352, 592, 597
Electrochemical fluorination: *see* Fluorination methods

Electronegativity, 57–58, 67–69, 509
Electronics applications, 95*ff*, 130*ff*, 153–154, 263*ff*, 348*ff*, 426, 453, 467, 586, 596
Electrophilic fluorinations, 42
Elimination reactions
  E1*cb*, 78
  of HF from HFCs, 72, 172
Embark®, 238
Emulsion No. II, 559
Emulsions: *see* Blood Substitutes; Surfactants
Enflurane, 69, 547
Environmental aspects, 159–161, 171, 187, 364, 456, 478
  of CFCs, HCFCs, and HFCs, 145, 160, 617
  surface fluorination, 478
Environmental conservation, application of PFPEs, 456
Epoxides: *see* Hexafluoropropene oxide; Oxiranes
Esbiol®, 238
Esbiothrin®, 238
ETFE, 352, 397
Ethers: *see under* Fluorinated derivatives; Perfluoropolyethers
Ethrane®, 547
Eulan®, 238
Evital®, 239
Exhaustive fluorinations of hydrocarbon-type feedstocks, 48–52, 92
  by ECF, *see above*
  by F_2, 48–49
  by high-valency metallic fluorides, 49, 92
  use of partially fluorinated feedstocks, 29, 129
Expanded poly(tetrafluoroethylene), 348

[18]F, x, 28, 42
FC-43®, 108, 112–113, 131
FC-75®, 108, 131
FCC®, 146
6-FDA, 420
FEP, 344
Fire-fighting agents
  extinguishants, 104, 133, 185–188
  foams, 130, 190
Flemion®, 403, 610
Flex®, 238
Florasan®, 238
Flugene®, 146
Fluon®, 19, 589; *see also* Poly(tetrafluoroethylene)
Fluorad®, 332, 592
Fluoride ion, role in fluorocarbon chemistry, 78

Fluorinated derivatives
  alcohols, 67–68, 183, 189, 588, 599
    as solvents, 69
  amides, 63, 67–68
  amines, 68–69, 126–127, 131
  aromatic compounds, 195*ff*, 221*ff*, 240*ff*, 271*ff*,
      592, 600
    early history, 8
    nuclear fluorinated, 8–10
    perfluorinated, 11
    with fluorinated side chains, 10, 179, 195
  bromides: *see* Perfluoroalkyl bromides
  carboxylic acids and derivatives, 12, 18, 62–4,
      67–68, 128–129, 139, 179, 181–185, 189,
      271, 597
  chlorides: *see* Chlorofluorocarbons
  ethers, 68–69, 72, 128–129, 131, 161, 544*ff*, 627
  heterocyclic compounds, 63, 126–127, 287*ff*, 598
  hydrides: *see* Hydrochlorofluorocarbons;
      Hydrofluorocarbons
  iodides, 18, 189, 325, 599; *see also*
      Perfluoroalkyl iodides
  ketones, 69, 139, 181
  nitriles, 225
  of main-group elements, 18
  oxetanes, 463*ff*
  oxiranes, 72, 328, 400, 405, 431*ff*, 452; *see also*
      Hexafluoropropene oxide
  pyridines, 199, 216, 290, 601, 605
  pyrimidines, 254, 287*ff*
  sulfonic and sulfinic acids and their derivatives,
      68, 128, 130,178,182–185, 189
  triazines, 199, 213, 287*ff*
  See also under individual compound names
Fluorinated fluids, 89*ff*, 145*ff*, 431*ff*, 463*ff*, 586,
    589, 592
Fluorinated membranes, 403*ff*
Fluorinated monomers
  acetylenes, 361
  carboxylated vinyl ethers, 405
  dienes, 349
  diglycidyl ethers, 399
  dioxoles, 349
  for addition polymers, 339
  for condensation polymers, 362
  hazards, 342
  hexafluoroacetone-based, 413*ff*
  oxetanes, 463
  oxiranes, 400, 405
  sulphonylated vinyl ethers, 405
  vinyl ethers, 389, 399
  vinyl monomers, 341, 399, 587, 596, 613

Fluorinating agents (sources of F)
  AgBF$_4$, 30
  AgF, 7, 31–32, 45–46
  AHF: *see* HF
  amine hydrofluorides, 37, 297, 299, 552; *see also*
      HF/pyridine
  AsF$_3$, 7
  BrF$_3$ + Br$_2$, 46
  BrF$_3$, 11, 550, 552
  CaF$_2$, 36, 199
  CF$_3$OF, 42
  CH$_3$CO$_2$F, 42
  (C$_2$H$_5$)$_2$NSF$_3$ (DAST), 35, 39, 169
  CoF$_3$, 11, 16, 29, 49, 92, 211
  COFCl, 36
  CsF, 36, 165, 199, 201, 297, 552
  CsSO$_4$F, 42
  CuF, 31
  DAST: *see* (C$_2$H$_5$)$_2$NSF$_3$
  electrophilic, 41*ff*
  F$_2$, 7, 15–16, 29, 40–43, 48–49, 91, 213, 463,
      464, 469*ff*, 483, 538, 552
  ^{18}F[F$_2$], 42
  FClO$_3$, 44
  fluoroalkylamine reagents (FAR), 34, 552
  HBF$_4$, 40, 196, 203–207
  HF (often with a catalyst), 10, 29–33, 37, 45–46,
      91, 146–149, 156, 162–170, 200, 212, 222–
      224, 229, 297, 299, 413, 550, 582, 591, 598
  HF/electrochemical, 18, 29, 44, 50, 92, 121*ff*,
      325, 600
  HF/NaNO$_2$, 8, 9, 39, 196, 207–211, 280, 592
  *HF/pyridine, 33, 36–37, 39, 40, 45–46, 280
  HF/SbCl$_5$, 11, 30–31, 46, 147–149, 162–163, 166,
      170, 546–547
  HgF$_2$, 30
  IF$_5$, 18
  IF$_5$ + I$_2$, 27, 46, 178, 188, 325, 588, 599
  KCoF$_4$, 50
  KF, 7, 10, 29, 31–32, 35–36, 46, 165, 196–197,
      199, 201, 212, 297, 299, 550, 602
  KHF$_2$, 7, 37
  MoF$_6$, 222
  MOST (morpholinosulfur trifluoride), 35
  NaBF$_4$, 9, 40, 203–207
  NaF, 33, 199, 297, 299, 550, 552
  *N–F reagents, 43
  SbF$_3$, 7, 10, 30–31, 224
  SbF$_3$/SbCl$_5$, 12–13, 30
  SbF$_5$, 11, 165, 230
  SF$_4$, 34, 38, 169, 222
  *Tetra-alkylammonium fluorides, 31, 36, 44

*Analogous species often react similarly.

Fluorinating agents (*cont.*)
  XeF$_2$, 43, 47, 550
  ZnF$_2$, 7
Fluorination methods
  background discussion, 6, 27
  [^{18}F] fluorination, 28, 42
  general processes
    electrochemical reactions, 50–51, 121*ff*
      anodic fluorination of arenes, 44
      Phillips (CAVE) process, 51, 133*ff*
      Simons (ECF) process, 50–51, 92, 121*ff*,
        325, 592, 600
    electrophilic fluorinations, 42
    exhaustive fluorinations of hydrocarbon-type
      feedstocks, 48–52, 92
      by ECF: *see* above
      by F$_2$, 48–49
      by high-valency metallic fluorides, 49, 92,
        211
      use of partially-fluorinated feedstocks, 29,
        129
  fluorination of carbon, 14–15, 91, 483
  halogen-exchange reactions
    in arenes and heteroarenes with nuclear
      halogen: *see* Halex reactions
    in arenes and heteroarenes with side-chain
      halogen, 10, 31, 221–225, 229
    in halogenocarbonyl compounds, 32, 413,
      598
    in monohalogeno-alkanes, 31
    in polyhalogeno-alkanes, 12, 30, 91,
      147–149, 156, 162–170
    in polyhalogeno-alkenes (allylic positions),
      31
  reactions of specific groups
    additions to C=C and to C≡C, 45–47, 162,
      165–170, 188
    conversions
      ArNH$_2$ → ArF: *see under* Balz–Schiemann
        reaction; HF-diazotization/fluorodedia-
        zoniation
      azirines → β-fluoro-amines, 40
      C(CO$_2$H)NH$_2$ → C(CO$_2$H)F, 39, 280
      C–H → C–F, 40 (aliphatic monofluorides),
        41–44 (arenes)
      C=O → CF$_2$, 38
      C–O–X → C–F, 33–37, 552
      CO$_2$H → CF$_3$, 38
      carbanions → C–F, 44
      epoxides → fluorohydrins, 37
Fluorination of carbon, 14–15, 91
Fluorine, 1, 15–17, 438, 464, 591
Fluorine "finishing" of polymers, 438, 464

Fluorinert®, 131, 592
Fluorinert liquid heat sink, 131
Fluoroalkenes
  π-bond dissociation energies, 73
  commercial aspects, 587
  cyclodimerization, 18, 73
  electrophilic reactions, 75
  free-radical reactions, 79
  nucleophilic reactions, 77, 547
  polymerization, 339*ff*, 373*ff*, 431*ff*
  synthesis via Wittig reagents, 181
  See also under individual compound names
Fluoroacetate, 20, 25, 506
Fluoroacetylenes, 73, 283, 361
Fluoroapatite, 579
Fluoroaromatics: *see* Fluorinated derivatives;
    Polyfluoroarenes; individual compound-
    names
Fluorobenzene, 8, 9, 27, 214–215, 281, 592, 604
Fluorobenzoic acid, 214–215
Fluorobenzonitriles, 198, 199, 216, 256, 273, 274,
    601
Fluorocarbon derivatives (with functional
    groups), micellization, 63; *see also under*
    Fluorinated derivatives; individual com-
    pound names
  physical properties: *see* Perfluorocarbons, physi-
    cal properties
  surface activities, 63
  surface energies, 63
Fluorochemicals industry, 579*ff* (U.S.A.), 595*ff*
    (Western Europe), 609*ff* (Japan)
Fluorocyclobutanes, formation of, 18, 73
Fluorodediazoniation: *see under* HF diazotization
Fluorodenitration, 603
Fluorodenitration (in aromatics), 603
1-Fluoro-2,4-dinitrobenzene, 10
Fluoroelastomers
  Aflas®, 387
  commercial aspects, 588*ff*, 607
  cure-site monomers, 383, 389
  curing (vulcanization) of, 380, 388, 390
  Dai-El®, 376
  Fluorel®, 376, 589
  history, 376
  Kalrez®, 389
  Kel-F®, 376
  molecular weight control, 378
  monomers, 341
  perfluorinated, 389
  polymerization techniques, 377, 387, 390
  polyphosphazenes, 373, 375
  processing, 385, 388, 393

Fluoroelastomers (*cont.*)
   properties, 383, 386, 388, 392
   silicones, 373, 375, 613
   SKF-32, 376
   Tecnoflon®, 376
   tetrafluoroethylene-perfluoro(alkyl vinyl ether)
      copolymers, 389
   tetrafluoroethylene-propylene copolymers, 387
   uses, 393
   vinylidene fluoride copolymers, 376, 589
Fluoroethers: *see* Fluorinated derivatives;
   Perfluoropolyethers
Fluoromar®, 544
Fluoronitrobenzenes, 199, 216
Fluoroolefins: *see* Fluoroalkenes
4-Fluorophenol, 215
Fluoroplastics, 339*ff*, 587*ff*, 606
   acryonyms, 340
   adhesion to (surface etching), 347
   Airopak®, 477
   amorphous, 349
   applications, 348, 352, 358, 360
   coatings, 397*ff*
   commercial aspects, 348, 350, 352, 354, 355, 358,
      360, 399, 587*ff*
   condensation (step-growth) types, 362, 399, 413*ff*
   curable coatings, 399
   degradation, 72
   electrical properties, 61, 347
   expanded ("porous") PTFE, 348
   fabrication, 343*ff*
   hazards, 342
   hexafluoroacetone-based, 362, 413*ff*
   history, 13, 19
   manufacturers, 349
   medical uses, 348
   membranes, 403*ff*
   monomers: *see* Fluorinated monomers
   perfluorinated, 342
   piezoelectricity, 61, 357, 358
   polymerization techniques, 342, 345, 349, 351,
      353, 355, 356, 361
   polymorphism, 357
   processing: *see* fabrication
   properties, 346, 350, 351, 353, 355, 357, 359
   pyroelectricity, 356
   sulfur-containing, 362
   surface-fluorinated, 363, 469*ff*
   tacticity, 359, 363
Fluoropolymers: *see* Fluoroelastomers; Fluoroplastics; Fluorinated fluids; Perfluoropolyethers
Fluoropyridazines, 290

Fluoropyridines, 197, 199, 216, 290, 601, 605
   2-Fluoropyridine, 9, 216
Fluoropyrimidines, 199, 254, 289*ff*, 297
Fluoropyridazinones, 290
Fluoroquinolines, 290
Fluoroquinolones: *see* Drugs
Fluorotoluenes, 215
Fluorotriazines, 295*ff*, 299
5-Fluorouracil, 20, 213, 591
Fluorspar, 5, 579
Fluosol®, 557*ff*
Fluothane®, 20, 69, 546, 598
Fluowet®, 332
Fluroxene, 19, 544
Flutec®, 90*ff*
Fomblin®, 431*ff*, 465, 589
Forafac®, 332
Forane®, 583
Forca®, 239
Force®, 239, 601
Forza®, 239
Free radicals (fluorinated) 1, 16, 78, 470–471
   fluorine atom migration in, 79
   See also Perfluorocarbon radicals

Freon®, 4, 146, 583
Friedel–Crafts reactions, 251, 253, 415
Frigen®, 146
Ftorosan®, 559*ff*
Fungaflor®, 238
Fungicides, 252*ff*
Fusilade®, 238

Galden®, 431, 465
Gallant®, 238
Gas solubilities: *see* Blood substitutes
Gas transport agents: *see* Blood substitutes
Gauntlet®, 239
Genetron®, 146, 583
Global warming potentials (CFCs and
   replacements), 627
Goal®, 239
Gore-Tex®, 348
Graphite fluoride, 14, 135, 483*ff*
Greases, 153–154, 466, 493
Grenade®, 238
Grignard reagents, 18, 180, 417

Hache UnoSuperOnecide®, 238
Halex reactions (exchange of F for nuclear halogen in arenes), 10, 32, 196–203, 212–214,
   297–299
   activation by substituents in, 197

Halex reactions (*cont.*)
  catalysis of, 201
  comparisons with alternative methods, 213, 602
  N-heteroarenes, 297*ff*
  industrial use and manufacturers, 216–218, 602
  plant procedures, 201–203
  solvents for, 200, 297
Halocarbon®, 147, 154
Halogen-exchange: *see* Fluorination methods
Halons [bromo(halogeno)fluorocarbons]
  120l (CHF₂Br), 187
  1211: *see* Bromochlorodifluoromethane
  1301: *see* Bromotrifluoromethane
  2402: *see* 1,2-dibromotetrafluoroethane
  alternatives, 587
  as fire extinguishants, 185–188
  environmental aspects, 187, 587, 625
  nomenclature, 5, 186, 587
  ODPs, 587
  physical properties, 186
  production figures, 186
  radical scavenging by, 187
  reactions with HO•, 187
  See also Perfluoroalkyl bromides
Halothane, 20, 69, 546, 598, 613
Hazards
  CFC alternatives, 171
  fluorine, 7, 14, 478
  fluoroacetate, 20, 506
  graphite fluoride, 495
  hydrogen fluoride, 6, 209–211
  perfluoroisobutene, 111
Health and safety aspects: *see* Hazards
Hemagen®, 560
Herbicides, 240*ff*
Heterocycles: *see* Fluorinated derivatives; specific compounds
Hexafluoroacetone, 74, 139, 362, 400, 413, 599, 611
Hexafluoroacetone hydrates, 414, 611
Hexafluorobenzene, 19, 27, 75, 78, 211–212, 511, 531
Hexafluorobut-2-yne, 73–74, 361
Hexafluoroethane, 15, 89, 92–93
Hexafluoroisopropanol, 550, 599, 611
Hexafluoroisopropylidene units, 362, 413*ff*
Hexafluoropropene, 75, 77–80, 341, 345, 374*ff*, 403–404, 432*ff*, 588–590, 596
Hexafluoropropene oxide, 27, 341, 405, 406, 413, 589, 596
Hexafluoroxylenes, 11–12, 223–224

HF-diazotization/fluorodediazoniation, 8, 9, 39, 196, 207–211, 280
  comparison with alternative methods, 213
  effect of substituents, 208
  industrial use and manufacturers, 214–215, 218, 592, 602
  plant procedures, 209–211
High-performance composites, 424
Historical aspects, 5–21
Hoelen®, 238
Hograss®, 238
Hostaflon®, 349
Hostinert®, 431, 607
Hydrochlorofluorocarbons (HCFCs)
  22 (CHF₂Cl), 19, 146, 150*ff*, 172–173, 245, 341, 585, 619, 622
  31 (CHFCl₂), 622
  123 (CF₃CHCl₂), 152, 155, 165–168, 172–173, 187, 585, 622
  123a (CF₂ClCHFCl), 166–167
  124 (CF₃CHFCl), 152, 156, 164–168, 172–173, 585, 622
  133a (CF₃CH₂Cl), 162, 165, 167, 546, 622
  141b (CFCl₂CH₃), 169, 172–173, 585, 622
  142b (CF₂ClCH₃), 169, 173, 585, 622
  225ca and 225cb, (C₃HF₅Cl₂), 170, 173
  applications, 162, 583, 595
  blowing agents, 172, 586
  eliminations from, 172; *see also* Carbenes(fluoro)
  environmental aspects, 171, 617*ff*
  fire extinguishants, 187–188, 587
  Montreal Protocol and GWPs, 618
  nomenclature, 4, 160, 583
  ODPs, 160, 587, 624
  physical properties, 159*ff*, 171, 627
  reactions with HO•, 79
  refrigerants, 172, 584
  solvents and cleaning fluids, 173, 586
  toxicities, 171
Hydrofluorocarbons (HFCs)
  32 (CH₂F₂), 585, 622
  125 (CF₃CHF₂), 16, 156, 166, 168–169, 172, 187, 585, 596, 622
  134 (CHF₂CHF₂), 622
  134a (CF₃CH₂F), 150, 156, 162–165, 172–173
  143a (CF₃CH₃), 164, 167, 585, 596, 622
  152a (CHF₂CH₃), 152, 169–170, 172–173, 585, 596, 622
  applications, 162, 583, 595
  blowing agents, 172, 586
  boiling points, 60, 171
  environmental aspects, 159, 171, 584, 617*ff*
  fire extinguishants, 187–188, 587

Hydrofluorocarbons (*cont.*)
  GWPs, 160, 585, 618, 627
  HF elimination from, 72, 172, 211
  nomenclature, 4, 160, 583
  ODPs, 160, 585, 624
  physical properties, 59, 171
  reactions with HO•, 79
  refrigerants, 172, 584
  solvents and cleaning fluids, 65, 173, 586
  stabilities, 72
  surface free energies, 62
  surface tensions, 62
  toxicities, 171
Hydrogen bonding, 69–70, 514*ff*
  and anesthetic activity, 69
  intramolecular, in FCH_2CH_2OH, 69
Hydrogen fluoride
  history, 5, 580
  manufacture, 580
  production statistics, 580
  uses, 582
  See also Fluorinating agents; HF-diazotization
Hydrogenation, 251, 274, 278, 298, 588, 603
Hydrogenolysis of C—Cl bonds in CFCs, 164–
     165, 167–168
Hydrophilicity, 67

I_π electron-pair repulsion, 76
ICI 195739, 535
Icon®, 238
Illoxan®, 238
Imagent®, 560
Imaging processes, 494
Impact®, 238
Imperator®, 238
Inductive effects, 57–58, 67–69, 509
Inhalation anaesthetics, 543*ff*, 598, 613
Inorganic polymers, 371, 375
Insecticides, 246*ff*
Intercalation compounds, 495
Iodine fluorides: *see* Fluorinating agents
Iodofluorocarbons: *see* Perfluoroalkyl iodides
Ion-exchange membranes, 403*ff*
Ionization potentials, 57–58
Ionomers, 403*ff*
Isceon®, 4, 146, 583
Isoflurane, 156, 547
Isotron®, 146, 583

Jaguar®, 238
Japanese fluorochemicals industry, 609*ff*
Javelin®, 604
Jupitor®, 238

Kafil®, 239
Kalrez®, 389
Kaltron®, 146
Karate®, 238
Kel-F®, 340, 376, 589
Ketones: *see under* Fluorinated derivatives
Khladon®, 146
Klea®, 585
Koltar®, 239
Krytox®, 43, 465, 589
Kynar®, 398, 590

Ladder polymers, 421
Larvakil®, 238
Leak testing, 133
Leather chemicals, 190
Ledon®, 146
Levafix®, 289*ff*
Light Water®, 130
Lipophilicity, 66, 539*ff*
Liquid crystals, 263*ff*
Lithium batteries, 483, 490
Lodyne®, 332
Lubricants, 115, 153–155, 454, 492
Lumiflon®, 399, 610

Manhattan Project, 17, 340
Manufacturers of fluorochemicals, 579*ff*
Mass spectrometric markers, 455
Matox®, 238
Mavril®, 238
Maxforce®, 238
Mechanisms, 70*ff*, 124–126, 180–181, 196–197,
    204–205, 208–209, 227, 231
Meisenheimer complexes, 197
Membranes, 403*ff*, 609
Methoxyflurane, 69, 546
Moissan, H., 1, 5–7
Moncut®, 238
Monomers: *see* Fluorinated monomers
Montreal Protocol, 146, 160, 187, 583, 596, 617,
    627
Mowdown®, 238
Multifluor®, 591
Murox®, 238

Nafion®, 403
Naturally occurring fluorocompounds, 20, 25
Negative hyperconjugation, 77
Neoflon®, 349, 612
Nerve gases, 20

Nomenclature (CFCs, fluorocarbon, Halon, HCFCs, HFCs, perfluorocarbon), 2–5, 160, 186, 583
Nomolt®, 239
NR-150, 424
Nucleocidin, 25
Nucleophilic substitution, effect of adjacent F, 71
Nustar®, 238
Nuva®, 332

Octafluoronaphthalene, 212
Octafluoropropane, 15, 89, 92–93
Oils, 115, 153–155, 454, 492
Oleophobal®, 332
Olymp®, 238
Oust®, 239
Outflank®, 239
Oxetanes, 463ff
Oxidation, 1,1,2,2-tetrafluoropropene, 464
Oxidative photopolymerization, 432
Oxiranes, 72, 328, 400, 431ff, 452; see also Hexafluoropropylene oxide
Oxygent®, 560
Oxypherol®, 559ff
Ozone depletion potential for CFCs, HCFCs, and HFCs (ODPs), 160, 587, 624
Ozone, stratospheric depletion by CFCs, 584, 624
Ozonolysis, 547

Paints, 399
Paper finishes, 190
Pay Off®, 238
PCTFE, 339, 350
PEEK, 604
Pentafluoroaniline, 27, 319
Pentafluorobenzene, 27, 75, 78, 211
Pentafluoroethyl iodide, 188–189, 325, 326, 558, 599; see also Perfluoroalkyl iodides
Pentafluorophenol, 27, 531
1H-Pentafluoropropene, 376
Pentafluoropyridine, 19, 69, 78, 199
Penthrane®, 546
Perflubron (PFOB), 560
Perflucol®, 560
Perfluidone, 246
Perfluorinated compounds
    alkanes: see Perfluorocarbons
    amines(tert.) 126–127, 131
    aromatics: see Polyfluoroarenes
    bromides: see Perfluoroalkyl bromides
    carboxylic acids and derivatives, 129, 182–185, 189, 598
    chlorides: see Chlorofluorocarbons

Perfluorinated compounds (cont.)
    cycloalkanes: see Perfluorocarbons
    definition, 2–4
    ethers, 128, 131
        formation by cyclization of carboxylic acids during ECF, 129
    iodides, 18; see also Perfluoroalkyl iodides
    molecular geometry, 81
    olefins: see Fluoroalkenes
    polymers: see Fluoroplastics; Fluoroelastomers; Perfluoropolyethers
    solubility of gases in, 65, 117, 555ff
    sulfonic acids and derivatives, 128, 130, 178, 182–185, 189, 592, 599
    uses: see Perfluorocarbons
    See also Fluorinated derivatives; individual compounds
Perfluoro compounds, general: see Fluorocarbon derivatives
Perfluoroadamantanes, 560
Perfluoroalkyl bromides, 177ff; see also Halons
    additions to C=C and C≡C, 178–179
    as "blood substitutes" (PFOB), 185, 556ff
    as radiopaques, 185, 561
    ionic reactions, 180–183
    preparation, 178
    radical nucleophilic reactions, 180–183
    reactions
        via ionic intermediates, 178
        via radical intermediates, 178
        with chlorine fluorosulfate, 185
        with sulfur-centered nucleophiles, 182
Perfluoroalkyl cadmium reagents, 184
Perfluoroalkyl copper reagents, 184, 279
Perfluoroalkyl effect, 74
β-(Perfluoroalkyl)ethyl iodides, 189
Perfluoroalkyl iodides, 18, 177ff
    additions to C=C and C≡C, 178–179
    additions to NC, 179
    electrochemical reactions, 185
    formation of Grignard reagents from, 18, 180
    formation of iodosobistrifluoroacetates, iodonium triflates and (perfluoroalkyl) arenes from, 185
    formation of lithium reagents from, 180
    ionic nucleophilic reactions of, 180–183
    manufacturers and uses, 188–190
    radical nucleophilic reactions of, 180–183
    reactions
        via ionic intermediates, 178
        via radical intermediates, 178
        with arenes, 179, 184, 232
        with carbon-centered nucleophiles, 181
        with copper, synthetic uses of, 184

Perfluoroalkyl iodides (*cont.*)
  reactions (*cont.*)
    with disulfides, 180
    with heteroarenes, 179, 184
    with nitrogen-centered nucleophiles, 183
    with oleum, 184, 189
    with sulfur-centered nucleophiles, 182
    with zinc, synthetic uses of, 183–184
    telomer iodides, 188, 325–327, 588, 599
    uses in synthesis, 178–190
    See also Trifluoromethyl iodide
Perfluoroalkyl zinc reagents, synthetic uses of,
    183–184
Perfluorocarbon radicals, 81, 178, 231; *see also*
    Free radicals
Perfluorocarbons (saturated; PFCs), 89*ff*, 596
  boiling points, 59
  bond strengths in, 72
  chemical and thermal stabilities, 72, 109
  cleaning fluids, 100
  coolants, 97–99, 131–132
  defluorination of, 72–73, 211
  etchants, 95–96
  fire extinguishants, 104, 133
  foam blowing, 103, 133
  heat transfer agents, 104
  history, 14
  leak testing, 133
  liquid barrier filters, 117
  lubricants, 115
  oxygen carriers, 65, 117, 555*ff*
  particle physics, 115–116
  physical properties, 58, 93–94, 556
  preparation, 89*ff*, 121*ff*
  refrigerants, 104
  solvents, 64
  sterilants, 112
  surface free energies, 62
  surface tensions, 61
  tracers, 114
  vapor phase soldering, 108–112, 132
  vapor-phase sterilization, 112–113
Perfluorocarbons (specific saturated compounds)
  perfluoro
    (cyclohexylmethyldecalin), 90
    decalin, 90, 556 *ff*
    (dimethylcyclohexanes), 90
    hexanes, 90, 97
    (methylcyclohexane), 90
    pentanes, 90
    perhydrofluoranthene, 90, 561
    perhydrofluorene, 90
    perhydrophenanthrene, 90

Perfluorochemical emulsions: *see* Blood substi-
    tutes
Perfluoro-(2,2-dimethyl-1,3-dioxole), 349
Perfluoroisobutene, 77–78, 111 (toxicity)
Perfluoro(methyloctahydroquinolidizine), 560
Perfluoronaphthalene, 212
Perfluoro-octyl bromide (PFOB) 185, 557*ff*
Perfluoro(perhydro-*N*-methylisoquinoline), 560
Perfluoropolyethers, 431*ff*, 463*ff*, 589
  applications, 453
  based on oxetanes, 463
  comparative properties, 453, 465
  degradation, 440
  end-group modification, 437, 450, 455
  functional, 450
  future for, 456
  mechanisms of formation, 435
  neutral, 437*ff*
  oxidative photopolymerization route, 432
  peroxide intermediates, 432*ff*
  properties, 440, 464
  stabilities, 72
Perfluoropropylene oxide: *see* Hexafluoropropene
    oxide
Perfluorotributylamine, 108, 131, 556*ff*
Perfluoro(trimethyladamantanes), 560
Perfluorotripropylamine, 556*ff*
Peroxidic polymers, 432
Persulan®, 238
Pesticides, 237*ff*
Petroleum alkylation, 590
PFA, 344
PFOB, 556*ff*
Picket Pounce®, 239
Piezoelectric effect, 61, 357–358
Pigments, 319
Polyfluoroarenes
  electrophilic attack on, 75
  nucleophilic reactions of, 77–78
  See also Fluoroaromatics
Polymers
  coatings, 413*ff*
  from hexafluoroacetone, 413*ff*
  membranes, 403*ff*
  polyacetylenes, 361
  polyamides, 422
  poly(carbon monofluoride), 483*ff*
  poly(chlorofluoroaldehydes), 362
  poly(chlorotrifluoroethylene), 339, 350
  polyesters, 421
  poly(ethylene-*alt*-tetrafluoroethylene), 61
  poly[fluoroalkyl (meth)acrylates], 361
  poly(hexafluorobut-2-yne), 361

Polymers (*cont.*)
    poly(hexafluoroisobutylene-*alt*-vinyl acetate), 361
    poly(hexafluoroisobutylene-*alt*-vinylidene fluoride), 361
    polyimides, 362, 422
    polymerizations, emulsifiers for, 190
    polymethine dyes, 319
    poly(*N*-perfluoroacylethylamines), 361
    polyperoxides, 432
    polystyrenes, 361
    poly(tetrafluoroethylene), 13, 19, 72, 79, 339, 342, 346, 397, 406, 589
    poly(tetrafluoroethylene-*alt*-isopropylene), 361, 387*ff*
    poly(tetrafluoroethylene-*co*-hexafluoropropylene), 344, 589
    poly[tetrafluoroethylene-*co*-perfluoro(methyl vinyl ether)], 344
    poly[tetrafluoroethylene-*co*-perfluoro(propyl vinyl ether)], 344, 389*ff*
    poly(tetrafluorothiirane), 362
    poly(thiocarbonyl fluoride), 362
    poly(trifluoroacetaldehyde), 362
    poly(vinyl fluoride), 358
    poly(vinylidene fluoride) (PVDF, PVF$_2$), 61, 397
    PTFE: *see* Poly(tetrafluoroethylene)
    See also Fluoroelastomers; Fluoroplastics; Perfluoropolyethers
Polytetrafluoroethylene
    chain length, 79, 346
    thermal stability, 72, 346–347
    See also under Polymers
Polytrin®, 238
Positron emission tomography (PET), 28
Pramex®, 238
Pride®, 238
Prime®, 238
Procure®, 239
Punch®, 238
Pydrin®, 238

Racer®, 238
Radical-ion chain process, $S_{RN}1$, 180–181
Radiopaque agents, 185, 570
Reactive dyes, 287*ff*
Reacton®, 291
Reaktofil®, 291
Reflex®, 238
Refrigerants
    CFC-phaseout, 145–147, 159–161
    CFC-substitutes, 160*ff*
    commercial, 583

Refrigerants (*cont.*)
    history, 13
    See also Chlorofluorocarbons; Hydrochlorofluorocarbons; Hydrofluorocarbons
Remazol®, 306
Resbuthrin®, 238
Responsar®, 238
Rifle®, 239
Ring-opening polymerizations, 349, 361, 362, 464
Ripcord®, 238
Rodenticide 1080, 20
Rodenticides, 257

Scotchgard®, 332, 592
Scotchrelease®, 332
Selectfluor™, 43*ff*
Semifluorinated alkanes, 59, 64, 271
Sevoflurane, 550
Silicones, 371, 375
Simons, J.H., 15, 17–18, 121
Single electron transfer (SET) reactions, 44, 47, 52, 71, 73, 180, 231, 347
S_N1 reactions, 70, 75, 178
S_N2 reactions, 70, 75, 178, 180
S_NAr reactions, 78, 196*ff*, 244, 288*ff*, 603
$S_{RN}1$ reactions, 71, 181
Solfac®, 238
Solicam®, 239
Solvent polarity, 64
Solvents (dipolar aprotic), 200–201
Spur®, 238
Starane®, 216, 238
Stereochemical effects, 536*ff*
Steric effects, 80, 346, 536
    size of CF$_3$ group, 81, 536, 537–538
    stabilization of PFC radicals by, 81
    van der Waals radius of F, 80, 536
    van der Waals volumes, 536
Stockade®, 238
Storm®, 238
Stratagem®, 238
Sumicidin®, 238
Sumifix®, 305
Sunfural®, 609
Super Blazer Complete®, 238
Suprane®, 551
Surface fluorination of polymers, 363, 469*ff*, 607
Surfactants, 190, 321*ff*, 546, 547, 561, 567, 588, 592
Surflon®, 611
Suva®, 583
Swarts, F., work on aliphatic fluorides, 12
Sylon®, 607

Tackle®, 238
Talcord®, 239
Talstar®, 238
Tecnoflon®, 376, 607, 610
Tedlar®, 358, 590
Teflon® AF, 349, 607
Teflon® FEP, 398
Teflon® PFA, 398
Teflon®, 14, 586; *see also* Poly(tetrafluoroethylene)
Tefzel®, 398, 589
Tegafur, 609
Tell®, 239
Telok®, 239
Telomer alcohols, 18, 188–190, 328, 588–589
Telomer iodides: *see* Fluorinated derivatives; Perfluoroalkyl iodides
Tetrafluoroethylene, 15, 18–19, 75, 77–80, 165, 169–170, 178, 188–189, 325, 341, 405, 433, 464, 558, 587–589, 599
    fluoropolymers from, 339*ff*, 589
    manufacture, 19, 156, 587
    telomers, 325, 588, 599
    uses in synthesis, 18, 165, 169–170, 178, 188–189
Tetrapion®, 613
Texgard®, 612
Textile chemicals, 190, 329
Thermosetting resins, 421
Tok®, 239
Tokurn®, 239
Transitional substances (HCFCs), 160, 625
Treflan®, 239
2,4,5-Trichloro-6-fluoropyrimidine, 199
Trifluoroacetaldehyde, 169, 362, 552
Trifluoroacetic acid and its derivatives, 12, 15, 18, 139, 178, 185, 230, 597
Trifluoroethanol, 65, 68, 548, 598
Trifluoroethylene, 155, 165
Trifluoroiodomethane: *see* Trifluoromethyl iodide
Trifluoromethanesulfonic (triflic) acid and its derivatives, 68, 182, 185, 245, 598, 612
Trifluoromethyl bromide: *see* Bromotrifluoromethane

Trifluoromethyl iodide, 18, 178
    additions to C=C and C≡C, 18, 178
    reactions with Hg, 18, 178
    reactions with NO, 178
    reactions with P, 18, 178
    reactions with S, 18, 178
Trifluoronitrosomethane, 180
Trifluoro-*s*-triazine, 212, 289, 295, 299
Trifmine®, 239
Trimidal®, 239
Triminol®, 239

Uranium hexafluoride, 15, 591

Van der Waals
    radii, 80, 536
    volumes, 536
Vapor-phase soldering, 108–112, 132
Vengeance®, 238
Verdict®, 238
Verofix®, 293
Victrex® PEEK, 604
Vistar®, 238
Viton®, 376*ff*, 589
Vulcanization: *see* Fluoroelastomers

Wettability, 61–64, 321–324
Winner®, 238
Wipeout®, 238
Wittig reagents containing F, 181

X-Ray contrast agents, 185, 570
Xenon difluoride: *see* Fluorinating agents
XR resin, 403

Zepel®, 332
"Zipper" mechanism of ECF, 125–126
Zisman plots, 321
Zonyl®, 332
Zorial®, 239